Alexander B. Borisov, Vladimir V. Kiselev

Quasi-One-Dimensional Magnetic Solitons

Also of Interest

Evolutionary Dynamic Equations
Stabilizability, Observability, and Controllability
Akram Ben Aissa, Khaled Zennir, 2025
ISBN 978-3-11-914539-8, e-ISBN (PDF) 978-3-11-221435-0

Error Freed CFD Mathematics
Stability, Monotonicity, Finite Element Theory
A. J. Baker, James D. Freels, 2026
ISBN 978-3-11-914776-7; e-ISBN 978-3-11-220643-0

Space-Time Algebra of Sedeons
Electrodynamics, Hydrodynamics, Field Theory, and Quantum Mechanics
Victor L. Mironov, Sergey V. Mironov, 2025
De Gruyter Series in Nonlinear Analysis and Applications
ISBN: 978-3-11-162519-5; e-ISBN 978-3-11-162517-1

Fixed Point Theory and Functional Analysis
Metric Spaces, Banach Spaces, Hilbert Spaces
Pradip Debnath, Hari Mohan Srivastava, Yeol Je Cho (eds.), 2026
ISBN 978-3-11-914503-9; e-ISBN 978-3-11-221576-0

Banach Algebras and Harmonic Analysis
Arens Products, Factorizations, and Bounded Operators
Mahmoud Filali (eds.), 2025
ISBN: 978-3-11-164324-3; e-ISBN 978-3-11-164358-8

Advanced Mathematical Models and Integration Methods
Seminar Proceedings on Applications to Mechanics and Physics
Oleg Viktorovich Kaptsov, Evgeniy I. Kaptsov (eds.), 2025
ISBN: 978-3-11-222692-6; e-ISBN 978-3-11-222693-3

Alexander B. Borisov, Vladimir V. Kiselev

Quasi-One-Dimensional Magnetic Solitons

Riemann Problem, Solitons, and Inverse Scattering Method

DE GRUYTER

Mathematics Subject Classification 2020
37K15, 35Q15, 35Q51, 37K40

Authors

An Associate Member Russian Academy of Sciences
Dr. Alexander B. Borisov
M. N. Mikheev Institute of Metal Physics
Ural Branch
Sofia Kovalevskaya str. 18
620108 Ekaterinburg
Russia
bor1947@gmail.com

Dr. Sci. (Phys.-Math) Vladimir V. Kiselev
Russian Academy of Sciences
M. N. Mikheev Institute of Metal Physics
Ural Branch
Sofia Kovalevskaya str. 18
620108 Ekaterinburg
Russia
kiseliev@imp.uran.ru

ISBN 978-3-11-223596-6
ISBN 978-3-11-223597-3 (PDF)
ISBN 978-3-11-223598-0 (E-PUB)
DOI https://doi.org/10.1515/9783112235973

Library of Congress Control Number: 2026938635

Bibliographic information published by the Deutsche Nationalbibliothek
The Deutsche Nationalbibliothek lists this publication in the Deutsche Nationalbibliografie; detailed bibliographic data are available on the Internet at https://dnb.dnb.de.

De Gruyter and Walter de Gruyter GmbH are part of De Gruyter Brill.
www.degruyterbrill.com

Questions about General Product Safety Regulation: productsafety@degruyterbrill.com

Cover illustration: ALIOUI Mohammed Elamine / iStock / Getty Images Plus
Printing: CPI books GmbH, Leck

MIX
Papier | Fördert gute Waldnutzung
FSC® C083411

With gratitude for our parents, teachers, and mentors

Contents

Preface

> So far as I remember, I have never seen an Author's Preface which had any purpose but one – to furnish reasons for the publication of the book.
>
> Mark Twain

To date, the monograph "Non-linear Magnetization Waves. Dynamic and topological solitons" by A. M. Kosevich, B. A. Ivanov, and A. S. Kovalev, 1983, is a sole research in the field of nonlinear dynamics of magnetically ordered media, which turned into a classic. Once published, the theory of solitons experienced a period of unusually rapid growth. Techniques developed as a wide variety of mathematical tools successfully solved an incredible number of problems. Studying so much materials demands a lot of effort. Meanwhile, the circle of people who use the methods of the theory of solitons is expanding. These circumstances prompted us to attempt a concise and understandable overview of the key concepts and outcomes of the theory of quasi-one-dimensional magnetic solitons, which serve as a "skeleton" for solving numerous specific problems.

Chapter 1, the longest chapter, bears an auxiliary nature. It briefly outlines the foundations of the phenomenological theory of magnetically ordered media, including a description of relaxation processes in magnets that takes the dynamical symmetry of the problem into account. Apart from the conventional approach that treats the dynamics of a magnetic crystal as a nonlinear dynamics of magnetic sublattices inserted into each other, the method of phenomenological Lagrangians of spin waves, based on the concept of spontaneous symmetry breaking of exchange interactions, is presented for the first time in the monographic literature. Both approaches examine the possibility of accounting for relativistic magneto-static and magneto-elastic interactions. When reading the book for the first time, a reader who is already familiar with the macroscopic theory of magnetism may skip this chapter and come back to it when needed.

Chapter 2 is the primary one. It deals with a Heisenberg ferromagnet as an example to explain methods for integrating quasi-one-dimensional models of magnetism. A simple scheme for constructing a $U - V$ pair for the nonlinear Landau–Lifshitz equation is presented. The $U - V$ pair provides a way of completely integrating nonlinear equations by the inverse scattering method or by a more universal method of "dressing" particular solutions using the Riemann problem. It is shown how to build the spectrum of elementary excitations of a strongly nonlinear integrable system. This chapter is concerned with a technique for constructing the $U - V$ pairs with a moving spectral parameter that allows us to compose new integrable equations with variable coefficients by employing established integrable nonlinear partial differential equations with constant coefficients. A theoretical description of axially symmetric waves of magnetization in an isotropic ferromagnet illustrates the effectiveness of the aforementioned method.

Chapters 3 and 4 describe a procedure for integrating the dynamic equations for a ferromagnet with magnetic anisotropy of the "easy-axis" and "easy-plane" types. Furthermore, the third chapter and its appendix explain a technique for working with el-

https://doi.org/10.1515/9783112235973-204

liptic functions as extremely useful for further expansions of the "dressing" method. All solitons with homogeneous asymptotics at spatial infinity are obtained and studied in detail. The interaction of large-amplitude nonlinear precession waves with magnetic solitons is described analytically. In an easy-axis ferromagnet, the velocity of motion of solitons, in particular of domain walls and magnetization reversal nuclei, can be regulated by manipulating a pumping wave amplitude and the wave number. It is demonstrated how the internal structure of solitons in an "easy-plane" anisotropy ferromagnet changes under an external magnetic field directed along the anisotropy axis. The magnetic field removes the degeneracy of the ground state and provokes a wide variety of solitons. In the case of an "easy-plane" ferromagnet, its integrable model is closely tied to that describing nonlinear magnetization waves and solitons in helicoidal magnetic structures.

A ferromagnet with biaxial magnetic anisotropy serves as an example of a highly dispersive system. For linear modes, the dependence of the frequency on the wave number has the form $\omega = \sqrt{P_4(k)}$, where P_4 is a fourth-degree polynomial. Since the dispersion law is not analytic in terms of anisotropy constants, a perturbation theory in terms of small anisotropy is impossible to construct. This manifests itself in the complication of the theory: a $U-V$ pair of a biaxial ferromagnet, in contrast to the previously considered problems, contains a spectral-parameter dependence of Jacobi elliptic functions rather than of rational functions. Chapter 5 focuses on a modification of the inverse scattering and "dressing" methods to fully investigate the features of the nonlinear dynamics of a ferromagnet with biaxial magnetic anisotropy. All solitons with homogeneous asymptotics at spatial infinity are found and analyzed. The spectrum of elementary excitations of strongly nonlinear states of a biaxial ferromagnet is deduced.

Chapters 6 and 7 dwell on multisublattice magnetic solitons in an isotropic ferrimagnet with two magnetic sublattices and in chiral models describing the nonlinear dynamics of noncollinear ferri- and antiferromagnets. It is worth noting that an integrable model of a ferrimagnet turns out to be universal. In particular, it offers a wide class of two-dimensional distributions of magnetization in isotropic ferro- and antiferromagnets. It is shown that elementary excitations of a ferrimagnet are nonlinear spin waves (two branches) and solitons, which make an additive contribution to the energy and momentum of any localized distribution of magnetization. For noncollinear ferri- and antiferromagnets, soliton-type solutions are found to describe a system of spatially localized magnetic excitations against the background of two nonlocalized magnetization waves running toward each other. It is established that magnetization waves produce the amplitude and phase modulation of localized excitations. At sufficiently large wave amplitudes, they destroy localized excitations.

Chapter 8 treats the reductive perturbation theory methods to correctly approximate the weakly nonlinear dynamics of a rather complicated system within simpler nonlinear models for a thorough exploration by means of soliton theory. Examples of exchange-magneto-static solitons in finite systems (ferro- and antiferromagnetic plates) and solitons associated with resonant energy exchange between different branches of the spectrum of magneto-elastic modes display the effectiveness of the nonlinear perturbation theory.

In Chapter 9 the nonlinear dynamics of solitons and spin waves in a helicoidal structure of inversion-center-free magnets is studied within the sine-Gordon model. The analysis of nonlinear excitations against the background of a spiral structure for arbitrary initial distributions of magnetization in the structure and preset boundary conditions is reduced to solving linear integral equations on the Riemann surface generated by the structure. Spectral expansions of the integrals of motion are obtained for the separated contributions of solitons and spin waves. Mutual soliton collisions are found to accompany not only a change in the coordinates of their centers and phases but also shifts in the structure itself. The chapter delineates the potential for solitons to be experimentally detected in a spiral structure.

The authors endlessly appreciate the participation of the scientific editor, Prof. A. P. Tankeev, in carefully reading the manuscript and making a number of insightful remarks. A. P. Tankeev and A. A. Raskovalov took part in the study of some of the issues outlined in the book. D. V. Dolgikh and A. A. Raskovalov afford relief in comprising the volume. The authors are very grateful to to prof. A. V. Mikhailov for his attention and constructive support of our work, to Yu. G. Gorelykh for the careful and creative translation of the manuscript into English and to the Administration of the M. N. Mikheev Institute of Metal Physics of the Ural Branch of the Russian Academy of Sciences for creating favorable conditions for work.

1 Phenomenological theory of magnetism

1.1 Phenomenological description of the nonlinear dynamics of magnets via magnetic sublattices

"In medius res" (From Latin: into the middle of things).

By Horace

1.1.1 Basic equations

The basic books on magnetism [67, 1, 68, 44, 16, 32, 80, 6] use the symmetrical CGS system of units, which defines the base units for length, mass, and time as centimeter (cm), gram (g), and second (s), respectively. This system also includes the permittivity ε_0 and magnetic μ_0 constants of a vacuum as being equal to unity. Therefore in a vacuum the magnetic field strength **H** and magnetic induction are the same, $\mathbf{B} = \mathbf{H}$, whereas in a magnetic medium, they differ by a term with the density of the magnetic moment **M**:

$$\mathbf{B} = \mathbf{H} + 4\pi\mathbf{M}.$$

In the SI system the magnetic field strength and magnetic induction differ from each other not only in a medium but also in a vacuum: $\mathbf{B} = \mu_0\mathbf{H}$.

The presence of two magnetic fields **H** and **B** in the SI system, even for empty space, and additional constants ε_0 and μ_0 complicates the presentation of magnetism theory. Furthermore, the dimensions of magnetic samples are typically of the order of a centimeter, and the width of domain walls, solitons, and domains is of the order of 10^{-6}–10^{-2} cm. For magnetism problems, the interpretation of experimental data in terms of meters is not appropriate. The SI system is mostly employed in nonlinear optics problems involving the propagation of information in the form of solitons via fibers tens and thousands of kilometers long.

The basic equations of the monograph on magnetic solitons are written in the symmetrical CGS system. The transition to the SI system encounters no essential difficulties because it involves only introducing additional factors and eliminating unnecessary ones. To clarify the statement, we recall the form of the laws of electromagnetism in the CGS and SI systems of units.

We begin with Maxwell's equations in terms of the true physical fields **B** and **E**. The latter are average values of microscopic fields over physically small times and macroscopic volumes of a medium. We call the fields **B** and **E** true because they determine the force **F** acting on a charge q moving in the medium with velocity **V**.

https://doi.org/10.1515/9783112235973-001

SI	CGS	
Lorentz force **F**:		
$\mathbf{F} = q\mathbf{E} + q[\mathbf{V} \times \mathbf{B}]$	$\mathbf{F} = q\mathbf{E} + \frac{q}{c}[\mathbf{V} \times \mathbf{B}]$	
Maxwell's laws:		
$\operatorname{rot} \mathbf{E} = -\frac{\partial \mathbf{B}}{\partial t}$	$\operatorname{rot} \mathbf{E} = -\frac{1}{c}\frac{\partial \mathbf{B}}{\partial t}$	(1.1.1a)
$\operatorname{rot} \mathbf{B} = \mu_0 \left\{\mathbf{j}_{ext} + \mathbf{j}_{pol} + \mathbf{j}_{mol} + \varepsilon_0 \frac{\partial \mathbf{E}}{\partial t}\right\}$	$\operatorname{rot} \mathbf{B} = \frac{4\pi}{c} \{\mathbf{j}_{ext} + \mathbf{j}_{pol} + \mathbf{j}_{mol}\} + \frac{1}{c}\frac{\partial \mathbf{E}}{\partial t}$	(1.1.1b)
$\operatorname{div} \mathbf{E} = \frac{1}{\varepsilon_0} \{\rho_{ext} + \rho_{pol}\}$	$\operatorname{div} \mathbf{E} = 4\pi \{\rho_{ext} + \rho_{pol}\}$	(1.1.1c)
$\operatorname{div} \mathbf{B} = 0$	$\operatorname{div} \mathbf{B} = 0$	(1.1.1d)

Here ε_0 and μ_0 are the electric and magnetic constants, respectively; $\varepsilon_0\mu_0 = 1/c^2$, where c is the speed of light in a vacuum.

The properties of a medium are characterized by the electric polarization vector **P** (dipole moment per unit volume of the medium) and the magnetization vector **M**. To simplify Maxwell's equations, two auxiliary vector fields **H** and **D** are defined; they are related to the vectors **M** and **P** and the physical fields **B** and **E** as follows:

SI	CGS
Electric displacement vector **D**:	
$\mathbf{D} = \varepsilon_0 \mathbf{E} + \mathbf{P}$	$\mathbf{D} = \mathbf{E} + 4\pi \mathbf{P}$
Magnetic field strength **H**:	
$\mathbf{H} = \frac{1}{\mu_0}\mathbf{B} - \mathbf{M}$	$\mathbf{H} = \mathbf{B} - 4\pi \mathbf{M}$

M = 0 in vacuum and nonmagnetic media. As a result, the fields **B** and **H** coincide in the CGS system but differ by the factor μ_0 in the SI system:

SI	CGS
$\mathbf{H} = \frac{1}{\mu_0}\mathbf{B}$	$\mathbf{H} = \mathbf{B}$

Therefore we can say that the intensity of the magnetic field **H** characterizes the magnetic field without considering the influence of a medium.

With the vectors **M** and **P**, the polarization charge density ρ_{pol}, the density of polarization currents $\mathbf{j}_{\text{pol}}$, and the density of the so-called molecular currents $\mathbf{j}_{\text{mol}}$ can be calculated in the both systems of units:

SI	CGS
$\rho_{\text{pol}} = -\operatorname{div}\mathbf{P}$	$\rho_{\text{pol}} = -\operatorname{div}\mathbf{P}$
$\mathbf{j}_{\text{pol}} = \dfrac{\partial \mathbf{P}}{\partial t}$	$\mathbf{j}_{\text{pol}} = \dfrac{\partial \mathbf{P}}{\partial t}$
$\mathbf{j}_{\text{mol}} = \operatorname{rot}\mathbf{M}$	$\mathbf{j}_{\text{mol}} = c\operatorname{rot}\mathbf{M}$

Then the system of Maxwell's equations for the true fields **B** and **E** (1.1.1) can be rewritten in another form:

SI	CGS
$\operatorname{rot}\mathbf{E} = -\dfrac{\partial \mathbf{B}}{\partial t}$	$\operatorname{rot}\mathbf{E} = -\dfrac{1}{c}\dfrac{\partial \mathbf{B}}{\partial t}$
$\operatorname{rot}\mathbf{H} = \mathbf{j}_{\text{ext}} + \dfrac{\partial \mathbf{D}}{\partial t}$	$\operatorname{rot}\mathbf{H} = \dfrac{4\pi}{c}\mathbf{j}_{\text{ext}} + \dfrac{1}{c}\dfrac{\partial \mathbf{D}}{\partial t}$
$\operatorname{div}\mathbf{D} = \rho_{\text{ext}}$	$\operatorname{div}\mathbf{D} = 4\pi\rho_{\text{ext}}$
$\operatorname{div}\mathbf{D} = 0$	$\operatorname{div}\mathbf{D} = 0$

As a consequence, these transformations of the Maxwell equations in the SI system eliminate the fundamental constants ε_0 and μ_0. Now the Maxwell equations (in both systems) include only the external charge density ρ_{ext} and the density of external currents $\mathbf{j}_{\text{ext}}$.

The conventional phenomenological theory of magnetic crystals treats complex magnetic structures as a set of magnetic sublattices embedded into each other [67, 1, 68, 44]. A continuum description characterizes each magnetic sublattice through the average density of the magnetic moment $M_{\nu i}(\mathbf{x}, t)$, where $\mathbf{x}$ and t are the spatial coordinates and time. The indices denoted in Latin letters take the values $i, j = 1, 2, 3$, and ν is the number of the sublattice.

The total energy of a finite sample includes bulk and surface energies. The latter specifies the boundary conditions for the field dynamics equations $M_{\nu i}(\mathbf{x}, t)$. An explicit form of the dynamic equations is determined only by the bulk part of the crystal magnet energy and is written as the expansion [67, 1]:

$$W = \int_{V_0} d^3\mathbf{x}\Big\{A_{ij}^{\nu_1\nu_2} M_{\nu_1 i} M_{\nu_2 j} \\ + B_{ij,s}^{\nu_1\nu_2}(M_{\nu_1 i}\partial_s M_{\nu_2 j} - M_{\nu_2 i}\partial_s M_{\nu_1 j}) \\ + C_{ij,sp}^{\nu_1\nu_2}\partial_s M_{\nu_1 i}\partial_p M_{\nu_2 j} + \cdots\Big\} + W^{(\mathrm{m})} - \int_{V_0} d^3\mathbf{x}(\mathbf{M}\cdot\mathbf{H}^{\mathrm{ext}}). \tag{1.1.3}$$

Here the summation is performed over all doubly appearing indices, and the integration is carried out over the region V_0 occupied by the magnetic material. In representation (1.1.3) the summand

$$-\int_{V_0} d^3\mathbf{x}(\mathbf{M}\cdot\mathbf{H}^{\mathrm{ext}})$$

corresponds to the energy of interaction between the magnet and an external magnetic field $\mathbf{H}^{\mathrm{ext}}$. Historically, the designation of $\mathbf{H}^{\mathrm{ext}}$ used herein was reserved for a magnetic field strength vector, although the true average macroscopic field in a medium (incorporated into the Lorentz force along with the electric field strength **E**) is certainly magnetic induction **B** [68]. For the CGS system, in which we work, this is nonessential because $\mathbf{H}^{\mathrm{ext}} = \mathbf{B}^{\mathrm{ext}}$ but should be taken into account when switching to the SI system, where $\mathbf{B}^{\mathrm{ext}} = \mu_0\mathbf{H}^{\mathrm{ext}}$.

The total local medium magnetization $\mathbf{M} = \sum_\nu \mathbf{M}_\nu$ generates a magnetic field $\mathbf{H}^{(\mathrm{m})}$ governed by the magnetostatics equations:

SI	CGS	
$\operatorname{rot}\mathbf{H}^{(\mathrm{m})} = 0,$ $\operatorname{div}(\mathbf{H}^{(\mathrm{m})} + \mathbf{M}) = 0$	$\operatorname{rot}\mathbf{H}^{(\mathrm{m})} = 0,$ $\operatorname{div}(\mathbf{H}^{(\mathrm{m})} + 4\pi\mathbf{M}) = 0$	(1.1.4)

with boundary conditions supporting the continuity of the tangential and normal components of the magnetic field strength and magnetic induction vectors, respectively, on the sample surface. In the CGS system, the solution to the magnetostatic problem (1.1.4) has the form

$$H_i^{(\mathrm{m})} = \mathrm{V.\,p.} \int_{V_0} M_k(\mathbf{x}') \frac{\partial}{\partial x_i' \partial x_k'} \frac{1}{|\mathbf{x}-\mathbf{x}'|} d^3\mathbf{x}'. \tag{1.1.5}$$

Letters “V. p.” indicate that the integral is evaluated in the sense of the Cauchy principal value, i. e., integration is to be carried out over the region excluding a small volume bounded by a sphere $|\mathbf{x}-\mathbf{x}'| = \rho$, and the limit as $\rho \to 0$ is taken at the end of the calculation. Relation (1.1.3) includes a nonlocal energy $W^{(\mathrm{m})}$ of interaction between the crystal magnetization $\mathbf{M}$ and the demagnetizing field $\mathbf{H}^{(\mathrm{m})}$:

$$W^{(\mathrm{m})} = -\frac{1}{2}\int\limits_{V_0} d^3\mathbf{x}(\mathbf{M}\cdot\mathbf{H}^{(\mathrm{m})}). \tag{1.1.6}$$

In the SI system the magneto-static energy (1.1.6) contains the additional multiplier μ_0 before the field $\mathbf{H}^{(\mathrm{m})}$ specified by eqs. (1.1.4)–SI.

The essence of the issues discussed stays the same, regardless of whether we use the CGS or SI system of units. We have opted for the symmetrical CGS system. It simplifies the basic formulas and clearly links our book to the foundational literature on magnetism.

For further analysis, it is advisable to introduce the magnetostatic potential Ψ as a solution to the following boundary value problem:

$$\begin{aligned}
&\Delta\Psi = 4\pi\,\mathrm{div}\,\mathbf{M} \quad (\mathbf{x}\in V_0);\\
&\Delta\Psi = 0 \quad (\mathbf{x}\in V_1);\\
&n_i[-\partial_i\Psi_- + 4\pi M_i]|_\sigma = -n_i\partial_i\Psi_+|_\sigma, \quad \Psi_-|_\sigma = \Psi_+|_\sigma,\\
&\Psi\to 0 \quad \text{as} \quad |x|\to\infty.
\end{aligned} \tag{1.1.7}$$

Here σ is the surface of a magnetic crystal. The indices "+" and "–" designate the fields outside and inside the crystal, n_i are the components of the unit outer normal to the surface σ, and V_1 is the region outside the magnetic crystal ($V_0 + V_1$ is the whole three-dimensional space with the coordinates x_i). The solution to problem (1.1.7) has the form [67, 1]

$$\Psi = -\int\limits_{V_0} d^3\mathbf{x}'\frac{\mathrm{div}\,\mathbf{M}(\mathbf{x}')}{|\mathbf{x}-\mathbf{x}'|} + \int\limits_{\sigma} d\sigma_k\frac{\mathbf{M}_k(\mathbf{x}')}{|\mathbf{x}-\mathbf{x}'|} = \int\limits_{V_0} d^3\mathbf{x}\mathbf{M}_k(\mathbf{x}')\frac{\partial}{\partial x_k'}\frac{1}{|\mathbf{x}-\mathbf{x}'|}.$$

Here $d\sigma_k$ is an element of the sample surface. All the integrals are assumed to be convergent and are calculated without constraints on $|\mathbf{x}-\mathbf{x}'|$. The potential Ψ governs a field $\tilde{H}_i^{(\mathrm{m})} = -\partial_i\Psi$, which slightly differs from $H_i^{(\mathrm{m})}$. Under the condition of continuity of the function $\mathbf{M}(\mathbf{x})$ at the point $\mathbf{x}=\mathbf{x}'$, we find

$$\begin{aligned}
\tilde{H}_i^{(\mathrm{m})} &= \int\limits_{V_0} d^3\mathbf{x}' M_k(\mathbf{x}')\frac{\partial^2}{\partial x_i'\partial x_k'}\frac{1}{|\mathbf{x}-\mathbf{x}'|}\\
&= -\frac{4\pi}{3}M_i(\mathbf{x}) + \lim_{\rho\to 0}\int\limits_{|\mathbf{x}-\mathbf{x}'|>\rho} M_k(\mathbf{x}')\frac{\partial^2}{\partial x_i'\partial x_k'}\frac{1}{|\mathbf{x}-\mathbf{x}'|}d^3\mathbf{x}'\\
&= -\frac{4\pi}{3}M_i(\mathbf{x}) + H_i^{(\mathrm{m})}.
\end{aligned}$$

In terms of $\tilde{\mathbf{H}}^{(\mathrm{m})}$, the magnetostatic energy (1.1.6) can be written as

$$W^{(\mathrm{m})} = -\frac{1}{2}\int_{V_0} \mathbf{M}\cdot\left(\tilde{\mathbf{H}}^{(\mathrm{m})} + \frac{4\pi}{3}\mathbf{M}\right)\mathrm{d}^3\mathbf{x}'. \tag{1.1.8}$$

The second summand in the right-hand side of eq. (1.1.8) can be conveniently combined with the first term in expansion (1.1.3) for the energy of a crystal magnet. Then the magneto-static energy of the sample is

$$\tilde{W}^{(\mathrm{m})} = -\frac{1}{2}\int_{V_0} (\mathbf{M}\cdot\tilde{\mathbf{H}}^{(\mathrm{m})})\mathrm{d}^3\mathbf{x} = \frac{1}{2}\int_{V_0} M_i\partial_i\Psi\mathrm{d}^3\mathbf{x}, \tag{1.1.9}$$

where there are no constraints on $|\mathbf{x}-\mathbf{x}'|$ when calculating Ψ and $\tilde{\mathbf{H}}^{(\mathrm{m})}$.

Let us give another useful representation for the energy $\tilde{W}^{(\mathrm{m})}$. By integrating by parts and taking into account relations (1.1.7) it is easy to derive the identities

$$\begin{aligned}
\int_{V_1} (\partial_i\Psi)^2\mathrm{d}^3\mathbf{x} &= -(\Psi n_i\partial_i\Psi)_+|_\sigma = \{\Psi n_i(-\partial_i\Psi + 4\pi M_i)\}_-|_\sigma,\\
\int_{V_0} M_i\partial_i\Psi\mathrm{d}^3\mathbf{x} &= \frac{1}{4\pi}\int_{V_0} \partial_i\Psi\{[-\partial_i\Psi + 4\pi M_i] + \partial_i\Psi\}\mathrm{d}^3\mathbf{x}\\
&= \frac{1}{4\pi}\{\Psi n_i(-\partial_i\Psi + 4\pi M_i)\}_-|_\sigma + \frac{1}{4\pi}\int_{V_0}(\partial_i\Psi)^2\mathrm{d}^3\mathbf{x}\\
&= \frac{1}{4\pi}\int_{V_0+V_1} (\partial_i\Psi)^2\mathrm{d}^3\mathbf{x}.
\end{aligned} \tag{1.1.10}$$

The first of eqs. (1.1.10) accounts for the equality of the outer normal to the region V_1 in the surface σ to $-\mathbf{n}$.

Using identities (1.1.10), energy (1.1.9) can be represented in the form

$$\tilde{W}^{(\mathrm{m})} = \int_{V_0} M_i\partial_i\Psi\mathrm{d}^3\mathbf{x} - \frac{1}{8\pi}\int_{V_0+V_1} (\partial_i\Psi)^2\mathrm{d}^3\mathbf{x}. \tag{1.1.11}$$

The last term in the right-hand side of eq. (1.1.11) contains the integration over the entire space $V_0 + V_1$. Expression (1.1.11) for the energy $\tilde{W}^{(\mathrm{m})}$ is preferable to formulate variational principles since it allows us to vary the fields Ψ and $\mathbf{M}_\nu$ independently. This brings not only to the equations needed to calculate these fields but also to the necessary boundary conditions for them.

The expression for the energy (1.1.3) must be invariant under transformations of a magnetic symmetry group of the crystal. This condition dictates a specific form of the phenomenological constants $A, B, \ldots$. Apart from spatial symmetry elements (rotations, reflections, and translations), the symmetry description of magnets requires to introduce another symmetry element, the time reversal operation. The latter is related to the invariance of the equations of motion for microparticles, which a magnetic crystal consists of, regarding a change in the sign of time. With that, the directions of the $\mathbf{H}^{\text{ext}}$ vector and magnetic moments of microparticles change simultaneously on reversal ones. Upon reversing time, we have

$$\mathbf{H}^{\text{ext}} \to -\mathbf{H}^{\text{ext}}, \quad \mathbf{M}_\nu \to -\mathbf{M}_\nu.$$

Therefore expansion (1.1.3) lacks terms containing products of an odd number of the multipliers $\mathbf{H}$ and $\mathbf{M}_\nu$.

In forming a magnetic structure, exchange interactions play the main role. The terms with coefficients $A, B, \ldots$, which are invariant with respect to the simultaneous rotation of all vectors by the same angle in the spin space, correspond to the exchange interactions in expansion (1.1.3). This rotation occurs in the spin space and therefore does not affect the Cartesian coordinates x_i of the crystal.

The rest of the summands appearing in the square bracket of equality (1.1.3) describe interactions that are less by a factor of v^2/c^2 in an order of magnitude than exchange ones (v has the order of magnitude of the velocities of atomic electrons, and c is the speed of light). In condensed matter, this magnitude is $v^2/c^2 \sim 10^{-4}$, and therefore such interactions are called relativistic. They lead to the dependence of the energy of the magnet on the orientation of the $\mathbf{M}_\nu$ magnetic vectors relative to the crystallographic axes.

The basic assumption of the phenomenological theory is that the modulus of the magnetization vector of each sublattice, even for excited states of the crystal, is thought to be constant:

$$(\mathbf{M}_\nu)^2 = M_{0\nu}^2 = \text{const.} \tag{1.1.12}$$

This geometric condition is quite well satisfied at low temperatures and by itself makes the problem essentially nonlinear. The evolution of the vector $\mathbf{M}_\nu(\mathbf{x}, t)$ is described as precession by the Landau–Lifshitz equations [44]

$$\partial_t \mathbf{M}_\nu = -\gamma_\nu [\mathbf{M}_\nu \times \mathbf{H}_\nu^{\text{eff}}], \quad \mathbf{M}_\nu^2 = M_{0\nu}^2, \tag{1.1.13}$$

where

$$\mathbf{H}_\nu^{\text{eff}} = -\frac{\delta W}{\delta \mathbf{M}_\nu}$$

is the effective field, $\gamma_\nu = \text{const}$ is the gyromagnetic ratio for the νth sublattice (in formulas (1.1.12) and (1.1.13), there is no summation over the index ν). For multisublattice magnets, direct integration of the equations meets formidable difficulties due to a huge number of the order-of-$\mathbf{M}_\nu(\mathbf{x}, t)$ vector parameters and a significant geometric and physical nonlinearity of the problem.

1.1.2 Connection with microtheory

The magnetic properties of matter exhibit themselves under the laws of relativistic quantum theory, and therefore they have no analogue in classical physics. Let us briefly explain the microscopic roots of the aforementioned macroscopic theory.

Due to the quantum mechanical concept of the indistinguishability of identical particles, the wave function of a crystal must be antisymmetric when the coordinates and projections of the electron spins are rearranged. This leads to a certain consistency and electron motion correlations that affect the crystal energy. Even in the scenario where an initial microscopic Hamiltonian of an aggregate of atoms is underlain by only Coulomb interactions and excludes any spin variables, taking into account the identity of electrons results in that the crystal energy becomes reliant on the spin moments of individual atoms. The final result can be achieved without resorting to antisymmetrization of the crystal wave function, but by adding to the initial microscopic Hamiltonian the effective (exchange) energy that depends on the spin mechanical moments $\hat{\mathbf{S}}_\nu^\alpha$ of individual crystal atoms [67, 1]:

$$\hat{H}_{\text{ex}} = -\frac{1}{2}\sum_{\nu_1,\nu_2,\alpha,\beta} J^{\nu_1\nu_2}(\mathbf{x}_{\nu_1}^\alpha - \mathbf{x}_{\nu_2}^\beta)(\hat{\mathbf{S}}_{\nu_1}^\alpha \cdot \hat{\mathbf{S}}_{\nu_2}^\beta). \tag{1.1.14}$$

Here the summation runs over the numbers of the sublattices ν_1, ν_2 and the numbers of atoms in them (α and β), and $J^{\nu_1\nu_2}(\mathbf{x}_{\nu_1}^\alpha - \mathbf{x}_{\nu_2}^\beta)$ is a function of the radius vector connecting the ν_1th and ν_2th sites of the magnetic sublattices in the crystal. Since the function $J^{\nu_1\nu_2}$ is defined by the degree of overlap of the wave functions of atoms, it decays exponentially with increasing distance $|\mathbf{x}_{\nu_1}^\alpha - \mathbf{x}_{\nu_2}^\beta|$ between atoms and is almost nonzero, but only when the atoms are neighbors. However, in an order of magnitude, we have

$$J^{\nu_1\nu_2} \sim \xi e^2/a,$$

where a is the lattice constant, $\xi \sim 0.1$ is a numerical factor reflecting the degree of overlap of the wave functions of atoms, and e is the charge of an electron [1].

A distinctive feature of the exchange energy is its invariance with respect to rotations of all the spin moments $\hat{\mathbf{S}}_\nu^\alpha$ through the same angle. The isotropy of the exchange energy with respect to rotations in spin space is Coulomb in nature. The name "exchange energy" comes from the effective interaction (1.1.14) as the result of the electron exchange between neighboring atoms of the crystal.

The electric fields of crystals cause different types of symmetry and different overlaps of the wave functions of the crystal atoms. Finally, a large number of exchange magnetic structures are formed, which can be described phenomenologically in terms of magnetic sublattices embedded in each other. Let us discuss this in more detail.

The operators of magnetic moments $\hat{\mathbf{M}}_\nu^\alpha$ of atoms are related to the operators of spin moments $\hat{\mathbf{S}}_\nu^\beta$ by the relation

$$\hat{\mathbf{M}}_\nu^a = -\hbar\gamma_\nu \hat{\mathbf{S}}_\nu^a. \tag{1.1.15}$$

Here there is no summation over ν; $\hbar\gamma_n = g_\nu \mu_B$ (g_ν is the g-factor, $\mu_B = |e|\hbar/(2mc)$ is the Bohr magneton, γ_ν is the gyromagnetic ratio for the νth sublattice, e is the charge of an electron, m is the mass of an electron, and $\hbar$ is Planck's constant). It should be underscored that in the general case, g_ν can be a tensor.

Let us define the density operators of the magnetic moments of the sublattices by the relations

$$\hat{\mathbf{M}}_\nu(\mathbf{x}, t) = \sum_a \hat{\mathbf{M}}_\nu^a \, \delta(\mathbf{x} - \mathbf{x}_\nu^a). \tag{1.1.16}$$

Then the exchange energy (1.1.14) can be written in the form

$$\hat{H}_{\text{ex}} = -\frac{1}{2} \iint E^{\nu_1\nu_2}(\mathbf{x} - \mathbf{x}')(\hat{\mathbf{M}}_{\nu_1}(\mathbf{x}, t) \cdot \hat{\mathbf{M}}_{\nu_2}(\mathbf{x}', t))\mathrm{d}^3\mathbf{x}\,\mathrm{d}^3\mathbf{x}', \tag{1.1.17}$$

where summation over twice repeated indices is implied. The coordinate-dependent functions $E^{\nu_1\nu_2}$ in (1.1.17) can be easily expressed through $J^{\nu_1\nu_2}$ using relation (1.1.15).

In magnets, due to the strong exchange interaction, a local quasiequilibrium distribution of magnetic moments is quickly established. Therefore the macroscopic density of the magnetic moments of the sublattices can be entered:

$$\mathbf{M}_\nu(\mathbf{x}, t) = \frac{1}{\Delta V} \int_{\Delta V} \mathrm{Sp}(\hat{\mathbf{M}}_\nu(\mathbf{x}', t)\hat{\rho}(\mathbf{x}', t))\mathrm{d}^3\mathbf{x}', \tag{1.1.18}$$

where $\hat{\rho}(\mathbf{x}', t)$ is the locally equilibrium density matrix of a magnet, and the integration is carried out within a physically small volume ΔV centered at a point with radius vector $\mathbf{x}$. The volume ΔV must satisfy the condition

$$a^3 \ll \Delta V \ll \lambda^3, \tag{1.1.19}$$

where λ is the characteristic size of magnetic inhomogeneities (which can be the length of a spin wave, the size of a soliton, etc.), and a is the interatomic distance.

When averaging the microscopic exchange energy (1.1.17), the product of the operators $(\hat{\mathbf{M}}_{\nu_1}(\mathbf{x}, t) \cdot \hat{\mathbf{M}}_{\nu_2}(\mathbf{x}', t))$ is replaced by the product of the average local densities (1.1.18), which are c-numbers. As a result, the Hamiltonian (1.1.14) meets the macroscopic exchange energy

$$W_{\text{ex}} = -\frac{1}{2} \iint E^{\nu_1\nu_2}(\mathbf{x} - \mathbf{x}')(\mathbf{M}_{\nu_1}(\mathbf{x}, t) \cdot \mathbf{M}_{\nu_2}(\mathbf{x}', t))\mathrm{d}^3\mathbf{x}\,\mathrm{d}^3\mathbf{x}'. \tag{1.1.20}$$

Recall that the function $E^{\nu_1\nu_2}(\mathbf{x})$ declines rapidly with increasing $|\mathbf{x}|$. Therefore in formula (1.1.20), $\mathbf{M}_{\nu_2}(\mathbf{x}', t)$ can be expanded in a series of powers $\mathbf{x} - \mathbf{x}'$:

$$\mathbf{M}_{\nu_2}(\mathbf{x}',t) \approx \mathbf{M}_{\nu_2}(\mathbf{x},t) + \partial_s \mathbf{M}_{\nu_2}(\mathbf{x},t)(x'-x)_s + \frac{1}{2}\partial_s\partial_p \mathbf{M}_{\nu_2}(\mathbf{x},t)(x'-x)_s(x'-x)_p. \tag{1.1.21}$$

Plugging (1.1.21) into (1.1.20) and integrating by parts, we arrive at the following expansion for the exchange energy of the crystal:

$$W_{\text{ex}} = -\frac{1}{2}\int [\alpha^{\nu_1\nu_2}(\mathbf{M}_{\nu_1}\cdot\mathbf{M}_{\nu_2}) + \beta_s^{\nu_1\nu_2}(\mathbf{M}_{\nu_1}\cdot\partial_s\mathbf{M}_{\nu_2} - \mathbf{M}_{\nu_2}\cdot\partial_s\mathbf{M}_{\nu_1}) - \gamma_{sp}^{\nu_1\nu_2}(\partial_s\mathbf{M}_{\nu_1}\cdot\partial_p\mathbf{M}_{\nu_2}) + \cdots]\mathrm{d}^3\mathbf{x}, \tag{1.1.22}$$

which agrees with (1.1.3). Note that passing from (1.1.21) to (1.1.22), we have omitted the surface integrals. For an infinite medium, such an approximation is justified. For finite samples, when deriving equations for calculating the fields $\mathbf{M}_\nu(\mathbf{x},t)$ and $\Psi(\mathbf{x},t)$ using the Hamilton variational principle, all the surface integrals with variations $\delta\mathbf{M}_\nu$, $\delta\Psi$, etc., arising from both the bulk and surface parts of the total energy of the magnet, must vanish. This requirement gives the boundary conditions for a particular problem [33]. The "delicate" procedure of integration by parts for finite samples should be understood through a concrete example.

The phenomenological constants $\alpha^{\nu_1\nu_2}$, $\beta_s^{\nu_1\nu_2}$, and $\gamma_{sp}^{\nu_1\nu_2}$ are the moments of the function $E^{\nu_1\nu_2}(\mathbf{x})$. In particular, we have

$$\gamma_{sp}^{\nu_1\nu_2} = \frac{1}{2}\int E^{\nu_1\nu_2}(\mathbf{x})x_s x_p \mathrm{d}^3\mathbf{x} \backsim \frac{J^{\nu_1\nu_2}a^5}{g_{\nu_1}g_{\nu_2}\mu_B^2},$$

where $J^{\nu_1\nu_2}$ is the quantity of the exchange integral between neighboring atoms of the ν_1th and ν_2th sublattices.

From a microscopic point of view, magnetoanisotropic (relativistic) interactions are related to spin-orbit and direct dipole–dipole couplings of the magnetic atoms of a crystal.

Let us first discuss the contribution of spin-orbit interactions to the magnetic anisotropy energy. The electric field of a crystal magnet orients the orbital motion of electrons in the atoms relative to the crystallographic axes in a certain way. The spin mechanical moments of the electrons in each atom interact with the orbital motion of the electrons of a given atom and its neighboring atoms. Whatever the nature of this coupling (**L** – **S**–coupling or **J** – **J**–coupling), according to quantum mechanical perturbation theory, it leads to an effective interaction of the magnetic moments of the crystal atoms, which in the quadratic approximation has the following form:

$$\hat{H}_{\text{so}} = \frac{1}{2}\iint k_{ij}^{\nu_1\nu_2}(\mathbf{x}-\mathbf{x}')\hat{M}_{\nu_1 i}(\mathbf{x},t)\hat{M}_{\nu_2 j}(\mathbf{x}',t)\mathrm{d}^3\mathbf{x}\,\mathrm{d}^3\mathbf{x}'. \tag{1.1.23}$$

This expression differs from the exchange Hamiltonian (1.1.17) only in that here the energy depends on the orientation of the magnetic moments $\hat{\mathbf{M}}_\nu$ with respect to the crystallographic axes. Since the function $k_{ij}^{\nu_1\nu_2}(\mathbf{x})$ decays rapidly with increasing $|\mathbf{x}|$, the energy (1.1.23) can be treated as the exchange energy in the same way. Thus the macroscopic energy produced by the spin-orbit interaction coincides with the magnetoanisotropic local part of formula (1.1.3).

It is intriguing that, generally speaking, the popular belief that magnetoanisotropic interactions are much weaker than exchange (isotropic) interactions is not always true. Recent advances have led to the discovery of an increasing number of quasi-one-dimensional and quasi-two-dimensional (layered) magnetic materials where spin-orbit interactions predominate over exchange interactions (see, for example, [60, 59]). The magnetic properties of such substances are special and are controlled, in particular, by exotic magnetic solitons, compactons and picons [41].

It should also be emphasized that only for inversion-center-free crystals, expansion (1.1.3) can contain the linear-in-gradient terms

$$B_{ij,s}^{\nu_1\nu_2}[M_{\nu_1 i}\,\partial_s M_{\nu_2 j} - M_{\nu_2 j}\,\partial_s M_{\nu_1 i}] \tag{1.1.24}$$

of an exchange and exchange-relativistic nature. The interactions of (1.1.24) are notable because they provoke spiral-shaped magnetic structures to arise, whose periods are incommensurable with the crystal–chemical ones and usually surpass the latter by a factor of many times.

Direct dipole–dipole interaction between the magnetic moments $\mathbf{M}_\nu^\alpha$ of crystal atoms corresponds to the microscopic Hamiltonian

$$\hat{H}_{\text{dip}} = \frac{1}{2}\sum_{\alpha,\nu,\alpha',\nu'}\left[\frac{(\hat{\mathbf{M}}_\nu^\alpha\cdot\hat{\mathbf{M}}_{\nu'}^{\alpha'})}{|\mathbf{r}_{\nu\nu'}^{\alpha\alpha'}|^3} - \frac{3(\hat{\mathbf{M}}_\nu^\alpha\cdot\mathbf{r}_{\nu\nu'}^{\alpha\alpha'})(\mathbf{r}_{\nu\nu'}^{\alpha\alpha'}\cdot\hat{\mathbf{M}}_\nu^\alpha)}{|\mathbf{r}_{\nu\nu'}^{\alpha\alpha'}|^5}\right], \tag{1.1.25}$$

where

$$\mathbf{r}_{\nu\nu'}^{\alpha\alpha'} = \mathbf{x}_\nu^\alpha - \mathbf{x}_{\nu'}^{\alpha'};$$

the summation is over all pairs of indices (α,ν), (α',ν') with a single constraint, $(\alpha,\nu)\neq(\alpha',\nu')$.

Using the expression for the operator of the microscopic magnetization of a medium

$$\hat{\mathbf{M}}(\mathbf{x},t) = \sum_{\alpha,\nu}\hat{\mathbf{M}}_\nu^\alpha\,\delta(\mathbf{x}-\mathbf{x}_\nu^\alpha) \equiv \sum_\nu \hat{\mathbf{M}}_\nu(\mathbf{x},t),$$

the discrete sum (1.1.25) can be rewritten as the integral

$$\hat{H}_{\text{dip}} = \frac{1}{2}\iint_{V_0}\hat{M}_i(\mathbf{x})\left(\frac{\partial^2}{\partial x_i\partial x_j'}\frac{1}{|\mathbf{x}-\mathbf{x}'|}\right)\hat{M}_j(\mathbf{x}')\mathrm{d}^3\mathbf{x}\,\mathrm{d}^3\mathbf{x}', \tag{1.1.26}$$

where V_0 is the volume of a crystal. However, in this case the formal transition from representation (1.1.26) to the continuum approximation along the scheme already described faces some difficulties. This is due to the divergence of the integrand in the region of small $|\mathbf{x} - \mathbf{x}'|$. At small distances $|\mathbf{x} - \mathbf{x}'|$, the discrete nature of the distribution of magnetic atoms in space should be borne in mind. For a proper transition to the continuum approximation, we break $\hat{H}_{\mathrm{dip}}$ into two summands:

$$\hat{H}_{\mathrm{dip}} = \hat{H}^{\mathrm{I}}_{\mathrm{dip}} + \hat{H}^{\mathrm{II}}_{\mathrm{dip}}.$$

Here $\hat{H}^{\mathrm{I}}_{\mathrm{dip}}$ is the value of the micro-Hamiltonian in the region $|\mathbf{x} - \mathbf{x}'| \geq \rho$:

$$\hat{H}^{\mathrm{I}}_{\mathrm{dip}} = \iint_{|\mathbf{x}-\mathbf{x}'|\geq\rho} \hat{M}_i(\mathbf{x})\left(\frac{\partial^2}{\partial x_i\,\partial x'_j}\frac{1}{|\mathbf{x}-\mathbf{x}'|}\right)\hat{M}_j(\mathbf{x}')\mathrm{d}^3\mathbf{x}\,\mathrm{d}^3\mathbf{x}',$$

where ρ is a sufficiently small macroscopic magnitude that satisfies the condition

$$a \ll \rho \ll \lambda.$$

For the region $|\mathbf{x} - \mathbf{x}'| \leq \rho$, we should return to the original form of writing the micro-Hamiltonian as the sum:

$$\hat{H}^{\mathrm{II}}_{\mathrm{dip}} = -\frac{1}{2}\sum_{i,j=1}^{3}\sum_{\alpha,\nu,\alpha',\nu'}\hat{M}^{\alpha}_{\nu i}\left[\frac{\partial^2}{\partial(\mathbf{r}^{\alpha\alpha'}_{\nu\nu'})_i\,\partial(\mathbf{r}^{\alpha\alpha'}_{\nu\nu'})_j}\frac{1}{|\mathbf{r}^{\alpha\alpha'}_{\nu\nu'}|}\right]\hat{M}^{\alpha'}_{\nu' j}. \tag{1.1.27}$$

Here the summation is carried out over all pairs of the indices (α, ν), (α', ν') satisfying the constraint

$$0 < |\mathbf{x}^{\alpha}_{\nu} - \mathbf{x}^{\alpha'}_{\nu'}| \leq \rho.$$

Let us dwell on the transition from the Hamiltonians $\hat{H}^{\mathrm{I}}_{\mathrm{dip}}$ and $\hat{H}^{\mathrm{II}}_{\mathrm{dip}}$ to the phenomenological magnet energy (1.1.3).

The expression for $\hat{H}^{\mathrm{I}}_{\mathrm{dip}}$ contains no singularity and yields a nonlocal macroscopic energy of a crystal magnet after averaging along the previous scheme. This energy coincides with the definition of $W^{(\mathrm{m})}$ in formulas (1.1.5) and (1.1.6).

Let us demonstrate that under physically justified approximations, the microscopic Hamiltonian $\hat{H}^{\mathrm{II}}_{\mathrm{dip}}$ in the phenomenological theory corresponds to the energy of the local magnetic anisotropy of the crystal. For this purpose, we regroup the terms in (1.1.27). Having taken an arbitrary atom of the νth sublattice with radius vector $\mathbf{x}^{\alpha}_{\nu}$, we first sum up over all atoms from the ν'th sublattice in its nearest neighborhood within a sphere of radius ρ. In other words, we sum up over all α' satisfying the condition $0 < |\mathbf{x}^{\alpha}_{\nu} - \mathbf{x}^{\alpha'}_{\nu'}| < \rho$ (the indices ν and ν' are still fixed). Suppose that such a sphere involves atoms of different sublattices. Then we calculate the sum over α, i. e., we take into account the contributions of all atoms of the νth sublattice (now there are no constraints on the values of α). In the long run, we sum up the contributions of the different ν and ν' sublattices. As a result, we come to the following representation for the Hamiltonian:

$$\hat{H}^{\mathrm{II}}_{\mathrm{dip}} = -\frac{1}{2}\sum_{\nu,\nu'}\sum_{\alpha}\sum_{\substack{\alpha' \\ 0<|\mathbf{x}^{\alpha}_{\nu}-\mathbf{x}^{\alpha'}_{\nu'}|<\rho}}\sum_{i,j=1}^{3}\hat{M}^{\alpha}_{\nu i}\left[\frac{\partial^2}{\partial(\mathbf{r}^{\alpha\alpha'}_{\nu\nu'})_i\,\partial(\mathbf{r}^{\alpha\alpha'}_{\nu\nu'})_j}\frac{1}{|\mathbf{r}^{\alpha\alpha'}_{\nu\nu'}|}\right]\hat{M}^{\alpha'}_{\nu' j}.$$

We introduce the energy operator $\hat{\tilde{H}}^{\mathrm{II}}_{\mathrm{dip}}$ for atoms enclosed within a physically small volume ΔV (1.1.19) centered at a point with radius vector $\mathbf{x} \approx \mathbf{x}^{\alpha}_{\nu}$:

$$\hat{\tilde{H}}^{\mathrm{II}}_{\mathrm{dip}} = -\frac{1}{2}\sum_{i,j=1}^{3}\sum_{\nu,\nu'}\sum_{\langle\alpha\rangle}\sum_{\substack{\alpha' \\ 0<|\mathbf{x}^{\alpha'}_{\nu'}-\mathbf{x}^{\alpha}_{\nu}|<\rho}}\hat{M}^{\alpha}_{\nu i}\left[\frac{\partial^2}{\partial(\mathbf{r}^{\alpha\alpha'}_{\nu\nu'})_i\,\partial(\mathbf{r}^{\alpha\alpha'}_{\nu\nu'})_j}\frac{1}{|\mathbf{r}^{\alpha\alpha'}_{\nu\nu'}|}\right]\hat{M}^{\alpha'}_{\nu' j}. \tag{1.1.28}$$

Here the symbol $\sum_{\langle\alpha\rangle}$ stands for the summation over the atoms of the νth sublattice that fall into the volume ΔV centered at the point with radius vector $\mathbf{x} \approx \mathbf{x}^{\alpha}_{\nu}$ (we assume that the atoms of all the sublattices fall into the volume ΔV).

Let us average expression (1.1.28) using the locally equilibrium density matrix $\hat{\rho}(\mathbf{x}, t)$ and the approximation

$$\mathrm{Sp}(\hat{M}^{\alpha}_{\nu i}\hat{M}^{\alpha'}_{\nu' j}\hat{\rho}) = \mathrm{Sp}(\hat{M}^{\alpha}_{\nu i}\hat{\rho})\mathrm{Sp}(\hat{M}^{\alpha'}_{\nu' j}\hat{\rho}).$$

Since the magnetic moments of crystal atoms have a close orientation within the physically small volume, their average values are independent of α and α' and can be ascribed to the point with radius vector $\mathbf{x}$:

$$\mathrm{Sp}(\hat{M}^{\alpha}_{\nu i}\hat{\rho}) = \mu_{\nu i}(\mathbf{x}, t), \quad \mathrm{Sp}(\hat{M}^{\alpha'}_{\nu' j}\hat{\rho}) = \mu_{\nu' j}(\mathbf{x}, t).$$

Upon averaging, the summation over the indices α and α' in eq. (1.1.28) can be performed explicitly, and we obtain the following expression for the macroscopic energy density of the crystal at the point with radius vector $\mathbf{x}$:

$$\frac{1}{\Delta V}\mathrm{Sp}(\hat{\tilde{H}}^{\mathrm{II}}_{\mathrm{dip}}\hat{\rho}) = \frac{1}{\Delta V}\sum_{i,j=1}\sum_{\nu,\nu'}\Delta N_{\nu}\beta^{\nu,\nu'}_{ij}\mu_{\nu i}\mu_{\nu' j}.$$

Here ΔN_{ν} is the number of magnetic moments of atoms of the νth sublattice in the volume ΔV. Since the quantity

$$\beta^{\nu\nu'}_{ij} = -\frac{1}{2}\sum_{\substack{\alpha' \\ 0<|\mathbf{r}^{\alpha,\alpha'}_{\nu,\nu'}|<\rho}}\frac{\partial^2}{\partial(\mathbf{r}^{\alpha\alpha'}_{\nu\nu'})_i\,\partial(\mathbf{r}^{\alpha\alpha'}_{\nu\nu'})_j}\frac{1}{|\mathbf{r}^{\alpha\alpha'}_{\nu\nu'}|}$$

weakly depends on α, α', and ρ, this dependence can be neglected.

Given that the ith projection of the magnetization of the νth sublattice is equal to

$$M_{\nu\, i}(\mathbf{x}, t) = \frac{\Delta N_\nu}{\Delta V}\, \mu_{\nu i}(\mathbf{x}, t) = \frac{1}{\mathrm{v}_\nu}\, \mu_\nu(\mathbf{x}, t),$$

where v_ν is the volume per atom of the νth sublattice, we obtain the following final expression for the macroscopic energy density of the magnet:

$$\sum_{i,j=1}^{3} \sum_{\nu,\nu'} \beta_{ij}^{\nu\nu'}\, \mathrm{v}_{\nu'} M_{\nu i}(\mathbf{x}, t)\, M_{\nu' j}(\mathbf{x}, t).$$

As we see, the discrete part of the dipole–dipole interaction micro-Hamiltonian also brings to a macroscopic energy consistent with expansion (1.1.3).

The phenomenological equations (1.1.13) for the nonlinear dynamics of magnets are the result of averaging the quantum mechanical Heisenberg equations for the sublattice micromagnetization operators (1.1.16):

$$\mathrm{i}\hbar\, \partial_t \hat{\mathbf{M}}_\nu = [\hat{\mathbf{M}}_\nu, \hat{H}], \qquad (1.1.29)$$

where $\hat{H}$ is the Hamiltonian of the crystal magnet, $[\hat{\mathbf{M}}_\nu, \hat{H}]$ is the commutator of the operators $\hat{\mathbf{M}}_\nu$ and $\hat{H}$, $\hbar$ is Planck's constant, and i is the imaginary unit. For this purpose, we have employed a locally equilibrium density matrix over physically small volumes.

If we regard sufficiently slow changes in the macroscopic magnetic moments of the sublattices as a characteristic of times that are much longer than that for the establishment of local quasiequilibrium in the magnet, then on the left side of (1.1.29), we arrive at the time derivative of the macroscopic magnetization density $\mathbf{M}_\nu(\mathbf{x}, t)$ of the νth sublattice. Suppose that quasiequilibrium distributions of the magnetic moments of atoms are quickly established in each of the sublattices. Then we neglect the higher-order correlation functions and replace the average values of the operators $\hat{M}_{\nu i}^{\alpha}$ by products of the averages. After physically justified expansions that are similar to those considered above, we derive macroscopic equations for the dynamics of magnetizations in the sublattices in the Landau–Lifshitz form (1.1.13).

Until now, it has actually been assumed that magnetism is associated with localized electrons of unfilled shells of crystal atoms, with no consideration given to the role of itinerant electrons responsible for electrical conductivity. Nevertheless, the phenomenological theory set forth here, by virtue of its symmetry basis, makes it possible to describe many phenomena in metals and, moreover, in magnetic semiconductors.

1.2 A phenomenological approach to relaxation processes in multisublattice magnets

...what is a theory? It strikes the eyes of a non-specialist that a theory is a pile of formulas that say nothing to the uninitiated. But these formulas do not constitute its essence.

L. Boltzmann

A macroscopic description of energy dissipation in magnets suggests the introduction of additional relaxation terms into the Landau–Lifshitz equations (1.1.13) in the form of Landau–Lifshitz, Bloch–Blombergen, Gilbert, and others [67, 42, 16, 32, 47]. However, all conventionally exploited dissipative terms are determined only by relativistic interactions. They leave aside the spatial dispersion characteristic of exchange interactions and lead to results that contradict the microtheory of spin waves. Bar'yakhtar was the first who noticed this [7–9]. He proposed a phenomenological approach for introducing dissipative terms into the Landau–Lifshitz equations to properly take into account the dynamic symmetry of relaxation processes in magnets. Furthermore, such a concept does not rule out the possibility of changing the nominal sublattice magnetizations due to relaxation and provides damping decrements of linear spin waves for consistency with microtheory. Therefore the relaxation of nonlinear magnetic excitations, in particular, the mobility of domain walls and other solitons in magnets, is expected to be more accurately described within the Bar'yakhtar approach. Let us focus on his main ideas.

To introduce dissipative terms into the equations of the dynamics of sublattice magnetizations, we consider the energy variation δW of a magnet. By the definition of the effective field we have

$$\delta W = -\int (\mathbf{H}_\nu^{\mathrm{eff}} \cdot \delta \mathbf{M}_\nu) \mathrm{d}^3\mathbf{x}. \tag{1.2.1}$$

Here and below, unless otherwise stated, a summation over doubly occurring indices is implied. From eq. (1.2.1) we find the change in the energy of the system with time (for simplicity of the analysis, we first assume that the field $\mathbf{H}^{\mathrm{ext}}$ is time-independent and the medium is unlimited):

$$\frac{\mathrm{d}W}{\mathrm{d}t} = -\int (\mathbf{H}_\nu^{\mathrm{eff}} \cdot \partial_t \mathbf{M}_\nu) \mathrm{d}^3\mathbf{x}.$$

It is not hard to check that the Landau–Lifshitz equations (1.1.13) provide the conservation of the energy W of a crystal magnet. If dissipation takes place, then the energy must decay over time:

$$\frac{\mathrm{d}W}{\mathrm{d}t} = -\int (\mathbf{H}_\nu^{\mathrm{eff}} \cdot \partial_t \mathbf{M}_\nu) \mathrm{d}^3\mathbf{x} = -T\frac{\mathrm{d}S}{\mathrm{d}t} < 0. \tag{1.2.2}$$

Here S is the entropy of a crystal magnet, and T is its temperature. The system entropy is maximum in the equilibrium state, and therefore

$$\mathbf{H}_\nu^{\text{eff}} = -\frac{\delta W}{\delta \mathbf{M}_\nu} = 0.$$

The phenomenological description of relaxation in magnets takes this circumstance into account, provided that we expand $\partial_t \mathbf{M}_\nu$ in the fields $\mathbf{H}_\nu^{\text{eff}}$ and their gradients:

$$\begin{aligned} \partial_t M_{\nu i} = &-\gamma_\nu \varepsilon_{ijk} M_{\nu j} H_{\nu k}^{\text{eff}} + \Gamma_{ik}^{\nu\nu_1} H_{\nu_1 k}^{\text{eff}} \\ &+ \Gamma_{ik,s}^{\nu\nu_1} \partial_s H_{\nu_1 k}^{\text{eff}} + \Gamma_{ik,sp}^{\nu\nu_1} \partial_s \partial_p H_{\nu_1 k}^{\text{eff}} + \cdots, \end{aligned} \tag{1.2.3}$$

where ε_{ijk} is an antisymmetric unit tensor ($\varepsilon_{123} = 1$). The first term in expansion (1.2.3) is chosen so that the Landau–Lifshitz equations (1.1.13) should be reproduced in the absence of energy dissipation. In this case, there is no summation over the index ν appearing more than twice. The coefficients $\Gamma_{ik}^{\nu\nu_1}$, etc., may depend on the magnetizations of the sublattices, their gradients, and an external magnetic field. Then they satisfy the Onsager symmetry relations [68, 8, 9]

$$\begin{aligned} \Gamma_{sk}^{\nu_1\nu_2}(\{\mathbf{M}_\nu\}, \mathbf{H}^{\text{ext}}) &= \Gamma_{ks}^{\nu_2\nu_1}(\{-\mathbf{M}_\nu\}, -\mathbf{H}^{\text{ext}}), \\ \Gamma_{sk,ij}^{\nu_1\nu_2}(\{\mathbf{M}_\nu\}, \mathbf{H}^{\text{ext}}) &= \Gamma_{ks,ij}^{\nu_2\nu_1}(\{-\mathbf{M}_\nu\}, -\mathbf{H}^{\text{ext}}). \end{aligned}$$

At the macroscopic level, they reflect the invariance of the microscopic equations of motion relative to a change in the sign of time, as well as to a change in the directions of the strength vector of the external magnetic field $\mathbf{H}^{\text{ext}}$ and of the magnetic moments of the crystal atoms. Due to the Onsager relations, the coefficients $\Gamma_{sk}^{\nu_1\nu_2}$, etc. can be represented as the sum of products of an even number of the factors $M_{\nu i}$ and H_j^{ext}. At the same time, due to the smallness of the relaxation terms, such expansions are often restricted to the first terms that are independent of the $\mathbf{H}^{\text{ext}}$ and $\mathbf{M}_\nu$ fields. According to microtheory, the following approximations correspond to higher orders of quantum mechanical perturbation theory. These orders take into account the contributions of spin–spin and spin–orbit interactions with small overlap integrals for the wave functions of neighboring crystal atoms and with small coefficients of the order of $(v^2/c^2)^n$. Besides, in formula (1.2.3), the terms $O((\mathbf{H}^{\text{eff}})^2)$ can be neglected in the main approximation. As a result, for the production of entropy in the system, we obtain the simple expression

$$\begin{aligned} T\frac{\mathrm{d}S}{\mathrm{d}t} \equiv 2Q = \int \Big[&\Gamma_{ik}^{\nu_1\nu_2} H_{\nu_1 i}^{\text{eff}} H_{\nu_2 k}^{\text{eff}} \\ &+ \Gamma_{ik,s}^{\nu_1\nu_2} (H_{\nu_1 i}^{\text{eff}} \partial_s H_{\nu_2 k}^{\text{eff}} - H_{\nu_2 k}^{\text{eff}} \partial_s H_{\nu_1 i}^{\text{eff}}) \\ &- \Gamma_{ik,sp}^{\nu_1\nu_2} \partial_s H_{\nu_1 i}^{\text{eff}} \partial_p H_{\nu_2 k}^{\text{eff}} \Big] \mathrm{d}^3\mathbf{x} > 0. \end{aligned} \tag{1.2.4}$$

Relation (1.2.4) determines the dissipative functional Q of the system. Noted that to simplify the analysis of general relations, we discuss here the case of an unbounded medium. Therefore, in eq. (1.2.4), we have omitted the surface integrals that appear as a result of the integration by parts when passing from eqs. (1.2.2) and (1.2.3) to eq. (1.2.4).

For finite samples, such integrals vanish under boundary conditions reflecting the absence of energy fluxes across the surface of a crystal magnet.

To completely define the specific form of the constants Γ in eq. (1.2.4), the following conditions should be added:

1. The functional Q is invariant with respect to transformations of the paramagnetic symmetry group of the crystal;
2. The sublattice magnetization components obey the laws of conservation as a consequence of the symmetry properties of the material;
3. The entropy production in the system is positive ($Q > 0$ for any $\mathbf{H}_\nu^{\text{eff}} \neq 0$).

In particular, the linear-in-gradients terms with coefficients $\Gamma_{ij,s}^{\nu_1\nu_2}$ always violate the positive definiteness of the functional Q and, most likely, should be absent even for crystals without a center of symmetry. It should also be underscored that building the invariants through the vector components of $\mathbf{H}_\nu^{\text{eff}}$ can be facilitated by transforming the latter under the action of crystal symmetry elements, in the same way as $\mathbf{M}_\nu$. This follows from definition (1.2.1) and the invariance of the energy W.

In the next section, we will clarify these statements using a simple example.

1.3 A ferroplate as an example of a finite sample

Let us take an old theory of magnetism, which treats a magnetized steel piece as a magnetic dipole consisting of two magnetic poles or "magnetic charges" (similar to an electric dipole). Using this concept, physicists created magnetostatics relied upon by the whole practice and technique of applying magnets. Meanwhile, it is well known that actually there exist no magnetic charges. And today, when it comes to magnetic poles, they have to necessarily stipulate that this **is a fictitious concept**.

I. K. Kikoin

1.3.1 Energy, a dissipative functional, and conditions for the existence of a homogeneous magnetization in a plate

Let us apply the above theory to the description of the nonlinear dynamics of magnetization in a plate made of a ferromagnetic material with uniaxial anisotropy. An idealized model of a uniaxial crystal with a symmetry axis C_∞ is used for solving a problem on real crystals of hexagonal, rhombohedral, and tetragonal systems, conditional upon neglecting a small anisotropy in the basal plane of the 4th and 6th orders in magnetization components. Suppose the plate has a thickness d, occupies a region $|x_3| \leq d/2$, is unlimited in directions Ox_α ($\alpha = 1, 2$), and the anisotropy axis coincides with the axis Ox_3.

With a single magnetic lattice, even for symmetry-center-free ferromagnetic crystals, expression (1.2.4) for the dissipative functional Q leaves out terms linear in the effective field gradients. In eq. (1.2.4), let us separate the main terms due to exchange interactions from the terms that are quadratic in the field $\mathbf{H}^{\text{eff}}$ gradients. A characteristic feature of exchange interactions is their invariance under rotations in spin space. At the macroscopic level, in eq. (1.2.4) for the dissipative functional Q, the terms unchanged under the rotation of the vector of $\mathbf{H}^{\text{eff}}$ through an arbitrary angle correspond to exchange relaxation. This condition imposes the following restriction on the choice of phenomenological constants $\Gamma_{ij,sp}$:

$$\Gamma_{ij,sp} = -\delta_{ij}\lambda^{\text{ex}}_{sp}.$$

The uniaxial symmetry of the crystal lattice, related to the transformations of the Cartesian coordinates x_i of the crystal magnet, yields the additional constraint

$$\lambda^{\text{ex}}_{sp} = \text{diag}(\lambda_1, \lambda_1, \lambda_3).$$

Other terms with gradients of the field $\mathbf{H}^{\text{eff}}$ are small because they contain a small parameter v^2/c^2 of relativistic interactions. Moreover, for the analysis of long-wavelength magnetization distributions, the gradients depreciate the importance of these terms. Therefore we leave aside the latter.

Relativistic interactions partially lift the degeneracy inherent to exchange relaxation. As a result of relativistic interactions only around the axis Ox_3, there are arbitrary rotations of spins (and thus the rotation of the vector of $\mathbf{H}^{\text{eff}}$) in a uniaxial crystal. Therefore the relativistic relaxation constants have the form

$$\Gamma_{sp} = \text{diag}(r_1, r_1, r_3).$$

The parameter r_3 should be set to zero. Otherwise, we will come into conflict with the conservation law

$$\int\limits_{V_0} M_3(\mathbf{x}, t)\text{d}^3\mathbf{x} = \text{const}, \tag{1.3.1}$$

which is a consequence of the symmetry of the micro-Hamiltonian of the system relative to the rotations of the spin moments of the crystal atoms around the axis Ox_3. This symmetry is violated by an external field when accounting for the anisotropy of the body shape or when more fully accounting for the real symmetry of the crystal.

Thus the symmetry properties of a uniaxial ferromagnetic crystal and the conservation law (1.3.1) give the following specific expression for the dissipative functional Q:

$$Q = \frac{1}{2}\int [\lambda^{\text{ex}}_{sp}(\partial_s\mathbf{H}^{\text{eff}} \cdot \partial_p\mathbf{H}^{\text{eff}}) + r_1(\mathbf{H}^{\text{eff}}_{\perp})^2]\text{d}^3\mathbf{x}, \tag{1.3.2}$$

where

$$\mathbf{H}^{\text{eff}}_{\perp} = (H^{\text{eff}}_1, H^{\text{eff}}_2, 0).$$

The requirement of the positive definiteness for Q is met if

$$\lambda_1, \lambda_3, r_1 > 0. \tag{1.3.3}$$

Bearing in mind the energy dissipation, the Landau–Lifshitz equation for a uniaxial anisotropy ferromagnet is written as

$$\begin{aligned} \partial_t M_\alpha &= -\gamma[\mathbf{M} \times \mathbf{H}^{\text{eff}}]_\alpha + r_1(H_\perp^{\text{eff}})_\alpha - \lambda_{sp}^{\text{ex}}\partial_s\partial_p H_\alpha^{\text{eff}}, \\ \partial_t M_3 &= -\gamma[\mathbf{M} \times \mathbf{H}^{\text{eff}}]_3 - \lambda_{sp}^{\text{ex}}\partial_s\partial_p H_3^{\text{eff}}, \end{aligned} \tag{1.3.4}$$

where

$$\lambda_{sp}^{\text{ex}} = \text{diag}(\lambda_1, \lambda_1, \lambda_3), \quad \alpha = 1, 2.$$

In eqs. (1.3.4) the terms $r_1(H_\perp^{\text{eff}})_u$ responsible for relativistic interactions trace out the relaxation of uniform oscillations of the magnetization. A characteristic feature of the resulting relaxation term $\lambda_{sp}^{\text{ex}}\partial_s\partial_p\mathbf{H}^{\text{eff}}$ of the exchange nature is that the components of the vector of the magnetization $\mathbf{M}(\mathbf{x}, t)$ begin diffusing from those regions where they are larger to those regions where their values are smaller. In this case, short-wavelength excitations dissipate fairly quickly, and long-wavelength excitations, which correspond to a quasihomogeneous distribution of magnetization in space, "survive". Accounting for relaxation reveals local changes in the length of the vector of $\mathbf{M}(\mathbf{x}, t)$ already in the leading exchange approximation.

In the general case the total energy of a finite magnetic crystal contains not only volume integrals of the type (1.1.3), but also integrals over the sample surface, which take into account the average macroscopic energy of spins pinning in the near-surface layers of the material [1]. Further, for simplicity, we assume that the spins are not fixed on the surface of the plate. Then the total energy includes no surface integrals and is written as [11, 27]

$$W = \int_{V_0} \left[f(|\mathbf{M}|) + \frac{\alpha_{ik}}{2}(\partial_i\mathbf{M}\cdot\partial_k\mathbf{M}) + \frac{\beta}{2}(M_1^2 + M_2^2) - (\mathbf{M}\cdot\mathbf{H}^{\text{ext}}) \right] \mathrm{d}^3\mathbf{x} + W^{(\text{m})}. \tag{1.3.5}$$

Here $\alpha_{ik} = \text{diag}(\alpha_1, \alpha_1, \alpha_3)$ and β are the constants of the exchange interaction and crystallographic magnetic anisotropy, respectively, and $W^{(\text{m})}$ is the magnetostatic energy (1.1.11). Then, for definiteness, the external magnetic field is thought to be directed along the axis Ox_3:

$$\mathbf{H}^{\text{ext}} = (0, 0, H_3^{\text{ext}}), \quad H_3^{\text{ext}} = \text{const} > 0.$$

The function $f(|\mathbf{M}|)$ will be specified later.

To separate the processes of changing the length of the vector of $\mathbf{M}(\mathbf{x}, t)$ from local rotations of $\mathbf{M}(\mathbf{x}, t)$ without changing its length, we use one of two parameterizations of the magnetization $\mathbf{M}(\mathbf{x}, t)$, either by angles $\theta(\mathbf{x}, t)$, $\varphi(\mathbf{x}, t)$ $(0 \le \theta < \pi, 0 \le \varphi < 2\pi)$ or by an auxiliary complex field $q(\mathbf{x}, t)$:

$$\begin{aligned} M_3 &= M\cos\theta = M\frac{1-|q|^2}{1+|q|^2}, \\ M_+ &= M_1 + \mathrm{i}M_2 = M\sin\theta\exp(\mathrm{i}\varphi) = M\frac{2q}{1+|q|^2}, \\ M_- &= M_1 - \mathrm{i}M_2 = M_+^*, \end{aligned} \tag{1.3.6}$$

where $M = M(\mathbf{x}, t)$ is the length of the magnetization vector (a function of coordinates x_i and time t), and i is the imaginary unit. The symbol "$*$" means a complex conjugation. We have introduced two parameterizations for the magnetization because the field variables θ and φ are often used in direct integrations of the Landau–Lifshitz equations, and the complex field $q(\mathbf{x}, t)$ is more convenient in analyzing weakly nonlinear magnetic solitons within models close to the nonlinear Schrödinger equation integrable by the inverse scattering method. In terms of M, θ, and φ, the total energy (1.3.5) has the form

$$\begin{aligned} W = \int\Big[f(M) &+ \frac{\alpha_{ik}}{2}(\partial_i M\,\partial_k M + M^2(\partial_i\theta\,\partial_k\theta + \sin^2\theta\,\partial_i\varphi\,\partial_k\varphi)) \\ &+ \frac{\beta}{2}M^2\sin^2\theta - MH_3^{\mathrm{ext}}\cos\theta\Big]\mathrm{d}^3\mathbf{x} + W^{(\mathrm{m})}. \end{aligned} \tag{1.3.7}$$

The magnetostatic energy $W^{(\mathrm{m})}$ will be expressed through the scalar potential $\Psi(\mathbf{x}, t)$ and fields M, θ, and φ if in formula (1.1.11), we put

$$\mathbf{M} = M(\sin\theta\cos\varphi, \sin\theta\,\sin\varphi, \cos\theta). \tag{1.3.8}$$

The ground state of a relaxation-free magnetic crystal corresponds to a minimum of the energy W. By varying the energy (1.3.7) over the fields M, θ, φ, and Ψ we obtain, due to the independence of volume and surface variations, equations that determine the possible equilibrium values of the fields M, θ, φ, and Ψ:

$$\begin{aligned} &\partial_M f + \alpha_{ik}[-\partial_i\partial_k M + M(\partial_i\theta\,\partial_k\theta + \sin^2\theta\,\partial_i\varphi\,\partial_k\varphi)] \\ &\quad + \beta M\sin^2\theta - (H_3^{\mathrm{ext}} - \partial_3\Psi)\cos\theta + (\mathbf{e}_1\cdot\nabla\Psi)\sin\theta = 0, \\ &\alpha_{ik}[-\partial_k(M^2\partial_i\theta) + M^2\partial_i\varphi\,\partial_k\varphi\sin\theta\cos\theta] + \beta M^2\sin\theta\cos\theta \\ &\quad + M(H_3^{\mathrm{ext}} - \partial_3\Psi)\sin\theta + M(\mathbf{e}_1\cdot\nabla\Psi)\cos\theta = 0, \\ &-\alpha_{ik}\partial_k(M^2\partial_i\varphi\,\sin^2\theta) + M\sin\theta(\mathbf{e}_2\cdot\nabla\Psi) = 0, \\ &\Delta\Psi = 4\pi\,\mathrm{div}\,\mathbf{M}\quad(\mathbf{x}\in V_0),\quad \Delta\Psi = 0\quad(\mathbf{x}\in V_1), \end{aligned} \tag{1.3.9}$$

and the boundary conditions necessary to solve them:

$$\begin{aligned}&\alpha_{ik}\partial_i M\, n_k|_\sigma = 0, \quad \alpha_{ik}\partial_i\theta\, n_k|_\sigma = 0, \quad \sin^2\theta\,\alpha_{ik}\,\partial_i\varphi\, n_k|_\sigma = 0,\\ &\Psi_-|_\sigma = \Psi_+|_\sigma, \quad n_k[-\partial_k\Psi + 4\pi M_k]_-|_\sigma = -n_k\partial_k\Psi_+|_\sigma,\\ &\partial_i\Psi \to 0 \quad \text{as} \quad |\mathbf{x}| \to \infty \quad (\mathbf{x} \in V_1).\end{aligned} \tag{1.3.10}$$

Here $\mathbf{n}$ is the unit vector of the outward normal to the plate surface σ, and V_0 (V_1) is the area of space inside (outside) the plate. Here and below, for simplicity of notation and calculations, it is reasonable to enter an orthonormal moving frame formed by the vectors $\mathbf{e}_i$:

$$\begin{aligned}&\mathbf{e}_1 = (\cos\varphi, \sin\varphi, 0), \quad \mathbf{e}_2 = (-\sin\varphi, \cos\varphi, 0),\\ &\mathbf{e}_3 = (0, 0, 1); \quad [\mathbf{e}_i \times \mathbf{e}_j] = \varepsilon_{ijk}\mathbf{e}_k.\end{aligned} \tag{1.3.11}$$

From eqs. (1.3.9) and (1.3.10) we conclude that the energy extremum conditions are satisfied, in particular, by the homogeneous state:

$$\begin{aligned}&\mathbf{H}_0^{(m)} = -\nabla\Psi_0 = (0, 0, -4\pi M_0) \quad (\mathbf{x} \in V_0),\\ &\mathbf{H}_0^{(m)} = 0 \quad (\mathbf{x} \in V_1),\\ &\theta_0 = 0, \quad \partial_{M_0} f - H_3^{\text{ext}} + 4\pi M_0 = 0,\end{aligned} \tag{1.3.12}$$

where $M_0 = \text{const}$ is the uniform magnetization. The energy W corresponding to the state (1.3.12) is degenerate relative to rotations around the axis Ox_3 ($\varphi = \varphi_0$ is an arbitrary constant angle).

Further, for the energy density $f(M)$, we accept the approximation [27]

$$\begin{aligned}f(M) &\approx f(M_0) + \partial_{M_0} f(M_0)(M - M_0)\\ &+ \frac{1}{2}\partial^2_{M_0} f(M_0)(M - M_0)^2 + \cdots = f(M_0)\\ &+ (H_3^{\text{ext}} - 4\pi M_0)(M - M_0) + \frac{1}{2\chi}(M - M_0)^2 + \cdots,\end{aligned} \tag{1.3.13}$$

where $\chi = (\partial^2_{M_0} f)^{-1}$ is the longitudinal susceptibility of a ferromagnet with uniform magnetization.

Let us look upon small deviations δM, $\delta\theta$, and $\delta\Psi$ from the extremal:

$$\begin{aligned}&M = M_0 + \delta M, \quad \theta = \delta\theta, \quad \Psi = \Psi_0 + \delta\Psi,\\ &\mathbf{H}^{(m)} = \mathbf{H}_0^{(m)} + \delta\mathbf{h}, \quad \delta\mathbf{h} = -\nabla\delta\Psi.\end{aligned}$$

The expansion of the energy W in the deviations from the extremal begins with the quadratic form $\delta^2 W$ over deviations. When the quadratic form is positive definite ($\delta^2 W > 0$ for any nonzero $\delta\Psi$, $\delta\theta$, and δM), the homogeneous state (1.3.12) corresponds to a local energy minimum and hence to the stable ground state of the magnetic crystal. To determine the stability limits of the solution (1.3.12), we calculate $\delta^2 W$. Making allowance for the relationships between variations $\delta\Psi$ and $\delta\mathbf{M}$,

$$\Delta\delta\Psi = 4\pi \operatorname{div} \delta\mathbf{M}, \quad n_k[\partial_k\delta\Psi - 4\pi\delta M_k]_-|_\sigma = n_k\partial_k\delta\Psi_+|_\sigma \quad (\mathbf{x} \in V_0),$$
$$\Delta\delta\Psi = 0, \quad \partial_i\delta\Psi \to 0 \quad \text{as} \quad |\mathbf{x}| \to \infty \quad (\mathbf{x} \in V_1),$$

after integration by parts, we arrive at the integrals

$$\delta^2 \int_{V_0} (\mathbf{M} \cdot \mathbf{H}^{(m)}) d^3\mathbf{x} = \int_{V_0} \left[(\delta_1\mathbf{M} \cdot \delta\mathbf{h}) - \frac{M_0}{2}(\delta\theta)^2 (\mathbf{H}_0^{(m)} \cdot \mathbf{e}_3) \right] d^3\mathbf{x},$$
$$\int_{V_0+V_1} \frac{(\delta\mathbf{h})^2}{8\pi} d^3\mathbf{x} = -\frac{1}{2} \int_{V_0} (\delta_1\mathbf{M} \cdot \delta\mathbf{h}) d^3\mathbf{x},$$

where

$$\delta_1\mathbf{M} = \delta\theta M_0\mathbf{e}_1 + \delta M\mathbf{e}_3, \quad \mathbf{H}_0^{(m)} = -4\pi M_0\mathbf{e}_3,$$

and, consequently,

$$\delta^2 W^{(m)} = \frac{M_0}{2} \int_{V_0} (\delta\theta)^2 (\mathbf{H}_0^{(m)} \cdot \mathbf{e}_3) d^3\mathbf{x} + \int_{V_0+V_1} \frac{(\delta\mathbf{h})^2}{8\pi} d^3\mathbf{x}.$$

As a result, the expression for the variation $\delta^2 W$ takes the form

$$\delta^2 W = \int_{V_0} \bigg(\frac{1}{2\chi}(\delta M)^2 + \frac{\alpha_{ik}}{2}[\partial_i\delta M\, \partial_k\delta M + M_0^2\partial_i\delta\theta\, \partial_k\delta\theta]$$
$$+ \frac{1}{2}M_0(\delta\theta)^2[H_3^{\text{ext}} - H_s] \bigg) d^3\mathbf{x} + \int_{V_0+V_1} \frac{(\delta\mathbf{h})^2}{8\pi} d^3\mathbf{x}.$$

Thus, as relaxation lacks, the uniform distribution of magnetization (see Figure 1.1) is stable under the following conditions:

$$\chi > 0, \quad \alpha_1, \alpha_3 > 0, \quad H_3^{\text{ext}} > H_s,$$

where $H_s = M_0(4\pi - \beta)$ is the saturation field. Recall that the constant $\beta > 0$ ($\beta < 0$) in the case of magnetic anisotropy of the "easy axis" ("easy plane") type.

When exposed to large in modulus negative fields H_3^{ext}, the crystal remagnetizes, and another homogeneous state will be stable:

$$\theta_0 = \pi, \quad \mathbf{H}^{(m)} = (0, 0, 4\pi M_0).$$

When $|H_3| < |H_s|$, the uniform distribution of magnetization is unstable. The sample is divided into macroscopic regions of self-arbitrary magnetization (ferromagnetic

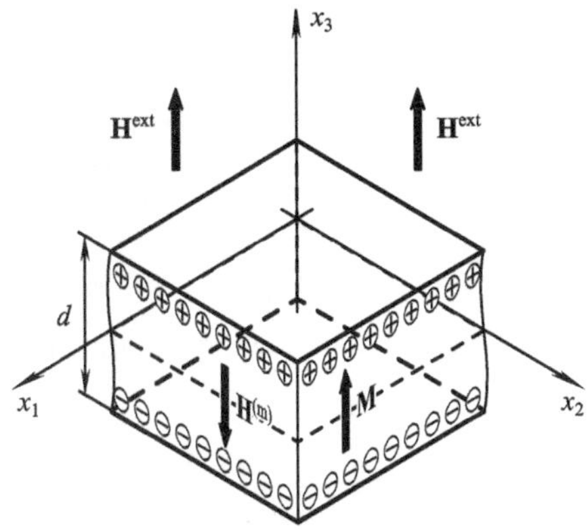

Figure 1.1: Formation of fictitious magnetic charges "+" and "–" in the surface of a uniformly magnetized plate and mutual orientation of the vectors **M**, $\mathbf{H}^{ext}$, and $\mathbf{H}^{(m)}$.

domains) with dimensions of the order of 10^{-3}–10^{-2} cm. The orientation of the magnetization in the domains contributes to the closure of the magnetic flux inside the sample. As a result, the magnetostatic field $\mathbf{H}^{(m)}$ outside the sample and hence the system energy are significantly reduced. In an unbounded medium, the domains are separated by narrow boundaries with a thickness of x_0: $x_0 \backsim \sqrt{\alpha/\beta}$ or $x_0 \backsim \sqrt{\alpha/(\beta+4\pi)}$, depending on the internal structure of the domain boundaries (Bloch- or Néel-type boundaries). Usually, the thickness of domain walls is of the order of $(10-10^3)\,a$, where a is the interatomic distance.

An expression for the energy of a ferromagnetic plate in terms of the complex field $q(\mathbf{x}, t)$ can be easily deduced using formulas (1.3.6), (1.3.7), and (1.3.13) and the relation

$$\partial_s\theta\,\partial_k\theta + \partial_s\varphi\,\partial_k\varphi\,\sin^2\theta = 2(\partial_s q\,\partial_k q^* + \partial_s q^*\,\partial_k q)(1+|q|^2)^{-2}.$$

1.3.2 Equations of dynamics

The total energy (1.3.5) and (1.3.13) of a ferromagnet corresponds to the effective field:

$$\mathbf{H}^{\text{eff}} = -\frac{\delta W}{\delta \mathbf{M}} = -\left(\mathbf{H}_3^{\text{ext}} - 4\pi M_0 + \frac{M-M_0}{\chi}\right)\frac{\mathbf{M}}{M} + \alpha_{ik}\partial_i\partial_k\mathbf{M} - \beta\mathbf{M}_\perp + \mathbf{H}^{\text{ext}} + \mathbf{H}^{(m)},$$

where $\mathbf{M}_\perp = (M_1, M_2, 0)$. Let us express $\mathbf{H}^{\text{eff}}$ in terms of the dynamic variables M and q:

$$H_+^{\text{eff}} = H_1^{\text{eff}} + iH_2^{\text{eff}} = -\left[H_3^{\text{ext}} - H_s + (M-M_0)\left(\frac{1}{\chi}+\beta\right)\right]\frac{2q}{1+|q|^2}$$

$$+ \alpha_{sp}\partial_s\partial_p\left(\frac{2Mq}{1+|q|^2}\right) + H_+^{(m)}, \quad H_-^{\text{eff}} = (H_+^{\text{eff}})^*,$$
$$H_3^{\text{eff}} = -\left(H_3^{\text{ext}} - 4\pi M_0 + \frac{M - M_0}{\chi}\right)\frac{1-|q|^2}{1+|q|^2}$$
$$+ \alpha_{sp}\partial_s\partial_p\left(M\frac{1-|q|^2}{1+|q|^2}\right) + H_3^{\text{ext}} + H_3^{(m)}.$$

Then the equations of magnetization dynamics (1.3.4) give the following equations for calculating M and q:

$$2M\partial_t q = \mathrm{i}\gamma M(2qH_3^{\text{eff}} - H_+^{\text{eff}} + q^2H_-^{\text{eff}}) + r_1(H_+^{\text{eff}} - q^2H_-^{\text{eff}})$$
$$+ \lambda_{sp}^{\text{ex}}(2q\,\partial_s\partial_p H_3^{\text{eff}} - \partial_s\partial_p H_+^{\text{eff}} + q^2\partial_s\partial_p H_-^{\text{eff}}),$$
$$\partial_t M = \frac{r_1}{1+|q|^2}(q^*H_+^{\text{eff}} + qH_-^{\text{eff}})$$
$$- \frac{\lambda_{sp}^{\text{ex}}}{1+|q|^2}(q^*\partial_s\partial_p H_+^{\text{eff}} + q\,\partial_s\partial_p H_-^{\text{eff}}) - \lambda_{sp}^{\text{ex}}\frac{1-|q|^2}{1+|q|^2}\partial_s\partial_p H_3^{\text{eff}},$$

where

$$2qH_3^{\text{eff}} - H_+^{\text{eff}} + q^2H_-^{\text{eff}} = \frac{2M}{1+|q|^2}[\alpha_{sp}(-\partial_s\partial_p q + 2q\,\partial_s q\,\partial_p q^*) + \beta q(1-|q|^2)]$$
$$+ 2q(H_3^{\text{ext}} + H_3^{(m)}) - H_+^{(m)} + q^2H_-^{(m)} - 4\alpha_{sp}\partial_p q\,\partial_s M.$$

The theory considered treats a change in the length of the magnetization vector in time and space as a consequence of only relaxation processes.

To rewrite the equations of magnetization dynamics (1.3.4) in terms of the fields M, θ, and φ, we expand the vectors $\mathbf{M}$ and $\mathbf{H}^{\text{eff}}$ (1.3.8) in the basis vectors (1.3.11):

$$\mathbf{M} = M\sin\theta\,\mathbf{e}_1 + M\cos\theta\,\mathbf{e}_3, \quad \mathbf{H}^{\text{eff}} = a_1\mathbf{e}_1 + a_2\mathbf{e}_2 + a_3\mathbf{e}_3,$$
$$a_1 = -\left[H_3^{\text{ext}} - H_s + (M - M_0)\left(\frac{1}{\chi} + \beta\right)\right]\sin\theta$$
$$+ \alpha_{ik}\hat{L}_{ik}^{(1)}(M\sin\theta) + (\mathbf{e}_1\cdot\mathbf{H}^{(m)}) \equiv -\sin\theta\frac{\delta W}{\delta M} - \frac{1}{M}\cos\theta\frac{\delta W}{\delta\theta},$$
$$a_2 = \alpha_{ik}\hat{L}_{ik}^{(2)}(M\sin\theta) + (\mathbf{e}_2\cdot\mathbf{H}^{(m)}) \equiv -\frac{1}{M\sin\theta}\frac{\delta W}{\delta\varphi},$$
$$a_3 = -\left(H_3^{\text{ext}} - 4\pi M_0 + \frac{M - M_0}{\chi}\right)\cos\theta + \alpha_{ik}\partial_i\partial_k(M\cos\theta)$$
$$+ H_3^{\text{ext}} + H_3^{(m)} \equiv -\cos\theta\frac{\delta W}{\delta M} + \frac{1}{M}\sin\theta\frac{\delta W}{\delta\theta}.$$

Here for shortness of notation, we introduce the differential operators

$$\hat{L}_{ik}^{(1)} = \partial_i\partial_k - (\partial_i\varphi)(\partial_k\varphi), \quad \hat{L}_{ik}^{(2)} = (\partial_i\varphi)\partial_k + (\partial_k\varphi)\partial_i + (\partial_i\partial_k\varphi).$$

Since the differentiation of the vectors $\mathbf{e}_\alpha$ ($\alpha = 1, 2$) follows the rules

$$\begin{aligned}\partial_i \mathbf{e}_\alpha &= \partial_i\varphi\, \varepsilon_{\alpha\beta}\, \mathbf{e}_\beta,\\ \partial_i(a_\alpha \mathbf{e}_\alpha + a_3\mathbf{e}_3) &= \mathbf{e}_\beta[\delta_{\beta\alpha}\partial_i - \varepsilon_{\beta\alpha}(\partial_i\varphi)]a_\alpha + \mathbf{e}_3\partial_i a_3,\\ \partial_i\partial_j(a_\alpha \mathbf{e}_\alpha + a_3\mathbf{e}_3) &= \mathbf{e}_\gamma(\delta_{\gamma\alpha}\hat{L}^{(1)}_{ij} - \varepsilon_{\gamma\alpha}\hat{L}^{(1)}_{ij})a_\alpha + \mathbf{e}_3\partial_i\partial_j a_3,\end{aligned}$$

where $\varepsilon_{\alpha\beta}$ is the antisymmetric unit tensor ($\varepsilon_{12} = 1$), all the derivatives in eqs. (1.3.4) are easily computed. By equating the coefficients of the linearly independent vectors $\mathbf{e}_i$ ($i = 1, 2, 3$) after differentiation, we find the equations of dynamics for the fields M, θ, and φ:

$$\begin{aligned}M\partial_t\theta &= \gamma M a_2 - \lambda^{\mathrm{ex}}_{sp}\cos\theta(\hat{L}^{(1)}_{sp}a_1 - \hat{L}^{(2)}_{sp}a_2)\\ &\quad + \lambda^{\mathrm{ex}}_{sp}\sin\theta\,\partial_s\partial_p a_3 + r_1 a_1\cos\theta,\\ M\sin\theta\,\partial_t\varphi &= \gamma M(a_3\sin\theta - a_1\cos\theta) + r_1 a_2\\ &\quad - \lambda^{\mathrm{ex}}_{sp}(\hat{L}^{(1)}_{sp}a_2 + \hat{L}^{(2)}_{sp}a_1),\\ \partial_t M &= r_1 a_1\sin\theta - \lambda^{\mathrm{ex}}_{sp}\sin\theta(\hat{L}^{(1)}_{sp}a_1 - \hat{L}^{(2)}_{sp}a_2)\\ &\quad - \lambda^{\mathrm{ex}}_{sp}\cos\theta\,\partial_s\partial_p a_3.\end{aligned} \tag{1.3.14}$$

When evaluating the expression $a_3\sin\theta - a_1\cos\theta$, the following identities are useful:

$$\begin{aligned}&\sin\theta\,\partial_i\partial_k(M\cos\theta) - \cos\theta\hat{L}^{(1)}_{ik}(M\sin\theta)\\ &\quad = -\partial_i M\,\partial_k\theta - \partial_k M\,\partial_i\theta - M\partial_i\partial_k\theta + \sin\theta\cos\theta\, M\,\partial_i\varphi\,\partial_k\varphi,\\ &a_{ik}M[\sin\theta\,\partial_i\partial_k(M\cos\theta) - \cos\theta\hat{L}^{(1)}_{ik}(M\sin\theta)]\\ &\quad = a_{ik}[-\partial_k(M^2\partial_i\theta) + M^2\sin\theta\cos\theta\,\partial_i\varphi\,\partial_k\varphi].\end{aligned}$$

A consequence of the first and third equations of system (1.3.14) is the simpler equation

$$\partial_t[M\cos\theta] = -\gamma M a_2\sin\theta - \lambda^{\mathrm{ex}}_{sp}\partial_s\partial_p a_3.$$

In calculating $Ma_2\sin\theta$, one more identity turns out to be effective:

$$M\sin\theta\, a_{ik}\hat{L}^{(2)}_{ik}(M\sin\theta) = a_{ik}\partial_k[M^2\sin^2\theta\,\partial_i\varphi].$$

Within the above approach, the methods of soliton perturbation theory are used for exploration of the features of the mobility of domain walls in infinite ferromagnets with “easy axis” anisotropy [35, 36]. Because magnetization rotations are significant in domain walls, the outcomes of the soliton perturbation theory for linear spin waves cannot be replicated in any finite order of the perturbation theory. The approach set forth above can quantitatively explain nontrivial relaxation phenomena.

1.3.3 Boundary conditions, Hamilton principle

To formulate the boundary conditions for the problem of the magnetization dynamics in a plate, we calculate the change in the plate energy over time (1.3.5). Using eqs. (1.3.4), upon integrating by parts, we get

$$\frac{\mathrm{d}W}{\mathrm{d}t} = -2Q - \int_{V_0} (\partial_t \mathbf{H}^{\mathrm{ext}} \cdot \mathbf{M})\mathrm{d}^3\mathbf{x}$$
$$+ \int_{\sigma} [\alpha_{ij}(\partial_i \mathbf{M} \cdot \partial_t \mathbf{M}) + (\mathbf{H}^{\mathrm{eff}} \cdot \lambda_{sj}^{\mathrm{ex}} \partial_s \mathbf{H}^{\mathrm{eff}})]\mathrm{d}S_j.$$

In the magnetostatic approximation the energy flux through the plate surface σ must be equal to zero. This is only possible under the following boundary conditions:

$$n_j \alpha_{ij} \partial_i \mathbf{M}|_\sigma = 0, \quad n_j \lambda_{ij}^{\mathrm{ex}} \partial_i \mathbf{H}^{\mathrm{eff}}|_\sigma = 0.$$

In virtue of the diagonal structure of the tensors α_{ij} and $\lambda_{ij}^{\mathrm{ex}}$, the given equalities are equivalent to the simpler ones:

$$\partial_n \mathbf{M}|_\sigma = \partial_n \mathbf{H}^{\mathrm{eff}}|_\sigma = 0,$$

where $\partial_n = n_j \partial_j$ is the derivative along the normal $\mathbf{n}$ to the surface σ. Here the plate is not assumed to be infinite in the directions along the axis Ox_α $(\alpha = 1, 2)$

In terms of the fields M, θ, and φ, the boundary conditions have the form

$$\partial_n M|_\sigma = \partial_n \theta|_\sigma = \partial_n \varphi \sin\theta|_\sigma = 0,$$
$$(\partial_n a_1 - a_2 \partial_n \varphi)|_\sigma = (\partial_n a_2 + a_1 \partial_n \varphi)|_\sigma = \partial_n a_3|_\sigma = 0.$$

In the case of the time-independent magnetic field $\mathbf{H}^{\mathrm{ext}}$, the plate energy decays with time until the dissipative functional Q vanishes:

$$\frac{\mathrm{d}W}{\mathrm{d}t} = -2Q \le 0.$$

Due to the positive definiteness of the functional Q in (1.3.2) and (1.3.3), this happens when the following equalities are fulfilled:

$$H_1^{\mathrm{eff}} = H_2^{\mathrm{eff}} = 0, \quad H_3^{\mathrm{eff}} = \mathrm{const}.$$

At $H_3^{\mathrm{eff}} = 0$, they coincide with those for the minimum energy of a crystal magnet:

$$\mathbf{H}^{\mathrm{eff}} = -\frac{\delta W}{\delta \mathbf{M}} = 0.$$

Let the dissipation of the system energy and the change in the length of the vector $\mathbf{M}(\mathbf{x}, t)$ be neglected. The equations for the magnetization dynamics in the plate, generated in this section, and the boundary conditions necessary for them to solve can then be derived from Hamilton's variational principle:

$$\delta \int_{t_1}^{t_2} L \, \mathrm{d}t = 0, \tag{1.3.15}$$

where $L = K - W$ is the Lagrange function of the plate. The functional K is an analogue of the kinetic energy of the system

$$K = \int_{V_0} \frac{1}{\gamma} M_0 (1 - \cos\theta) \partial_t \varphi \, \mathrm{d}^3\mathbf{x} = \frac{M_0}{\mathrm{i}\gamma} \int_{V_0} \frac{q^* \partial_t q - q \partial_t q^*}{1 + |q|^2} \mathrm{d}^3\mathbf{x}.$$

The variation of the action of the plate (1.3.15) is carried out along the fields θ, φ, and Ψ (or q, q^*, and Ψ), provided that $\delta\varphi = 0$ ($\delta q = \delta q^* = 0$) at the ends of the time interval $[t_1, t_2]$.

1.4 Magnetoelastic interactions in multisublattice magnets

Let a man who does not know geometry will not enter here.

Plato's words on the pediment of his Academy

Under deformations of a crystal magnet, the distances between magnetic atoms change. This means that the energies of their exchange (1.1.17), spin–orbit (1.1.23), and dipole–dipole (1.1.25) interactions change too. In contrast, interactions between magnetic atoms can cause alterations in the body size and shape throughout the process magnetization. Although the connection between the crystal magnetic and elastic subsystems is tenuous, it nonetheless produces some fascinating phenomena and effects. So, spontaneous deformations of the crystal break the translational symmetry of the total micro-Hamiltonian of the system. As a consequence, extra modes of magneto-elastic oscillations with an energy gap in the spectrum of "quasi-magnons" arise [69, 70]. In magneto-elastic media, rotations of a polarization plane, birefringence of quasi-sonic waves, and renormalization of their velocity propagation, etc. are possible [68].

There are two main ways to theoretically describe changes in the positions of material particles in a deformed body. As a "mark" of the material point of the body, the Lagrangian approach uses Cartesian coordinates of the position of this point in the undeformed state of the body. Then we can watch the evolution of the marked material points of the body.

In the Euler approach, we observe what happens in a fixed area of space as time goes by. "Such a description is ideally suitable for studying liquids, where often a rapidly deforming mass comes from nowhere and leaves for nowhere, so it would be preferable to consider what is happening here and now, before our eyes. However, being convenient kinematically, the Eulerian description becomes clumsy when it comes to the laws of dynamics and leads to difficulties. The point is that the dynamic equations include quantities related to material particles of a body rather than to spatial regions occupied temporarily by these particles. Some relations as being obvious under the Lagrangian approach require complicated reasoning when using Eulerian coordinates" [66]. Further, we will adhere to the Lagrangian description when discussing magneto-elastic interactions [65, 1].

Let x_k be the Cartesian coordinates of a material point of an undeformed body, let $X_k = x_k + u_k(\mathbf{x}, t)$ be the coordinates of the same point after deformation, let $\mathbf{u}(\mathbf{x}, t)$ be a vector of displacements of the medium, and let $\mathbf{M}_\nu(\mathbf{x}, t)$ denote the magnetic moment of a νth sublattice of the magnet per unit volume of the undeformed crystal (we assume that the medium density is $\rho_0 = \text{const}$ in the undeformed state).

Being bound by magnetic anisotropy to certain crystallographic axes, the macroscopic magnetizations $\mathbf{M}_\nu(\mathbf{x}, t)$ of the magnet sublattices tend to rotate after these axes when the sample is deformed. As a result, the total angular momentum is conserved in the system at $\mathbf{H}^{\text{ext}} = 0$, rather than the angular momentum of the magnetic and elastic subsystems separately. This means that the phenomenological expansion of the magnetic, elastic, and magneto-elastic energies of the crystal magnet can only contain quantities that are invariant under simultaneous global rotations of the deformable body and sublattice magnetizations:

$$X_k \to D_{ks} X_s, \quad M_{\nu k} \to D_{ks} M_{\nu s}. \tag{1.4.1}$$

Here D is the radius-vector-$\mathbf{x}$ independent matrix of the rotation group SO(3) ($D_{is} D_{ik} = \delta_{sk}$, $\det D = 1$).

Next, we need to compose combinations of the quantities $M_{\nu k}$, $\partial M_{\nu k}/\partial x_i$, and $\partial X_k/\partial x_i$ as they are invariant under the transformations (1.4.1). Note that only the vector index k is subjected to the transformation (the Lagrangian coordinates x_i do not react to transformation (1.4.1), since they serve as a "mark" of the material point of the body). Only 18 independent invariants can be made. We choose the following ones from them [65, 1]:

$$\begin{aligned} \eta_{ij} &= \frac{1}{2}\left(\frac{\partial X_k}{\partial x_i} \frac{\partial X_k}{\partial x_j} - \delta_{ij} \right), \\ K_{\nu i} &= M_{\nu k} \frac{\partial X_k}{\partial x_i}, \quad K_{\nu ij} = \frac{\partial M_{\nu k}}{\partial x_i} \frac{\partial X_k}{\partial x_j}. \end{aligned} \tag{1.4.2}$$

Other invariants can be expressed through invariants (1.4.2); for example,

$$(\mathbf{M}_{\nu_1} \cdot \mathbf{M}_{\nu_2}) = K_{\nu_1 i} [2\eta + E]^{-1}_{ij} K_{\nu_2 j},$$

$$\left(\frac{\partial \mathbf{M}_{\nu_1}}{\partial x_i} \cdot \frac{\partial \mathbf{M}_{\nu_2}}{\partial x_j}\right) = K_{\nu_1 is}[2\eta + E]^{-1}_{sp} K_{\nu_2 jp},$$

where E is the identity matrix.

Invariants (1.4.2) correspond to local characteristics of the deformed medium. To clarify their physical sense, let us pass to a curvilinear coordinate system connected to the deformed body. In curvilinear coordinates, we should tell apart the upper (contravariant) and lower (covariant) indices for vectors and tensors. In the Cartesian coordinate system used above, the covariant and contravariant vectors and their components coincide: $X^k = X_k$, $x^k = x_k$, $\mathbf{i}_k = \mathbf{i}^k$ ($\mathbf{i}_k$ are the unit vectors of the Cartesian coordinate system connected to the undeformed body). We should introduce a local curvilinear coordinate system with basis vectors $\mathbf{e}_i$, attached to a point of the deformed medium with the Lagrangian coordinates x_p:

$$\mathbf{e}_s = \frac{\partial X^k}{\partial x^s}\mathbf{i}_k. \tag{1.4.3}$$

When the body is deformed, each of the frames $\{\mathbf{e}_i\}$ changes, but the Lagrangian coordinates of the material point of the medium, to which the frame is assigned, remain intact. In this sense, the curvilinear coordinate system (1.4.3) is, as it were, "frozen" into the medium and "traces" the shifts and rotations of small areas of the crystal (see Figures 1.2 and 1.3).

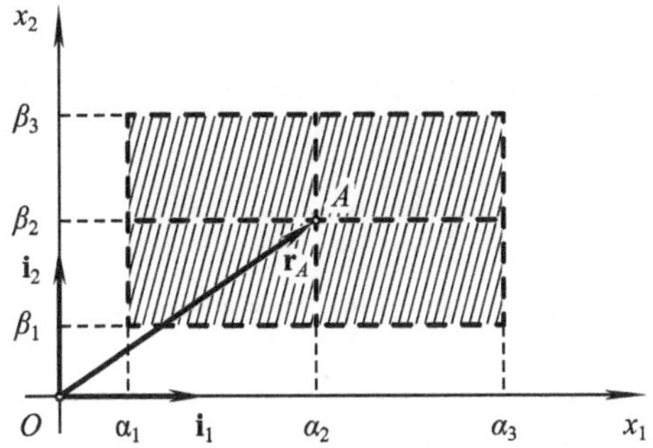

Figure 1.2: An undeformed body (dashed) and coordinate lines $x_1 = \alpha =$ const and $x_2 = \beta =$ const in it. Radius vector $\mathbf{r}_A = x_1^{(A)}\mathbf{i}_1 + y_1^{(A)}\mathbf{i}_2$ defines the position of a material point A.

For the undeformed state of the body, the vector connecting two close particles of the medium can be written as $\mathbf{dx} = \mathrm{d}x^k\mathbf{i}_k$. The square of the distance between the particles is determined by the relation

$$(\mathbf{dx})^2 = \mathrm{d}x^k\mathrm{d}x^m(\mathbf{i}_k \cdot \mathbf{i}_m) = \mathrm{d}x^k\mathrm{d}x^m\delta_{km}.$$

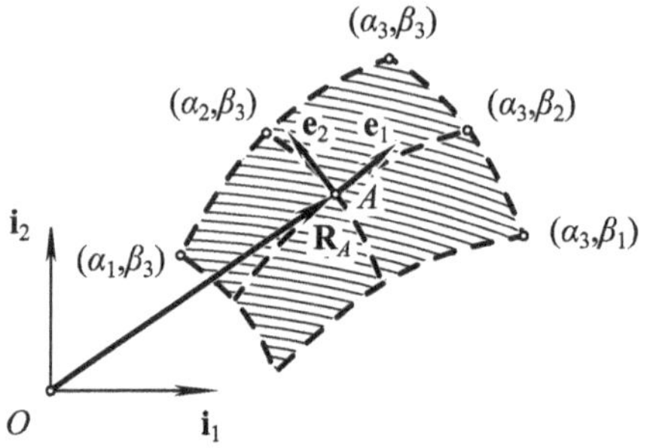

Figure 1.3: In a deformed body a particle A is in the point with radius vector $\mathbf{R}(x_1^{(A)}, x_2^{(A)}) = X_1(x_1^{(A)}, x_2^{(A)})\mathbf{i}_1 + X_2(x_1^{(A)}, x_2^{(A)})\mathbf{i}_2$. Coordinate lines $\mathbf{R}(\alpha, x_2)$ and $\mathbf{R}(x_1, \beta)$ become curved. The vectors $\mathbf{e}_1$ and $\mathbf{e}_2$ tangential to the coordinate lines make up a local basis.

After deformation of the body, the vector connecting the same close particles becomes different: $\mathrm{d}\mathbf{X} = \mathrm{d}X^k \mathbf{i}_k$. The distance between the particles changes:

$$(\mathrm{d}\mathbf{X})^2 = \frac{\partial X^k}{\partial x^s}\frac{\partial X^m}{\partial x^p}\delta_{km}\,\mathrm{d}x^s\mathrm{d}x^p = (\mathbf{e}_s \cdot \mathbf{e}_p)\,\mathrm{d}x^s\mathrm{d}x^p.$$

The quantity

$$g_{sp} = \frac{\partial X^k}{\partial x^s}\frac{\partial X^m}{\partial x^p}\delta_{km}$$

characterizes the metric properties of the deformed medium. This quantity is invariant under rotations of the body as a whole (1.4.1).

Since the change in the distance between close particles of the deformed body can be written as

$$(\mathrm{d}\mathbf{X})^2 - (\mathrm{d}\mathbf{x})^2 = (g_{is} - \delta_{is})\,\mathrm{d}x^i\mathrm{d}x^s,$$

as a measure of the nonlinear elastic deformation of the medium, usually, not the quantity g_{is} is chosen, but rather the Lagrangian strain tensor

$$\eta_{ij} = \frac{1}{2}(g_{ij} - \delta_{ij}).$$

In terms of Cartesian displacements of the medium, the components of the tensor $\boldsymbol{\eta}$ have the form

$$\eta_{ij} = \frac{1}{2}\left(\frac{\partial u_i}{\partial x^j} + \frac{\partial u_j}{\partial x^i} + \frac{\partial u_k}{\partial x^i}\frac{\partial u_k}{\partial x^j}\right).$$

For further analysis, the tensor g^{kl} inverse to g_{km} is useful:

$$g^{kl}g_{lm} = \delta^k_m.$$

It is obvious that

$$g^{sp} = \frac{\partial x^s}{\partial X^l}\frac{\partial x^p}{\partial X^m}\delta^{lm}.$$

The metric tensors g^{sp} and g_{km} can be used to lower and raise the indices of tensors and vectors in a curvilinear coordinate system. In particular, we can define contravariant basis vectors of the local coordinate system:

$$\mathbf{e}^m = g^{mn}\mathbf{e}_n = \frac{\partial x^m}{\partial X^p}\,\mathbf{i}^p. \tag{1.4.4}$$

The orthogonality property of the vectors $\mathbf{e}^m$ and $\mathbf{e}_s$ turns out to be useful:

$$(\mathbf{e}^m \cdot \mathbf{e}_s) = \delta^m_s. \tag{1.4.5}$$

Let us calculate the derivatives of the vectors $\mathbf{e}_s$ and $\mathbf{e}^s$ over the Lagrangian variables x_p. Given (1.4.3) and (1.4.4), we get

$$\begin{aligned}\frac{\partial\,\mathbf{e}_s}{\partial x^p} &= \frac{\partial^2 X^k}{\partial x^p \partial x^s}\,\mathbf{i}_k = \frac{\partial^2 X^n}{\partial x^p \partial x^s}\frac{\partial x^k}{\partial X^n}\mathbf{e}_k,\\ \frac{\partial\,\mathbf{e}^s}{\partial x^p} &= \frac{\partial}{\partial x^p}\left(\frac{\partial x^s}{\partial X^n}\right)\mathbf{i}^n = \left[\frac{\partial}{\partial x^p}\left(\frac{\partial x^s}{\partial X^n}\right)\right]\frac{\partial X^n}{\partial x^m}\mathbf{e}^m.\end{aligned} \tag{1.4.6}$$

In virtue of the identities

$$\frac{\partial x^s}{\partial X^n}\frac{\partial X^n}{\partial x^m} = \delta^s_m, \qquad \frac{\partial}{\partial x^p}\left(\frac{\partial x^s}{\partial X^n}\right)\frac{\partial X^n}{\partial x^m} + \frac{\partial x^s}{\partial X^n}\frac{\partial^2 X^n}{\partial x^p \partial x^m} = 0,$$

relations (1.4.6) can be rewritten as follows:

$$\frac{\partial\,\mathbf{e}_s}{\partial x^p} = \Gamma^k_{ps}\,\mathbf{e}_k, \qquad \frac{\partial\,\mathbf{e}^s}{\partial x^p} = -\Gamma^k_{pm}\,\mathbf{e}^m, \tag{1.4.7}$$

where

$$\Gamma^k_{ps} = \frac{\partial^2 X^n}{\partial x^p \partial x^s}\frac{\partial x^k}{\partial X^n}$$

are the connection coefficients (the second-kind Christoffel symbols). They determine the operation of parallel translation of vectors along curves in a deformed body [50, 61] and can be expressed via the metric tensor (or via the tensor $\boldsymbol{\eta}$):

$$\Gamma^k_{ij} = \frac{1}{2}g^{kl}\left(\frac{\partial g_{li}}{\partial x^j} + \frac{\partial g_{lj}}{\partial x^i} - \frac{\partial g_{ij}}{\partial x^l}\right).$$

When theoretically describing the magneto-elastic interactions, it would be more appropriate to use the νth sublattice magnetization vector components $\mu_{\nu i}$ calculated relative to the local basis e^k instead of the components $M_{\nu i}$ of the same vector in the basis $\{\mathbf{i}^k\}$ associated with an undeformed body:

$$M_{\nu k}\,\mathbf{i}^k = \mu_{\nu p}\,\mathbf{e}^p. \tag{1.4.8}$$

Using relation (1.4.5), we notice that the local magnetizations $\mu_{\nu p}$ of the sublattices coincide with the invariants $K_{\nu p}$ in eq. (1.4.2):

$$\mu_{\nu p} = M_{\nu k}(\mathbf{i}^k \cdot \mathbf{e}_p) = M_{\nu k}\frac{\partial X^k}{\partial x^p} = K_{\nu p}.$$

Similarly, instead of the derivatives $\partial M_{\nu i}/\partial x^s$ of the magnetization vector components for the νth sublattice, which are defined with respect to the basis $\{\mathbf{i}^k\}$, it would be better to use the absolute derivatives of the components of the same vector calculated with respect to the local frame $\{\mathbf{e}^k\}$. Differentiating equality (1.4.8) over x_s and accounting for (1.4.7), we arrive at

$$\frac{\partial M_{\nu k}}{\partial x^s}\,\mathbf{i}^k = \left(\frac{\partial \mu_{\nu p}}{\partial x^s}\right)\mathbf{e}^p + \mu_{\nu p}\frac{\partial\,\mathbf{e}^p}{\partial x^s} = \left(\frac{\partial \mu_{\nu j}}{\partial x^s} - \Gamma^n_{sj}\,\mu_{\nu n}\right)\mathbf{e}^j \equiv \nabla_s\mu_{\nu j}\,\mathbf{e}^j.$$

Here $\nabla_s\,\mu_{\nu j}$ is the absolute (covariant) derivative of the local magnetization $\mu_{\nu i}$ of the νth sublattice. The covariant derivatives $\nabla_s\mu_{\nu j}$ coincide with the invariants $K_{\nu sj}$:

$$\nabla_s\,\mu_{\nu j} = \left(\frac{\partial M_{\nu k}}{\partial x^s}\,\mathbf{i}^k \cdot \mathbf{e}_j\right) = \frac{\partial M_{\nu k}}{\partial x^s}\frac{\partial X^k}{\partial x^j} = K_{\nu sj}.$$

Under arbitrary continuous transformations of the Lagrangian coordinates

$$x^s \to \tilde{x}^s = \tilde{x}^s(\mathbf{x}),$$

the quantities g_{is}, $\mu_{\nu i}$, and $\nabla_s\,\mu_{\nu j}$ are transformed as tensors

$$\tilde{g}_{km} = \frac{\partial x^s}{\partial \tilde{x}^k}g_{sp}\frac{\partial x^p}{\partial \tilde{x}^m}, \quad \tilde{\nabla}_k\tilde{\mu}_{\nu m} = \frac{\partial x^s}{\partial \tilde{x}^k}\nabla_s\mu_{\nu p}\frac{\partial x^p}{\partial \tilde{x}^m}, \quad \tilde{\mu}_{\nu m} = \frac{\partial x^s}{\partial \tilde{x}^m}\mu_{\nu s}.$$

The thermodynamic potential of a magnet, including the energies of magnetic, elastic, and magneto-elastic interaction, is given as

$$W = \int_{V_0}[f(\eta_{ij}, K_{\nu i}, K_{\nu ij}) - (\mathbf{M}\cdot\mathbf{H}^{\text{ext}})]\mathrm{d}^3\mathbf{x}, \tag{1.4.9}$$

where the undeformed body volume V_0 is the domain of the integration, and the energy density f is an expansion in terms of invariants (1.4.2). The energy invariance requirement (1.4.9) controls the coefficients in the expanded f relative to transformations of the symmetry group of the undeformed magnetic crystal, which affect the Lagrangian coordinates x_i.

In the general case the potential W contains not only bulk but also surface energies. The bulk energy specifies the form of the dynamic equations of a crystal magnet. The surface energy defines the boundary conditions of the problem. In the absence of medium deformations ($u_i = 0$), the potential (1.4.9) coincides with the previously considered energy of a perfect magnetic crystal (1.1.3). When $\mathbf{M}_\nu = 0$, the potential (1.4.9) boils down to the nonlinear elastic energy $W^{(e)}$ of the sample. This energy can be represented in the form of expansion [54, 29]:

$$W^{(e)} = \int_{V_0} [\lambda_{ik,sp}\, \eta_{ik}\eta_{sp} + \lambda_{ik,sp,mn}\, \eta_{ik}\eta_{sp}\eta_{mn} + \cdots] \mathrm{d}^3\mathbf{x}, \tag{1.4.10}$$

where $\lambda_{ik,sp}$, etc. are elastic moduli of the material. On the right side of eq. (1.4.10), we have omitted the linear-in-η_{ij} terms because they meet the initial loading of the sample.

At first glance, it may seem that the energy of magnetic and magneto-elastic interactions can be derived by formally substituting

$$M_{\nu i} \to \mu_{\nu i}, \quad \frac{\partial M_{\nu i}}{\partial x^s} \to \nabla_s \mu_{\nu i} \tag{1.4.11}$$

in the expression for the energy of a perfect undeformed magnetic crystal (1.1.3). However, this is untrue, though, as the magnet symmetry allows for more complex interactions as well. Moreover, minimal interactions are not basic, as the gauge theories of elementary particles always suggest [34, 40, 53, 20]. For example, we cannot get the correct magneto-elastic interaction energy if we make the formal substitution (1.4.11) in the energy (1.3.5) of a uniaxial ferromagnet. First, this is because the model of a uniaxial crystal is an approximation that denies magnetic anisotropy terms in the basal plane, which are terms of the same order of smallness as the magneto-elastic interactions. Secondly, even if we take into account the terms of the magnetic anisotropy in the basal plane of the crystal, with the formal replacement (1.4.11), they still will not reproduce the necessary magneto-elastic interactions (of course, this entails the main magneto-elastic interactions of the same order of magnitude).

Let the energy dissipation in the system and changes in the lengths of the sublattice magnetization vectors be neglected. Then the nonlinear dynamics of a magneto-elastic medium is described by the following system of the Landau–Lifshitz equations and equations of elasticity theory:

$$\begin{aligned} \partial_t \mathbf{M}_\nu &= -\gamma_\nu [\mathbf{M}_\nu \times \mathbf{H}_\nu^{\mathrm{eff}}], \\ \mathbf{M}_\nu^2 &= M_{0\nu}^2 = \mathrm{const}, \\ \rho_0\, \partial_t^2 u^i &= \partial_s P^{is}, \end{aligned} \tag{1.4.12}$$

where

$$\mathbf{H}_\nu^{\text{eff}} = -\frac{\delta W}{\delta \mathbf{M}_\nu}$$

is the effective field, and P^{is} is the Piola–Kirchhoff tensor (generally speaking, it is not symmetrical in the indices i and s):

$$P^{is} = \frac{\partial W}{\partial[\partial_s u_i]}.$$

In the case of a ferromagnetic crystal with one magnetic lattice, the dynamic equations (1.4.12) and the boundary conditions necessary for them to solve can be derived from Hamilton's variational principle

$$\delta \int_{t_1}^{t_2} L \, \mathrm{d}t + \int_{t_1}^{t_2} \delta A \, \mathrm{d}t = 0. \tag{1.4.13}$$

Here $L = K - W$ is the Lagrange function of a magnetoelastic sample, and the functional K has the meaning of the kinetic energy of the medium:

$$K = \int_{V_0} \left[\frac{1}{2}\rho_0 (\partial_t^2 \mathbf{u})^2 + \frac{1}{\gamma} M_0 (1 - \cos\theta)\, \partial_t \varphi \right] \mathrm{d}^3\mathbf{x}.$$

We have used the parameterization of the medium magnetization $\mathbf{M}(\mathbf{x}, t)$ by the angles θ, φ (1.3.6) ($|\mathbf{M}| = M_0 = \text{const}$).

The work of external surface forces has the form

$$\delta A = \int_S \delta u_i \, T^{ij} \, \mathrm{d}S_j. \tag{1.4.14}$$

It is worth emphasizing that the external stresses T^{ij} are applied to the surface S of the deformed body. The force $T^{ij}\,\mathrm{d}S_j$ acts on the surface element $\mathrm{d}S_j$ of the deformed body. Therefore, to use the Hamilton principle, in (1.4.14), we should replace the integration over the surface S by integration over the surface σ of an undeformed body. The required transformation

$$\int_S \delta u_i \, T^{ij} \, \mathrm{d}S_j = \int_\sigma \delta u_i \, T^{ij} \frac{\partial \det A}{\partial A_s^j} \mathrm{d}\sigma_s \tag{1.4.15}$$

is achieved by means of the identities

$$\begin{aligned}
\mathrm{d}S_j &= \frac{1}{2} \varepsilon_{jkp} \, \mathrm{d}X^k \wedge \mathrm{d}X^p \\
&= \frac{1}{2} \left(\varepsilon_{j'kp} \frac{\partial X^{j'}}{\partial x^s} \frac{\partial X^k}{\partial x^m} \frac{\partial X^p}{\partial x^n} \right) \frac{\partial x^s}{\partial X^j} \mathrm{d}x^m \wedge \mathrm{d}x^n
\end{aligned}$$

$$= \det\left\|\frac{\partial \mathbf{X}}{\partial \mathbf{x}}\right\| \frac{\partial x^s}{\partial X^j} \frac{1}{2} \varepsilon_{smn} \mathrm{d}x^m \Lambda \, \mathrm{d}x^n$$

$$= \det\left\|\frac{\partial \mathbf{X}}{\partial \mathbf{x}}\right\| \frac{\partial x^s}{\partial X^j} \mathrm{d}\sigma_s = \frac{\partial \det A}{\partial A_s^j} \mathrm{d}\sigma_s,$$

where Λ is the symbol of the outer product, and

$$A_l^k = \frac{\partial X^k}{\partial x^l} = \delta_l^k + \frac{\partial u^k}{\partial x^l}.$$

Here we have employed the representation of the inverse matrix

$$(A^{-1})_j^s \equiv \frac{\partial x^s}{\partial X^j} = \frac{\partial \ln \det A}{\partial A_s^j}.$$

In formulas (1.4.14) and (1.4.15), it would be possible not to distinguish between the upper and lower indices for the tensors since in this case, all the tensor quantities refer to the Cartesian coordinate system connected to the nondeformable body.

The variation of the action functional (1.4.13) is carried out over the fields θ, φ, and u_i, provided that $\delta\theta = 0$, $\delta\varphi = 0$, and $\delta u_s = 0$ at the ends of the time interval $[t_1, t_2]$. In addition, $\delta u_s = 0$ on that part of the surface σ where the displacements u_s^{ext} are given rather than the stresses T^{ij}.

In this case the total energy E of the magnet does not coincide with the potential energy W:

$$E = \int_{V_0} \frac{1}{2} \rho_0 \, (\partial_t \mathbf{u})^2 \mathrm{d}^3\mathbf{x} + W.$$

Note also that the unharmonicity induced into elatic subsystem of the crystal by the magnetic subsystem is more significant than the intrinsic unharmonicity of the elastic subsystem. Therefore, when analyzing nonlinear phenomena, the magnetoelastic energy of a crystal is typically written in a linear approximation, and the elastic energy is written in approximation that is quadratic in displacement gradients $\partial_s u_k$. Furthermore, it would be possible not to distinguish between the surfaces of the body before and after deformation ($\mathrm{d}S_i \approx \mathrm{d}\sigma_i$). In such approximations, the Lagrangian and Euler descriptions coincide.

We have deduced an expression for the energy of a crystal magnet relying on the condition of invariance of the energy (when $\mathbf{H}^{\mathrm{ext}} = 0$) with respect to global rotations of the deformable body (1.4.1). We can assume that rotational defects of the crystal lattice (disclinations), if any, preserve the energy invariance solely with respect to local rotations (1.4.1) with the matrix D depending on the Lagrangian coordinates x_s. In turn, translational lattice defects (dislocations), if any, retain the invariance of the system energy only for local translations $X_k \to X_k + \eta_k(\mathbf{x}, t)$. The requirement of local symmetry of the system energy can be satisfied by including additional gauge fields into the theory, which by their physical meaning should theoretically describe the magnitude of the relative shift or rotation of neighboring areas of the medium, i. e., plastic deformation of the material or a disordered state of the medium, locally resembling to a perfect crystal. The principle of local invariance of the system energy, which reflects the preservation of the ideal order in small areas of the medium, is responsible for the form of differential invariants that are the energy of interaction between the magnetic and elastic subsystems and the gauge fields. Within this approach, geometry no longer serves as a tool of description. The local symmetry of the medium becomes dynamical symmetry in the sense that the features of the field dynamics turn out to be determined by the geometric properties of the medium.

Unfortunately, the concept set forth in [21, 79, 57, 38, 56, 24, 55] faces serious difficulties and contains controversial statements [48, 49]. Choosing a Lagrangian for free gauge fields is one of the most difficult aspects. This manifests itself, for example, in the discrepancy between the conclusions of the gauge theory of dislocations and disclinations in crystals [38] and the results of the less general but reliably established inconsistent theory of elasticity [48]. The considered example of magneto-elastic interactions shows that in defect-free crystals, the invariance of the geometric characteristics of the medium is not enough. To formulate the theory, it is necessary to write down the total invariant energy of the system.

The application of the gauge approach to constructing field models of defects and theoretically describing disordered magnets that have a local crystalline order still remains an open problem.

1.5 Exchange symmetry and the method of phenomenological Lagrangians

Symmetry, no matter how broadly or narrowly we understand this word, is an idea with the help of which a person tried to explain and create order, beauty and perfection.

Hermann Weyl, Symmetry

The hypothesis of the existence of magnetic sublattices is basic for the conventional macroscopic theory of crystal magnets. Despite the correct conclusions that come from this model assumption, there is, strictly speaking, no convincing microscopic justification for it. Andreev, Volkov, Marchenko, and Zheltukhin [77, 75, 76, 3, 2, 4] proposed an alternative phenomenological theory underlain by general symmetry considerations. It excludes the concept of sublattices, which in many cases is redundant and complicates a theoretical description of the nonlinear dynamics of magnetic media. The new approach is equally applicable to both crystals and disordered magnets, conditional upon magnetic structures arising in these media due to the fact that exchange interactions are mainly stronger as compared to magnetoanisotropic (relativistic) ones. It turned out that any exchange magnetic structure can be characterized by no more than three mutually orthogonal magnetic vectors. In the case of spontaneously broken symmetry of the system as a result of forming an equilibrium magnetic structure, the group-theoretic analysis of the interaction of Goldstone modes that tend to restore the system symmetry relative to rotations in the spin space of the spin moments of crystal atoms gives equations of nonlinear dynamics for spin waves with the lowest excitation energy. Although these equations are simpler than the Landau–Lifshitz equations for sublattice magnetizations, the new phenomenological approach is self-consistent with the previous one in its range of applicability. Moreover, it makes it possible to quite simply reveal a number of general regularities derived only under the extremely stringent limitations of the microscopic theory.

1.5.1 Exchange symmetry of magnets

The conventional phenomenological theory describes the symmetry properties of crystal magnets via magnetic space groups. Groups as elements contain purely spatial transformations (rotations, reflections, and translations) combined with the time reversal operation R. Under transformations of the magnetic symmetry group, a charge density $\rho(\mathbf{x})$ and a semiclassical density of the magnetic moment $\mathbf{m}(\mathbf{x})$ of the medium remain invariant ($\rho(\mathbf{x})$ and $\mathbf{m}(\mathbf{x})$ are the quantities derived through quantum mechanical and thermodynamic averaging over small volumes of the medium). Despite the fact that such a description of the symmetry of the magnetic medium is complete, anyway, it loses the most essential information. The point is that a crystal's equilibrium magnetic structure to be established should be exposed mainly to exchange forces whose symmetry is higher than that of weak relativistic interactions. Therefore formulating the equations of dynamics of magnetic media requires a more correct approach concerning a consideration of the approximate symmetry of the exchange forces rather than an analysis of the exact magnetic symmetry of the crystal [4].

Let us neglect all relativistic interactions and discuss the exchange symmetry of a magnet. Since the exchange interactions depend only on the relative orientations of the spin moments of neighboring crystal atoms, the general orientation of the spin moments of atoms with respect to the crystallographic axes becomes conditional. As a result, an infinite aggregate of new symmetry elements appears, consisting of rotations U of the spin space. In other words, this set comprises arbitrary rotations of all the crystal atoms' spin moments through the same angle. The exchange symmetry of crystal magnets is determined by assigning the exchange space groups, each of which contains combinations of the time reversal operation R with purely spatial transformations and spin space rotations U as its elements. If in the elements of the exchange group of a crystal magnet, we formally identify the R- and U-transformations with the identity one, then we get the usual space group G of the crystal, with respect to which the charge density and the densities of all spin scalars in the system remain invariant. During rotations in spin space, the quantity $\mathbf{m}(\mathbf{x})$ behaves like a vector. The time reversal operation corresponds to the inversion of the vector of $\mathbf{m}(\mathbf{x})$ in spin space:

$$R\mathbf{m} = -\mathbf{m}.$$

To characterize the equilibrium exchange structure of a magnet, we represent the density of the magnetic moment $\mathbf{m}(\mathbf{x})$ as an expansion in a complete system of functions $\{\varphi_{\nu\alpha}^{(a)}(\mathbf{x})\}$. The latter are basic for irreducible representations of the space group G:

$$\mathbf{m}(\mathbf{x}) = \sum_{a=1}^{3} \sum_{\nu,\alpha} \mathbf{M}_{\nu\alpha}^{(a)} \varphi_{\nu\alpha}^{(a)}(\mathbf{x}). \tag{1.5.1}$$

The functions $\varphi_{\nu\alpha}^{(a)}$ with given a are transformed according to the νth irreducible representation of the group G. The index α enumerates the functions belonging to this representation. For each ν, there are, generally speaking, three sets of independent functions labeled by the index a. Since $\mathbf{m}(\mathbf{x})$ is real valued, the expansion coefficients $\mathbf{M}_{\nu\alpha}^{(a)}$ and functions $\varphi_{\nu\alpha}^{(a)}(\mathbf{x})$ can be chosen to be real as well. The quantities $\mathbf{M}_{\nu\alpha}^{(a)}$ are transformed as vectors during spin rotations U and change their sign under the time reversal operation R. Since the group is finite, we can construct a representation to express the action of an element of the symmetry $g \in G$ on any of the coordinate functions $\varphi_{\nu\alpha}^{(a)}(\mathbf{x})$ as

$$\varphi_{\nu\alpha}^{(a)}(\mathbf{x}) \xrightarrow{g} \sum_{\beta} U_{\nu\beta\alpha}^{(a)}(g)\, \varphi_{\nu\beta}^{(a)}(\mathbf{x}),$$

where the real matrices $U_{\nu\beta\alpha}^{(a)}(g)$ satisfy the orthogonality conditions of [58, 64] ($n(G)$ is the number of elements in the group G, and l_ν is the dimension of the νth representation):

$$\sum_{g} U_{\nu\alpha\beta}^{(a)}(g)\, U_{\nu'\alpha'\beta'}^{(a')}(g) = \frac{n(G)}{l_\nu} \delta_{aa'} \delta_{\nu\nu'} \delta_{\alpha\alpha'} \delta_{\beta\beta'}. \tag{1.5.2}$$

The quantity $\mathbf{m}^2$ is a spin scalar: it must remain invariant under the G-group transformations in the crystal's equilibrium state. By the orthogonality conditions (1.5.2) this is possible only if the following equalities are fulfilled:

$$(\mathbf{M}_{\nu\alpha}^{(a)} \cdot \mathbf{M}_{\nu'\alpha'}^{(a')}) = C(a,\nu)\, \delta_{aa'} \delta_{\nu\nu'} \delta_{\alpha\alpha'}. \tag{1.5.3}$$

Relations (1.5.3) indicate that the different magnetic vectors $\mathbf{M}_{\nu\alpha}^{(a)}$ are perpendicular to each other, and those of them that correspond to one irreducible representation have the same length. The maximum number of the mutually orthogonal three-dimensional vectors with a given ν is equal to three. Consequently, the dimension of an irreducible representation cannot be greater than three.

For further analysis, it is more convenient to assume that under the action of the G-group elements, it is not the functions $\varphi^{(a)}_{\nu\alpha}$ that are transformed, but the magnetic vectors themselves:

$$\mathbf{M}^{(a)}_{\nu\beta} \xrightarrow{g} \sum_{\alpha} U^{(a)}_{\nu\beta\alpha}(g)\,\mathbf{M}^{(a)}_{\nu\alpha}.$$

The magnetic vectors $\mathbf{M}^{(a)}_{\nu\alpha}$ transforming according to nonidentity representations of the group G make no contribution to the total magnetization of the medium. Therefore they are antiferromagnetic vectors. The latter always can be achieved equal to unit length by changing the normalization of the functions $\varphi^{(a)}_{\nu\alpha}$. We denote them as $\mathbf{l}_a$ (the index a now enumerates different antiferromagnetic vectors and takes no more than three values $a = 1,2,3$). When the vectors $\mathbf{M}^{(a)}_{\nu\alpha}$ are transformed according to the identity representation of the group G,

$$\mathbf{M}^{(a)}_{\nu\alpha} \xrightarrow{g} \mathbf{M}^{(a)}_{\nu\alpha},$$

one of them can be made coinciding with the spontaneous magnetization $\mathbf{M}$ of the medium by means of a suitable orthogonal transformation and normalization of the functions $\varphi^{(a)}_{\nu\alpha}$. The rest of the identity representation vectors, if any, are antiferromagnetic and make no contribution to the average crystal magnetization.

There are only four different cases that may arise (Figure 1.4).

1. There is a single magnetic vector that transforms according to the identity representation and coincides with the equilibrium magnetization $\mathbf{M}$ of the crystal as a ferromagnet or a collinear ferrimagnet.
2. There is a single vector $\mathbf{l}$ transformed according to a nonidentity one-dimensional representation. The body is a collinear antiferromagnet.
3. There is a vector of the total magnetization $\mathbf{M}$ of the medium and one or two antiferromagnetic vectors $\mathbf{l}_a$ ($a = 1$ or $a = 1, 2$) orthogonal to it. The body is a noncollinear ferrimagnet.
4. There are two or three mutually perpendicular vectors $\mathbf{l}_a$. The crystal is a noncollinear antiferromagnet.

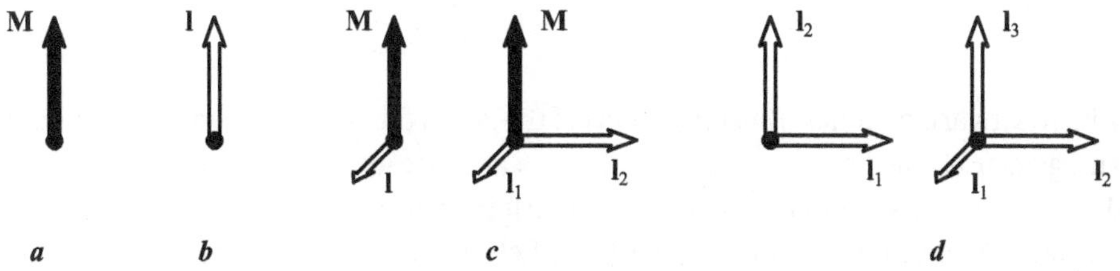

Figure 1.4: Vectors of ferromagnetism (the dark arrow) and antiferromagnetism (the light arrow) for different magnetic crystals: (a) a ferromagnet or a collinear ferrimagnet; (b) a collinear antiferrromagnet; (c) a noncollinear ferrimagnet; (d) a noncollinear antiferromagnet.

As is known, the translational symmetry of a crystal is described by functions of the form

$$u_{\mathbf{k}}(\mathbf{x})\ \exp(i\mathbf{k}\cdot\mathbf{x}),$$

where $\mathbf{k}$ is a wave vector. The function $u_{\mathbf{k}}(\mathbf{x})$ has the periodicity of the Bravais lattice of the crystal. We cover one-, two-, or three-dimensional real irreducible representations of the group G. Such representations can be characterized by wave vectors occupying a general position in the reciprocal cell only for crystals of the triclinic system (in more symmetric crystals, the combined star of the vectors $\mathbf{k}$ and $-\mathbf{k}$ contains more than three vectors). In crystals of a monoclinic system, representations with a wave vector occupying a general position in the symmetry plane of the reciprocal lattice are possible. Representations with a wave vector occupying a general position on the axis of symmetry are possible for all noncubic crystals. In the cases listed above, the irreducible representation of the group G is always two-dimensional [4], with a pair of the mutually orthogonal vectors $\mathbf{l}_1$ and $\mathbf{l}_2$ transformed under translations as a pair of the functions $\sin(\mathbf{k}\cdot\mathbf{x})$ and $\cos(\mathbf{k}\cdot\mathbf{x})$. Then from expansion (1.5.1) it follows that the density of the magnetic moment of the medium is transformed under translations as

$$\mathbf{m}(\mathbf{x}) \backsim \mathbf{l}_1 \sin(\mathbf{k}\cdot\mathbf{x}) + \mathbf{l}_2 \cos(\mathbf{k}\cdot\mathbf{x}).$$

We have come to the conclusion that in a periodic crystal, the only exchange structures whose periods are incommensurable with the lattice ones are helicoidal magnetic structures.

Nonhelicoidal magnetic structures are theoretically described by wave vectors that correspond to selected points of the reciprocal lattice. In particular, if all nonzero magnetic vectors are transformed according to the representation with $\mathbf{k} = 0$, then the magnetic and spatial elementary cells coincide.

In reality, not all possible magnetic structures from the point of view of exchange symmetry are observed, but only those that satisfy the energy criterion of stability. For example, the summand

$$K_i^{ab}\left(\mathbf{l}_a \cdot \frac{\partial \mathbf{l}_b}{\partial x_i} - \mathbf{l}_b \cdot \frac{\partial \mathbf{l}_a}{\partial x_i}\right), \tag{1.5.4}$$

which is invariant under transformations of the group G in the expansion of the system energy, causes the magnetic structure to be instable because it can be negative at small deviations of the vectors $\mathbf{l}_a$ from their equilibrium values.

Meanwhile, an important peculiarity of helicoidal magnetic structures with wave vectors occupying a general position in the plane (on the axis) of reciprocal cell symmetry is as follows: Invariants (1.5.4) result in instabilities of the structures only when expression (1.5.4) includes a derivative over a spatial coordinate along the direction characterized by a wave vector orthogonal to the symmetry plane (axis) of the reciprocal cell.

There exist no invariants of the type (1.5.4) for representations with wave vectors whose components are equal to 1/2, 1/3, or 1/4 of the reciprocal lattice periods [4]. This means that the periods of a magnetic cell are greater by a factor of 2, 3, or 4 than those of a spatial cell. The tripling of magnetic periods compared to spatial ones is possible only in hexagonal spatial lattices, and the quadrupling is possible in face-centered and body-centered cubic lattices.

The exchange magnetic structures classified above can also be extended to crystalline disordered magnets. The macroscopic symmetry of such media can be characterized using multipole moments [4]:

$$\begin{aligned} \mathbf{M} &= \langle \mathbf{m}(\mathbf{x}) \rangle, \\ \mathbf{M}_i &= \langle x_i\, \mathbf{m}(\mathbf{x}) \rangle, \\ \mathbf{M}_{ik} &= \left\langle \left(x_i x_k - \frac{1}{3}\, \delta_{ik} x_s^2 \right) \mathbf{m}(\mathbf{x}) \right\rangle, \\ &\dots, \end{aligned} \tag{1.5.5}$$

where $\mathbf{m}(\mathbf{x})$ is the semiclassical density of the magnetic moment. The angle brackets stand for averaging over all configurations of the disordered system.

For the nonzero multipole moments to be invariant, the exchange symmetry group of a crystal-disordered medium consists of all combinations of spatial rotations, inversion, spin rotations U, and time reversal transformations R. The purely spatial symmetry group G of the medium comes from the exchange group by formally identifying the elements R and U under the identity transformation. The multipole moments (1.5.5) are transformed by irreducible representations of the group G. Therefore the same phenomenological approach [4] might be applied for classifying the exchange structures in disordered magnetic media and theoretically describing their noninear dynamics as well.

1.5.2 Dynamical symmetries, method of phenomenological Lagrangians for Goldstone particles

It is well known that the conservation laws for a system of fields are intimately related to the invariance of the Lagrange function under transformations that reflect the algebraic symmetry of a problem (Noether's theorem) [13]. Since linear field-gradients-independent transformations yield algebraic symmetries on field variables, they appear to be insufficient for choosing the system Hamiltonian writing format or for specifying the Lagrange function. The explicit form of the Lagrange function can be required through its invariance with respect to a wider group of transformations and be implemented on field variables inhomogeneously and nonlinearly. Besides, it characterizes the so-called dynamic symmetry of the problem. The principles of dynamical symmetry cannot be deduced by logical reasoning, as Einstein emphasized, but "intuition based on insight into the essence of experience" leads to them. The hypothesis we have already discussed about the invariance of the Lagrange function of a disordered medium with respect to local gauge transformations can serve as an example of dynamic symmetry. This hypothesis reflects the preservation of a perfect crystalline order in small volumes of the medium. Building phenomenological Lagrangians to describe the nonlinear dynamics of spin waves with the lowest excitation energy is a further dramatic example of the successful application of dynamical symmetry principles in the theory of magnetism. Progress in this direction becomes achievable through a deep group-theoretical analysis of the following general property of magnetically ordered states. The latter always arise accompanied by the phenomenon of spontaneous breaking of the symmetry of exchange interactions: an exchange-interaction Hamiltonian is invariant under any turn of all atomic spins in a crystal by the same angle. However, no magnetic structure is invariant under such rotations. The magnetic vectors $\mathbf{M}_{\nu a}^{(a)}$ that specify the equilibrium exchange magnetic structure are changed when the spin moments of all atoms of the crystal rotate by the same angle. The spin moments' coordinate- and time-dependent local rotations that alter the equilibrium values of the structure's magnetic vectors $\mathbf{M}_{\nu a}^{(a)}$ involve elementary magnetic excitations of the system. Due to the short-range nature of the exchange forces, the energy of these elementary excitations (magnons) vanishes when the momentum approaches zero. The last claim comes from the general Goldstone theorem [81, 45, 31], which is formulated as follows.

Let G be a continuous symmetry group whose generators are written as integrals over the entire space of local densities, and let the Hamiltonian of a system with short-range forces be invariant with respect to this group. If the system ground (vacuum) state is not invariant with respect to this symmetry group, then the Hamiltonian spectrum necessarily contains a branch of elementary excitations, with their energy tending to zero as the excitation momentum tends to zero.

From a qualitative point of view, in every equilibrium magnetic structure, one of the possible vacuum states of a magnet prevails, and thereby the algebraic symmetry of the system's initial micro-Hamiltonian with respect to rotations in spin space is violated. As a consequence, Goldstone excitations (magnons) appearing in the system tend to restore the broken symmetry. Since magnons are Goldstone quasiparticles, they can be described by applying a universal group-theoretic method for the nonlinear dynamics of Goldstone fields in systems with spontaneous vacuum symmetry breaking. This technique is frequently employed in elementary particle physics [19, 18, 72, 73, 78]. Such an approach allows us to reveal the transformation properties of Goldstone fields by means of inhomogeneous nonlinear transformations that take into account the dynamic symmetry of the problem. Ultimately, the method provides an explicit expression for the Lagrange function of Goldstone particles tending to restore the primary symmetry of interactions, spontaneously broken by nonzero vacuum averages arisen in the system. In addition, it should be stressed that the internal mechanism of emerging Goldstone particles, as well as the mechanism of restoring the symmetry, remain beyond the scope of the method.

The above approach enables us to:

- Successfully describe processes involving soft pions [78, 63];
- Regard a neutrino as a Goldstone fermion in the theory for a supersymmetry group to be nonlinearly realized [74];
- Consider the theory of gravity as a theory of joint nonlinear realizations of affine and conformal groups [15];
- Treat the theory of gauge fields as a theory of nonlinear realization of the group of gauge transformations [37].

It is worth underscoring that the extension of the method of phenomenological Lagrangians to the theoretical description of nonlinear Goldstone excitations in condensed matter physics is extremely important. We first discuss the main postulates and mathematical technique of the method, followed by the construction of phenomenological Lagrangians for spin waves.

Let G be a continuous symmetry group of interactions in a system or a symmetry group of some initial Lagrangian that causes vacuum degeneration and the appearance of Goldstone particles, and let H be its maximum subgroup that keeps the vacuum invariant. The method for constructing the effective Lagrangian of Goldstone particles is based on the following assumptions [72, 73]:

1. In the long-wavelength limit, Goldstone particles are described by local fields $A(\mathbf{x}, t)$ whose quantum characteristics and transformation properties are determined by the unique correspondence of these fields to the parameters a of the left (right) cosets of the group G by the subgroup H.
2. The effective Lagrangian for the interaction of Goldstone particles is invariant under transformations of the group G.

Further, when discussing the transformation properties of the fields, we will demonstrate that the subgroup H in the approach at hand stands out through saving the Lagrangian's algebraic symmetry and characterizing it by homogeneous linear representations. Therefore the transformations of the subgroup H are unrelated to Goldstone particles.

To establish the transformation law for local Goldstone fields under the operations of the group G, we write down an arbitrary element g of the group as a product

$$g = K(a)\,H(b), \tag{1.5.6}$$

where $H(b)$ is a transformation of the subgroup H, $K(a)$ is a representative of the left coset G/H of the group G by the subgroup H, and a and b are real parameters of the group space (in the general case, each of the a and b quantities has a finite number of components).

After multiplying from the left by any element $g_1 \in G$, an element $K(a) \in G$ passes into some element g' of the group G to be decomposed again in accordance with (1.5.6):

$$g_1 K(a) = g' = K(a')\, h(b'). \tag{1.5.7}$$

According to (1.5.7), under the operation $g_1 \in G$, the coordinates (parameters) a of the factor-space G/H are transformed according to the nonlinear law:

$$a \xrightarrow{g_1} a' = a'(a, g_1). \tag{1.5.8}$$

This induces a left shift of $h(b')$ on the vacuum symmetry group of the system with parameters b' depending on a and g_1:

$$b \xrightarrow{g_1} b' = b'(a, g_1). \tag{1.5.9}$$

Combining eqs. (1.5.6) and (1.5.7), we have

$$g_1 g = K(a')\, h(b')\, H(b).$$

For further analysis, it is important that under factorization (1.5.6), transformation (1.5.8) contains no parameters b. We also note that each specific parameterization of the G-group elements corresponds to a certain choice of the coordinates a in the factor-space G/H. It can be shown that the arbitrariness in the choice of coordinate systems in the factor-space G/H is not essential for describing the interaction of Goldstone particles.

The method of phenomenological Lagrangians claims that each parameter a meets a local Goldstone field $A(\mathbf{x}, t)$ so that under the operations of the group G, the Goldstone fields are transformed according to the law

$$A(\mathbf{x}, t) \xrightarrow{g_1} A'(\mathbf{x}, t) = a'(A(\mathbf{x}, t), g_1). \tag{1.5.10}$$

Fields $\Psi(\mathbf{x}, t)$ that are not Goldstone ones must be transformed by linear representations of the subgroup H:

$$\Psi(\mathbf{x}, t) \xrightarrow{h(b')} \Psi'(\mathbf{x}, t) = D(h(b'))\Psi(\mathbf{x}, t). \tag{1.5.11}$$

Here $D(h)$ are matrices that make the representation of the subgroup H in the space of the functions Ψ. Using representation (1.5.11) and transformation law (1.5.9) for the H-subgroup parameters under the action of the elements of the group G, we can construct the so-called induced representation of the group G in the space of the functions Ψ:

$$\Psi(\mathbf{x}, t) \xrightarrow{g_1} \Psi'(\mathbf{x}, t) = D\{h(b'(A, g_1))\}\Psi(\mathbf{x}, t). \tag{1.5.12}$$

It is clear that the transformations on $\Psi(\mathbf{x}, t)$ make sense only together with the transformations on $A(\mathbf{x}, t)$ because b' depends on A. Relations (1.5.10) and (1.5.12) determine the G-group nonlinear realization on the local fields A and Ψ of the physical system at hand.

Given the known transformational properties of the functions A and Ψ, the phenomenological Lagrangian of fields to be constructed in the long-wave approximation reduces to finding the G-group transformation invariants that contain a minimum number of derivatives of the local fields $A(\mathbf{x}, t)$. The heterogeneity of transformations (1.5.10) brings about an important consequence: there are no invariants composed only of the derivative-free Goldstone fields $A(\mathbf{x}, t)$. This means, in particular, that the phenomenological Lagrangian of Goldstone particles does not have a mass term.

As an illustration of the mathematical technique of the method, we resort to a specific example [78].

Let the symmetry group G of primary interactions in a system be determined by generators X_l ($l = 1, 2, \ldots, n$) and Y_α ($\alpha = 1, 2, \ldots, m$), and let only the generators Y_α of the subgroup H satisfy the relations $Y_\alpha|0\rangle = |0\rangle$, where the ket-vector $|0\rangle$ describes the vacuum state of the system. This implies that H is a vacuum stability group. The commutation relations of the G-group algebra have the form

$$[X_l, X_m] = \mathrm{i}C^{\alpha}_{lm} Y_\alpha, \quad [X_l, Y_\alpha] = \mathrm{i}C^{m}_{l\alpha} X_m, \quad [Y_\alpha, Y_\beta] = \mathrm{i}C^{\gamma}_{\alpha\beta} Y_\gamma. \tag{1.5.13}$$

When permutating the inferior indices, the structure constants C^{α}_{lm}, $C^{m}_{l\alpha}$, and $C^{\gamma}_{\alpha\beta}$ become antisymmetric and are related to each other by the equalities that follow from (1.5.13) and the Jacobi identities. In some vicinity of the unit element of the group G, any element of $g \in G$ uniquely decomposes into a product of the form

$$g = \exp(\mathrm{i}\, a^l X_l)\, \exp(\mathrm{i}\, b^\alpha\, Y_\alpha),$$

where a^l and b^α are real parameters. The aggregate of the elements $K(a) = \exp(\mathrm{i}\, a^l X_l)$ originates the factor-space G/H, whereas the set of the elements $\exp(\mathrm{i}\, b^\alpha\, Y_\alpha)$ produces the subgroup $H(b)$.

Given the commutator algebra (1.5.13), the specific form of formulas (1.5.10) and (1.5.12) for determining the nonlinear realization of the group G on the system's Goldstone fields $A^l(\mathbf{x}, t)$ and on the non-Goldstone fields $\Psi(\mathbf{x}, t)$ can be derived from eq. (1.5.7).

In particular, at $g_1 = h = \exp(\mathrm{i}\, b^\alpha\, Y_\alpha) \in H$, the left side of relation (1.5.7) can be written as

$$h \exp(\mathrm{i}\, a^l X_l) = (h\, \exp(\mathrm{i}\, a^l X_l)\, h^{-1})\, h. \tag{1.5.14}$$

Further calculations are simplified by an artificial technique. Let us introduce the auxiliary operator function of the parameter s:

$$F(s) = h \exp(\mathrm{i}\, a^l X_l\, s) h^{-1},$$

which is a solution of the differential equation

$$\frac{\mathrm{d}F}{\mathrm{d}s} = \mathrm{i}\, a^l h X_l h^{-1} F(s) \tag{1.5.15}$$

with the initial condition $F(s = 0) = I$. Upon finding $F(s)$, we also calculate the product of the operators in (1.5.14):

$$F(s = 1) = h \exp(\mathrm{i}\, a^l X_l)\, h^{-1}.$$

Using both the identity

$$e^A B e^{-A} = B + \frac{1}{1!}[A, B] + \frac{1}{2!}[A[A, B]] + \cdots, \tag{1.5.16}$$

which is valid for any operators A and B, and the permutation relations (1.5.13), it is not hard to show that

$$h X_l h^{-1} = X_p S(h)^p_l, \tag{1.5.17}$$

where $S(h)$ is an $n \times n$ matrix dependent on the parameters b^a:

$$S^k_l = \sum_{n=0}^{\infty} \frac{1}{n!} (P^n)^k_l = (\exp P)^k_l, \tag{1.5.18}$$

where $P^k_l = C^k_{la}\, b^a$, $(P^2)^k_l = P^k_s\, P^s_l$, etc. With formula (1.5.17), we can integrate eq. (1.5.15):

$$F(s) = \exp(\mathrm{i} X_m S^m_l a^l s).$$

In the long run, we find

$$F(s = 1) = h \exp(\mathrm{i} a^l X_l)\, h^{-1} = \exp(\mathrm{i} X_m S^m_l a^l).$$

This means that representation (1.5.14) gives the required factorization (1.5.7):

$$h \exp(\mathrm{i} a^l X_l) = \exp(\mathrm{i} a^l S^m_l(h) X_m) \exp(\mathrm{i} b^a Y_a).$$

Hence we can establish the law of transformation of the local fields

$$A^m(\mathbf{x}, t) \xrightarrow{h} S^m_l(h) A^l(\mathbf{x}, t), \quad \Psi(\mathbf{x}, t) \xrightarrow{h} D(h)\Psi(\mathbf{x}, t).$$

Here $h = \exp(\mathrm{i}\, b^a\, Y_a)$ and b^a are independent of A^l.

Thus we have shown that the subgroup H is realized uniformly and linearly on all the fields and thereby describes the algebraic symmetry of the problem.

When $g_1 \notin H$, the transformation properties of the local fields are far much complicated, so we restrict ourselves to considering infinitesimal transformations with $g_1 = I + \mathrm{i}\varepsilon^l X_l$ ($|\varepsilon^l| \ll 1$). Then eq. (1.5.7) appears as

$$(I + \mathrm{i}\varepsilon^l X_l)\exp(\mathrm{i}a^l X_l) = \exp(\mathrm{i}(a^l + \delta a^l)X_l)\exp(\mathrm{i}\delta b^\alpha Y_\alpha). \tag{1.5.19}$$

Next, we find an explicit form of the parameters δa^l and δb^α. Since $\delta a^l, \delta b^\alpha = O(\varepsilon^l)$, the exponential functions of noncommuting operators on the right side of eq. (1.5.19) need to be expanded into a series up to linear terms in the parameters δa^l and δb^α for collating the order-of-magnitude terms.

We first expand the operator function

$$\exp(\mathrm{i}\,(a^l + \delta a^l)X_l).$$

To achieve this goal, it is also useful to introduce the auxiliary operator function of the parameter s:

$$F(s) = \exp(\mathrm{i}(a^l + \delta a^l)X_l s), \tag{1.5.20}$$

which is the solution of the differential equation

$$\frac{\mathrm{d}F}{\mathrm{d}s} = \mathrm{i}(a^l + \delta a^l)X_l F \tag{1.5.21}$$

with initial condition $F(s=0) = I$. We represent $F(s)$ as

$$F(s) = \exp(\mathrm{i}a^l X_l s)F_1(s).$$

Then from eq. (1.5.21) we find that the new operator function $F_1(s)$ is the solution of the problem

$$\frac{\mathrm{d}F_1}{\mathrm{d}s} = \mathrm{i}\delta a^k X_k(s)F_1, \quad F_1(s=0) = I, \tag{1.5.22}$$

where

$$X_k(s) = \exp(-\mathrm{i}a^l X_l s)X_k \exp(\mathrm{i}a^p X_p s).$$

Integrating (1.5.22), we arrive at

$$F_1(s) = I + \mathrm{i}\delta a^k \int_0^s X_k(s')F_1(s')\mathrm{d}s'. \tag{1.5.23}$$

The integral equation (1.5.23) can be solved by the method of successive iterations in the form of a series in powers of δa^k. With an accuracy to terms linear in δa^l, we have

$$F_1(s) = I + \mathrm{i}\delta a^k \int_0^s X_k(s')\mathrm{d}s'.$$

Finally, we have deduced $F(s)$ and hence the required expansion of the right-hand side of equality (1.5.20):

$$\begin{aligned} F(s=1) &= \exp(\mathrm{i}(a^l + \delta a^l)X_l) \\ &= \exp(\mathrm{i}a^l X_l)\left[I + \mathrm{i}\delta a^k \int_0^1 X_k(s)\mathrm{d}\,s + \cdots\right]. \end{aligned} \tag{1.5.24}$$

The expansion of the second exponential function is trivial:

$$\exp(\mathrm{i}\delta b^\alpha Y_\alpha) = I + \mathrm{i}\delta b^\alpha Y_\alpha + \cdots. \tag{1.5.25}$$

Plugging eqs. (1.5.24) and (1.5.25) into eq. (1.5.19) and equating terms of the same order of smallness, we bring eq. (1.5.19) to the form

$$\varepsilon^l X_l(s=1) = \delta a^k \int_0^1 X_k(s)\mathrm{d}\,s + \delta b^\alpha Y_\alpha. \tag{1.5.26}$$

To go further, note that the differentiation of the operator function $X_k(s)$ over the parameter s yields the new function

$$Y_\alpha(s) = \exp(-\mathrm{i}a^l X_l s)\, Y_\alpha \exp(\mathrm{i}a^p X_p s).$$

However, after differentiating the function $Y_\alpha(s)$ by s, $X_l(s)$ appears again. As a result, we come up with the closed set of differential equations

$$\frac{\mathrm{d}X_k(s)}{\mathrm{d}s} = -C^\alpha_{kp} a^p Y_\alpha(s), \quad \frac{\mathrm{d}Y_\alpha(s)}{\mathrm{d}s} = -C^l_{\alpha p} a^p X_l(s)$$

with initial conditions

$$X_k(s=0) = X_k, \quad Y_\alpha(s=0) = Y_\alpha.$$

This set of equations is equivalent to the following problem for determining $X_k(s)$:

$$\begin{aligned} &\frac{\mathrm{d}^2 X_k(s)}{\mathrm{d}s^2} = -m^l_k X_l(s), \\ &\left.\frac{\mathrm{d}X_k(s)}{\mathrm{d}s}\right|_{s=0} = -C^\alpha_{kp} a^p Y_\alpha, \quad X_k(s)|_{s=0} = X_k, \end{aligned} \tag{1.5.27}$$

where $m^l_k = -C^\alpha_{kp} C^l_{\alpha s} a^p a^s$. An explicit form of $X_k(s)$ can be found from eq. (1.5.27) in the form of a series that corresponds to the symbolic notation

$$X_k(s) = (\cos\sqrt{m}s)_k^p X_p - \left(\frac{\sin\sqrt{m}s}{\sqrt{m}}\right)_k^l C_{lp}^{\alpha} a^p Y_\alpha. \tag{1.5.28}$$

Substituting eq. (1.5.28) into eq. (1.5.26) and equating the coefficients for the identical generators, we get the formulas necessary for the infinitesimal transformation:

$$\begin{aligned}
\delta a^k &= (\sqrt{m}\cot\sqrt{m})_i^k \varepsilon^i,\\
\delta b^\alpha &= -\left([\sin\sqrt{m} + (\cos\sqrt{m} - 1)\cot\sqrt{m}]\frac{1}{\sqrt{m}}\right)_i^l C_{lp}^{\alpha}\, a^p\, \varepsilon^i.
\end{aligned}$$

To construct a phenomenological Lagrangian, it is necessary to know covariant derivatives of the fields A^l and Ψ with respect to the spatial coordinates $\mathbf{x}$ and time t. Next, for simplification of the notation, it is convenient to introduce the four-vector

$$x^\mu \quad (\mu = 0, 1, 2, 3); \quad \mathbf{x} = (x^1, x^2, x^3), \quad t = x^0.$$

The covariant derivatives of the fields A^l and Ψ are related to the differential Cartan forms $\omega^l(A, \partial_\mu A)$ and $\theta^\alpha(A, \partial_\mu A)$ [72, 73, 78] defined by the relations

$$K^{-1}\partial_\mu K = \mathrm{i}\,\omega^l(A, \partial_\mu A)X_l + \mathrm{i}\,\theta^\alpha(A, \partial_\mu A)\,Y_\alpha, \tag{1.5.29}$$

where $K = \exp(\mathrm{i}A^l X_l)$, and only the fields A^l depend on the coordinates x^μ.

Let us show that the specific form of the left side of eq. (1.5.29) and the compatibility conditions for the derivatives $\partial_\mu\partial_\nu K = \partial_\nu\partial_\mu K$ give equations that specify the structure of the functions ω^l and θ^α. For this, the identities

$$\begin{aligned}
&\partial_\nu K^{-1} = -K^{-1}(\partial_\nu K)\,K^{-1},\\
&(K^{-1}\partial_\nu K)\,(K^{-1}\partial_\mu K) - (K^{-1}\partial_\mu K)\,(K^{-1}\partial_\nu K)\\
&\quad = \mathrm{i}\,\big[(\theta^\alpha(A, \partial_\mu A)\,\omega^s(A, \partial_\nu A) - \theta^\alpha(A, \partial_\nu A)\,\omega^s(A, \partial_\mu A))\\
&\qquad \cdot C_{\alpha s}^{l} X_l + \big(\omega^k(A, \partial_\mu A)\,\omega^s(A, \partial_\nu A)\,C_{ks}^{\alpha}\\
&\qquad + \theta^\gamma(A, \partial_\mu A)\,\theta^\beta(A, \partial_\nu A)\,C_{\gamma\beta}^{\alpha}\big)\,Y_\alpha\big]
\end{aligned} \tag{1.5.30}$$

are useful. The former is derived by differentiating the equality $K^{-1}K = I$, the latter is a consequence of formulas (1.5.13) and (1.5.29). Taking into account (1.5.30), we easily obtain the Maurer–Cartan integrability conditions by differentiating relations (1.5.29):

$$\begin{aligned}
&\partial_\nu\omega^l(A, \partial_\mu A) - \partial_\mu\omega^l(A, \partial_\nu A) + [\theta^\alpha(A, \partial_\mu A)\,\omega^s(A, \partial_\nu A)\\
&\quad - \theta^\alpha(A, \partial_\nu A)\,\omega^s(A, \partial_\mu A)]\,C_{\alpha s}^{l} = 0,\\
&\partial_\nu\theta^\alpha(A, \partial_\mu A) - \partial_\mu\theta^\alpha(A, \partial_\nu A) + \omega^k(A, \partial_\mu A)\,\omega^s(A, \partial_\nu A)\,C_{ks}^{\alpha}\\
&\quad + \theta^\gamma(A, \partial_\mu A)\,\theta^\beta(A, \partial_\nu A)\,C_{\gamma\beta}^{\alpha} = 0.
\end{aligned}$$

They are necessary and sufficient for the field K to be reconstructed on the field $K^{-1}\partial_\mu K$.

We will find representations for the Cartan forms ω^l and θ^α in terms of the Goldstone fields A^l by applying a technique similar to that considered above. We incorporate the parameter s into eq. (1.5.29) through the substitution $A^l \to s A^l$:

$$\begin{aligned}&\exp(-\mathrm{i}A^l X_l s)\,\partial_\mu \exp(\mathrm{i}A^k X_k s)\\&\quad= \mathrm{i}\,\omega^l(sA,\, s\,\partial_\mu A)X_l + \mathrm{i}\,\theta^\alpha(sA,\, s\,\partial_\mu A)Y_\alpha.\end{aligned} \tag{1.5.31}$$

Differentiating the left and right sides of equality (1.5.31) over s, using the commutation relations (1.5.13), and equating the coefficients for identical generators, we obtain a closed set of ordinary differential equations for computing the functions $\omega^l(sA,\, s\,\partial_\mu A)$ and $\theta^\alpha(sA,\, s\,\partial_\mu A)$:

$$\frac{\mathrm{d}\omega^l}{\mathrm{d}s} = \partial_\mu A^l + A^p\theta^\beta C^l_{p\beta}, \qquad \frac{\mathrm{d}\theta_\beta}{\mathrm{d}s} = A^k\omega^p C^\beta_{kp}. \tag{1.5.32}$$

According to eq. (1.5.31), the following initial conditions hold for system (1.5.32):

$$\omega^l(sA,\, s\,\partial_\mu A)|_{s=0} = 0, \quad \theta^\alpha(sA,\, s\,\partial_\mu A)|_{s=0} = 0. \tag{1.5.33}$$

The Cartan forms (1.5.29) coincide with the values of the solutions (1.5.32) and (1.5.33) of the problem for $s = 1$. Here is the final result:

$$\begin{aligned}\omega^l(A,\,\partial_\mu A) &= \left(\frac{\sin\sqrt{m}}{\sqrt{m}}\right)^l_p \partial_\mu A^p,\\ \theta^\beta(A,\,\partial_\mu A) &= \left(\frac{1-\cos\sqrt{m}}{\sqrt{m}}\right)^p_s \partial_\mu A^s\, C^\beta_{kp}\, A^k.\end{aligned} \tag{1.5.34}$$

Let us disclose the transformation properties of Cartan's forms under the operations of the group G. As we have already discussed, under the operation $g \in G$, an element of the factor-space $K(A) = \exp(\mathrm{i}A^l X_l)$ changes, and the result can be represented in the form

$$g\,K(A) = K(A')\,h(b). \tag{1.5.35}$$

Here $K(A') = \exp(\mathrm{i}A'^s(A, g)X_s)$, $h(b) = \exp(\mathrm{i}\,b^\alpha(A, g)\,Y_\alpha)$, and the quantities A^l are functions of the variables x^μ. Now we calculate the derivative of the left and right sides of equality (1.5.35) over x^μ. Since g, X_l, and Y_α are x^μ-independent, we have

$$g\,\partial_\mu K(A) = \partial_\mu K(A')\,h(b) + K(A')\,\partial_\mu h(b). \tag{1.5.36}$$

Combining the relation

$$[g\,K(A)]^{-1} = K^{-1}(A)\,g^{-1} = h^{-1}(b)\,K^{-1}(A')$$

with formula (1.5.36), we find

$$K^{-1}(A)\,\partial_\mu K(A) = h^{-1}(b)(K^{-1}(A')\,\partial_\mu K(A'))h(b) + h^{-1}(b)\,\partial_\mu h(b). \tag{1.5.37}$$

Using eq. (1.5.29), we pass in eq. (1.5.37) to Cartan's forms to transform them through the original ones:

$$\omega^l(A',\partial_\mu A')\,X_l + \theta^\alpha(A',\partial_\mu A')\,Y_\alpha = h(b)\,[\omega^s(A,\partial_\mu A)X_s + \theta^\beta(A,\partial_\mu A)\,Y_\beta]h^{-1}(b) + h(b)\,\partial_\mu h^{-1}(b).$$

Identity (1.5.16) and the commutation relations (1.5.13) make it possible to verify that the following partition is valid (we equate the terms linear in the generators X_n and Y_α, respectively):

$$\omega^l(A',\partial_\mu A')\,X_l = h(b)\,X_s h^{-1}(b)\,\omega^s(A,\partial_\mu A), \tag{1.5.38}$$

$$\theta^\alpha(A',\partial_\mu A')\,Y_\alpha = h(b)\,Y_\beta\,h^{-1}(b)\,\theta^\beta(A,\partial_\mu A) + h(b)\,\partial_\mu h^{-1}(b). \tag{1.5.39}$$

From eq. (1.5.38) we can infer that the forms of ω^l under the operations $g \in G$ are transformed according to the previously introduced representation (1.5.17) of the subgroup H, with the parameters b^α depending, in the general case, on the Goldstone fields:

$$\omega^l(A,\partial_\mu A) \xrightarrow{g} S(h)^l_k\,\omega^k(A,\partial_\mu A), \quad h = \exp(\mathrm{i}\,b^\alpha(A,g)\,Y_\alpha). \tag{1.5.40}$$

Notably, the operators Y_α and $h(b)$ are not assigned in the space of the functions $\Psi(\mathbf{x},t)$ in the general case. Therefore we should proceed from relation (1.5.39) to its implementation on the functions Ψ to expose the transformation features of the Ψ-field derivatives:

$$\theta^\alpha(A',\partial_\mu A')\,\lambda_\alpha = D(h)\,\lambda_\beta\,D^{-1}(h)\,\theta^\beta(A,\partial_\mu A) + D(h)\,\partial_\mu D^{-1}(h). \tag{1.5.41}$$

Here λ_α is a necessary representation of the generators Y_α, and $D(h)$ are the previously discussed matrices of the linear representation of the subgroup H on the local fields Ψ. Using these matrices, we can build the induced representation of the group G on the same fields:

$$\Psi \to D(h)\,\Psi, \quad h = \exp(\mathrm{i}\,b^\alpha(A,g)\,Y_\alpha).$$

Covariant derivatives of the local fields A^l and Ψ are defined by the relations

$$D_\mu A^l = \omega^l(A,\partial_\mu A), \quad \nabla_\mu\Psi = \partial_\mu\Psi + \theta^\beta(A,\partial_\mu A)\,\lambda_\beta\,\Psi. \tag{1.5.42}$$

With formulas (1.5.40) and (1.5.41) for the Cartan forms to be transformed, we can make sure that the covariant derivatives (1.5.42) are transformed under the operations of the group G by linear representations of the subgroup H,

$$D_\mu A^l \xrightarrow{g} S^l_k(h)\, D_\mu A^k, \quad \nabla_\mu \Psi \xrightarrow{g} D(h)\, \nabla_\mu \Psi,$$

where $h = \exp(\mathrm{i}\, b^\alpha(A, g)\, Y_\alpha)$ is an element of the subgroup H with parameters b^α, which nontrivially depend both on a specific operation $g \in G$ and the Goldstone fields A^l.

Hence we can draw an important conclusion about the necessary and sufficient invariance of the Lagrangian $L(\Psi, \nabla_\mu \Psi, D_\mu A^l)$ with respect to the subgroup H for the invariance with respect to the whole group G.

Now we can formulate simple working rules for constructing phenomenological Lagrangians for Goldstone particles in any physical systems with spontaneous symmetry breaking:

1. In the subgroup-invariant Lagrangian of the free non-Goldstone fields Ψ, ordinary derivatives of the fields Ψ are replaced by covariant ones. The resulting Lagrangian is invariant with respect to the group G and describes the interaction between these fields and Goldstone particles.
2. Such a Lagrangian is supplemented by a Lagrangian of free Goldstone fields to be an invariant expression of the covariant derivatives $D_\mu A^l$.

Constructing the Lagrangian of free Goldstone particles in the long-wavelength approximation, it suffices to restrict ourselves to invariants containing a minimum number of derivatives of the fields A^l over the spatial coordinates x and time t.

In the example considered above, we have the identity

$$S^m_k\, g_{mr}(0)\, S^r_n = g_{kn}(0). \tag{1.5.43}$$

Here $g_{mr}(0) = C^\rho_{ml}\, C^l_{\rho r}$, and therefore the minimal interactions of Goldstone particles are theoretically described by the invariants

$$g_{ij}(0)\, \omega^i(A, \partial_\mu A)\, \omega^j(A, \partial_\mu A).$$

Formula (1.5.43) is a consequence of the algebraic structure of the matrices S^k_n (1.5.18) and simpler relations

$$P^m_k\, g_{mn} = -g_{km} P^m_n, \quad C^m_{k\sigma} C^\rho_{ml} C^l_{\rho n} + C^\gamma_{kl} C^l_{\gamma m} C^m_{n\sigma} = 0. \tag{1.5.44}$$

To verify eq. (1.5.44), note that the Jacobi identities for the generators of the group G

$$[X_l\, [X_k, Y_\sigma]] + [X_k\, [Y_\sigma, X_l]] + [Y_\sigma\, [X_l, X_k]] = 0,$$
$$[Y_\sigma\, [Y_\gamma, X_n]] + [Y_\gamma\, [X_n, Y_\sigma]] + [X_n\, [Y_\sigma, Y_\gamma]] = 0$$

give important restrictions on the structural constants of the group

$$\begin{aligned} C^m_{k\sigma}\, C^\rho_{lm} + C^m_{\sigma l} C^\rho_{km} + C^\gamma_{lk}\, C^\rho_{\sigma\gamma} &= 0, \\ C^m_{\gamma n}\, C^l_{\sigma m} + C^m_{n\sigma} C^l_{\gamma m} + C^\rho_{\sigma\gamma}\, C^l_{n\rho} &= 0. \end{aligned} \tag{1.5.45}$$

Next, we multiply the first of equalities (1.5.45) by $C^{l}_{\rho n}$, sum up the result over the indices l and ρ, and use the second equality. Then we obtain identity (1.5.44):

$$C^{m}_{k\sigma}C^{\rho}_{ml}C^{l}_{\rho n} + C^{\gamma}_{kl}C^{l}_{\gamma m}C^{m}_{n\sigma} = C^{\rho}_{km}C^{m}_{\sigma l}C^{l}_{\rho n} - C^{\gamma}_{kl}C^{l}_{\sigma m}C^{m}_{\gamma n} = 0.$$

There is a deep connection between the above formalism and the concepts of differential geometry. Let us define the metric $g_{ij}(A)$ of the factor-space G/H by the relation

$$g_{ij}(A)\,\mathrm{d}A^{i}\,\mathrm{d}A^{j} = g_{sp}(0)\,\omega^{s}(A,\,\mathrm{d}A)\,\omega^{p}(A,\mathrm{d}A), \tag{1.5.46}$$

where

$$\omega^{i}(A,\,\mathrm{d}A) = \omega^{i}(A,\,\partial_{\mu}A)\,\mathrm{d}x^{\mu}, \quad \mathrm{d}A^{l} = \partial_{\mu}A^{l}\,\mathrm{d}x^{\mu}.$$

From (1.5.46) and (1.5.34) we find

$$g_{ij}(A) = \left(\frac{\sin\sqrt{m}}{\sqrt{m}}\right)^{s}_{i} g_{sk}(0)\left(\frac{\sin\sqrt{m}}{\sqrt{m}}\right)^{k}_{j}. \tag{1.5.47}$$

The metric $g_{ij}(A)$ allows us to introduce the notion of a moving orthogonal frame in the group space. In this case the forms of $\omega^{i}(A,\,\mathrm{d}A)$ can be thought of as components of an infinitely small displacement of the origin of the frame from a point with local coordinates of A^{l} to a neighboring point in the group space. The forms of

$$\omega^{i}_{k}\,(A,\,\mathrm{d}A) = C^{i}_{k\beta}\,\theta^{\beta}(A,\,\partial_{\mu}A)\,\mathrm{d}x^{\mu}$$

govern the rotation of the basis vectors of the shifted frame at the point with coordinates of A^{l}. With this interpretation, the Mauer–Cartan equations define the structure of an n-dimensional Riemannian space with zero torsion but nonzero curvature [78]. The origin of the group space remains immobile under the transformations of the subgroup H. Therefore these transformations correspond to rotations in the group space, whereas the transformations $g \notin H$ meet shifts along with rotations. From this point of view, the definition of the covariant differential $\nabla_{\mu}\Psi\,\mathrm{d}x^{\mu}$, introduced previously, coincides with the usual definition of an absolute differential in Riemannian spaces and spaces with affine connection.

1.6 Phenomenological Lagrangians of nonlinear dynamics of Goldstone spin waves

It always seems odd to me that the fundamental laws of physics, when discovered, can appear in so many different forms that are not apparently identical at first, but, with a little mathematical fiddling you can show the relationship. I think it is somehow a representation of the simplicity of nature. Perhaps a thing is simple if you can describe it fully in several different ways without immediately knowing that you are describing the same thing.

R. Feynman

In this section, we show how the concept of spontaneous breaking of the SO(3)-symmetry of exchange interactions makes it possible to construct phenomenological Lagrangians for spin waves in magnets. The spontaneous high-symmetry breaking exhibits itself as the emergence of Goldstone fields. The realization of spontaneous symmetry breaking by nonlinear transformations (the so-called nonlinear realizations of the SO(3)-symmetry) of Goldstone fields underlies the method of phenomenological Lagrangians. In the long-wavelength limit, the approach accurately describes the spectrum of Goldstone spin waves and all their potential nonlinear interactions. Depending on the invariance subgroup of the magnet's ground state, one, two, or three Goldstone branches of spin waves can be observed in the excitation spectrum. The first two cases include a group SO(2) as a subgroup of the ground state invariance. These cases fit simple ferro- and antiferromagnets. In such spin systems, tiny departures from the ground state are described by two angles (by coordinates of the SO(3)/ SO(2) factor-space) that meet one or two branches of the spectrum of spin excitations, depending on whether the equations of dynamics contain time derivatives of the first or second order. Three branches of spin waves appear under maximum symmetry-breaking of the ground state of a magnet when the subgroup of its invariance is only an identical transformation. Then local deviations from the magnet's equilibrium state are described by three angles (parameters of the group SO(3)) whose space-time dependence specifies the dynamics of the Goldstone modes.

Resorting to [77, 75, 76], we construct the most general expression for the phenomenological Lagrangian of spin waves, which is invariant under transformations of the group SO(3) – symmetry of exchange interaction for the ground state invariant only under an identical SO(3)-transformation.

As is known, the group SO(3) consists of 3×3 orthogonal matrices D with determinant equal to unit:

$$D_{sk}\, D_{pk} = D_{ks}\, D_{kp} = \delta_{sp}, \quad \det D = 1. \tag{1.6.1}$$

The last of equalities (1.6.1) implies that

$$\varepsilon_{ijk}\, D_{is}\, D_{jp} D_{kq} = \varepsilon_{spq}, \tag{1.6.2}$$

where ε_{spq} is an absolutely antisymmetric unit tensor ($\varepsilon_{123} = 1$). The indices designated by the Latin letters take the values 1, 2, and 3. For specific calculations, the following identities turn out to be useful:

$$\varepsilon_{ijk}\, \varepsilon_{ksp} = \delta_{is}\, \delta_{jp} - \delta_{ip}\, \delta_{js}, \quad \varepsilon_{ijk}\, \varepsilon_{jkp} = 2\, \delta_{ip}. \tag{1.6.3}$$

For further analysis, the matrices D can be conveniently parameterized by the Euler angles θ, φ, and ψ [71]:

$$D = \begin{pmatrix} \cos\psi\cos\varphi - \cos\theta\sin\varphi\sin\psi & -\sin\psi\cos\varphi - \cos\theta\sin\varphi\cos\psi & \sin\theta\sin\varphi \\ \cos\psi\sin\varphi + \cos\theta\cos\varphi\sin\psi & -\sin\psi\sin\varphi + \cos\theta\cos\varphi\cos\psi & -\sin\theta\cos\varphi \\ \sin\psi\sin\theta & \cos\psi\sin\theta & \cos\theta \end{pmatrix}. \tag{1.6.4}$$

Let us introduce the notation

$$A_1 = \theta, \quad A_2 = \varphi, \quad A_3 = \psi.$$

Then $D_{ij} = D_{ij}(A)$.

Next, we define transformations of the group SO(3) by means of left shifts on the group:

$$D(A) \overset{D_0}{\rightarrow} D_0 D(A) = D(A'(A, D_0)), \tag{1.6.5}$$

where D_0 is an arbitrary element of the group SO(3). The method of phenomenological Lagrangians establishes the transformation properties of the magnon Goldstone fields $A_k(\mathbf{x}, t)$ $(k = 1, 2, 3)$ by specifically matching these fields with the parameters of the group SO(3), thereby satisfying the transformation law for the local fields $A_k(\mathbf{x}, t)$:

$$A_k(\mathbf{x}, t) \overset{D_0}{\rightarrow} A'_k(A, D_0). \tag{1.6.6}$$

As a consequence of the first of equalities (1.6.1), the product $D_{ks}\partial_\mu D_{kp}$ is antisymmetric with respect to the permutation of the indices s and p. We express this by means of the relation

$$D_{ks}\partial_\mu D_{kp} = -\varepsilon_{spn}\, \omega_n(A, \partial_\mu A). \tag{1.6.7}$$

Here $\mu = 0, 1, 2, 3$, $x_0 = t$ is time, $\mathbf{x} = (x_1, x_2, x_3)$ is spatial coordinates, and

$$\partial_\mu = \frac{\partial}{\partial x_\mu}.$$

Furthermore, we do not distinguish between superscripts and subscripts until otherwise stated. The matrices $(J_n)_{sp} = -\mathrm{i}\varepsilon_{spn}$ use the commutation relations

$$[J_s, J_p] = \mathrm{i}\,\varepsilon_{spk} J_k \tag{1.6.8}$$

to implement the algebra of the group SO(3) generators. Equality (1.6.7) can be rewritten as

$$D^{-1}\partial_\mu D = -\mathrm{i}\,\omega_s(A,\,\partial_\mu A)\,J_s. \tag{1.6.9}$$

It shows that the quantities ω_s coincide with the left Cartan forms.

The Cartan forms are invariants under transformation (1.6.6):

$$\begin{aligned}
-\mathrm{i}\omega_s(A',\partial_\mu A')J_s &= D^{-1}(A')\,\partial_\mu D(A') \\
&= \left[D_0 D(A)\right]^{-1}\partial_\mu\left[D_0 D(A)\right] = D^{-1}(A)\,D_0^{-1}\,D_0\,\partial_\mu D(A) \\
&= D^{-1}(A)\,\partial_\mu D(A) = -\mathrm{i}\,\omega_s(A,\partial_\mu A)\,J_s.
\end{aligned}$$

As in the previous section, it is easy to demonstrate that the form of writing the left side of equality (1.6.9) and the permutation relations (1.6.8) bring to the Maurer–Cartan equations for specifying the structure of the functions ω_s:

$$\partial_\nu\omega_s(A,\,\partial_\mu A) - \partial_\mu\omega_s(A,\,\partial_\nu A) + \varepsilon_{snp}\,\omega_n(A,\,\partial_\nu A)\,\omega_p(A,\,\partial_\mu A) = 0. \tag{1.6.10}$$

Using eq. (1.6.3), from (1.6.9) we find

$$\omega_n(A,\,\partial_\mu A) = -\frac{1}{2}\,\varepsilon_{nsp}\,D_{ks}\,\partial_\mu D_{kp}. \tag{1.6.11}$$

To avoid tedious calculations for the Cartan forms (1.6.11) in terms of the fields A_l, we give a mechanical analogy. The functions $\omega_s(A,\,\partial_\mu A)$ are linear in the derivatives $\partial_\mu A_l$. Moreover, when s is fixed, the coefficients for derivatives with the same indices l but different μ are the same for all Cartan's forms. Calculating three functions $\omega_s(A,\,\partial_\mu A)$ is sufficient to find the remaining ones. It is not difficult to deduce an explicit form of the quantity $\omega_s(A,\,\partial_\mu A)$ by noting that their mathematical definition (1.6.7) or (1.6.11) coincides with the definition of the vector ω_s-components of the angular velocity of a rigid body rotating around a fixed point in a Cartesian coordinate system whose axes are strongly connected to the body, with the origin of the system coordinates and the center of rotation of the body being merged [43, 30, 5]. In this case the matrices $D(A)$ make the transition from a fixed Cartesian frame of reference x_1, x_2, x_3 to a frame of reference x_1', x_2', x_3' associated with the rotating body: $\mathbf{x} = D\mathbf{x}'$ (see Figure 1.5).

The well-known results of classical mechanics [43, 30, 5] allow us to express the forms of $\omega_s(A,\,\partial_\mu A)$ in terms of the Euler angles

$$A_1 = \theta, \quad A_2 = \varphi, \quad A_3 = \psi$$

for describing here certainly not rotation, but local spin turns in small volumes of a magnetic-ordered medium:

$$\begin{aligned}
\omega_1(A,\,\partial_\mu A) &= \partial_\mu A_2 \sin A_1 \sin A_3 + \partial_\mu A_1 \cos A_3, \\
\omega_2(A,\,\partial_\mu A) &= \partial_\mu A_2 \sin A_1 \cos A_3 - \partial_\mu A_1 \sin A_3, \\
\omega_3(A,\,\partial_\mu A) &= \partial_\mu A_2 \cos A_1 + \partial_\mu A_3.
\end{aligned} \tag{1.6.12}$$

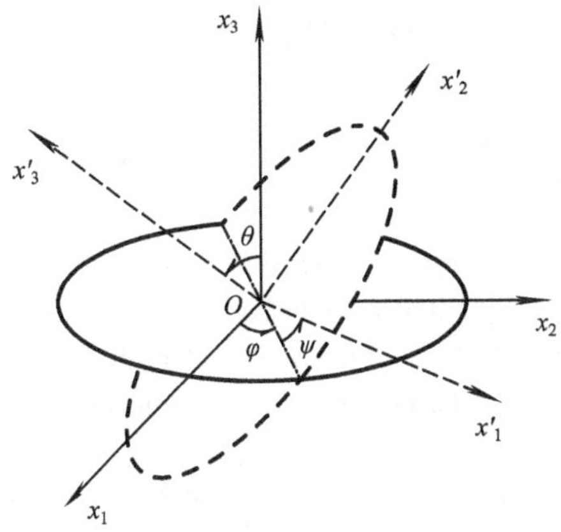

Figure 1.5: Euler angles: $\theta = A_1$, $\varphi = A_2$, $\psi = A_3$.

It is crucial to emphasize that there are no invariants under transformations (1.6.5) and (1.6.6), which are composed only of the matrices $D(A)$ or magnon fields $A_l(\mathbf{x}, t)$. Additionally, we can demonstrate that the linear independence and completeness of the form $\omega_s(A, \partial_\mu A)$ makes it possible to represent any differential form $\Xi(A, \partial_\mu A)$ as

$$\Xi(A, \partial_\mu A) = a_i\, \omega_i(A, \partial_\mu A),$$

where a_i are constant coefficients. Since the Cartan forms (1.6.12) involve a complete set of invariants, the most general SO(3)-invariant expression for the phenomenological Lagrangian of spin waves, which contains derivatives of quasiclassical magnon fields not higher than the second degree, has the form

$$Z = \int L\, \mathrm{d}^3\mathbf{x},$$

where L is the density of the Lagrange function,

$$\begin{aligned} L = \frac{1}{2}\, &[a_{ik}\, \omega_i(A, \partial_t A)\, \omega_k(A, \partial_t A) + 2b_i \omega_i(A, \partial_t A) \\ &- c_{ik,lm}\, \omega_i(A, \partial_l A)\, \omega_k(A, \partial_m A) - 2d_{i,l}\, \omega_i(A, \partial_l A)], \end{aligned} \qquad (1.6.13)$$

and a, b, c, and d are phenomenological constants. Without loss of generality, we can suppose that $a_{ik} = a_{ki}$ and $c_{ik,lm} = c_{ki,ml}$. From a microscopic point of view, the parameters a, b, c, and d are nonzero vacuum averages of certain combinations of the spin operators of the system. Further, we will see that the constants b_i are proportional to the vector components of equilibrium spontaneous magnetization of a medium. Therefore they alter the sign under the time-reversal operation R:

$$b_i \xrightarrow{R} -b_i.$$

As a result, the Lagrange function (1.6.13) is invariant under the transformation R. It should be stressed that interactions of the type

$$d_{i,l}\,\omega_i(A,\,\partial_l A) \tag{1.6.14}$$

are only capable of arising in inversion-center-free crystals and are associated with the existence of helicoidal magnetic structures. The explicit form of the constants a, b, c, and d arises from the requirement of the invariance of the Lagrangian (1.6.13) under transformations of the medium's spatial symmetry group. Later, we will cover some potential restrictions on the choice of constants.

It is interesting that the Lagrangian (1.6.13) for $b_i = 0$, regardless the spatial coordinate dependence of the angles A_l (for $\partial_k A_l = 0$, $k = 1, 2, 3$), literally coincides with the Lagrangian for rotation of an asymmetric rigid body relative to a fixed point [43]. In this case, $\omega_s(A,\,\partial_t A)$ is the angular velocity of the body in the reference frame rigidly connected to the body, and the constants a_{ij} are the tensor components of the moments of inertia of the body.

When $b_j \neq 0$ and $\partial_k A_l = 0$, expression (1.6.13) coincides with the Lagrange function of a rotating charged body with its own magnetic moment [26] (a magnetic top).

In the general case, the Lagrangian (1.6.13) describes a system of interacting mechanical tops located at different points in space. The quantities $\omega_s(A,\,\partial_k A)$ characterize the relative turns of the tops at neighboring points of the system. The constants $c_{ik,lm}$ and $d_{i,l}$ are responsible for the rigidity of the connections of the tops between themselves and the system ground state.

In a similar way, in magnets the forms $\omega_s(A,\,\partial_\mu A)$ characterize local turns of spins in small volumes of a medium. These turns take place in the spin space with respect to the basis associated with the magnetic vectors of the exchange structure. It is significant to notice that in tiny volumes of the medium, a perfect exchange structure is preserved and characterized by no more than three unchanged-length magnetic vectors rigidly aligned toward each another. In the excited states of a magnet, when passing from one small volume of the medium to another one, these vectors that create a local reference system rotate as a whole with respect to the fixed coordinate system in the spin space. Ultimately, the SO(3)-symmetry of the phenomenological Lagrangian (1.6.13) in a fixed spin space is provided by expressing the relative displacements and rotations of spins through a moving frame. Below we will exemplify these statements.

The parameterization of the matrices D by Euler angles is convenient for interpreting the analogy between spin dynamics and the rotation of a rigid body. It can be shown [72, 73] that other SO(3)-group parameterizations used reproduce physically equivalent outcomes. In this case the theory accounts for all interactions between magnons, corresponding to tree-type Feynman diagrams (diagrams without closed loops) that are most significant in the long-wavelength limit.

By using the well-known Lagrangian we find the exchange energy of the magnet:

$$E = \int\left(\frac{\partial L}{\partial(\partial_t A_l)}\partial_t A_l - L\right)\mathrm{d}^3\mathbf{x}$$
$$= \frac{1}{2}\int\left[a_{ik}\,\omega_i(A, \partial_t A)\,\omega_k(A, \partial_t A) + 2d_{i,l}\,\omega_i(A, \partial_l A)\right.$$
$$\left. + c_{ik,lm}\,\omega_i(A, \partial_l A)\omega_k(A, \partial_m A)\right]\mathrm{d}^3\mathbf{x}. \tag{1.6.15}$$

The equations of spin wave dynamics can be derived by varying the action corresponding to the Lagrangian

$$Z = \int L\mathrm{d}^3\mathbf{x}$$

over the fields A_l or D_{ik}. Further, to simplify the analysis, we consider the case of an unbounded medium, assuming that all the surface integrals vanish at infinity. When choosing the fields $D_{ij}(\mathbf{x}, t)$ as dynamic variables, connections (1.6.1) should be taken into account. In doing so, we exploit the Lagrange method of indefinite multipliers [30]. Let us enter the auxiliary Lagrangian

$$\tilde{Z} = Z + \frac{1}{2}\int\lambda_{ij}(D_{ki}D_{kj} - \delta_{ij})\mathrm{d}^3\mathbf{x}, \tag{1.6.16}$$

where the fields $\lambda_{ij}(\mathbf{x}, t) = \lambda_{ji}(\mathbf{x}, t)$ play the role of Lagrange multipliers. The original variational problem is equivalent to the variational problem with the Lagrangian $\tilde{Z}$, and the variations of the fields λ_{ij} and D_{ij} can now be thought to be independent.

The Lagrange equations for λ_{ij} and D_{ij} have the form

$$\partial_\mu\frac{\tilde{\delta}Z}{\tilde{\delta}(\partial_\mu D_{sj})} - \frac{\tilde{\delta}Z}{\tilde{\delta}D_{sj}} - \lambda_{ij}D_{si} = 0, \quad D_{ki}D_{kj} = \delta_{ij}. \tag{1.6.17}$$

Here and below, the quantity $\tilde{\delta}F/\tilde{\delta}u$ means the "partial" variational derivative of the functional F over the field variable u, namely, in formula (1.6.17),

$$\frac{\tilde{\delta}Z}{\tilde{\delta}(\partial_\mu D_{sj})} = -\frac{\partial L}{\partial\omega_n(A, \partial_\mu A)}\frac{1}{2}\varepsilon_{nrj}D_{sr},$$
$$\frac{\tilde{\delta}Z}{\tilde{\delta}D_{sj}} = \frac{\partial L}{\partial\omega_n(A, \partial_\mu A)}\frac{1}{2}\varepsilon_{nrj}\,\partial_\mu D_{sr}. \tag{1.6.18}$$

The partial variational derivatives (1.6.18) are related to the "total" variational derivative of the functional Z over the field variable D_{sj} as follows:

$$\frac{\delta Z}{\delta D_{sj}} = \frac{\tilde{\delta}Z}{\tilde{\delta}D_{sj}} - \partial_\mu\frac{\tilde{\delta}Z}{\tilde{\delta}(\partial_\mu D_{sj})}.$$

To exclude the Lagrange multipliers λ_{ij} from the first equality in (1.6.17), we multiply it by $D_{sk}\,\varepsilon_{kjl}$ and sum up the result over the indices s and j. Using the second equality and formulas (1.6.7) and (1.6.3), we arrive at the equation

$$\partial_\mu\left(\frac{\partial L}{\partial\omega_l(A,\partial_\mu A)}\right)-\varepsilon_{lnm}\left(\frac{\partial L}{\partial\omega_n(A,\partial_\mu A)}\right)\omega_m(A,\partial_\mu A)=0.$$

Thus the evolution equations for quasiclassical magnon fields are given by

$$\begin{aligned}&a_{lk}\,\partial_t\omega_k(A,\partial_t A)-c_{lr,nm}\,\partial_n\omega_r(A,\partial_m A)\\&\quad-\varepsilon_{lnm}\left[a_{nk}\,\omega_k(A,\partial_t A)+b_n\right]\omega_m(A,\partial_t A)\\&\quad+\varepsilon_{lnm}\left[c_{nr,pq}\,\omega_r(A,\partial_q A)+d_{n,p}\right]\omega_m(A,\partial_p A)=0.\end{aligned}\tag{1.6.19}$$

The invariance of the Lagrange function $\tilde{Z}$ (or Z) with respect to rotations in the spin space is related to the law of conservation of the mechanical spin moment of a magnet. The infinitesimal transformation corresponding to such a symmetry is

$$D\to D+\mathrm{i}\,\varepsilon_k J_k\,D.\tag{1.6.20}$$

The generators J_k control the rotation around the kth axis in a fixed spin space, and ε_k are constant parameters (small angles of a turn).

There is a simple way proposed by Gell-Man and Levy for constructing conserved currents and their divergences [78, 28]. We formally think that the parameters ε_k under the infinitesimal transformation (1.6.20) are coordinate-and-time dependent. Consequently, under transformation (1.6.20), the Lagrangian (1.6.16) of the system boils down to

$$\delta\tilde{Z}=\frac{\tilde{\delta}\tilde{Z}}{\tilde{\delta}\varepsilon_k}\,\varepsilon_k+\frac{\tilde{\delta}\tilde{Z}}{\tilde{\delta}(\partial_\mu\varepsilon_k)}\,\partial_\mu\varepsilon_k,$$

where

$$\begin{aligned}\frac{\tilde{\delta}\tilde{Z}}{\tilde{\delta}\varepsilon_k}&=\left[\left(\frac{\tilde{\delta}Z}{\tilde{\delta}D_{sj}}+\lambda_{ij}\,D_{si}\right)\varepsilon_{ksp}\,D_{pj}+\frac{\tilde{\delta}Z}{\tilde{\delta}(\partial_\mu D_{sj})}\,\varepsilon_{ksp}\,\partial_\mu D_{pj}\right],\\\frac{\tilde{\delta}\tilde{Z}}{\tilde{\delta}(\partial_\mu\varepsilon_k)}&=\frac{\tilde{\delta}Z}{\tilde{\delta}(\partial_\mu D_{sj})}\,\varepsilon_{ksp}\,D_{pj}.\end{aligned}\tag{1.6.21}$$

By definition the current associated with transformation (1.6.20) is the quantity

$$S_{k\mu}\equiv-\frac{\tilde{\delta}\tilde{Z}}{\tilde{\delta}(\partial_\mu\varepsilon_k)}=-\frac{\tilde{\delta}Z}{\tilde{\delta}(\partial_\mu D_{sj})}\,\varepsilon_{ksp}\,D_{pj}.$$

Resorting to formulas (1.6.18) and (1.6.3) and identity

$$\varepsilon_{nrj} D_{sr} D_{pj} = \varepsilon_{isp} D_{in},$$

which follows from eqs. (1.6.1) and (1.6.2), we come to the following relations between the current components and the Cartan forms:

$$\begin{aligned} S_{k\mu} &= \frac{\partial L}{\partial \omega_n(A, \partial_\mu A)} D_{kn}, \\ S_{k0} &= D_{kn}\left[b_n + a_{ni}\, \omega_i(A, \partial_t A)\right], \\ S_{kr} &= -D_{kn}\left[c_{ni,rm}\, \omega_i(A, \partial_m A) + d_{n,r}\right]. \end{aligned} \tag{1.6.22}$$

Bearing in mind the Lagrange equations (1.6.17), we can rewrite the first of eqs. (1.6.21) as

$$\frac{\tilde{\delta} \tilde{Z}}{\tilde{\delta} \varepsilon_k} - -\partial_\mu S_{k\mu}.$$

Since the Lagrangian $\tilde{Z}$ does not depend on the parameters ε_k, and it is just the case, we have $\partial_\mu S_{k\mu} = 0$. This is because $\tilde{Z}$ is invariant under transformations (1.6.20) with constants ε_k. The quantities S_{k0} have a sense of the vector components of the spin moment density of the magnet. The explicit form of the conservation law for the mechanical spin moment of the system is

$$\begin{aligned} &\partial_t\left[D_{kn}\left(b_n + a_{ni}\, \omega_i(A, \partial_t A)\right)\right] \\ &\quad - \partial_r\left[D_{kn}\left(d_{n,r} + c_{ni,rm}\, \omega_i(A, \partial_m A)\right)\right] = 0. \end{aligned} \tag{1.6.23}$$

By the relation

$$\partial_\mu D_{kn} = -\varepsilon_{lnq}\, \omega_q(A, \partial_\mu A)\, D_{kl}, \tag{1.6.24}$$

which comes from (1.6.7), the conservation law (1.6.23) and the magnon field dynamics equations (1.6.19) are equivalent.

Here it is worth elucidating this finding within a mechanical model. Expression (1.6.22) for S_{k0} formally coincides with the definition of the angular momentum of a rotating body in a fixed coordinate system, with its center and the rotation center of the body being the same [5]. It is important to underscore that the quantities

$$\sigma_{n0} = b_n + a_{ni}\, \omega_i(A, \partial_t A) \tag{1.6.25}$$

are also components of the angular momentum of the body, but calculated in a mobile coordinate system whose axes are tightly connected to the body. From this point of view, eqs. (1.6.19) and (1.6.23) for $\partial_k A = 0$ $(k = 1, 2, 3)$ are nothing else than the law of conservation of angular momentum for a rotating body, but written in different frames of reference, namely, in a frame associated with a rotating body and in a fixed frame of reference.

In magnets the situation is similar. As for spin space, expression (1.6.22) for S_{k0} yields the system's spin moment relative to a fixed coordinate system, whereas σ_{k0} in eq. (1.6.25) stands for the same characteristic, but written in a mobile frame rigidly connected to the magnetic vectors of the exchange structure. As already mentioned, small regions of a magnet keep the exchange structure perfect on the mobile basis in spite of continuously rotating the magnetic vectors in the fixed frame of spin space as a whole when passing from one region of the coordinate space to another. Let us now explain what has been said using the formulas for S_{k0} and σ_{k0} as examples. Equations (1.6.22) and (1.6.25) imply that the vector $\mathbf{b}$ defines the equilibrium spontaneous spin moment of a medium. When the magnet is in the excited state characterized by angles $A_l \neq 0$, the vector $\mathbf{b}$ remains unchanged in the mobile frame (see (1.6.25)), whereas it rotates according to the law $B_k = D_{kn}\, b_n$, as by eq. (1.6.22) (see Figure 1.6) in the fixed basis of the spin space. As in mechanics, when going from a mobile frame to a fixed one, the forms $\omega_s(A, \partial_t A)$ are transformed as components of the vector:

$$\omega_k(A, \partial_t A) \to \Omega(A, \partial_t A) = D_{kn}\, \omega_n(A, \partial_t A).$$

As a result, the energy and Lagrangian of the system remain unchanged. For example, the summand $b_n\, \omega_n(A, \partial_t A) = B_k \Omega_k(A, \partial_t A)$ remains invariant. From this point of view, for the matrix elements D_{ij}, the left index refers to the fixed frame, and the right index refers to the mobile frame in the spin space.

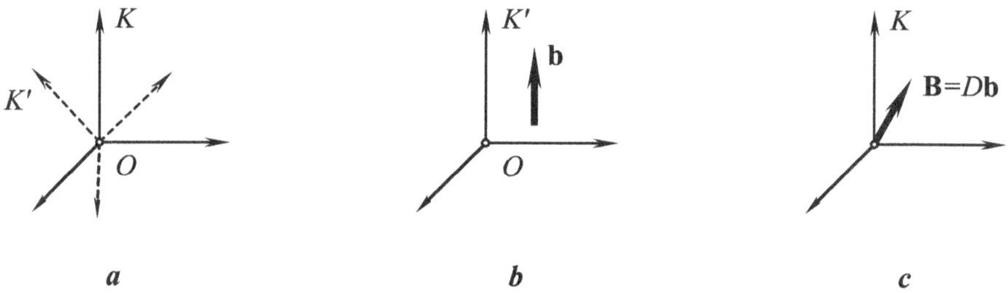

Figure 1.6: Fixed and mobile Cartesian coordinate systems in spin space (a). The K' system is rigidly connected to the spontaneous spin moment of a medium; therefore $\mathbf{b} = \text{const}$ in it (b). In a laboratory reference frame K, the spin moment in excited states is $\mathbf{B} = D(\mathbf{x}, t)\mathbf{b} \neq \text{const}$ (c).

Obeying the laws of quantum mechanics, the spin moment vector of a system differs from the medium magnetization vector only by the magnetomechanical ratio γ:

$$M_k(\mathbf{x}, t) = -\hbar\gamma\, S_{k0}(\mathbf{x}, t) = -\hbar\gamma\, D_{kn}\left[b_n + a_{ni}\, \omega_i(A, \partial_t A)\right]. \tag{1.6.26}$$

In inversion-center-free magnets, the density of the Lagrange function, apart from the term (1.6.14), may contain terms $\mu_{m,pq}\,\omega_p(A,\partial_m A)\,\omega_q(A,\partial_t A)$. Among them, the following are of interest:

$$\mu_{ms}\varepsilon_{pqs}\,\omega_p(A,\partial_m A)\,\omega_q(A,\partial_t A). \tag{1.6.27}$$

In antiferromagnets, the coefficient μ_{ms} can be given by $\mu_{ms} = \mu_m^\nu l_{s\nu}^{(0)}$, μ_m^ν are the exchange constants, and the constant vectors $\mathbf{l}_\nu^{(0)} = (l_{1\nu}^{(0)}, l_{2\nu}^{(0)}, l_{3\nu}^{(0)})$ define the equilibrium structure. Vectors labeled by different numbers ν are orthogonal (ν runs over no more than three values). Under the time-reversal operation, we have

$$\mu_m^\nu \to \mu_m^\nu, \quad \mathbf{l}_\nu^{(0)} \to -\mathbf{l}_\nu^{(0)}, \quad \omega_q(A,\partial_t A) \to -\omega_q(A,\partial_t A).$$

Therefore expression (1.6.27) is invariant. The additional terms in (1.6.27) change the components of the conserved current:

$$S_{k0} = D_{kn}\left[b_n + a_{ni}\,\omega_i(A,\partial_t A) + \mu_m^\nu l_{s\nu}^{(0)}\varepsilon_{qns}\,\omega_q(A,\partial_m A)\right],$$
$$S_{kr} = -D_{kn}\left[c_{ni,rm}\,\omega_i(A,\partial_m A) + d_{n,r} + \mu_r^\nu l_{s\nu}^{(0)}\varepsilon_{qns}\,\omega_q(A,\partial_t A)\right],$$

and together with them, the magnetization of the medium will be other:

$$M_k(\mathbf{x},t) = -\hbar\gamma D_{kn}\big[b_n + a_{ni}\,\omega_i(A,\partial_t A) + \mu_m^\nu l_{s\nu}^{(0)}\varepsilon_{qns}\,\omega_q(A,\partial_m A)\big]. \tag{1.6.28}$$

However, the Goldstone fields $A_n(\mathbf{x},t)$ can be evaluated by the same equations. By relations (1.6.3), (1.6.10), and (1.6.28) the additional terms in the equations $\partial_\mu S_{k\mu} = 0$ mutually cancel out.

From (1.6.28) we can infer that in antiferromagnets ($b_n = 0$), magnetization can be generated not only by the dynamics of the fields A_n but also by static defects and textures: $A_n = A_n(\mathbf{x})$.

The spatial symmetry of a magnetically ordered medium leads to restrictions on the parameters a, b, c, and d of the phenomenological Lagrangian of spin waves (1.6.13). As an example, consider a noncollinear antiferromagnet UO_2 [4].

A four-sublattice antiferromagnet UO_2 has a cubic face-centered spatial lattice (group O_h^5). The magnetic ions of uranium occupy positions coinciding with the sites of the FCC lattice (see Figure 1.7). The spin moments of these atoms are directed along the spatial diagonals of the cube, i. e., along the directions $[1, 1, 1]$, $[1, \bar{1}, \bar{1}]$, $[\bar{1}, 1, \bar{1}]$, $[\bar{1}, \bar{1}, 1]$. The quasiclassical spin momentum density is invariant under translations into doubled elementary vectors of translation $\mathbf{a}_\nu$. The exchange structure is described by three mutually orthogonal antiferromagnetic vectors $\mathbf{l}_\nu^{(0)}$ directed along the edges of the cube:

$$\mathbf{l}_1^{(0)} = (1,0,0), \quad \mathbf{l}_2^{(0)} = (0,1,0), \quad \mathbf{l}_3^{(0)} = (0,0,1).$$

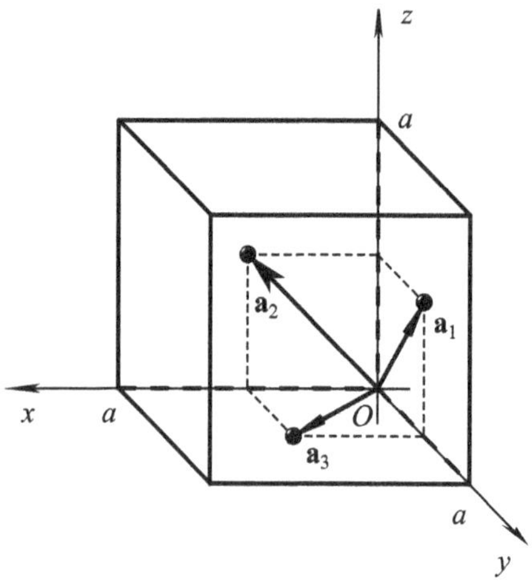

Figure 1.7: Magnetic atom positions in the compound UO_2 (black balls). The arrows show the translation vectors of the spatial lattice of the compound UO_2: UO_2: $\mathbf{a}_1 = (a/2)(0,1,1)$, $\mathbf{a}_2 = (a/2)(1,0,1)$, $\mathbf{a}_3 = (a/2)(1,1,0)$, where a is the length of the cube edge (the cube is a crystallographic cell of the magnetic structure. Passing to dimensionless variables, we set $a = 1$).

Here $\nu = 1,2,3$ is the number of the vector. In excited states of a magnet characterized by the angles $A_l(\mathbf{x}, t)$, the magnetic vectors $\mathbf{l}_\nu = D\mathbf{l}_\nu^{(0)}$ have the components $l_{s\nu} = D_{sk}\, l_{k\nu}^{(0)}$. The magnetic ordering in UO_2 corresponds to an irreducible star with rays

$$\mathbf{k}_1 = \frac{1}{2}(\mathbf{b}_2 + \mathbf{b}_3), \quad \mathbf{k}_2 = \frac{1}{2}(\mathbf{b}_1 + \mathbf{b}_3), \quad \mathbf{k}_3 = \frac{1}{2}(\mathbf{b}_1 + \mathbf{b}_2),$$

where $\mathbf{b}_\nu$ are reciprocal lattice vectors. When translated into $\mathbf{a}_\mu$, the vector $\mathbf{l}_\mu^{(0)}$ does not change, whereas the other two antiferromagnetic vectors $\mathbf{l}_\mu^{(0)}$ $(\mu \neq \nu)$ change sign. The vectors $\mathbf{l}_\nu^{(0)}$ are transformed according to one three-dimensional representation of the space group G of the magnet. This representation corresponds to the cubic symmetry of the spin space with the invariants

$$\begin{aligned}
\omega_i^2(A, \partial_t A) &= \Omega_i^2(A, \partial_t A), \\
\omega_i(A, \partial_k A)\, \omega_i(A, \partial_k A) &= \Omega_i(A, \partial_k A)\, \Omega_i(A, \partial_k A), \\
\sum_{\nu=1}^{3} \left(\mathbf{l}_\nu^{(0)} \cdot \boldsymbol{\omega}(A, \partial_\nu A)\right)^2 &= \sum_{\nu=1}^{3} \left(\mathbf{l}_\nu \cdot \boldsymbol{\Omega}(A, \partial_\nu A)\right)^2 \\
&= \omega_1^2(A, \partial_1 A) + \omega_2^2(A, \partial_2 A) + \omega_3^2(A, \partial_3 A),
\end{aligned}$$

where x_i are the Cartesian coordinates along the principal axes of the crystal, and

$$\boldsymbol{\omega} = (\omega_1, \omega_2, \omega_3), \quad \boldsymbol{\Omega} = (\Omega_1, \Omega_2, \Omega_3), \quad \Omega_i = D_{ik}\omega_k.$$

Therefore, for the compound UO_2, the phenomenological constants in the exchange Lagrangian (1.6.13) are

$$\begin{aligned} &b_n = 0, \quad d_{n,k} = 0, \quad a_{ij} = R\delta_{ij}, \\ &c_{ik,lm} = P\delta_{ik}\delta_{lm} + Q[\delta_{i1}\delta_{k1}\delta_{l1}\delta_{m1} + \delta_{i2}\delta_{k2}\delta_{l2}\delta_{m2} + \delta_{i3}\delta_{k3}\delta_{l3}\delta_{m3}], \end{aligned} \tag{1.6.29}$$

where $R, P > 0, P > Q$.

It should be emphasized that the method of phenomenological Lagrangians of spin waves is based on accounting for the general properties of the symmetry of magnets. Therefore its single limitation consists of the fact that the characteristic frequencies of long-wavelength Goldstone excitations must be less than the exchange frequencies of spin waves separated by energy gaps from the Goldstone modes.

Note that eqs. (1.6.19) find their application in various realms of physics. In particular, it can be shown that for a certain choice of constants a, b, c, and d, they coincide with the equations proposed in [16, 17] for describing the dynamics of the A- and B-phases of superfluid ^{3}He. Equations (1.6.19) are universal due to the general symmetry requirements laid down for many physical systems.

The description of the nonlinear dynamics of Goldstone excitations based on models of magnetic sublattices and the method of phenomenological Lagrangians of spin waves are consistent with each other, at least in the exchange approximation. Below we will illustrate the validity of this assertion with specific examples.

Finally, it is worth underscoring that a local isomorphism of the groups SO(3) and SU(2) simplifies specific calculations. Recall that SU(2) is a group of two-dimensional unitary unimodular matrices g:

$$g g^\dagger = g^\dagger g = I, \quad \det g = 1. \tag{1.6.30}$$

Here and in what follows, the sign "†" stands for the Hermitian conjugation.

There is a biunique correspondence between vicinities of identity elements of the groups SU(2) and SO(3), which is expressed by the following working formula (see [71, 62]):

$$g^\dagger \sigma_p g = \sigma_k D_{kp}, \quad g \sigma_p g^\dagger = D_{pk}\sigma_k. \tag{1.6.31}$$

Here σ_i are Pauli matrices for the algebra of generators of the group SU(2) to be realized:

$$\begin{aligned} &\sigma_1 = \begin{pmatrix} 0 & 1 \\ 1 & 0 \end{pmatrix}, \quad \sigma_2 = \begin{pmatrix} 0 & -\mathrm{i} \\ \mathrm{i} & 0 \end{pmatrix}, \\ &\sigma_3 = \begin{pmatrix} 1 & 0 \\ 0 & -1 \end{pmatrix}, \quad I = \begin{pmatrix} 1 & 0 \\ 0 & 1 \end{pmatrix}, \\ &\sigma_s\sigma_p = I\delta_{sp} + \mathrm{i}\,\varepsilon_{spq}\sigma_q, \quad \sigma_k^\dagger = \sigma_k, \quad \mathrm{Sp}\,\sigma_k = 0. \end{aligned} \tag{1.6.32}$$

It is easy to check that if the matrices $g_1, g_2 \in \mathrm{SU}(2)$ correspond to the matrices D_1, $D_2 \in \mathrm{SO}(3)$, then the product of the matrices $g_1 g_2 \in \mathrm{SU}(2)$ corresponds to the product of the matrices $D_2 D_1 \in \mathrm{SO}(3)$ [30]. The relationship between the groups allows us to replace the construction of the SO(3)-group representations by a similar and simpler problem for the group SU(2). In particular, the Cartan forms can be determined from the expansion

$$(\partial_\mu g)\, g^\dagger = \frac{\mathrm{i}}{2}\sigma_k\, \omega_k(A, \partial_\mu A).$$

Then the result is

$$\omega_k(A, \partial_\mu A) = -\mathrm{i}\,\mathrm{Sp}\,((\partial_\mu g)\, g^\dagger\, \sigma_k). \tag{1.6.33}$$

Here A_l are the SU(2)-group parameters. Using the parameterization of the matrices $g \in \mathrm{SU}(2)$ by Euler angles, we have

$$g(A) = \begin{pmatrix} \cos\frac{A_1}{2}\, \exp(\frac{\mathrm{i}}{2}\,(A_2 + A_3)) & \mathrm{i}\,\sin\frac{A_1}{2}\, \exp(\frac{\mathrm{i}}{2}\,(A_3 - A_2)) \\ \mathrm{i}\,\sin\frac{A_1}{2}\, \exp(-\frac{\mathrm{i}}{2}\,(A_3 - A_2)) & \cos\frac{A_1}{2}\, \exp(-\frac{\mathrm{i}}{2}\,(A_2 + A_3)) \end{pmatrix}.$$

The functions $A_l = A_l(\mathbf{x}, t)$ correspond to quasiclassical magnon fields.

Using the possibility of cyclic permutation of matrices under the sign of the operation “Sp” (“trace”), the identity

$$(\partial_\mu g^\dagger)\, g = -g^\dagger \partial_\mu\, g,$$

and relations (1.6.30), (1.6.31), and (1.6.32), it is easy to make sure by direct verification that the previous definition of Cartan forms (1.6.11) coincides with that derived above (1.6.33):

$$\begin{aligned} \omega_n(A, \partial_\mu A) &= -\frac{1}{2}\,\varepsilon_{nsp}\, D_{ks}\, \partial_\mu D_{kp} \\ &= -\frac{1}{4}\,\varepsilon_{nsp}\,\mathrm{Sp}\,[(g^\dagger \sigma_s\, g)\, \partial_\mu (g^\dagger \sigma_p\, g)] \\ &= -\frac{\mathrm{i}}{2}\varepsilon_{nsp}\,\varepsilon_{spm}\mathrm{Sp}[(\partial_\mu\, g)\, g^\dagger \sigma_m] = -\mathrm{i}\,\mathrm{Sp}[(\partial_\mu\, g)\, g^\dagger \sigma_n]. \end{aligned} \tag{1.6.34}$$

The previous definition (1.6.11) includes the invariance of Cartan forms with respect to turns through an arbitrary angle in spin space as their invariance under left shifts on the group SO(3): $D(A) \to D_0 D(A)$. It is mandatory to stress that the invariance of Cartan forms under rotations in spin space corresponds to their invariance under right shifts on the group SU(2), as seen in representation (1.6.33): $g(A) \to g(A) g_0$. Such a choice of two different representations is needed to achieve the consistency of the formulas in this book.

The Goldstone field transformation law (1.6.6) can be found in explicit form using the group isomorphism

$$g(A) \xrightarrow{g_0} g(A)g_0 = g(A'(A, g_0)).$$

1.7 Accounting for an external magnetic field, relativistic interactions, and relaxation processes in the method of phenomenological Lagrangians for spin waves

If a scientist does not understand a problem, he writes down many formulae, but as soon as he has understood it, there remain at most two.

N. Bohr

In the theory set forth above, the energy of interaction between a magnet and an external magnetic field $\mathbf{H}^{\text{ext}}$ should be taken into account. In doing so, it suffices to add the term $-\int(\mathbf{M}\cdot\mathbf{H}^{\text{ext}})d^3\mathbf{x}$ to the system energy (1.6.15). Moreover, the Lagrange function (1.6.14) must include the same term but with the opposite sign.

The magnetization distribution (1.6.26) or (1.6.28) creates its own magnetic field $\mathbf{H}^{(\text{m})}$ in the sample. The interaction between the magnetization and the field $\mathbf{H}^{(\text{m})}$ can be accounted for in the same way by incorporating the magnetostatic energy (1.1.6) into the system energy (1.6.15).

Let us consider whether the theory could include local relativistic interactions. Relativistic interactions break the symmetry of the Lagrangian under rotations in spin space; its invariance is maintained only when the spin and space variables are rotated simultaneously. In a crystal lattice the spatial variables are the vectors connected to the crystallographic axes. The matrices D and vectors oriented along the crystallographic axes make up the invariants describing the relativistic interactions. The left index of the matrix elements D_{ij} is responsible for the orientation of the magnetic structure relative to the crystallographic axes.

For an antiferromagnet UO_2, there is only one quadratic-in-D_{ij} relativistic invariant

$$\sum_{\nu=1}^{3}(\mathbf{l}_\nu^{(0)}\cdot\mathbf{l}_\nu)^2 = D_{11}^2 + D_{22}^2 + D_{33}^2.$$

Therefore the terms

$$W_\text{a} = -\frac{\beta}{2}\int[D_{11}^2 + D_{22}^2 + D_{33}^2]d^3\mathbf{x}$$

should be added to the energy of UO_2 ($\beta > 0$).

Relativistic interactions cause a variety of effects, including spatial anisotropy in spin wave propagation, correlations between spin wave propagation and its direction, the emergence of a specific spin wave polarization during interspin scattering, and more.

Since the magnetic field and relativistic interactions break the phenomenological Lagrangian's invariance with respect to spin rotations, the continuity equations for the spin current are invalid.

An intriguing attempt to include relaxation processes in the macroscopic theory discussed was undertaken in [10]. When energy losses take place, the equations for the nonlinear dynamics of Goldstone excitations can be formulated as follows:

$$\partial_\mu \frac{\tilde{\delta} Z}{\tilde{\delta}(\partial_\mu A_l)} - \frac{\tilde{\delta} Z}{\tilde{\delta} A_l} + \frac{\delta Q}{\delta(\partial_t A_l)} = 0, \tag{1.7.1}$$

where

$$\begin{aligned} \frac{\tilde{\delta} Z}{\tilde{\delta}(\partial_\mu A_l)} &= \frac{\partial L}{\partial \omega_n(A, \partial_\nu A)} \frac{\partial \omega_n(A, \partial_\nu A)}{\partial(\partial_\mu A_l)}, \\ \frac{\tilde{\delta} Z}{\tilde{\delta} A_l} &= \frac{\partial L}{\partial \omega_n(A, \partial_\nu A)} \frac{\partial \omega_n(A, \partial_\nu A)}{\partial A_l}, \end{aligned} \tag{1.7.2}$$

Z is the previously defined phenomenological Lagrangian of spin waves, $\mu,\nu = 0, 1, 2, 3$, and

$$Q = \int R \, \mathrm{d}^3\mathbf{x}$$

is the dissipative functional of the system with $R = R(A, \partial_t A, \partial_t \partial_l A, \ldots)$. As opposed to the partial variational derivatives (1.7.2), formula (1.7.1) contains the quantity $\delta Q/\delta(\partial_t A_l)$ as the functional total variational derivative of Q over the generalized velocities $(\partial_t A_l)$. This quantity is derived after integrating by parts the integrands with factors of the type $\partial_n[\delta\, \partial_t A_l]$.

In the general case, Q is a positive definite functional of the same symmetry as the Lagrangian Z of a magnet. In the exchange approximation, a dissipative functional Q_{ex} must meet the following requirement: the total spin momentum of the system must be conserved, i. e.,

$$\int S_{k0}(\mathbf{x}, t) \mathrm{d}^3\,\mathbf{x} = \mathrm{const.}$$

Exchange relaxation has no influence on the total spin momentum of the medium but merely causes the spin momentum density to diffuse from higher spin density regions of the coordinate space to lower spin density ones. The simplest quadratic-in-Cartan's form exchange dissipative functional that satisfies the outlined criteria and contains the minimum number of derivatives has the form

$$Q_{\mathrm{ex}} = \frac{1}{2} \int r_{ij,nm}\, \partial_t \omega_i(A, \partial_n A)\, \partial_t \omega_j(A, \partial_m A) \mathrm{d}^3\,\mathbf{x} > 0,$$

where without loss of generality, we can assume that $r_{ij,nm} = r_{ji,mn}$. The spatial symmetry of the medium imposes restrictions on the choice of the phenomenological parameters $r_{ij,nm}$. For example, for a noncollinear antiferromagnet UO_2, we have

$$r_{ij,mn} = r_1 \delta_{ij} \delta_{mn} + r_2 [\delta_{i1} \delta_{j1} \delta_{n1} \delta_{m1} \\ + \delta_{i2} \delta_{j2} \delta_{m2} \delta_{n2} + \delta_{i3} \delta_{j3} \delta_{m3} \delta_{n3}]; \quad r_1 > 0, \ r_1 > r_2.$$

To write down the dynamic equations (1.7.1) in a more precise manner, we use the following representation of the Cartan forms:

$$\omega_n(A, \partial_\mu A) = K_{nl}(A) \, \partial_\mu A_l. \tag{1.7.3}$$

The functions $K_{nl}(A)$ in (1.7.3) are dependent only on the angles A_l. Although their form may be easily found from formulas (1.6.12), we will not offer it here because only a 3×3 inverse matrix is required for calculations:

$$K_{il}(K^{-1})_{lp} = \delta_{ip}.$$

Its matrix elements satisfy the identities

$$\frac{\partial K_{nl}}{\partial A_s} - \frac{\partial K_{ns}}{\partial A_l} + \varepsilon_{npq} K_{ps} K_{ql} = 0,$$

which are consequences of the Maurer–Cartan structural equations (1.6.10). These properties of the functions $K_{nl}(A)$ allow us to exclude them from the final equations (1.7.1):

$$\partial_\mu \left(\frac{\partial L}{\partial \omega_l(A, \partial_\mu A)} \right) - \varepsilon_{lnm} \frac{\partial L}{\partial \omega_n(A, \partial_\mu A)} \omega_m(A, \partial_\mu A) \\ = \partial_n \left(\frac{\partial R}{\partial [\partial_t \omega_l(A, \partial_n A)]} \right) - \varepsilon_{liq} \frac{\partial R}{\partial [\partial_t \omega_i(A, \partial_n A)]} \omega_q(A, \partial_n A).$$

Thus we arrive at the equations for the dynamics of magnon fields, which take into account energy losses:

$$a_{lk} \, \partial_t \omega_k(A, \partial_t A) - c_{lr,nm} \, \partial_n \omega_r(A, \partial_m A) \\ + \varepsilon_{lnm} [c_{nr,pq} \, \omega_r(A, \partial_q A) + d_{n,p}] \, \omega_m(A, \partial_p A) \\ - \varepsilon_{lnm} \, [a_{nk} \, \omega_k(A, \partial_t A) + b_n] \omega_m(A, \partial_t A) \\ = -\varepsilon_{liq} \, r_{ij,nm} \, \partial_t \omega_j(A, \partial_m A) \, \omega_q(A, \partial_n A) \\ + r_{lj,nm} \, \partial_n \partial_t \omega_j(A, \partial_m A). \tag{1.7.4}$$

Using (1.6.24) and (1.7.4), it is easy to show that the equation for the system spin moment density obeys, as it should, the law of conservation of the total mechanical spin moment:

$$\partial_t S_{k0} = -\partial_n [\Pi_{kn} + S_{kn}].$$

Here the dynamic spin current $S_{k\mu}$ is the same as in formula (1.6.22). The flux Π_{kn} determines the exchange diffusion of magnetization in the sample:

$$\Pi_{kn} = -D_{kl}\, r_{lj,nm}\, \partial_t \omega_j(A, \partial_m A).$$

The exchange diffusion of magnetization is accompanied by energy losses

$$\frac{\mathrm{d}E}{\mathrm{d}t} = -\int \frac{\delta Q_{\mathrm{ex}}}{\delta(\partial_t A_s)}\, \partial_t A_s \mathrm{d}^3\mathbf{x} = -2Q_{\mathrm{ex}} < 0.$$

Relativistic interactions reduce the symmetry of the dissipative functional. The relativistic dissipative functional Q_{a} is invariant only when the magnetic vectors of the exchange structure and the lattice rotate concurrently. The most general expression for Q_{a} containing the minimum number of derivatives is

$$Q_{\mathrm{a}} = \frac{1}{2}\int \lambda_{ijkl}\, \partial_t D_{ij}\, \partial_t D_{kl} \mathrm{d}^3\mathbf{x} > 0. \tag{1.7.5}$$

The functional (1.7.5) describes the relaxation of a homogeneous spin precession in a magnetically ordered medium. For an antiferromagnet UO_2, we have

$$Q_{\mathrm{a}} = \frac{\lambda}{2}\int [(\partial_t D_{11})^2 + (\partial_t D_{22})^2 + (\partial_t D_{33})^2] \mathrm{d}^3\mathbf{x},$$

where $\lambda > 0$.

Next, we demonstrate that the phenomenological Lagrangians of magnets, when relaxation processes are ignored, result in completely integrable nonlinear models, which are physically significant and admit exact solutions defining certain varieties of magnetic solitons in multisublattice magnets. In magnetically ordered media, energy losses always occur. At low energy losses, the above strategy allows for a detailed exploration of the evolution of magnetic solitons using the methods of nonlinear perturbation theory.

1.8 Discussion of the equivalence of the Landau–Lifshitz approach to the description of the dynamics of magnets and the method of phenomenological Lagrangians of spin waves

Similis similiti gaudet (in Latin).

Like takes pleasure in like

To date, there are two macroscopic theories used for theoretically describing the nonlinear dynamics of magnets: a magnetic-sublattice-based approach and the method of phenomenological Lagrangians. Further, we consider a single-sublattice ferromagnet, a ferrimagnet with two magnetic sublattices, and a four-sublattice noncollinear antiferromagnet UO_2. Using these examples, we discuss the question of the equivalence of these phenomenological theories.

In the exchange approximation, the Landau–Lifshitz equation for a single-sublattice ferromagnet has the form

$$\partial_t \mathbf{M} = -\gamma[\mathbf{M} \times \alpha_{rm} \partial_r \partial_m \mathbf{M}], \quad \mathbf{M}^2 = M_0^2 = \text{const}, \tag{1.8.1}$$

where $\alpha_{mn} = \alpha_{nm}$ are the constants of the exchange interaction, M_0 is the magnitude of the nominal magnetization, and γ is the magnetomechanical ratio.

After passing from the medium's magnetization vector $\mathbf{M}$ to its spin momentum density vector $\mathbf{S}$, eq. (1.8.1) can be rewritten as

$$\partial_t \mathbf{S} = \tilde{\alpha}_{rm}\, \partial_r [\mathbf{S} \times \partial_m \mathbf{S}], \quad \mathbf{M} = -\gamma\hbar \mathbf{S}, \quad \tilde{\alpha}_{mn} = \alpha_{mn}\, \hbar\gamma^2. \tag{1.8.2}$$

Let us show that eqs. (1.8.2) coincide with eqs. (1.6.23) for the following choice of phenomenological constants:

$$\begin{aligned} &a_{ni} = 0, \quad d_{n,r} = 0, \quad b_n = -\delta_{n3}\frac{M_0}{\gamma\hbar}, \\ &c_{ni,rm} = (\delta_{ni} - \delta_{n3}\,\delta_{i3})\,\tilde{\alpha}_{rm}\, b_3^2. \end{aligned} \tag{1.8.3}$$

Such a choice of the parameters produces the following forms of eqs. (1.6.23) and the density vector of the medium's magnetic moment:

$$\begin{aligned} &\partial_t [D_{k3}\, b_3] - \tilde{\alpha}_{rm}\, \partial_r \big[D_{k1}\, \omega_1(A, \partial_m A) \\ &\quad + D_{k2}\, \omega_2(A, \partial_m A)\big]\, b_3^2 = 0; \end{aligned} \tag{1.8.4}$$

$$M_k = -\gamma\,\hbar\, b_3 D_{k3} \equiv D_{k3} M_0, \quad S_k = D_{ks} b_s. \tag{1.8.5}$$

Due to the orthogonality condition of the matrices D, the length of the vector $\mathbf{M}$ is unchanged: $\mathbf{M}^2 = M_0^2$. From (1.6.4) we can infer that the vector $\mathbf{M}$ is parameterized by two Euler angles:

$$\mathbf{M} = M_0\left(\sin\theta\cos\left(\varphi - \frac{\pi}{2}\right),\ \sin\theta\sin\left(\varphi - \frac{\pi}{2}\right),\ \cos\theta\right).$$

To transform the second summand in eq. (1.8.4), we need an auxiliary relation, which can be easily obtained by multiplying eq. (1.6.24)

$$\partial_m D_{ji} = -\varepsilon_{liq}\, \omega_q(A, \partial_m A)\, D_{jl}$$

by $\varepsilon_{knj} D_{nr}$, followed by summation over twice occurring indices:

$$\partial_m D_{ji}\, \varepsilon_{knj}\, D_{nr} = -\omega_q(A, \partial_m A)\, \varepsilon_{knj}\, D_{nr}\, D_{jl}\, \varepsilon_{liq}$$

$$
\begin{aligned}
&= -\omega_q(A, \partial_m A)\, \varepsilon_{srl}\, D_{ks}\, \varepsilon_{liq} \\
&= -\omega_q(A, \partial_m A)\, D_{ks}\, [\delta_{si}\delta_{rq} - \delta_{sq}\delta_{ri}].
\end{aligned} \tag{1.8.6}
$$

Performing the transformations, we have used the formula

$$
\varepsilon_{knj} D_{nr} D_{jl} = \varepsilon_{srl} D_{ks},
$$

derived from (1.6.1), (1.6.2), and identities (1.6.3). Assuming that $r = i = 3$ in eq. (1.8.6), we come up with

$$
\varepsilon_{knj}\, D_{n3}\, \partial_m D_{j3} = D_{k1}\, \omega_1(A, \partial_m A) + D_{k2}\, \omega_2(A, \partial_m A). \tag{1.8.7}
$$

Equation (1.8.4) perfectly matches the Landau–Lifshitz equations (1.8.2) for a basic ferromagnet when relations (1.8.5) and (1.8.7) are taken into account. Additionally, the example at hand demonstrates how altering the values of the constants a, b, c, and d can diminish the number of Goldstone fields in eqs. (1.6.23) or (1.6.19). In this instance, two local fields, $A_1 = \theta$ and $A_2 = \varphi$, are sufficient for eqs. (1.6.23) to be parameterized. To interpret the findings, recall that in the exchange approximation, a single macroparameter, spontaneous magnetization $-\gamma \hbar \mathbf{b}$ characterizes the equilibrium state of a single-sublattice ferromagnet. The direction of spontaneous magnetization distinguishes one of the potential vacuum states of the ferromagnet. Since the vector $\mathbf{b}$ remains unchanged and coincides in direction with the axis, around which all spins synchronously turn at the same angle, the given vacuum state possesses the nontrivial invariance subgroup SO(2). Goldstone fields describe the ferromagnet's excited states that only correspond to those local spin turns that modify the spontaneous magnetization $-\gamma \hbar \mathbf{b}$. It is sufficient to assign two dynamic variables to describe such turns.

As shown in Chapter 6, in the exchange approximation, the effective equations of motion

$$
[\mathbf{l} \times (c^{-2}\, \partial_t^2 \mathbf{l} - \Delta \mathbf{l})] + \beta c^{-1} \partial_t \mathbf{l} = 0, \quad \mathbf{l}^2 = 1, \tag{1.8.8}
$$

can be derived from the Landau–Lifshitz equations for an isotropic ferrimagnet with two crystallographically nonequivalent magnetic sublattices.

If we put

$$
\begin{aligned}
&d_{n,r} = 0, \quad a_{ni} = c^{-2}\left(\delta_{ni} - \frac{b_n\, b_i}{\mathbf{b}^2}\right), \\
&c_{ni,rm} = \delta_{rm}\left(\delta_{ni} - \frac{b_i\, b_n}{\mathbf{b}^2}\right), \\
&\mathbf{b} = (0,\, 0,\, \beta c^{-1}), \quad l_k = D_{k3},
\end{aligned} \tag{1.8.9}
$$

in eqs. (1.6.23), they will coincide with eqs. (1.8.8). The proof of this statement is similar to that given for a ferromagnet.

Let us focus on the phenomenological equations describing a noncollinear antiferromagnet UO_2. In [23, 22, 4] the Landau–Lifshitz approach covers the subject of potential exchange magnetic structure types and the range of spin waves in UO_2. The magnetic ions of this crystal (uranium ions) are assembled into a face-centered cubic lattice. The particular arrangement of the nonmagnetic ions is irrelevant. The only thing that matters is that the latter does not violate the symmetry of the lattice created by the magnetic ions, i. e., the symmetry of the group O_h^5. In the context of magnetism, a crystal is a system of four magnetic sublattices embedded into one another, with each being formed by uranium ions in the positions $\mathbf{a}_0 = (0, 0, 0)$, $\mathbf{a}_1 = (0, 1/2, 1/2)$, $\mathbf{a}_2 = (1/2, 0, 1/2)$, $\mathbf{a}_3 = (1/2, 1/2, 0)$ (see Figure 1.7). The spins of the ions are directed along the cube's spatial diagonals (along the directions $[1, 1, 1]$, $[1, \bar{1}, \bar{1}]$, $[\bar{1}, 1, \bar{1}]$, and $[\bar{1}, \bar{1}, 1]$, respectively).

For describing the magnetic structure of UO_2 in terms of the magnetic sublattice model, it is convenient to use the following linear combinations of the sublattice spin moments:

$$\begin{aligned}
\mathbf{m} &= \mathbf{S}_1 + \mathbf{S}_2 + \mathbf{S}_3 + \mathbf{S}_4, & \mathbf{k} &= 0;\\
\mathbf{l}_1 &= \mathbf{S}_1 + \mathbf{S}_2 - \mathbf{S}_3 - \mathbf{S}_4, & \mathbf{k} &= \frac{\mathbf{b}_2 + \mathbf{b}_3}{2};\\
\mathbf{l}_2 &= \mathbf{S}_1 - \mathbf{S}_2 + \mathbf{S}_3 - \mathbf{S}_4, & \mathbf{k} &= \frac{\mathbf{b}_1 + \mathbf{b}_3}{2};\\
\mathbf{l}_3 &= \mathbf{S}_1 - \mathbf{S}_2 - \mathbf{S}_3 + \mathbf{S}_4, & \mathbf{k} &= \frac{\mathbf{b}_1 + \mathbf{b}_2}{2}.
\end{aligned}$$

On the right, there is the wave vector $\mathbf{k}$ that controls the translational symmetry corresponding to the magnetic vector ($\mathbf{b}_s$ are the vectors of the lattice inverse relative to that formed by the elementary translation vectors $\mathbf{a}_s$).

The spin moments of the ions have the same length,

$$\mathbf{S}_1^2 = \mathbf{S}_2^2 = \mathbf{S}_3^2 = \mathbf{S}_4^2 = S^2 = \text{const}, \tag{1.8.10}$$

and can be expressed through magnetic vectors $\mathbf{m}$, $\mathbf{l}_\nu$ ($\nu = 1, 2, 3$):

$$\begin{aligned}
\mathbf{S}_1 &= \frac{1}{4}(\mathbf{m} + \mathbf{l}_1 + \mathbf{l}_2 + \mathbf{l}_3), & \mathbf{S}_2 &= \frac{1}{4}(\mathbf{m} + \mathbf{l}_1 - \mathbf{l}_2 - \mathbf{l}_3),\\
\mathbf{S}_3 &= \frac{1}{4}(\mathbf{m} - \mathbf{l}_1 + \mathbf{l}_2 - \mathbf{l}_3), & \mathbf{S}_4 &= \frac{1}{4}(\mathbf{m} - \mathbf{l}_1 - \mathbf{l}_2 + \mathbf{l}_3).
\end{aligned}$$

In terms of $\mathbf{m}$ and $\mathbf{l}_\nu$, the normalization conditions (1.8.10) take the form

$$(\mathbf{l}_1)^2 + (\mathbf{l}_2)^2 + (\mathbf{l}_3)^2 + \mathbf{m}^2 = (4S)^2; \tag{1.8.11}$$

$$\begin{aligned}
(\mathbf{l}_1 \cdot \mathbf{l}_2) + (\mathbf{l}_3 \cdot \mathbf{m}) &= (\mathbf{l}_1 \cdot \mathbf{l}_3) + (\mathbf{l}_2 \cdot \mathbf{m})\\
&= (\mathbf{l}_2 \cdot \mathbf{l}_3) + (\mathbf{l}_1 \cdot \mathbf{m}) = 0.
\end{aligned} \tag{1.8.12}$$

In the equilibrium antiferromagnetic state of the crystal, $\mathbf{m} = 0$, and the vectors $\mathbf{l}_\nu$ have an identical length and are directed along the cube's edges $[1, 0, 0]$, $[0, 1, 0]$, $[0, 0, 1]$.

In [23, 22] an invariant expansion of the UO_2 exchange energy in the powers of the order parameters was obtained. The expressions for the energy include the terms of an inhomogeneous bilinear and a homogeneous biquadratic exchange coupling. Accounting for the biquadratic exchange is of fundamental importance for a rigid mutual orientation of the antiferromagnetic vectors $\mathbf{l}_\nu$ of the magnetic structure to be realized. In [25] the estimates yield the same-order-magnitudes of the bilinear and biquadratic exchange in compounds of the UO_2 type. In addition, in the main approximation, the dynamics of Goldstone magnons in UO_2 are only driven by the bilinear exchange and reduce to local turns of the vectors $\mathbf{l}_\nu$, retaining their length and mutual orientation as dictated by the homogeneous biquadratic exchange. Let us focus on this discussion.

The bilinear exchange energy of UO_2 is given by [22]

$$\begin{aligned} w_2 = \frac{1}{2} \sum \bar{J}_{ij} \, [(\mathbf{S}_1^i \cdot \mathbf{S}_1^j) + (\mathbf{S}_2^i \cdot \mathbf{S}_2^j) + (\mathbf{S}_3^i \cdot \mathbf{S}_3^j) \\ + (\mathbf{S}_4^i \cdot \mathbf{S}_4^j)] + \sum J_{ij} \, [(\mathbf{S}_1^i \cdot \mathbf{S}_2^j) + (\mathbf{S}_1^i \cdot \mathbf{S}_3^j) + (\mathbf{S}_1^i \cdot \mathbf{S}_4^j) \\ + (\mathbf{S}_2^i \cdot \mathbf{S}_4^j) + (\mathbf{S}_2^i \cdot \mathbf{S}_3^j) + (\mathbf{S}_3^i \cdot \mathbf{S}_4^j)] + \cdots . \end{aligned} \tag{1.8.13}$$

Here $\bar{J}_{ij}$ and J_{ij} are the exchange integrals, and the indices i and j denote the numbers of the sublattices. All of the sums are taken over the nearest neighbors. In the nearest-neighbor approximation, the Fourier components of exchange integrals have the form

$$\begin{aligned} \bar{J}_k &= 2\bar{J}(\cos k_1 + \cos k_2 + \cos k_3), \\ J_{12} = J_{34} &= 2J\left(\cos \frac{k_2 + k_3}{2} + \cos \frac{k_2 - k_3}{2}\right), \\ J_{13} = J_{24} &= 2J\left(\cos \frac{k_1 + k_3}{2} + \cos \frac{k_1 - k_3}{2}\right), \\ J_{14} = J_{23} &= 2J\left(\cos \frac{k_1 + k_2}{2} + \cos \frac{k_1 - k_2}{2}\right). \end{aligned} \tag{1.8.14}$$

To make writing easier, we have used dimensionless variables ($\mathbf{k}$ measured in units of the reciprocal length of a crystallographic cell).

Passing in expansions (1.8.13) and (1.8.14) to the continuum limit gives the following expression for the exchange energy density [23]:

$$\begin{aligned} w_2 = \frac{A}{2}[(\mathbf{l}_1)^2 + (\mathbf{l}_2)^2 + (\mathbf{l}_3)^2] + \frac{B}{2}\mathbf{m}^2 \\ + \frac{\alpha}{2} \sum_{\nu,j=1}^{3} \left(\frac{\partial \mathbf{l}_\nu}{\partial x_j} \cdot \frac{\partial \mathbf{l}_\nu}{\partial x_j}\right) + \frac{\beta}{2}\left[\left(\frac{\partial \mathbf{l}_1}{\partial x_1}\right)^2 + \left(\frac{\partial \mathbf{l}_2}{\partial x_2}\right)^2 + \left(\frac{\partial \mathbf{l}_3}{\partial x_3}\right)^2\right], \end{aligned} \tag{1.8.15}$$

where $B, \alpha > 0$, $\alpha > \beta$. We have omitted the terms $\propto (\partial_i \mathbf{m})^2$ because they are small in the long-wavelength limit compared to the term $\propto \mathbf{m}^2$.

The summands $(\mathbf{m}\cdot\mathbf{l}_\nu)^2$ and $\mathbf{m}^4$ can be excluded from the biquadratic exchange energy density w_4 due to the smallness of the terms containing $\mathbf{m}$ compared to the expressions incorporating $\mathbf{l}_\nu$ ($\mathbf{m}^2 \ll (\mathbf{l}_\nu)^2$) Consequently, we have

$$\begin{aligned} w_4 = {} & \frac{C}{4}[(\mathbf{l}_1)^2 + (\mathbf{l}_2)^2 + (\mathbf{l}_3)^2]^2 + \frac{D}{4}[(\mathbf{l}_1)^4 + (\mathbf{l}_2)^4 + (\mathbf{l}_3)^4] \\ & + \frac{E}{2}[(\mathbf{l}_1\cdot\mathbf{l}_2)^2 + (\mathbf{l}_1\cdot\mathbf{l}_3)^2 + (\mathbf{l}_2\cdot\mathbf{l}_3)^2] \\ & + F[(\mathbf{m}\cdot\mathbf{l}_1)(\mathbf{l}_2\cdot\mathbf{l}_3) + (\mathbf{m}\cdot\mathbf{l}_2)(\mathbf{l}_1\cdot\mathbf{l}_3) + (\mathbf{m}\cdot\mathbf{l}_3)(\mathbf{l}_1\cdot\mathbf{l}_2)], \end{aligned}$$

where $3C + D > 0$ and $E > 0$ (the choice of the inequalities will be clear from further analysis).

Furthermore, to deduce the equilibrium values of $\mathbf{l}_\nu$, it is sufficient to minimize the homogeneous part of the crystal's total exchange energy density, neglecting the dependence of the latter on the vector $\mathbf{m}$, since $\mathbf{m}^2 \ll \mathbf{l}_\nu^2$ [23]:

$$\begin{aligned} w = {} & \frac{A}{2}[(\mathbf{l}_1)^2 + (\mathbf{l}_2)^2 + (\mathbf{l}_3)^2] \\ & + \frac{C}{4}[(\mathbf{l}_1)^2 + (\mathbf{l}_2)^2 + (\mathbf{l}_3)^2]^2 + \frac{D}{4}[(\mathbf{l}_1)^4 + (\mathbf{l}_2)^4 + (\mathbf{l}_3)^4] \\ & + \frac{E}{2}[(\mathbf{l}_1\cdot\mathbf{l}_2)^2 + (\mathbf{l}_1\cdot\mathbf{l}_3)^2 + (\mathbf{l}_2\cdot\mathbf{l}_3)^2]. \end{aligned} \tag{1.8.16}$$

In this approximation, coupling (1.8.11) should be replaced by the simpler restriction

$$(\mathbf{l}_1)^2 + (\mathbf{l}_2)^2 + (\mathbf{l}_3)^2 \approx (4S)^2,$$

which can be easily taken into account by the method of indefinite Lagrange multipliers.

Under the conditions $3C + D > 0$, $E > 0$, and $A < 0$, the minimum energy (1.8.16) corresponds to the following magnetic structure:

$$\mathbf{l}_1 \perp \mathbf{l}_2 \perp \mathbf{l}_3, \quad (\mathbf{l}_1)^2 = (\mathbf{l}_2)^2 = (\mathbf{l}_3)^2 \equiv l^2 = \frac{(4S)^2}{3} = -\frac{A}{3C + D}. \tag{1.8.17}$$

Returning to the complete expression for the exchange energy density $w = w_2 + w_4$ containing $\mathbf{m}$, we conclude that the total spin moment $\mathbf{m}$ is equal to zero in the ground state (1.8.17) when $B > 0$. The values found for the vectors $\mathbf{m}$ and $\mathbf{l}_\nu$ exactly satisfy restrictions (1.8.12) and are consistent with the experiment.

In the context of Landau's theory of phase transitions, the parameter A is assumed to be temperature-dependent, obeying the law $A = -r(T_c - T)$, where $r > 0$, and T_c is the transition temperature. Therefore in the temperature range $T > T_c$, $A > 0$, the magnetic structure (1.8.17) is not achieved, and the crystal enters the paramagnetic ground state: $\mathbf{l}_\nu = \mathbf{m} = 0$.

Restrictions (1.8.17) on the mutual orientation of the vectors $\mathbf{l}_\nu$ can be represented in the concise mathematical form:

$$l_{i\nu}\, l_{j\nu} = \delta_{ij}\, l^2, \quad l_{i\nu}\, l_{i\mu} = \delta_{\nu\mu}\, l^2. \tag{1.8.18}$$

Here the Greek indices enumerate various vectors $\mathbf{l}_\nu$, and the Latin indices enumerate the projections of these vectors onto the coordinate axes: $\mathbf{l}_\nu = (l_{1\nu}, l_{2\nu}, l_{3\nu})$.

In the excited states of the UO_2 crystal, the components of the vector $\mathbf{m}$ are not equal to zero, but nevertheless, they remain negligibly small compared to the components of the vector $\mathbf{l}_\nu$: $(\mathbf{l}_\nu)^2 \gg \mathbf{m}^2$. We assume that in excited states the relative orientation of the vectors $\mathbf{l}_\nu$, as before, is determined by relations (1.8.18). By comparing formulas (1.8.18) and (1.6.1) we can match the components of the vector $\mathbf{l}_\nu$ and the elements of the orthogonal matrices $D \in SO(3)$ [39]:

$$l_{i\nu} = D_{i\nu}\, l. \tag{1.8.19}$$

The stated approximation theoretically describes the UO_2 nonlinear dynamics by the following summands of the bilinear exchange energy:

$$W_{\mathrm{ex}} = \int \Bigg(\frac{\tilde{B}}{2}\mathbf{m}^2 + \frac{\alpha}{2} \sum_{\nu,j=1}^{3} \left(\frac{\partial \mathbf{l}_\nu}{\partial x_j} \cdot \frac{\partial \mathbf{l}_\nu}{\partial x_j} \right) + \frac{\beta}{2} \left[\left(\frac{\partial \mathbf{l}_1}{\partial x_1} \right)^2 + \left(\frac{\partial \mathbf{l}_2}{\partial x_2} \right)^2 + \left(\frac{\partial \mathbf{l}_3}{\partial x_3} \right)^2 \right] \Bigg) \mathrm{d}^3\mathbf{x}. \tag{1.8.20}$$

Going from expression (1.8.15) to (1.8.20), we have used the first of the normalization conditions (1.8.11) to eliminate the invariant $(\mathbf{l}_1)^2 + (\mathbf{l}_2)^2 + (\mathbf{l}_3)^2$. Then we have $\tilde{B} = B - A$ in (1.8.20) and can exploit approximation (1.8.19) for the vectors $\mathbf{l}_\nu$. Further, when estimating, we assume that $\alpha, \beta \sim a^2 \tilde{B}$ (a is the interatomic distance).

Using (1.8.20), it is easy to rewrite the Landau–Lifshitz equations (1.1.13) in terms of the vectors $\mathbf{m}$ and $\mathbf{l}_\nu$. In this case, it comes to the evolution of the spin moments of sublattices, so we should formally set $\gamma_\nu = -1$ in (1.1.13) (all γ_ν are identical, since the uranium ions are in crystallographically equivalent positions). Therefore the equations of motion for the vectors $\mathbf{m}$ and $\mathbf{l}_\nu$ appear as follows:

$$\partial_t \mathbf{m} = [\mathbf{m} \times \mathbf{H}_{\mathrm{m}}] + \sum_\nu [\mathbf{l}_\nu \times \mathbf{H}_{\mathbf{l}_\nu}],$$

$$\partial_t \mathbf{l}_\nu = [\mathbf{l}_\nu \times \mathbf{H}_{\mathrm{m}}] + [\mathbf{m} \times \mathbf{H}_{\mathbf{l}_\nu}] + \sum_{\chi,\lambda} |\varepsilon_{\nu\chi\lambda}| [\mathbf{l}_\nu \times \mathbf{H}_{\mathbf{l}_\lambda}],$$

where

$$\mathbf{H}_{\mathrm{m}} = -\frac{\delta W_{\mathrm{ex}}}{\delta \mathbf{m}}, \quad \mathbf{H}_{\mathbf{l}_\nu} = -\frac{\delta W_{\mathrm{ex}}}{\delta \mathbf{l}_\nu}$$

are the effective fields corresponding to the vectors $\mathbf{m}$ and $\mathbf{l}_\nu$, $\varepsilon_{\nu\chi\lambda}$ is an antisymmetric unit tensor. In more detailed form,

$$\begin{aligned}\partial_t \mathbf{m} &= \sum_\nu [\mathbf{l}_\nu \times (\alpha\Delta\mathbf{l}_\nu + \beta\partial_\nu^2\mathbf{l}_\nu)],\\ \partial_t \mathbf{l}_\nu &= -\tilde{B}[\mathbf{l}_\nu \times \mathbf{m}]\\ &\quad + [\mathbf{m} \times (\alpha\Delta\mathbf{l}_\nu + \beta\partial_\nu^2\mathbf{l}_\nu)]\\ &\quad + \sum_{\chi,\lambda} |\varepsilon_{\nu\chi\lambda}|[\mathbf{l}_\nu \times (\alpha\Delta\mathbf{l}_\lambda + \beta\partial_\lambda^2\mathbf{l}_\lambda)].\end{aligned} \tag{1.8.21}$$

Next, we believe that the characteristic size of magnetic inhomogeneities d is much larger than the interatomic distance a. Therefore, against the terms $\tilde{B}[\mathbf{m} \times \mathbf{l}_\nu]$ in eqs. (1.8.21), we can discard the terms $\alpha[\mathbf{m} \times \Delta\mathbf{l}_\nu]$ and $\beta[\mathbf{m} \times \partial_\nu^2\mathbf{l}_\nu]$:

$$\frac{\alpha|[\mathbf{m} \times \Delta\mathbf{l}_\nu]|}{\tilde{B}|[\mathbf{m} \times \mathbf{l}_\lambda]|} \backsim \frac{\beta|[\mathbf{m} \times \partial_\nu^2\mathbf{l}_\nu]|}{\tilde{B}|[\mathbf{m} \times \mathbf{l}_\lambda]|} \backsim \left(\frac{a}{d}\right)^2 \ll 1.$$

Under these assumptions, we can express the vector $\mathbf{m}$ through the components $l_{i\nu}$ and their derivatives from the second equation in (1.8.21). After substituting such a representation for $\mathbf{m}$ into the first equation, we obtain a closed equation for calculating $l_{i\nu}$. However, before doing so, it is worth paying attention to the fact that the terms of the form $\alpha[\mathbf{l}_\nu \times \Delta\mathbf{l}_\lambda]$ and $\beta[\mathbf{l}_\nu \times \partial_\lambda^2\mathbf{l}_\lambda]$ on the right-hand side of the last equation of system (1.8.21) bring, after substituting the expression for $\mathbf{m}$ into the first equation, to terms containing third derivatives of the components of the vectors $\mathbf{l}_\nu$. In the long-wavelength limit, the latter are small compared to the terms containing second derivatives of these vectors. Consequently, the terms $\alpha[\mathbf{l}_\nu \times \Delta\mathbf{l}_\lambda]$ and $\beta[\mathbf{l}_\nu \times \partial_\lambda^2\mathbf{l}_\lambda]$ on the right-hand side of the last equation of system (1.8.21) should likewise be neglected. In the long run, this equation becomes substantially more straightforward and takes the form

$$\partial_t D_{i\nu} = -\tilde{B}\varepsilon_{ijk} D_{j\nu} m_k. \tag{1.8.22}$$

Using formulas (1.6.1), (1.6.2), (1.6.3), and (1.6.11), from eq. (1.8.22), it is easy to express the components of the vector $\mathbf{m}$ in terms of the elements of the matrix $D \in SO(3)$ and Cartan forms [39]:

$$\begin{aligned}m_k &= -\frac{1}{2\tilde{B}}\varepsilon_{kij} D_{j\nu}\partial_t D_{i\nu} = -\frac{1}{2\tilde{B}}\varepsilon_{\lambda\mu\nu} D_{k\lambda} D_{i\mu}\partial_t D_{i\nu}\\ &= \frac{1}{\tilde{B}} D_{k\lambda}\omega_\lambda(A, \partial_t A).\end{aligned} \tag{1.8.23}$$

Plugging eq. (1.8.23) into the first equation of system (1.8.21), we obtain

$$\frac{1}{\tilde{B}}\partial_t[D_{k\lambda}\omega_\lambda(A, \partial_t A)] = \sum_\nu [\mathbf{l}_\nu \times (\alpha\Delta\mathbf{l}_\nu + \beta\partial_\nu^2\mathbf{l}_\nu)]_k. \tag{1.8.24}$$

To transform the right side of eq. (1.8.24), we need two auxiliary relations

$$[\mathbf{l}_\nu \times \partial_\nu \mathbf{l}_\nu]_k = -D_{k\mu}\omega_\lambda(A, \partial_\nu A)[\delta_{\lambda\nu}\delta_{\nu\mu} - \delta_{\mu\lambda}]l^2,$$
$$[\mathbf{l}_\nu \times \partial_q \mathbf{l}_\nu]_k = 2l^2 D_{k\mu}\omega_\mu(A, \partial_q A). \tag{1.8.25}$$

They are consequences of formula (1.8.6). It should be emphasized that the first of eqs. (1.8.25) does not execute the summation over the index ν. The second equality implies the summation over all twice occurring indices, including ν. Formulas (1.8.25) allow us to easily verify that the effective equations (1.8.24) for the nonlinear dynamics of the UO_2-antiferromagnet exactly coincide with eqs. (1.6.23) and (1.6.29). The phenomenological parameters of the two approaches are related by the relationships

$$R = \frac{1}{\tilde{B}}, \quad P = (2\alpha + \beta)l^2, \quad Q = -\beta l^2.$$

Expressions (1.8.23) and (1.6.26) also coincide for the total spin moment of the system.

To describe the magnetic structure and nonlinear dynamics of the UO_2-compound, the Landau–Lifshitz theory calls for the setting of four vectors $\mathbf{S}_1$, $\mathbf{S}_2$, $\mathbf{S}_3$, and $\mathbf{S}_4$ (or $\mathbf{m}$, $\mathbf{l}_\nu$, $\nu = 1,2,3$), which are parameterized by means of eight separate fields.

An important advantage of the method of phenomenological Lagrangians is the efficient reduction of the number of field variables needed for a theoretical description of Goldstone excitations connected to local rotations of the vectors of a magnet's exchange structure. There are only three Goldstone magnon fields in total.

The examined cases reveal a relationship between the mathematical structures of the Landau–Lifshitz equations and the Andreev–Volkov–Marchenko–Zheltukhin equations: they are equivalent in physical consequences in the region characteristic of Goldstone excitations.

The consistency in the exchange approximation of the two macroscopic theories established in this section can be traced further. It is simple to check, in particular, that the UO_2-compound's relativistic and magnetoelastic effects manifest in the sublattice model in precisely the same way as in the phenomenological Lagrangian method.

1.9 Phenomenological Lagrangian of magnetoelastic interactions

What we have to look for are not fundamental particles, but fundamental symmetries.

W. Heisenberg

The Lagrange function of any closed system of particles is always invariant under a continuous translation group of the coordinate space. With that, in a crystal's ground state, atoms form a lattice with symmetry lower than the initial symmetry of the Lagrangian or Hamiltonian of the system of microparticles forming the crystal. The physical characteristics of an equilibrium crystal are invariant only with respect to a discrete translation group. In accordance with the Goldstone theorem, this means that gapless modes of elastic lattice vibrations (acoustic phonons) are easily excited in the crystal to restore the broken translational symmetry The number of broken elements of the initial system symmetry group of the micro-Hamiltonian determines how many branches there exist in the spectrum of acoustic phonons. The equilibrium state of the crystal breaks the symmetry with respect to continuous translations in the three independent directions. These translations generate the three momentum components P_i ($i = 1, 2, 3$). Consequently, three Goldstone branches of lattice vibrations are excited in the crystal.

In magnetically ordered media, along with acoustic phonons, there are three branches of Goldstone magnons. Let us discuss a theoretical description of magnetoelastic interactions within the method of phenomenological Lagrangians [14, 52, 51].

Suppose that a crystal magnet's micro-Hamiltonian is invariant under the spin rotation group SO(3), which, as well as a continuous translation group $T(3)$ of the coordinate space, is completely broken by the ground state of the crystal. The transformations of the group SO(3) and of the translation group $T(3)$ in the posed problem commute with each other, because they are carried out in different spaces (spin and coordinate, respectively). Next, we represent an arbitrary element g of the group $T(3) \otimes \mathrm{SO}(3)$ in the form

$$g(X, A) = t(X)\, D(A),$$

where $D(A) \in \mathrm{SO}(3)$. We use the parameterization for an element of $t(X) \in T(3)$:

$$t(X) = \exp(\mathrm{i}\, X_s P_s).$$

The shift generators P_s commute with each other but do not commute with spatial coordinates. It should be underscored here that there is an important difference between the above problem and those previously considered. The point is that the parameters X_s and A_i must be coordinate- and time-independent at the first stages of the calculations for determining the transformation properties of the parameters of the group $T(3) \otimes \mathrm{SO}(3)$ and constructing its differential invariants. Let us explain the features of the calculations.

Suppose that $g_0 = t(x_0)\, D(a_0)$ is another element of the group $T(3) \otimes \mathrm{SO}(3)$. Then

$$g_0 g = t(X + x_0) D(A'(A, a_0)). \tag{1.9.1}$$

The differential invariants of the group are defined by the equality

$$g^{-1} \mathrm{d}g = \mathrm{i}\, \mathrm{d}X_s P_s - \mathrm{i}\, \omega_s(A, \mathrm{d}A) J_s. \tag{1.9.2}$$

To derive eqs. (1.9.1) and (1.9.2), the group parameters X_i and A_i should be thought of as independent of the space-time coordinates x_i and t. Meanwhile, in accordance with the axiomatics of the method of phenomenological Lagrangians, relation (1.9.1) regulates the transformation properties of local Goldstone fields:

$$X_i(\mathbf{x},t) \xrightarrow{g_0} X_i(\mathbf{x},t) + x_{0i}, \quad A_i(\mathbf{x},t) \xrightarrow{g_0} A_i'(A,a_0).$$

The transformational properties of the magnon fields $A_i(\mathbf{x},t)$ in this problem coincide with those considered earlier. The fields $X_i(\mathbf{x},t)$ matching phonons can be rewritten as follows:

$$X_i(\mathbf{x},t) = x_i + u_i(\mathbf{x},t) \quad (i = 1,2,3).$$

Then, as will be clear from further analysis, the variables x_i acquire the sense of the Cartesian coordinates of material particles in a body's undeformed state, whereas the quantities $u_i(\mathbf{x},t)$ are components of a medium's displacement vector.

Relation (1.9.2) allows us to introduce local differential forms. The magnon fields $A_i(\mathbf{x},t)$ correspond to Cartan's forms

$$\omega_s(A,\mathrm{d}A) = \omega_s(A,\partial_\mu A)\,\mathrm{d}x_\mu, \quad \mathrm{d}A_s = \partial_\mu A_s \mathrm{d}x_\mu, \quad \mu = 0,1,2,3,$$
$$x_0 = t, \quad \mathbf{x} = (x_1, x_2, x_3),$$

which are the same as discussed earlier. Acoustic phonons are described by local differential invariants that characterize the distortion of the medium:

$$\partial_\mu X_s = \delta_{\mu s} + \partial_\mu u_s(\mathbf{x},t), \quad \mathrm{d}X_s = \partial_\mu X_s\,\mathrm{d}x_\mu.$$

The most general expression for the density of the phonon Lagrange function, which is invariant under the rotations of the crystal as a whole and contains the minimum number of derivatives, is

$$L_\mathrm{e} = \frac{1}{2}\rho_0(\partial_t u_s)^2 - w_\mathrm{e}(\eta_{ij}), \tag{1.9.3}$$

where $\rho_0 = \mathrm{const}$ is the density of the medium in the undeformed state. We assume that the medium's nonlinear elastic energy w_e depends only on the Lagrangian strain tensor

$$\eta_{ij} = \frac{1}{2}\left[\frac{\partial X_s}{\partial x_i}\frac{\partial X_s}{\partial x_j} - \delta_{ij}\right] = \frac{1}{2}[\partial_i u_j + \partial_j u_i + \partial_i u_s\,\partial_j u_s]. \tag{1.9.4}$$

The tensor η_{ij} is invariant under global rotations of the deformable body:

$$X_i \to \tilde{D}_{ik} X_k.$$

In this case the constant orthogonal matrix $\tilde{D}$ describes the rotation of the crystal as a whole in the usual coordinate space.

The density of the phonon Lagrangian (1.9.3) perfectly corresponds to the conventional description of the nonlinear elastic dynamics of crystals (see Section 1.4).

The density of the Lagrange function L_m of Goldstone magnons coincides with that in eq. (1.6.13).

The Lagrangian $\int L_{\mathrm{me}} d^3\mathbf{x}$ of magnetoelastic interactions breaks the symmetry of the system containing the fields u_i and A_j with respect to rotations in spin space. However, the total angular momentum of the crystal's elastic and spin subsystems must also be preserved when accounting for the magnetoelastic coupling. Consequently, arbitrary rotations of the magnetic crystal as a whole must not change the density of the Lagrange function $L = L_{\mathrm{m}} + L_{\mathrm{e}} + L_{\mathrm{me}}$. The derivatives $\partial X_k/\partial x_s$ and vector functions depending on magnon fields and defined with respect to a fixed coordinate system in spin space should be exploited to construct an expression that satisfies the above requirement for the density of the Lagrange function L_{me}. The most important quantities are

$$\begin{aligned} S_{i0}(\mathbf{x}, t) &= D_{in}(\mathbf{x}, t)[b_n + a_{np}\omega_p(A, \partial_t A)], \\ l_{i\nu}(\mathbf{x}, t) &= D_{in}(\mathbf{x}, t) l_{n\nu}^{(0)}. \end{aligned} \tag{1.9.5}$$

They coincide with projections both of the vector of the spin moment density

$$\mathbf{S}_0 = (S_{10}, S_{20}, S_{30})$$

and of the antiferromagnetic vectors

$$\mathbf{l}_\nu = (l_{1\nu}, l_{2\nu}, l_{3\nu})$$

on the axes of the fixed coordinate system in the spin space. The vectors

$$\mathbf{l}_\nu^{(0)} = (l_{1\nu}^{(0)}, l_{2\nu}^{(0)}, l_{3\nu}^{(0)}) = \mathrm{const}$$

define the exchange structure of a magnet in the unexcited state in the same coordinate system.

The combinations

$$\begin{aligned} K_j &= \frac{\partial X_i}{\partial x_j} S_{i0} = \frac{\partial X_i}{\partial x_j} D_{in}[b_n + a_{np}\omega_p(A, \partial_t A)], \\ L_{j\nu} &= \frac{\partial X_i}{\partial x_j} l_{i\nu} = \frac{\partial X_i}{\partial x_j} D_{in}\, l_{n\nu}^{(0)} \end{aligned} \tag{1.9.6}$$

remain invariant under simultaneous global rotations of the magnetic crystal:

$$D_{in} \to \tilde{D}_{ik} D_{kn}, \quad S_{i0} \to \tilde{D}_{ik} S_{k0}, \quad X_i \to \tilde{D}_{ik} X_k. \tag{1.9.7}$$

The same constant orthogonal matrix $\tilde{D}$ governs the rotations of the spin and space vectors of a deformed body with regard to fixed reference systems in the spin and coordinate spaces, respectively.

Recall that in eqs. (1.9.5) and (1.9.6), only the index i is the vector index. The coordinates x_j serve as "marks" of material particles of an undeformed body. They do not respond to rotations (1.9.7) affecting the coordinates $X_i(\mathbf{x}, t)$ of the deformed crystal particles.

Due to the weakness of magnetoelastic interactions, invariants containing derivatives of the matrices D are less important in the long-wavelength limit. Such invariants, for example, are

$$K_{jp} = \frac{\partial X_i}{\partial x_j}\frac{\partial S_{i0}}{\partial x_p}, \quad L_{jp\nu} = \frac{\partial X_i}{\partial x_j}\frac{\partial l_{i\nu}}{\partial x_p}.$$

Other invariants can be built from the derivatives $\partial X_i/\partial x_j$ and forms

$$\Omega_m(A, \partial_j A) = D_{mn}\omega_n(A, \partial_j A),$$

which determine local rotations of spins with respect to a fixed coordinate system in spin space. These will be invariants

$$P_{ij} = \frac{\partial X_m}{\partial x_i}\Omega_m(A, \partial_j A).$$

The spin momentum flux density $\mathbf{S}_r = (S_{1r}, S_{2r}, S_{3r})$ (see (1.6.22)) is directly related to the spatial variables and, consequently, should also be used to construct invariants in the expansion of the crystal's magnetoelastic energy:

$$T_{ij} = \frac{\partial X_m}{\partial x_i} S_{mj} = -\frac{\partial X_m}{\partial x_i} D_{mn}\big[c_{np,js}\omega_p(A, \partial_s A) + d_{n,j}\big].$$

The invariants

$$\eta_{ij}, \quad K_j, \quad L_{j\nu}, \quad K_{jp}, \quad L_{jp\nu}, \quad P_{ij}, \quad T_{ij} \tag{1.9.8}$$

including the minimum number of derivatives are basic. They can be dependent only under special choices of the phenomenological constants:

$$c_{np,js} \backsim \delta_{np}c_{js}$$

or

$$c_{np,js} \propto (\delta_{np}\mathbf{b}^2 - b_n b_p)c_{js}, \quad c_{np,js} \propto \big[\delta_{np}(\mathbf{l}_\nu^{(0)})^2 - l_{n\nu}^{(0)} l_{p\nu}^{(0)}\big]c_{js}$$

(there is no summation over ν). We have already met the last two variants listed above (see (1.8.3) and (1.8.9)).

In the general case the density of the Lagrangian L_{me} of magnetoelastic interactions is a function of the basis invariants (1.9.8). This statement can be verified [14] using the following theorem.

A rotation-invariant function $f(\mathbf{V}^1, \mathbf{V}^2, \ldots, \mathbf{V}^n)$ of n vectors $\mathbf{V}^A = (V_1^A, V_2^A, V_3^A)$ can only be a function $f(\pi^{AB}, \ldots, \Delta^{ABC})$ of the quantities $\pi^{AB} = V_i^A V_i^B$ and $\Delta^{ABC} = \varepsilon_{ijk} V_i^A V_j^B V_k^C$.

Our problem includes all possible combinations of this type, which are expressed in terms of the basic invariants (1.9.8). For example, using the ratios

$$\varepsilon_{ijk} \det\left\| \frac{\partial \mathbf{X}}{\partial \mathbf{x}} \right\| = \varepsilon_{lmn} \frac{\partial X_i}{\partial x_l} \frac{\partial X_j}{\partial x_m} \frac{\partial X_k}{\partial x_n}, \quad \det \|2\eta + E\| = \det^2 \left\| \frac{\partial \mathbf{X}}{\partial \mathbf{x}} \right\|,$$

it is easy to show that

$$\varepsilon_{ijk} \frac{\partial X_i}{\partial x_p} S_{jr} S_{k0} = \frac{1}{\sqrt{\det \|2\eta + E\|}} \varepsilon_{lmn} (2\eta_{pl} + \delta_{pl}) T_{mr} K_n,$$

where E is the unit 3×3 matrix.

The theory outlined above is consistent with the description of magnetoelastic interactions within the magnetic sublattice model. So, in [12] the quadratic-in-spins magnetostriction density for UO_2 is given inside the Landau–Lifshitz approach using the exchange approximation

$$w_{\mathrm{me}}^{\mathrm{ex}} = -\frac{\Lambda}{2} (\mathbf{l}_1^2 u_{11} + \mathbf{l}_2^2 u_{22} + \mathbf{l}_3^2 u_{33}) - \frac{\Lambda_1}{2} \mathbf{m}^2 u_{ss}. \tag{1.9.9}$$

Due to the smallness of the anharmonicity of the elastic subsystem, expression (1.9.9) is written in a linear-in-displacements approximation;

$$u_{si} = (\partial_s u_i + \partial_i u_s)/2$$

is the linear part of the tensor η_{si}, and Λ and Λ_1 are the magnetoelastic coupling constants.

Using the outcomes of the previous section, it is easy to verify that expression (1.9.9), in the context of the Andreev–Marchenko–Volkov–Zheltukhin approach, should match the invariant

$$w_{\mathrm{me}}^{\mathrm{ex}} = -\frac{\Lambda}{2} ((\mathbf{L}_1)^2 \eta_{11} + (\mathbf{L}_2)^2 \eta_{22} + (\mathbf{L}_3)^2 \eta_{33}) - \frac{\Lambda_1}{2\tilde{B}^2} \eta_{kk} K_s^2, \tag{1.9.10}$$

where $\mathbf{L}_\nu = (L_{1\nu}, L_{2\nu}, L_{3\nu})$ (see (1.8.23), (1.9.4), and (1.9.6)).

Formulas (1.9.9) and (1.9.10) coincide in the main linear-in-displacements approximation. Expression (1.9.10) corresponds to the density of the Lagrange function

$$L_{\mathrm{me}}^{\mathrm{ex}} = \sum_{\nu=1}^{3} \left[\frac{\Lambda_1}{2\tilde{B}^2} \eta_{kk} \left(\frac{\partial X_i}{\partial x_\nu} D_{in}\, \omega_n(A, \partial_t A) \right)^2 + \frac{\Lambda l^2}{2} \left(\frac{\partial X_i}{\partial x_\nu} D_{i\nu} \right)^2 \eta_{\nu\nu} \right]. \tag{1.9.11}$$

We have taken into account that

$$l_{nv}^{(0)} = l\delta_{nv}, \quad b_n = 0, \quad a_{np} = \frac{\delta_{np}}{\tilde{B}}$$

(see the previous section).

Since $K_s^2 \approx \omega_s^2(A, \partial_t A)/\tilde{B}^2$, the first term on the right side of formula (1.9.11) leads to a renormalization of the exchange constants in the crystal's Lagrangian $\int(L_\mathrm{e} + L_\mathrm{me} + L_\mathrm{m})\mathrm{d}^3\mathbf{x}$ due to a change in the spacings between magnetic ions of the deformed medium:

$$a_{ij} = \frac{\delta_{ij}}{\tilde{B}} \to \tilde{a}_{ij} = \frac{\delta_{ij}}{\tilde{B}}\left(1 + \frac{\Lambda_1}{\tilde{B}}\eta_{kk}\right).$$

The first term in (1.9.11) can be neglected in the main approximation because the magnetoelastic deformations $|\Lambda_1\, \eta_{kk}/\tilde{B}| \ll 1$ are small.

With accuracy up to $O(\eta^2)$, the energy density of elastic deformations for cubic system crystals, one of which is UO_2, is

$$\begin{aligned} w_\mathrm{e} = {} & \frac{c_{11}}{2}\left[\eta_{11}^2 + \eta_{22}^2 + \eta_{33}^2\right] + c_{12}\left[\eta_{11}\eta_{22} + \eta_{11}\eta_{33} + \eta_{22}\eta_{33}\right] \\ & + 2c_{44}\left[\eta_{23}^2 + \eta_{13}^2 + \eta_{12}^2\right], \end{aligned}$$

where c_{ij} are the elastic moduli of a crystal. The inequalities

$$c_{11} > 0, \quad c_{11} > \max(c_{12}, -2c_{12}), \quad c_{44} > 0$$

guarantee the positive definiteness of the energy w_e.

2 Nonlinear dynamics of an isotropic ferromagnet. Methods of integration in the theory of solitons

2.1 The inverse scattering method and solitons

> The discovery by Gardner, Green, Kruskal, and Miura in 1967 that there was an analytical method for solving the initial value problem for this particular equation, and the subsequent discovery that similar methods could be applied to the sine-Gordon equation and other non-linear equations, have created a revolution in nonlinear physics which has changed our views and understanding of many physical problems. There is no sign that this revolution subsiding and what it has already achieved seems to be of permanent value.
>
> R. K. Bullough, P. J. Caudrey, Solitons

The earliest studies of collective excitations in magnets have been connected to the concept of low-amplitude spin waves formed under weak external influences on a medium being in a close-to-stable equilibrium state, for example, a state with uniform distribution of magnetization. The method of linearizing equations of the nonlinear dynamics of magnets near the ground state of the medium, followed by accounting for the anharmonics of perturbation theory, was an effective (and, in fact, the only) method for analyzing spin waves. As is known, linear equations satisfy the superposition principle. The latter greatly simplifies an analysis of linear wave processes: an arbitrarily complex wave field can be represented as the sum of the simplest harmonic waves through the Fourier transform. This is a powerful mathematical tool. Any change in the shape of an arbitrary wave packet can be described by finding a change in the characteristics of the simplest harmonic waves that comprise the packet.

With the quantum description of low-amplitude oscillations of magnetization, a complete set of linear spin-wave modes is collated with a gas of noninteracting quasiparticles known as magnons. The quantum numbers of magnons characterize the degree of excitation of one or another branch of the spectrum of spin waves.

The Fourier procedure and the related concept of quasiparticles cannot always be applied. Under strong external influences on a medium, magnetization oscillations of a crystal are described by nonlinear differential equations without adhering to the superposition principle. Magnetization distributions of a real crystal include such strongly nonlinear objects as domain walls, ring- and spiral-shaped domains, and regularly ordered domain structures. Numerous nonlinear defects, such as magnetic dislocations, disclinations, vortices, Bloch lines, Bloch points, and other nontrivial nonlinear objects, can be uncovered in patterns comprising magnetic domains and separate domain walls. Some of them are prototypes of defects in the linear theory of elasticity. At the same time, nonlinear defects, being richer in their internal structure than those of the linear theory of elasticity, have specific interaction peculiarities.

Nonlinear magnetic excitations and defects have a significant impact on a variety of processes, including magnetization reversal, energy exchange between various subsys-

https://doi.org/10.1515/9783112235973-002

tems of a crystal, transfer phenomena, etc. They also convey vital information about the structure and dynamics of a medium. None of the finite orders of linear perturbation theory can provide the physical properties of such states.

A distinctive feature of even weakly nonlinear magnetization waves is their mutual interaction. In this case the nonlinear waves can change beyond recognition. For example, two nonlinear waves having met can disappear giving rise to a new wave with frequency equal to the sum or difference of the original wave frequencies.

This monograph covers magnetic media with spatially dispersed nonlinear magnetization waves, conditional upon neglecting energy losses due to wave relaxation.

It turns out that when exposed to strong external impacts in any dispersed condensed medium with low energy losses, a complete set of nonlinear normal modes involves, apart from waves similar to those arising under weak influences, entirely new particle-like waves called solitons. Moreover, solitons determine the main physical properties of condensed matter under strong external influences.

To explain the mechanism of how solitons emerge, we should recall that in media with dispersion, different wavelength waves propagate with different phase velocities. Therefore the linear theory claims that any wave packet of harmonic waves always decomposes over time into its component waves. It is necessary to stress that in this case the wave packet energy is, of course, conserved, although the wave packet itself spreads out, and its amplitude diminishes. However, even a weak nonlinear relationship between the harmonic waves that make up the wave packet provokes interactions to compress the packet or increase the steepness of the resulting wave front. The dispersion-induced spreading out of the wave packet is balanced by a nonlinear pumping of energy from some of its harmonic components to others.

Because of the leveling of two competing effects in condensed media, and in particular in magnets, special nonlinear waves arise, called solitons. A distinctive feature of solitons is that they are like particles. Similar to particles, solitons are concentrated in space and keep their shape and velocity intact both during their own travel and after colliding with other solitons. After a mutual collision, the solitons fly apart to restore the shape and velocity they had before the collision. The result of the collision looks outwardly as if the "center of gravity" and the phase of the internal oscillations of each of the solitons have gained an additional permanent shift.

Since the properties inherent in any condensed matter govern the formation of solitons, universal models make it possible to explore nonlinear systems of various physical natures. It turns out that nonlinear phenomena in actual complicated systems can be theoretically described by constructing their simplified models in each area of characteristic spatio-temporal scales that correctly account for the main nonlinear interactions and admit exact solutions. For building universal models, special versions of nonlinear perturbation theory have been developed [62, 107].

Finally, the most important thing is that scientists have succeeded in masterminding effective methods of integrating a large number of universal models. In some sense the novel methods generalize the Fourier transform for linear systems to analytically describe strongly nonlinear states of a medium. The cornerstone of the mathematical technique for integrating nonlinear partial differential equations is known as the "inverse scattering method" [62, 107, 72, 97, 130, 3, 108, 96, 68, 1].

The classical version of the inverse scattering method is widely used for integrating evolutionary (wave) nonlinear partial differential equations that contain only one spatial coordinate x and time t.

To apply the inverse scattering method, we should represent an initial nonlinear equation in the form of the compatibility condition

$$[\partial_x - U(\lambda), \partial_t - V(\lambda)] \equiv \partial_t U(\lambda) - \partial_x V(\lambda) + [U(\lambda), V(\lambda)] = 0 \tag{2.1.1}$$

for some set of linear differential equations for the vector function $\chi(\lambda)$:

$$\partial_x \chi = U(\lambda)\chi, \quad \partial_t \chi - V(\lambda)\chi. \tag{2.1.2}$$

The matrix elements U and V contain field variables and their derivatives over the coordinates x and t, in terms of which the initial nonlinear differential equation is written. It is worth noting that the quantities U, V, and χ depend not only on the coordinates x and t but also on an additional complex parameter λ. The fact that U and V are meromorphic functions of the parameter λ is crucial for further analysis. Recall that a function is said to be meromorphic on some bounded domain if it is an analytic function there everywhere except for a finite number of poles. We also point out that after performing all the operations in formula (2.1.1), the latter becomes identical to the initial nonlinear equation for the fields under consideration for arbitrary values of the parameter λ.

System (2.1.2) and its compatibility condition (2.1.1) can be geometrically interpreted as the equality of the covariant derivatives of the vector-function χ to zero on some manifold with the connectivity coefficients U and V (Christoffel symbols). Then the compatibility condition (2.1.1) is equivalent to the statement that the manifold has zero curvature. Therefore the initial nonlinear equation (2.1.1) is sometimes referred to as the zero-curvature representation. The latter appeared for the first time in the works of Lax [97], and then it was generalized by Zakharov and Shabat [130]. Therefore the matrix functions U and V are also called the generalized Lax pair, the Zakharov–Shabat representation, or simply a $U - V$ pair [62, 107, 3, 108, 96, 68, 1, 112].

The representation of the initial nonlinear differential equation (2.1.1) is more universal than system (2.1.2). The fact is that eqs. (2.1.2) are slightly modified during calculations in each specific problem, for example, to preserve the invariant-in-time asymptotics of the fundamental solutions of system (2.1.2), prescribed for $x \to \pm\infty$.

Representations (2.1.1) and (2.1.2) are used for passing from the initial dynamic variables (fields) to new generalized variables (scattering data). In the new variables the nonlinear equation splits into a series of noninterconnected linear differential equations. Their solution suggests solving the initial nonlinear equation through inverting the change of variables. Thus all steps of integrating a nonlinear equation via the inverse scattering method are associated with seeking a solution to certain linear problems. This provides the possibility of comprehensively describing essentially nonlinear phenomena and processes in as much depth as the pertinent linear problems are researched.

The generalized variables used for integrating meet the admissible types of excitations in the system and are naturally divided into two groups.

The former covers a discrete set of variables that correspond to stable, mutually interacting excitations of the soliton type. Examples of soliton-type excitations in magnetic media include domain walls, solitary domains, and self-localized magnetization waves. The latter deals with variables that continuously depend on the spectral parameter λ. In a physical space, they describe nonlinear, localized waves with the predominant influence of dispersion. Such waves do not have a stationary shape and fade away over time.

Although the solution types listed above can be clearly told apart in the language of generalized variables, it is more challenging to separate them in the space of observed variables. Nevertheless, if an initial wave packet is examined for a long time, then its "radiative" components spread out due to dispersion, but the solitons continue to exist as localized excitations of an unchanged shape against the background of small-amplitude oscillating waves.

The division of an arbitrary initial wave packet into solitons is similar in significance to such distinctive properties as the maintenance of solitons' shape and velocity throughout their motion and mutual interaction.

The classical version of the inverse scattering method makes it possible to trace in detail how an arbitrary initial pulse decays into a set of solitons and an oscillating tail. However, this method explores the analyticity properties of partial solutions of the auxiliary linear system (2.1.2). Solving the latter is a daunting task, and its relatively simple solution is possible only for solutions to the initial nonlinear equation that rapidly decrement as $x \to \pm\infty$. Meanwhile, it turns out that there is no need for a complete investigation of the spectral problem (2.1.2) to construct particular solutions to the equations of motion. When $U-V$ pairs are rationally λ-dependent, exact solutions to nonlinear models were derived by direct methods. Among them are the polynomial closure method [98, 50], the "dressing" procedure [129, 131, 108], the Hirota method [75, 74], the Bäcklund (Darboux) transformation, and others [62, 107, 96, 1, 112, 99].

Of all the above modifications of the conventional inverse scattering method scheme, the "dressing" procedure is more universal and technically transparent. The "dressing" procedure is based on the use of the matrix Riemann problem in the complex λ-plane and allows us to construct new solutions knowing a particular solution to the initial nonlinear equation (2.1.1). Unlike the classical version of the inverse scattering method, the "dressing" method for particular solutions has the drawback that the Cauchy problem for the initial evolutionary non-linear equation is usually not specified.

The advantage of the "dressing" procedure is that new solutions to the equations of motion are built locally, in the vicinity of each point x, t, to study any solutions, not just rapidly decreasing ones for $x \to \pm\infty$. Among other things, this method provides non-localized, soliton-type magnetic excitations in space (solitons against the background of nonlinear spin waves) [37, 41, 126]. The main feature of such solitons is their stability when they interact.

It should be emphasized that soliton-type solutions are found explicitly by purely algebraic calculations. The study of waves with a dispersion predominance necessitates a solution to linear integral equations and, in general, the use of numerical methods.

In analytical mechanics, the Liouville–Arnold theorem [8] states that equations of motion for a Hamiltonian system with N degrees of freedom in a $2N$-dimensional phase space can be integrated in quadratures only if the system has N (half the dimension of the phase space) first integrals of motion in involution. Moreover, the phase space itself is stratified into N-dimensional submanifolds, the motion along which is linear. From a formal point of view, any nonlinear partial differential equation has an infinite number of degrees of freedom and therefore can be integrated in quadratures only if there are an infinite number of integrals of motion. Such a situation has been thought to be unreal for a long time. Nevertheless, it has been turned out that the complete integrability of a large number of universal models is associated with the presence of an infinite series of conservation laws [127, 128, 108]. In [128, 68] a number of examples show that from the point of view of analytical mechanics, the transformation of the inverse scattering method is a canonical transformation to the action-angle variables to "diagonalize" the energy functional of a system at hand. Here and below, the integrability of equations implies that they have a hidden algebraic symmetry that allows us to find all their exact solutions for a certain class of functions (for example, decaying functions at infinity). We can say that an infinite series of conservation laws guarantees the dynamic stability of solitons.

Among the variety of solitons, there are those whose dynamic stability is enhanced by topological factors. Recall that topology studies the properties of mappings that are preserved under continuous deformations. From a mathematical point of view, vector fields describing topological solitons have features that cannot be eliminated by continuous deformations of vector fields. From a physical point of view, to eliminate a topological soliton arising in a system, enormous energy is required to break the geometric bonds characterizing the soliton or to destroy the ordered state in a vast volume of matter. This circumstance makes topological solitons the most nonlinear and stable of all those available.

The generalized Lax representation (2.1.1) can also be used to calculate quasi-periodic-in-x ("finite-gap") solutions of nonlinear equations (see, for example, [65, 93, 63, 64, 14]).

Of course, not all nonlinear systems have representation (2.1.1), and hence not all nonlinear partial differential equations can be integrated using the inverse scattering method or its modifications, just as the Fourier method does not enable integrating all linear differential equations. Thus the question arises about the practical significance of the inverse scattering method. In this connection, first and foremost, we should recall the universality property of exactly integrable equations. The universality of the integrable equations is explained by the fact that they combine the most spread types of dispersion with the characteristic types of nonlinearities and share the symmetry requirements for many physical systems. The number of nonlinear phenomena that can be described (in a certain approximation) by a given equation is unusually large. For example, the "sine-Gordon" equation describes the propagation of dislocations in crystals, the motion of domain walls in ferromagnets, the propagation of magnetic flux along the Josephson line, the behavior of charge density waves, and commensurability–incommensurability phase transitions [112, 113, 24, 23] and can be exploited for researching a series of magnetic excitations in weak ferromagnets and antiferromagnets [132, 12, 10, 11].

In addition, as stated in [86, 112, 100, 89, 96], it is conceivable to develop a perturbation theory apparatus based on the inverse scattering method for studying close-to-exactly integrable equations. Solutions found within such a perturbation theory cannot be deduced in any finite order of the perturbation theory for linear modes. The perturbation theory significantly broadens the scope of the inverse scattering method's applicability and adds even more value to the study of exactly integrable models.

It is important to highlight that Sklyanin, Takhtadzhan, and Faddeev have proposed a generalization of the inverse method to precisely solve numerous quantum mechanical models designed by quantizing integrable classical systems [115, 116, 29]. It was discovered that the quantum inverse scattering method is closely related to methods for exactly solving two-dimensional lattice models of statistical physics [13, 73]. Furthermore, it is also a natural generalization of the algebraic Bethe ansatz, which was earlier used for theoretically describing the Heisenberg chain with exchange interaction between nearest neighbors and atomic spin $S = 1/2$ [15].

The quantum inverse scattering method and generalizations of the Bethe method have been applied for a full-fledged investigation [116, 13, 117, 69, 123–125] of the ground state and excitation spectrum of a one-dimensional chain of spins with the Hamiltonian

$$H = -\frac{1}{2}\sum_{l=1}^{N}[J_x\hat{\sigma}_l^x\hat{\sigma}_{l+1}^x + J_y\hat{\sigma}_l^y\hat{\sigma}_{l+1}^y + J_z\hat{\sigma}_l^z\hat{\sigma}_{l+1}^z],$$

where $\hat{\sigma}_l^\alpha$ are the operators of spin $S = 1/2$, related to a particular chain site. The constant parameters J_x, J_y, J_z control the type of spin ordering in the ground state of the chain (ferro- or antiferromagnetic ordering) and its magnetic anisotropy.

A striking illustration of the generality and fruitfulness of the quantum inverse scattering method is its application to the construction of new analytical solutions for the one-dimensional Hubbard model in the theory of magnetism. The Hubbard model describes a strongly correlated system of electrons. It relies on both the lattice's site-to-site electron transfer and the Coulomb repulsion of two electrons with opposite spins at the same site. Fresh findings are also obtained for the problem of a gas of electrons resonantly interacting with an impurity magnetic center (the Kondo problem). Kondo magnets are interesting in that with a relatively small change in their chemical composition due to specific many-particle effects, a transition from completely "localized" magnetism to a behavior reminiscent of "collectivized" magnets with a low saturation magnetization is possible (see [82] and references therein).

The quantum inverse scattering method underlies effective methods put in place for evaluating the asymptotic behavior of correlation functions in quantum mechanical models [29]. The asymptotic behavior of the spin correlator can be calculated for large spacings between spins, in particular, for an anisotropic Heisenberg chain.

Let us list the main theoretical outcomes of studying classical magnetic solitons.

The papers [95, 118] first have proved the possibility of integrating the Landau–Lifshitz equations for an isotropic quasi-one-dimensional ferromagnet via the inverse scattering method. In [68, 118] a procedure for computing all exact solutions to the model with homogeneous magnetization asymptotics was formulated for $x \to \pm\infty$.

The manuscripts [49, 51, 52, 55] have established that the one-dimensional Landau–Lifshitz equations for an uniaxial ferromagnet also comprise a $U - V$ pair. To construct exact solutions to these equations, a simplified version of the general integration technique, the polynomial closure method, is proposed in [49, 52]. The authors of [51, 54, 53] have advanced the classical version of the inverse scattering method for an uniaxial anisotropy ferromagnet, which makes it possible to calculate all solutions with uniform asymptotics for $x \to \pm\infty$.

Direct integration [80] has found many particular solutions of the type of solitary magnetization waves to the equations of nonlinear dynamics of magnets. It is shown that some of them can be treated as a bound state of a great number of magnons. Such research plays a crucial role in understanding the physical underpinnings of nonlinear collective excitations and provides guidance in the search for integrable systems.

Being integrated directly, the Landau–Lifshitz equations for an "easy plane" anisotropy ferromagnet exposed to a constant magnetic field aligned along the anisotropy axis demonstrated that there are several types of spatially localized magnetization waves. In a magnetic field below a certain critical value, there exist turning waves, algebraic solitons, and pulsating solitons [81, 111, 92, 9]. At a high field strength, precessional solitons were revealed [111].

A procedure proposed in [37, 41] for integrating the equations for an "easy plane" anisotropy ferromagnet is based on the matrix Riemann problem. This method is a mathematically quite simple instrument for studying new types of nonlinear collective excitations. In particular, this method can be used for exploring magnetic excitations with inhomogeneous asymptotics as $x \to \pm\infty$ under an external magnetic field oriented along the anisotropy axis.

A generalized Lax pair for one-dimensional equations for a biaxial ferromagnet has been derived in [55, 114]. Instead of being in the complex plane, as was the case in all previously known problems, the matrices $U(\lambda)$ and $V(\lambda)$ in representation (2.1.1) in this instance turned out to be meromorphic functions of the parameter λ on a Riemann surface topologically equivalent to a torus. The biaxial ferromagnet model has led to the need to study a fundamentally novel class of $U - V$ pairs whose λ-dependence is expressed in terms of Jacobi elliptic functions. Because of the aforementioned constraints, the inverse scattering method could not be literally applied but required a generalization. An important class of soliton-type solutions to the equations for a biaxial ferromagnet was obtained in [6, 7, 79] by direct integration of the Landau–Lifshitz equations and also by the Hirota method in [27, 28]. The necessary generalization of the algorithm for “dressing” particular solutions has been done independently in [102, 103, 36, 30, 31, 109, 110, 25]. The “dressing” method-based procedure for constructing multisoliton excitations in a biaxial ferromagnet is covered in [102, 103, 31, 109, 110, 25].

For a biaxial ferromagnet, Borisov has put forward a classical version of the inverse scattering method for the Cauchy problem to be solved for the Landau–Lifshitz equation with initial distributions of magnetization with uniform asymptotics at infinity [32].

Finite-gap solutions for uniaxial and biaxial ferromagnets to describe periodically modulated-in-x magnetization waves are given in [20, 21, 19, 18, 26, 106].

Since the systems mentioned above loss the integrability property when passing to a larger number of dimensions, two- and three-dimensional magnetization waves require special treatment. Although it allows the investigation of some particular classes of non-one-dimensional magnetization waves, the classical version of the inverse scattering method becomes ineffective when applied to a larger number of dimensions. In this context, [103–105] and [40, 39] envisage analytical descriptions of cylindrically symmetric magnetic excitations, two-dimensional soliton-like precession waves, and magnetic disclinations in isotropic ferro- and antiferromagnets, respectively.

References [90, 91] present the key concepts, methods, and conclusions of the theory of nonlinear integrable equations in $(2+1)$-dimensions. A vast class of exact solutions to the effective equations for an isotropic antiferromagnet in $(3+1)$-space can be found in [34, 35]. All of them exhibit typical effects of dispersion-free media, such as steepening of the front and ambiguity of solutions.

Unlike domain walls and localized waves in an ordinary ferromagnet with one sublattice, domain boundaries and nonlinear magnetization waves in magnets with several magnetic sublattices are richer in their internal structure, have a number of peculiarities, and are yet far from being studied in depth. The Landau–Lifshitz equations for magnets with several sublattices constitute a more complicated system of nonlinear equations, which makes them challenging to integrate.

As was noted in [67, 121], the Landau–Lifshitz equations for media with several magnetic sublattices can be a new source of physically meaningful integrable models. However, the integrability of complete dynamical equations, even for two-sublattice magnets, has not yet been proven. By directly integrating the Landau–Lifshitz equations some types of solitary magnetization waves have been studied only in two-sublattice magnets [132, 12, 10, 11, 7, 66, 122].

Andreev, Marchenko, Volkov, and Zheltukhin are the first to propose phenomenological equations based on the concept of spontaneous symmetry breaking of the spin rotation group, regarding which exchange interactions are invariant. We believe that these equations can play a significant role in studying the dynamics of magnets with several magnetic sublattices.

Within this approach, physically meaningful integrable axisymmetric and asymmetric chiral models have been revealed [36, 58, 59, 87, 88, 38, 45], which are effective for analyzing the nonlinear dynamics of multisublattice noncollinear ferro- and antiferromagnets. For example, the axisymmetric model [38] offers magnetic excitations in six-sublattice antiferromagnets with a structure of the $YMnO_3$ type. "Multisublattice" solitons were obtained, and their interaction was studied [36, 38, 45]. In [38, 44], it was established that the Landau–Lifshitz equations for a two-sublattice isotropic ferrimagnet, under physically justified approximations, can also be reduced to an integrable model. These results will be presented in Chapters 6 and 7.

This chapter clarifies the general technique of the inverse scattering method using the example of integrating the Landau–Lifshitz equations for an isotropic quasi-one-dimensional ferromagnet (see Sections 2.2–2.8). Let us consider a simple method for constructing a necessary $U - V$ pair to demonstrate how the problem of the evolution of the initial magnetization distribution is reduced to solving a number of linear problems. We will illustrate a general algorithm for constructing an infinite series of conservation laws for integrable models and explain the connection between soliton states of the semiclassical model of a ferromagnet and Bethe spin complexes in a quantum chain of spins. Using the model of an isotropic ferromagnet as an example, in Sections 2.9 and 2.10, we will introduce the universal procedure for "dressing" particular solutions to integrable equations first suggested by Zakharov and Shabat. Section 2.11 deals with a method for constructing $U - V$ pairs with a moving spectral parameter [56, 57]. It allows for constructing new integrable equations with variable coefficients using existing integrable nonlinear partial differential equations with fixed coefficients. In addition, we will display the utility of this method to theoretically describe ring-shaped solitons in an isotropic ferromagnet [103, 104].

Due to the importance of models of a quasi-one-dimensional ferromagnet with uniaxial and biaxial magnetic anisotropy, the next three chapters (Chapters 3–5) discuss the specific features of the inverse scattering method and "dressing" procedure for the appropriate Landau–Lifshitz equations. For ferromagnets with uniaxial and biaxial magnetic anisotropy, we will completely classify all possible soliton states, show that the "dressing" procedure makes it possible to analytically describe multisoliton excitations against the background of nonlocalized nonlinear magnetization waves, and examine mutual soliton interactions and interactions between solitons and a nontrivial background.

2.2 Basic equations of quasi-one-dimensional dynamics of an isotropic ferromagnet

One cannot escape the feeling that these mathematical formulae have an independent existence and an intelligence of their own, that they are wiser than we are, wiser even than their discoverers, that we get more out of them than was originally put into them.

H. Hertz

2.2.1 A continuum model. Constructing *U* – *V* pair

Let us consider the one-dimensional Landau–Lifshitz equations for an isotropic ferromagnet with one magnetic sublattice:

$$\partial_t \mathbf{M} = -\alpha\gamma[\mathbf{M} \times \partial_x^2 \mathbf{M}], \quad \mathbf{M}^2 = M_0^2 = \text{const}. \tag{2.2.1}$$

Here $\alpha > 0$ is the inhomogeneous exchange interaction constant, M_0 is the nominal magnetization per unit length of a medium, and $\gamma > 0$ is the magneto-mechanical ratio. Equations (2.2.1) include only the main (exchange) interactions. The free energy of an isotropic quasi-one-dimensional ferromagnet has the form

$$W = \frac{\alpha}{2} \int (\partial_x \mathbf{M})^2 \mathrm{d}x.$$

This model can be used, for example, to describe nonlinear excitations in structures consisting of a collection of chains of magnetic atoms when the interactions between atoms in different chains and the relativistic and magneto-static interactions of atoms within each chain can be neglected.

By means of scaling transformations, we introduce dimensionless variables

$$\mathbf{S} = -\frac{\mathbf{M}}{M_0}, \quad t' = \gamma M_0 t, \quad x' = \frac{x}{\sqrt{\alpha}}$$

and rewrite eqs. (2.2.1) in a form convenient for analysis:

$$\partial_{t'} \mathbf{S} = [\mathbf{S} \times \partial_{x'} \mathbf{S}], \quad \mathbf{S}^2 = 1. \tag{2.2.2}$$

Next, we will omit the "stroke" over the variables x' and t'.

According to the general scheme of the inverse scattering method, it is necessary to find matrices $U(\lambda)$ and $V(\lambda)$ such that eq. (2.2.2) should be equivalent to the commutation representation

$$[\partial_x - U, \partial_t - V] \equiv \partial_t U - \partial_x V + [U, V] = 0. \tag{2.2.3}$$

The symmetry of the initial nonlinear partial differential equation, i. e., in this case, the invariance of eqs. (2.2.2) with respect to rotations in the three-dimensional spin space, determines the explicit shape of the matrices $U(\lambda)$ and $V(\lambda)$ in representation (2.2.3). If the matrices $U(\lambda)$ and $V(\lambda)$ are transformed according to the linear representation of the group SO(3) as components of a three-dimensional vector, the written form (2.2.3) of eqs. (2.2.2) will be invariant under transformations of the spin rotation group. To satisfy the specified requirement, we will search for U and V as an expansion in terms of generators σ_s $(s = 1, 2, 3)$ of the rotation group:

$$U = -\mathrm{i}\, u_s\, \sigma_s, \quad V = -\mathrm{i}\, v_s\, \sigma_s. \tag{2.2.4}$$

We choose a two-dimensional representation of the generators σ_s, implemented by the Pauli matrices:

$$\sigma_1 = \begin{pmatrix} 0 & 1 \\ 1 & 0 \end{pmatrix}, \quad \sigma_2 = \begin{pmatrix} 0 & -\mathrm{i} \\ \mathrm{i} & 0 \end{pmatrix}, \quad \sigma_3 = \begin{pmatrix} 1 & 0 \\ 0 & -1 \end{pmatrix}, \quad I = \begin{pmatrix} 1 & 0 \\ 0 & 1 \end{pmatrix};$$

$$\sigma_s\, \sigma_p = I\, \delta_{sp} + \mathrm{i}\, \varepsilon_{spq}\, \sigma_q.$$

Here δ_{sp} is the Kronecker symbol, and ε_{spq} is the antisymmetric unit tensor ($\varepsilon_{123} = 1$).

Let SU(2) be the group of unitary unimodular 2×2 matrices. We use the local isomorphism of the groups SU(2) and SU(3), which is expressed by the formula

$$g^{\dagger} \sigma_p g = \sigma_k D_{kp}, \tag{2.2.5}$$

where

$$g \in \mathrm{SU}(2) \quad (g^{\dagger} g = I,\ \det g = 1),$$
$$D \in \mathrm{SO}(3) \quad (D^{\mathrm{T}} D = E,\ \det D = 1).$$

The superscripts "†" and "T" indicate the operations of Hermitian conjugation and transposition of the corresponding matrices, and E is the identity 3×3 matrix.

Using formula (2.2.5), it is easy to verify that the transformations

$$U \to g^{\dagger} U g = -\mathrm{i}\, \sigma_k\, D_{kp}\, u_p, \quad V \to g^{\dagger} V g = -\mathrm{i}\, \sigma_k\, D_{kp}\, v_p$$

define a linear representation of the group SO(3) on the manifold of the matrices (2.2.4), so that, under spin rotations, the quantities u_p (v_p) in expansion (2.2.4) are transformed as vector components.

Relation (2.2.3) in terms of the variables u_p and v_s takes the form of the vector equation

$$\partial_t \mathbf{u} - \partial_x \mathbf{v} + 2[\mathbf{u} \times \mathbf{v}] = 0, \tag{2.2.6}$$

which is invariant under rotations in the spin space and must be equivalent to the equation of motion for an isotropic ferromagnet (2.2.2). Since eq. (2.2.2) includes the vector **S** and its derivatives, the vectors **u** and **v** need to be sought as an expansion in linearly independent vectors formed from the vector **S** and its derivatives:

$$\begin{aligned}\mathbf{u} &= \delta_1\mathbf{S} + \delta_2\,\partial_x\mathbf{S} + \delta_3\,\partial_t\mathbf{S} + \delta_4[\mathbf{S}\times\partial_x\mathbf{S}] + \delta_5[\mathbf{S}\times\partial_t\mathbf{S}] + \cdots,\\ \mathbf{v} &= \gamma_1\mathbf{S} + \gamma_2\,\partial_x\mathbf{S} + \gamma_3\,\partial_t\mathbf{S} + \gamma_4[\mathbf{S}\times\partial_x\mathbf{S}] + \gamma_5[\mathbf{S}\times\partial_t\mathbf{S}] + \cdots.\end{aligned} \tag{2.2.7}$$

We fit the constants δ and γ and the number of terms in expansion (2.2.7) so that eqs. (2.2.2) and (2.2.6) should be equivalent. Omitting algebraic calculations, we present one of the possible options for choosing vectors **u** and **v**:

$$\mathbf{u} = \delta_1\mathbf{S}, \quad \mathbf{v} = \gamma_1\mathbf{S} + \gamma_4[\mathbf{S}\times\partial_x\mathbf{S}]. \tag{2.2.8}$$

Here the coefficients δ_1, γ_1, and γ_4 are related as follows:

$$\delta_1 = \gamma_4, \quad \gamma_1 = -2\,\delta_1\gamma_4. \tag{2.2.9}$$

To use the inverse scattering method, relations (2.2.9) must be parameterized in such a way that all coefficients in expansion (2.2.8) should be meromorphic functions of an independent parameter. In the present case, this requirement can easily be satisfied by setting

$$\delta_1 = \gamma_4 = \frac{\lambda}{2}, \quad \gamma_1 = -\frac{\lambda^2}{2},$$

where λ is an arbitrary complex parameter (the factor 1/2 is introduced to simplify the algebraic expressions).

Thus the $U - V$ pair for eqs. (2.2.2) has the form

$$\begin{aligned}U(\lambda) &= \frac{\lambda}{2\mathrm{i}}\,S_k\,\sigma_k \equiv \frac{\lambda}{2\mathrm{i}}\,S,\\ V(\lambda) &= \frac{1}{2\mathrm{i}}(-\lambda^2 S_p + \lambda[\mathbf{S}\times\partial_x\mathbf{S}]_p)\,\sigma_p \equiv -\frac{1}{2\mathrm{i}}\lambda^2 S - \frac{\lambda}{2}\,S\,\partial_x S,\end{aligned} \tag{2.2.10}$$

where $S = S_p\sigma_p$.

The symmetry of the exchange interaction group SO(3) is decisive; therefore a simple generalization of the above scheme allows us to reproduce the generalized Lax pairs for all known integrable equations of the theory of magnetism, including models with the energy of relativistic magneto-anisotropic interactions.

2.2.2 Auxiliary linear equations

Auxiliary linear equations, for which (2.2.3) is a compatibility condition, play an important role in the inverse scattering method. For model (2.2.2), we choose

$$\partial_x \chi = U(\lambda)\chi \tag{2.2.11}$$

as one of the auxiliary equations where the matrix $U(\lambda)$ is defined by formula (2.2.10). The second equation will be discussed a little later.

It is known from the theory of differential equations that the general solution of the linear system (2.2.11) can be represented as a superposition of any two linearly independent solutions (vector functions $\chi^{(1)}$ and $\chi^{(2)}$):

$$\chi = c_1\chi^{(1)} + c_2\chi^{(2)}, \tag{2.2.12}$$

where c_1 and c_2 are arbitrary x-independent factors. For further analysis, it is convenient to combine the column vectors $\chi^{(a)}$ ($a = 1, 2$) into the 2×2 matrix of fundamental solutions,

$$\varphi = (\chi^{(1)}, \chi^{(2)}).$$

Then the general solution (2.2.12) of system (2.2.11) can be written as

$$\chi = \varphi\,\mathbf{c},$$

where $\mathbf{c}$ is an arbitrary complex two-dimensional vector that can depend only on the variables λ and t.

Let us examine magnetization distributions with uniform asymptotics at infinity:

$$\mathbf{S} \to (0,0,1) \quad \text{as} \quad |x| \to \infty.$$

Then, as $|x| \to \infty$, eq. (2.2.11) becomes more simplified:

$$\partial_x \chi \cong \frac{\lambda}{2\mathrm{i}}\sigma_3 \chi. \tag{2.2.13}$$

System (2.2.13) has two linearly independent solutions that produce the 2×2 matrix

$$\varphi_0 = \exp\left(\frac{\lambda}{2\mathrm{i}}\sigma_3 x\right). \tag{2.2.14}$$

It is easy to verify with a simple check that the matrix columns φ_0 are linearly independent:

$$\det \varphi_0 = 1 \neq 0.$$

Recall that any particular solution of system (2.2.11) can be fixed by its asymptotic behavior as $x \to -\infty$ (or as $x \to +\infty$). We define two partial solutions of the auxiliary system (2.2.11) by their asymptotic behavior as $x \to -\infty$:

$$\varphi_- \to \varphi_0 = \exp\left(\frac{\lambda}{2\mathrm{i}}\sigma_3 x\right) \quad \text{as} \quad x \to -\infty. \tag{2.2.15}$$

Here we have also combined the column vectors of the solutions into a 2×2 matrix function (the Jost function) that satisfies the differential equation

$$\partial_x \varphi_- = U(\lambda)\varphi_-. \tag{2.2.16}$$

The column vectors of the matrix φ_- are linearly independent in virtue of the linear independence of their asymptotics as $x \to -\infty$.

Given that $\operatorname{Sp} U(\lambda) = 0$, from eq. (2.2.16) we find

$$\operatorname{Sp}[\partial_x \varphi_- \varphi_-^{-1}] \equiv \partial_x \ln \det \varphi_- = \operatorname{Sp} U = 0,$$

i. e., $\det \varphi_-$ does not depend on x. Therefore the determinants of the matrices φ_- and φ_0 coincide:

$$\det \varphi_- = \lim_{x \to -\infty} \det \varphi_- = \det \varphi_0 = 1.$$

Further analysis requires one more Jost function φ_+ to be the solution of a matrix equation that coincides with eq. (2.2.16),

$$\partial_x \varphi_+ = U(\lambda)\varphi_+,$$

but with asymptotic behavior given at opposite infinity:

$$\varphi_+ \to \varphi_0 = \exp\left(\frac{\lambda}{2\mathrm{i}}\sigma_3 x\right) \quad \text{as} \quad x \to +\infty.$$

The columns of the matrix φ_- constitute a basis in the space of all solutions of system (2.2.11) exactly as the columns of the matrix φ_+. This suggests that the Jost functions are connected algebraically:

$$\varphi_- = \varphi_+ T. \tag{2.2.17}$$

Here T is the 2×2 transition matrix whose elements depend only on the spectral parameter λ and time t. Due to the unimodularity of the matrices φ_+ and φ_-, the transition matrix also becomes unimodular:

$$\det T = 1. \tag{2.2.18}$$

Let us now dwell on the dependence of the matrix Jost functions on time. To obtain the evolution equation for φ_-, we differentiate eq. (2.2.16) over t:

$$\partial_t \partial_x \varphi_- = U \partial_t \varphi_- + \partial_t U \varphi_-.$$

Rewrite the result using the commutation relation (2.2.3) as

$$\begin{aligned}\partial_t\partial_x\varphi_- &= U\partial_t\varphi_- + (\partial_x V + VU - UV)\varphi_- \\ &= U\partial_t\varphi_- + \partial_x V\varphi_- + V\partial_x\varphi_- - UV\varphi_- = U(\partial_t\varphi_- - V\varphi_-) + \partial_x(V\varphi_-).\end{aligned}$$

Bearing in mind that

$$\partial_t\partial_x\varphi_- = \partial_x\partial_t\varphi_-,$$

we find

$$(\partial_x - U)[\partial_t\varphi_- - V\varphi_-] = 0.$$

We have shown that the columns of the matrix $\partial_t\varphi_- - V\varphi_-$ satisfy the eqs. (2.2.16). Therefore they can be expanded in fundamental solutions of system (2.2.16):

$$(\partial_t - V)\varphi_- = \varphi_- C. \tag{2.2.19}$$

Here C is a λ- and t-dependent matrix. Now we rely on the fact that C does not depend on x and pass in eq. (2.2.19) to the limit as $x \to -\infty$. As a result, using (2.2.10) and (2.2.15), we have

$$C = \lim_{x\to-\infty} \varphi_-^{-1}(\partial_t - V)\varphi_- = \frac{\lambda^2}{2\mathrm{i}}\sigma_3.$$

Thus the differential equation in the variable t for the Jost function φ_- appears as

$$\partial_t\varphi_- = V\varphi_- + \frac{\lambda^2}{2\mathrm{i}}\varphi_-\sigma_3. \tag{2.2.20}$$

When deriving the auxiliary linear equation (2.2.20) supplementary to eq. (2.2.16), we have employed the commutation representation (2.2.3). By direct verification we can make sure that the converse statement is valid: the compatibility condition for the system of auxiliary linear eqs. (2.2.16) and (2.2.20), that is, the requirement $\partial_t\partial_x\varphi_- = \partial_x\partial_t\varphi_-$, is equivalent to the zero curvature representation (2.2.3) for the initial nonlinear equation (2.2.2).

Further, we will focus on a differential equation in a variable t for the matrix function φ_-^{-1}. We can easily derive it from eq. (2.2.20) using the relation

$$\partial_t\varphi_-^{-1} = -\varphi_-^{-1}\partial_t\varphi_-\varphi_-^{-1},$$

which is a consequence of the identity $\varphi_-^{-1}\varphi_- = \varphi_-\varphi_-^{-1} = I$. Overall, we see the outcome

$$\partial_t\varphi_-^{-1} = -\varphi_-^{-1}V - \frac{\lambda^2}{2\mathrm{i}}\sigma_3\varphi_-^{-1}. \tag{2.2.21}$$

In a similar way, we can show that

$$\begin{aligned}\partial_t \varphi_+ &= V\varphi_+ + \frac{\lambda^2}{2\mathrm{i}}\varphi_+\sigma_3,\\ \partial_t \varphi_+^{-1} &= -\varphi_+^{-1}V - \frac{\lambda^2}{2\mathrm{i}}\sigma_3\varphi_+^{-1}.\end{aligned} \tag{2.2.22}$$

To obtain an evolution equation for the transition matrix T (see (2.2.17)), we should differentiate the relation

$$T = \varphi_+^{-1}\varphi_-$$

over t. Using (2.2.22) and (2.2.20), we arrive at the simple matrix equation

$$\partial_t T = \frac{\mathrm{i}\lambda^2}{2}[\sigma_3, T]. \tag{2.2.23}$$

Further, we will demonstrate that the elements of the transition matrix T and some parameters of the Jost functions $\varphi_\pm$ are new generalized variables, which can be used to linearize the initial nonlinear partial differential equation (2.2.2). To pass to the new generalized variables, it is mandatory to study the analytical properties of the Jost functions and the transition matrix. This is the content of the so-called direct scattering problem.

2.3 Direct scattering problem

> The most powerful method of advance that can be suggested at present is to employ all the resources of pure mathematics in attempts to perfect and generalize the mathematical formalism that forms the existing basis of theoretical physics.
>
> P. A. M. Dirac

2.3.1 Integral representations of Jost functions

In the previous section, we showed that the generalized Lax representation (2.2.3) of the equations of dynamics (2.2.2) for an isotropic ferromagnet is the compatibility condition for the auxiliary linear system of differential equations for the Jost functions $\varphi_\pm$:

$$\partial_x \varphi_\pm = \frac{\lambda}{2\mathrm{i}} S\,\varphi_\pm, \tag{2.3.1}$$

$$\partial_t \varphi_\pm = -\frac{1}{2}\left(\frac{\lambda^2}{\mathrm{i}}S + \lambda\, S\,\partial_x S\right)\varphi_\pm + \frac{\lambda^2}{2\mathrm{i}}\varphi_+\sigma_3 \tag{2.3.2}$$

with asymptotic conditions at infinity

$$\varphi_\pm \to \varphi_0(x,\lambda) = \exp\left(\frac{\lambda}{2i}\sigma_3 x\right) \quad \text{as} \quad x \to \pm\infty. \tag{2.3.3}$$

Equations (2.3.1) and conditions (2.3.3) can be specified alternatively using the integral equations

$$\begin{aligned}\varphi_-(x,t,\lambda) =&\varphi_0(x,\lambda)\\ &+\frac{\lambda}{2i}\int_{-\infty}^{x}\varphi_0(x-y,\lambda)[S(y,t)-\sigma_3]\varphi_-(y,\lambda)\,dy,\end{aligned} \tag{2.3.4}$$

$$\begin{aligned}\varphi_+(x,t,\lambda) =&\varphi_0(x,\lambda)\\ &-\frac{\lambda}{2i}\int_{x}^{+\infty}\varphi_0(x-y,\lambda)[S(y,t)-\sigma_3]\varphi_+(y,\lambda)\,dy.\end{aligned} \tag{2.3.5}$$

For real values of the parameter λ, the integral equations (2.3.4) and (2.3.5) are Volterra, and their iterations (under certain restrictions on $S(y,\ t)$) converge absolutely. Analyzing the iterations, after performing simple transformations and integrations by parts, we make sure that the matrices $\varphi_\pm(x,t,\lambda)$ admit the representations

$$\begin{aligned}\varphi_- &= \varphi_0(x,\lambda) + \frac{\lambda}{2i}\int_{-\infty}^{x}\Gamma_-(x,y,t)\,\varphi_0(y,\lambda)\,dy,\\ \varphi_+ &= \varphi_0(x,\lambda) + \frac{\lambda}{2i}\int_{x}^{+\infty}\Gamma_+(x,y,t)\,\varphi_0(y,\lambda)\,dy,\end{aligned} \tag{2.3.6}$$

where the kernels $\Gamma_\pm$ do not depend on λ.

To justify the validity of representations (2.3.6), we substitute them into eqs. (2.3.1). The latter turn into identities when the matrices $\Gamma_\pm$ meet the Goursat problems, that is, the partial differential equations

$$\partial_x\Gamma_\pm(x,y) + S(x)\,\partial_y\Gamma_\pm(x,y)\,\sigma_3 = 0 \tag{2.3.7}$$

for $\pm(y-x) > 0$ and under the boundary conditions

$$\lim_{y\to\pm\infty}\Gamma_\pm(x,y) = 0, \tag{2.3.8}$$

$$\mp[\Gamma_\pm(x,x) - S(x)\,\Gamma_\pm(x,x)\,\sigma_3] = S(x) - \sigma_3. \tag{2.3.9}$$

Formulas (2.3.8), (2.3.9), and below, where this does not cause confusion, indicate no dependence of the functions $\Gamma_\pm$ on the parameter t.

The Goursat problem (2.3.7)–(2.3.9) always has a solution for the rapidly decreasing magnetization distributions $\mathbf{S}(x,t)$ as $|x| \to \infty$. Therefore representations (2.3.6) do exist.

It follows from relation (2.3.9) that the evolution of magnetization distributions in an isotropic ferromagnet is theoretically described if we succeed in calculating the matrices $\Gamma_{\pm}(x, y, t)$:

$$S(x,t) = [I \mp \Gamma_{\pm}(x,x,t)\,\sigma_3]\sigma_3[I \mp \Gamma_{\pm}(x,x,t)\,\sigma_3]^{-1}. \tag{2.3.10}$$

2.3.2 Reductions to the Jost functions and a transition matrix

Using the relations

$$\sigma_2\,\sigma_a\,\sigma_2 = -\sigma_a, \quad a = 1,3,$$

it is easy to check that for real values of λ, the matrices $U(\lambda)$, $V(\lambda)$ and $\varphi_0(\lambda)$ satisfy the involutions (the symbol "$*$" stands for complex conjugation)

$$U(\lambda) = \sigma_2\,U^*(\lambda)\,\sigma_2, \quad V(\lambda) = \sigma_2\,V^*(\lambda)\,\sigma_2, \quad \varphi_0(\lambda) = \sigma_2\,\varphi^*(\lambda)\,\sigma_2,$$

and the functions $\varphi_{\pm}(\lambda)$ and $\sigma_2\,\varphi_{\pm}^*(\lambda)\,\sigma_2$ are solutions to the same differential equations (2.3.1) and (2.3.2) under the same asymptotic conditions (2.3.3). By the uniqueness theorem for solutions to differential equations the latter means that for real values of λ, the Jost functions obey the restriction

$$\varphi_{\pm}(\lambda) = \sigma_2\,\varphi_{\pm}^*(\lambda)\,\sigma_2, \tag{2.3.11}$$

which carries over to the kernels in the integral representations (2.3.5) and (2.3.6), and

$$\Gamma_{\pm} = -\sigma_2\,\Gamma_{\pm}^*\,\sigma_2. \tag{2.3.12}$$

Formulas (2.3.11) and (2.3.12) establish a connection between the columns of the Jost functions

$$\varphi_{\pm} = (\varphi_{\pm}^{(1)}, \varphi_{\pm}^{(2)})$$

for real values of λ,

$$\varphi_{\pm}^{(1)}(\lambda) = \mathrm{i}\sigma_2[\varphi_{\pm}^{(2)}(\lambda)]^*,$$

and specify the kernels' type

$$\Gamma_{\pm} = \begin{pmatrix} \alpha_{\pm} & \beta_{\pm}^* \\ \beta_{\pm} & -\alpha_{\pm}^* \end{pmatrix}. \tag{2.3.13}$$

Since the Jost functions $\varphi_{\pm}$ are related by the algebraic relation (2.2.17), reduction (2.3.11) is transferred to the transition matrix

$$T(\lambda) = \sigma_2 T^*(\lambda) \sigma_2. \tag{2.3.14}$$

For real λ, relations (2.2.18) and (2.3.14) define concretely the algebraic structure of the matrix $T(\lambda)$:

$$T(\lambda) = \begin{pmatrix} a(\lambda) & -b^*(\lambda) \\ b(\lambda) & a^*(\lambda) \end{pmatrix}, \quad |a(\lambda)|^2 + |b(\lambda)|^2 = 1. \tag{2.3.15}$$

To simplify formula (2.3.10), which determines the magnetization distribution, we note that the matrices $I \mp \Gamma_\pm \sigma_3$ are the first terms of the Jost functions $\varphi_\pm$ expanded in inverse powers of λ:

$$\varphi_\pm = \left(I \mp \Gamma_\pm(x,x)\sigma_3 + O\left(\frac{1}{\lambda}\right)\right)\varphi_0(x,t). \tag{2.3.16}$$

It is easy to verify the validity of the asymptotic expansion (2.3.16) by integrating by parts representations (2.3.6), taking into account the relation

$$\varphi_0(y,\lambda) = \left(\frac{\lambda}{2\mathrm{i}}\right)^{-1} \sigma_3 \partial_y \varphi_0(y,\lambda).$$

Since the matrices $\varphi_\pm$ are unimodular, the matrices $I \mp \Gamma_\pm \sigma_3$ must also be unimodular. Using the general formula

$$A^{-1} = \sigma_2 A^{\mathrm{T}} \sigma_2, \tag{2.3.17}$$

which is valid for any unimodular 2×2 matrix A (the index T means transposition), and reduction (2.3.12), it is easy to check that

$$(I \mp \Gamma_\pm \sigma_3)^{-1} = (I \mp \Gamma_\pm \sigma_3)^\dagger.$$

Finally, relation (2.3.10) responsible for the magnetization distribution can be rewritten in a form convenient for calculations:

$$S(x,t) = [I \mp \Gamma_\pm(x,x,t)\sigma_3]\sigma_3[I \mp \Gamma_\pm(x,x,t)\sigma_3]^\dagger. \tag{2.3.18}$$

2.3.3 Analytic properties of the Jost functions and transition matrix elements

Let us now focus on the analytic properties of the Jost solutions $\varphi_\pm(x,t,\lambda)$ as functions of λ for fixed x and t. As before, it is convenient to use special notation for the first and second columns of matrices $\varphi_\pm(\lambda)$:

$$\varphi_\pm(\lambda) = (\varphi_\pm^{(1)}(\lambda),\ \varphi_\pm^{(2)}(\lambda)).$$

From eq. (2.3.6) we find integral representations for the first column of the matrix φ_- and the second column of the matrix φ_+:

$$\begin{aligned} \varphi_-^{(1)}(\lambda) &= \begin{pmatrix}1\\0\end{pmatrix}\exp\left(\frac{\lambda}{2i}x\right) + \frac{\lambda}{2i}\int_{-\infty}^{x}\Gamma_-(x,y)\begin{pmatrix}1\\0\end{pmatrix}\exp\left(\frac{\lambda}{2i}y\right)dy;\\ \varphi_+^{(2)}(\lambda) &= \begin{pmatrix}0\\1\end{pmatrix}\exp\left(-\frac{\lambda}{2i}x\right) + \frac{\lambda}{2i}\int_{x}^{+\infty}\Gamma_+(x,y)\begin{pmatrix}0\\1\end{pmatrix}\exp\left(-\frac{\lambda}{2i}y\right)dy, \end{aligned} \tag{2.3.19}$$

which indicate that the functions $\varphi_-^{(1)}(\lambda)$ and $\varphi_+^{(2)}(\lambda)$ continue analytically from the real λ-axis into the upper half-plane of the complex variable λ. Indeed, for the indicated values of λ, the factors $\exp(\pm\frac{\lambda}{2i}y)$ appearing in the integrands decay as the variable y goes to $-\infty$ or $+\infty$, respectively.

Similarly, it can be shown that the column vectors $\varphi_-^{(2)}(\lambda)$ and $\varphi_+^{(1)}(\lambda)$ continue analytically from the real λ-axis into the lower half-plane of the variable λ.

The connection of the Jost functions (2.2.17) and formula (2.3.15) leads to the equalities

$$\varphi_-^{(1)}(\lambda) = a(\lambda)\,\varphi_+^{(1)}(\lambda) + b(\lambda)\,\varphi_+^{(2)}(\lambda), \tag{2.3.20}$$

$$\varphi_-^{(2)}(\lambda) = -b^*(\lambda)\,\varphi_+^{(1)}(\lambda) + a^*(\lambda)\,\varphi_+^{(2)}(\lambda). \tag{2.3.21}$$

Given that the matrix functions $\varphi_\pm(\lambda)$ are unimodular, these equalities yield expressions for the transition matrix elements:

$$\begin{aligned} a(\lambda) &= \det(\varphi_-^{(1)}(\lambda),\ \varphi_+^{(2)}(\lambda)),\\ b(\lambda) &= \det(\varphi_+^{(1)}(\lambda),\ \varphi_-^{(1)}(\lambda)),\\ b^*(\lambda) &= \det(\varphi_+^{(2)}(\lambda),\ \varphi_-^{(2)}(\lambda)),\\ a^*(\lambda) &= \det(\varphi_+^{(1)}(\lambda),\ \varphi_-^{(2)}(\lambda)). \end{aligned} \tag{2.3.22}$$

From the analyticity properties of the columns $\varphi_-^{(1)}(\lambda)$ and $\varphi_+^{(2)}(\lambda)$ it follows that the function $a(\lambda)$ continues analytically from the real λ-axis into the upper half-plane $\operatorname{Im}\lambda \geq 0$. The coefficient $a^*(\lambda)$ continues analytically from the real axis into the lower half-plane. Expressions (2.3.22) for $b(\lambda)$ and $b^*(\lambda)$, generally speaking, admit no analytic continuations from the real λ-axis into the complex plane.

The procedure for integrating the initial nonlinear equation can be formulated using either eq. (2.3.20) or eq. (2.3.21). Further analysis will be carried out on the example of eq. (2.3.20), and therefore we will only discuss the properties of two elements $a(\lambda)$ and $b(\lambda)$ of the transition matrix.

We can infer from formulas (2.3.6) and (2.3.22) that

$$\varphi_\pm(x,\lambda=0) = \varphi_0(x,0) \equiv I; \quad a(\lambda=0) = 1. \tag{2.3.23}$$

Hence, due to the connection

$$|a(\lambda)|^2 + |b(\lambda)|^2 = 1,$$

we have

$$|b(\lambda)| < 1, \quad b(\lambda = 0) = 0. \tag{2.3.24}$$

In the next section, we will calculate the asymptotics of the function $a(\lambda)$ as $|\lambda| \to \infty$ and show that

$$a(\lambda) = \omega_0 + O\left(\frac{1}{|\lambda|}\right), \tag{2.3.25}$$

where $|\omega_0| = 1$. The property of analyticity of the function $a(\lambda)$ and its asymptotic behavior as $|\lambda| \to \infty$ imply that the zeros of the function $a(\lambda)$ are concentrated in the finite part of the half-plane $\operatorname{Im}\lambda \geq 0$ and can have condensation points only on the real axis. To facilitate further research, we assume that the function $a(\lambda)$ has no zeros on the real axis and that its zeros in the region $\operatorname{Im}\lambda > 0$ are simple. Under the stated conditions, the total number of zeros of the function $a(\lambda)$ is finite.

Suppose $\lambda_1, \lambda_2, \ldots, \lambda_n$ are a complete set of zeros of the function $a(\lambda)$ ($\operatorname{Im}\lambda_j > 0$, $j = 1, 2, \ldots, n$). Since $a(\lambda = \lambda_j) = 0$, due to the first equality (2.3.22), the columns $\varphi_-^{(1)}(\lambda_j)$ and $\varphi_+^{(2)}(\lambda_j)$ are proportional for $\lambda = \lambda_j$:

$$\varphi_-^{(1)}(x, t, \lambda_j) = \gamma_j(t)\, \varphi_+^{(2)}(x, t, \lambda_j). \tag{2.3.26}$$

The set of the complex-valued time-dependent functions $\gamma_j(t)$ ($j = 1, 2, \ldots, n$) characterizes the auxiliary linear problem and is crucial in further analysis.

For both $x \to +\infty$ and $x \to -\infty$, eq. (2.3.26) defines a column-solution to eq. (2.3.1) with exponentially decreasing asymptotics. The above statement is a consequence of formulas (2.2.14) and (2.3.6) and the condition $\operatorname{Im}\lambda_j > 0$. This means that the zeros of the function $a(\lambda)$ correspond to the discrete spectrum points of the spectral problem.

The Jost solutions $\varphi_\pm(\lambda)$ have oscillating asymptotics on the real axis λ. In particular, from (2.3.3) and (2.3.20) we find

$$\begin{aligned} &\varphi_-^{(1)} \to \begin{pmatrix}1\\0\end{pmatrix} \exp\left(\frac{\lambda}{2i}x\right) \quad \text{as} \quad x \to -\infty; \\ &\varphi_-^{(1)} \to a(\lambda) \begin{pmatrix}1\\0\end{pmatrix} \exp\left(\frac{\lambda}{2i}x\right) \\ &\qquad + b(\lambda) \begin{pmatrix}1\\0\end{pmatrix} \exp\left(-\frac{\lambda}{2i}x\right) \quad \text{as} \quad x \to +\infty. \end{aligned} \tag{2.3.27}$$

Therefore the real values of λ constitute a continuous spectrum of the linear problem (2.3.1).

The name "scattering method" comes from nonrelativistic quantum mechanics, where solutions to the Schrödinger equation with oscillating asymptotics similar to (2.3.27) describe the scattering of a microparticle by a spatially localized potential. The solutions of the type (2.3.26), exponentially decreasing as $|x| \to \infty$, are responsible for bound states of the particle in this potential. It turns out that the explicit form of the scattering potential can be restored if a set of so-called scattering data is known.

The problem at hand includes the variables

$$\{\lambda_j, \gamma_j(t), b(\lambda, t)\} \tag{2.3.28}$$

as scattering data, and the role of the potential is played by the magnetization distribution $\mathbf{S}(x, t)$.

From a mathematical point of view, using the auxiliary linear system (2.3.1), we have constructed the mapping

$$\mathbf{S}(x, t) \to \{\lambda_j, \gamma_j(t), b(\lambda, t)\}. \tag{2.3.29}$$

2.3.4 Evolution of scattering data

Let us show that for an isotropic ferromagnet, the Landau–Lifshitz equations (2.2.2) with the variables λ_j, γ_j, and b are substantially simplified.

At $\lambda = \lambda^*$, the evolution equation for the transition matrix (2.2.23) gives

$$\partial_t a(\lambda, t) = 0, \quad \partial_t b(\lambda, t) = -\mathrm{i}\lambda^2 b(\lambda, t). \tag{2.3.30}$$

In particular, we can see that for $\operatorname{Im}\lambda = 0$, the coefficient $a(\lambda)$ is time-independent. By the analyticity of the function $a(\lambda)$ this is also true for $\operatorname{Im}\lambda > 0$, whence it implies that the zeros of the function $a(\lambda)(\lambda = \lambda_j)$ also do not depend on time: $\lambda_j = \text{const}$ $(j = 1, 2, \ldots, n)$.

Now we establish the evolution of the transition coefficients $\gamma_j(t)$ for the discrete spectrum. In doing so, we use eqs. (2.3.2) for the columns $\varphi_-^{(1)}$ and $\varphi_+^{(2)}$:

$$\begin{aligned} \partial_t \varphi_-^{(1)} &= V(\lambda)\, \varphi_-^{(1)} + \frac{\lambda^2}{2\mathrm{i}}\, \varphi_-^{(1)}, \\ \partial_t \varphi_+^{(2)} &= V(\lambda)\, \varphi_+^{(2)} - \frac{\lambda^2}{2\mathrm{i}}\, \varphi_+^{(2)}. \end{aligned} \tag{2.3.31}$$

Equations (2.3.31) are also valid for $\operatorname{Im}\lambda > 0$, and in the case $\lambda = \lambda_j$, they are consistent with eq. (2.3.26) only when

$$\partial_t \gamma_j = -\mathrm{i}\lambda_j^2\, \gamma_j. \tag{2.3.32}$$

Equations (2.3.30) and (2.3.32) can be easily integrated. The time dependence of the transition coefficients is given by simple formulas

$$a(\lambda, t) = a(\lambda, t = 0), \quad b(\lambda, t) = b(\lambda, 0) \exp(-i\lambda^2 t);$$
$$\gamma_j(t) = \gamma_j(0) \exp(-i\lambda_j^2 t), \quad j = 1, 2, \ldots, n. \tag{2.3.33}$$

Thus the nonlinear Landau–Lifshitz equations (2.2.2) in variables (2.3.28) are solved explicitly. The values of the integration constants λ_j, γ_j $(t = 0)$ and $a(\lambda, t = 0)$, $b(\lambda, t = 0)$ can be found from the linear equation (2.3.1) for a given initial magnetization distribution $\mathbf{S}(x, t = 0)$. For a theoretical description of the evolution of the magnetization $\mathbf{S}(x, t)$ at $t > 0$, it is only necessary to invert the mapping (2.3.29). This is the content of the inverse scattering method.

To construct the inversion procedure, we need information about the asymptotic behavior of the function $a(\lambda)$ as $|\lambda| \to \infty$. This information is also useful for finding an infinite series of integral of motion for an isotropic ferromagnet.

2.4 Asymptotics of transition matrix elements as $|\lambda| \to \infty$. A generating function of integrals of motion

The Great Book of Nature is written in the language of mathematics.

G. Galileo

In view of the ratio

$$T(\lambda) = \lim_{x \to +\infty} \varphi_+^{-1}(x, \lambda)\varphi_-(x, \lambda) = \lim_{x \to +\infty} \varphi_0^{-1}(x, \lambda)\varphi_-(x, \lambda), \tag{2.4.1}$$

obtaining an asymptotic expansion of the transition matrix elements as $|\lambda| \to \infty$ reduces to constructing an expansion of the Jost function $\varphi_-(x, \lambda)$ as $|\lambda| \to \infty$ to be represented as [68]

$$\varphi_-(x, \lambda) = (I + W(x, \lambda)) \exp\left(\frac{\lambda}{2i} x \sigma_3 + Z(x, \lambda)\right). \tag{2.4.2}$$

Here the matrix $W(x, \lambda)$ is antidiagonal, and the matrix $Z(x, \lambda)$ is diagonal. The asymptotic conditions (2.3.3) imply that

$$W(x, \lambda) \to 0, \quad Z(x, \lambda) \to 0 \quad \text{as} \quad x \to -\infty.$$

To prove formula (2.4.2), we show that the functions W and Z are uniquely determined by the differential equation (2.3.1). Plugging expansion (2.4.2) into eq. (2.3.1) and separating the diagonal and antidiagonal parts produce the following system of equations:

$$\partial_x W + W\left(\frac{\lambda}{2i}\sigma_3 + \partial_x Z\right) = \frac{\lambda}{2i}(S_1\sigma_1 + S_2\sigma_2 + S_3\sigma_3 W),$$
$$\frac{\lambda}{2i}\sigma_3 + \partial_x Z = \frac{\lambda}{2i}[S_3\sigma_3 + (S_1\sigma_1 + S_2\sigma_2)W]. \tag{2.4.3}$$

Further, we demonstrate that the matrices W and Z as $|\lambda| \to \infty$ admit the asymptotic representations

$$W = \sum_{n=0}^{\infty} \frac{W_n}{\lambda^n} + O(|\lambda|^{-\infty}); \quad Z = \sum_{n=0}^{\infty} \frac{Z_n}{\lambda^n} + O(|\lambda|^{-\infty}).$$

Here we have denoted by $O(|\lambda|^{-\infty})$ a function that has a vanishing asymptotic series in powers of λ^{-1}.

From the second equation of system (2.4.3) we find an equation for calculating W_0:

$$(S_1\sigma_1 + S_2\sigma_2)W_0 = \sigma_3(1 - S_3).$$

Given the identity

$$(S_2\sigma_1 - S_1\sigma_2)(S_1\sigma_1 + S_2\sigma_2) = (1 - S_3^2)\mathrm{i}\sigma_3,$$

we arrive at

$$W_0(x) = \frac{\mathrm{i}[S_2(x)\sigma_1 - S_1(x)\sigma_2]}{1 + S_3(x)}. \tag{2.4.4}$$

After that, the second equation of system (2.4.3) boils down to the relations

$$Z_n(x) = -\frac{\mathrm{i}}{2}\int_{-\infty}^{x} [S_1(x')\sigma_1 + S_2(x')\sigma_2]W_{n+1}(x')\mathrm{d}x'. \tag{2.4.5}$$

The first equation, after excluding $\partial_x Z$ from it and transforming the terms containing W_0, gives recurrent relations for calculating the coefficients W_{n+1} $(n = 0, 1, \ldots)$:

$$W_{n+1} = \mathrm{i}\,\sigma_3\partial_x W_n - \frac{\mathrm{i}}{2}\sum_{k=1}^{n} W_k(S_1\sigma_2 - S_2\sigma_1)W_{n+1-k}. \tag{2.4.6}$$

For $n = 0$, eq. (2.4.6) losses the last sum.

Analyzing formulas (2.4.4) and (2.4.6), we can infer that the coefficients $W_n(x)$ are anti-Hermitian matrices of the form

$$W_n(x) = \begin{pmatrix} 0 & -w_n^*(x) \\ w_n(x) & 0 \end{pmatrix}. \tag{2.4.7}$$

In terms of the functions $w_n(x)$, the recurrence relation (2.4.6) is

$$\begin{aligned} w_{n+1}(x) &= -\mathrm{i}\,\partial_x w_n(x) \\ &\quad - \frac{1}{2}[S_1(x) - \mathrm{i}\,S_2(x)] \sum_{k=1}^{n} w_k(x)\, w_{n+1-k}(x), \quad n = 0, 1, \ldots, \\ w_0 &= \frac{S_1(x) + \mathrm{i}\,S_2(x)}{1 + S_3(x)}. \end{aligned} \tag{2.4.8}$$

From relations (2.4.5) and (2.4.7) we obtain the explicit shape of the matrices Z_n:

$$Z_n(x) = \begin{pmatrix} z_n(x) & 0 \\ 0 & z_n^*(x) \end{pmatrix},$$

$$z_n(x) = -\frac{\mathrm{i}}{2} \int_{-\infty}^{x} [S_1(x') - \mathrm{i}\,S_2(x')]\, w_{n+1}(x')\mathrm{d}x'.$$

The asymptotic expansion of the transition matrix $T(\lambda)$ as $\lambda \to \infty$ can be deduced by a limit passage in accordance with formula (2.4.1). Bearing in mind that the matrix $W(x)$ vanishes as $x \to +\infty$, we arrive at the representation

$$T(\lambda) = \exp Q(\lambda) + O(|\lambda|^{-\infty}), \tag{2.4.9}$$

where the diagonal matrix $Q(\lambda)$ appears as:

$$\begin{aligned} Q(\lambda) &= \mathrm{i}\begin{pmatrix} q(\lambda) & 0 \\ 0 & -q^*(\lambda) \end{pmatrix}, \quad q(\lambda) = \sum_{n=0}^{\infty} \frac{I_n}{\lambda^n}, \\ I_n &= \lim_{x \to +\infty} \frac{1}{\mathrm{i}} z_n(x) = -\frac{1}{2} \int_{-\infty}^{+\infty} [S_1(x) - \mathrm{i}\,S_2(x)]\, w_{n+1}(x)\mathrm{d}x. \end{aligned} \tag{2.4.10}$$

It is worth emphasizing the fact that the matrix $Q(\lambda)$, expanded in eq. (2.4.9), is diagonal, and the element $b(\lambda)$ of the transition matrix makes the contribution $O(|\lambda|^{-\infty})$ as $|\lambda| \to \infty$.

The unimodularity of the matrix $T(\lambda)$ signifies that

$$\operatorname{Sp} Q(\lambda) = \mathrm{i}(q(\lambda) - q^*(\lambda)) = 0.$$

Therefore the quantities $q(\lambda)$ in (2.4.10) and, consequently, the coefficients I_n are real.

So we have shown that the decomposition

$$\frac{1}{\mathrm{i}} \ln a(\lambda) = \sum_{n=0}^{\infty} \frac{I_n}{\lambda^n} \tag{2.4.11}$$

is valid as $|\lambda| \to \infty$, where the real functionals I_n given by formulas (2.4.8) and (2.4.10) are the integrals of motion for the model of an isotropic ferromagnet because of the time-independence of the coefficients $a(\lambda)$. In this sense the function $a(\lambda)$ is a generating function of an infinite series of local integrals of motion.

After simple transformations, the first conservation integrals coincide with the field momentum P and energy E of an isotropic ferromagnet:

$$\begin{aligned} P = -2I_0 &= \int \left(\frac{S_1\,\partial_x S_2 - S_2\,\partial_x S_1}{1+S_3} \right) \mathrm{d}x, \\ E = -2I_1 &= \frac{1}{2} \int (\partial_x \mathbf{S})^2 \mathrm{d}x. \end{aligned} \tag{2.4.12}$$

From the Landau–Lifshitz equation for an isotropic ferromagnet,

$$\partial_t \mathbf{S} = \partial_x [\mathbf{S} \times \partial_x \mathbf{S}], \quad \mathbf{S}^2 = 1,$$

it directly follows that the components of the total spin are also integrals of motion. However, the family of $\{I_n\}$ does not contain these integrals. Next, we show that the most important of aforementioned integrals, which determines the number of “spin deviations” in the model’s semiclassical quantization [80],

$$N = \int_{-\infty}^{+\infty} (1 - S_3(x))\mathrm{d}x,$$

is expressed in terms of the λ-derivative of the function $a(\lambda)$.

For this aim, we examine again the function

$$\theta(x,\lambda) = \varphi_0^{-1}(x,\lambda)\,\varphi_-(x,\lambda),$$

whose limit yields the transition matrix $T(\lambda)$ as $x \to +\infty$ (see (2.4.1)). The matrix function $\theta(x,\lambda)$ is the solution to the differential equation

$$\partial_x \theta = -\frac{\lambda}{2\mathrm{i}}\,\sigma_3\,\theta + \varphi_0^{-1}\frac{\lambda}{2\mathrm{i}}\,S\,\varphi_0\,\theta \tag{2.4.13}$$

under the asymptotic condition

$$\theta(x,\lambda) \to I \quad \text{as} \quad x \to -\infty.$$

Now we differentiate eq. (2.4.13) over λ and set $\lambda = 0$. Since

$$\varphi_-(x,\lambda=0) = \varphi_0(x,\lambda=0) = I,$$

we derive the equation

$$\partial_x n = \frac{1}{2\mathrm{i}}(S - \sigma_3)$$

for the matrix function $n(x) = \partial_\lambda \theta(x,\lambda)|_{\lambda=0}$ with the latter's asymptotic behavior $n(x) \to 0$ as $x \to -\infty$. A solution of this equation is

$$n(x) = \frac{1}{2\mathrm{i}} \int_{-\infty}^{x} [S(x') - \sigma_3]\mathrm{d}x'. \tag{2.4.14}$$

Then the limit passage as $x \to +\infty$ in accordance with formula (2.4.1) gives

$$\partial_\lambda T(\lambda)|_{\lambda=0} = \frac{1}{2\mathrm{i}} \int_{-\infty}^{+\infty} [S(x) - \sigma_3]\mathrm{d}x.$$

Hence we have the value of the desired integral of motion:

$$N = -2\mathrm{i}\,\partial_\lambda a(\lambda)|_{\lambda=0}. \tag{2.4.15}$$

In the next section, we will show that the explicit form of the function $a(\lambda)$ is entirely determined by its zeros λ_j $(j = 1, 2, \ldots, n)$ and the quantity $b(\lambda)$. This will enable us to specify the values of the integrals of motion for an isotropic ferromagnet.

2.5 Dispersion relations. Relationship between the results of a quasiclassical model and quantum states of an isotropic chain of spins

It is more important to have beauty in the equations of physics than to have them agree with experiments.

P. A. M. Dirac

Let us show that the analyticity of the function $a(\lambda)$ and the normalization relation $a(\lambda = 0) = 1$ (2.3.23) allow us to express the coefficient $a(\lambda)$ in terms of its zeros, if any, and the function $b(\lambda)$; namely, we have the relations

$$\begin{aligned} a(\lambda) &= \omega_0 \prod_{j=1}^{n} \frac{\lambda - \lambda_j}{\lambda - \lambda_j^*} \exp\left(\frac{1}{2\pi\mathrm{i}} \int_{-\infty}^{+\infty} \frac{\ln(1 - |b(\mu)|)}{\mu - \lambda} \mathrm{d}\mu \right), \\ \omega_0(\lambda) &= \prod_{j=1}^{n} \frac{\lambda_j^*}{\lambda_j} \exp\left(-\frac{1}{2\pi\mathrm{i}} \int_{-\infty}^{+\infty} \frac{\ln(1 - |b(\mu)|)}{\mu} \mathrm{d}\mu \right) \end{aligned} \tag{2.5.1}$$

for $\operatorname{Im}\lambda > 0$. It allows for the transition to the real λ-axis according to the rule (the Sokhotski–Plemelj formula)

$$\frac{1}{\mu - \lambda} \to \frac{1}{\mu - \lambda - \mathrm{i}\,0} = \mathrm{V.p.}\,\frac{1}{\mu - \lambda} + \pi\mathrm{i}\,\delta(\mu - \lambda). \tag{2.5.2}$$

The letters "V. p." in formula (2.5.2) indicate that the integration over the variable μ of the singular term must be carried out in the sense of the principal Cauchy value.

To prove representation (2.5.1), we consider the analytic function

$$\tilde{a}(\lambda) = \omega_0^{-1} a(\lambda) \prod_{j=1}^{n} \frac{\lambda - \lambda_j^*}{\lambda - \lambda_j},$$

in the region $\operatorname{Im} \lambda \geq 0$. We have entered the numerical factor ω_0 ($|\omega_0| = 1$) for the function $\tilde{a}(\lambda)$ to have the asymptotic behavior (see (2.3.25))

$$\tilde{a}(\lambda) = 1 + O\left(\frac{1}{\lambda}\right) \quad \text{as} \quad |\lambda| \to \infty.$$

By its definition the function $\tilde{a}(\lambda)$ no longer has any zeros in the region $\operatorname{Im} \lambda \geq 0$. On the real axis, its modulus is expressed by the function $b(\lambda)$:

$$|\tilde{a}(\lambda)|^2 = |a(\lambda)|^2 = 1 - |b(\lambda)|^2.$$

Now we look at the function $\eta(\lambda) = \ln \tilde{a}(\lambda)$, which behaves also analytically for $\operatorname{Im} \lambda > 0$ and vanishes as $|\lambda| \to \infty$. Moreover, the function $a(\lambda)$ is presumed to have no zeros on the real λ-axis; therefore the function $\eta(\lambda) = \ln \tilde{a}(\lambda)$ is continuous up to the real axis. The function $\eta(\mu)/(\mu - \lambda)$ is an analytic function with regard to the variable μ inside the contour Γ depicted in Figure 2.1 if and only if λ lies on the real axis. Then the Cauchy theorem claims that

$$\int_{\Gamma} \frac{\eta(\mu)}{\mu - \lambda}\, d\mu = 0. \tag{2.5.3}$$

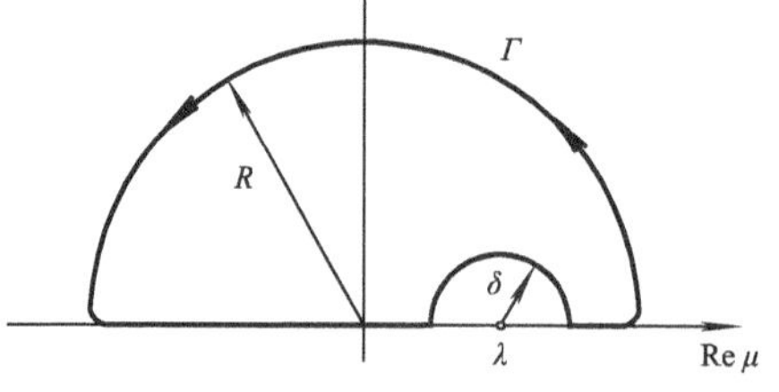

Figure 2.1: Contour Γ includes a part of the real axis and semicircles C_R and C_δ.

The function $\eta(\mu)$ rapidly decays as $|\mu| \to \infty$, and thus the integral over the semicircle C_R with radius R cancels out in the limit as $R \to \infty$. As a consequence, the left side of formula (2.5.3) contains two terms. The former is an integral in the sense of the principal value along the real μ-axis, excluding the neighborhood of the point $\mu = \lambda$; the latter is an integral over a small semicircle C_δ with radius δ, going around the point $\mu = \lambda$ in the upper half-plane $\mathrm{Im}\,\mu > 0$. We introduce a new variable $\mu - \lambda = \delta \exp(\mathrm{i}\varphi)$ $(0 \le \varphi \le \pi)$. Then integrating along the arc C_δ, it is easy to show that in the limit as $R \to \infty$ and $\delta \to 0$, formula (2.5.3) gives

$$\mathrm{V.p.} \int_{-\infty}^{+\infty} \frac{\eta(\mu)}{\mu - \lambda}\,\mathrm{d}\mu - \pi\,\mathrm{i}\,\eta(\lambda) = 0. \tag{2.5.4}$$

Equality (2.5.4) enables us to relate the real and imaginary parts of the function $\eta(\lambda)$:

$$\mathrm{Im}\,\eta(\lambda) = -\frac{1}{\pi}\mathrm{V.p.} \int_{-\infty}^{+\infty} \frac{\mathrm{Re}\,\eta(\mu)}{\mu - \lambda}\,\mathrm{d}\mu. \tag{2.5.5}$$

Using the Sokhotski–Plemelj formula (2.5.2), we can rewrite the dispersion relation (2.5.5) as

$$\eta(\lambda) = \mathrm{Re}\,\eta(\lambda) + \mathrm{i}\,\mathrm{Im}\,\eta(\lambda) = \frac{1}{\pi\mathrm{i}} \int_{-\infty}^{+\infty} \frac{\mathrm{Re}\,\eta(\mu)}{\mu - \lambda - \mathrm{i}\,0}\,\mathrm{d}\mu.$$

The fresh writing admits an analytic continuation from the real λ-axis to the upper half-plane:

$$\eta(\lambda) = \frac{1}{\pi\mathrm{i}} \int_{-\infty}^{+\infty} \frac{\mathrm{Re}\,\eta(\mu)}{\mu - \lambda}\,\mathrm{d}\mu, \quad \mathrm{Im}\,\lambda > 0. \tag{2.5.6}$$

Representation (2.5.1) now follows from (2.5.6), the obvious relation

$$\mathrm{Re}\,\eta(\mu) = |\ln \tilde{a}(\mu)| = \frac{1}{2}\ln(1 - |b(\mu)|),$$

and the normalization condition $a(\lambda = 0) = 1$.

Further, we employ representation (2.5.1) for a more detailed analysis of the basic integrals of motion of an isotropic ferromagnet, such as an integral for field momentum P, an integral for energy E, and an integral for the "number of spin deviations" N. Explicit expressions for them are given by formulas (2.4.11), (2.4.12), (2.4.15), and (2.5.1):

$$\begin{aligned} N &= \int_{-\infty}^{+\infty} \rho(\lambda)\mathrm{d}\lambda + \sum_{j=1}^{n} \rho_j; \\ P &= \int_{-\infty}^{+\infty} \lambda\rho(\lambda)\mathrm{d}\lambda - 4\sum_{j=1}^{n} \arctan\frac{\rho_j}{s_j}; \\ E &= \int_{-\infty}^{+\infty} \lambda^2\rho(\lambda)\mathrm{d}\lambda + 16\sum_{j=1}^{n} \arctan\frac{\rho_j}{s_j^2 + \rho_j^2}. \end{aligned} \tag{2.5.7}$$

Here we have introduced the notations

$$\rho(\lambda) = -\frac{1}{\pi\lambda^2}\ln\left(1 - |b(\lambda)|^2\right),$$
$$s_j = -\frac{4\mathrm{Re}\,\lambda_j}{|\lambda_j|^2}, \quad \rho_j = \frac{4\mathrm{Im}\,\lambda_j}{|\lambda_j|^2}.$$

They, as we will see further, make it possible to more fully disclose the physical meaning of the above findings.

First and foremost, it should be pointed out that the right parts of formulas (2.5.7) involve the sums of two independent types of collective excitations of the system.

The first terms in formulas (2.5.7) describe mode packets of the continuous spectrum of spin waves with density $\rho(\lambda)$. Due to the inequality $|b(\lambda)| < 1$ and equality $|b(\lambda = 0)| = 0$ (2.3.24), the density $\rho(\lambda)$ is nonnegative and nonsingular. Each value of λ of the continuous spectrum of the scattering method corresponds to a quasiparticle ("magnon") with momentum $p = \lambda$ and frequency (energy) $\omega = \lambda^2$. Thus the momentum and energy of the magnon satisfy the dispersion relation

$$\omega(p) = p^2.$$

Further, we show that the second terms in formulas (2.5.7), related to the discrete spectrum of the scattering method, describe the contributions of particle-like magnetization waves called solitons. The momentum of an individual soliton has the form

$$P_1 = -4\arctan\frac{\rho_1}{s_1}$$

and varies in the range of $|P_1| \le 2\pi$. Its energy is defined by the formula

$$E_1 = \frac{16\rho_1}{s_1^2 + \rho_1^2},$$

which is useful to rewrite in terms of the soliton's total momentum P_1 and related number of spin deviations $N_1 = \rho_1$:

$$E_1 = \frac{16}{N_1}\sin^2\frac{P_1}{4}. \tag{2.5.8}$$

Recall that long before the soliton theory was proposed, Bethe developed a consistent quantum mechanical theory of excitations of an isotropic ferromagnetic chain of spins $S = 1/2$ with an interaction between nearest neighbors [15]. He established all states of the system with a one-dimensional Hamiltonian

$$\hat{H} = -\frac{J}{2}\sum_l (\hat{\sigma}_l \cdot \hat{\sigma}_{l+1}),$$

where $\hat{\sigma}_l = (\sigma_l^x, \sigma_l^y, \sigma_l^z)$ are the Pauli operators acting on the lth site of the chain of spins, and $J > 0$ is the exchange constant. Apart from the states that can be represented as a superposition of independent spin waves, Bethe predicted the so-called spin complexes. Their energy lies outside the continuous spectrum of spin waves. With interrelated a well-defined energy E_1 and a total momentum P_1, each spin complex moves as a whole along the ferromagnetic chain. Remarkably, the Bethe quantum result coincides with formula (2.5.8) up to notations, where the parameter N_1 characterizes the total number of magnons in the complex (the number of flipped spins in a chain with parallel spins). Bethe discovered that for the same total momentum, the energy of a complex of N_1 bound magnons is lower than that of N_1 free magnons. Thus Bethe applied the quantum language to describe localized spin excitations corresponding to magnetic solitons in the quasiclassical approach (for $N_1 \gg 1$). Several other dynamic magnetic solitons can also be interpreted as bound states of numerous magnons [80].

When the points λ_j ($j = 1, 2, \dots, n$; $\operatorname{Im} \lambda_j > 0$) are concentrated toward the real λ-axis, discrete soliton modes of an isotropic ferromagnet pass into continuous (spin-wave) ones. In this case the dispersion law (2.5.8) of magnetic solitons turns into the dispersion law of N_1 noninteracting magnons with a quasimomentum p under the natural linearization of formula (2.5.8) and the condition $P_1 = N_1 p$.

2.6 The inverse scattering method

Of all the services that can be rendered to science, introducing new ideas is the most important.

J. J. Thompson

The present section deals with the Gel'fand–Levitan–Marchenko formalism that allows us to return from the generalized coordinates $\{\lambda_j, \gamma_j(t), b(\lambda, t)\}$ to the magnetization distribution $\mathbf{S}(x, t)$ of an isotropic ferromagnet. The inversion method relies on the analytical properties of the functions $\varphi_\pm(\lambda)$ and elements of the transition matrix $T(\lambda)$.

To formulate the inverse method, we use relation (2.3.20) to write it as

$$\frac{\varphi_-^{(1)}(x,\lambda)}{\lambda\, a(\lambda)} = \frac{\varphi_+^{(1)}(x,\lambda)}{\lambda} + \frac{r(\lambda)}{\lambda}\varphi_+^{(2)}(x,\lambda), \tag{2.6.1}$$

where $r(\lambda) = b(\lambda)/a(\lambda)$.

To start with, we multiply eq. (2.6.1) by $\exp(-\frac{\lambda}{2i}y)$, replace the functions $\varphi_+^{(1)}$ and $\varphi_+^{(2)}$ by their integral representations (2.3.6), and rewrite the result in a format suitable for further analysis:

$$\frac{\varphi_-^{(1)}(x,\lambda)\exp(i\lambda y/2)}{\lambda\, a(\lambda)} - \frac{1}{\lambda}\begin{pmatrix}1\\0\end{pmatrix}\exp\left(\frac{\lambda i}{2}(y-x)\right)$$

$$= \frac{1}{2\mathrm{i}} \int_{x}^{+\infty} \begin{pmatrix} \alpha_{+}(x,z) \\ \beta_{+}(x,z) \end{pmatrix} \exp\left(\frac{\lambda\,\mathrm{i}}{2}(y-z)\right) dz$$

$$+ \frac{r(\lambda)}{\lambda} \left[\begin{pmatrix} 0 \\ 1 \end{pmatrix} \exp\left(\frac{\lambda\,\mathrm{i}}{2}(y+x)\right) \right.$$

$$\left. + \frac{\lambda}{2\mathrm{i}} \int_{x}^{+\infty} \begin{pmatrix} \beta_{+}^{*}(x,z) \\ -\alpha_{+}^{*}(x,z) \end{pmatrix} \exp\left(\frac{\lambda\,\mathrm{i}}{2}(y+z)\right) dz \right]. \tag{2.6.2}$$

Let $y > x$. Then the left side of formula (2.6.2) continues analytically from the real axis into the upper half-plane of the complex parameter λ, except for the points $\lambda = \lambda_j$ ($j = 1, 2, \dots, n$), where it has simple poles $a(\lambda_j) = 0$. Due to formulas (2.3.23) and (2.3.24), the left and right sides of eq. (2.6.2) are regular at the point $\lambda = 0$.

Now we integrate eq. (2.6.2) along the real λ-axis. According to the results of Section 2.4, the left-hand side of eq. (2.6.2) tends to zero rather quickly as $|\lambda| \to \infty$; therefore the integral can be evaluated using the Jordan lemma and the Cauchy residue theorem. Taking relation (2.3.26)

$$\varphi_{-}^{(1)}(x,\lambda_j) = \gamma_j\, \varphi_{+}^{(2)}(x,\lambda_j)$$

and the integral representation (2.3.19) for the column-vector $\varphi_{+}^{(2)}$ into account, we obtain

$$\int_{-\infty}^{+\infty} \left[\frac{\varphi_{-}^{(1)}(x,\lambda)\exp(\lambda\,\mathrm{i}\,y/2)}{\lambda\, a(\lambda)} - \frac{1}{\lambda} \begin{pmatrix} 1 \\ 0 \end{pmatrix} \exp\left(\frac{\lambda\,\mathrm{i}}{2}(y-x)\right) \right] d\lambda$$

$$= 2\pi\mathrm{i} \sum_{j=1}^{n} \operatorname*{res}_{\lambda=\lambda_j} \left[\frac{\varphi_{-}^{(1)}(x,\lambda)}{\lambda\, a(\lambda)} \exp\left(\frac{\lambda\mathrm{i}}{2} y\right) \right]$$

$$= 2\pi\mathrm{i} \sum_{j=1}^{n} \frac{m_j}{\lambda_j} \varphi_{+}^{(2)}(x,\lambda) \exp\left(\frac{\mathrm{i}\lambda_j}{2} y\right)$$

$$= 2\pi\mathrm{i} \sum_{j=1}^{n} \frac{m_j}{\lambda_j} \exp\left(\frac{\mathrm{i}\lambda_j}{2} y\right) \left[\begin{pmatrix} 0 \\ 1 \end{pmatrix} \exp\left(\frac{\mathrm{i}\lambda_j}{2} x\right) \right.$$

$$\left. + \frac{\lambda_j}{2\mathrm{i}} \int_{x}^{+\infty} \begin{pmatrix} \beta_{+}^{*}(x,z) \\ -\alpha_{+}^{*}(x,z) \end{pmatrix} \exp\left(\frac{\mathrm{i}\lambda_j}{2} z\right) dz \right],$$

where

$$m_j = \gamma_j [\partial_\lambda a(x)]^{-1} |_{\lambda=\lambda_j}.$$

When integrating over λ the right side of eq. (2.6.2), we use the formula

$$\int_{-\infty}^{+\infty} \exp\left(\frac{\lambda \mathrm{i}}{2}(y-x)\right) \mathrm{d}\lambda = 4\pi\,\delta(y-x).$$

Combining the results of the calculations, we get

$$\begin{pmatrix} \alpha_+(x,y,t) \\ \beta_+(x,y,t) \end{pmatrix} - \begin{pmatrix} 0 \\ k(x+y,t) \end{pmatrix} + \int_x^{+\infty} \begin{pmatrix} \beta_+^*(x,z,t) \\ -\alpha_+^*(x,y,t) \end{pmatrix} \partial_y k(z+y,t)\,\mathrm{d}z = 0, \tag{2.6.3}$$

where

$$k(x,t) = \frac{1}{2\pi \mathrm{i}} \int_{-\infty}^{+\infty} \frac{r(\lambda,t)}{\lambda} \exp\left(\frac{\mathrm{i}\lambda}{2}x\right) \mathrm{d}\lambda - \sum_{j=1}^{n} \frac{m_j(t)}{\lambda_j} \exp\left(\frac{\mathrm{i}\lambda_j}{2}x\right).$$

Equation (2.6.3) and their complex conjugated constitute a closed system of integral equations that defines the functions $\alpha_+(x,y,t)$, $\alpha_+^*(x,y,t)$, $\beta_+(x,y,t)$, and $\beta_+^*(x,y,t)$. Taking into account the explicit shape of the matrix $\Gamma_+(x,y,t)$ (2.3.13),

$$\Gamma_+ = \begin{pmatrix} \alpha_+ & \beta_+^* \\ \beta_+ & -\alpha_+^* \end{pmatrix},$$

we represent the indicated system in the form of an equation for the matrix function Γ_+:

$$\Gamma_+(x,y,t) + \sigma_3 K(x,y,t) + \int_x^{+\infty} \Gamma_+(x,z,t)\,\partial_y K(z+y,t)\,\mathrm{d}z = 0, \tag{2.6.4}$$

where

$$K(x,t) = \begin{pmatrix} 0 & -k^*(x,t) \\ k(x,t) & 0 \end{pmatrix}.$$

Knowing a given set of the generalized coordinates $\{\lambda_j, \gamma_j(t), b(\lambda,t)\}$, we can restore the shape of the matrix $\Gamma_+(x,y,t)$ using eq. (2.6.4). As a result, connection (2.3.18) between the functions $\mathbf{S}(x,t)$ and $\Gamma_+(x,x,t)$ opens up the possibility of a detailed theoretical description of the nonlinear dynamics of an isotropic ferromagnet.

System (2.6.4) is called the Gel'fand–Levitan–Marchenko equations for the right end. Analogously, we can derive integral equations for defining the matrix $\Gamma_-(x,y,t)$ using eq. (2.3.21). In these equations the integration over the variable z is performed from $-\infty$ to x, so they are called the Gel'fand–Levitan–Marchenko equations for the left end. It is worth noting that solving the inverse problem along the above scheme, generally speaking, requires both systems of integral equations for the left and right ends. The magnetization distribution in the vicinities of $-\infty$ and $+\infty$ can be theoretically described using the first and second methods, respectively. The claim that the magnetization distributions calculated in two different methods coincide calls for a unique proof [68], which we will not dwell on here.

The stated way for solving the initial problem for the Landau–Lifshitz equation (2.2.2) can be rendered as a diagram:

$$\begin{array}{ccc} \mathbf{S}(x,t=0) & \xrightarrow{F} & \{\lambda_j, \gamma_j(t=0), b(\lambda, t=0)\}; \\ \tau_1 \downarrow & & \downarrow \tau_2 \\ \mathbf{S}(x,t) & \xleftarrow{F^{-1}} & \{\lambda_j, \gamma_j(t), b(\lambda, t)\}. \end{array} \tag{2.6.5}$$

Here the mapping F is a nonlinear substitution of variables that linearizes the Landau–Lifshitz equation (2.2.2) for an isotropic ferromagnet. The mappings τ_1 and τ_2 produce a time shift according to the initial nonlinear equation (2.2.2) and the explicit formulas (2.3.33), respectively.

It is instructive to explore the mapping F for a linearized problem when the magnetization distribution has the form

$$\mathbf{S}(x,t) \approx \tilde{\mathbf{S}}(x,t) = (\tilde{S}_1(x,t), \tilde{S}_2(x,t), 1), \quad |\tilde{S}_\alpha| \ll 1, \quad \alpha = 1, 2,$$

and the Landau–Lifshitz equation is simplified to

$$\partial_t \tilde{\mathbf{S}} = [\mathbf{n} \times \partial_x^2 \tilde{\mathbf{S}}], \tag{2.6.6}$$

where $\mathbf{n} = (0,0,1)$.

From the integral equation (2.3.4) for the Jost solution $\varphi_-(x,\lambda)$ in the linear-in-$\tilde{S}_\alpha$ approximation, we have

$$\begin{aligned} \varphi_-(x,\lambda) &\cong \varphi_0(x,\lambda) \\ &\quad + \frac{\lambda}{2\mathrm{i}} \int_{-\infty}^{x} \varphi_0(x-y,\lambda)[\tilde{S}_1(y)\sigma_1 + \tilde{S}_2(y)\sigma_2]\varphi_0(y)\mathrm{d}y \\ &= \varphi_0(x,\lambda)\left[I + \frac{\lambda}{2\mathrm{i}} \int_{-\infty}^{x} [\tilde{S}_1(y)\sigma_1 + \tilde{S}_2(y)\sigma_2]\varphi_0(2y_1\lambda)\,\mathrm{d}y\right]. \end{aligned} \tag{2.6.7}$$

Here we have resorted to the identity

$$\exp(\beta\sigma_3)\,\sigma_\alpha = \sigma_\alpha \exp(-\beta\sigma_3) \quad (\alpha = 1, 2).$$

Formulas (2.4.1) and (2.6.7) furnish the transition matrix's elements

$$a(\lambda) \cong 1, \quad b(\lambda,t) \cong \frac{\lambda}{2\mathrm{i}} \int_{-\infty}^{+\infty} [\tilde{S}_1(x,t) + \mathrm{i}\,\tilde{S}_2(x,t)] \exp(-\mathrm{i}\lambda x)\,\mathrm{d}x.$$

We can see that the discrete spectrum for the linearized problem (2.6.6) fades and the mapping F boils down to the usual Fourier transform. Evidently, the temporal dynamics of the coefficient $b(\lambda)$, assigned by formula (2.3.33), coincides with that of the Fourier transform of the function $\tilde{\mathbf{S}}(x,t)$ that meets eq. (2.6.6).

The above considerations make it possible, in general, to interpret the mapping F as a nonlinear analogue of the Fourier transform. Then the scheme (diagram (2.6.5)) for integrating the Landau–Lifshitz equations for an isotropic ferromagnet by the inverse method is a nonlinear analogue of the Fourier method.

2.7 Soliton solutions

Abstract ideas are very useful tools, and sometimes it is almost impossible to do without them. However, it should never be forgotten that they have no physical reality. Only the movements of localized elements over time possess physical reality.

Louis de Broglie

The formalism set forth above gives a complete solution to the Cauchy problem for the Landau–Lifshitz equation for an isotropic ferromagnet (2.2.2). In other words, it captures the evolution of the initial distribution of magnetization. At the same time, another possibility also opens up within the formalism. Namely, wide classes of partial solutions to the Landau–Lifshitz equation (2.2.2) can be extracted and thoroughly studied by appropriately selecting the initial values of the generalized coordinates $\{\lambda_j, \gamma_j, b(\lambda)\}$. Among the multitude of exact solutions for integrable models, soliton solutions for a ferromagnet model at hand occupy a special place to meet the case:

$$b(\lambda, t=0) = b(\lambda, t) = 0. \tag{2.7.1}$$

Note that in the scattering theory for an auxiliary linear problem, the ratio $b(\lambda)/a(\lambda)$ for the transition coefficients plays the role of a reflection coefficient. Such an interpretation comes from the shape of the asymptotics of the function $\varphi_-^{(1)}(x,\lambda)$ (2.3.27). Therefore the case (2.7.1) and all its characteristics are commonly referred to as reflectionless. It is important that such scattering data convert the Gel'fand–Levitan–Marchenko equations (2.6.4) into clear linear algebraic equations to be solved explicitly.

As an example, we consider a one-soliton state ($n = 1$). For it, the coefficient $a(\lambda)$ has one zero $\lambda = \lambda_1$ ($\operatorname{Im}\lambda_1 > 0$), and therefore the discrete spectrum of the auxiliary linear problem (2.3.1) is characterized by one transition coefficient

$$\gamma_1(t) = \gamma_1(t=0)\exp(-\mathrm{i}\lambda_1^2 t).$$

For $b = 0$, the kernel of the integral equation (2.6.4) becomes degenerate:

$$k(z+y,t) = -\frac{m_1(t)}{\lambda_1}\exp\left(\frac{\mathrm{i}\lambda_1}{2}z\right)\exp\left(\frac{\mathrm{i}\lambda_1}{2}y\right),$$

where

$$m_1(t) = m_0 \exp(-\mathrm{i}\lambda_1^2 t), \quad m_0 = \gamma_1(t=0)[\partial_\lambda a(\lambda)]^{-1}|_{\lambda=\lambda_1}.$$

As is known from the theory of integral equations, in this case the functions $\alpha_+(x,y,t)$, and $\beta_+(x,y,t)$ should be sought as

$$\begin{aligned} \alpha_+(x,y,t) &= a(x,t)\exp\left(\frac{\mathrm{i}\lambda_1 y}{2}\right), \\ \beta_+(x,y,t) &= b(x,t)\exp\left(\frac{\mathrm{i}\lambda_1 y}{2}\right). \end{aligned} \tag{2.7.2}$$

After substituting formulas (2.7.2) into eq. (2.6.4), all the integrations are performed explicitly. The matrix integral equation (2.6.4) is reduced to an algebraic system for the quantities a, b, a^*, and b^*:

$$\begin{aligned} a(x,t) &= \frac{|m_1(t)|^2}{\lambda_1^*(\lambda_1-\lambda_1^*)\Delta}\exp\left(-\frac{\mathrm{i}\lambda_1^* x}{2} + \frac{\mathrm{i}}{2}(\lambda_1-\lambda_1^*)x\right), \\ b(x,t) &= -\frac{m_1(t)}{\lambda_1\Delta}\exp\left(\frac{\mathrm{i}\lambda_1 x}{2}\right), \end{aligned}$$

where

$$\Delta = 1 + \left|\frac{m_1(t)}{\lambda_1-\lambda_1^*}\right|^2 \exp(\mathrm{i}(\lambda_1-\lambda_1^*)x).$$

Using this result, we calculate the matrix

$$\begin{aligned} I - \Gamma_+(x,x,t)\sigma_3 &= \begin{pmatrix} A(x,t) & B^*(x,t) \\ -B(x,t) & A^*(x,t) \end{pmatrix}; \\ A(x,t) &= \frac{1}{\Delta}\left[1 + \left|\frac{m_1(t)}{\lambda_1-\lambda_1^*}\right|^2 \frac{\lambda_1}{\lambda_1^*}\exp(\mathrm{i}(\lambda_1-\lambda_1^*)x)\right], \\ B(x,t) &= -\frac{m_1(t)}{\lambda_1\Delta}\exp(\mathrm{i}\lambda_1 x). \end{aligned}$$

After having done that, from formula (2.3.18) we obtain an analytical expression for solving the Landau–Lifshitz equation (2.2.2):

$$\begin{aligned} S_3(x,t) &= |A(x,t)|^2 - |B(x,t)|^2 = 1 - 2|B(x,t)|^2, \\ S_+(x,t) &\equiv S_1(x,t) + \mathrm{i}\,S_2(x,t) = -2A^*(x,t)B(x,t). \end{aligned} \tag{2.7.3}$$

Introducing real parameters u and v and notations

$$\lambda_1 = \frac{v}{2} + \frac{\mathrm{i}u}{2} \quad (u>0), \quad x_0 = \frac{2}{u}\ln\left(\frac{|m_0|}{u}\right), \quad \alpha_0 = \arg\frac{m_0}{(v+\mathrm{i}u)^2},$$

we write down solution (2.7.3) explicitly:

$$
\begin{aligned}
S_3(x,t) &= 1 - \frac{2u^2}{(u^2+v^2)\cosh^2[u\,(x-vt-x_0)/2]},\\
S_1(x,t) + \mathrm{i}\,S_2(x,t) &= \frac{2u\exp[\mathrm{i}\,(\alpha_0 + v\,x/2 + (u^2-v^2)\,t/4)]}{(u^2+v^2)\cosh^2[u\,(x-vt-x_0)/2]}\\
&\quad\times\left[\mathrm{i}\,u\sinh\left(\frac{u}{2}(x-vt-x_0)\right) + v\cosh\left(\frac{u}{2}(x-vt-x_0)\right)\right].
\end{aligned}
\tag{2.7.4}
$$

These formulas show that the solution $\mathbf{S}(x,t)$ is a solitary wave localized in a region with a characteristic size $\sim u^{-1}$, centered at the point

$$x(t) = x_0 + v\,t.$$

The latter moves at a constant speed v (see Figure 2.2). The wave has a finite energy, and, moreover, the values of all the integrals of motion (2.4.11) over the solution (2.7.4) are finite. In the strict sense, waves with such properties are traditionally called solitons, although in the physical literature, a soliton is frequently understood as any particle-like solution, i. e., a localized solution with a finite energy.

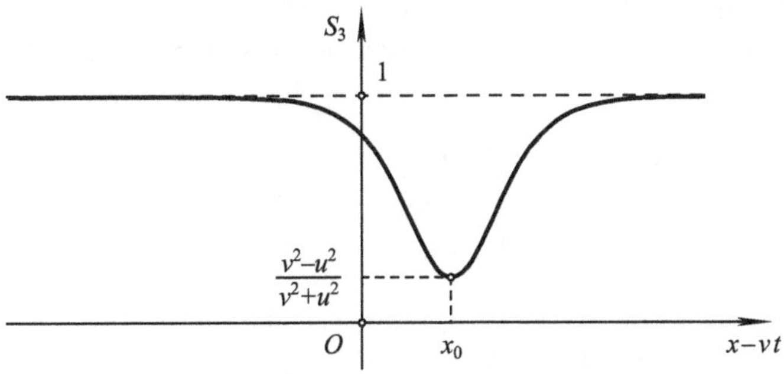

Figure 2.2: Component S_3 of a precession soliton in a ferromagnet.

Since an $\mathbf{S}$-vector's small-amplitude wave with wave vector $k = v/2$ and frequency $\omega = \frac{(u^2-v^2)}{4}$ (in a laboratory coordinate system), precessing around the Ox_3-axis, runs through the soliton localization region, magnetic excitation of (2.7.4) not only moves forward but also oscillates. The wave originates at one edge of the soliton and disappears at the opposite one. In the coordinate system connected to the soliton, the precession frequency $\omega' = \omega + kv = (u^2+v^2)/4$ is always positive. The choice of a well-defined ground state of $\mathbf{M} = M_0(0,0,-1)$ matches the lack of symmetry of the expression for the frequency ω' with regard to the point $\omega' = 0$. The precession of the vector's magnetization $\mathbf{M} = -M_0\,\mathbf{S}$ around this state is not physically equivalent in different directions.

The continuum approximation is true if the low-amplitude wavelength and the soliton width in the initial dimensional variables are greater than the lattice constant a. Hence we arrive at the constraint on the parameters of the solution (2.7.4):

$$v^2, u^2 < \alpha/a^2,$$

where α is the exchange interaction constant.

Multisoliton solutions to the equations of nonlinear dynamics for an isotropic ferromagnet (2.2.2) describe elastic pair collisions $n > 1$ of solitons of the type (2.7.4). The simplest way of theoretically describing the soliton collisions is the Zakharov–Shabat "dressing" method, to which we now proceed.

2.8 The Riemann matrix problem

> The roots of any discovery lie far in the depths of the past, and like waves smashing against a shore, human thought splashes many times around a prepared discovery until the ninth shaft comes.
>
> V. I. Vernadsky

The present section deals with the classical matrix Riemann problem of the theory of functions of a complex variable, which underlies the above algorithm for integrating the equation of nonlinear dynamics of an isotropic ferromagnet.

Recall that the starting point of the inverse scattering method is the connection between the Jost matrix solutions

$$\begin{aligned} \varphi_-(x,\lambda) &= \varphi_+(x,\lambda)\,T(\lambda), \\ T(\lambda) &= \begin{pmatrix} a(\lambda) & -b^*(\lambda) \\ b(\lambda) & a^*(\lambda) \end{pmatrix} \end{aligned} \tag{2.8.1}$$

for real values of the spectral parameter λ and the analyticity of the functions $\varphi_\pm(\lambda)$ and $T(\lambda)$ regarding it. To more fully establish the role of the analytic properties, we enter the matrix functions

$$\begin{aligned} F_+(x,\lambda) &= \frac{1}{a(\lambda)}(\varphi_-^{(1)}(x,\lambda),\ \varphi_+^{(2)}(x,\lambda)), \\ F_-(x,\lambda) &= (\varphi_+^{(1)}(x,\lambda),\ \varphi_-^{(2)}(x,\lambda)), \end{aligned}$$

as the solutions to the auxiliary linear problem (2.3.1), which continue analytically from the real λ-axis into the upper and lower half-planes, respectively.

The matrix relation (2.8.1) is equivalent to the connections between the column vectors:

$$\begin{aligned} \varphi_-^{(1)}(x,\lambda) &= a(\lambda)\,\varphi_+^{(1)}(x,\lambda) + b(\lambda)\,\varphi_+^{(2)}(x,\lambda), \\ \varphi_-^{(2)}(x,\lambda) &= -b^*(\lambda)\,\varphi_+^{(1)}(x,\lambda) + a^*(\lambda)\,\varphi_+^{(2)}(x,\lambda). \end{aligned} \tag{2.8.2}$$

Since the matrix inverse with respect to $T(\lambda)$ has the form

$$T^{-1}(\lambda) = \begin{pmatrix} a^*(\lambda) & b^*(\lambda) \\ -b(\lambda) & a(\lambda) \end{pmatrix},$$

along with (2.8.2), we have the following equalities:

$$\begin{aligned} \varphi_+^{(1)}(x,\lambda) &= a^*(\lambda)\,\varphi_-^{(1)}(x,\lambda) - b(\lambda)\,\varphi_-^{(2)}(x,\lambda), \\ \varphi_+^{(2)}(x,\lambda) &= b^*(\lambda)\,\varphi_-^{(1)}(x,\lambda) + a(\lambda)\,\varphi_-^{(2)}(x,\lambda). \end{aligned} \tag{2.8.3}$$

Using formulas (2.8.2) and (2.8.3) and relation (2.8.1), we write the matrix functions $F_\pm(x,\lambda)$ for real values of λ as follows:

$$\begin{aligned} F_+(x,\lambda) &= \frac{1}{a(\lambda)}\,\varphi_+(x,\lambda)M_{++}(\lambda) = \frac{1}{a(\lambda)}\,\varphi_-(x,\lambda)M_{-+}(\lambda), \\ F_-(x,\lambda) &= \varphi_+(x,\lambda)M_{+-}(\lambda) = \varphi_-(x,\lambda)M_{--}(\lambda), \end{aligned} \tag{2.8.4}$$

where the matrices $M_{\pm\pm}(\lambda)$ are of a special triangular shape,

$$\begin{aligned} M_{++}(\lambda) &= \begin{pmatrix} a(\lambda) & 0 \\ b(\lambda) & 1 \end{pmatrix}, \quad M_{-+}(\lambda) = \begin{pmatrix} 1 & b^*(\lambda) \\ 0 & a(\lambda) \end{pmatrix}, \\ M_{+-}(\lambda) &= \begin{pmatrix} 1 & -b^*(\lambda) \\ 0 & a^*(\lambda) \end{pmatrix}, \quad M_{--}(\lambda) = \begin{pmatrix} a^*(\lambda) & 0 \\ -b(\lambda) & 1 \end{pmatrix}. \end{aligned} \tag{2.8.5}$$

Formulas (2.8.4) and (2.8.1) are compatible only when two equivalent-to-factorization transition matrices are valid:

$$T(\lambda) = M_{+-}(\lambda)M_{--}^{-1}(\lambda) = M_{++}(\lambda)M_{-+}^{-1}(\lambda).$$

The triangular shape of the matrices $M_{\pm\pm}(\lambda)$ allows them to be uniquely restored by a given unimodular matrix $T(\lambda)$.

As follows from eqs. (2.8.4), relation (2.8.1) in terms of the matrices $F_\pm(x,\lambda)$ appears as

$$F_-(x,\lambda) = F_+(x,\lambda)\,G_0(\lambda), \tag{2.8.6}$$

where

$$\begin{aligned} G_0(\lambda) &= a(\lambda)M_{++}^{-1}(\lambda)M_{+-}(\lambda) \\ &= a(\lambda)M_{-+}^{-1}(\lambda)M_{--}(\lambda) = \begin{pmatrix} 1 & -b^*(\lambda) \\ -b(\lambda) & 1 \end{pmatrix}. \end{aligned} \tag{2.8.7}$$

For real values of λ, the matrix $G_0(\lambda)$ is Hermitian:

$$G_0^\dagger(\lambda) = G_0(\lambda).$$

Given the unimodularity of the matrices $\varphi_{\pm}(x,\lambda)$, relations (2.8.4) for real values of λ provide

$$\det F_+^{-1}(x,\lambda) = a(\lambda), \quad \det F_-(x,\lambda) = a^*(\lambda). \tag{2.8.8}$$

From the real axis these relations can be analytically continued to the upper and lower half-planes of the complex variable λ.

Formulas (2.8.8) imply that the matrix functions $F_+^{-1}(\lambda)$ and $F_-(\lambda)$ have no singularities in their analyticity domains, except for zeros of the coefficients $a(\lambda)$ and $a^*(\lambda^*)$.

Recall that the real parameter λ is sensible for a magnon's quasi-momentum. With a large quasi-momentum (as $|\lambda| \to \infty$, the magnon "does not feel the scattering potential $\mathbf{S}(x,t)$"). The asymptotic behavior of the functions $F_{\pm}(x,\lambda)$ as $|\lambda| \to \infty$ and $|x| \to \infty$ confirms this assertion. According to formulas (2.8.4) and the outcomes of Section 2.4 on exploring the asymptotic behavior of the function $\varphi_-(x,\lambda)$ for large values of $|\lambda|$, we have

$$F_-(x,\lambda) \sim F_+(x,\lambda) \sim \varphi_0(x,\lambda) = \exp\left(\frac{\lambda}{2\mathrm{i}} x\sigma_3\right)$$

as $|\lambda| \to \infty$ and $|x| \to \infty$. Now we extract these factors with an essential singularity as $\lambda = \infty$ from the matrices $F_+(x,\lambda)$ and $F_-(x,\lambda)$ and introduce new matrix functions

$$\begin{aligned} G_-(x,\lambda) &= F_-(x,\lambda)\,\varphi_0^{-1}(x,\lambda), \\ G_+(x,\lambda) &= \varphi_0(x,\lambda)\,F_+^{-1}(x,\lambda). \end{aligned} \tag{2.8.9}$$

They also continue analytically from the real λ-axis into the upper and lower half-planes of the complex variable λ.

Formulas (2.8.6) and (2.8.9) generate an equality valid on the real λ-axis:

$$G_+(x,\lambda)\,G_-(x,\lambda) = G(x,\lambda), \tag{2.8.10}$$

where

$$G(x,\lambda) = \varphi_0(x,\lambda)\,G_0(\lambda)\,\varphi_0^{-1}(x,\lambda). \tag{2.8.11}$$

Relation (2.8.10) displays that one of the classical problems of the theory of functions of a complex variable, known as the Riemann problem, plays a crucial role in formulating the method of integrating the Landau–Lifshitz equations for an isotropic ferromagnet (2.2.2). It is typically stated as follows: Let a closed contour Γ, which may pass through an infinitely distant point, be assigned on the plane of a complex variable λ, and let $G(x,\lambda)$ be a matrix function on it. It is required to build a matrix function $G_+(\lambda)$ analytic inside the contour Γ and a matrix function $G_-(\lambda)$ analytic outside this contour that factorize the matrix $G(\lambda)$ on the contour itself:

$$G_+(\lambda)\,G_-(\lambda) = G(\lambda), \quad \lambda \in \Gamma. \tag{2.8.12}$$

For the model of an isotropic ferromagnet, the real axis

$$\Gamma = \{\lambda : \operatorname{Im}\lambda = 0,\ -\infty < \operatorname{Re}\lambda < \infty\}$$

acts as a contour. The variables x and t are parameters.

Obviously, the Riemann problem can be solved, though not uniquely. So, if $G_+(\lambda)$ and $G_-(\lambda)$ are some solutions, and g is an arbitrary nondegenerate matrix λ-independent function, then the pair of the matrix functions $G_+(\lambda)g$ and $g^{-1}G_-(\lambda)$ is also a solution to the Riemann problem (2.8.12) with the same function $G(\lambda)$. The ambiguity can be eliminated by setting a value for one of the functions $G_+(\lambda)$ or $G_-(\lambda)$ at some point in the analyticity domain. Different choices of this point and different values of the function correspond to different normalizations of the Riemann problem. The most commonly used normalization is that on the identity matrix at an infinitely distant point. Such a normalization is called canonical. It is not always feasible to choose the canonical normalization because it needs to be compatible with the auxiliary linear system for an integrable model. In particular, for an isotropic ferromagnet, according to the results obtained previously, we have

$$b(\lambda = 0) = 0, \quad a(\lambda = 0) = 1;$$
$$\varphi_\pm(x, \lambda = 0) = \varphi_0(x, \lambda = 0) = I.$$

Therefore it is natural to normalize the Riemann problem (2.8.10) at the point $\lambda = 0$ rather than at the point $\lambda = \infty$:

$$G_\pm(x, t, \lambda = 0) = G(x, t, \lambda = 0) = I. \tag{2.8.13}$$

Normalization (2.8.13) is consistent both with the auxiliary linear system (2.3.1)–(2.3.2) and equality (2.8.10).

It is worth emphasizing that, when posted, specific problems may impose additional constraints (reductions on solution of a Riemann problem) on the solutions to the Riemann problem (2.8.12). As for an isotropic ferromagnet, such a reduction is the connection between the functions $G_+(\lambda)$ and $G_-(\lambda)$ on the real λ-axis:

$$G_+(\lambda) = G_-^\dagger(\lambda), \tag{2.8.14}$$

which continues to the analyticity domain of the function

$$G_+(\lambda) = G_-^\dagger(\lambda^*), \quad \operatorname{Im}\lambda > 0. \tag{2.8.15}$$

Reduction (2.8.14) is a consequence of formulas (2.3.11), (2.3.17), (2.8.4), and (2.8.9). However, the easiest way to prove this is as follows: Since the matrices $U(\lambda)$ and $V(\lambda)$ (2.2.10) are anti-Hermitian on the real λ-axis,

$$V^{\dagger}(\lambda) = -V(\lambda), \quad U^{\dagger}(\lambda) = -U(\lambda),$$

the differential equations (2.2.21) and (2.2.22) for the functions $\varphi_{\pm}^{-1}(x,t,\lambda)$ coincide with the differential equations for the functions $\varphi_{\pm}^{\dagger}(x,t,\lambda)$ for real values of λ, just as the asymptotic conditions corresponding to them coincide as $x \to \pm\infty$:

$$\varphi_0^{\dagger}(x,\lambda) = \varphi_0^{-1}(x,\lambda).$$

By the uniqueness theorem for solutions to differential equations for the functions $\varphi_{\pm}^{-1}(x,t,\lambda)$ and $\varphi_{\pm}^{\dagger}(x,t,\lambda)$, this means that on the real λ-axis, we have

$$\varphi_{\pm}^{-1}(x,t,\lambda) = \varphi_{\pm}^{\dagger}(x,t,\lambda).$$

Bearing in mind formulas (2.8.4) and (2.8.9), we immediately arrive at eq. (2.8.14).

It is essential that the anti-Hermitian matrices $U(\lambda)$ and $V(\lambda)$ create many generalized Lax pairs. Therefore reduction (2.8.15) becomes universal for a large portion of integrable models. Within the Gel'fand–Levitan–Marchenko formalism, reduction (2.8.15) allows us to confine ourselves to constructing the integral equations of the inverse problem only for one of the ends (either the right or left one); strictly speaking, this statement requires a special proof.

In addition to reduction (2.8.15), other limitations are possible for models different from those of an isotropic ferromagnet [101]. All of them are extremely significant because they characterize the algebraic structure of the $U-V$ pairs, establish the properties of the matrices $G_{\pm}(\lambda)$ in the Riemann problem (2.8.12), and, ultimately, determine the explicit form of the solutions to the initial nonlinear equation.

Note that there is one more fascinating and paramount aspect of the Riemann problem (2.8.10) for an isotropic ferromagnet. The shape of the matrix $G(x,\lambda)$ (2.8.11) is defined by the function $\varphi_0(x,\lambda)$, which is related to the trivial but exact solution to the Landau–Lifshitz equations (2.2.2):

$$\mathbf{S}(x,t) = (0,0,1).$$

As a result, this manifests itself in the fact that the new solutions to the Landau–Lifshitz equation constructed using the Riemann problem have the asymptotics

$$\mathbf{S}(x,t) \to (0,0,1) \quad \text{as} \quad |x| \to \infty.$$

The next section illustrates a general algorithm for constructing new solutions for a nonlinear integrable model by "dressing" known particular solutions of the model using the Riemann problem. The procedure will preserve the aforementioned regularities. In other words, a chosen particular solution to the nonlinear equation specifies the matrix solution of the auxiliary linear system $\varphi_0(x,t,\lambda)$, which must be included in the Riemann problem formulation. All newly built soliton solutions to the initial nonlinear equation have a selected "seeding" solution as the asymptotics as $|x| \to \infty$.

2.9 Integration of nonlinear equations using the matrix Riemann problem

In sciences and arts ... everything should make noise with new research to move forward.

F. Bacon

2.9.1 "Dressing" particular solutions of integrable models using the regular Riemann problem

Let us look at the lowdown of a direct method for calculating exact solutions to nonlinear equations based on the matrix Riemann problem [130, 129, 68, 131].

Suppose an initial nonlinear equation is equivalent to the compatibility condition

$$\partial_t U(\lambda) - \partial_x V(\lambda) + [U(\lambda), V(\lambda)] = 0 \tag{2.9.1}$$

for some auxiliary system of linear differential equations for the matrix function $\chi(x, t, \lambda)$:

$$\partial_x \chi = U(\lambda)\chi, \quad \partial_t \chi = V(\lambda)\chi. \tag{2.9.2}$$

In general, $U(\lambda)$ and $V(\lambda)$ can be meromorphic functions of the parameter λ on a Riemann surface of kind g. The case $g = 1$ will be discussed separately in Chapters 5 and 7. At the same time, the case $g = 0$ is a typical property of many integrable models with rational λ-dependence of the U and V matrices. The latter can be represented as expansions into simple fractions:

$$\begin{aligned} U &= \sum_{i=1}^{m} \sum_{r=1}^{n_i} \frac{U_{i,r}(x,t)}{(\lambda - a_i)^r} + \sum_{k=0}^{n_\infty} \lambda^k U_k(x,t), \\ V &= \sum_{j=1}^{\tilde{m}} \sum_{s=1}^{\tilde{n}_j} \frac{V_{j,s}(x,t)}{(\lambda - b_j)^s} + \sum_{l=0}^{\tilde{n}_\infty} \lambda^l V_l(x,t). \end{aligned} \tag{2.9.3}$$

The coefficients $U_{i,r}$, $V_{j,s}$, U_k, and V_l are generally $n \times n$ matrices in the space C^n. Each specific problem furnishes the commutation relation (2.9.1) as a separate system of nonlinear differential equations for these matrix coefficients. In turn, each system inhibits itself as a dependence only on the pole divisors of the $U(\lambda)$ and $V(\lambda)$ matrices. Here recall that by definition pole divisors set the position, multiplicity, and number of poles on the right-hand sides of the eqs. (2.9.3). Formally, a pole at the point $\lambda = \infty$ is thought to correspond to an expression λ^k with a natural exponent k. Its multiplicity is determined after the mapping $\lambda \to 1/\lambda$, i. e., it is equal to k.

The pole divisors of the matrices $U(\lambda)$ and $V(\lambda)$ are written as

$$D_U = \{(a_i, n_i), i = 1, 2, \dots, m; \quad (\lambda = \infty, n_\infty)\},$$
$$D_V = \{(b_i, \tilde{n}_i), i = 1, 2, \dots, \tilde{m}; \quad (\lambda = \infty, \tilde{n}_\infty)\}.$$

As a rule, it is necessary to diminish the amount of matrix coefficients in expansion (2.9.3) for meaningful physical applications. Specific nonlinear models come from a general system of the kinds (2.9.1) and (2.9.3) as a result of reductions, namely, by imposing constraints compatible with the problem at hand on the matrices $U(\lambda)$ and $V(\lambda)$.

To simplify the further presentation, we assume that the matrices $U(\lambda)$ and $V(\lambda)$ do not have poles at the point $\lambda = \infty$ and satisfy the additional constraints

$$U^\dagger(\lambda^*) = -U(\lambda), \quad V^\dagger(\lambda^*) = -V(\lambda); \tag{2.9.4}$$
$$U(\lambda = \infty) = V(\lambda = \infty) = I. \tag{2.9.5}$$

Other and/or additional reductions are also possible. A complete description of them is an important problem of classifying integrable models [101].

Under condition (2.9.5), the matrices $U(\lambda)$ and $V(\lambda)$ have a more specific shape:

$$U = I + \sum_{i=1}^{m} \sum_{r=1}^{n_i} \frac{U_{i,r}(x,t)}{(\lambda - a_i)^r}, \quad V = I + \sum_{j=1}^{\tilde{m}} \sum_{s=1}^{\tilde{n}_j} \frac{V_{j,s}(x,t)}{(\lambda - b_j)^s}. \tag{2.9.6}$$

Their pole divisors are

$$\begin{aligned} D_U &= \{(a_i, n_i), i = 1, 2, \dots, m\}, \\ D_V &= \{(b_i, \tilde{n}_i), i = 1, 2, \dots, \tilde{m}\}. \end{aligned} \tag{2.9.7}$$

For integrable models of magnets, the well-known shapes of the matrices $U(\lambda)$ and $V(\lambda)$ are written as expansions in terms of the SU(2)-group generators, the group of unitary unimodular 2×2 matrices. The group SU(2) is the most natural for describing the main (exchange) interactions in magnets. Therefore, in what follows, for simplicity, the matrices $U(\lambda)$ and $V(\lambda)$ are assumed to be of order 2×2.

Let $\chi(\lambda)$ be a solution to eq. (2.9.2). Then, due to eq. (2.9.4), the function $[\chi^\dagger(\lambda^*)]^{-1}$ is a solution as well. We put

$$\chi^{-1}(\lambda) = \chi^\dagger(\lambda^*). \tag{2.9.8}$$

Making allowance for eq. (2.9.5), from system (2.9.2) we derive

$$\chi(\lambda = \infty) = \text{const}.$$

Due to the linearity of eqs. (2.9.2), without loss of generality, we can look upon that

$$\chi(\lambda = \infty) = I. \tag{2.9.9}$$

Now we demonstrate how having one of the solutions of system (2.9.1)–(2.9.2) (for example, $U_0(\lambda)$, $V_0(\lambda)$, and $\chi_0(\lambda)$), other its solutions can be constructed. Constructing the new solutions is reduced to finding the 2×2 matrix functions $G_+(\lambda)$ and $G_-(\lambda)$. The former is analytic inside a closed contour Γ in the complex λ-plane, and the latter is analytic outside the contour Γ. Further, for definiteness, the real axis of the complex λ-plane is used as a contour:

$$\Gamma = \{\lambda : \operatorname{Im}\lambda = 0, \ -\infty < \operatorname{Re}\lambda < \infty\}.$$

Such a contour is formally closed at an infinitely distant point. On the contour, the functions $G_+(\lambda)$ and $G_-(\lambda)$, which are analytical in the upper and lower half-planes of the complex λ-plane, satisfy the condition

$$G_+(x,t,\lambda)G_-(x,t,\lambda) = G(x,t,\lambda), \tag{2.9.10}$$

where

$$G(x,t,\lambda) = \chi_0(x,t,\lambda)\, G_0(\lambda)\chi_0^{-1}(x,t,\lambda).$$

The formulation of the matrix Riemann problem (2.9.10) includes the variables x and t as parameters. The matrix $G_0(\lambda)$ does not depend on the variables x and t. The constraints on the matrix function $G_0(\lambda)$, necessary for solving the problem, will be formulated below.

Next, we should get rid of the ambiguity of the solution to the Riemann problem using the canonical normalization

$$G_+(\lambda = \infty) = G_-(\lambda = \infty) = I. \tag{2.9.11}$$

Then the matrix function $G_0(\lambda)$ must satisfy the condition

$$G_0(\lambda = \infty) = I. \tag{2.9.12}$$

Further, we will see that constraints (2.9.11) and (2.9.12) are consistent both with the $U-V$ pair form (2.9.6) and with the asymptotic behavior of new solutions $U(\lambda)$, $V(\lambda)$, and $\chi(\lambda)$ of system (2.9.1)–(2.9.2) as $|\lambda| \to \infty$ (2.9.5), (2.9.9).

Our immediate goal is to prove that the new solution of system (2.9.1)–(2.9.2) for real values of the parameter λ is given by the formulas

$$\begin{aligned} U(\lambda) &= -G_+^{-1}(\lambda)\big[\partial_x G_+(\lambda) - U_0(\lambda)\, G_+(\lambda)\big] \\ &= \big[\partial_x G_-(\lambda) + G_-(\lambda)\, U_0(\lambda)\big] G_-^{-1}(\lambda), \\ V(\lambda) &= -G_+^{-1}(\lambda)\big[\partial_t G_+(\lambda) - V_0(\lambda)\, G_+(\lambda)\big] \\ &= \big[\partial_t G_-(\lambda) + G_-(\lambda)\, V_0(\lambda)\big] G_-^{-1}(\lambda); \end{aligned} \tag{2.9.13}$$

$$\begin{aligned} \chi_+(x,t,\lambda) &= G_+^{-1}(x,t,\lambda)\chi_0(x,t,\lambda), \\ \chi_-(x,t,\lambda) &= G_-(x,t,\lambda)\chi_0(x,t,\lambda). \end{aligned} \tag{2.9.14}$$

Here it is worth highlighting a remarkable fact, related to formulas (2.9.13) and (2.9.14). The point is that according to a known particular solution $U_0(\lambda)$, $V_0(\lambda)$ and $\chi_0(\lambda)$ of system (2.9.1)–(2.9.2), they generate a new solution of system (2.9.1)–(2.9.2) via the matrices of the Riemann problem (2.9.10). The double form of writing the right-hand sides of eqs. (2.9.13) is because of the relations

$$\begin{aligned} &\partial_x G_+(\lambda)\, G_-(\lambda) + G_+(\lambda)\, \partial_x G_-(\lambda) \\ &\quad = U_0(\lambda)\, G_+(\lambda)\, G_-(\lambda) - G_+(\lambda)\, G_-(\lambda)\, U_0(\lambda), \\ &\partial_t G_+(\lambda)\, G_-(\lambda) + G_+(\lambda)\, \partial_t G_-(\lambda) \\ &\quad = V_0(\lambda)\, G_+(\lambda)\, G_-(\lambda) - G_+(\lambda)\, G_-(\lambda)\, V_0(\lambda). \end{aligned}$$

They are obtained by differentiating conditions (2.9.10) over x and t.

To prove this claim, we start with finding out how the matrices $U(\lambda)$ and $V(\lambda)$ defined by expressions (2.9.13) depend on the parameter λ.

In this section, we discuss the case of the regular Riemann problem when the matrices $U(\lambda)$ and $V(\lambda)$ are not degenerate in their analyticity domains:

$$\det G_+(\lambda) \neq 0, \quad \det G_-(\lambda) \neq 0.$$

Then from formulas (2.9.13) it follows that the functions $U(\lambda)$ and $V(\lambda)$ admit an analytic continuation from the real axis over the entire complex λ-plane, except for the poles of the “seeding” functions $U_0(\lambda)$ and $V_0(\lambda)$ characterized by the divisors D_U and D_V (2.9.7).

Suppose there is an additional limitation on the matrix function $G_0(\lambda)$ that controls the right side of eq. (2.9.10) of the Riemann problem. If a point (λ_0, n_0) of the divisor D_U (or D_V) falls on the contour Γ, then the function $G_0(\lambda)$ is assumed to appear as

$$G_0(\lambda) = I + O(|\lambda - \lambda_0|^{n_0}) \tag{2.9.15}$$

in a small neighborhood of this point.

Let us show that under condition (2.9.15), the matrix functions $U(x,t,\lambda)$ and $V(x,t,\lambda)$ defined by formulas (2.9.13) for any values of the variables x and t are rational functions of the parameter λ with the pole divisors (2.9.7), i. e., they have the same algebraic-in-λ structure (2.9.6) as the “seeding” matrices $U_0(\lambda)$ and $V_0(\lambda)$.

Consider the first version: The point (λ_0, n_0) from the set (2.9.7) does not lie on the contour Γ. Then the matrix function $U(x, t, \lambda)$ defined by formula (2.9.13) has a pole of the same order n_0 at the point $\lambda = \lambda_0$ as the matrix function $U_0(x, t, \lambda)$ has.

The second version involves the following: The divisor's point (λ_0, n_0) lies on the contour Γ. Then the "seeding" solution of the auxiliary linear system (2.9.2) has an essential singularity on the contour Γ. Nevertheless, condition (2.9.15) guarantees the regularity of both functions $G(x, t, \lambda)$ and $G_\pm(x, t, \lambda)$ and of the functions $\partial_x G_\pm(x, t, \lambda)$ at the point $\lambda = \lambda_0$ (see (2.9.10)).

Thus both cases provide for the matrix function $U(x, t, \lambda)$ given by formulas (2.9.14) to have a pole of order n_0 at the point $\lambda = \lambda_0$.

By Liouville's theorem the matrix function $U(x, t, \lambda)$ analytic over the entire complex λ-plane except for the poles with the divisor D_U and the asymptotic behavior (2.9.5) as $\lambda \to \infty$ is a rational function of the type (2.9.6).

Analogously, we can prove that the matrix function $V(x, t, \lambda)$ settled by formulas (2.9.13) is a rational λ-dependent function of the type (2.9.6) with the pole divisor D_V (2.9.7).

Thus the "seeding" rationally λ-dependent matrix functions $U_0(\lambda)$ and $V_0(\lambda)$ of the type (2.9.6) produce new functions $U(\lambda)$ and $V(\lambda)$ plotted by formulas (2.9.13). The λ-dependence pattern persists. Only the coefficients $U_{i,r}$ and $V_{j,s}$ in formulas (2.9.6) change.

The canonical normalization (2.9.11) of the Riemann problem ensures the fulfillment of conditions (2.9.5) and (2.9.9) for the new matrix functions $U(\lambda)$, $V(\lambda)$ (2.9.13), and $\chi(\lambda)$ (2.9.14).

The latter satisfy reductions (2.9.4) and (2.9.8) if we impose additional constraints on the matrices $G_+(\lambda)$, $G_-(\lambda)$, and $G_0(\lambda)$ (2.9.10) of the Riemann problem:

$$\begin{aligned} &G_+(\lambda) = G_-^\dagger(\lambda), \quad G_0^\dagger(\lambda) = G_0(\lambda), \quad \det G_0 > 0; \\ &\lambda \in \Gamma = \{\lambda : \operatorname{Im}\lambda = 0, -\infty < \operatorname{Re}\lambda < \infty\}. \end{aligned} \tag{2.9.16}$$

The regular Riemann problem that includes the matrices $G_+(\lambda)$ and $G_-(\lambda)$ nondegenerate in their analyticity domains enables us to persuade oneself by direct verification that the matrix functions $\chi_\pm(x, t, \lambda)$, $U(x, t, \lambda)$, and $V(x, t, \lambda)$ driven by formulas (2.9.13) and (2.9.14) give an exact solution to the system of auxiliary differential equations (2.9.1)–(2.9.2). The stated procedure for integrating system (2.9.1)–(2.9.2) can be called the method of "dressing" of the "seeding" solutions $U_0(x, t, \lambda)$, $V_0(x, t, \lambda)$, and $\chi_0(x, t, \lambda)$ using the Riemann matrix problem combining the contour Γ and the function $G_0(\lambda)$.

We have designed the matrix functions $U(x, t, \lambda)$ and $V(x, t, \lambda)$ satisfying the differential equation (2.9.1). Their algebraic-in-λ structure coincides with that of the "seeding" functions $U_0(x, t, \lambda)$ and $V_0(x, t, \lambda)$. Recall that the initial nonlinear partial differential equation is equivalent to the commutation representation (2.9.1) for all values of the parameter λ. Therefore the difficulties met in constructing a new solution to the initial nonlinear equation via a known particular solution have been smoothed away to a simple algebraic task: The matrix coefficients $U_{i,r}$ and $V_{j,s}$ in formulas (2.9.6) need to be converted back to the field variables that make up the initial nonlinear equation.

The solutions to the wave equations of motion (2.9.1), constructed using the regular Riemann problem, describe waves with a predominance of the dispersion effect over the nonlinearity.

Let us show that the regular solution of the Riemann problem is boiled down to integrating a linear system of singular integral equations on the contour Γ.

We explore the function

$$\Phi(\lambda) = \begin{cases} G_+^{-1}(\lambda), & \operatorname{Im}\lambda > 0, \\ G_-(\lambda), & \operatorname{Im}\lambda < 0. \end{cases}$$

By constructing the function $\Phi(\lambda)$ is analytic in the upper and lower half-planes of the complex variable λ, and on the real axis, it has a jump that, as follows from (2.9.10), is equal to

$$G_+^{-1}(\lambda) - G_-(\lambda) = G_+^{-1}(\lambda)[I - G(\lambda)], \tag{2.9.17}$$

where $G(\lambda) = \chi_0(\lambda)G_0(\lambda)\chi_0^{-1}(\lambda)$. According to eq. (2.9.11), the function $\Phi(\lambda) \to I$ as $|\lambda| \to \infty$. Next, we resort to the Cauchy formula

$$\Phi(\lambda) - I = \frac{1}{2\pi i}\int_\gamma \frac{[\Phi(\lambda') - I]}{\lambda' - \lambda}\, d\lambda',$$

where the integration contour γ is chosen as shown in Figure 2.3. Further, given that the jump magnitude of the function $\Phi(\lambda)$ on the real axis is governed by expression (2.9.17), we can represent the function $\Phi(\lambda)$ as

$$\Phi(\lambda) = I + \frac{1}{2\pi i}\int_{-\infty}^{+\infty} \frac{G_+^{-1}(\lambda')[I - G(\lambda')]}{\lambda' - \lambda}\, d\lambda'.$$

By definition, $\Phi(\lambda) = G_-(\lambda)$ at $\operatorname{Im}\lambda < 0$. This equality holds, in particular, when λ tends to the real axis from the lower half-plane:

$$G_-(\lambda) = I + \frac{1}{2\pi i}\int_{-\infty}^{+\infty} \frac{G_+^{-1}(\lambda')[I - G(\lambda')]}{\lambda' - \lambda + i0}\, d\lambda'. \tag{2.9.18}$$

In relation (2.9.18) the variable λ is real; the integral is evaluated according to the Sokhotski–Plemelj formulas:

$$\frac{1}{\lambda' - \lambda \pm i0} = \text{V. p.}\frac{1}{\lambda' - \lambda} \mp \pi i\, \delta(\lambda' - \lambda), \tag{2.9.19}$$

where the letters “V. p.” are the symbol of the Cauchy principal value.

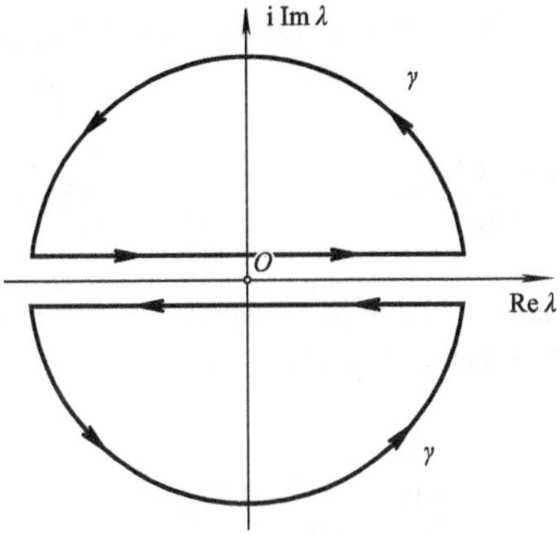

Figure 2.3: Choice of a contour γ in the complex λ-plane.

For an arbitrary nondegenerate 2×2 matrix G_+, we have the identity

$$G_+^{-1}(\lambda) = \frac{\sigma_2 G_+^{\mathrm{T}}(\lambda) \sigma_2}{\det G_+(\lambda)}. \tag{2.9.20}$$

The superscript “T” means the transposition operation. Using reduction (2.9.16), we transform eq. (2.9.20) to the form

$$G_+^{-1}(\lambda) = \frac{\sigma_2 G_-^*(\lambda) \sigma_2}{\det G_+(\lambda)}, \quad \lambda \in \Gamma.$$

The values of $|\det G_+(\lambda)|$ and $\det G_0$ are mutually tied. Indeed, from (2.9.10) and (2.9.16) we find

$$|\det G_+| = [\det G_0]^{1/2}.$$

The argument of the function $\det G_+(\lambda)$ can be restored modulo using the dispersion relation for the function $\ln \det G_+(\lambda)$:

$$\arg \det G_+(\lambda) = -\frac{1}{\pi} \mathrm{V.p.} \int_{-\infty}^{+\infty} \frac{\ln|\det G_+(\lambda')|\,\mathrm{d}\lambda'}{\lambda' - \lambda}, \quad \lambda \in \Gamma. \tag{2.9.21}$$

In the case of an isotropic ferromagnet, formula (2.9.21) coincides with relation (2.5.5), which relates the real and imaginary parts of the function $\ln a(\lambda)$. Let us by analogy denote

$$\begin{aligned} a(\lambda) &\equiv \det G_+(x,t,\lambda) \\ &= [\det G_0]^{1/2} \exp\left(\frac{1}{2\pi \mathrm{i}} \mathrm{V.p.} \int_{-\infty}^{+\infty} \frac{\ln \det G_0(\lambda')}{\lambda' - \lambda}\,\mathrm{d}\lambda' \right), \quad \lambda \in \Gamma. \end{aligned} \tag{2.9.22}$$

Finally, to calculate the matrix function $G_-(\lambda)$, we derive the closed singular integral equation

$$G_-(x,t,\lambda) = I + \frac{1}{2\pi\mathrm{i}} \int\limits_{-\infty}^{+\infty} \frac{\sigma_2\, G_-^*(x,t,\lambda')\,\sigma_2\,[I - G(x,t,\lambda')]}{(\lambda' - \lambda + \mathrm{i}\,0)\, a(\lambda')}\,\mathrm{d}\lambda', \quad \lambda \in \Gamma. \tag{2.9.23}$$

From eq. (2.9.23) we calculate $G_-(\lambda)$ for real values of λ and then analytically continue the result to the region $\operatorname{Im}\lambda < 0$. We restore the function $G_+(\lambda)$ from the function $G_-(\lambda)$ using the analytic continuation of reduction (2.9.16):

$$G_+(\lambda) = G_-^\dagger(\lambda^*), \quad \operatorname{Im}\lambda > 0.$$

Formulas (2.9.22) and (2.9.23) indicate that the convergence of the integrals, in particular, the absence of zeros and poles of the function $\det G_0(\lambda)$ is a necessary condition for the regular Riemann problem to solve. Simultaneously, it should be stressed that the requirement of the convergence of all the integrals may be insufficient for the solvability of system (2.9.23). A sufficient condition is, for example, the requirement for the matrix $G(x,t,\lambda) + G^\dagger(x,t,\lambda)$ to be positively defined for real values of λ [108, 68]. Everywhere below, the regular Riemann problem will be thought of being solvable. Moreover, when a normalization is imposed, it is uniquely solvable.

When analyzing specific problems, the above algorithm for solving the regular Riemann problem can be modified. As an example, we may examine the model of an isotropic ferromagnet. Since the Riemann problem (2.8.10) in this case is normalized to the identity matrix at the point $\lambda = 0$ (2.8.13), the shape of matrices $G_+^{-1}(\lambda)$ and $G_-(\lambda)$ needs to be sought as

$$G_+^{-1}(\lambda) = I + \lambda \int\limits_{-\infty}^{+\infty} \frac{\varphi(\mu)}{\mu - \lambda}\,\mathrm{d}\mu, \quad \operatorname{Im}\lambda > 0;$$

$$G_-(\lambda) = I + \lambda \int\limits_{-\infty}^{+\infty} \frac{\varphi(\mu)}{\mu - \lambda}\,\mathrm{d}\mu, \quad \operatorname{Im}\lambda < 0.$$

On the real λ-axis, by the Sokhotski–Plemelj formulas (2.9.19) we have

$$\begin{aligned} G_+^{-1}(\lambda) &= I + \lambda\left[\mathrm{V.p.}\int\limits_{-\infty}^{+\infty} \frac{\varphi(\mu)}{\mu - \lambda}\,\mathrm{d}\mu + \pi\,\mathrm{i}\,\varphi(\lambda)\right], \\ G_-(\lambda) &= I + \lambda\left[\mathrm{V.p.}\int\limits_{-\infty}^{+\infty} \frac{\varphi(\mu)}{\mu - \lambda}\,\mathrm{d}\mu - \pi\,\mathrm{i}\,\varphi(\lambda)\right]. \end{aligned} \tag{2.9.24}$$

Plugging eqs. (2.9.24) into eq. (2.9.17), we find

$$\varphi(\lambda) = \frac{1}{2\pi i \lambda} G_+^{-1}(\lambda) [I - G(\lambda)].$$

After repeating the calculations previously outlined above for the second equation from (2.9.24), we arrive at a closed system of singular integral equations for calculating the matrix function $G_-(\lambda)$:

$$G_-(\lambda) = I + \frac{\lambda}{2\pi i} \int_{-\infty}^{+\infty} \frac{\sigma_2 G_-^*(\mu) \sigma_2 [I - G(\mu)]}{a(\mu) \mu (\mu - \lambda + i0)} d\mu, \tag{2.9.25}$$

where the matrix function $G(\mu)$ is controlled by formulas (2.8.7) and (2.8.11). Although eq. (2.9.25) differs in form from the Gel'fand–Levitan–Marchenko equations, for example, from eq. (2.6.4) "for the right end", it is completely equivalent to them as long as that the functions $a(\lambda)$ and $a^*(\lambda)$ do not have zeros in their analyticity domains. A simple check using formulas (2.3.6), (2.8.4), (2.8.5), and (2.8.9) can verify the validity of the statement.

2.9.2 The Riemann problem with zeros

Within the formalism stated above, soliton-type solutions are called the solutions of system (2.9.1)–(2.9.2) constructed using the Riemann problem (2.9.10) with the function

$$G_0(\lambda) = I.$$

In this case, the Riemann problem appears as

$$\begin{aligned} &G_-(\lambda) G_+(\lambda) = G_+(\lambda) G_-(\lambda) = I, \\ &\lambda \in \Gamma = \{\lambda : \operatorname{Im}\lambda = 0, -\infty < \operatorname{Re}\lambda < \infty\}, \end{aligned} \tag{2.9.26}$$

with the normalization $G_+(\lambda = \infty) = I$. A nontrivial solution to the Riemann problem (2.9.26) is possible only under the condition that the matrices $G_+(\lambda)$ and $G_-(\lambda)$ are degenerate in their analyticity domains at the points λ_i and μ_j, respectively:

$$\det G_+(\lambda_i) = 0, \quad \det G_-(\mu_j) = 0, \quad i, j = 1, 2, \dots, n.$$

Otherwise, the matrix functions $G_\pm(\lambda)$ continue analytically from the contour Γ over the entire complex plane and, consequently, are constant as Liouville's theorem claims (from the normalization condition $G_+(\lambda) = G_-(\lambda) = I$). In what follows, for definiteness, the zeros of the functions $\det G_+(\lambda)$ and $\det G_-(\lambda)$ will be assumed to be simple. Here it should be pointed out that the zeros of the functions $G_+(\lambda)$ and $G_-(\lambda)$ coincide with their mutually opposite poles because $G_-(\lambda) = G_+^{-1}(\lambda)$ and $G_+(\lambda) = G_-^{-1}(\lambda)$.

Let the matrices of the Riemann problem obey the additional constraint (2.9.16). Reduction (2.9.16) continues from the contour Γ to the analyticity domains of the functions $G_\pm(\lambda)$, where it takes the form

$$G_+^\dagger(\lambda^*) = G_-(\lambda), \quad \operatorname{Im}\lambda < 0; \quad G_-^\dagger(\lambda^*) = G_+(\lambda), \quad \operatorname{Im}\lambda > 0. \tag{2.9.27}$$

Hence the poles of the matrices $G_+(\lambda)$ and $G_-(\lambda)$ are complex conjugate to each other:

$$\mu_i = \lambda_i^*, \quad \operatorname{Im}\lambda_i > 0, \quad i = 1, 2, \dots, n.$$

The connection of matrices (2.9.26) on the contour Γ can be rewritten as the equalities

$$G_+^\dagger(\lambda^*)G_+(\lambda) = G_+(\lambda)G_+^\dagger(\lambda^*) = I, G_+(\lambda = \infty) = I, \tag{2.9.28}$$

which define the meromorphic function $G_+(\lambda)$ in the complex plane of λ. The poles of this function are located at points $\lambda = \lambda_i^*$, $i = 1, 2, \dots, n$. Therefore its expansion into simple fractions has the form

$$G_+(\lambda) = I + \sum_{k=1}^{n} \frac{A_k}{\lambda - \lambda_k^*}. \tag{2.9.29}$$

The first term in the right-hand side of formula (2.9.29) is equal to the identity matrix because

$$G_+(\lambda = \infty) = I.$$

The vanishing-residues requirement on the left-hand side of the first of relations (2.9.28) leads to a system of n independent matrix equations:

$$A_k^\dagger G_+(\lambda_k) = 0, \quad k = 1, 2, \dots, n. \tag{2.9.30}$$

A nontrivial solution of eqs. (2.9.30) is possible only under the condition

$$\det A_k = 0, \quad \det G_+(\lambda_k) = 0.$$

It is known that any two-dimensional degenerate matrix can be written in the form

$$(A_k)_{\alpha\beta} = (\mathbf{m}^k)_\alpha (\mathbf{X}^k)_\beta \quad (\alpha, \beta = 1, 2), \tag{2.9.31}$$

where $\mathbf{X}^k$ and $\mathbf{m}^k$ are two-dimensional complex vectors. After substituting eq. (2.9.31) into eq. (2.9.30), we obtain a linear system of algebraic equations for determining the vectors $\mathbf{X}^k$:

$$(\mathbf{m}^k)_\alpha^* + \sum_{s=1}^{n} \frac{1}{\lambda_k - \lambda_s^*} (\mathbf{m}^{k*} \cdot \mathbf{m}^s)(\mathbf{X}^s)_\alpha = 0, \quad k = 1, 2, \dots, n, \tag{2.9.32}$$

where

$$(\mathbf{m}^{k*} \cdot \mathbf{m}^s) = \sum_{\alpha=1,2} (\mathbf{m}^k)^*_\alpha (\mathbf{m}^s)_\alpha$$

is the scalar product of the two-dimensional vectors $\mathbf{m}^{k*}$ and $\mathbf{m}^s$. Then the solution to system (2.9.32) is given by

$$(\mathbf{X}^s)_\alpha = -\sum_{p=1}^{n} \frac{\partial \ln \det A}{\partial A_{ps}} (\mathbf{m}^p)^*_\alpha. \tag{2.9.33}$$

Here we have incorporated the matrix A with elements

$$A_{ij} = \frac{(\mathbf{m}^{i*} \cdot \mathbf{m}^j)}{\lambda_i - \lambda_j^*}$$

and used the identity

$$(A^{-1})_{ij} = \frac{\partial \ln \det A}{\partial A_{ji}}.$$

Thus the general solution to the Riemann problem (2.9.27)–(2.9.28) is

$$[G_+(\lambda)]_{\alpha\beta} = \delta_{\alpha\beta} - \sum_{i,j}^{n} \frac{1}{\lambda - \lambda_i^*} (\mathbf{m}^i)_\alpha (\mathbf{m}^j)_\beta \frac{\partial \ln \det A}{\partial A_{ji}}. \tag{2.9.34}$$

The integration of system (2.9.1)–(2.9.2) using the function (2.9.34) yields extra poles for new matrices $U(\lambda)$ and $V(\lambda)$ at the points $\lambda = \lambda_i$ and $\lambda = \lambda_i^*$ $(i = 1, 2, \dots, n)$ on the right-hand sides of eqs. (2.9.13). The poles are associated with zeros of the Riemann problem. It is necessary to require the residues at these poles to be equal to zero. Only then the λ-dependence pattern of the new matrix functions $U(\lambda)$ and $V(\lambda)$ (2.9.13) will have that as (2.9.6). Taking into account representation (2.9.29), we bring about the formulated requirement to the following system of n independent equations:

$$G_+^\dagger(\lambda_k)[\partial_x A_k - U_0(\lambda_k^*) A_k] = 0, \quad G_+^\dagger(\lambda_k)[\partial_t A_k - V_0(\lambda_k^*) A_k] = 0, \\ k = 1, 2, \dots, n. \tag{2.9.35}$$

It should be underscored that the differentiation of the matrix function A_k with the elements $(A_k)_{\alpha\beta} = (\mathbf{m}^k)_\alpha (\mathbf{X}^k)_\beta$ over the coordinates x and t produces two additional terms. For example,

$$\partial_x (A_k)_{\alpha\beta} = [\partial_x (\mathbf{m}^k)_\alpha] (\mathbf{X}^k)_\beta + (\mathbf{m}^k)_\alpha \partial_x (\mathbf{X}^k)_\beta. \tag{2.9.36}$$

By the choice of the vectors $\mathbf{m}^k$ we have the equalities

$$G_+^\dagger(\lambda_k^*)\mathbf{m}^k = 0$$

(there is no summation over the index k). Therefore a number of terms in eq. (2.9.35) vanishes. The remaining terms contain the multipliers

$$G_-(\lambda_k^*)(\partial_x \mathbf{m}^k - U_0(\lambda_k^*)\mathbf{m}^k) \quad \text{and} \quad G_-(\lambda_k^*)(\partial_t \mathbf{m}^k - V_0(\lambda_k^*)\mathbf{m}^k)$$

and are equal to zero only when the vectors $\mathbf{m}^k$ satisfy the equations

$$\begin{aligned} \partial_t \mathbf{m}^k &= V_0(x,t,\lambda_k^*)\mathbf{m}^k, \\ \partial_x \mathbf{m}^k &= U_0(x,t,\lambda_k^*)\mathbf{m}^k, \end{aligned} \tag{2.9.37}$$

whose solutions are

$$\mathbf{m}^k = \chi_0(x,t,\lambda_k^*)\mathbf{c}^k. \tag{2.9.38}$$

Here $\mathbf{c}^k$ are arbitrary constant two-dimensional complex vectors. Thus the vectors $\mathbf{m}^k$ are completely determined by the matrices $U_0(\lambda)$, $V_0(\lambda)$, and $\chi_0(\lambda)$ of the seeding solution.

Let us show that the vectors $\mathbf{X}^k$, on the contrary, are determined by the matrices $U(\lambda)$ and $V(\lambda)$ of the "dressing" solution (2.9.13). We use the consequences of formulas (2.9.13),

$$\begin{aligned} \partial_x G_+(\lambda) - U_0(\lambda)G_+(\lambda) &= -G_+(\lambda)U(\lambda), \\ \partial_t G_+(\lambda) - V_0(\lambda)G_+(\lambda) &= -G_+(\lambda)V(\lambda). \end{aligned} \tag{2.9.39}$$

The condition of equality of residues at the poles $\lambda = \lambda_k^*$ of relations (2.9.39) leads to the equations

$$\begin{aligned} &\partial_x A_k - U_0(\lambda_k^*)A_k = -A_k U(\lambda_k^*), \quad \partial_t A_k - V_0(\lambda_k^*)A_k = -A_k V(\lambda_k^*), \\ &k = 1, 2, \ldots, n. \end{aligned} \tag{2.9.40}$$

Hence eqs. (2.9.37) immediately imply consistent equations for the vectors $\mathbf{X}^k$:

$$\partial_t \mathbf{X}^k = -\mathbf{X}^k U(x,t,\lambda_k^*), \quad \partial_x \mathbf{X}^k = -\mathbf{X}^k V(x,t,\lambda_k^*). \tag{2.9.41}$$

The procedure for constructing the multisoliton solutions to equations (2.9.1) and (2.9.2) has been accomplished totally. The application of a particular solution U_0, V_0, and χ_0 to system (2.9.1)–(2.9.2) causes formulas (2.9.13), (2.9.14), (2.9.34), and (2.9.38) to generate a series of new solutions.

It is interesting that formulas (2.9.29), (2.9.31), (2.9.38), and (2.9.41) also become effective also in the integrable nonlinear model "multiplication" procedure, which allows us to comprise new integrable equations from a given integrable equation [46, 33, 48, 47].

From a mathematical point of view, the matrices A_k with the elements $(\mathbf{m}^k)_\alpha(\mathbf{X}^k)_\beta$ are connected to projection operators. The language of linear algebra in the space $\mathbb{C}^2$ of two-dimensional complex vectors is convenient to disclose this connection. Let a linear operator U be given in the space $\mathbb{C}^2$. Recall that the kernel $\operatorname{Ker} U$ of operator U is an aggregate of those vectors from $\mathbb{C}^2$ that the operator sets to zero. Denote the image of the operator U by $\operatorname{Im} U$, i. e., the result of applying the operator U to the entire space $\mathbb{C}^2$. Obviously, $\operatorname{Ker} U$ and $\operatorname{Im} U$ are linear spaces, and their direct sum is the entire space $\mathbb{C}^2$:

$$\operatorname{Ker} U \otimes \operatorname{Im} U = \mathbb{C}^2.$$

Now we return to the operator A_k to arrive at the equality

$$A_k G_-(\lambda_k^*) = 0$$

by the no-residue condition at the point $\lambda = \lambda_k^*$ in the product $G_+(\lambda)G_+^\dagger(\lambda^*)$. In other words, $\operatorname{Ker} A_k = \operatorname{Im} G_+^\dagger(\lambda_k)$ or $\operatorname{Ker} A_k = \operatorname{Im} \chi_-(\lambda_k^*)$, provided that formula (2.9.14) is used.

At the same time, the structure of the matrix A_k (2.9.31) makes it clear that applying the operator A_k to the entire space $\mathbb{C}^2$ provides a space spanned by the vectors $\mathbf{m}^k$ of the form (2.9.38), i. e., $\operatorname{Im} A_k = \operatorname{Im} \chi_0(\lambda_k^*)$.

The construction process set forth above claims that the kernel and image of the matrix A_k completely determine its structure (this is a specific property of projection matrices). It is curious that the Riemann problem creates a connection between $\operatorname{Im} A_k$ and the "seeding" solution $\chi_0(x, t, \lambda)$ to the auxiliary linear system (2.9.2), and between the new solution $\chi_-(x, t, \lambda)$ of the latter built by the "dressing" of $\chi_0(x, t, \lambda)$ and $\operatorname{Ker} A_k$.

The matrix function (2.9.34) can be sometimes represented as a product of matrices to highlight its dependence on projection operators more clearly [108]:

$$G_+(\lambda) = \prod_{k=1}^{n} \left(I - \frac{\lambda_k - \lambda_k^*}{\lambda - \lambda_k^*} P_k \right), \tag{2.9.42}$$

where P_k is a projection operator matrix:

$$\det P_k = 0, \quad \operatorname{Sp} P_k = 1, \quad P_k^2 = P_k, \quad P_k^+ = P_k. \tag{2.9.43}$$

Direct verification enables us to make sure that the matrix function (2.9.42) satisfies conditions (2.9.27) and (2.9.28). Representations (2.9.34) and (2.9.42) of the function $G_+(\lambda)$ are equivalent since the solution to the Riemann problem with given zeros, reductions, and normalization is unique.

Using expressions (2.9.42) and (2.9.43), it is easy to show that

$$\det G_+(\lambda) = \prod_{k=1}^{n} \frac{\lambda - \lambda_k}{\lambda - \lambda_k^*}. \tag{2.9.44}$$

The matrix $G_+(\lambda)$ factorized into the product (2.9.43) is convenient for multisoliton solutions to be recursively built. In this case, an nth soliton solution is calculated by the same formulas as a one-soliton solution. Then an $(n-1)$th soliton solution ($n > 1$) acts as a "seeding" one. Below, using specific examples, we will show that a recurrent procedure allows us to relatively simply investigate the asymptotics of multisoliton solutions as $t \to \pm\infty$.

2.9.3 The scheme of "dressing" in the general case

Let us now dive into finding solutions to system (2.9.1)–(2.9.2) in the most general case where the Riemann problem is posed as follows:

$$\begin{aligned} G_+(x,t,\lambda)\,G_-(x,t,\lambda) &= G(x,t,\lambda), \\ G(x,t,\lambda) &= \chi_0(x,t,\lambda)G_0(\lambda)\chi_0^{-1}(x,t,\lambda), \quad \lambda \in \Gamma, \end{aligned} \tag{2.9.45}$$

the matrix $G_0(\lambda) \neq I$, and the matrix functions $G_+(\lambda)$ and $G_-(\lambda)$ have zeros at points λ_i and λ_i^* in the complex λ-plane:

$$\begin{gathered} \det G_+(\lambda = \lambda_i) = 0, \quad \det G_-(\lambda = \lambda_i^*) = 0 \\ (\operatorname{Im}\lambda_i > 0, \quad i = 1, 2, \ldots, n). \end{gathered} \tag{2.9.46}$$

Seeking the solution to this problem using one of the indicated methods begins with constructing a solution $G^0_\pm(\lambda)$ to the Riemann problem (2.9.26)–(2.9.27) with zeros (2.9.46). We represent the desired matrices $G_\pm(\lambda)$ as

$$G_+(\lambda) = G^0_+(\lambda)\,\tilde{G}_+(\lambda), \quad G_-(\lambda) = \tilde{G}_-(\lambda)\,G^0_-(\lambda). \tag{2.9.47}$$

Obviously, $\tilde{G}_\pm(\lambda)$ are analytic and nonsingular matrices in the upper and lower half-planes of the complex variable λ, respectively.

Thus the problem of determining the functions $\tilde{G}_\pm(\lambda)$ is reduced to the already considered regular Riemann problem

$$\tilde{G}_+(\lambda)\,\tilde{G}_-(\lambda) = \tilde{G}(\lambda), \quad \lambda \in \Gamma, \tag{2.9.48}$$

with the reductions

$$\tilde{G}^\dagger_-(\lambda) = \tilde{G}_+(\lambda), \quad \tilde{G}(\lambda = \infty) = I, \quad \lambda \in \Gamma.$$

The formulation of the regular Riemann problem (2.9.48) contains the matrix function $\tilde{G}(\lambda) \neq G(\lambda)$:

$$\tilde{G}(\lambda) = [G^0_+(\lambda)]^{-1} G(\lambda)\,G^0_+(\lambda), \quad [G^0_+(\lambda)]^{-1} = G^0_+(\lambda)^\dagger, \quad \operatorname{Im}\lambda = 0.$$

The solutions to the equations of motion (2.9.1), built using the general Riemann problem (2.9.45), describe both a set of solitons that retain their shape over time and decreasing-in-amplitude nonlinear waves that spread over time.

Reducing the general Riemann problem (2.9.45) to the regular case allows for changing the order of the comultipliers in formula (2.9.47):

$$G_+(\lambda) = \tilde{G}_+(\lambda)\, G_+^0(\lambda), \quad G_-(\lambda) = G_-^0(\lambda)\, \tilde{G}_-(\lambda).$$

Then the factorable matrix $G(\lambda)$ stays the same. Relation (2.9.10) reduces to the regular Riemann problem that differs from (2.9.48):

$$\tilde{G}_+(\lambda)\, \tilde{G}_-(\lambda) = G(\lambda), \quad \lambda \in \Gamma.$$

Meanwhile, such an approach defines the x- and t-dependence of the matrices $G_\pm^0(x,t,\lambda)$ of the Riemann problem with zeros not only through the "seeding" function $\chi_0(x,t,\lambda)$ but also through the solution $\tilde{G}_+(x,t,\lambda)$ to the regular Riemann problem. The vectors $\mathbf{m}^k$ of the Riemann problem with zeros are

$$\mathbf{m}^k = \tilde{G}_-(x,t,\lambda_k^*)\chi_0(x,t,\lambda_k^*)\,\mathbf{c}^k,$$

where $\mathbf{c}^k$ are arbitrary constant two-dimensional complex vectors.

The general Riemann problem (2.9.45) can be also directly reduced to a system of algebraic and integral equations. For this, we represent the functions $G_+(\lambda)$ and $G_-(\lambda)$ in the form [108]

$$\begin{aligned} G_+^{-1}(\lambda) &= I + \sum_{k=1}^{n} \frac{A_k^\dagger}{\lambda - \lambda_k} + \frac{1}{2\pi \mathrm{i}} \int_{-\infty}^{+\infty} \frac{\varphi(\mu)}{\mu - \lambda}\, \mathrm{d}\mu \quad (\operatorname{Im}\lambda > 0), \\ G_-(\lambda) &= I + \sum_{k=1}^{n} \frac{A_k^\dagger}{\lambda - \lambda_k} + \frac{1}{2\pi \mathrm{i}} \int_{-\infty}^{+\infty} \frac{\varphi(\mu)}{\mu - \lambda}\, \mathrm{d}\mu \quad (\operatorname{Im}\lambda < 0). \end{aligned} \tag{2.9.49}$$

Here, as before, the elements of the degenerate matrix A_k are given by

$$(A_k)_{\alpha\beta} = (m^k)_\alpha (X^k)_\beta.$$

Due to reduction (2.9.16), on the real λ-axis, we have

$$G_-(\lambda) = G_+^\dagger(\lambda) = I + \sum_{k=1}^{n} \frac{A_k^\dagger}{\lambda - \lambda_k} + \frac{1}{2\pi \mathrm{i}} \int_{-\infty}^{+\infty} \frac{\varphi(\mu)}{\mu - \lambda + \mathrm{i}\,0}\, \mathrm{d}\mu. \tag{2.9.50}$$

This relation continues from the real axis into the upper half-plane and gives the following representation for the function $G_+(\lambda)$:

$$G_+(\lambda) = I + \sum_{k=1}^{n} \frac{A_k}{\lambda - \lambda_k^*} - \frac{1}{2\pi i} \int_{-\infty}^{+\infty} \frac{\varphi^\dagger(\mu)}{\mu - \lambda} \, d\mu \quad (\operatorname{Im} \lambda > 0). \tag{2.9.51}$$

Formulas (2.9.49) and (2.9.51) imply that the requirement for poles to be nonexistent in the product $G_+^{-1}(\lambda)\, G_+(\lambda)$ results in the following system of independent algebraic equations for the components $(X^k)_\beta$:

$$\begin{aligned} &\sum_{\alpha=1,2} (m^k)_\alpha^* \left(\delta_{\alpha\beta} - \frac{1}{2\pi i} \int_{-\infty}^{+\infty} \frac{\varphi_{\beta\alpha}^*(\mu)}{\mu - \lambda_k} \, d\mu \right) \\ &\quad + \sum_{q=1}^{n} \frac{(\mathbf{m}^{k*} \cdot \mathbf{m}^q)}{\lambda_k - \lambda_q^*} (X^q)_\beta = 0, \end{aligned} \tag{2.9.52}$$

where

$$(\mathbf{m}^{k*} \cdot \mathbf{m}^q) = \sum_{\alpha=1,2} (m^k)_\alpha^* (m^q)_\alpha.$$

Further, plugging eq. (2.9.49) into relation (2.9.17) and repeating the arguments already given, we find

$$\varphi(\lambda) = G_+^{-1}(\lambda)[I - G(\lambda)] = \frac{1}{a(\lambda)} \sigma_2\, G_-^*(\lambda)\, \sigma_2\, [I - G(\lambda)]. \tag{2.9.53}$$

Then from eq. (2.9.50) we obtain an integral representation for the matrix function $G_-(\lambda)$ for real values of λ:

$$[G_-(\lambda)]_{\alpha\beta} = \delta_{\alpha\beta} + \sum_{k=1}^{n} \frac{(X^k)_\alpha^* (m^k)_\beta^*}{\lambda - \lambda_k} + \frac{1}{2\pi i} \int_{-\infty}^{+\infty} \frac{\varphi_{\alpha\beta}(\mu)}{\mu - \lambda + i0} \, d\mu. \tag{2.9.54}$$

Equalities (2.9.50)–(2.9.54) constitute a closed system of equations for solving the Riemann problem in general form.

As before, the requirement for extra poles to be nonexistent in the "dressing" relations (2.9.13) leads to dependence (2.9.38) of the vectors $\mathbf{m}^k$ on coordinates and time.

It is worth pointing out that the procedure set forth above for integrating nonlinear equations dwells on no special properties of the matrix function $G(x, t, \lambda)$, except for the condition for the unique solvability of the Riemann problem (2.9.45), reductions, and the explicit dependence of the "seeding" solution χ_0 on the variables x, t, and λ.

It is also important to note that the connection between the Riemann problem and the commutation representation (2.9.1) is not only very general but also a local one (the Gel'fand–Levitan–Marchenko approach loses this locality). This circumstance opens up the possibility of constructing a wide class of new solutions within the framework of the above formalism, including spatially nonlocalized solutions. The latter are extremely difficult to describe theoretically applying the Gel'fand–Levitan–Marchenko approach.

The advantages of the procedure of “dressing” particular solutions of integrable models will be illustrated with instances in the section below.

The Riemann problem also turns out to be effective in calculating the asymptotics of exact solutions of integrable models as $t \to +\infty$ [68, 76, 77, 22, 60, 61].

Meanwhile, the analytical properties of the “seeding” functions $\chi_0(\lambda)$ of the auxiliary linear system (2.9.2) can be formulated on multisheeted Riemann surfaces even in the presence of a $U - V$ pair with a rational dependence on the spectral parameter λ. Such situations allow for no obvious selection of new (“dressed”) solutions of the original nonlinear model (2.9.1) with the properties required by the physical sense of the problem. Moreover, in this case the data of the continuous and discrete spectra of the auxiliary linear system (2.9.2) turn out to be interconnected. The edges in the continuous spectrum additionally complicate the analytical properties of the transition matrix coefficients [68]. Therefore an exhaustive research of the Riemann problem is very cumbersome and unproductive. In addition, the normalization conditions for the Riemann problem may turn out to be nontrivial. Sometimes it is easier to return to the Gel’fand–Levitan–Marchenko universal formalism to overcome such challenges.

The two-sheeted Riemann surface of the function $\chi_0(\lambda)$ appears in Chapters 3 and 9 in analyzing quasi-one-dimensional solitons against the background of periodic waves and domain structures and in describing theoretically two-dimensional magnetic dislocations in incommensurate magnetic and crystalline structures [42, 43]. It turns out that magnetic dislocations (vortices) with a minimum topological charge are not soliton-like objects.

2.10 Analysis of soliton states of an isotropic ferromagnet with the “dressing” method

In my current understanding, I came to the thought that in order for the “wave-particle” dualism to be kept in mind it is necessary to develop wave mechanics based on non-linear equations, regarding to those linear equations have only approximate forms being valid under certain conditions.

Louis de Broglie

This section illustrates the effectiveness of the “dressing” method for constructing soliton solutions of the Landau–Lifshitz model (2.2.2) for an isotropic ferromagnet, which were previously sought within the Gel’fand–Levitan–Marchenko formalism. The “dressing” method makes it easy to show that in the general case, an nth soliton solution breaks down into a sum of n spatially separated single solitons as $t \to \pm\infty$.

Recall that the “dressing” procedure described in the previous section was developed for nonlinear differential equations. They can be represented in the form of a compatibility condition

$$\partial_t U(\lambda) - \partial_x V(\lambda) + [U(\lambda), V(\lambda)] = 0 \tag{2.10.1}$$

of the auxiliary line system

$$\partial_x \chi = U(\lambda)\chi, \quad \partial_t \chi = V(\lambda)\chi,$$

where the matrices $U(\lambda)$ and $V(\lambda)$ are rational functions of the parameter λ.

In spite of representing eqs. (2.2.2) of the nonlinear dynamics of an isotropic ferromagnet as (2.10.1), the auxiliary linear system was written in a different form within the Gel'fand–Levitan–Marchenko method:

$$\partial_x \varphi = U(\lambda)\,\varphi, \quad \partial_t \varphi = V(\lambda)\,\varphi + \frac{\lambda^2}{2\mathrm{i}}\,\varphi\,\sigma_3. \tag{2.10.2}$$

The reason for choosing the auxiliary equations as (2.10.2) was that the Gel'fand–Levitan–Marchenko formalism includes the Jost functions $\varphi_\pm(x,t,\lambda)$ whose asymptotics must remain unchanged over time as $x \to \pm\infty$.

It is easy to check that the matrix functions $\chi(x,t,\lambda)$ and $\varphi(x,t,\lambda)$ are related by the simple relationship

$$\chi(x,t,\lambda) = \varphi(x,t,\lambda)\exp\left(\frac{\mathrm{i}\lambda^2}{2}\sigma_3 t\right).$$

Therefore, in the "dressing" method, the trivial "seeding" solution

$$\mathbf{S}^{(0)} = (0,0,1) \tag{2.10.3}$$

of the Landau–Lifshitz equations corresponds to the matrix function (see (2.2.10))

$$\chi^{(0)}(x,t,\lambda) = \exp\left(\frac{\lambda}{2\mathrm{i}}\,x\,\sigma_3 - \frac{\lambda^2}{2\mathrm{i}}\,t\,\sigma_3\right).$$

Recall also that the Gel'fand–Levitan–Marchenko formalism contains the time-dependent transition matrix coefficient $b(\lambda)$ (2.3.33):

$$b(x,t) = b(x,t=0)\exp(-\mathrm{i}\lambda^2 t).$$

Therefore the Riemann problem (2.8.10) for an isotropic ferromagnet can be rewritten as coinciding in form with problem (2.9.10):

$$G_+(x,t,\lambda)\,G_-(x,t,\lambda) = G(x,t,\lambda),$$

where

$$G(x,t,\lambda) = \chi_0(x,t,\lambda)G_0(\lambda)\chi_0^{-1}(x,t,\lambda),$$

$$G_0(\lambda) = \begin{pmatrix} 1 & -b^*(\lambda,0) \\ -b(\lambda,0) & 1 \end{pmatrix}.$$

Thus two methods of integrating the equations for an isotropic ferromagnet (the Gel'fand–Levitan–Marchenko and “dressing” methods) are self-consistent at the level of constitutive relations.

The general scheme for integrating nonlinear equations, presented in Section 2.9, uses the reduction to solutions to the Riemann problem. For real values of the parameter λ, this reduction has the form

$$G_+(\lambda) = G_-^\dagger(\lambda), \quad G_0^\dagger(\lambda) = G_0(\lambda). \tag{2.10.4}$$

The same reduction is valid for an isotropic ferromagnet (see (2.8.14)). The difference lies in the choice of the normalization point for the Riemann problem. The Riemann problem for an isotropic ferromagnet should be normalized to the identity matrix at the point $\lambda = 0$ rather than at the point $\lambda = \infty$ (see (2.8.13)):

$$G_\pm(x,t,\lambda = 0) = G(x,t,\lambda = 0) = I, \quad G_0(\lambda = \infty) = I. \tag{2.10.5}$$

This circumstance will slightly change the way of constructing the matrices $G_\pm(x,t,\lambda)$. As for the rest, the scheme for integrating the equations for an isotropic ferromagnet does not differ from that considered in Section 2.9.

As an example, let us construct the already known one-soliton solution (2.7.4) to the Landau–Lifshitz equations by the “dressing” method. For a one-soliton state, each matrix functions $G_+(\lambda)$ and $G_-(\lambda)$ in the Riemann problems have one degeneracy point in their analyticity domains: $\lambda = \lambda_1$ and $\lambda = \lambda_1^*$, respectively, $\operatorname{Im}\lambda_1 > 0$. Given reduction (2.10.4) and normalization (2.10.5), the one-soliton matrices $G_\pm^{(1)}$ should be sought as

$$G_+^{(1)}(\lambda) = I + \frac{\lambda}{\lambda - \lambda_1^*} A, \quad G_-^{(1)}(\lambda) = I + \frac{\lambda}{\lambda - \lambda_1} A^\dagger.$$

Repeating the speculations of Section 2.9, we conclude that the elements of the matrix A can be written as follows:

$$A_{\alpha\beta} = m_\alpha X_\beta \quad (\alpha, \beta = 1, 2),$$

where the components of the two-dimensional vector $\mathbf{X}$ are determined by the algebraic equations

$$m_\beta^* + \frac{\lambda_1}{\lambda_1 - \lambda_1^*} |\mathbf{m}|^2 X_\beta = 0.$$

Hence we find

$$X_\beta = -\frac{(\lambda_1 - \lambda_1^*)}{\lambda_1} \frac{m_\beta^*}{|\mathbf{m}|^2},$$

and therefore

$$[G_+^{(1)}(\lambda)]_{\alpha\beta} = \delta_{\alpha\beta} - \frac{(\lambda_1 - \lambda_1^*)\,\lambda}{\lambda_1\,(\lambda - \lambda_1^*)}\frac{m_\alpha m_\beta^*}{|\mathbf{m}|^2}. \tag{2.10.6}$$

The matrix P with elements

$$P_{\alpha\beta} = \frac{m_\alpha m_\beta^*}{|\mathbf{m}|^2}$$

is Hermitian and projection one:

$$P^\dagger = P, \quad P^2 = P, \quad \operatorname{Sp} P = 1, \quad \det P = 0.$$

The mapping $\lambda \to \lambda^{-1}$ carries the point $\lambda = 0$ into the point $\lambda = \infty$, and formula (2.10.6) coincides with the more general formula (2.9.42) ($n = 1$) given above:

$$G_+^{(1)}(\lambda) = I - \frac{\lambda_1^{-1} - \lambda_1^{*-1}}{\lambda^{-1} - \lambda_1^{*-1}}\,P.$$

The outcomes of Section 2.9 provide the dependence of the vector $\mathbf{m}$ on the variables x and t as

$$\mathbf{m} = \chi^{(0)}(x, t, \lambda_1^*)\,\mathbf{c}, \tag{2.10.7}$$

where $\mathbf{c} = (c_1, c_2)$ is an arbitrary constant complex vector.

The “seeding” solution (2.10.3) corresponds to the matrices (see (2.2.10))

$$U^{(0)} = \frac{\lambda}{2\mathrm{i}}\,\sigma_3, \quad V^{(0)} = -\frac{1}{2\mathrm{i}}\lambda^2\sigma_3.$$

The matrix functions $U^{(1)}(\lambda)$ and $V^{(1)}(\lambda)$ defining a new solution to the Landau–Lifshitz equation (2.2.2) are calculated by formulas (2.9.13).

In particular, taking into account reduction (2.10.4), we have

$$U^{(1)}(x,t,\lambda) = \frac{\lambda}{2\mathrm{i}}S_k^{(1)}\sigma_k = -[G_+^{(1)}]^\dagger(x,t,\lambda)\Big[\partial_x G_+^{(1)}(x,t,\lambda) - \frac{\lambda}{2\mathrm{i}}\sigma_3 G_+^{(1)}(x,t,\lambda)\Big],$$

from which we deduce a general formula that generates a new solution to the Landau–Lifshitz equations (2.2.2) using the chosen “seeding” one:

$$S_k^{(1)}\sigma_k = [G_+^{(1)}]^\dagger(x,t,\lambda=\infty)\,\sigma_3\,G_+^{(1)}(x,t,\lambda=\infty). \tag{2.10.8}$$

It is easy to check that the solution $\mathbf{S}^{(1)}(x,t)$ built by formula (2.10.8) using the matrix functions (2.10.6) and (2.10.7) coincides with the one-soliton solution of Landau–Lifshitz equations previously derived by the more tedious Gel’fand–Levitan–Marchenko procedure:

$$S_3^{(1)}(x,t) = 1 - \frac{2\,u_1^2}{(u_1^2 + v_1^2)\,\cosh^2[u_1\,(x - v_1\,t - x_{01})/2]},$$
$$S_1^{(1)}(x,t) + \mathrm{i}\,S_2^{(1)}(x,t) = \frac{2\,u_1\,\exp(\mathrm{i}\,[\varphi_{01} + v_1\,x/2 + (u_1^2 - v_1^2)\,t/4])}{(u_1^2 + v_1^2)\,\cosh^2[u_1\,(x - v_1\,t - x_{01})/2]}$$
$$\times\left(-u_1\,\sinh\left[\frac{u_1}{2}(x - v_1\,t - x_{01})\right] + \mathrm{i}\,v_1\,\cosh\left[\frac{u_1}{2}(x - v_1\,t - x_{01})\right]\right). \tag{2.10.9}$$

Here we have employed the notations

$$\lambda_1 = \frac{1}{2}(v_1 + \mathrm{i}\,u_1) \quad (u_1 > 0), \quad x_{01} = \frac{2}{u_1}\ln\left|\frac{c_1}{c_2}\right|, \quad \varphi_{01} = \arg\frac{c_1^*}{c_2^*}.$$

The soliton (2.10.9) is characterized by four real parameters: velocity v_1, a coordinate x_{01} of its "center of gravity" at $t = 0$, an initial phase φ_{01}, and a projection value of $S_3^{(1)}(x,t)$ in the soliton's "center of gravity":

$$A_1 = \frac{v_1^2 - u_1^2}{v_1^2 + u_1^2}. \tag{2.10.10}$$

The "dressing" procedure gives an algorithm for exact solutions to an isotropic ferromagnet equations to "multiplicate". In particular, the method makes it possible to build recursively multisoliton solutions. Using this circumstance, we show that n-soliton solutions describe elastic pair collisions in a system of n self-localized magnetization waves of the type (2.10.9).

For simplicity, we confine ourselves to the analysis of a two-soliton solution to the equations for an isotropic ferromagnet. The "dressing" procedure can construct a two-soliton solution through choosing the one-soliton excitation matrix functions as new "seeding" functions:

$$\begin{aligned}
U^{(1)}(\lambda) &= -[G_+^{(1)}(\lambda)]^{-1}(\partial_x G_+^{(1)}(\lambda) - U^{(0)}G_+^{(1)}(\lambda)) \\
&= (\partial_x G_-^{(1)}(\lambda) + G_-^{(1)}U^{(0)}(\lambda))[G_-^{(1)}(\lambda)]^{-1}, \\
V^{(1)}(\lambda) &= -[G_+^{(1)}(\lambda)]^{-1}(\partial_t G_+^{(1)}(\lambda) - V^{(0)}G_+^{(1)}(\lambda)) \\
&= (\partial_t G_-^{(1)}(\lambda) + G_-^{(1)}V^{(0)}(\lambda))[G_-^{(1)}(\lambda)]^{-1}; \\
\chi_+^{(1)}(x,t,\lambda) &= [G_+^{(1)}(x,t,\lambda)]^{-1}\chi^{(0)}(x,t,\lambda), \\
\chi_-^{(1)}(x,t,\lambda) &= G_-^{(1)}(x,t,\lambda)\chi^{(0)}(x,t,\lambda).
\end{aligned} \tag{2.10.11}$$

On the other hand, due to factorization (2.9.42), formulas obtained from relations (2.10.11) after the formal replacement

$$G_\pm^{(1)}(\lambda) \to \tilde{G}_\pm^{(2)}(\lambda), \quad \{\chi_\pm^{(0)},\, U^{(0)},\, V^{(0)}\} \to \{\chi_\pm^{(1)},\, U^{(1)},\, V^{(1)}\},$$

ascertain the two-soliton excitation. In this case the matrices $\tilde{G}_+^{(2)}(\lambda)$ and $\tilde{G}_-^{(2)}(\lambda)$, as before, retain the shape (2.10.6). However, their zeros are located at other points, $\lambda = \lambda_2$ and $\lambda = \lambda_2^*$, respectively ($\operatorname{Im}\lambda_2 > 0$). In addition, when calculating $\tilde{G}_\mp^{(2)}(\lambda)$, the previous vectors $\mathbf{m}$ should be replaced by the vectors $\tilde{\mathbf{m}}$ built using the matrix functions of the one-soliton state:

$$\tilde{\mathbf{m}} = \chi_-^{(1)}(x,t,\lambda_2^*)\,\tilde{\mathbf{c}} = \left(I + \frac{(\lambda_1 - \lambda_1^*)\,\lambda_2^*}{\lambda_1^*\,(\lambda_2^* - \lambda_1)}\,P\right)\chi^{(0)}(x,t,\lambda_2^*)\,\tilde{\mathbf{c}}, \tag{2.10.12}$$

where $\tilde{\mathbf{c}} = (\tilde{c}_1, \tilde{c}_2)$ is an arbitrary constant complex vector. In particular, the matrix $\tilde{G}_+^{(2)}(\lambda)$ has the shape

$$\tilde{G}_+^{(2)}(\lambda) = I - \frac{(\lambda_2 - \lambda_2^*)\,\lambda}{\lambda_2\,(\lambda - \lambda_2^*)}\,\tilde{P}. \tag{2.10.13}$$

Formulas (2.10.12) and (2.10.13) contain the projection matrices P and $\tilde{P}$ with elements

$$P_{\alpha\beta} = \frac{m_\alpha m_\beta^*}{|\mathbf{m}|^2}, \qquad \tilde{P}_{\alpha\beta} = \frac{\tilde{m}_\alpha \tilde{m}_\beta^*}{|\tilde{\mathbf{m}}|^2}.$$

The magnetization distribution corresponding to the two-soliton state is calculated by the formula

$$\begin{aligned} S_k^{(2)}(x,t)\sigma_k &= \left[\tilde{G}_+^{(2)}(x,t,\lambda=\infty)\right]^\dagger S_k^{(1)}(x,t)\sigma_k\,\tilde{G}_+^{(2)}(x,t,\lambda=\infty) \\ &= \left[\tilde{G}_+^{(2)}(x,t,\lambda=\infty)\right]^\dagger \left[G_+^{(1)}(x,t,\lambda=\infty)\right]^\dagger \sigma_3 \\ &\quad \times G_+^{(1)}(x,t,\lambda=\infty)\tilde{G}_+^{(2)}(x,t,\lambda=\infty). \end{aligned} \tag{2.10.14}$$

We do not give here an explicit form of the solution $\mathbf{S}^{(2)}(x,t)$ but show only that the two-soliton magnetization distribution (2.10.14) tends to a one-soliton one and decreases exponentially in all other directions of the (x,t)-plane on the trajectories of the second soliton as $t \to \pm\infty$, when

$$x - v_2\,t = \text{const} \quad (\lambda_2 = (v_2 + \mathrm{i}\,u_2)/2, \quad u_2 > 0).$$

For this, we draw attention to the fact that the components of the vector $\mathbf{m}$ are expressed as

$$\mathbf{m} = \left(A(t)\exp\left[\frac{u_1}{4}(v_1 - v_2)\,t\right],\ B(t)\exp\left[\frac{u_1}{4}(v_2 - v_1)\,t\right]\right)$$

in the limit

$$x - v_2\,t = \text{const}, \quad t \to \pm\infty. \tag{2.10.15}$$

Here $u_1 > 0$, and $A(t)$ and $B(t)$ are bounded functions of the variable t. Let $v_1 > v_2$ for definiteness. Then in the limit (2.10.15) we have

$$\lim_{t\to+\infty} G_+^{(1)}(\lambda) = \begin{pmatrix} \frac{\lambda_1^*(\lambda-\lambda_1)}{\lambda_1(\lambda-\lambda_1^*)} & 0 \\ 0 & 1 \end{pmatrix} = d_+(\lambda),$$
$$\lim_{t\to-\infty} G_+^{(1)}(\lambda) = \begin{pmatrix} 1 & 0 \\ 0 & \frac{\lambda_1^*(\lambda-\lambda_1)}{\lambda_1(\lambda-\lambda_1^*)} \end{pmatrix} = d_-(\lambda); \tag{2.10.16}$$
$$\lim_{t\to\pm\infty} \tilde{\mathbf{m}} = \chi^{(0)}(x,t,\lambda_2^*)\, d_\pm^*(\lambda_2)\, \tilde{\mathbf{c}}.$$

Formulas (2.10.14) and (2.10.16) imply that the two-soliton solution to the Landau–Lifshitz equation reduces to a one-soliton one in the limit (2.10.15). In other words, in the coordinate system moving with the second soliton's velocity v_2, formulas (2.10.9) govern the two-soliton magnetization distribution (2.10.14) in the asymptotics as $t \to \pm\infty$. In the process, we should perform the formal replacement

$$v_1 \to v_2, \qquad u_1 \to u_2$$

and change the parameters x_{01} and φ_{01}. It is worth noting that the asymptotics differ from each other only in a shift of the second soliton's gravity center coordinate and in a change in the initial phase of its precession:

$$\Delta x_{02} = -\frac{2}{\operatorname{Im}\lambda_2} \ln\left|\frac{\lambda_2-\lambda_1^*}{\lambda_2-\lambda_1}\right|,$$
$$\Delta\varphi_{02} = -2\left(\arg\frac{\lambda_2-\lambda_1^*}{\lambda_2-\lambda_1} + 2\arg\lambda_1\right). \tag{2.10.17}$$

The above recursive algorithm enables us to conduct a similar treatment, beginning with a one-soliton solution tied with the Riemann problem with the zero of the function $G_+^{(1)}(\lambda)$ at the point $\lambda = \lambda_2$ ($\operatorname{Im}\lambda_2 > 0$) and adding one more zero at the point $\lambda = \lambda_1$ ($\operatorname{Im}\lambda_1 > 0$). By the unique solvability of the Riemann problem we again arrive at the same two-soliton excitation of an isotropic ferromagnet. Its new notation is useful for studying the asymptotics of the two-soliton state in another limiting case

$$x - v_1 t = \text{const}, \qquad t \to \pm\infty. \tag{2.10.18}$$

Repeating the above calculation, it is easy to show that the two-soliton solution reduces to the one-soliton solution for $v_1 > v_2$ in the limit (2.10.18). The asymptotic states as $t \to +\infty$ and $t \to -\infty$ differ only in a shift of the first soliton's gravity center coordinate and a change in the initial phase of its internal precession around the direction $(0, 0, 1)$:

$$\Delta x_{01} = \frac{2}{\operatorname{Im}\lambda_1}\ln\left|\frac{\lambda_1 - \lambda_2^*}{\lambda_1 - \lambda_2}\right|,$$
$$\Delta\varphi_{01} = 2\left(\arg\frac{\lambda_1 - \lambda_2^*}{\lambda_1 - \lambda_2} + 2\arg\lambda_2\right). \tag{2.10.19}$$

The findings secured can be illustrated in terms of the general theory of scattering. Recall that against the linear theory, the one-soliton solution (2.10.9) matches a particle-like formation rather than a wave packet. A soliton is characterized by a velocity v_j, a coordinate of its center of gravity, $x_j(t) = v_j t + x_{0j}$, a projection value of A_j (2.10.10) in the soliton's "center of gravity", and an initial phase φ_{0j}. As $t \to \pm\infty$, the two-soliton solution $\mathbf{S}^{(2)}(x,t)$ describes the free motion of two solitons with parameters $(v_j, x_{0j}^{(\pm)}, A_j, \varphi_{0j}^{(\pm)})$, $j = 1, 2$, $v_1 > v_2$. As $t \to -\infty$, their "centers of gravity" are separated by a large distance of the order of $(v_2 - v_1)\,t$, and the fastest first soliton is located to the left of the second soliton.

Thus, as $t \to -\infty$, the asymptotic state abided by the two-soliton solution reflects the motion of two spatially separated solitons moving toward one another. At finite times, the picture of spatially separated solitons disappears, and the two-soliton solution describes a nontrivial interaction of solitons. However, as $t \to +\infty$, the distance between the solitons increases over time so that spatially separated solitons reemerge. In this case the first soliton that is the fastest of them "overtakes" the second one and happens to be to its right side. The picture depicted is typical of scattering theory, which deals with asymptotic states in terms of free particles. Only the characteristics of the particles and, in the general case, their number change during the scattering process.

This problem covers a unique scattering mechanism. Namely, the number of particles and their amplitudes and velocities remain unchanged during scattering. In this case, only the parameters that vary are the solitons' "centers of gravity" and their internal motion phases.

For n-soliton excitation, the increments of coordinates and phases of solitons after each of their pair collisions are generally depicted additively through two-particle shifts. The latter are computed by formulas (2.10.17) and (2.10.19) for the case $v_1 > v_2$ or for the case $v_2 > v_1$ after the replacement $1 \leftrightarrow 2$. Then the sum of the increments of the coordinate x_{0j} and phase φ_{0j} for a soliton labeled j is taken over all its successive collisions with other solitons [68]. The sequence of the collisions between a single soliton and others has no impact on their mutual interaction. This specific property of scattering, which lowers n-particle scattering to two-particle one, is usually called factorization. When it comes to solitons in the narrow sense of the word, the concept of "soliton" includes the term factorization.

From the point of view of scattering theory, the situation where all soliton velocities v_j are different is essential for an n-soliton solution to interpret. However, even if two or more velocities coincide, the solution itself retains its meaning. In this case, solitons initiate bound states rather than diverge.

The remark made also addresses restrictions on the initial parameters λ_j and c_k for the multisoliton state. The general formula for deducing an n-soliton solution makes it possible to pass to the limits where some of the complex numbers of λ_j become coincident and even go out on the real axis, and the numbers of c_k turn out to be tending to zero. With such a degeneration, the solutions $\mathbf{S}(x,t)$ found are polynomial or exponential-polynomial ones [53, 17, 16].

Due to the algebraic nature of the initial formula for computing $\mathbf{S}(x,t)$, all degenerate solutions satisfy the Landau–Lifshitz equations.

It should be stressed that the Riemann problem is convenient for analyzing asymptotics for any, not only soliton, solutions of integrable models as $t \to \pm\infty$. As $t \to +\infty$, the general Riemann problem proves to be greatly simplified on straight lines $x - vt = \text{const}$, where v is the group velocity of a nonlinear wave. In particular, it reduces to a problem with the λ-independent factorable matrix $G_0(\lambda)$. In this case, we can trace connections with the so-called isomonodromic solutions, self-similar solutions, and Painlevé-type equations [1, 94]. From an applied perspective, the solutions of Painlevé equations can be used as stitching functions in describing various transition regimes in nonlinear condensed matter physics. This extensive theme stays beyond the present book; we can only refer the reader to the original literature [1, 76, 77, 22, 60, 61, 4, 5, 2, 70, 85, 83, 84, 119, 120, 78, 71].

2.11 Radial waves in an isotropic ferromagnet

Throwing stones into water, look at the circles they form; otherwise, such throwing will be empty fun.

Kozma Prutkov

A wide class of integrable nonlinear partial differential equations with variable coefficients can be derived from well-known integrable equations with constant coefficients. Each integrable equation with constant coefficients is described by a $U - V$ pair with a constant spectral parameter. Integrable models with variable coefficients are designed by a regular procedure [56, 57] involving the passage from known $U - V$ pairs with a constant spectral parameter to $U - V$ pairs with a moving spectral parameter. Let us illustrate the idea of the method.

A quasi-one-dimensional equation for an isotropic ferromagnet,

$$\partial_t \mathbf{S} = [\mathbf{S} \times \partial_r^2 \mathbf{S}], \quad \mathbf{S}^2 = 1, \tag{2.11.1}$$

is equivalent to the compatibility condition for the following auxiliary linear system with constant spectral parameter λ:

$$\begin{aligned} \partial_r \chi &= \frac{\lambda}{2\mathrm{i}} S_p \sigma_p \chi, \\ \partial_t \chi &= \frac{1}{2\mathrm{i}} (-\lambda^2 S_p + \lambda [\mathbf{S} \times \partial_r \mathbf{S}]_p) \sigma_p \chi. \end{aligned} \tag{2.11.2}$$

To obtain a new integrable equation with variable coefficients, we assume that representation (2.11.2) contains the spectral parameter λ as a function of the variables r and t. Then the system compatibility condition (2.11.2) leads to the equality

$$\lambda\,\partial_t \mathbf{S} + \partial_t \lambda\,\mathbf{S} + (\partial_r \lambda^2)\,\mathbf{S} - \partial_r \lambda [\mathbf{S} \times \partial_r \mathbf{S}] - \lambda [\mathbf{S} \times \partial_r^2 \mathbf{S}] = 0. \tag{2.11.3}$$

We require the function $\lambda(r, t)$ to satisfy the following consistent system of differential equations:

$$\partial_r \lambda = \frac{\lambda}{r}, \quad \partial_t \lambda = -\frac{2\lambda^2}{r}. \tag{2.11.4}$$

Then

$$\partial_r \lambda^2 = \frac{2\lambda^2}{r}. \tag{2.11.5}$$

System (2.11.4) can be integrated:

$$\lambda = -\alpha(t)\,r, \quad \alpha(t) = -\frac{1}{2\,(t + \kappa)}.$$

Here κ is the integration constant; it plays the role of a new spectral parameter in further analysis.

Using formulas (2.11.4) and (2.11.5) and using eq. (2.11.3), we arrive at a new differential equation with variable coefficients [104]:

$$\partial_t \mathbf{S} = [\mathbf{S} \times \partial_r^2 \mathbf{S}] + \frac{1}{r}[\mathbf{S} \times \partial_r^2 \mathbf{S}], \quad \mathbf{S}^2 = 1. \tag{2.11.6}$$

The new integrable model (2.11.6) was found and studied in [103, 104]. It describes the distributions of magnetization in an isotropic ferromagnet, which depend only both on time t and on the distance $r = \sqrt{x_1^2 + x_2^2}$ from the origin of the Cartesian coordinate system.

For further analysis, it is convenient to pass from the radial variable r to the new coordinate

$$\xi = \frac{r^2}{4} = \frac{x_1^2 + x_2^2}{4}.$$

Then the auxiliary linear system (2.11.2) appears as

$$\begin{aligned} \partial_\xi \chi &= \mathrm{i}\,\alpha(t)\,S_p \sigma_p \chi, \\ \partial_t \chi &= \mathrm{i}\,\xi(2\,\alpha^2(t)\,S_p + \alpha(t)[\mathbf{S} \times \partial_r \mathbf{S}]_p)\sigma_p \chi. \end{aligned} \tag{2.11.7}$$

Due to the specifics of system (2.11.7), the matrix function χ is thought to satisfy the canonical normalization condition for $\kappa = \infty$ ($\alpha(t) = 0$):

$$\chi(\xi, t, \kappa = \infty) = I.$$

Let us analyze the nature of axially symmetric excitations of an isotropic ferromagnet with homogeneous asymptotic behavior at spatial infinity:

$$\mathbf{S} = (0, 0, 1) \quad \text{as} \quad \xi \to \infty.$$

For $\xi = 0$ from the second equation in (2.11.7), we have

$$\partial_t \chi(\xi = 0, t, \kappa) = 0. \tag{2.11.8}$$

Therefore the matrix function $\chi(\xi = 0, t, \kappa)$ is a generating function of the conservation laws.

Using the first equation in (2.11.7), we find the asymptotic expansion of the matrix solution $\chi(\xi, t, \kappa)$ into a series in the parameter $|\kappa| \ll 1$:

$$\chi = \left(I + \sum_{n=1}^{\infty} \chi_n(\xi, t) \kappa^{-n} \right) \exp(\mathrm{i}\, a\, \xi\, \sigma_3).$$

The first expansion terms are

$$\chi_1(\xi, t) = \frac{\mathrm{i}}{2} \int_{\xi}^{+\infty} [S_p(\xi', t)\sigma_p - \sigma_3] \mathrm{d}\,\xi',$$

$$\begin{aligned}\chi_2(\xi, t) = {} & \frac{\mathrm{i}\,t}{2} \int_{\xi}^{+\infty} [\sigma_3 - S_p(\xi', t)\sigma_p] \mathrm{d}\,\xi' \\ & - \frac{1}{4} \int_{\xi}^{+\infty} \int_{\xi'}^{+\infty} (S_k(\xi', t)\, S_p(\xi'', t)\, \sigma_k\, \sigma_p \\ & - [S_p(\xi'', t) + S_p(\xi', t)]\sigma_p \sigma_3 + I) \mathrm{d}\,\xi'\, \mathrm{d}\,\xi''.\end{aligned}$$

According to relation (2.11.8), the values of the coefficients $\chi_n(\xi, t)$ for $\xi = 0$ are time-independent. Hence we deduce the first two integrals of motion:

$$I_1 = 2\mathrm{i}\chi_1(\xi = 0, t) = \int_0^{+\infty} [\sigma_3 - S_p(\xi', t)\sigma_p] \mathrm{d}\,\xi',$$

$$\begin{aligned}I_2 = -2\mathrm{i}\chi_2(\xi = 0, t) = t I_1 + \frac{\mathrm{i}}{2} \int_0^{+\infty} \int_{\xi'}^{+\infty} & (S_k(\xi', t) S_p(\xi'', t)\, \sigma_k \sigma_p \\ & - [S_p(\xi', t) + S_p(\xi'', t)]\sigma_p \sigma_3 + I) \mathrm{d}\,\xi'' \mathrm{d}\,\xi'.\end{aligned}$$

The integral I_1 is responsible for the total magnetic moment of the system.

Let us examine the conservation law

$$J = \frac{1}{2}\,\mathrm{Sp}\,[I_2\sigma_3] = t\int\limits_0^{+\infty}[1 - S_3(\xi,t)]\,\mathrm{d}\,\xi - \frac{1}{2}\int\limits_0^{+\infty}\int\limits_{\xi'}^{+\infty}[\mathbf{S}(\xi',t)\times\mathbf{S}(\xi'',t)]_3\mathrm{d}\,\xi''\mathrm{d}\,\xi'. \quad (2.11.9)$$

The first term in formula (2.11.9) is positive definite and grows linearly with time. Hence the localization radius of the solution must also enlarge with time. Otherwise, the second term in formula (2.11.9) would stay limited, which would imply an unlimited growth of the quantity J as $t \to \infty$. In other words, there are no solutions like "magnetic drops". Any magnetization fluctuation localized at an initial moment of time, dependent only on the distance to the origin, spreads out [104]. It is self-evident that the aforementioned proof holds only when the integrals I_1 and J remain finite as $t \to \infty$.

It is easy to check that the "dressing" method allows for constructing soliton solutions to models (2.11.6) and (2.11.7) with the variable spectral parameter $\alpha(t)$ using almost the same formulas as for models (2.11.1) and (2.11.2) with constant spectral parameter λ. Namely, the known particular solution $\mathbf{S} = (0,0,1)$ contributes to seeking new solutions to model (2.11.6) by the formula

$$S_k\sigma_k = G_+^{\dagger}(\alpha = \infty)\,\sigma_3\,G_+(\alpha = \infty).$$

For multisoliton states, the matrix $G_+(\alpha)$ of the Riemann problem is a meromorphic function of the spectral parameter $\alpha = \alpha(t)$, normalized to the identity matrix at the point $\alpha = 0$ and satisfying the condition

$$G_+^{\dagger}(\alpha^*)\,G_+(\alpha) = I.$$

Formally, the matrix function $G_+(\alpha)$ has the same form as for model (2.11.1). So, for single-soliton excitation, we have

$$G_+(\alpha) = I - \frac{\alpha_1^{-1} - (\alpha_1^*)^{-1}}{\alpha^{-1} - \alpha_1^*}P,$$

where P is a projection matrix with elements

$$P_{\alpha\beta} = \frac{m_\alpha m_\beta^*}{|\mathbf{m}|^2}.$$

The peculiarity of this problem is only that the spectral parameter α and zero $\alpha = \alpha_1$ ($\mathrm{Im}\alpha_1 > 0$) of the matrix $G_+(\alpha)$ depend on time:

$$\alpha_1 = -\frac{1}{2\,(t + \kappa_1)}, \qquad \mathrm{Im}\,\kappa_1 > 0.$$

As before, the vector $\mathbf{m}$ is expressed by the "seeding" solution $\chi_0(\alpha) = \exp(\mathrm{i}\,\alpha\,\xi\,\sigma_3)$ of the auxiliary linear system (2.11.7):

$$\mathbf{m} = \chi_0(a_1^*)\,\mathbf{c} = \exp(\mathrm{i}\,a_1^*\,\xi\,\sigma_3)\,\mathbf{c},$$

where $\mathbf{c} = (c_1, c_2)$ is an arbitrary constant complex vector.

The magnetization distribution corresponding to the one-soliton solution to model (2.11.6) can be represented as

$$\begin{aligned} S_3 &= 1 - \frac{2a^2}{[a^2 + (t+b)^2]\cosh^2\varphi}, \qquad \varphi = \frac{a\xi}{a^2 + (t+b)^2} + \ln\left|\frac{c_1}{c_2}\right|, \\ S_1 + \mathrm{i}\,S_2 &= \frac{2a\exp(\mathrm{i}X)}{[a^2 + (t+b)^2]\cosh\varphi}(-\mathrm{i}\,(t+b) + a\,\mathrm{th}\,\varphi), \\ X &= \frac{\xi(t+b)}{a^2 + (t+b)^2} + \arg\frac{c_1^*}{c_2^*}. \end{aligned} \tag{2.11.10}$$

To derive formula (2.11.10), we have switched from the complex number $\kappa_1 = b + \mathrm{i}\,a$ to two real parameters a and b.

Expression (2.11.10) confirms the conclusion about the spreading out of localized initial distributions of magnetization (see Figure 2.4). The soliton (2.11.10) is a ring-shaped wave whose radius and thickness increase with time along a linear law (at large t) and whose amplitude decays as t^{-1} at a fixed ξ [104].

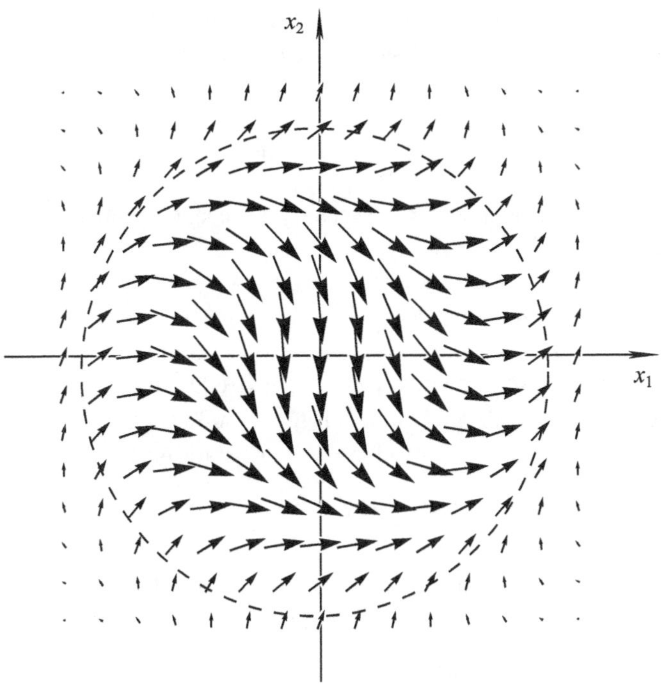

Figure 2.4: Magnetization distribution in the Ox_1x_2-plane for a ring-shaped soliton (2.11.10).

3 Nonlinear dynamics of a ferromagnet with "easy-axis" anisotropy

3.1 Quasi-one-dimensional dynamics of a ferromagnet with allowance for demagnetizing fields and magnetic anisotropy energy

Split off each problem you study into as many parts as you can to solve them easily.

R. Descartes

3.1.1 Basic integrable models

The present section extends the previously proposed ferromagnet model involving only the main exchange interactions to include models that additionally account for weaker magneto-anisotropic (relativistic) and magneto-static interactions. It is important that when subjected to the aforementioned interactions, physically meaningful quasi-one-dimensional models of a ferromagnet also turn out to be completely integrable by the inverse scattering method (or its modifications). Let us focus on this issue.

In the general case, the dynamics of a single-sublattice ferromagnet is described by the Landau–Lifshitz equations

$$\partial_t \mathbf{M} = -\gamma[\mathbf{M} \times \mathbf{H}^{\text{eff}}], \quad \mathbf{H}^{\text{eff}} = -\frac{\delta W}{\delta \mathbf{M}}, \quad \mathbf{M}^2 = M_0^2 = \text{const}, \tag{3.1.1}$$

where $\mathbf{M}$ is the medium's magnetic moment density vector, γ is the magneto-mechanical ratio, and W is the ferromagnet's free energy,

$$W = \int \left[\frac{1}{2} \alpha_{ik} (\partial_i \mathbf{M} \cdot \partial_k \mathbf{M}) + w^{(a)} + w^{(m)} - (\mathbf{M} \cdot \mathbf{H}_0) \right] \mathrm{d}^3\mathbf{x},$$

where α_{ik} are the constants of the exchange interaction, $\mathbf{H}_0 = \mathbf{H}_0(t)$ is an external magnetic field, assumed to be slowly time-dependent, $w^{(a)}$ is the crystal's energy density of the magnetic anisotropy, and $w^{(m)}$ is the density of the magneto-static energy:

$$w^{(m)} = -\frac{1}{2}(\mathbf{M} \cdot \mathbf{H}^{(m)}). \tag{3.1.2}$$

The crystal's intrinsic magnetic field $\mathbf{H}^{(m)}$ generated by the distribution of magnetization $\mathbf{M}(\mathbf{x}, t)$ is determined from the equations of magneto-statics

$$\operatorname{rot} \mathbf{H}^{(m)} = 0, \quad \operatorname{div}(\mathbf{H}^{(m)} + 4\pi\mathbf{M}) = 0. \tag{3.1.3}$$

Equations (3.1.1)–(3.1.3), together with appropriate boundary conditions, completely describe the dynamics of a ferromagnet. In general, solving system (3.1.1)–(3.1.3) faces

https://doi.org/10.1515/9783112235973-003

serious analytical difficulties. The situation is greatly simplified in studying quasi-one-dimensional problems, for example, when exploring the magnetic dynamics of highly elongated ellipsoid-shaped ferromagnetic samples or nonlinear excitations in magnetic structures consisting of parallel chains of magnetic atoms, provided that mutual interactions between the various magnetic chains are neglected and that there is a dominant direction (along the chains) in which the exchange interaction constants can be nonzero.

For one-dimensional magnetization waves propagating along the axis given by a unit vector **e**, the magneto-static equations (3.1.3) have an explicit solution

$$\mathbf{H}^{(m)} = -4\pi(\mathbf{M}\cdot\mathbf{e})\,\mathbf{e}. \tag{3.1.4}$$

In the simplest case, the anisotropy energy density $w^{(a)}$ is a quadratic form in the components of the vector **M**. Jointly with eq. (3.1.4), the energy $w^{(m)}$ also turns out to be a quadratic form in the components of the vector **M**. The expression $w^{(a)} + w^{(m)}$ can be reduced by a suitable choice of axes of the coordinate system to

$$w^{(a)} + w^{(m)} = -\frac{1}{2}(J_1 M_1^2 + J_2 M_2^2 + J_3 M_3^2). \tag{3.1.5}$$

The constants J_1, J_2, and J_3 in formula (3.1.5) are now responsible for not only the magnetic anisotropy energy but also for the contribution of demagnetizing fields.

We introduce the dimensionless variables

$$\mathbf{M} = -M_0\mathbf{S}, \quad \mathbf{H}_0 = -M_0\mathbf{H}, \quad \xi' = \alpha^{-1/2}\xi, \quad t' = \gamma M_0 t, \tag{3.1.6}$$

where ξ is a coordinate along the direction of the magnetization wave propagation,

$$\xi = (\mathbf{e}\cdot\mathbf{r}),$$

$\mathbf{M}_0$ is the nominal magnetization of a ferromagnet, and α is an exchange constant.

In view of eqs. (3.1.5) and (3.1.6), eqs. (3.1.1) take the form

$$\partial_{t'}\mathbf{S} = [\mathbf{S}\times\partial^2_{\xi'}\mathbf{S}] + [\mathbf{S}\times J\,\mathbf{S}] + [\mathbf{S}\times\mathbf{H}], \quad \mathbf{S}^2 = 1, \tag{3.1.7}$$

where $J = \mathrm{diag}(J_1, J_2, J_3)$.

The dimensionless variables allow us to rewrite the free energy of a ferromagnet as

$$W = M_0^2\alpha^{1/2}\int\left[\frac{1}{2}(\partial_{\xi'}\mathbf{S})^2 - \frac{1}{2}(\mathbf{S}\cdot J\,\mathbf{S}) - (\mathbf{S}\cdot\mathbf{H})\right]d\xi'. \tag{3.1.8}$$

Further, wherever this will not cause misunderstandings, we will write the variables ξ' and t' without "primes", ξ and t.

Let us itemize specific special cases of quadratic magnetic anisotropy:

1. When $J_1 = J_2 = J_3$, eqs. (3.1.7) are reduced to the model of an isotropic ferromagnet in an external magnetic field $\mathbf{H} = \mathbf{H}(t)$:

$$\partial_t\mathbf{S} = [\mathbf{S}\times\partial^2_{\xi}\mathbf{S}] + [\mathbf{S}\times\mathbf{H}], \quad \mathbf{S}^2 = 1.$$

2. When $J_1 = J_2 \neq J_3$, we examine a ferromagnet with a distinguished effective anisotropy axis $\mathbf{n} = (0, 0, 1)$:

$$\partial_t \mathbf{S} = [\mathbf{S} \times \partial_\xi^2 \mathbf{S}] + \tilde{\beta}(\mathbf{S} \cdot \mathbf{n})[\mathbf{S} \times \mathbf{n}] + [\mathbf{S} \times \mathbf{H}], \quad \mathbf{S}^2 = 1, \tag{3.1.9}$$

where $\tilde{\beta} = J_3 - J_2$. It is easy to check that the complete Landau–Lifshitz equations (3.1.7) can be reduced to the model (3.1.9) only when the direction of propagation of magnetization waves and the direction of the medium axis of the uniaxial crystallographic anisotropy coincide (for $\mathbf{e} = \mathbf{n}$, see (3.5.12)). Next, we construct exact solutions to the simplified model (3.1.9) without imposing the condition $\mathbf{e} = \mathbf{n}$. For $\mathbf{e} \neq \mathbf{n}$, these solutions are relevant only conditional upon the smallness of the magnetostatic interaction energy (3.1.2) as compared to the energy of the medium's uniaxial crystallographic anisotropy.

Interestingly, when the directions of the external magnetic field and the effective anisotropy axis coincide, i. e., when $\mathbf{H} = H(t)\mathbf{n}$, the vector field $\mathbf{S}(\xi, t)$ can be transformed:

$$\begin{aligned} S_1(\xi, t) + \mathrm{i}\, S_2(\xi, t) &= [S_1'(\xi, t) + \mathrm{i}\, S_2'(\xi, t)] e^{-\mathrm{i}\int^t H(t')\,\mathrm{d}t'}, \\ S_3(\xi, t) &= S_3'(\xi, t). \end{aligned} \tag{3.1.10}$$

After that, the external magnetic field is excluded from system (3.1.9):

$$\partial_t \mathbf{S}' = [\mathbf{S}' \times \partial_\xi^2 \mathbf{S}'] + \tilde{\beta}(\mathbf{S}' \cdot \mathbf{n})[\mathbf{S}' \times \mathbf{n}], \quad \mathbf{S}'^2 = 1. \tag{3.1.11}$$

In the absence of an external magnetic field, the uniform magnetization distribution in a ferromagnet corresponds to the energy density

$$w = -\frac{1}{2}\tilde{\beta} M_3^2 + \text{const} = -\frac{1}{2}\tilde{\beta} M_0^2 S_3^2 + \text{const}. \tag{3.1.12}$$

For $\tilde{\beta} > 0$, energy (3.1.12) is minimal if the vector $\mathbf{S}$ is codirectional or antiparallel to the anisotropy axis:

$$\mathbf{S} = (0, 0, \pm 1). \tag{3.1.13}$$

For $\tilde{\beta} < 0$, energy (3.1.12) is minimal when the vector $\mathbf{S}$ lies in a plane perpendicular to the anisotropy axis $\mathbf{n} = (0, 0, 1)$.

Thus the uniaxial magnetic anisotropy implies two physically different situations. A ferromagnet with $\tilde{\beta} > 0$ ($\tilde{\beta} < 0$) is customarily called a ferromagnet with an “easy-axis” (“easy-plane”) anisotropy. The methods for integrating eqs. (3.1.11) are different in each of the two cases of uniaxial anisotropy. The properties of particular solutions to the Landau–Lifshitz equations are also different. Moreover, against isotropic ferromagnets, uniaxial ferromagnets exhibit new nonlinear magnetic formations, topological solitons of the domain wall type. An “easy-axis” anisotropy ferromagnet's domain walls separate regions with two different directions of spontaneous magnetization (3.1.13). It can be put otherwise: domain walls separate two nonequivalent ground states of a magnetic medium with the same energy density. Since the equations for "easy-axis" and "easy-plane" anisotropy ferromagnets correspond to rationally λ-dependent $U-V$ pairs, the methods of their integration represent a natural generalization of the scheme for integrating the equations of an isotropic ferromagnet described in the previous section.

3. When $J_1 < J_2 < J_3$ and $\mathbf{H} = 0$, the quasi-one-dimensional model of a ferromagnet with biaxial magnetic anisotropy (3.1.7) is also completely integrable. However, it fundamentally differs from the models of an isotropic ferromagnet and a uniaxial magnetic anisotropy ferromagnet. The point is that under the condition $J_1 < J_2 < J_3$ and $\mathbf{H} = 0$, the generalized Lax pair for eqs. (3.1.8) becomes meromorphic on a Riemann surface topologically equivalent to a torus rather than in the complex plane of the spectral parameter λ. The inverse scattering method and the "dressing" procedure in such a situation require special discussion.

The problems considered below suggest that the direction of propagation of the magnetization wave coincides either with the axis Ox_1 of the Cartesian coordinate system, that is, $\xi = x$ (a general case of biaxial anisotropy and an "easy-axis" anisotropy ferromagnet), or with the direction of the anisotropy axis, that is, $\xi = z$ (an "easy-plane" anisotropy ferromagnet).

3.1.2 Zero-curvature representations

Zero-curvature representations for quasi-one-dimensional Landau–Lifshitz models with quadratic magnetic anisotropy were first found by Borovik and Robuk [9] and Sklyanin [29]. The latter in his works showed that in the presence of biaxial magnetic anisotropy, the Landau–Lifshitz equations

$$\partial_t \mathbf{S} = [\mathbf{S} \times \partial_\xi^2 \mathbf{S}] + [\mathbf{S} \times J\mathbf{S}], \quad \mathbf{S}^2 = 1, \quad J = \operatorname{diag}(J_1, J_2, J_3) \tag{3.1.14}$$

are equivalent to the commutation relation

$$\begin{aligned} &\partial_t U - \partial_\xi V + [U, V] = 0; \quad U(z, t, \lambda) = -\mathrm{i} \sum_{\alpha=1}^{3} w_\alpha(\lambda)\, S_\alpha \sigma_\alpha, \\ &V(z, t, \lambda) = -\mathrm{i} \sum_{\alpha=1}^{3} \{ w_\alpha(\lambda)\, [\mathbf{S} \times \partial_\xi \mathbf{S}]_\alpha + 2\, a_\alpha(\lambda)\, S_\alpha \} \sigma_\alpha, \end{aligned} \tag{3.1.15}$$

where σ_α are Pauli matrices, and $a_1(\lambda) = -w_2(\lambda)\, w_3(\lambda)$ (and cyclic permutation of indices 1, 2, and 3 for other coefficients $a_\alpha(\lambda)$). The functions $w_\alpha(\lambda)$ satisfy the quadratic relations

$$w_\alpha^2(\lambda) - w_\beta^2(\lambda) = -\frac{1}{4}(J_\alpha - J_\beta), \quad \alpha, \beta = 1, 2, 3. \tag{3.1.16}$$

The complex spectral parameter λ should be chosen so that eqs. (3.1.16) are uniformized by the meromorphic functions $w_\alpha(\lambda)$. Representation (3.1.15) is, in a certain sense, universal. Exploiting it, we can build all the above models of the quasi-one-dimensional dynamics of a ferromagnet through various limit transitions.

When $J_1 = J_2 = J_3$, formulas (3.1.14) boil down to the commutation representation of the equations of an isotropic ferromagnet ($w_1 = w_2 = w_3 = \lambda/2$).

Assume that $J_1 = J_2$ and $J_3 - J_1 = \beta^2 > 0$. Then in the case of “easy axis” anisotropy, the Landau–Lifshitz equations (3.1.14) can be written as ($\xi = x$)

$$\partial_t \mathbf{S} = \left[\mathbf{S} \times \partial_x^2 \mathbf{S}\right] + \beta^2 (\mathbf{n} \cdot \mathbf{S})[\mathbf{S} \times \mathbf{n}], \quad \mathbf{S}^2 = 1, \tag{3.1.17}$$

where $\mathbf{n} = (0, 0, 1)$. In this context, the formulas

$$w_1 = w_2 = \frac{\beta}{4}\left(\lambda + \lambda^{-1}\right), \quad w_3 = \frac{\beta}{4}\left(\lambda - \lambda^{-1}\right) \tag{3.1.18}$$

give a possible parameterization of relations (3.1.16).

For an “easy-plane” anisotropy ferromagnet, we have $J_1 = J_2$ and $J_1 - J_3 = \rho^2 > 0$, and the Landau–Lifshitz equations (3.1.14) are also simplified ($\xi = z$):

$$\partial_t \mathbf{S} = \left[\mathbf{S} \times \partial_z^2 \mathbf{S}\right] - \rho^2 (\mathbf{n} \cdot \mathbf{S})[\mathbf{S} \times \mathbf{n}], \quad \mathbf{S}^2 = 1. \tag{3.1.19}$$

The equalities

$$w_1 = w_2 = \frac{\rho\lambda}{\lambda^2 - 1}, \quad w_3(\lambda) = \frac{\rho(\lambda^2 + 1)}{2(\lambda^2 - 1)} \tag{3.1.20}$$

are the essence of the possible parameterization of the functions $w_\alpha(\lambda)$.

In general, the quadratic relations (3.1.16) define a complex elliptic curve $w_\alpha = w_\alpha(\lambda)$ in the space with coordinates (w_1, w_2, w_3). Under the assumption $J_3 > J_2 > J_1$, the functions $w_\alpha(\lambda)$ admit the uniformization

$$w_1(\lambda) = \frac{\rho}{\mathrm{sn}(\lambda, k)}, \quad w_2(\lambda) = \rho\,\frac{\mathrm{dn}(\lambda, k)}{\mathrm{sn}(\lambda, k)}, \quad w_3(\lambda) = \rho\,\frac{\mathrm{cn}(\lambda, k)}{\mathrm{sn}(\lambda, k)}. \tag{3.1.21}$$

Here $\mathrm{sn}(\lambda, k)$, $\mathrm{cn}(\lambda, k)$, and $\mathrm{dn}(\lambda, k)$ are the Jacobi elliptic functions with modulus $k = [(J_2 - J_1)/(J_3 - J_1)]^{1/2}$ and $2\rho = (J_3 - J_1)^{1/2}$.

In contrast to the previous cases, the functions $w_\alpha(\lambda)$ turn out to be doubly periodic regarding the complex parameter λ. In the fundamental parallelogram of their periodicity on the λ-plane, they are analytic functions of the parameter λ everywhere except for a finite number of poles.

All the above models tested for a ferromagnet's nonlinear dynamics admit representations of zero curvature with 2×2 matrices $U(\xi, t, \lambda)$ and $V(\xi, t, \lambda)$. Their expansion in terms of the Pauli matrices σ_α ($\alpha = 1, 2, 3$) is natural when theoretically describing the symmetry of the main exchange interactions. Concurrently, taking into account the magnetic anisotropy, in contrast to the $U-V$ pair for an isotropic ferromagnet, representation (3.1.15) contains nonequivalently different components of the vector $\mathbf{S}$.

3.2 Jost solutions, reductions for a ferromagnet with "easy-axis" anisotropy and equilibrium values of magnetization at infinity

Ne quid nimis...

(Latin) Nothing in excess

The inverse scattering method-based procedure for integrating the equations of nonlinear dynamics of a ferromagnet with "easy-axis" anisotropy was first proposed in [8, 7]. However, these papers use a $U-V$ pair more complicated than representations (3.1.15) and (3.1.18). In addition, [8, 7] do not consider the dynamics of a ferromagnet with a domain wall. The present section explains the procedure for integrating the equations of an "easy-axis" anisotropy ferromagnet, which is underlain by the inverse scattering method and the representation of zero curvature (3.1.15) and (3.1.18).

Suppose that the order parameter $\mathbf{S}(x, t)$ tends rather quickly to the equilibrium values as $|x| \to \infty$:

$$\begin{aligned} &\mathbf{S}(x, t) \to (0, 0, 1) \quad \text{as} \quad x \to +\infty, \\ &\mathbf{S}(x, t) \to (0, 0, \varepsilon) \quad \text{as} \quad x \to -\infty, \end{aligned} \tag{3.2.1}$$

where $\varepsilon = \pm 1$. The quantity ε is equal to minus unity only if there is one uncompensated domain wall in the medium.

Let us discuss a spectral problem for the auxiliary linear system

$$\partial_x \varphi = U(\lambda)\varphi, \tag{3.2.2}$$

where the matrix $U(\lambda)$ is defined by formulas (3.1.15) and (3.1.18).

Due to conditions (3.2.1), system (3.2.2) has two types of asymptotic solutions:

$$\begin{aligned} \chi_0(x,\lambda) &= \exp\left[\frac{\varepsilon}{\mathrm{i}} w_3(\lambda)\, \sigma_3 x\right] \quad \text{as} \quad x \to -\infty, \\ \varphi_0(x,\lambda) &= \exp\left[\frac{1}{\mathrm{i}} w_3(\lambda)\, \sigma_3 x\right] \quad \text{as} \quad x \to +\infty. \end{aligned} \tag{3.2.3}$$

We fix two sets of basic solutions $\varphi_\pm(x,t,\lambda)$ of system (3.2.2) (Jost solutions) by choosing the following asymptotic conditions as $|x| \to \infty$:

$$\begin{aligned} \varphi_-(x,t,\lambda) &\to \chi_0(x,\lambda) \quad \text{as} \quad x \to -\infty, \\ \varphi_+(x,t,\lambda) &\to \varphi_0(x,\lambda) \quad \text{as} \quad x \to +\infty. \end{aligned} \tag{3.2.4}$$

The evolution of the functions $\varphi_\pm$ is described by the equations

$$\begin{aligned}\partial_t\varphi_+ &= V(\lambda)\,\varphi_+ - \varphi_+ 2\mathrm{i}w_1^2(\lambda)\,\sigma_3,\\ \partial_t\varphi_- &= V(\lambda)\,\varphi_- - \varphi_- 2\mathrm{i}w_1^2(\lambda)\,\varepsilon\sigma_3,\end{aligned} \tag{3.2.5}$$

where the matrix $V(\lambda)$ is defined in (3.1.15) and (3.1.18).

The consistency condition

$$\partial_t U - \partial_x V + [U, V] = 0 \tag{3.2.6}$$

for system (3.2.2),(3.2.5) is equivalent to the Landau–Lifshitz equations for an “easy-axis” anisotropy ferromagnet:

$$\partial_t \mathbf{S} = [\mathbf{S}\times\partial_x^2\mathbf{S}] + \beta^2(\mathbf{n}\cdot\mathbf{S})[\mathbf{S}\times\mathbf{n}], \quad \mathbf{S}^2 = 1, \tag{3.2.7}$$

where $\mathbf{n} = (0,\ 0,\ 1)$ and $\beta^2 > 0$. The last summands on the right-hand sides of eqs. (3.2.5) do not violate the consistency conditions (3.2.6) and are introduced to keep asymptotics (3.2.4) unchanged in time.

Since $\operatorname{Sp} U(\lambda) = 0$, eqs. (3.2.2) imply that $\det\varphi_\pm$ is x-independent. In the limit as $x \to \pm\infty$, we find that

$$\det\varphi_+ = \det\varphi_0 = 1, \quad \det\varphi_- = \det\chi_0 = 1. \tag{3.2.8}$$

It is easy to check that system (3.2.2),(3.2.5) and the choice of conditions (3.2.4) suggest constraints on the Jost solutions. For definiteness, we formulate them only for real values of λ:

$$\varphi_\pm(-\lambda^{-1}) = \sigma_3\varphi_\pm(\lambda)\sigma_3, \tag{3.2.9}$$

$$\varphi_\pm^*(\lambda) = \sigma_2\varphi_\pm(\lambda)\sigma_2. \tag{3.2.10}$$

Further, we see that different columns of the Jost matrices $\varphi_\pm$ turn out to be analytic functions regarding the parameter λ in different half-planes of the complex λ-plane. Reduction (3.2.9) leaves the positions of the columns unchanged; therefore it is preserved under the analytical continuation of the solution columns from the real axis into the appropriate half-planes. In contrast to condition (3.2.9), reduction (3.2.10) ties the solution columns on the real λ-axis. They continue analytically regarding the parameter λ from the real axis into different half-planes of the complex variable λ.

For real values of λ, the solutions of one of the bases are a linear combination of the solutions of the other basis, and therefore

$$\varphi_-(x,t,\lambda) = \varphi_+(x,t,\lambda)\,T(\lambda,t), \tag{3.2.11}$$

where the transition matrix T can depend only on the variables λ and t.

Reduction (3.2.10) is transferred to the transition matrix and specifies its algebraic structure:

$$T(\lambda) = \begin{pmatrix} a(\lambda) & -b^*(\lambda) \\ b(\lambda) & a^*(\lambda) \end{pmatrix}.$$

Due to reduction (3.2.9), we have

$$a(\lambda) = a(-\lambda^{-1}), \quad b(\lambda) = -b(-\lambda^{-1}). \tag{3.2.12}$$

Formulas (3.2.8) and (3.2.11) confirm the unimodularity of the transition matrix:

$$\det T(\lambda) = |a(\lambda)|^2 + |b(\lambda)|^2 = 1.$$

The elements of matrix $T(\lambda, t)$ are part of the new variables for ease of integrating the initial nonlinear equation (3.2.7). Using eq. (3.2.5), we arrive at an equation that determines the dependence of the matrix T on time:

$$\partial_t T = -2\mathrm{i}\, w_1^2(\lambda)(\varepsilon T\sigma_3 - \sigma_3 T). \tag{3.2.13}$$

From eq. (3.2.13) it follows that

$$a(\lambda, t) = a(\lambda) = \text{const}, \quad b(\lambda, t) = b(\lambda, t = 0)\exp[-4\mathrm{i}\, w_1^2(\lambda)\, t] \tag{3.2.14}$$

for $\varepsilon = 1$ and

$$a(\lambda, t) = a(\lambda, t = 0)\exp[4\mathrm{i}\, w_1^2(\lambda)\, t], \quad b(\lambda, t) = b(\lambda) = \text{const} \tag{3.2.15}$$

for $\varepsilon = -1$.

As with an isotropic ferromagnet, the t-independent elements of the transition matrix can be used for constructing an infinite series of integrals of motion for an "easy-axis" anisotropy ferromagnet's model.

To reveal a full set of new dynamic variables for the initial nonlinear model (3.2.7) to integrate, we need to explore the analytical properties of the Jost functions $\varphi_\pm(\lambda)$.

3.3 Direct scattering method

Never set a task you fail to solve.

Berke's rule

An important step in studying the direct scattering method is to construct integral operators to transform the asymptotic solutions $\varphi_0(x, \lambda)$ and $\chi_0(x, \lambda)$ (3.2.3) of system (3.2.2) as $|x| \to \infty$ into its solutions for arbitrary x.

The specificity of the spectral problem (3.2.2) for a uniaxial ferromagnet is that the integral representations of the Jost functions $\varphi_\pm(x,t,\lambda)$ have a more complex form

$$\begin{aligned}\varphi_+(x,t,\lambda) &= R(x,t)\,\varphi_0(x,t)\\ &\quad+\int_x^{+\infty}[\Gamma_+^{(1)}(x,y,t)+w_1(\lambda)\Gamma_+^{(2)}(x,y,t)]\varphi_0(y,\lambda)dy,\\ \varphi_-(x,t,\lambda) &= K(x,t)\chi_0(x,t)\\ &\quad+\int_{-\infty}^{x}[\Gamma_-^{(1)}(x,y,t)+w_1(\lambda)\Gamma_-^{(2)}(x,y,t)]\chi_0(y,\lambda)dy,\end{aligned}\tag{3.3.1}$$

compared to that encountered in the model of an isotropic ferromagnet. Further, where it does not cause misunderstandings, we will omit the dependence of functions on the parameter t.

Let us explain a simple algorithm for finding such integral representations using the example of the function φ_+. The differential equation (3.2.2) for the function φ_+ and the asymptotic condition (3.2.4) are equivalent to the integral equation

$$\begin{aligned}\varphi_+(x,\lambda) &= \varphi_0(x,\lambda)\\ &\quad+\mathrm{i}\int_x^{+\infty}\varphi_0(x-y,\lambda)[w_1(\lambda)(S_1(y)\sigma_1+S_2(y)\sigma_2)\\ &\quad+w_3(\lambda)(S_3(y)-1)\sigma_3]\varphi_+(y,\lambda)dy.\end{aligned}\tag{3.3.2}$$

Assume that the field $\mathbf{S}(x)$ differs slightly from its equilibrium value $\mathbf{S} = (0, 0, 1)$. Then the solution to eq. (3.3.2) should be sought by the method of successive approximations. To unify calculations, it is useful to introduce a parameter δ by replacing

$$\begin{aligned}&S_{1,2}\to\delta S_{1,2},\quad S_3\to\sqrt{1-\delta^2(S_1^2+S_2^2)}=1-\frac{\delta^2}{2}(S_1^2+S_2^2)+\cdots,\\ &\varphi_+=\varphi_0+\delta\varphi_1+\delta^2\varphi_2+\cdots.\end{aligned}$$

The parameter δ serves for a visual grouping of the same-order-smallness terms in eqs. (3.3.2). At the end of the calculations, we set $\delta = 1$. The range of applicability of a series in the powers of δ is determined by the requirement that each subsequent expansion term must be less than the previous one (for $\delta = 1$). Due to the connection $w_1^2 = w_3^2 + \beta^2/4$, when constructing the expansion, all λ-dependent multipliers of the type $w_1^n(\lambda)$ with a degree of n higher than the first one can be reduced to the coefficients $1, w_3^{2k}, w_1 w_3^{2k}$ ($k = 1, 2, \ldots$). After the replacement

$$w_3(\lambda)\,\varphi_0(y,\lambda)=\mathrm{i}\sigma_3\partial_y\varphi_0(y,\lambda),$$

followed by integration by parts, the multipliers $w_3(\lambda)$ can be eliminated from the terms containing the integrals of the function $\varphi_0(y,\lambda)$ over the variable y. As a result, such terms are converted to the form

$$\int_x^{+\infty} [\Gamma_n^{(1)}(x,y) + w_1(\lambda)\Gamma_n^{(2)}(x,y)]\varphi_0(y,\lambda)\mathrm{d}y,$$

where $n = 1, 2, \dots$ is the perturbation theory order. It turns out that in any finite order of the perturbation theory, the nonintegral terms can be written as

$$R_n(x)\,\varphi_0(x,\lambda),$$

where the matrices R_n do not depend on the parameter λ.

These calculations enable us to reveal the general structure of the integral representations (3.3.1) for the Jost solutions $\varphi_\pm(x,\lambda)$. By reductions (3.2.9) and (3.2.10) the matrices $R(x)$, $K(x)$, and $\Gamma_\pm^{(1,2)}(x,y)$ in formulas (3.3.1) must appear as

$$\begin{gathered} R = \begin{pmatrix} r & 0 \\ 0 & r^* \end{pmatrix}, \quad K = \begin{pmatrix} k & 0 \\ 0 & k^* \end{pmatrix}, \\ \Gamma_\pm^{(1)} = \begin{pmatrix} \alpha_\pm & 0 \\ 0 & \alpha_\pm^* \end{pmatrix}, \quad \Gamma_\pm^{(2)} = \begin{pmatrix} 0 & -\beta_\pm^* \\ \beta_\pm & 0 \end{pmatrix}. \end{gathered} \tag{3.3.3}$$

The asymptotic behavior (3.2.4) holds if

$$\begin{aligned} R(x) &\to I \quad \text{as} \quad x \to +\infty, \\ K(x) &\to I \quad \text{as} \quad x \to -\infty. \end{aligned}$$

We insert representation (3.3.1) into system (3.2.2) for a mere rigorous justification of their validity. Next, under the already formed rules, we transform the y-variable integrals of the functions $\varphi_0(y,\lambda)$ and $\chi_0(y,\lambda)$ with the coefficients w_1^2, w_1w_3, w_3. Then we equate the coefficients of the linearly independent functions 1, $w_1(\lambda)$, $w_3(\lambda)$ in nonintegral terms to zero. Tracking the completeness conditions for the functions $\exp[\mathrm{i}w_3(\lambda)x]$

$$\begin{gathered} \int_{-\infty}^{+\infty} \exp[\mathrm{i}w_3(\lambda)x]\mathrm{d}\lambda = \int_{-\infty}^{+\infty} \frac{1}{\lambda^2}\exp[\mathrm{i}w_3(\lambda)x]\mathrm{d}\lambda = \frac{8\pi}{\beta}\delta(x), \\ \int_{-\infty}^{+\infty} \frac{1}{\lambda}\exp[\mathrm{i}w_3(\lambda)x]\mathrm{d}\lambda = 0, \\ \int_{-\infty}^{+\infty} \frac{1}{\lambda}w_1(\lambda)\exp[\mathrm{i}w_3(\lambda)x]\mathrm{d}\lambda = 4\pi\delta(x), \end{gathered} \tag{3.3.4}$$

and considering terms containing integrals of the functions $\varphi_0(y,\lambda)$ and $\chi_0(y,\lambda)$, we compose differential equations to determine the quantities $R(x)$, $K(x)$, and $\Gamma_\pm^{(1,2)}(x,y)$. These equations yield an explicit form of the matrix elements $r(x)$ and $k(x)$:

$$\begin{aligned} r &= \sqrt{\frac{1+S_3(x)}{2}}\exp\left(\frac{\mathrm{i}}{2}\int_x^{+\infty}[1-S_3(y)]\partial_y\varphi(y)\mathrm{d}y\right),\\ k &= \sqrt{\frac{1+\varepsilon S_3(x)}{2}}\exp\left(-\frac{\mathrm{i}}{2}\int_{-\infty}^{x}[1-\varepsilon S_3(y)]\partial_y\varphi(y)\mathrm{d}y\right). \end{aligned} \tag{3.3.5}$$

Here and further,

$$\varphi(x) = \arctan\frac{S_2(x)}{S_1(x)}.$$

The integral operator kernels $\Gamma_\pm^{(1)}(x,y)$ are defined in terms of $\Gamma_\pm^{(2)}(x,y)$. So, for the kernel $\Gamma_-^{(1)}(x,y)$, we have

$$\begin{aligned} \Gamma_-^{(1)}(x,y) = {}& \frac{\mathrm{i}\,(S_1(x)\sigma_1 + S_2(x)\sigma_2)}{1-S_3^2(x)}\\ & \times(\partial_x\Gamma_-^{(2)}(x,y) - \varepsilon\,S_3(x)\partial_y\Gamma_-^{(2)}(x,y)). \end{aligned} \tag{3.3.6}$$

An expression for the function $\Gamma_+^{(1)}(x,y)$ can be deduced from eq. (3.3.6) by formally replacing

$$\varepsilon = 1, \quad \Gamma_-^{(1,2)} \to \Gamma_+^{(1,2)}. \tag{3.3.7}$$

The kernel $\Gamma_-^{(2)}(x,y)$ satisfies the differential equation of the hyperbolic type:

$$\begin{aligned} &\partial_x^2\Gamma_-^{(2)}(x,y) - \partial_y^2\Gamma_-^{(2)}(x,y)\\ &\quad + \left(\frac{S_3(x)\partial_x S_3(x)}{1-S_3^2(x)} + \partial_x\varphi(x)\,\mathrm{i}\sigma_3\right)\partial_x\Gamma_-^{(2)}(x,y)\\ &\quad - \varepsilon\left(\frac{\partial_x S_3(x)}{1-S_3^2(x)} + S_3(x)\partial_x\varphi(x)\,\mathrm{i}\sigma_3\right)\partial_y\Gamma_-^{(2)}(x,y)\\ &\quad + \frac{\beta^2}{4}(1-S_3^2(x))\Gamma_-^{(2)}(x,y) = 0 \end{aligned} \tag{3.3.8}$$

with the boundary condition

$$\Gamma_-^{(2)}(x,x) = -\frac{\mathrm{i}(S_1\sigma_1 + S_2\sigma_2)}{1+\varepsilon S_3}K. \tag{3.3.9}$$

A differential equation for the function $\Gamma_+^{(2)}(x,y)$ differs from eq. (3.3.9) by the formal change (3.3.7). Here the boundary condition can be stated as

$$\Gamma_+^{(2)}(x,x) = \frac{\mathrm{i}\,(S_1\sigma_1 + S_2\sigma_2)}{1+S_3} R. \tag{3.3.10}$$

The relationship between the integral representations (3.3.1) and the differential equations (3.3.6)–(3.3.10) is single-valued. By the theory of characteristics the aforementioned differential equations and boundary conditions for them can be easily replaced by Voltaire integral equations, which are uniquely solvable. Thus the validity of the integral representations (3.3.1) is completely substantiated.

From formulas (3.3.9) and (3.3.10) it follows that the dynamics of the magnetization of an "easy-axis" anisotropy ferromagnet can be theoretically described if we succeeds in calculating the functions $\Gamma_\pm^{(2)}(x,y,t)$.

Further analysis requires the asymptotics of the Jost solutions $\varphi_\pm(x,\lambda)$ near singular points $\lambda = 0$ and $\lambda = \infty$ of eqs. (3.2.2). These asymptotics can be found directly from system (3.2.2). However, in this case the final result is easier to obtain using the substitutions

$$\varphi_0(y,\lambda) = \frac{\mathrm{i}\,\sigma_3}{w_3(\lambda)}\partial_y\varphi_0(y,\lambda), \quad \chi_0(y,\lambda) = \frac{\mathrm{i}\,\varepsilon\sigma_3}{w_3(\lambda)}\partial_y\chi_0(y,\lambda)$$

and integrating by parts in representations (3.3.1).

The asymptotic behavior of the Jost functions as $|\lambda| \to \infty$ has the form

$$\begin{aligned}\varphi_+ &= \left[\left(I + \frac{S_1\sigma_1 + S_2\sigma_2}{1+S_3}\sigma_3\right)R + O\left(\frac{1}{|\lambda|}\right)\right]\varphi_0(x,\lambda),\\ \varphi_- &= \left[\left(I + \frac{S_1\sigma_1 + S_2\sigma_2}{1+\varepsilon S_3}\varepsilon\sigma_3\right)K + O\left(\frac{1}{|\lambda|}\right)\right]\chi_0(x,\lambda).\end{aligned} \tag{3.3.11}$$

The asymptotics of the Jost solutions as $\lambda \to 0$ can be extracted from formulas (3.3.11) and reduction (3.2.9):

$$\begin{aligned}\varphi_+ &= \left[\left(I - \frac{S_1\sigma_1 + S_2\sigma_2}{1+S_3}\sigma_3\right)R + O(\lambda)\right]\varphi_0(x,\lambda),\\ \varphi_- &= \left[\left(I - \frac{S_1\sigma_1 + S_2\sigma_2}{1+\varepsilon S_3}\varepsilon\sigma_3\right)K + O(\lambda)\right]\chi_0(x,\lambda).\end{aligned} \tag{3.3.12}$$

As with in an isotropic ferromagnet's model, we arrive at the following representations:

$$a(\lambda) = \det[\varphi_-^{(1)}(\lambda),\, \varphi_+^{(2)}(\lambda)], \quad b(\lambda) = \det[\varphi_+^{(1)}(\lambda),\, \varphi_-^{(1)}(\lambda)] \tag{3.3.13}$$

for elements of the transition matrix. To study the analytic properties of the functions $a(\lambda)$ and $b(\lambda)$ and their asymptotic behavior in the neighborhood of the singular points $\lambda = 0, \infty$ of the integral representations (3.3.1), we should further exploit representations (3.3.13).

Analytical properties of the Jost solutions and the elements of the transition matrix are different, depending on the value of the parameter $\varepsilon = \pm 1$.

Let us first focus on the case $\varepsilon = +1$. The integral representations (3.3.1) indicate that for $\varepsilon = +1$, the functions $\varphi_-^{(1)}(x,\lambda)$ and $\varphi_+^{(2)}(x,\lambda)$ continue analytically from the real λ-axis to the upper half-plane of the complex variable λ, whereas the functions $\varphi_-^{(2)}(x,\lambda)$ and $\varphi_+^{(1)}(x,\lambda)$ continue analytically from the real λ-axis to the lower half-plane. According to formulas (3.3.13), this means that the function $a(\lambda)$ may be analytically continued into the upper half-plane of the variable λ, whereas the function $b(\lambda)$, generally speaking, possesses no certain analyticity properties.

For real values of the parameter λ, the Jost solutions $\varphi_\pm(x,\lambda)$ exhibit oscillatory behavior at infinity (see (3.2.3) and (3.2.11)). Therefore the real λ-axis corresponds to the continuous spectrum of problem (3.2.2), (3.2.4).

The zeros of the function $a(\lambda)$ resting in the domain $\mathrm{Im}\,\lambda > 0$ coincide with the discrete spectrum of problem (3.2.2),(3.2.4). In other words, these zeros coincide with those values of $\lambda = \lambda_k$ ($\mathrm{Im}, \lambda_k > 0$) for which the solutions to system (3.2.2) tend to zero as $|x| \to \infty$. Indeed, at $a(\lambda = \lambda_k) = \det[\varphi_-^{(1)}(x,\lambda_k), \varphi_+^{(2)}(x,\lambda_k)] = 0$ the solution columns become linearly dependent,

$$\varphi_-^{(1)}(x,\lambda_k) = c_k\, \varphi_+^{(2)}(x,\lambda_k), \tag{3.3.14}$$

and, due to their definition (3.2.4), exponentially disappear as $x \to \pm\infty$.

Further, for simplicity, we assume that the function $a(\lambda)$ has no zeros on the real axis and that its zeros in the domain $\mathrm{Im}\,\lambda > 0$ are simple. Under the formulated conditions, the total number of zeros of the function $a(\lambda)$ is finite.

In contrast to an isotropic ferromagnet's model, this problem contains the additional reduction (3.2.12) $a(\lambda) = a(-\lambda^{-1})$, which leads to the fact that the zeros of the function $a(\lambda)$ enter in pairs. If $\lambda = \lambda_a \neq \mathrm{i}$ is the zero of the function $a(\lambda)$, then $\lambda = -\lambda_a^{-1}$ is also its zero ($\mathrm{Im}\,\lambda_a > 0$). The special case $\lambda = \mathrm{i}$ is excluded. Further, we see that the presence of a single simple zero $\lambda = \mathrm{i}$ of the function $a(\lambda)$ contradicts the dispersion relation. The appearance of multiple zeros, including a zero of even multiplicity at the point $\lambda = \mathrm{i}$, is technically possible. Soliton solutions to the Landau–Lifshitz equations with “easy-axis” anisotropy, associated with multiple zeros of the coefficient $a(\lambda)$, were discussed in [8, 7]. Such solutions are exponential-polynomial rather than exponential ones. They correspond to weakly bound soliton states.

Thus, for $\varepsilon = 1$, the function $a(\lambda)$ has an even number of simple zeros $\lambda = \lambda_n$ ($\lambda_n \neq \mathrm{i}$, $\mathrm{Im}\,\lambda_n > 0$, $n = 1, 2, \ldots, 2N$) to be detailed, when necessary, in accordance with the equality:

$$\{\lambda_n,\ n = 1, 2, \ldots, 2N\} \equiv \{\lambda_a, \lambda_{a+N} = -\lambda_a^{-1},\ a = 1, 2, \ldots, N\}. \tag{3.3.15}$$

The coefficient c_n in formula (3.3.14) is time-dependent only. The equation of its evolution arises from eqs. (3.2.5) and (3.3.14):

$$\partial_t c_n = -4\mathrm{i}\, w_1(\lambda_n) c_n.$$

Hence we find

$$c_n(t) = c_n(t=0)\exp[-4\mathrm{i}\, w_1(\lambda_n)\, t]. \tag{3.3.16}$$

Due to reduction (3.2.9), the coefficients $c_n(t)$ come in pairs:

$$\{c_n(t),\ n = 1, 2, \ldots, 2N\} \equiv \{c_a(t),\ c_{a+N}(t) = -c_a(t), a = 1, 2, \ldots, N\}.$$

Formulas (3.3.11) and (3.3.13) describe the asymptotics of the function $a(\lambda)$ as $\lambda \to \infty$:

$$a(\lambda) = \exp(-\mathrm{i}\beta_0) + O(1/|\lambda|), \quad \beta_0 = \frac{1}{2}\int_{-\infty}^{+\infty} (1 - S_3(y))\partial_y \varphi\, \mathrm{d}y, \tag{3.3.17}$$

whereas the asymptotic behavior of the coefficient $a(\lambda)$ as $\lambda \to 0$ follows from the reduction $a(-\lambda^{-1}) = a(\lambda)$:

$$a(\lambda) = \exp(-\mathrm{i}\beta_0) + O(\lambda). \tag{3.3.18}$$

Since $|a(\lambda)|^2 + |b(\lambda)|^2 = 1$, the coefficient $b(\lambda)$ at the points $\lambda = 0$ and $\lambda = \infty$ vanishes (for $\lambda = 0$, the function $b(\lambda)$ disappears along with all its derivatives).

As with for an isotropic ferromagnet, the analyticity of the function $a(\lambda)$ and relation (3.3.17) allow us to express the coefficient $a(\lambda)$ in terms of its zeros (if any) and the function $b(\lambda)$. Namely, for $\operatorname{Im}\lambda \geq 0$, the dispersion relation takes place:

$$a(\lambda) = \exp(-\mathrm{i}\beta_0)\prod_{a=1}^{N}\frac{(\lambda - \lambda_a)(\lambda + \lambda_a^{-1})}{(\lambda - \lambda_a^*)(\lambda + \lambda_a^{*-1})}$$
$$\times \exp\left(\frac{1}{2\pi \mathrm{i}}\int_{-\infty}^{+\infty}\frac{\ln(1 - |b(\mu)|^2)}{\mu - \lambda}\mathrm{d}\mu\right). \tag{3.3.19}$$

It is easy to verify by direct verification that the dispersion relation (3.3.19) meets the asymptotic behavior (3.3.18) as $\lambda \to 0$ and the reduction $a(\lambda) = a(-\lambda^{-1})$. For this goal, it suffices to use the equalities

$$b(\lambda = 0) = 0, \qquad \int_{-\infty}^{+\infty}\frac{1}{\lambda}\ln(1 - |b(\lambda)|^2)\mathrm{d}\lambda = 0.$$

The second of them is a consequence of symmetry $|b(\lambda)|^2 = |b(-\lambda^{-1})|^2$.

It is worth pointing out that, in contrast to an isotropic ferromagnet's model, the problem at hand includes no normalization condition for the coefficient $a(\lambda)$ at a point $\lambda = 0$ or any other point in the λ-plane. Therefore the phase multiplier $\exp(-\mathrm{i}\beta_0)$ is left undefined in the dispersion relation.

Summing up the preliminary result, we can say that we have moved from the field $\mathbf{S}(x,t)$ to a new set of variables:

$$\mathbf{S}(x,t) \to \{\lambda_n,\ c_n(t),\ b(\lambda,t)\} \equiv \{g\}_{\varepsilon=1}. \qquad (3.3.20)$$

In terms of the generalized variables (3.3.20), the model of an "easy-axis" anisotropy ferromagnet can be easily integrated for the same equilibrium values of the field $\mathbf{S}(x,t)$ at spatial infinity. The next task is to return from variables (3.3.20) to the solution of $\mathbf{S}(x,t)$ to the Landau–Lifshitz equations.

Let us now discuss the analytic properties of the Jost functions when $\varepsilon = -1$. In this case the functions $\varphi_-^{(1)}(\lambda)$ and $\varphi_+^{(1)}(\lambda)$ continue analytically from the real λ-axis to the region $\operatorname{Im}\lambda < 0$, whereas the functions $\varphi_-^{(2)}(\lambda)$ and $\varphi_+^{(2)}(\lambda)$ can be analytically continued from the real λ-axis to the region $\operatorname{Im}\lambda > 0$. Therefore, when $\varepsilon = -1$, in contrast to the problem with $\varepsilon = 1$, the function $b(\lambda)$ can be analytically continued from the real λ-axis into the lower half-plane, and the function $a(\lambda)$ has no specific analyticity properties.

For distributions of the field $\mathbf{S}(x,t)$ with different equilibrium states as $x \to -\infty$ and $x \to +\infty$, the discrete spectrum of problem (3.2.2), (3.2.4) corresponds to the zeros $\lambda = \mu_n$ ($\operatorname{Im}\mu_n < 0$) of the function $b(\lambda)$. At the points $\lambda = \mu_n$, the solution-columns $\varphi_-^{(1)}(x,\mu_n)$ and $\varphi_+^{(1)}(x,\mu_n)$ turn out to be connected:

$$\varphi_-^{(1)}(x,t,\mu_n) = \gamma_n(t)\varphi_+^{(1)}(x,t,\mu_n).$$

The time-dependence of the coefficient $\gamma_n(t)$ is given by

$$\gamma_n(t) = \gamma_n(t=0)\exp\big[4\mathrm{i}\, w_1^2(\mu_n)\, t\big]. \qquad (3.3.21)$$

The reduction $b(\lambda) = -b(-\lambda^{-1})$ means that if $\lambda = \mu_a \neq -\mathrm{i}$ is a zero of the function $b(\lambda)$, then $\lambda = -\mu_a^{-1}$ is also a zero ($\operatorname{Im}\mu_a < 0$). Further, for simplicity, we assume that all zeros of the coefficient $b(\lambda)$ are simple and none of them lies on the real axis. In contrast to the case $\varepsilon = 1$, the case $\varepsilon = -1$ implies the existence of a single simple zero at the point $\lambda = -\mathrm{i}$. Moreover, this zero is necessary to lead to a soliton solution of the domain wall type.

Thus, for $\varepsilon = -1$, the coefficient $b(\lambda)$ has an odd number of simple zeros, which can be conveniently ordered as follows:

$$\{\mu_n,\ n = 0,1,\dots,2N\} \equiv \{\mu_0 = -\mathrm{i},\ \mu_a,\ \mu_{a+N} = -\mu_a^{-1},\ a = 1,2,\dots,N\},$$

where $\mu_a \neq -\mathrm{i}$.

Due to reduction (3.2.9), the coefficients $\gamma_n(t)$ corresponding to these zeros turn out to be dependent:

$$\begin{aligned}&\{\gamma_n(t),\ n = 0,1,2,\dots,N\}\\ &\quad\equiv \{\gamma_0(t),\ \gamma_a(t),\ \gamma_{a+N}(t) = \gamma_a(t),\ a = 1,2,\dots,N\}.\end{aligned}$$

The function $b(\lambda)$ has the asymptotics

$$b(\lambda) = -\exp(\mathrm{i}\alpha_0) + O(1/|\alpha|) \quad \text{as} \quad |\lambda| \to \infty \tag{3.3.22}$$

and

$$b(\lambda) = \exp(\mathrm{i}\alpha_0) + O(\lambda) \quad \text{as} \quad \lambda \to 0,$$

where

$$\alpha_0 = \frac{1}{2}[\varphi(x=\infty) + \varphi(x=-\infty)] - \frac{1}{2}\int_{-\infty}^{+\infty} S_3(y)\partial_y\varphi(y)\,\mathrm{d}y.$$

At the points $\lambda = 0$ and $\lambda = \infty$ the coefficient $a(\lambda)$ vanishes (at the point $\lambda = 0$ the function $a(\lambda)$ is equal to zero together with all its derivatives).

The coefficient $b(\lambda)$ can be expressed in terms of $a(\lambda)$ and zeros μ_n ($\operatorname{Im}\mu_n < 0$). The corresponding dispersion relation has the form

$$b(\lambda) = -\exp(\mathrm{i}\alpha_0)\left(\frac{\lambda+\mathrm{i}}{\lambda-\mathrm{i}}\right)\prod_{\alpha=1}^{N}\frac{(\lambda-\mu_\alpha)(\lambda+\mu_\alpha^{-1})}{(\lambda-\mu_\alpha^*)(\lambda+\mu_\alpha^{*-1})}$$
$$\times\exp\left(-\frac{1}{2\pi\mathrm{i}}\int_{-\infty}^{+\infty}\frac{\ln(1-|a(\mu)|^2)}{\mu-\lambda}\,\mathrm{d}\mu\right), \quad \operatorname{Im}\lambda \le 0.$$

So we have described the mapping

$$\mathbf{S}(x,t) \to \{\mu_n,\ \gamma_n(t),\ a(\lambda,t)\} \equiv \{g\}_{\varepsilon=-1}. \tag{3.3.23}$$

It assigns the transition from the field $\mathbf{S}(x,t)$ to the new dynamic variables $\{g\}_{\varepsilon=-1}$ to integrate an "easy-axis" anisotropy ferromagnet's model with different magnetization equilibrium values as $x \to +\infty$ and $x \to -\infty$.

In the following sections, we will show that mappings (3.3.20) and (3.3.23) are invertible.

3.4 The inverse scattering problem in the case of initial magnetization distributions with asymptotics $\mathbf{S} \to (0, 0, 1)$ as $|x| \to \infty$. Precessing solitons

We must concentrate our mind's eye totally upon the most insignificant and easiest of matters, and dwell on them long enough to acquire the habit of intuiting the truth distinctly and clearly.

R. Descartes

Let $\{g\}_{\varepsilon=1}$ be the scattering data (3.3.20) for obtaining the Gel'fand–Levitan–Marchenko-type equations for "the right end" to restore the field $\mathbf{S}(x,t)$. We only derive the latter assuming that those for the “left end” give equivalent results. Then we need only the first column of the matrix equality (3.2.11),

$$\varphi_-^{(1)} = a\,\varphi_+^{(1)} + b\,\varphi_+^{(2)}, \tag{3.4.1}$$

as well as integral representations of the column-solutions $\varphi_+^{(1)}(x,\lambda)$ and $\varphi_+^{(2)}(x,\lambda)$:

$$\begin{aligned}\varphi_+^{(1)} &= r(x)\begin{pmatrix}1\\0\end{pmatrix}\exp\left[\frac{1}{\mathrm{i}}w_3(\lambda)x\right]\\ &+\int_x^{+\infty}\begin{pmatrix}\alpha_+(x,y)\\ w_1(\lambda)\beta_+(x,y)\end{pmatrix}\exp\left[\frac{1}{\mathrm{i}}w_3(\lambda)y\right]\mathrm{d}y,\end{aligned} \tag{3.4.2}$$

$$\begin{aligned}\varphi_+^{(2)} &= r^*(x)\begin{pmatrix}0\\1\end{pmatrix}\exp\left[-\frac{1}{\mathrm{i}}w_3(\lambda)x\right]\\ &+\int_x^{+\infty}\mathrm{d}y\begin{pmatrix}-w_1(\lambda)\beta_+^*(x,y)\\ \alpha_+^*(x,y)\end{pmatrix}\exp\left[-\frac{1}{\mathrm{i}}w_3(\lambda)y\right]\mathrm{d}y.\end{aligned} \tag{3.4.3}$$

To construct one of the equations of the inverse scattering problem, we multiply the first component of the vector equality (3.4.1) by the function

$$a^{-1}(\lambda)\exp\left[-\frac{1}{\mathrm{i}}w_3(\lambda)z\right] \quad (z > x)$$

and rewrite the resulting formula as follows:

$$\begin{aligned}&\exp\left[-\frac{1}{\mathrm{i}}w_3(\lambda)(z-x)\right]\left(\frac{(\varphi_-^{(1)}(x,\lambda))_1}{a(\lambda)}\exp\left[-\frac{1}{\mathrm{i}}w_3(\lambda)x\right]-r(x)\right)\\ &=\int_x^{+\infty}\alpha_+(x,y)\exp\left[\frac{1}{\mathrm{i}}w_3(\lambda)(y-z)\right]\mathrm{d}y\\ &\quad-\frac{b(\lambda)}{a(\lambda)}w_1(\lambda)\int_x^{+\infty}\beta_+^*(x,y)\exp\left[-\frac{1}{\mathrm{i}}w_3(\lambda)(y+z)\right]\mathrm{d}y.\end{aligned} \tag{3.4.4}$$

The left side of formula (3.4.4) is an analytic function in the domain $\operatorname{Im}\lambda \geq 0$ everywhere except for simple poles at the points $\lambda = \lambda_n$ ($\operatorname{Im}\lambda_n > 0$, $n = 1, 2, \dots, 2N$). By relations (3.3.14) and (3.4.3) we have

$$\begin{aligned}&\operatorname*{res}_{\lambda=\lambda_n}\frac{1}{a(\lambda)}[\varphi_-^{(1)}(x,\lambda)]_1\exp\left[-\frac{1}{\mathrm{i}}w_3(\lambda)z\right]\\ &=\frac{c_n}{a'(\lambda_n)}[\varphi_+^{(2)}(x,\lambda_n)]_1\exp\left[-\frac{1}{\mathrm{i}}w_3(\lambda_n)z\right]\end{aligned}$$

$$= -\frac{c_n w_1(\lambda_n)}{a'(\lambda_n)} \int_x^{+\infty} \beta_+^*(x,y) \exp\left[-\frac{1}{\mathrm{i}} w_3(\lambda_n)(y+z)\right] \mathrm{d}y, \tag{3.4.5}$$

where $a'(\lambda_n) = \partial_\lambda a(\lambda)|_{\lambda=\lambda_n}$.

Bearing in mind the asymptotics of the functions $[\varphi_-^{(1)}(x,\lambda)]_1$ (3.3.11) and $a(\lambda)$ (3.3.17), it is easy to check that

$$\frac{[\varphi_-^{(1)}(x,\lambda)]_1}{a(\lambda)} \exp\left[-\frac{1}{\mathrm{i}} w_3(\lambda) x\right] - r(x) = O\left(\frac{1}{|\lambda|}\right) \tag{3.4.6}$$

as $|\lambda| \to \infty$.

Next, we integrate eq. (3.4.4) over the parameter λ along the real axis. The integral of the left-hand side of the equality converges as $|\lambda| \to \infty$ (see (3.4.6)) and can be evaluated using the Jordan lemma and formula (3.4.5). When integrating the right-hand side, we should recall the completeness relation

$$\int_{-\infty}^{+\infty} \exp\left[\frac{1}{\mathrm{i}} w_3(\lambda)(y-z)\right] \mathrm{d}\lambda = \frac{8\pi}{\beta} \delta(y-z)$$

for the functions $\exp[\mathrm{i} w_3(\lambda) x]$ (3.3.4). As a result, we obtain one of the integral equations of the inverse scattering problem

$$\alpha_+(x,z,t) - \int_x^{+\infty} \beta_+^*(x,y,t) f_1(y+z,\, t)\, \mathrm{d}y = 0, \tag{3.4.7}$$

where

$$f_1(y,t) = \frac{1}{4\pi} \int_{-\infty}^{+\infty} \frac{b(\lambda,t)}{a(\lambda)} w_1^2(\lambda) \exp[\mathrm{i} w_3(\lambda) y] \mathrm{d}\lambda \\ - \mathrm{i} \sum_{\alpha=1}^{N} \frac{c_\alpha(t)}{a'(\lambda_\alpha)\lambda_\alpha} w_1^2(\lambda_\alpha) \exp[\mathrm{i} w_3(\lambda_\alpha) y]. \tag{3.4.8}$$

Here we have simplified expression (3.4.8) for $f_1(y,t)$ using the equalities

$$\frac{c_\alpha}{a'(\lambda_\alpha)\,\lambda_\alpha} = \frac{c_{\alpha+N}}{a'(\lambda_{\alpha+N})\,\lambda_{\alpha+N}},$$
$$w_1(\lambda_{\alpha+N}) = -w_1(\lambda_\alpha), \quad w_3(\lambda_{\alpha+N}) = w_3(\lambda_\alpha)$$

and the symmetry properties (3.2.12) of the coefficients $a(\lambda)$ and $b(\lambda)$.

The procedure for composing another integral equation for the inverse scattering problem slightly differs from that already considered and consists of the following. We multiply the second component of the vector equality (3.4.1) by the function

$$\frac{1}{\lambda\, a(\lambda)} \exp\left[-\frac{1}{\mathrm{i}} w_3(\lambda) z\right] \quad (z > x)$$

and transform the finding into the form ($S_+ = S_1 + \mathrm{i}S_2$)

$$\begin{aligned}
&\frac{1}{\lambda} \exp\left[-\frac{1}{\mathrm{i}} w_3(\lambda)(z-x)\right] \\
&\quad\times\left(\frac{(\varphi_-^{(1)}(x,\lambda))_2}{a(\lambda)} \exp\left[-\frac{1}{\mathrm{i}} w_3(\lambda) x\right] - \frac{S_+(x) r(x)}{(1+S_3(x))} \frac{w_1(\lambda)}{w_3(\lambda)}\right) \\
&\quad= -\frac{S_+(x) r(x)}{(1+S_3(x))} \frac{w_1(\lambda)}{\lambda\, w_3(\lambda)} \exp\left[-\frac{1}{\mathrm{i}} w_3(\lambda)(z-x)\right] \\
&\quad+ \frac{w_1(\lambda)}{\lambda} \int_x^{+\infty} \beta_+(x,y) \exp\left[-\frac{1}{\mathrm{i}} w_3(\lambda)(z-y)\right] \mathrm{d}y \\
&\quad+ \frac{b(\lambda)}{\lambda\, a(\lambda)} \left(r^*(x) \exp\left[-\frac{1}{\mathrm{i}} w_3(\lambda)(x+z)\right]\right. \\
&\quad+ \left.\int_x^{+\infty} \alpha_+^*(x,y) \exp\left[-\frac{1}{\mathrm{i}} w_3(\lambda)(y+z)\right] \mathrm{d}y \right).
\end{aligned} \tag{3.4.9}$$

Note that we have incorporated the same summand into the left-and right-hand sides of the equality

$$-\frac{S_+ r}{(1+S_3)} \frac{w_1(\lambda)}{\lambda\, w_3(\lambda)} \exp\left[-\frac{1}{\mathrm{i}} w_3(\lambda)(z-x)\right]. \tag{3.4.10}$$

At first glance, this summand worsens the equality because additional singularities appear at the points $\lambda = 0, \pm 1$. The purpose of the operation is to regularize divergent integrals. Further calculations will make them completely clear.

Let us integrate relation (3.4.9) over the variable λ along the real axis. Because of the singularities at the points $\lambda = 0, \pm 1$, the integrals must be understood in the sense of the Cauchy principal value.

The left-hand side of eq. (3.4.9) is an analytic function in the domain $\operatorname{Im} \lambda \geq 0$ except for simple poles at the points $\lambda = \lambda_n$ ($\operatorname{Im} \lambda_n > 0$, $n = 1, 2, \ldots, 2N$) and $\lambda = \pm 1$. At the point $\lambda = 0$, it is regular because

$$\left(\frac{(\varphi_-^{(1)}(x,\lambda))_2}{a(\lambda)} \exp\left[-\frac{1}{\mathrm{i}} w_3(\lambda) x\right] - \frac{S_+(x) r(x)}{(1+S_3(x))} \frac{w_1(\lambda)}{w_3(\lambda)}\right)\Bigg|_{\lambda=0} = 0.$$

This is a consequence of formulas (3.3.12) and (3.3.18). Moreover, in view of asymptotics (3.3.11) and (3.3.17), we can easily show that

$$\frac{(\varphi_-^{(1)}(\lambda))_2}{a(\lambda)} \exp\left[-\frac{1}{\mathrm{i}} w_3(\lambda) x\right] - \frac{S_+ r}{(1+S_3)} \frac{w_1(\lambda)}{w_3(\lambda)} = O\left(\frac{1}{|\lambda|}\right)$$

as $|\lambda| \to \infty$. Therefore the integral of the left-hand side of eq. (3.4.9) is well defined. We represent it as an integral along the contour Γ shown in Figure 3.1, minus the integrals along the small semicircles C_1 and C_2 that bypass the poles at the points $\lambda = -1$ and $\lambda = 1$ in the upper half-plane of the complex variable. The integral along the contour Γ can be calculated using Jordan's lemma, whereas the semicircle integrals can be calculated directly. Finally, we arrive at

$$\begin{aligned} \text{V.p.} \int_{-\infty}^{+\infty} \frac{1}{\lambda} \exp\left[-\frac{1}{\mathrm{i}} w_3(\lambda)(z-x)\right] \\ \times \left(\frac{(\varphi_-^{(1)}(x,\lambda))_2}{a(\lambda)} \exp\left[-\frac{1}{\mathrm{i}} w_3(\lambda) x\right] - \frac{S_+(x) r(x)}{(1+S_3(x))} \frac{w_1(\lambda)}{w_3(\lambda)} \right) \mathrm{d}\lambda \\ = 2\pi\mathrm{i} \left(\sum_{n=1}^{2N} \frac{1}{\lambda_n} \frac{(\varphi_-^{(1)}(x,\lambda_n))_2}{a'(\lambda_n)} \exp\left[-\frac{1}{\mathrm{i}} w_3(\lambda_n) z\right] - \frac{S_+(x) r(x)}{1+S_3(x)} \right). \end{aligned} \tag{3.4.11}$$

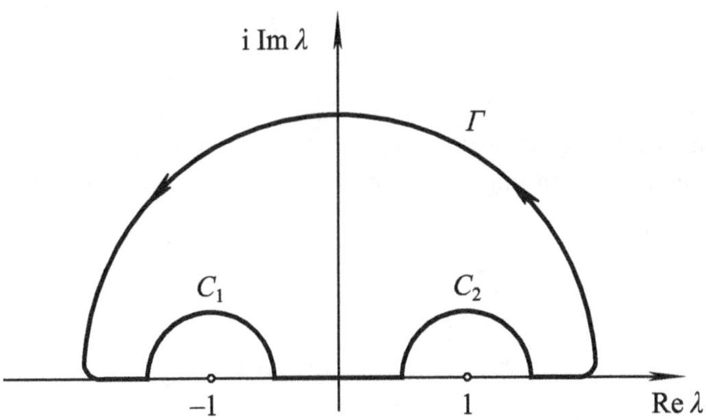

Figure 3.1: Selection of the integration contour.

Further, as before, we modify formula (3.4.11) using connection (3.3.14)

$$(\varphi_-^{(1)}(x,\lambda_n))_2 = c_n (\varphi_+^{(2)}(x,\lambda))_2$$

and the integral representation (3.4.3).

When integrating the right-hand side of eq. (3.4.9) over λ, we keep in mind the relation

$$\int_{-\infty}^{+\infty} \frac{1}{\lambda} w_1(x) \exp[\mathrm{i} w_3(\lambda) x] \, \mathrm{d}\lambda = 4\pi\delta(x).$$

Interestingly, the integral of the added function (3.4.10) for $z > x$ gives the expression

$$-\mathrm{V.p.}\int_{-\infty}^{+\infty}\frac{1}{\lambda}\frac{w_1(\lambda)}{w_3(\lambda)}\exp[iw_3(\lambda)(z-x)]\frac{S_+(x)\,r(x)}{(1+S_3(x))}\,\mathrm{d}\lambda=-2\pi\mathrm{i}\,\frac{S_+(x)\,r(x)}{1+S_3(x)}.$$

The latter fully compensates for the contribution of the last term in formula (3.4.11).

As a result of simple transformations, we obtain one more equation of the inverse scattering problem:

$$\beta_+(x,z,t)+r^*(x,t)f_2^*(x+z,\ t)+\int_x^{+\infty}\alpha_+^*(x,y,t)f_2^*(y+z,\ t)\,\mathrm{d}y=0, \tag{3.4.12}$$

where

$$f_2^*(y,t)=\frac{1}{4\pi}\int_{-\infty}^{+\infty}\frac{b(\lambda,t)}{\lambda\,a(\lambda)}\exp[iw_3(\lambda)y]\,\mathrm{d}\lambda-\mathrm{i}\sum_{\alpha=1}^{N}\frac{c_\alpha(t)}{\lambda_\alpha a'(\lambda_\alpha)}\exp[iw_3(\lambda_\alpha)y]. \tag{3.4.13}$$

Equalities (3.4.7) and (3.4.12) and their complex conjugates form a closed system of integral equations for calculating the functions $\alpha_+(x,y,t)$ and $\beta_+(x,y,t)$. Recollecting formulas (3.3.3), we write this system in the form of matrix equations for calculating $\Gamma_+^{(1,2)}$:

$$\begin{aligned}&R(x,t)\,F_2(x+z,\ t)+\Gamma_+^{(2)}(x,z,t)\,\mathrm{i}\sigma_2+\int_x^{+\infty}\Gamma_+^{(1)}(x,y,t)F_2(y+z,\ t)\mathrm{d}y=0,\\&\Gamma_+^{(1)}(x,z,t)+\int_x^{+\infty}\Gamma_+^{(2)}(x,y,t)\,F_1(y+z,\ t)\mathrm{d}y=0,\end{aligned} \tag{3.4.14}$$

where

$$F_1=\begin{pmatrix}0 & -f_1^*\\ f_1 & 0\end{pmatrix},\quad F_2=\begin{pmatrix}f_2 & 0\\ 0 & f_2^*\end{pmatrix}.$$

The scheme for integrating the Landau–Lifshitz equations for an “easy-axis” anisotropy ferromagnet and the same equilibrium values of the magnetization as $x\to\pm\infty$ is as follows.

1. Having solved system (3.2.2), we find a set of scattering data for a given initial order-parameter distribution $\mathbf{S}(x,t=0)$ at $t=0$:

$$\{g(t=0)\}_{\varepsilon=1}\equiv\{\lambda_n,\ c_n(t=0),\ b(\lambda,t=0)\}.$$

2. Using formulas (3.2.14) and (3.3.16), we understand the evolution of the scattering data. Then the functions $f_1(y,t)$ and $f_2(y,t)$ are defined by formulas (3.4.8) and (3.4.13).
3. We solve linear integral equations of the Gel'fand–Levitan–Marchenko type (3.4.14).
4. Applying formula (3.3.10), we obtain the solution to the Landau–Lifshitz equations at an arbitrary moment of time $t > 0$:

$$\frac{S_+(x,t)}{1+S_3(x,t)} = \frac{\beta_+(x,x,t)}{\mathrm{i}\,r(x,t)}. \tag{3.4.15}$$

Here it is necessary to note an important feature of relation (3.4.15). According to equations (3.4.7) and (3.4.12), the function $\beta_+(x,x,t)$ is proportional to the quantity $r^*(x,t)$. Therefore the right-hand side of the calculation formula (3.4.15) includes the $\mathbf{S}(x,t)$-dependent phase multiplier $r^*(x,t)/r(x,t)$ (see (3.3.5)). The integration scheme depicted, in contrast to a similar scheme for an isotropic ferromagnet, contains a new point: in searching for the field $\mathbf{S}(x,t)$ at $t > 0$, not only algebraic calculations are mandatory to perform but also one more additional integration. In a specific example, we will show that the indicated obstacle causes no great difficulties.

The simplest class of exact solutions of an "easy-axis" anisotropy ferromagnet's model can be compiled when $b(\lambda) = 0$ and the function $a(\lambda)$ has $2N$ simple zeros (3.3.15) (an N-soliton solution). For an N-soliton state, the integral equations (3.4.14) have degenerate kernels F_1 and F_2 and are solved explicitly. As in the case of an isotropic ferromagnet, the N-soliton solution to the equations of an "easy-axis" ferromagnet describes elastic pair collisions of N particle-like nonlinear magnetization waves. Since solitons retain their individuality during interactions, it suffices to analyze the internal structure of a single-soliton excitation.

For a single-soliton state, the functions $f_1(y,t)$ and $f_2^*(y,t)$ are

$$f_1(y,t) = -\mathrm{i}\,w_1^2(\lambda_1)\,\gamma(t)\,\exp[\mathrm{i}w_3(\lambda_1)y],$$
$$f_2^*(y,t) = -\mathrm{i}\,\gamma(t)\,\exp[\mathrm{i}w_3(\lambda_1)y],$$

where

$$\gamma(t) = \gamma_0\exp[-4\mathrm{i}\,w_1^2(\lambda_1)\,t], \quad \gamma_0 = \frac{c_1(t=0)}{\lambda_1 a'(\lambda_1)} = \text{const.}$$

The solution to the equations to the inverse scattering problem (3.4.14) is sought as

$$\alpha_+(x,z,t) = \alpha(x,t)\exp[\mathrm{i}w_3(\lambda_1)z], \quad \beta_+(x,z,t) = \beta(x,t)\exp[\mathrm{i}w_3(\lambda_1)z].$$

Then, after simple calculations, we find

$$\beta(x,t) = \frac{\mathrm{i}\,r^*(x,t)\,\gamma(t)\,\exp[\mathrm{i}w_3(\lambda_1)x]}{1-\left[\frac{|\gamma(t)|w_1^*(\lambda_1)}{w_3(\lambda_1)-w_3^*(\lambda_1)}\right]^2\exp[2\mathrm{i}\,x\,(w_3(\lambda_1)-w_3^*(\lambda_1))]}. \tag{3.4.16}$$

Further analysis is simplified by the parameterization:

$$\mathbf{S} = (\sin\theta\cos\varphi,\ \sin\theta\sin\varphi,\ \cos\theta). \tag{3.4.17}$$

Formulas (3.4.15)–(3.4.17) imply

$$\tan\frac{\theta}{2}\exp(\mathrm{i}\varphi) = \frac{r^*}{r}\gamma\exp[2\mathrm{i}\,w_3 x] \times\left(1-\frac{|\gamma|^2 w_1^{*2}}{(w_3-w_3^*)^2}\exp[2\mathrm{i}(w_3-w_3^*)\,x]\right)^{-1}. \tag{3.4.18}$$

In what follows, $w_{1,3} = w_{1,3}(\lambda_1)$. Hence we explicitly define $\theta(x,t)$:

$$\begin{aligned}\tan^2\frac{\theta}{2} &= \frac{|w_3-w_3^*|^2}{4|w_1|^2\cosh^2\eta - |w_1-w_1^*|^2},\\ \eta &= \mathrm{i}\,(w_3-w_3^*)[x-2(w_3+w_3^*)t] + \ln\left|\frac{\gamma_0 w_1}{w_3-w_3^*}\right|\end{aligned} \tag{3.4.19}$$

and, consequently, the component $S_3(x,t)$ of the vector $\mathbf{S}(x,t)$:

$$S_3 = 1 - \frac{|w_3-w_3^*|^2}{2|w_1|^2}\left(\cosh^2\eta + \frac{|w_3-w_3^*|^2-|w_1-w_1^*|^2}{4|w_1|^2}\right)^{-1}. \tag{3.4.20}$$

Equating the phases of the left- and right-hand sides of eq. (3.4.18), we deduce equations for $\varphi(x,y)$:

$$\begin{aligned}\varphi(x,y) &+ \int_x^{+\infty}(1-S_3(y,t))\,\partial_y\varphi(y,t)\,\mathrm{d}y = (w_3+w_3^*)\,x\\ &- 2(w_1^2+w_1^{*2})\,t + \delta - \arctan\left(\frac{\mathrm{i}(w_1^2-w_1^{*2})\exp 2\eta}{2|w_1|^2+(w_1^2+w_1^{*2})\exp 2\eta}\right),\end{aligned} \tag{3.4.21}$$

where $\delta = \arg\gamma_0$. Differentiation over x by formula (3.4.21) brings about an explicit expression for $\partial_x\varphi$ to be integrated. Then we find

$$\begin{aligned}\varphi &= \frac{w_3+w_3^*}{\mathrm{i}\,(w_3-w_3^*)}\eta + \arctan\left[\frac{\mathrm{i}\,(w_1-w_1^*)}{w_1+w_1^*}\tanh\eta\right]\\ &+ (4|w_3|^2-\beta^2)\,t + \text{const}.\end{aligned} \tag{3.4.22}$$

The one-soliton solution (3.4.19), (3.4.22) describes a solitary magnetization wave that moves as a whole with speed

$$v = 2(w_3 + w_3^*). \tag{3.4.23}$$

In the coordinate system connected to the soliton, the medium's magnetization vector precesses with constant frequency

$$\omega' = 4|w_3|^2 - \beta^2 \tag{3.4.24}$$

around the Ox_3-axis. The region of a sharp change in the magnetization l_0 extends within the following order of magnitude:

$$l_0 \sim \frac{1}{|w_3 - w_3^*|}. \tag{3.4.25}$$

The value of the projection S_3 in the soliton's "center of gravity" is

$$A = \frac{(w_1 + w_1^*)^2 + (w_3 - w_3^*)^2}{(w_1 + w_1^*)^2 - (w_3 - w_3^*)^2}. \tag{3.4.26}$$

Let us pass to the limit as

$$\beta \to 0, \quad |\lambda_1| \to \infty, \quad \beta|\lambda_1| = \text{const}. \tag{3.4.27}$$

In this limit, $w_1 \cong w_3 = (v + iu)/4$, where u and v are real parameters ($u > 0$). As a result, as $\beta \to 0$, solution (3.4.20), (3.4.22) is reduced to the previously found precessing soliton (2.10.9) of the model of an isotropic ferromagnet.

To analyze the features of the solution at finite values of the anisotropy constant β, it is useful to switch from the complex parameter λ_1 to two real ones: $\lambda_1 = \exp(\rho + i\sigma)$ with $-\infty < \rho < \infty$ and $0 < \sigma < \pi$.

Next, we express the soliton's physical characteristics (3.4.23)–(3.4.26) in terms of the parameters ρ and σ:

$$\begin{aligned} v &= 2\beta \sinh\rho \cos\sigma, \quad \omega' = \beta^2(\sinh^2\rho - \cos^2\sigma), \\ l_0 &\backsim (\beta\, \mathrm{ch}\rho \sin\sigma)^{-1}, \quad A = \cos 2\sigma. \end{aligned} \tag{3.4.28}$$

We rewrite the obvious inequality

$$\cosh^2\rho \sin^2\sigma > 0 \tag{3.4.29}$$

in terms of the soliton's velocity and precession frequency

$$1 + \frac{\omega'}{\beta^2} - \frac{v^2}{4\beta^2} > 0. \tag{3.4.30}$$

Recall that for an isotropic ferromagnet, the precession frequency ω' is always positive at the same equilibrium values of the order parameter $\mathbf{S}(x,t)$ at infinity. For an "easy-axis" anisotropy ferromagnet, the precession frequency ω' can vanish, be either positive or negative.

At a frequency $\omega' = 0$ ($\sinh^2\rho = \cos^2\sigma$), the speed of the magnetic soliton lies in finite limits

$$0 < v^2 < 4\beta^2.$$

Negative values of the frequency ω' are also limited:

$$-\beta^2 < \omega' < 0.$$

Infinitely large positive values of the frequency ω' are possible only under the condition $|\sinh\rho| \gg 1$ and, consequently, for $\cosh\rho \gg 1$ (see (3.4.28)). However, at high values of $\cosh\rho$, the width l_0 of the domain of a sharp change in the magnetization becomes small. Therefore the condition for applicability of the long-wavelength approximation may be violated. The values of the quantity $\cosh\rho$ and frequency $\omega' > 0$ must be bounded from above by the condition

$$l_0 \backsim \sqrt{\alpha}\,[\beta\,\cosh\rho\,\sin\sigma]^{-1} \gg a, \tag{3.4.31}$$

where a is the crystal lattice constant. In formula (3.4.31), we have returned to dimensional variables.

Near the soliton existence region's boundaries (3.4.29) and (3.4.30), the parameter $\sigma \approx 0, \pi$. As $\sigma \to 0$ ($\sigma \to \pi$), the deviations of the fields $\mathbf{S}(x,t)$ from the equilibrium value $\mathbf{S} \approx (0, 0, 1)$ diminish, and the soliton lengthens. The magnetic soliton (3.4.19), (3.4.22) becomes like a spin wave with the wave number $k = \beta\,\sinh\rho$ and the dispersion law $\omega = \beta^2 + k^2$, as in the linear theory (in a laboratory coordinate system). For example, when $\sigma \approx 0$, the solution $\mathbf{S}(x,t)$ can be approximately rewritten in the form (see Figure 3.2)

$$S_3 = 1 + O(\sigma^2), \quad S_1 + iS_2 \approx \frac{2\sigma\,\exp(i\beta x\,\sinh\rho - i\beta^2 t\,\cosh^2\rho)}{\cosh[\beta\sigma\,\cosh\rho\,(x - 2\beta t\,\sinh\rho)]}.$$

In the limit as $\sigma \to 0$, the length of the soliton is $(\beta\,\sigma\,\cosh\rho)^{-1} \to \infty$.

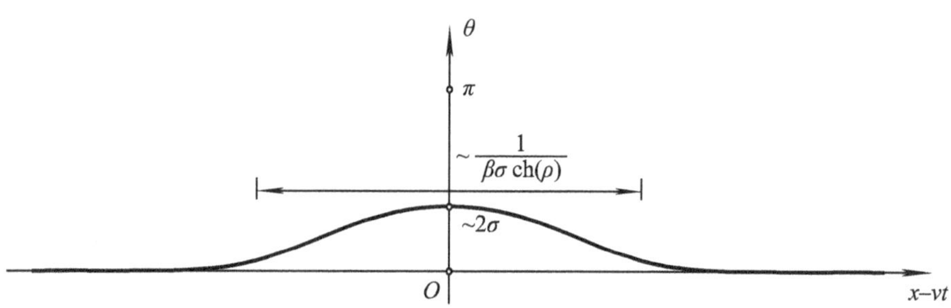

Figure 3.2: A small-amplitude soliton. When $\sigma \ll 1$, a precessing soliton moves with a speed $v = 2\beta\sinh\rho$ and resembles a “cut-off” spin wave with an azimuthal angle $\theta \ll 1$, a wavenumber $k = \beta\sinh\rho$, and a frequency $\omega = \beta^2\cosh^2\rho = \beta^2 + k^2$, as in the linear theory.

Compared to moving solitons, motionless solitons [12] are theoretically described by simpler analytical expressions. The different functional forms of solutions (3.4.19) and (3.4.22) for motionless and moving solitons are a consequence of the lack of Galilean invariance of the original dynamic equations (3.2.7). Formula (3.4.28) states that there are two types of motionless solitons: the former is possible when $\rho = 0, \sigma \neq 0$, and the latter exists when $\rho \neq 0, \sigma = \pi/2$.

When $\rho = 0$, solution (3.4.19), (3.4.22) takes the form

$$S_3 = 1 - \frac{2\tan^2\sigma}{\cosh^2\eta + \tan^2\sigma}, \quad S_1 + iS_2 = \frac{2\tan\sigma\cosh\eta}{\cosh^2\eta + \tan^2\sigma}\exp i\varphi; \qquad (3.4.32)$$
$$\eta = \beta x\sin\sigma, \quad \varphi = -\beta^2 t\cos^2\sigma.$$

The internal precession frequency of soliton (3.4.32) is always negative: $\omega = -\beta^2\cos^2\sigma$. At low precession frequencies ($\sigma \approx \pi/2$), the soliton is a wide region of almost over-magnetized material, where $S_3 \approx \cos 2\sigma \approx -1$. Two areas of width l_0, in which the medium's magnetization abruptly shifts from the equilibrium value $S_3 \cong -1$ to the equilibrium value $S_3 = 1$ divide this region from the rest of the ferromagnet. In dimensional variables, the width

$$l_0 \sim \frac{\sqrt{\alpha}}{\beta}$$

of the region of a sharp change in the magnetization coincides with the thickness of an "easy-axis" ferromagnet's domain wall (see the next section). The boundaries of the overmagnetized region are determined by those values of x where the projection of S_3 vanishes. In dimensional variables, the length Δ of the overmagnetized region is greater than the magnetic length l_0:

$$\Delta \sim \frac{\sqrt{\alpha}}{\beta\sin\sigma}\ln(4\tan^2\sigma) \gg l_0$$

for $\sigma \approx \pi/2$.

Thus at finite values of the parameter β and for $\sigma \approx \pi/2$, the motionless soliton (3.4.32) can be interpreted as a bound state of two domain walls with opposite chiralities [12], namely, as the medium's "magnetization-reversal nucleus". The tendency to the medium's magnetization reversal is accompanied by the precession of the vector **S**. In this case the vector **S** rotates synchronously at every point of the soliton localization area (see Figure 3.3).

The distribution of the field $\mathbf{S}(x,t)$ for another motionless soliton ($\rho \neq 0, \sigma = \pi/2$) has the form

$$S_3 = 1 - \frac{2\coth^2\rho}{\cosh^2\eta + \sinh^{-2}\rho}, \quad S_1 + iS_2 = \frac{2\coth\rho\,\sinh\eta\,\exp i\varphi}{\cosh^2\eta + \sinh^{-2}\rho}; \qquad (3.4.33)$$
$$\eta = \beta x\cosh\rho, \quad \varphi = \beta^2 t\sinh^2\rho.$$

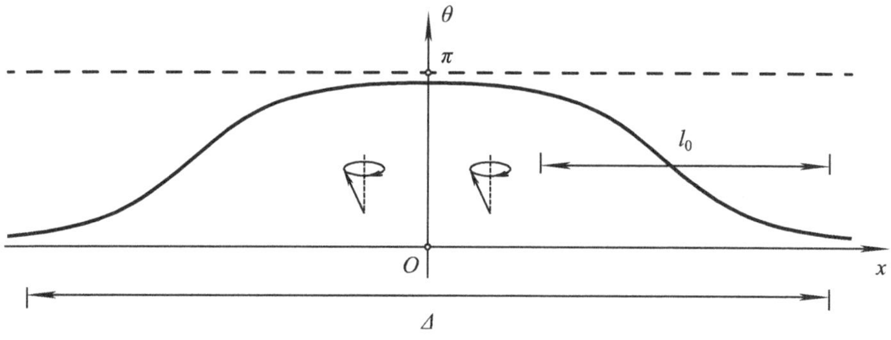

Figure 3.3: Changing the azimuth angle θ for a motionless soliton (3.4.32). The precession of the vector **S** in an overmagnetized region is shown schematically.

The soliton precession frequency (3.4.33) is always positive: $\omega = \beta^2 \sinh^2 \rho$. At low frequencies (for $\rho \approx 0$), the "shapes" of solitons (3.4.33) and (3.4.32) differ little. A main distinguishing feature of soliton (3.4.33) is that although the vector **S** of the localized soliton rotates in phase around the Ox_3-axis, the rotation phases in the left and right "halves of the soliton" (in the regions $x > 0$ and $x < 0$) differ by π (see Figure 3.4).

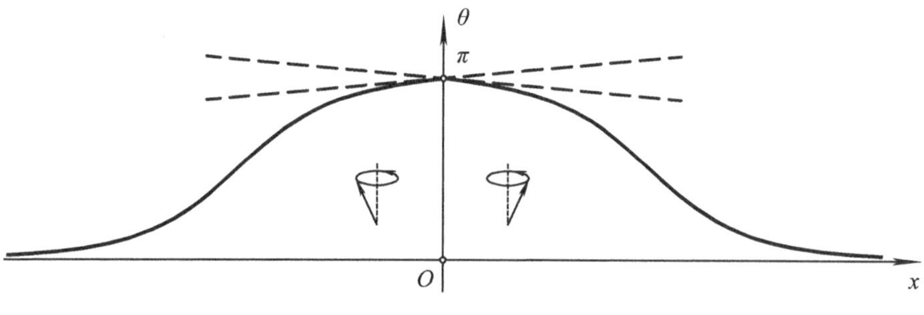

Figure 3.4: Changing the azimuthal angle θ and the nature of the precession of the vector **S** for a motionless soliton (3.4.33).

In general, in the reference frame connected to the soliton, the moving soliton's vector **S** precesses around the Ox_3-axis with unchanged frequency. However, within the region of the overmagnetized material, the initial phase of the precession is spatially inhomogeneous. The latter represents the sum of a linear function and an inflection-type function of the variable η (see (3.4.22)). At $\sigma \approx \pi/2$ the moving soliton resembles a solitary domain with magnetization precession inside the domain. Nevertheless, as the parameter σ changes, the difference between the solitary domain and the moving soliton grows. At $0 < \sigma < \pi$ the projection value at the center of the soliton, equal to $\cos 2\sigma$, may significantly differ from the equilibrium value $S_3 \cong -1$.

Summarizing, we can claim that against an "easy-axis" ferromagnet's soliton, the previously considered precessing soliton of an isotropic ferromagnet has no pronounced localization of magnetization reversals during passing from the equilibrium values of the magnetization at spatial infinity to its internal precession and contains no extended overmagnetized region with a slightly changing magnetization projection $M_3 = -M_0 S_3$.

The aforesaid difference between the solitons can be abolished by continuously deforming the field $\mathbf{S}(x,t)$, which we can observe in the limit (3.4.27).

3.5 The inverse scattering problem in the case of initial magnetization distributions with asymptotic behavior $\mathbf{S} \to (0, 0, +1)$ as $x \to \pm\infty$. Domain walls

Each problem that I solved became a rule, which served afterwards to solve other problems.

R. Descartes

The scheme for constructing the inverse mapping of scattering data $\{g\}_{\varepsilon=-1}$ (3.3.23) into the solution $\mathbf{S}(x,t)$ according to the model of an "easy-axis" ferromagnet with different equilibrium values of magnetization at infinity is largely similar to the procedure outlined in the previous section. The differences are related to other analytic properties of the functions $\varphi_-^{(1)}(\lambda)$, $\varphi_+^{(1,2)}(\lambda)$, $a(\lambda)$, and $b(\lambda)$ in eq. (3.4.1). Let us highlight the key points.

To compose the Gel'fand–Levitan–Marchenko equations, we proceed as follows. We multiply the first component of the vector equality (3.4.1) by the function $\frac{1}{\lambda\, b(\lambda)} \exp[\frac{1}{\mathrm{i}} w_3(\lambda) z]$ $(z > x)$ and rewrite the obtained outcome in the form ($S_- = S_1 - \mathrm{i}S_2$):

$$
\begin{aligned}
&\frac{1}{\lambda}\exp\left[\frac{1}{\mathrm{i}} w_3(\lambda)(z-x)\right] \\
&\quad\times\left(\frac{(\varphi_-^{(1)}(x,\lambda))_1}{b(\lambda)}\exp\left[\frac{1}{\mathrm{i}} w_3(\lambda)\, x\right] + \frac{S_-(x)\, r^*(x)}{(1+S_3(x))}\frac{w_1(\lambda)}{w_3(\lambda)}\right) \\
&\quad= \frac{S_-(x) r^*(x)}{(1+S_3(x))}\frac{w_1(\lambda)}{w_3(\lambda)}\exp\left[\frac{1}{\mathrm{i}} w_3(\lambda)(z-x)\right] \\
&\quad\quad - \frac{w_1(\lambda)}{\lambda}\int_x^{+\infty} \beta_+^*(x,y)\exp\left[\frac{1}{\mathrm{i}} w_3(\lambda)(z-y)\right] \mathrm{d}y \\
&\quad\quad + \frac{a(\lambda)}{\lambda\, b(\lambda)}\left(r(x)\exp\left[\frac{1}{\mathrm{i}} w_3(\lambda)(z+x)\right]\right. \\
&\quad\quad \left. + \int_x^{+\infty} \alpha_+(x,y)\exp\left[\frac{1}{\mathrm{i}} w_3(\lambda)(z+y)\right] \mathrm{d}y\right).
\end{aligned}
\tag{3.5.1}
$$

Next, we multiply the second component of eq. (3.4.1) by the function

$$\frac{1}{b(\lambda)}\exp\left[\frac{1}{\mathrm{i}}w_3(\lambda)\,z\right]\quad(z>x)$$

and modify the result:

$$\begin{aligned}&\exp\left[\frac{1}{\mathrm{i}}w_3(\lambda)(z-x)\right]\left(\frac{(\varphi_-^{(1)}(x,\lambda))_2}{b(\lambda)}\exp\left[\frac{1}{\mathrm{i}}w_3(\lambda)\,x\right]-r^*(x)\right)\\&=\int_x^{+\infty}\alpha_+^*(x,y)\exp\left[\frac{1}{\mathrm{i}}w_3(\lambda)(y-z)\right]\mathrm{d}y\\&+\frac{a(\lambda)}{b(\lambda)}w_1(\lambda)\int_x^{+\infty}\beta_+(x,y)\exp\left[\frac{1}{\mathrm{i}}w_3(\lambda)(z+y)\right]\mathrm{d}y.\end{aligned}\tag{3.5.2}$$

The functions in the left-hand sides of eqs. (3.5.1) and (3.5.2) are analytic in the domain $\operatorname{Im}\lambda\le 0$, except for some number of simple poles, and rather quickly tend to zero as $|\lambda|\to\infty$.

The left side of eq. (3.5.1) has simple poles at the points $\lambda=\mu_n$ ($\operatorname{Im}\mu_n<0$, $n=0,1,2,\dots,2N$) and $\lambda=\pm1$. It is worth underscoring that the additional term

$$\frac{S_-r^*}{(1+S_3)}\frac{w_1}{w_3}\exp\left[\frac{1}{\mathrm{i}}w_3(z-x)\right]$$

implies the regularity of the left-hand side of relation (3.5.1) at the point $\lambda=0$.

Interestingly, the left-hand side of eq. (3.5.2) does not have a pole at $\lambda=-\mathrm{i}$, because

$$(\varphi_-^{(1)}(x,\lambda=-\mathrm{i}))_2=\gamma_0(\varphi_+^{(1)}(x,\lambda=-\mathrm{i}))_2=0.$$

This is a consequence of integral representations for Jost solutions (see (3.4.2)). As a result, the left-hand side of formula (3.5.2) contains simple poles only at the points $\mu_1,\mu_2,\dots,\mu_{2N}$.

Further, we integrate relations (3.5.1) and (3.5.2) over the parameter λ along the real axis. The singularities at the points $\lambda=0,\pm1$ stand for the fact that the integration of relation (3.5.1) should be carried out in the sense of the principal Cauchy value. Calculations similar to those performed in the previous section yield

$$\begin{aligned}&-\beta_+^*(x,z,t)+r(x,t)f_1(x+z,\,t)\\&\quad+\int_x^{+\infty}\alpha_+(x,y,t)f_1(y+z,\,t)\,\mathrm{d}y=0,\\&\alpha_+^*(x,z,t)+\int_x^{+\infty}\beta_+(x,y,t)f_2(y+z,\,t)\,\mathrm{d}y=0,\end{aligned}\tag{3.5.3}$$

where

$$
\begin{aligned}
f_1(y,t) &= \frac{1}{4\pi}\int_{-\infty}^{+\infty} \frac{a(\lambda,t)}{\lambda\, b(\lambda)} \exp\left[\frac{1}{\mathrm{i}} w_3(\lambda) y\right] \mathrm{d}\lambda \\
&\quad + \mathrm{i}\sum_{a=0}^{N} \frac{\gamma_a(t)}{\mu_a b'(\mu_a)} \exp\left[\frac{1}{\mathrm{i}} w_3(\mu_a) y\right], \\
f_2(y,t) &= \frac{1}{4\pi}\int_{-\infty}^{+\infty} \frac{a(\lambda,t)}{\lambda\, b(\lambda)} w_1^2(\lambda) \exp\left[\frac{1}{\mathrm{i}} w_3(\lambda) y\right] \mathrm{d}\lambda \\
&\quad + \mathrm{i}\sum_{a=1}^{N} \frac{\gamma_a(t)}{\mu_a b'(\mu_a)} w_1^2(\mu_a) \exp\left[\frac{1}{\mathrm{i}} w_3(\mu_a) y\right].
\end{aligned}
\tag{3.5.4}
$$

Equalities (3.5.3) and their complex conjugates form a closed set of equations for the inverse scattering problem to compute the functions $\alpha_+(x,z,t)$ and $\beta_+(x,z,t)$ or, equivalently, to calculate the matrix functions $\Gamma_+^{(1,2)}(x,z,t)$ from the integral equations

$$
\begin{aligned}
&\Gamma_+^{(2)}(x,z,t)\sigma_1 + R(x,t)F_1(x+z,\ t) \\
&\quad + \int_x^{+\infty} \mathrm{d}y\, \Gamma_+^{(1)}(x,y,t) F_1(y+z,\ t) = 0, \\
&\Gamma_+^{(1)}(x,z,t) + \int_x^{+\infty} \mathrm{d}y\, \Gamma_+^{(2)}(x,y,t)\, \mathrm{i}\sigma_2 F_2^*(y+z,\ t) = 0.
\end{aligned}
\tag{3.5.5}
$$

The kernels in eqs. (3.5.5) have the shape

$$
F_1 = \begin{pmatrix} f_1 & 0 \\ 0 & f_1^* \end{pmatrix}, \quad F_2 = \begin{pmatrix} f_2 & 0 \\ 0 & f_2^* \end{pmatrix}.
$$

The algorithm for integrating the Landau–Lifshitz equations for an “easy-axis” anisotropy ferromagnet with different equilibrium magnetization values at $x \to \pm\infty$ is as follows:

1. Using a known initial field distribution, from system (3.2.2) we find the scattering data $\{\mu_n,\ \gamma_n(0),\ a(\gamma,0)\}$ for $t = 0$.
2. We calculate the scattering data for $t > 0$ with by formulas (3.2.15) and (3.3.21). Then formulas (3.5.4) define the functions $f_1(y,t)$ and $f_2(y,t)$.
3. We search for solutions to the linear integral equations (3.5.5).
4. Applying relation (3.3.10), we obtain the field $\mathbf{S}(x,t)$ for $t > 0$:

$$
\frac{\mathrm{i}\, S_+(x,t)}{1+S_3(x,t)} = \frac{\beta_+(x,x,t)}{\mathrm{i}\, r(x,t)}.
\tag{3.5.6}
$$

The multisoliton solutions to the model should be sought under the condition $a(\lambda) = 0$ and can be written down explicitly. They are set by the zeros of the function $b(\lambda)$:

$$\{\mu_n\} = \{\mu_0 = -\mathrm{i},\ \mu_\alpha,\ \mu_{\alpha+N} = -\mu_\alpha^{-1},\ \alpha = 1, 2, \ldots, N\}.$$

The paired zeros $\mu_\alpha \neq -\mathrm{i}$ and $-\mu_\alpha^{-1}$ ($\operatorname{Im}\mu_\alpha < 0$) parameterize the already known precessing solitons of types (3.4.19) and (3.4.22).

The zero $\mu_0 = -\mathrm{i}$ gives a new type of magnetic soliton. Let us analyze the internal structure and features of such a soliton. The latter corresponds to the coefficient

$$b(\lambda) = -\exp(\mathrm{i}\alpha_0)\frac{\lambda + \mathrm{i}}{\lambda - \mathrm{i}}$$

and the function $f_2(y, t) = 0$. Moreover, the variable γ_0 is time-independent because $w_1(\lambda = -\mathrm{i}) = 0$ (see (3.3.21)). Solving the equations of the inverse scattering problem (3.5.3) is trivial:

$$\alpha_+(x, z, t) \equiv 0, \quad \beta_+(x, z, t) = \mathrm{i}\gamma_0^* r^*(x, t) \exp\left[-\frac{\beta}{2}(x + z) + \mathrm{i}\alpha_0\right].$$

Having chosen the real and positive variable γ_0, from eq. (3.5.6) we find the distribution of the order parameter $\mathbf{S}(x, t)$:

$$S_3 = \tanh[\beta(x - x_0)], \quad S_1 + \mathrm{i}S_2 = \frac{\exp(\mathrm{i}\alpha_0)}{\cosh[\beta(x - x_0)]}, \tag{3.5.7}$$

where $\beta x_0 = \ln \gamma_0$.

Since the medium's magnetization is $\mathbf{M}(x, t) = -M_0\,\mathbf{S}(x, t)$, the solution (3.5.7) describes a 180-degree domain wall as a spatially localized layer of magnetization parallel to the x_2x_3-plane centered at the point $x_1 \equiv x = x_0$ (see Figure 3.5).

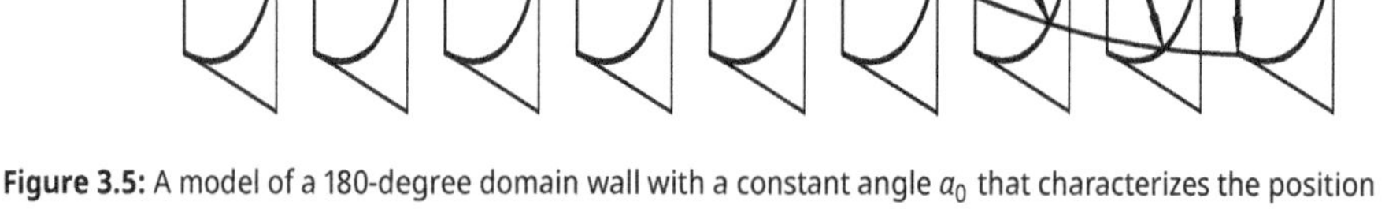

Figure 3.5: A model of a 180-degree domain wall with a constant angle α_0 that characterizes the position of the reversal plane of the wall's magnetic moments. The wall lies in the x_2x_3-plane. The medium's magnetization vector is parallel to the P-plane, making up the angle α_0 with the x_1x_3-plane ($0 \le \alpha_0 < 2\pi$; in the figure the angle α_0 is obtuse). In the P-plane the vector **M** rotates from the position $(0, 0, M_0)$ at $x = -\infty$ to the position $(0, 0, -M_0)$ at $x = +\infty$.

The domain wall (3.5.7) delimits two semiinfinite spaces, with the left one ($x < x_0$) being an upward-magnetization domain (along the positive Ox_3-axis) and the right one ($x > x_0$) involving a downward-magnetization domain.

As noted earlier, the energy of the uniaxial anisotropy (3.1.12) is minimal if there is an abrupt change in the direction of magnetization from orientation $\mathbf{M} = M_0(0, 0, 1)$ to position $\mathbf{M} = M_0(0, 0, -1)$ at the point $x = x_0$ on the Ox_1-axis. On the other hand, the exchange energy is minimal if the orientation of the vector $\mathbf{M}(x)$ alters gradually. Ultimately, the width of the threshold region is a consequence of a compromise between the two competing effects.

According to solution (3.5.7), the width of the transition region is $\backsim \beta^{-1}$. In dimensional variables, this value corresponds to the characteristic magnetic length

$$l_0 = \sqrt{\frac{\alpha}{\tilde{\beta}_0}}, \tag{3.5.8}$$

where $\alpha > 0$ is the inhomogeneous exchange interaction constant, and $\tilde{\beta}_0 = J_3 - J_1 = \beta^2 > 0$ is the easy-axis anisotropy constant.

Typical values of α are $\alpha \backsim (10^4–10^6)\, a^2$, where a is the crystal lattice constant. The anisotropy constant $\tilde{\beta}$ varies over a wide range for different materials. Assuming that $\tilde{\beta} \backsim 10–10^2$, we arrive at the following estimate of the domain wall thickness:

$$l_0 \backsim (10–10^3)a \gg a.$$

The value obtained for the magnetic length turns out to be much larger than that of the crystal lattice constant. Therefore the semiclassical approach used here is well justified. Quantum mechanical calculations of the structure of domain walls are redundant for most magnetic materials.

Note that the deviations of the magnetization from the limiting values in domains are small, even at distances $\backsim 10\, l_0$ from the center of the domain wall. Therefore we should often confine ourselves to evaluating the structure of domain walls in an unbounded medium. In an unbounded medium, interactions between fairly distant flat domain walls are exponentially small.

Meanwhile, in thin magnetic films, demagnetizing fields can lead to strong long-range interactions.

In an unbounded medium, the total energy per unit wall area is determined in dimensional variables by the expression

$$\sigma = \frac{1}{2} \int_{-\infty}^{+\infty} [\alpha(\partial_x \mathbf{M})^2 + \beta^2 (M_0^2 - M_3^2)]\mathrm{d}x, \tag{3.5.9}$$

where $\mathbf{M} = -M_0\, \mathbf{S}$ (here M_0 is the nominal magnetization per unit volume of the medium), and $\tilde{\beta}_0 = \beta^2 > 0$ is the crystallographic magnetic anisotropy constant of a material. Using formulas (3.1.6), (3.5.7), and (3.5.9), we find

$$\sigma = 2M_0^2\alpha^{1/2}\beta = 2M_0^2\sqrt{\alpha\tilde{\beta}_0}. \tag{3.5.10}$$

The contributions of the exchange and anisotropy to energy (3.5.10) are identical.

It is worth recalling that when deducing solution (3.5.7), we have neglected the contribution of the magneto-static interactions (3.1.2) compared to that of the energy of the medium’s crystallographic magnetic anisotropy. Let us discuss the influence of magnetic fields produced by the distribution of magnetization in a medium on the parameters and properties of domain walls in an “easy-axis” anisotropy ferromagnet.

Given the geometry of the problem at hand, the component M_1 of the medium magnetization vector, perpendicular to the x_2x_3-plane of the domain wall, creates magnetic charges on the wall surface with density

$$\rho_{\mathrm{m}} = -4\pi\,\partial_x M_1.$$

Here we have again returned to dimensional variables. Magnetic charges cause the magneto-static field (3.1.4) to emerge and, consequently, the demagnetization energy density (3.1.2). The latter can be written as the additional anisotropy energy density $\tilde{w}^{(\mathrm{a})}$:

$$w^{(\mathrm{m})} = -\frac{1}{2}(\mathbf{M}\cdot\mathbf{H}^{(\mathrm{m})}) = 2\pi M_1^2 \equiv \tilde{w}^{(\mathrm{a})}. \tag{3.5.11}$$

As a result, the equations of magnetization dynamics are modified in comparison with eqs. (3.1.17) and take the form (3.1.14):

$$\partial_t\mathbf{M} = -\gamma\big[\mathbf{M}\times(\alpha\,\partial_x^2\mathbf{M} + \tilde{\beta}_0(\mathbf{n}\cdot\mathbf{M})\,\mathbf{n} - 4\pi(\mathbf{e}\cdot\mathbf{M})\,\mathbf{e})\big], \tag{3.5.12}$$

where $\mathbf{n} = (0,\,0,\,1)$, $\mathbf{e} = (1,\,0,\,0)$, and $\mathbf{M}^2 = M_0^2$. Next, we present a general algorithm for integrating eqs. (3.5.12) by the inverse scattering method and show that their solution corresponding to the domain wall is

$$M_1 + \mathrm{i}M_2 = -\frac{M_0\,\exp\mathrm{i}\alpha_0}{\cosh(\xi\,x + \tau\,t + \delta)}, \quad M_3 = -M_0\,\tanh(\xi\,x + \tau\,t + \delta), \tag{3.5.13}$$

where α_0 and δ are real parameters, and

$$\xi = \sqrt{(\tilde{\beta}_0 + 4\pi\cos^2\alpha_0)/\alpha}, \quad \tau = 2\pi\gamma M_0\sin(2\alpha_0).$$

Formulas (3.5.13) can be verified by a simple check.

According to formulas (3.5.13), the magneto-static interactions provoke the motion of domain walls with velocity

$$v(\alpha_0) = 2\pi\gamma M_0\sin(2\alpha_0)\sqrt{\alpha}(\tilde{\beta}_0 + 4\pi\cos^2\alpha_0)^{-1/2}.$$

The maximum value of the velocity $v(\alpha_0)$ matches the value of $\alpha_0 = \alpha_{\max}$ to be the root of the equation

$$\frac{4\pi}{\beta}\cos^2\alpha_{max} = \sqrt{1+\frac{4\pi}{\tilde{\beta}_0}} - 1.$$

Hence we arrive at the expression for the limiting velocity of a domain boundary in a uniaxial ferromagnet (the Walker limit):

$$v_{max} = \gamma M_0\sqrt{\alpha\tilde{\beta}_0}\left(\sqrt{1+\frac{4\pi}{\tilde{\beta}_0}} - 1\right) \approx 2\pi\gamma M_0\sqrt{\frac{\alpha}{\tilde{\beta}_0}} \quad \text{at} \quad \tilde{\beta}_0 \gg 4\pi.$$

The value of v_{max} for different ferromagnets is about 100–1000 m/s. In reality, a lower value of the limiting velocity is typical for ferromagnetic plates. The point is that due to "twisting", the domain boundary in the plate cannot always be described by the simple dependence $\mathbf{M} = \mathbf{M}(x + vt)$.

Let us rewrite expressions (3.5.13) for the thickness of the domain wall and its energy σ through relations (3.5.8) and (3.5.9):

$$l_0 = \sqrt{\frac{\alpha}{\tilde{\beta}}}, \quad \sigma = \frac{1}{2}\int_{-\infty}^{+\infty}[\alpha(\partial_x\mathbf{M})^2 + \tilde{\beta}(M_0^2 - M_3^2)]\mathrm{d}x = 2M_0^2\sqrt{\alpha\tilde{\beta}}.$$

Here $\tilde{\beta} = \tilde{\beta}_0 + 4\pi\cos^2\alpha_0$ is the effective anisotropy parameter of the medium, including the contribution of magneto-static interactions ($\tilde{\beta}_0$ is the constant of crystallographic magnetic "easy-axis" anisotropy).

Formulas (3.5.11) and (3.5.13) mean that as the value of $\cos^2\alpha_0$ increases, the demagnetization energy, the effective magnetic anisotropy constant, and the domain wall's energy σ also increase, with the diminishing domain wall thickness l_0.

The energy of the domain wall turns out to be minimal when $\cos^2\alpha_0 = 0$, that is, for $\alpha_0 = \pi/2, 3\pi/2$. For such values of the angle α_0, the magnetization vector $\mathbf{M}$ lies in the domain boundary x_2x_3-plane. Therefore no magnetic charges appear on its surface $M_1 = 0$. The vector $\mathbf{M}$ is orthogonal to the Ox_1-axis, and its rotation along the domain wall thickness can compared to a screw rotation. Such domain walls are usually called "Bloch walls". When not exposed to additional perturbations and interactions, they can only be motionless. The symmetry allows two possible directions of a magnetization turn in the plane of the Bloch domain boundary. Therefore there are two types of domain walls with the same energy (see Figure 3.6). The former are right-handed domain walls ($\alpha_0 = \pi/2$):

$$\mathbf{M} = -M_0\left(0, \operatorname{sech}\sqrt{\frac{\tilde{\beta}_0}{\alpha}}x, \tanh\sqrt{\frac{\tilde{\beta}_0}{\alpha}}x\right).$$

The latter meet left-handed ones $\alpha_0 = 3\pi/2$:

$$\mathbf{M} = -M_0\left(0, -\operatorname{sech}\sqrt{\frac{\tilde{\beta}_0}{\alpha}}x, \tanh\sqrt{\frac{\tilde{\beta}_0}{\alpha}}x\right).$$

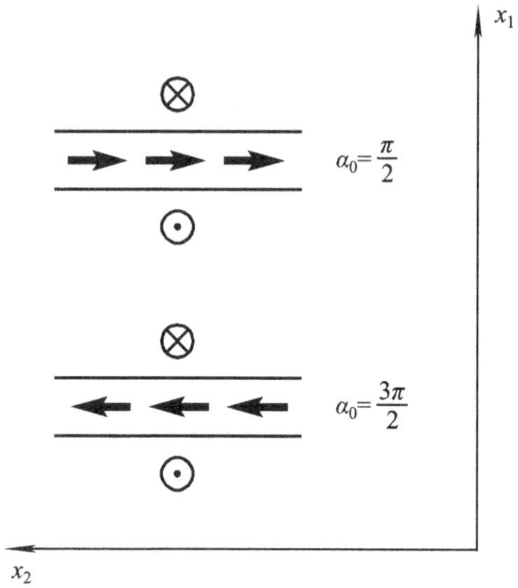

Figure 3.6: Bloch domain walls. The Ox_3-axis side view.

In experiments, domain walls having sections with opposite directions of rotation of the magnetization vector can be implemented under certain conditions. A finite-thickness region passing along the Bloch line separates these sections from each other. The distribution of magnetization gains a vortex-like appearance around the Bloch line. The Bloch line cannot terminate at any point of the domain wall; it either closes into a ring or emerges on the crystal surface together with the domain wall. The Bloch lines are topological defects in the fine structure of domain walls.

The magneto-static energy (3.5.11) is maximum when $\alpha_0 = 0, \pi$. In this case, the domain wall (3.5.13) is also immobile. A turn of the vector **M** along the wall thickness is carried out in the x_1x_3-plane perpendicular to the wall plane (x_2x_3-plane). The values of $\alpha_0 = 0$ and $\alpha_0 = \pi$ correspond to two types of domain walls, which differ in the directions of rotation of the vector **M**. They are referred to as Néel walls (see Figure 3.7).

Here we need to draw attention to an important circumstance. Despite the fact that the Bloch walls minimize the demagnetization energy, the Néel walls or intermediate configurations with $\alpha_0 \neq 0, \pi$ may become energetically more favorable than the Bloch walls. This happens when an external magnetic field directed at an angle to the domain wall plane or additional magneto-anisotropic interactions are taken into account. Under altered external fields, the Bloch domain wall may turn out to be unstable. Then its magnetization spontaneously reorients to another configuration. This is especially crucial for finite-size samples.

In any case, the domain wall (3.5.7) or (3.5.13) separates two different ground states of a ferromagnet. Therefore topological factors enhance the stability of soliton (3.5.7). Un-

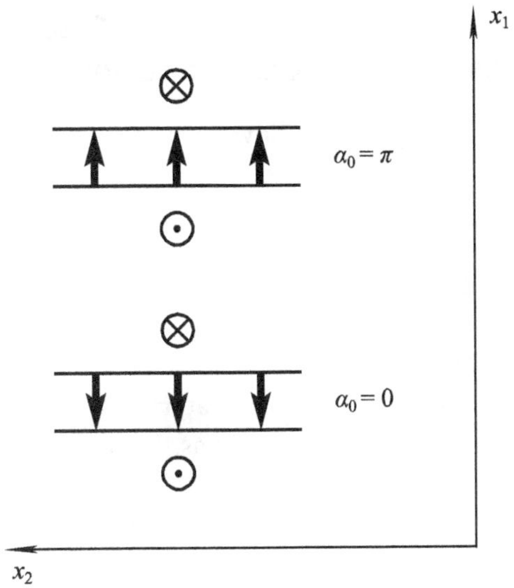

Figure 3.7: Néel domain walls.

der fixed boundary conditions ($\mathbf{S} \to (0, 0, \pm 1)$ as $x \to \pm\infty$), the magnetization distribution corresponding to the domain wall (3.5.7) cannot be boiled down to a uniform ground state by continuously deforming the field $\mathbf{S}(x)$. When the coordinate x runs through the interval from $-\infty$ to $+\infty$, the end of the vector $\mathbf{S}(x)$ on the surface of the sphere $\mathbf{S}^2 = 1$ describes an arc that connects the opposite poles of the sphere, namely, points with radius vectors $\mathbf{S} = (0, 0, 1)$ and $\mathbf{S} = (0, 0, -1)$. Continuously deforming the field $\mathbf{S}(x)$ can never tighten such a trajectory into a point that would satisfy the uniform magnetization distribution of the medium.

These findings show that when not subjected to an external magnetic field or nonlinear spin waves, the domain wall in a boundless easy-axis crystallographic anisotropy ferromagnet is always immobile (see (3.5.7)). If a ferromagnet's internal magnetic field along the Ox_3-axis $\mathbf{H} = (0, 0, H')$ ($H' \neq 0$) is nonzero, then there exist no immobile domain boundaries. The point is that when $H' \neq 0$, the energy densities of domains with the magnetizations $\mathbf{M} = (0, 0, -M_0)$ and $\mathbf{M} = (0, 0, M_0)$ differ by the magnitude of $2M_0H'$. Therefore the domain wall cannot be in equilibrium. A profitable domain with $\mathbf{M} \uparrow\uparrow \mathbf{H}$ expands at the expense of a nonprofitable one, that is, the domain boundary moves. Section 3.7.1 concerns the issue in more detail.

It stands to reason that solution (3.5.7) would be easier to deduce by directly integrating the Landau–Lifshitz equations (3.2.7). However, the advantage of the proposed method over others is that it allows us to trace the evolution of any initial magnetization distributions with asymptotics $\mathbf{S} \to (0, 0, \pm 1)$ as $x \to \pm\infty$. In this case, all calculations are reduced to solving linear problems.

3.6 The “dressing” method for trivial solutions $\mathbf{S} = (0,\ 0,\ \pm 1)$ to the Landau–Lifshitz equations with “easy-axis” anisotropy

The entire method consists in the order and arrangement of things to which the mind’s eye must turn so that we can discover some truth.

R. Descartes

Soliton solutions to the model of an easy-axis ferromagnet with equilibrium values of the field $\mathbf{S}(x, t)$ as $x \to \pm\infty$, constructed in the previous sections, are easier to seek by the “dressing” method for trivial solutions to the model $\mathbf{S} = (0,\ 0,\ \pm 1)$ using the matrix Riemann problem. Since this procedure contains several technical features outlying the isotropic ferromagnet model and repeated in analyzing other problems, the present section examines it in-depth.

Recall that the “dressing” procedure includes an auxiliary linear system not in the form (3.2.2), (3.5.5), but in the form:

$$\partial_x \chi = U(\lambda)\chi, \quad \partial_t \chi = V(\lambda)\chi. \tag{3.6.1}$$

For an “easy-axis” anisotropy ferromagnet’s model, the matrices $U(\lambda)$ and $V(\lambda)$ are determined by formulas (3.1.15) and (3.1.18). In contrast to the model of an isotropic ferromagnet, they do not simultaneously vanish neither at the point $\lambda = 0$ nor at any other point of the λ-plane. This peculiarity leads to the fact that the matrix solutions of system (3.6.1) with given analyticity properties cannot be normalized to an identity matrix for any value of λ. For an easy-axis ferromagnet, the singular points of system (3.6.1) are $\lambda = 0$ and $\lambda = \infty$. In what follows, we fix no numerical values of the matrix solutions to eqs. (3.6.1); we stipulate only their asymptotic properties. Ultimately, the problem of conjugation of analytic functions underlying the “dressing” method is nevertheless reduced to the Riemann problem normalized to an identity matrix at the point $\lambda = \infty$ through a gauge transformation. However, the gauge transformation transfers the initial linear reductions to solutions of system (3.6.1) into nonlinear reductions to the Riemann problem’s matrices. Fortunately, it turns out that even under nonlinear reductions, the Riemann problem with zeros has an explicit solution.

We will be concerned with solutions to the Landau–Lifshitz equations (3.1.17) with the following boundary conditions:

$$\mathbf{S} \to (0,\ 0,\ 1) \quad \text{as} \quad x \to +\infty,$$
$$\mathbf{S} \to (0,\ 0,\ \varepsilon) \quad \text{as} \quad x \to -\infty,$$

where $\varepsilon = \pm 1$. Under these conditions, the auxiliary linear system (3.6.1) has the asymptotic solutions

$$\chi_2^{(0)}(x,t,\lambda) = \exp\left[\frac{1}{\mathrm{i}}(w_3(\lambda)x - 2w_1^2(\lambda)\,t)\sigma_3\right] \quad \text{as} \quad x \to +\infty,; \tag{3.6.2}$$
$$\chi_1^{(0)}(x,t,\lambda) = \exp\left[\frac{\varepsilon}{\mathrm{i}}(w_3(\lambda)x - 2w_1^2(\lambda)\,t)\sigma_3\right] \quad \text{as} \quad x \to -\infty.$$

Next, we fix two sets of basic solutions of system (3.6.1) by the asymptotics

$$\chi_{1,2}(x,t,\lambda) \to \chi_{1,2}^{(0)}(x,t,\lambda) \quad \text{as} \quad x \to \mp\infty.$$

Using the columns of the matrices $\chi_{1,2}(x,t,\lambda)$, we need to build two matrix solutions $\chi_\pm(x,t,\lambda)$ of system (3.6.1). The former $\chi_+(\lambda)$ and the latter $\chi_-(\lambda)$ are analytic functions in the domains $\operatorname{Im}\lambda \geq 0$ and $\operatorname{Im}\lambda \leq 0$, respectively. The problem of conjugation of the analytic functions $\chi_\pm(\lambda)$ on the real λ-axis provides the basis for the “dressing” method. The analysis can be simplified by noting that the solutions $\chi_{1,2}(x,t,\lambda)$ of system (3.6.1) are expressed through the Jost solutions $\varphi_\pm(x,t,\lambda)$ of the related systems (3.2.2) and (3.2.5):

$$\chi_2(\lambda) = \varphi_+(\lambda)\exp[2\mathrm{i}w_1^2(\lambda)\sigma_3\,t], \tag{3.6.3}$$
$$\chi_1(\lambda) = \varphi_-(\lambda)\exp[2\mathrm{i}\varepsilon w_1^2(\lambda)\sigma_3\,t].$$

We have already described the analytic properties of the Jost functions $\varphi_\pm(x,t,\lambda)$. Therefore we formulate all the necessary statements in terms of the Jost solutions. Only at the final stage of calculations, we will pass from the Jost functions to the functions $\chi_\pm(\lambda)$ with the required analyticity properties.

Recall that the Jost solutions on the real λ-axis are related by the matrix equalities

$$\varphi_-(\lambda) = \varphi_+(\lambda)\,T(\lambda), \quad \varphi_+(\lambda) = \varphi_-(\lambda)\,T^{-1}(\lambda), \tag{3.6.4}$$

where

$$T(\lambda) = \begin{pmatrix} a(\lambda) & -b^*(\lambda) \\ b(\lambda) & a^*(\lambda) \end{pmatrix}, \quad T^{-1}(\lambda) = \begin{pmatrix} a^*(\lambda) & b^*(\lambda) \\ -b(\lambda) & a(\lambda) \end{pmatrix},$$
$$|a(\lambda)|^2 + |b(\lambda)|^2 = 1.$$

Equalities (3.6.4) should be detailed:

$$\varphi_-^{(1)} = a\,\varphi_+^{(1)} + b\,\varphi_+^{(2)}, \quad \varphi_-^{(2)} = -b^*\,\varphi_+^{(1)} + a^*\,\varphi_+^{(2)}; \tag{3.6.5}$$
$$\varphi_+^{(1)} = a^*\,\varphi_-^{(1)} - b\,\varphi_-^{(2)}, \quad \varphi_+^{(2)} = b^*\,\varphi_-^{(1)} + a\,\varphi_-^{(2)}.$$

The final form of the functions $\chi_\pm(\lambda)$ with the prescribed analyticity properties depends on the parameter $\varepsilon = \pm 1$.

Let us first discuss the problem with $\varepsilon = -1$. As shown earlier, for $\varepsilon = -1$, the functions $\varphi_+^{(1)}(\lambda)$, $\varphi_-^{(1)}(\lambda)$, and $b(\lambda)$ turn out to be analytic in the lower half-plane of the complex variable λ, whereas the functions $\varphi_+^{(2)}(\lambda)$, $\varphi_-^{(2)}(\lambda)$, and $b^*(\lambda)$ continue analytically from the real λ-axis into the upper half-plane. We define matrix the solutions $\Phi_-(\lambda)$ and $\Phi_+(\lambda)$ of systems (3.2.2) and (3.2.5), which are analytic in the domains $\operatorname{Im}\lambda \le 0$ and $\operatorname{Im}\lambda \ge 0$, respectively:

$$\Phi_- = (\varphi_+^{(1)}, \varphi_-^{(1)}), \quad \Phi_+ = (\varphi_-^{(2)}, \varphi_+^{(2)}). \tag{3.6.6}$$

Exploiting eqs. (3.6.5) and (3.6.6), it is easy to check that the functions $\Phi_\pm(\lambda)$ on the real λ-axis are expressed through the Jost solutions:

$$\begin{aligned} \Phi_-(\lambda) &= \varphi_+(\lambda)\, M_{+-}(\lambda) = \varphi_-(\lambda)\, M_{--}(\lambda)\, \mathrm{i}\sigma_2, \\ \Phi_+(\lambda) &= -\varphi_+(\lambda)\sigma_3\, M_{++}(\lambda) = \varphi_-(\lambda)\, M_{-+}(\lambda)\sigma_1, \end{aligned} \tag{3.6.7}$$

where the matrices M are triangular,

$$M_{+-} = \begin{pmatrix} 1 & a \\ 0 & b \end{pmatrix}, \quad M_{--} = \begin{pmatrix} 1 & -a^* \\ 0 & b \end{pmatrix},$$
$$M_{++} = \begin{pmatrix} b^* & 0 \\ a^* & 1 \end{pmatrix}, \quad M_{-+} = \begin{pmatrix} b^* & 0 \\ a & 1 \end{pmatrix},$$

and satisfy the identity

$$T(\lambda) = -\sigma_3 M_{++}(\lambda)\sigma_1 M_{-+}^{-1}(\lambda) = -M_{+-}(\lambda)\, \mathrm{i}\sigma_2 M_{--}^{-1}(\lambda).$$

Formulas (3.6.7) make it possible to reveal all necessary properties of the functions $\Phi_\pm(\lambda)$ knowing the previous properties of the Jost solutions.

First and foremost, we note that although the matrices $\varphi_\pm(\lambda)$ are unimodular, the matrix solutions $\Phi_\pm(\lambda)$ do not satisfy this property. On the real λ-axis the determinant of the matrix $\Phi_-(\lambda)$ is

$$\det \Phi_-(\lambda) = -\left[\det \Phi_+(\lambda)\right]^* = b(\lambda).$$

Reductions (3.2.9) and (3.2.10), to which the functions $\varphi_\pm(\lambda)$ obey, are transferred to the matrix functions $\Phi_\pm(\lambda)$. On the real λ-axis, we have

$$\Phi_\mp(-\lambda^{-1}) = \pm\sigma_3\Phi_\mp(\lambda), \tag{3.6.8}$$

$$\Phi_\mp^*(\lambda) = \pm\mathrm{i}\,\sigma_2\Phi_\pm(\lambda)\sigma_1. \tag{3.6.9}$$

Reduction (3.6.8) can be continued from the real λ-axis into the analyticity domain of the functions $\Phi_+(\lambda)$ and $\Phi_-(\lambda)$, whereas reduction (3.6.9) holds only on the real λ-axis. Using the identity

$$\Phi^{-1} = \frac{1}{\det\Phi}\sigma_2\Phi^{\mathrm{T}}\sigma_2$$

for arbitrary matrices of 2×2, we rewrite reduction (3.6.9) as

$$[\Phi_{\mp}(\lambda)]^{\dagger} = \pm\sigma_3 \det \Phi_{\pm}(\lambda)\, \Phi_{\pm}^{-1}(\lambda).$$

The asymptotics of the functions $\Phi_{\pm}(\lambda)$ as $|\lambda| \to \infty$ arise from the previously obtained asymptotics for the functions $\varphi_{\pm}(\lambda)$ and $b(\lambda)$, (3.3.11) and (3.3.22):

$$\begin{aligned} \Phi_{-}(\lambda) &= \left[g_{-1} + O\left(\frac{1}{|\lambda|}\right)\right] \exp\left[\frac{1}{\mathrm{i}} w_3(\lambda)\sigma_3 x\right], \\ \Phi_{+}(\lambda) &= \exp(-\mathrm{i}\alpha_0)\left[g_{-1}\sigma_3 + O\left(\frac{1}{|\lambda|}\right)\right] \exp\left[\frac{1}{\mathrm{i}} w_3(\lambda)\sigma_3 x\right], \end{aligned} \tag{3.6.10}$$

where g_{-1} is a λ-independent unitary matrix,

$$\begin{aligned} g_{-1} &= \begin{pmatrix} 1 & -S_{-}(1+S_3)^{-1} \\ S_{+}(1+S_3)^{-1} & 1 \end{pmatrix} \begin{pmatrix} r & 0 \\ 0 & -r^* \exp \mathrm{i}\alpha_0 \end{pmatrix}, \\ g_{-1}^{\dagger} g_{-1} &= I. \end{aligned}$$

Relations (3.6.7) generate the conjugation condition for the analytic functions $\Phi_{\pm}(\lambda)$ on the real λ-axis, which underlies the inverse scattering method and “dressing” method:

$$\Phi_{-}(x,t,\lambda) = -\frac{1}{b^*(\lambda)} \Phi_{+}(x,t,\lambda)\,\sigma_3 \begin{pmatrix} 1 & a(\lambda,t) \\ a^*(\lambda,t) & 1 \end{pmatrix}. \tag{3.6.11}$$

Recall that for $\varepsilon = -1$, the coefficient $b(\lambda)$ is time-independent, and the coefficient $a(\lambda,t)$ evolves along eq. (3.2.15):

$$a(\lambda,t) = a(\lambda, t=0) \exp[4\mathrm{i}\, w_1^2(\lambda)\, t].$$

To pass from eq. (3.6.11) to the standard formulation of the Riemann problem, we extract the multiplier $\exp[\frac{1}{\mathrm{i}} w_3(\lambda)\sigma_3 x]$ from the functions $\Phi_{\pm}(\lambda)$ (see (3.6.10)) to represent them in the form

$$\begin{aligned} \Phi_{-}(x,t,\lambda) &= G_{-}(x,t,\lambda) \exp\left[\frac{1}{\mathrm{i}} w_3(\lambda)\sigma_3 x\right], \\ \Phi_{+}(x,t,\lambda) &= -b^*(\lambda) G_{+}^{-1}(x,t,\lambda)\sigma_3 \exp\left[\frac{1}{\mathrm{i}} w_3(\lambda)\sigma_3 x\right]. \end{aligned} \tag{3.6.12}$$

As a result, we rewrite eq. (3.6.11) as

$$\begin{aligned} &G_{+}(x,t,\lambda)\, G_{-}(x,t,\lambda) \\ &\quad = \chi_2^{(0)}(x,t,\lambda) \begin{pmatrix} 1 & a(\lambda,0) \\ a^*(\lambda,0) & 1 \end{pmatrix} [\chi_2^{(0)}(x,t,\lambda)]^{-1}, \end{aligned} \tag{3.6.13}$$

where $\chi_2^{(0)}(x,t,\lambda)$ is the "seeding" solution (3.6.2). The matrix functions $G_\pm(\lambda)$ satisfy the conditions

$$\det G_-(\lambda) = -[\det G_+(\lambda^*)]^* = b(\lambda) \quad (\operatorname{Im}\lambda \le 0);$$
$$G_+(\lambda) = G_-^\dagger(\lambda^*) \quad (\operatorname{Im}\lambda \ge 0);$$
$$G_-(-\lambda^{-1}) = \sigma_3 G_-(\lambda), \quad G_+(-\lambda^{-1}) = G_+(\lambda)\sigma_3; \tag{3.6.14}$$
$$G_-(\lambda) = g_{-1} + O\left(\frac{1}{|\lambda|}\right) \quad \text{as} \quad |\lambda| \to \infty. \tag{3.6.15}$$

The required solutions of the auxiliary linear system (3.6.1), $\chi_-(\lambda)$ and $\chi_+(\lambda)$, which are analytic in the domains $\operatorname{Im}\lambda \le 0$ and $\operatorname{Im}\lambda \ge 0$, respectively, take the form typical of the "dressing" scheme of particular solutions:

$$\chi_+(\lambda) = G_+^{-1}(\lambda)\chi_2^{(0)}(\lambda), \quad \chi_-(\lambda) = G_-(\lambda)\chi_2^{(0)}(\lambda). \tag{3.6.16}$$

When passing from relations (3.6.3) and (3.6.12) to formulas (3.6.16), we have eliminated insignificant factors.

Multisoliton states of the model of an "easy-axis" anisotropy ferromagnet with different equilibrium values of magnetization at infinity are associated with solutions to the Riemann problem (3.6.13) for $a(\lambda, t = 0) = 0$. In this case, on the real λ-axis, we have

$$G_+(\lambda)\, G_-(\lambda) = I. \tag{3.6.17}$$

It is worth pointing out that nontrivial solutions to the Riemann problem (3.6.17) are possible only when the functions $G_-(\lambda)$ have zeros in their domains of analyticity. Further, for simplicity, we assume that all zeros of the function $G_-(\lambda)$ are simple and do not lie on the real axis. In virtue of the reduction (3.6.14), they have the form

$$\{\mu_n,\ n = 0, 1, 2, \dots, 2N\}$$
$$\equiv \{\mu_0 = -\mathrm{i},\ \mu_a \ne -\mathrm{i},\ \mu_{a+N} = -\mu_a^{-1},\ a = 1, 2, \dots, N\}.$$

Since $G_+(\lambda) = [G_-(\lambda^*)]^\dagger$ $(\operatorname{Im}\lambda \ge 0)$, the zeros of the function $G_+(\lambda)$ are $\{\mu_n^*,\ n = 0, 1, 2, \dots, 2N\}$.

Equality (3.6.17) implies that the zeros of the functions $G_\pm(\lambda)$ coincide with the poles of the functions $G_\mp(\lambda)$.

As is known, the solution to the Riemann problem (3.6.17) is determined up to an arbitrary nondegenerate λ-independent matrix. We use the condition (3.6.15) to abolish the arbitrariness. We put that

$$G_-(\lambda) = g_{-1}\Psi_-(\lambda), \quad G_+(\lambda) = \Psi_+(\lambda)g_{-1}^\dagger, \tag{3.6.18}$$

where $\Psi_+(\lambda) = \Psi_-^\dagger(\lambda)$ for real values of the parameter λ. Then the matrix function $\Psi_-(\lambda)$ will have the canonical normalization

$$\Psi_-(\lambda = \infty) = I.$$

Plugging (3.6.18) into formula (3.6.14), we find

$$\Psi_-(-\lambda^{-1}) = g_{-1}^\dagger \sigma_3 g_{-1} \Psi_-(\lambda).$$

Hence

$$\begin{aligned} g_{-1}^\dagger \sigma_3 g_{-1} &\equiv \begin{pmatrix} S_3 & S_-(\frac{r^*}{r}) \exp(i\alpha_0) \\ S_+(\frac{r}{r^*}) \exp(-i\alpha_0) & -S_3 \end{pmatrix} \\ &= \Psi_-(\lambda = 0) = \Psi_-^\dagger(\lambda = 0). \end{aligned} \tag{3.6.19}$$

As a result, the Riemann problem (3.6.13) with linear reduction (3.6.14) is reduced to constructing the matrices $\Psi_\pm(\lambda)$ satisfying the conditions

$$\begin{aligned} &\Psi_+(\lambda)\Psi_-(\lambda) = I, \quad \Psi_+(\lambda) = \Psi_-^\dagger(\lambda), \quad \operatorname{Im}\lambda = 0, \\ &\Psi_-(\lambda = \infty) = I \end{aligned} \tag{3.6.20}$$

and the nonlinear reduction

$$\Psi_-(-\lambda^{-1}) = \Psi_-^\dagger(\lambda = 0)\, \Psi_-(\lambda). \tag{3.6.21}$$

The general solution to problem (3.6.20) without imposing the additional reduction (3.6.21) has been obtained earlier:

$$\begin{aligned} [\Psi_+(\lambda)]_{\sigma\sigma'} &= \delta_{\sigma\sigma'} - \sum_{p,s=0}^{2N} \frac{(m_p)_\sigma (m_s)_{\sigma'}^*}{\lambda - \mu_p} (A^{-1})_{ps}, \\ [\Psi_-(\lambda)]_{\sigma\sigma'} &= \delta_{\sigma\sigma'} + \sum_{k,n=0}^{2N} \frac{(m_k)_\sigma (m_n)_{\sigma'}^*}{\lambda - \mu_n^*} (A^{-1})_{kn}, \end{aligned} \tag{3.6.22}$$

where $\sigma, \sigma' = 1, 2$, and $(A^{-1})_{ij}$ are elements of the $2N \times 2N$ matrix inverse to the matrix A with elements

$$A_{ij} = \frac{(\mathbf{m}_i^* \cdot \mathbf{m}_j)}{\mu_i^* - \mu_j} = -A_{ji}^*. \tag{3.6.23}$$

When deriving formulas (3.6.22), we have taken into account the relations

$$(A^{-1})_{ij} = -(A^{-1})_{ji}^*, \tag{3.6.24}$$

which are a consequence of eq. (3.6.23).

The validity of solution (3.6.22) can be verified by a simple check using the identity

$$\sum_{s,k=0}^{2N} (A^{-1})_{ps}(\mathbf{m}_s^* \cdot \mathbf{m}_k)(A^{-1})_{kn} = (A^{-1})_{pn}\mu_n^* - \mu_p(A^{-1})_{pn}. \tag{3.6.25}$$

Here formula (3.6.25) contains no summation over the indices p and n.

Identity (3.6.25) also turns out to be useful for justifying the nonlinear reduction (3.6.21). The latter provides simple evidence that

$$\begin{aligned}[\Psi_-^\dagger(\lambda = 0)\Psi_-(\lambda)]_{\sigma\sigma'} &= \delta_{\sigma\sigma'} + \sum_{p,n=0}^{2N} (m_p)_\sigma \frac{1}{\mu_p}(A^{-1})_{pn}(m_n)_{\sigma'}^* \\ &\quad + \sum_{p,n=0}^{2N} (m_p)_\sigma \frac{\mu_n^*}{\mu_p}\frac{(A^{-1})_{pn}}{(\lambda - \mu_n^*)}(m_n)_{\sigma'}^*. \end{aligned} \tag{3.6.26}$$

At the same time, we get

$$\begin{aligned}[\Psi_-(-\lambda^{-1})]_{\sigma\sigma'} &= \delta_{\sigma\sigma'} - \sum_{n,p=0}^{2N} (m_p)_\sigma \frac{1}{\mu_n^*}(A^{-1})_{pn}(m_n)_{\sigma'}^* \\ &\quad + \sum_{n,p=0}^{2N} (m_p)_\sigma \frac{(A^{-1})_{pn}(\mu_n^*)^{-2}}{(\lambda + \mu_n^{*-1})}(m_n)_{\sigma'}^*. \end{aligned} \tag{3.6.27}$$

Expressions (3.6.26) and (3.6.27) may coincide conditional upon coinciding the matrix coefficients in the same poles. This happens under the condition

$$m_{\alpha+N} = m_\alpha, \quad \mu_{\alpha+N} = -\mu_\alpha^{-1}, \tag{3.6.28}$$

when the elements of the matrix A^{-1} acquire additional symmetry properties:

$$\begin{aligned}(A^{-1})_{\alpha+N\,0} &= \frac{1}{i\mu_\alpha}(A^{-1})_{\alpha\,0}, \\ \mu_\alpha(A^{-1})_{N+\alpha\,\beta}\frac{1}{\mu_\beta^*} &= -(A^{-1})_{\alpha\,\beta+N}, \\ (A^{-1})_{\alpha+N\,\beta+N} &= \frac{1}{\mu_\alpha}(A^{-1})_{\alpha\,\beta}\frac{1}{\mu_\beta^*}. \end{aligned}$$

The remaining connections between the elements of the matrix A^{-1} follow from the anti-Hermicity requirement (3.6.24).

Expressions (3.6.22) and (3.6.28) give a complete and explicit solution to the Riemann problem (3.6.20) with nonlinear reduction (3.6.21).

As a result, the newly derived solution $\chi_\pm(x,t,\lambda)$ of the auxiliary system is obtained by “dressing” the seeding solution $\chi_2^{(0)}(x,t,\lambda)$ using the matrix functions g_{-1} and $\Psi_\pm(\lambda)$:

$$\chi_+(x,t,\lambda) = g_{-1}(x,t)\Psi_+^{-1}(x,t,\lambda)\chi_2^{(0)}(x,t,\lambda),$$
$$\chi_-(x,t,\lambda) = g_{-1}(x,t)\Psi_-(x,t,\lambda)\chi_2^{(0)}(x,t,\lambda).$$

The vectors $\mathbf{m}_n$ stay x- and t-dependent as before due to the requirement of lacking “extra” poles at the points $\{\lambda_n \equiv \mu_n^*, n = 0,1,2,\ldots,2N\}$ and $\{\mu_n, n = 0,1,2,\ldots,2N\}$ in the “dressing” formulas

$$U = \partial_x\chi_\pm\chi_\pm^{-1}, \quad V = \partial_t\chi_\pm\chi_\pm^{-1},$$

which can be written in more detail:

$$\begin{aligned} U &= \partial_x g_{-1} g_{-1}^{-1} - g\,\Psi_+^{-1}(\partial_x\Psi_+ - U_0\Psi_+)g_{-1}^{-1} \\ &= \partial_x g_{-1} g_{-1}^{-1} + g_{-1}(\partial_x\Psi_- + \Psi_- U_0)\Psi_-^{-1}g_{-1}^{-1}, \\ V &= \partial_t g_{-1} g_{-1}^{-1} - g_{-1}\Psi_+^{-1}(\partial_t\Psi_+ - V_0\Psi_+)g_{-1}^{-1} \\ &= \partial_t g_{-1} g_{-1}^{-1} + g_{-1}(\partial_t\Psi_- + \Psi_- V_0)\Psi_-^{-1}g_{-1}^{-1}. \end{aligned} \tag{3.6.29}$$

Here

$$U_0 = -\mathrm{i}\sigma_3 w_3(\lambda), \quad V_0(\lambda) = 2\mathrm{i}\, w_1^2(\lambda)\sigma_3.$$

“The dressing” formulas (3.6.29) now include the matrix function $g_{-1}(x,t)$, thereby differing from those used earlier (2.9.13). This matrix performs the so-called gauge transformation. The peculiarity of the transformation is that the function $g_{-1}(x,t)$ is λ-independent; therefore the requirement of lacking the “extra” poles in relations (3.6.29) is boiled down to the previous equations:

$$\partial_x\mathbf{m}_n = U_0(\mu_n)\mathbf{m}_n, \quad \partial_t\mathbf{m}_n = V_0(\mu_n)\mathbf{m}_n. \tag{3.6.30}$$

Equations (3.6.30) have trivial solutions consistent with requirement (3.6.28):

$$\mathbf{m}_\alpha = \chi_2^{(0)}(x,t,\mu_\alpha)\,\mathbf{c}_\alpha, \quad \mathbf{m}_{\alpha+N} = \mathbf{m}_\alpha,$$

where $\mathbf{c}_\alpha$ are arbitrary constant two-dimensional complex vectors.

Equality (3.6.19) is the main one for constructing multisoliton solutions for the model of an “easy-axis” anisotropy ferromagnet with different equilibrium values of magnetization at infinity. As in the Gel’fand–Levitan–Marchenko approach, the component S_3 of the order parameter **S** is calculated explicitly:

$$S_3(x,t) = \left[\Psi_-(x,t,\lambda=0)\right]_{11}. \tag{3.6.31}$$

Calculation of the field phase $S_+ = S_1 + \mathrm{i}S_2$ involves additional integration using the equality

$$S_+(x,t)\frac{r(x,t)}{r^*(x,t)}\exp(-i\alpha_0) = [\Psi_-(x,t,\lambda=0)]_{21}. \tag{3.6.32}$$

As an illustration of the effectiveness of the proposed method, let us again obtain the soliton solution corresponding to the domain boundary. In this case the function $\Psi_-(\lambda)$ has a single pole at the point $\lambda = \mathrm{i}$:

$$\Psi_-(\lambda) = I + \frac{2\mathrm{i}}{\lambda - \mathrm{i}}P, \tag{3.6.33}$$

where P is a projection matrix with elements

$$P_{\sigma\sigma'} = \frac{(m_0)_\sigma (m_0)^*_{\sigma'}}{|\mathbf{m}_0|^2}, \quad \mathbf{m}_0 = \chi_2^{(0)}(x,t,\lambda=-\mathrm{i})\,\mathbf{c} = \exp\left(-\frac{\beta\sigma_3}{2}x\right)\mathbf{c}.$$

For definiteness, we suppose that $\mathbf{c} = (c_1, c_2)$, $c_2c_1^* = -|c_1c_2|$. Then, bearing in mind eqs. (3.6.31)–(3.6.33), we can reproduce the solution corresponding to the domain wall:

$$S_3 = \tanh[\beta(x-x_0)], \quad S_+ = \frac{\exp(i\alpha_0)}{\cosh[\beta(x-x_0)]}, \quad x_0 = \frac{1}{\beta}\ln\left|\frac{c_1}{c_2}\right|.$$

Using the explicit formulas (3.6.31) and (3.6.32), we can verify that the interaction between a domain wall and precessing magnetic solitons is reduced to mutual elastic pair collisions. After collisions, the phases of the solitons' internal precession and the coordinates of their “centers of gravity” gain additional constant shifts.

For $\varepsilon = 1$, all the constructions are analogous. Solutions of systems (3.2.2) and (3.2.5), $F_+(x,t,\lambda)$ and $F_-(x,t,\lambda)$, which are analytic in the domains $\mathrm{Im}\,\lambda \geq 0$ and $\mathrm{Im}\,\lambda \leq 0$, respectively, have the form

$$F_+ = (\varphi_-^{(1)}, \varphi_+^{(2)}), \quad F_- = (\varphi_+^{(1)}, \varphi_-^{(2)}).$$

On the real λ-axis, the Jost solutions and the functions $F_\pm(\lambda)$ are tied by the relations

$$\begin{aligned} F_+(\lambda) &= \varphi_+(\lambda)M_{++}(\lambda) = \varphi_-(\lambda)M_{-+}(\lambda),\\ F_-(\lambda) &= \varphi_+(\lambda)M_{+-}(\lambda) = \varphi_-(\lambda)M_{--}(\lambda), \end{aligned} \tag{3.6.34}$$

where

$$M_{++} = \begin{pmatrix} a & 0 \\ b & 1 \end{pmatrix}, \quad M_{-+} = \begin{pmatrix} 1 & b^* \\ 0 & a \end{pmatrix},$$

$$M_{+-} = \begin{pmatrix} 1 & -b^* \\ 0 & a^* \end{pmatrix}, \quad M_{--} = \begin{pmatrix} a^* & 0 \\ -b & 1 \end{pmatrix};$$

$$T(\lambda) = M_{++}(\lambda)M_{-+}^{-1}(\lambda) = M_{+-}(\lambda)M_{--}^{-1}(\lambda).$$

For $\varepsilon = 1$, the time-independent coefficient $a(\lambda)$ continues analytically to the domain $\mathrm{Im}\,\lambda \geq 0$, whereas the coefficient $b(\lambda, t)$ evolves in the following manner:

$$b(\lambda, t) = b(\lambda, t = 0) \exp[-4\mathrm{i} w_1^2(\lambda)\, t].$$

The appropriate results for the functions $\varphi_\pm(\lambda)$ allow us to obtain the necessary reductions of the functions $F_\pm(\lambda)$ and their asymptotic behavior in the vicinity of the singular points $\lambda = 0$ and $\lambda = \infty$:

$$\begin{aligned}
F_\pm^*(\lambda) &= \sigma_2 F_\mp(\lambda)\sigma_2,\\
\det F_+(\lambda) &= [\det F_-(\lambda)]^* = a(\lambda), \quad \mathrm{Im}\lambda = 0;\\
F_\pm(-\lambda^{-1}) &= \sigma_3 F_\pm(\lambda)\sigma_3;\\
F_+(\lambda) &= \exp(-\mathrm{i}\beta_0)\left[g_1 + O\left(\frac{1}{|\lambda|}\right)\right] \exp\left[\frac{1}{\mathrm{i}} w_3(\lambda)\sigma_3 x\right],\\
F_-(\lambda) &= \left[g_1 + O\left(\frac{1}{|\lambda|}\right)\right] \exp\left[\frac{1}{\mathrm{i}} w_3(\lambda)\sigma_3 x\right] \quad \text{as} \quad |\lambda| \to \infty.
\end{aligned}$$

Here

$$g_1 = \begin{pmatrix} 1 & -S_-(1+S_3)^{-1} \\ S_+(1+S_3)^{-1} & 1 \end{pmatrix} \begin{pmatrix} r & 0 \\ 0 & r^* \exp(\mathrm{i}\beta_0) \end{pmatrix}, \quad g_1^\dagger g_1 = I.$$

The problem of conjugation of the analytic functions $F_\pm(\lambda)$ on the real λ-axis can be posed as

$$F_-(x,t,\lambda) = \frac{1}{a(\lambda)} F_+(x,t,\lambda) \begin{pmatrix} 1 & -b^*(\lambda,t) \\ -b(\lambda,t) & 1 \end{pmatrix}. \tag{3.6.35}$$

Employing the transformations

$$\begin{aligned}
\frac{1}{a(\lambda)} F_+(x,t,\lambda) &= g_1(x,t)\Psi_+^{-1}(x,t,\lambda) \exp\left[\frac{1}{\mathrm{i}} w_3(\lambda) x\sigma_3\right],\\
F_-(x,t,\lambda) &= g_1(x,t)\Psi_-(x,t,\lambda) \exp\left[\frac{1}{\mathrm{i}} w_3(\lambda) x\sigma_3\right],
\end{aligned}$$

we reduce relation (3.6.35) to the following formulation of the Riemann problem:

$$\begin{aligned}
&\Psi_+(x,t,\lambda)\Psi_-(x,t,\lambda)\\
&\quad = \chi_2^{(0)}(x,t,\lambda) \begin{pmatrix} 1 & -b^*(\lambda, t=0) \\ -b(\lambda, t=0) & 1 \end{pmatrix} [\chi_2^{(0)}(x,t,\lambda)]^{-1},\\
&\Psi_-(\lambda) = \Psi_+^\dagger(\lambda) \quad (\mathrm{Im}\lambda = 0)
\end{aligned} \tag{3.6.36}$$

with canonical normalization

$$\Psi_\pm(\lambda = \infty) = I$$

and nonlinear reduction

$$\Psi_+(-\lambda^{-1}) = \sigma_3 \Psi_+(\lambda) \Psi_+^\dagger(\lambda = 0)\sigma_3,$$
$$\Psi_-(-\lambda^{-1}) = \sigma_3 \Psi_-^\dagger(\lambda = 0) \Psi_-(\lambda)\sigma_3. \tag{3.6.37}$$

Here we have taken into account the equality

$$g_1^\dagger \sigma_3 g_1 = \Psi_+^{-1}(\lambda = 0)\sigma_3 = \Psi_-(\lambda = 0)\sigma_3.$$

The latter, in turn, implies that

$$\Psi_\pm^\dagger(\lambda = 0) = \Psi_\pm^{-1}(\lambda = 0), \quad \Psi_\pm(\lambda = 0) = \sigma_3 \Psi_\pm^\dagger(\lambda = 0)\sigma_3.$$

We choose the functions

$$\chi_+(\lambda) = g_1 \Psi_+^{-1}(\lambda) \chi_2^{(0)}(\lambda), \quad \chi_-(\lambda) = g_1 \Psi_-(\lambda) \chi_2^{(0)}(\lambda) \tag{3.6.38}$$

as "dressed" solutions of the auxiliary linear system (3.6.1) with the desired analytic properties.

For $\varepsilon = 1$, the working formula for calculating new solutions of the Landau–Lifshitz equations (3.1.17) is given by

$$\sigma_3 \Psi_+(\lambda = 0) = g_1^\dagger \sigma_3 g_1 = \begin{pmatrix} S_3 & -S_-(\frac{r^*}{r}) \exp(\mathrm{i}\beta_0) \\ -S_+(\frac{r}{r^*}) \exp(-\mathrm{i}\beta_0) & -S_3 \end{pmatrix}.$$

For the multisoliton states ($b(\lambda, t = 0) \equiv 0$), the Riemann problem (3.6.36) is solved explicitly:

$$[\Psi_+(\lambda)]_{\sigma\sigma} = \delta_{\sigma\sigma'} - \sum_{p,s=1}^{2N} \frac{(m_p)_\sigma (m_s)^*_{\sigma'}}{\lambda - \lambda_p^*} (A^{-1})_{ps}.$$

Here $\sigma, \sigma' = 1, 2$, and $(A^{-1})_{ij}$ are the elements of the inverse $2N \times 2N$ matrix with respect to the matrix with elements A_{ij}:

$$A_{ij} = \frac{(\mathbf{m}_i^* \cdot \mathbf{m}_j)}{\lambda_i - \lambda_j^*}; \quad (A^{-1})_{ij} = \frac{\partial \ln \det A}{\partial A_{ji}};$$
$$\mathbf{m}_\alpha = \chi_2^{(0)}(x, t, \lambda_\alpha^*) \mathbf{c}_\alpha, \quad \mathbf{m}_{\alpha+N} = \sigma_3 \mathbf{m}_\alpha,$$
$$\lambda_{\alpha+N} = -\lambda_\alpha^{-1} \quad (\lambda_\alpha \neq \mathrm{i}, \quad \mathrm{Im}\lambda_\alpha > 0, \quad \alpha = 1, 2, \dots, N),$$

where $\mathbf{c}_\alpha$ are arbitrary constant two-dimensional complex vectors.

The simplest of the multisoliton states corresponding to the pair of poles of the function $\Psi_+(\lambda)$ at the points $\lambda = \lambda_1^*$ and $\lambda = -(\lambda_1^*)^{-1}$ coincides with the precessing magnetic soliton (3.4.19), (3.4.22) found previously by a more tedious procedure.

3.7 Forced motion of a domain wall in the field of a nonlinear spin wave

Any complicated problem always makes a researcher want to solve it. It seems to us that the pleasure of solving such a problem grows much faster than its complexity. Here, the non-linearity typical of all laws of nature is also revealed.

V. G. Bar'yakhtar, B. A. Ivanov, "In the world of magnetic domains"

3.7.1 Setting up a problem

An easy-axis ferromagnet model contains a prohibition on the motion of a domain wall. It is easy to see this by rewriting the Landau–Lifshitz equations (3.1.17) as follows:

$$\begin{aligned} \partial_t S_3 &= \frac{\mathrm{i}}{2}\partial_x [S_+\partial_x S_- - S_-\partial_x S_+], \\ \partial_t S_+ &= \mathrm{i}\partial_x [S_3\partial_x S_+ - S_+\partial_x S_3] - \mathrm{i}\beta^2 S_3 S_+, \end{aligned} \tag{3.7.1}$$

where $S_\pm = S_1 \pm \mathrm{i}S_2$. According to the first equation in (3.7.1), the projection of the medium's total magnetic moment on the anisotropy axis must remain unchanged:

$$I = \int_{-L}^{L} S_3(x,t)\mathrm{d}x = \mathrm{const}, \tag{3.7.2}$$

where $2L$ is the sample size. As the domain wall moves, the value of I inevitably changes. This circumstance leads to a prohibition on its motion. The conservation law (3.7.2) also holds for an external magnetic field oriented along the anisotropy axis. This means that the wall motion is also impossible.

The most well-known mechanism of domain wall motion in an external magnetic field directed along the anisotropy axis suggests that noncircular magnetization precession around the external magnetic field generates a magneto-static field [18, 11]. In the process the latter, which is orthogonal to the wall surface, breaks the law of conservation of the magnetization projection on the anisotropy axis, and the prohibition on the domain wall's motion is lifted. During stationary motion of the domain wall, the wall energy gained from the external field compensates for dissipative losses.

The present section delves deeper into another mechanism of motion of a nonfixed-by-defects domain wall interacting with a running nonlinear spin wave. The considered scenario of a stationary motion of the wall is realized even under the neglect of magnetostatic interactions and relaxation processes. Analytical solutions deduced in this chapter do not contradict the qualitative conclusions [18, 11] and are consistent with the results of [2] on the excitation and control of the low-amplitude solitons' motion in the field of a circularly polarized wave. Let us give arguments [26] in favor of the fact that when interacting with a packet of spin waves, a domain wall displaces and, under the field of a running spin wave, must move toward the wave.

Let a localized packet of spin waves move from left to right and make a positive contribution I_0 to the conservation law (3.7.2) before colliding with a domain wall. The area hatched in Figure 3.8a is equal to the value of I_0. Then the packet partially passes through the wall and partially reflects off of it (Figure 3.8b). The contribution of the wave transmitted through the wall to the conservation law (3.7.2) is negative and equals $-I_1$. The contribution of the reflected wave, equal to I_2, is positive. The appropriate areas are hatched in Figure 3.8b. The amplitude of the reflected wave cannot be too large (this contradicts the law of conservation of energy). Therefore an imbalance arises:

$$\Delta I = I_0 - (I_2 - I_1) = I_0 + I_1 - I_2 > 0.$$

It can only be compensated by shifting the domain wall to the left. Indeed, when the wall is shifted to the left by Δx, the integral (3.7.2) changes by a magnitude of $\Delta I \cong 2\Delta x$ approximately equal to the area of the hatched parallelogram in Figure 3.8b. Hence the value of the shift of the domain wall is equal to

$$\Delta x = \frac{1}{2}(I_0 + I_1 - I_2).$$

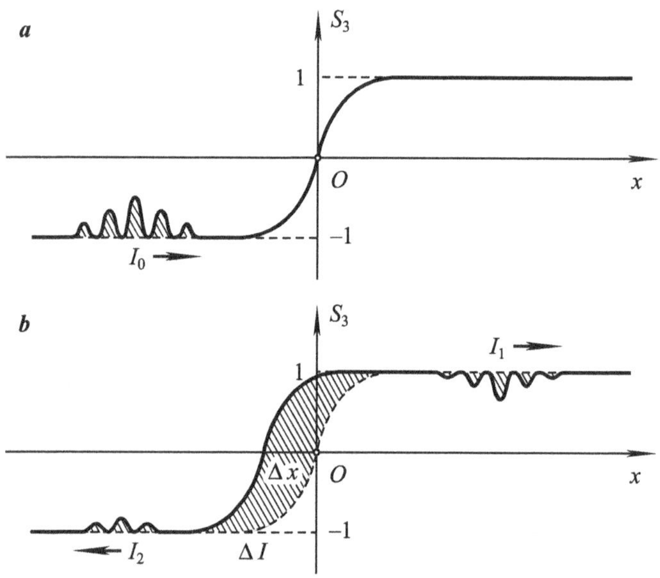

Figure 3.8: Interaction between a domain wall and a spin-wave pulse in an “easy-axis” anisotropy ferromagnet: (a) the position of the spin-wave packet before the collision with the domain wall, (b) after the collision.

The above speculation is general and does not require the spin-wave amplitude to be small. In real crystals, anisotropy is more complicated. Energy dissipation processes, lattice defects, and surface and size effects can destroy the integral (3.7.2). However, if a localized spin-wave pulse passes through the domain wall far faster than over the time required for the destruction of the integral (3.7.2), the above considerations remain valid.

The paper [30] uses the "dressing" method to find an explicit solution of model (3.7.1), which describes the interaction between a domain wall and a nonlinear precession wave of large amplitude. In this and the next section, following [15–17, 14], we modify the calculations in [30] to disclose the analytic properties of the matrix functions of the Riemann problem. For this goal, an apparatus based on the theory of elliptic functions is required. It is extremely useful for further generalizations of the "dressing" method through analytically describing not only spatially localized solitons but also more general nonlocalized soliton-like states.

Kuznetsov and Mikhailov [22] have used for the first time the "dressing" through Shabat's integral operators to study soliton-like excitations on the pedestal of a cnoidal wave in media within the Korteweg–de Vries model. Based on the Riemann problem solution, the paper [6] analytically describes solitons in noncollinear ferri- and antiferromagnets against the background of nonlocalized magnetization waves running toward each other. The paper [5] analyzes the interaction between running precession waves and solitons in a quasione-dimensional "easy-plane" anisotropy ferromagnet. It has been established that the waves give rise to the amplitude and phase modulation of solitons and, under certain conditions, destroy them. Chapters 4 and 7 of this book deal with these outcomes. The Darboux transformation related to the "dressing" method has been used for constructing solutions to the sine-Gordon wave equation to describe domain boundaries moving between two spatially homogeneous but time-periodic states [10]. The same method within the sh-Gordon wave equation has been applied in [21, 20] to explore interaction between "dark" solitons and homogeneous oscillations of nonzero vacuum. The paper [23] investigates the features of interaction between solitons and radiation using analytical and numerical methods for media whose dynamics is described by the nonlinear Schrödinger equation.

Let us examine a sample having two uniformly magnetized regions separated by a domain wall. The wall thickness is small, so we can assume that $\mathbf{S} \to (0,0,1)$ as $x \to +\infty$ and $\mathbf{S} \to (0,0,-1)$ as $x \to -\infty$. If a spin wave is excited in the sample, then these asymptotics change as follows:

$$\mathbf{S} \to \mathbf{S}_0^{(\pm)} \quad \text{as} \quad x \to \pm\infty, \tag{3.7.3}$$

where

$$\mathbf{S}_0^{(\pm)} = (\sin\theta_0 \cos\varphi_\pm,\ \pm\sin\theta_0 \sin\varphi_\pm,\ \pm\cos\theta_0). \tag{3.7.4}$$

The angle θ_0 characterizes the wave amplitude, $\varphi_\pm = px - \omega t + \delta_\pm$, p is the wavenumber, $\omega = \cos\theta_0(p^2 + \beta^2)$ is the precession frequency, and the phase shifts δ_+ and δ_-, as we will see later, are related to the domain wall topology.

Here the wave (3.7.4) is modulationally unstable. The initial development stage of instability can be analyzed through stability in the linear approximation regarding a small periodic modulation. Small perturbations of the pumping wave (3.7.4) are $\sim \exp(\mathrm{i}\tilde{\omega}t + \mathrm{i}\tilde{p}x)$, where $\tilde{p}$ is a real number, and obey the dispersion law

$$\tilde{\omega}(\tilde{p}) = -2p\tilde{p}\cos\theta_0 \pm \tilde{p}\sqrt{\tilde{p}^2 - \nu^2}, \quad \nu^2 = \sin^2\theta_0(p^2 + \beta^2). \tag{3.7.5}$$

The perturbation growth increment $\delta = |\tilde{p}|\sqrt{\nu^2 - \tilde{p}^2}$ is real in the range of the wave numbers $-\nu < \tilde{p} < \nu$. The maximum value $\delta_{\max}$ of the increment is related to the minimum time $\tau_{\min}$, during which the wave perturbation increases by a factor of e: $\delta_{\max} = \tau_{\min}^{-1} = \nu^2/2$.

The wave (3.7.4) passes through the sample of length $2L$ over a time $T = 2Lp/\omega$. Due to the instability of the wave, its interaction with solitons can only be considered under the condition $\tau_{\min} > T$, which imposes constraints on the problem parameters:

$$\frac{pL\sin^2\theta_0}{\cos\theta_0} < 1.$$

From a physical point of view, an external source generates the wave (3.7.4) at the boundary of a ferromagnet. This wave interacts with domain walls and magnetization-reversal nuclei (or breathers) initially present in the sample. On the one hand, the sample length must be large enough to neglect wave excitation at the sample boundary, and on the other hand, it must be small enough so that the wave instability has no time to develop.

It should be stressed that the nonlinear waves (3.7.4) themselves are exact solutions to the Landau–Lifshitz equations (3.1.17). Our goal is to construct a new solution to the Landau–Lifshitz equations by the "dressing" method. This solution should satisfy the asymptotic boundary conditions (3.7.3) and thus describe the interaction of the domain wall with a large-amplitude precession wave.

For the magnetization distributions (3.7.4), the auxiliary linear system

$$\begin{aligned}
\partial_t\chi &= -\mathrm{i}\{w_1([\mathbf{S}\times\partial_x\mathbf{S}]_1\sigma_1 + [\mathbf{S}\times\partial_x\mathbf{S}]_2\sigma_2) \\
&\quad + w_3[\mathbf{S}\times\partial_x\mathbf{S}]_3\sigma_3 - 2w_1^2 S_3\sigma_3 \\
&\quad - 2w_1w_3(S_1\sigma_1 + S_2\sigma_2)\}\chi \equiv V\chi, \\
\partial_x\chi &= -\mathrm{i}\{w_1(S_1\sigma_1 + S_2\sigma_2) + w_3S_3\sigma_3\}\chi \equiv U\chi,
\end{aligned} \tag{3.7.6}$$

where

$$w_1 = \frac{\beta}{4}\left(\lambda + \frac{1}{\lambda}\right), \quad w_3 = \frac{\beta}{4}\left(\lambda - \frac{1}{\lambda}\right),$$

has the exact solutions

$$\chi_{1,2}^{(0)}(\lambda, x, t) = a(\lambda, \tau) \exp\left(\pm \frac{\mathrm{i}\sigma_3 \varphi_{\mp}}{2} \right) \times \begin{pmatrix} 1 & \pm\Delta(\lambda, \tau) \\ \mp\Delta(\lambda, \tau) & 1 \end{pmatrix} \exp\left(\mp \frac{\tau}{\mathrm{i}} \sigma_3 \eta \right). \tag{3.7.7}$$

Here we have introduced the notations

$$\Delta(\lambda, \tau) = \frac{\tau - (w_3 \cos\theta_0 - p/2)}{w_1 \sin\theta_0} = \frac{w_1 \sin\theta_0}{\tau + w_3 \cos\theta_0 - p/2},$$
$$\tau = \left[(w_3 \cos\theta_0 - p/2)^2 + w_1^2 \sin^2\theta_0 \right]^{1/2}, \quad \eta = x - (2w_3 + p\cos\theta_0)t.$$

In what follows, we choose the multiplier $a(\lambda, \tau)$ in a way that simplifies the calculations.

3.7.2 Properties of the Jost solutions. The "dressing" method

Let us define the Jost functions χ_1 and χ_2 for a problem of domain wall-spin wave interaction as the solutions of system (3.7.6) with the following asymptotic behavior:

$$\chi_{1,2} \to \chi_{1,2}^{(0)} \quad \text{as} \quad x \to \mp\infty. \tag{3.7.8}$$

We pass from problems (3.7.6) and (3.7.8) to equivalent integral equations for calculating the Jost functions:

$$\begin{aligned} \chi_1(\lambda, x) &= \chi_1^{(0)}(\lambda, x) \left(I + \int_{-\infty}^{x} [\chi_1^{(0)}(\lambda, x')]^{-1} \times [U(\lambda, x') - U_1(\lambda, x')] \chi_1(\lambda, x') \mathrm{d}x' \right), \\ \chi_2(\lambda, x) &= \chi_2^{(0)}(\lambda, x) \left(I - \int_{x}^{+\infty} [\chi_2^{(0)}(\lambda, x')]^{-1} \times [U(\lambda, x') - U_2(\lambda, x')] \chi_2(\lambda, x') \mathrm{d}x' \right), \end{aligned} \tag{3.7.9}$$

where $U_{1,2}$ are the values of the U-operator for $\mathbf{S} = \mathbf{S}_0^{(\mp)}$. By means of iterations solutions to equations (3.7.9) can be treated as formal series in powers of the functions $\chi_{1,2}^{(0)}$. Analyzing the series, we can infer that the analytic properties of the Jost solutions are formulated on the two-sheeted Riemann surface of the function $\tau = \tau(\lambda)$. The Riemann surface arises as a natural domain of defining for the "seeding" matrix functions $\chi_{1,2}^{(0)}(\lambda)$

(3.7.7). As in other problems, the analytic properties of the Jost solutions are simple consequences of the analytic properties of the functions $\chi_{1,2}^{(0)}(\lambda)$ and the arrangement of the x-integration limits in representations (3.7.9).

Further, we see that for a domain boundary to exist, the Riemann problem's functions must have poles at four points in the Riemann surface, lying above the values of the spectral parameter $\lambda = \pm\mathrm{i}$. The situation becomes entangled because the "seeding" functions (3.7.7) also have poles at the same points.

To construct a solution that describes the domain wall-spin wave interaction, it is necessary to outline rules for transforming points in the Riemann surface both under operations related to complex conjugation and under transformations that leave the auxiliary linear system (3.7.6) intact. In [30], before formulating such rules, the spectral parameter had been changed. After that, instead of the function $\tau = \tau(\lambda)$, another function appears. The latter still represents the square root of the fourth-degree polynomial over other spectral parameter. Of course, such a substitution for the spectral parameters remains some questions unanswered: which signs should be chosen in front of the square root after symmetry transformations of the auxiliary linear system or after complex conjugation operations.

To avoid similar difficulties, we choose a new spectral parameter in other way. Consider any rational function $f(\lambda, y)$ of variables λ and y, where $y = \sqrt{R(\lambda)}$ with $R(\lambda)$ a third- or fourth-degree polynomial over λ. This function can always be represented parametrically in terms of doubly periodic Jacobi analytic functions having only pole singularities. Formulas for connecting the initial variables λ and y to the "uniformizing" variable u,

$$\lambda = \lambda(u), \quad y = y(u),$$

are classified in reference books. In our problem, we have [3]

$$\begin{aligned} \lambda &= (k')^{-1/2} \tan\frac{\theta_0+\theta_4}{2}\left[\frac{\mathrm{i}\,\mathrm{dn}(u+\mathrm{i}K') + k'\cot((\theta_0+\theta_4)/2)}{\mathrm{i}\,\mathrm{dn}(u+\mathrm{i}K') - \tan((\theta_0+\theta_4)/2)}\right], \\ \tau &= \frac{\mathrm{i}\beta k^2 \sin\theta_0 [\sin(\theta_0-\theta_4)\sin(\theta_0+\theta_4)]^{-1/2}}{[\mathrm{i}\,\mathrm{dn}(u+\mathrm{i}K') + k'\cot((\theta_0+\theta_4)/2)]} \\ &\quad \times \frac{\mathrm{cn}(u+\mathrm{i}K')\mathrm{sn}(u+\mathrm{i}K')}{[\mathrm{i}\,\mathrm{dn}(u+\mathrm{i}K') - \tan((\theta_0+\theta_4)/2)]}. \end{aligned} \tag{3.7.10}$$

In formulas (3.7.10) the angle $0 \le \theta_4 \le \theta_0$ is determined by the relation

$$\tan\theta_4 = \kappa(1+\kappa^2)^{-1/2}\tan\theta_0, \quad \kappa = p/\beta.$$

The modulus k of the Jacobi elliptic functions $\mathrm{sn}\,u$, $\mathrm{cn}\,u$, and $\mathrm{dn}\,u$ has the form

$$k = \frac{2\sqrt{\cos\theta_0\cos\theta_4}}{\cos\theta_0+\cos\theta_4}.$$

Meanwhile, an additional module is

$$k' = \sqrt{1-k^2} = \tan\frac{\theta_0+\theta_4}{2}\tan\frac{\theta_0-\theta_4}{2}.$$

In the complex u-plane the functions $\lambda = \lambda(u)$ and $\tau = \tau(u)$ (3.7.10) are meromorphic doubly periodic functions with periods $2K$ and $4\mathrm{i}K'$, where $K = K(k)$ and $K' = K(k')$ are the complete elliptic integrals of the first kind.

Among elliptic functions, the Jacobi functions are often used; however, they are not practical for precise calculations. It is easier to carry out all calculations using Weierstrass' functions and to write down the findings in terms of the well-researched and tabulated Jacobi functions only at the very end [1]. A technique for exploiting elliptic functions is explained in the Appendix.

It is simpler to disclose the analytical properties of Jost solutions through choosing $\alpha(\lambda(u), \tau(u)) = \beta(-u)$ as a common multiplier in formulas (3.7.7). The representation $\beta(u)$ in terms of the Weierstrass zeta functions and the Jacobi elliptic functions is obtained in Appendix:

$$\begin{aligned}\beta(u) &= \zeta(u+\mathrm{i}K'+K/2) - \zeta(u-3\mathrm{i}K'+K/2) - \eta_3\\ &= \frac{1}{2}\left(\frac{\operatorname{cn}(v,k)}{\operatorname{sn}(v,k)} - k'\frac{\operatorname{sn}(v,k)}{\operatorname{cn}(v,k)}\right)\bigg|_{v=(u+\mathrm{i}K'+K/2)/2}.\end{aligned}$$

Here the periods of the zeta functions are equal to $2K$ and $8\mathrm{i}K'$, and $\eta_3 = \zeta(4\mathrm{i}K'|K, 4\mathrm{i}K')$. After the vertical bar, it is customary to indicate halves of primitive periods [3, 1]. Appendix contains identities useful for the functions $\beta(-u)$, $\beta^*(u^*)$, and $\beta(u)$ to compute. In terms of the spectral parameter u, the "seeding" matrix functions hold:

$$\chi_{1,2}^{(0)}(u) = \exp\left(\pm\frac{\mathrm{i}\sigma_3}{2}\varphi_\mp\right)\begin{pmatrix}\beta(-u) & \pm\beta(u)\\ \mp\beta(u) & \beta(-u)\end{pmatrix}\exp\left(\mp\frac{\tau}{\mathrm{i}}\sigma_3\eta\right). \tag{3.7.11}$$

Functions (3.7.11) and hence the Jost solutions $\chi_{1,2}(u)$ are doubly periodic functions with periods $2K$ and $8\mathrm{i}K'$. Since $\chi_{1,2}(u + 4\mathrm{i}K') = -\chi_{1,2}(u)$, it suffices to explore the analytical properties of Jost solutions within a periodicity cell of the function $\tau = \tau(u)$, i. e., within a rectangle with sides $2K$ and $4\mathrm{i}K'$.

Constructing the Jost solutions and hence integrating the initial nonlinear Landau–Lifshitz equation is further reduced to solving the Riemann problem for doubly periodic functions of the complex variable u. Such a problem is called the Riemann problem on a torus. Let us highlight the fundamental difference between the considered equation (3.2.7) and those integrated using the Riemann problem on a torus (see [25, 4, 28]; the results are presented in Chapters 5 and 7). The papers [25, 4, 28] investigate solitons against a homogeneous background but in strong dispersion media. For linear modes, the dependence of a frequency ω on a wavenumber k had the form $\omega = \sqrt{P_4(k)}$, where P_4 is a fourth-degree polynomial. This circumstance manifested itself in a complication

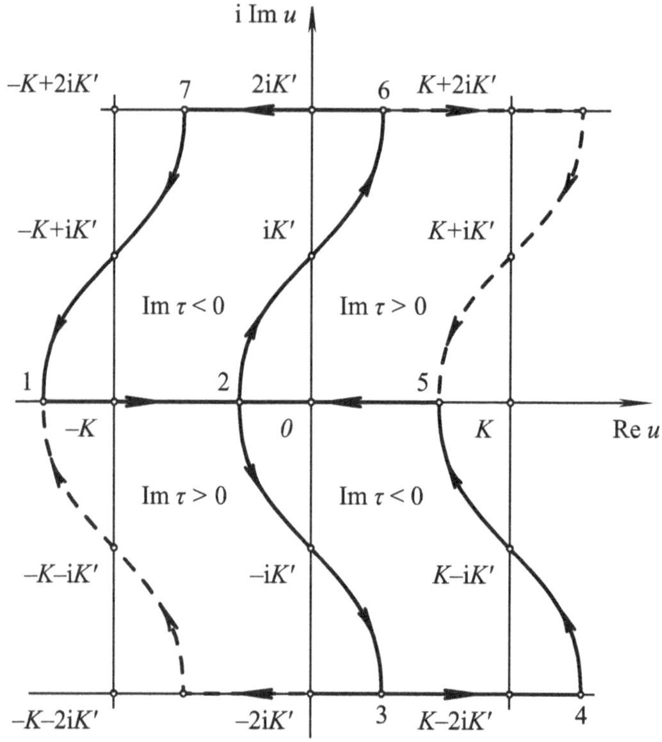

Figure 3.9: Analyticity domain of the Jost functions. The solid line denotes the contour Γ, to the left and right of which are $\operatorname{Im}\tau(u) < 0$ and $\operatorname{Im}\tau(u) > 0$, respectively; the dashed line represents a contour equivalent to the contour Γ.

of U–V pairs: the matrices U and V were λ-dependent through doubly periodic elliptic Jacobi functions rather than through rational functions. This peculiarity was brought about by the Riemann problem on a torus. In the case under consideration, the U–V operators (3.1.15) and (3.1.18) have a rational dependence on the parameter λ. When integrating eq. (3.2.7), the Riemann problem on a torus has a different origin: it is related to the accounting of an inhomogeneous periodic background as a pumping wave.

As $|x| \to \infty$, the oscillating Jost solutions exist only when the values of the spectral parameter u lie on the contour $\Gamma = \{u : \operatorname{Im}\tau(u) = 0\}$. The set Γ matches the continuous spectrum of the scattering problem connected to the Jost solutions.

In Figure 3.9, angular points labeled by $1, 2 \ldots, 7$ in the contour Γ have the coordinates

$$-\tilde{u} - K, \quad -\tilde{u}, \quad \tilde{u} - 2iK', \quad \tilde{u} + K - 2iK',$$
$$-\tilde{u} + K, \quad \tilde{u} + 2iK', \quad \tilde{u} - K + 2iK',$$

where $0 < \tilde{u} < K$ is the root of the equation

$$\frac{\mathrm{cn}\tilde{u}}{\mathrm{sn}\tilde{u}} = \frac{1}{1-\gamma}\left[\tan\frac{\theta_0-\theta_4}{2} + \gamma\tan\frac{\theta_0+\theta_4}{2}\right],$$

$$\gamma = \frac{\kappa\cos\theta_0 + \sqrt{\kappa^2\cos^2\theta_0 + 1}}{\kappa + \sqrt{1+\kappa^2}}, \qquad \kappa = \frac{p}{\beta}.$$

The segments of the contour Γ being beyond the boundaries of the fundamental rectangle can be brought back inside the rectangle by means of translations to the primitive periods of the lattice. Note that any contour that coincides with that shown in the figure after the translations $2Km + 4\mathrm{i}K'm'$ can be taken as the contour Γ in the u-plane (here m and m' are integers).

The functions $\lambda(u)$, $\tau(u)$, $\beta(u)$, and $\beta(-u)$ have the following transformation properties:

$$\begin{aligned}
&\lambda(-u+2\mathrm{i}K') = \lambda(u), \quad \tau(-u+2\mathrm{i}K') = -\tau(u);\\
&\lambda(u+K) = -\lambda^{-1}(u), \quad \tau(u+K) = -\tau(u);\\
&\lambda^*(u^*) = \lambda(u), \quad \tau^*(u^*) = \tau(u),\\
&2\beta^*(\pm u^*)\beta(\mp u) = (1+k')\sqrt{k'};\\
&\beta(-u+2\mathrm{i}K') = \beta(u), \quad \beta[-(-u+2\mathrm{i}K')] = -\beta(-u);\\
&\beta^*[(-u+2\mathrm{i}K')^*] = \beta(-u), \quad \beta^*[-(-u+2\mathrm{i}K')^*] = -\beta(u).
\end{aligned} \tag{3.7.12}$$

Using relations (3.7.12), it is easy to prove the following reductions to the Jost solutions:

$$\begin{aligned}
\chi_{1,2}(-u+K+2\mathrm{i}K') &= -\sigma_3\chi_{1,2}(u)\sigma_3;\\
\chi^*_{1,2}[(-u+2\mathrm{i}K')^*] &= \pm\mathrm{i}\sigma_2\chi_{1,2}(u);\\
\chi^*_{1,2}(u^*) &= \frac{(1+k')\sqrt{k'}}{2\beta(u)\beta(-u)}\sigma_2\chi_{1,2}(u)\sigma_2 \quad (u\in\Gamma).
\end{aligned} \tag{3.7.13}$$

As before, formulas (3.7.13) are proven using the uniqueness theorem for solutions of linear differential equations under the same asymptotic conditions.

In the contour Γ the basic Jost solutions must be interconnected:

$$\chi_1(u) = \chi_2(u)T(u). \tag{3.7.14}$$

Reductions (3.7.13) specify the shape of the transition matrix:

$$T(u) = \begin{pmatrix} a(u) & -b^*(u^*)\\ b(u) & a^*(u^*)\end{pmatrix}, \quad a(u)a^*(u^*) + b(u)b^*(u^*) = 1. \tag{3.7.15}$$

The translations $2Km + 4iK'm'$ (m and m' are arbitrary integers) leave the functions $a(u)$ and $b(u)$ unaltered. The latter satisfy the conditions

$$\begin{aligned} &a(-u + K + 2iK') = a(u), \quad b(-u + K + 2iK') = -b(u); \\ &a^*[(-u + 2iK')^*] = -a(u), \quad b^*[(-u + 2iK')^*] = -b(u). \end{aligned} \tag{3.7.16}$$

Analyzing the formal series obtained by iterations from the integral equations (3.7.9), we can conclude that the matrix functions

$$\chi_+(u) = (\chi_1^{(2)}(u), \chi_2^{(2)}(u)), \quad \chi_-(u) = (\chi_2^{(1)}(u), \chi_1^{(1)}(u))$$

admit analytic continuation from the contour Γ to the domains $\Gamma_+ = \{u : \operatorname{Im}\tau > 0\}$ and $\Gamma_- = \{u : \operatorname{Im}\tau < 0\}$, respectively. Here the superscripts (1) and (2) enumerate the columns of the Jost matrix functions.

The first two reductions of (3.7.13) do not change the columns of the Jost matrices and therefore are transferred from the contour Γ into the analyticity domain of the function $\chi_\pm(u)$:

$$\begin{aligned} \chi_\pm(-u + K + 2iK') &= \pm\sigma_3\chi_\pm(u), \\ \chi_\pm^*[(-u + 2iK')^*] &= \pm i\sigma_2\chi_\pm(u)\sigma_3. \end{aligned} \tag{3.7.17}$$

The last relation of (3.7.13) can be rewritten in terms of $\chi_\pm(u)$ only on the contour Γ:

$$\chi_\pm^*(u^*) = \mp\frac{(1 + k')\sqrt{k'}}{2\beta(u)\beta(-u)} i\sigma_2\chi_\mp(u)\sigma_1 \quad (u \in \Gamma). \tag{3.7.18}$$

On the contour Γ, all basic solutions are interconnected:

$$\chi_- = \chi_2 M_{+-} = \chi_1 M_{--} i\sigma_2, \quad \chi_+ = -\chi_2\sigma_3 M_{++} = \chi_1 M_{-+}\sigma_1. \tag{3.7.19}$$

The matrices M have the same shape as with a domain wall not subject to a spin wave:

$$\begin{aligned} M_{+-} &= \begin{pmatrix} 1 & a \\ 0 & b \end{pmatrix}, \quad M_{--} = \begin{pmatrix} 1 & -a^* \\ 0 & b \end{pmatrix}, \\ M_{++} &= \begin{pmatrix} b^* & 0 \\ a^* & 1 \end{pmatrix}, \quad M_{-+} = \begin{pmatrix} b^* & 0 \\ a & 1 \end{pmatrix}. \end{aligned}$$

The relationship between the Jost solutions (3.7.14) on the contour Γ is equivalent to the condition of conjugation of the analytic functions $\chi_\pm(u)$ on the same contour:

$$\chi_-(u) = -\frac{1}{b^*(u^*)}\chi_+(u)\sigma_3 \begin{pmatrix} 1 & a(u) \\ a^*(u^*) & 1 \end{pmatrix}. \tag{3.7.20}$$

It is easy to show (see (3.7.19)) that

$$b(u) = \frac{\det\chi_-(u)}{\det\chi_2^{(0)}(u)}, \quad b^*(u^*) = -\frac{\det\chi_+(u)}{\det\chi_1^{(0)}(u)}. \tag{3.7.21}$$

Consequently, the elements $b(u)$ and $b^*(u^*)$ of the transition matrix can be analytically continued from the contour Γ to the domains Γ_- and Γ_+, respectively, with the exception perhaps of the points where

$$\det\chi_1^{(0)}(u) = \det\chi_2^{(0)}(u) = \beta^2(u) + \beta^2(-u) \tag{3.7.22}$$

goes to zero.

The zeros of the coefficients $b(u)$ and $b^*(u^*)$ meet the discrete spectrum of the auxiliary linear system (3.7.6). Due to reductions (3.7.16), the function $b(u)$ has two special zeros

$$u = \mathrm{i}K' - K/2, \quad -\mathrm{i}K' + K/2 \quad (\mathrm{mod}(2K, 4\mathrm{i}K')) \tag{3.7.23}$$

and a set of zeros of the form

$$\begin{aligned} u = \mu_s, \quad -\mu_s^* - 2\mathrm{i}K', \quad \mu_s^* + K, \\ -\mu_s - K + 2\mathrm{i}K' \quad (\mathrm{mod}(2K, 4\mathrm{i}K')), \end{aligned} \tag{3.7.24}$$

where $\mu_s \in \Gamma_-$, $s = 1, 2, \dots, N$, are complex parameters. The special zeros (3.7.23) are always present because they guarantee the existence of a domain wall. Additional zeros (3.7.24) may not exist because they are associated with breathers excited against the background of a spin wave.

The zeros of the functions $b^*(u^*)$ and $b(u)$ turn out to be complex conjugate.

Next, we are concerned with soliton-like excitations for which the reflection coefficients are equal to zero,

$$a(u) = a^*(u^*) = 0.$$

In this case the condition for the unimodularity of the transition matrix takes a simple form

$$b(u)b^*(u^*) = 1. \tag{3.7.25}$$

Among a set of soliton solutions to the Landau–Lifshitz equations, of special interest is a solution describing the domain wall-spin wave interaction without breathers. As already noted, such a solution is determined by the specific zeros (3.7.23) of the function $b(u)$. For soliton solutions, the function $b(u)$ becomes meromorphic in the u-plane. Formula (3.7.25) implies that its poles coincide with the zeros of the function $b^*(u^*)$,

$$u = \pm(\mathrm{i}K' + K/2) \quad (\mathrm{mod}(2K, 4\mathrm{i}K')).$$

Using the theorem on the factorization of an elliptic function (see Appendix), we restore the explicit form of the function $b(u)$:

$$b(u) = \theta \exp(8\mathrm{i}K'\eta_1)\frac{\sigma(u - \mathrm{i}K' + K/2)\sigma(u + \mathrm{i}K' - K/2)}{\sigma(u + \mathrm{i}K' + K/2)\sigma(u - \mathrm{i}K' - K/2)}.$$

Here $\theta = \pm 1$, the Weierstrass sigma-functions have periods $2K$ and $4\mathrm{i}K'$, and $\eta_1 = \zeta(K|K, 2\mathrm{i}K')$

By direct verification we can verify the validity of equivalent representations (see Appendix):

$$\begin{aligned} b(u) &= \mathrm{i}\theta(\mathrm{sn}v - \mathrm{i}\,\mathrm{cn}v)(\mathrm{sn}w - \mathrm{i}\,\mathrm{cn}w)|_{v=u+K/2, w=-u+K/2} \\ &= -\theta\left(\frac{\mathrm{i}+1}{\mathrm{i}-1}\right)\left(\frac{\mathrm{dn}u - \mathrm{i}\sqrt{k'}}{\mathrm{dn}u + \mathrm{i}\sqrt{k'}}\right) = -\frac{2\theta\beta(u)\beta(-u)}{\sqrt{k'}(1+k')}. \end{aligned} \tag{3.7.26}$$

Constructing the desired solution to the Landau–Lifshitz equations requires applying the "dressing" method for the function

$$\exp(-\mathrm{i}\tau\sigma_3\eta), \tag{3.7.27}$$

rather than for an exact solution to the U–V equations (3.7.6) [30]. Here $\eta = x - (2w_3 + p\cos\theta_0)t$. Further, we show that function (3.7.27) contains all the essential features necessary for recovering the U–V pair and the solutions $\chi_\pm(u, x, t)$.

The solution $\chi_-(u)$ should be sought as

$$\chi_-(u) = \Psi(u)\exp(-\mathrm{i}\tau\sigma_3\eta). \tag{3.7.28}$$

The only singularities of the function $\Psi(u)$ in the u-plane are the poles at the points $u = \pm(K/2 + \mathrm{i}K')$ and $\mathrm{mod}(2K, 4\mathrm{i}K')$

According to formulas (3.7.12) and (3.7.18), the matrix function $\Psi(u)$ must satisfy the reductions

$$\begin{aligned} \Psi(-u + K + 2\mathrm{i}K') &= -\sigma_3\Psi(u), \\ \Psi^*[(-u + 2\mathrm{i}K')^*] &= -\mathrm{i}\sigma_2\Psi(u)\sigma_3. \end{aligned} \tag{3.7.29}$$

Let us show that the conjugation condition for the analytic functions $\chi_\pm(u)$ (3.7.20) on the contour Γ can be rewritten in terms of only one function $\Psi(u)$. For this purpose, using the identity

$$\chi_+^{-1} = \frac{1}{\det\chi_+}\sigma_2\chi_+^{\mathrm{T}}\sigma_2,$$

valid for any nondegenerate 2×2-matrix χ_+, and relations (3.7.18), (3.7.21), and (3.7.22), we should establish a connection between the functions $\chi_+^{-1}(u)$ and $\chi_-^\dagger(u^*)$ on the contour Γ:

$$\chi_+^{-1}(u) = -\frac{1}{b^*(u^*)}\frac{2\beta(u)\beta(-u)\chi_-^\dagger(u^*)}{(1+k')\sqrt{k'}[\beta^2(u) + \beta^2(-u)]}. \tag{3.7.30}$$

Then, for $a = a^* = 0$, connection (3.7.30) and formulas (3.7.20) and (A.17) yield

$$\Psi^\dagger(u^*)\Psi(u) = [\beta(u)\beta^*(u^*) + \beta(-u)\beta^*(-u^*)]I \quad (u \in \Gamma). \tag{3.7.31}$$

Further analysis requires knowing the asymptotic behavior of the function $\chi_-(x, t, u)$ for $u \in \Gamma$ as $x \to \pm\infty$. Formulas (3.7.19) prompt how this function behaves:

$$\begin{aligned} \chi_- &\to \exp\left(\frac{\mathrm{i}\sigma_3}{2}\varphi_-\right)\begin{pmatrix} -\beta(u)b(u) & \beta(-u) \\ -\beta(-u)b(u) & -\beta(u) \end{pmatrix} \\ &\quad\times \exp(\mathrm{i}\tau\sigma_3\eta), \quad x \to -\infty; \\ \chi_- &\to \exp\left(-\frac{\mathrm{i}\sigma_3}{2}\varphi_+\right)\begin{pmatrix} \beta(-u) & -\beta(u)b(u) \\ \beta(u) & \beta(-u)b(u) \end{pmatrix} \\ &\quad\times \exp(-\mathrm{i}\tau\sigma_3\eta), \quad x \to +\infty. \end{aligned} \tag{3.7.32}$$

Let us show that constructing the matrix function $\Psi(u)$ is reduced to algebraic calculations. Comparison of formulas (3.7.28) and (3.7.32) makes it possible to guess an u-dependence of the matrix elements of the function $\Psi(u)$ to additionally specify the form of the latter through reductions (3.7.29):

$$\Psi(u) = \frac{1}{\sqrt{\gamma}}\begin{pmatrix} \beta(-u)a_1 - \beta(u)b(u)a_2 & -\beta(u)b(u)b_1 + \beta(-u)b_2 \\ \beta(u)a_1^* - \beta(-u)b(u)a_2^* & \beta(-u)b(u)b_1^* - \beta(u)b_2^* \end{pmatrix},$$

where $a_{1,2}$, $b_{1,2}$, and $\sqrt{\gamma}$ depend only on x and t. Unlike the previous problems, the matrix function $\Psi(u)$ has double poles instead of simple ones at the points $u = \pm(\mathrm{i}K' + K/2)$ $(\mathrm{mod}(2K, 4\mathrm{i}K'))$.

We can arrive at additional connections between the parameters $a_{1,2}$, $b_{1,2}$, and $\sqrt{\gamma}$ on the contour Γ under condition (3.7.31) and the requirement (see (3.7.21) and (3.7.22))

$$\det\Psi(u) = b(u)[\beta^2(u) + \beta^2(-u)].$$

These connections can be presented as

$$a_i = b_i \quad (i = 1, 2), \quad \gamma = |a_1|^2 + |a_2|^2 - 2\theta\sqrt{k'}(1 + k')^{-1}(a_1a_2^* + a_2a_1^*).$$

This can be directly verified using formulas (A.17), (3.7.25), and (3.7.26) and the identities

$$\begin{aligned} \beta^2(\pm u) &= -k'\left[\frac{\tau + \mathrm{i}(\kappa w_3 + \beta\cos\theta_0/2)}{\tau \pm (w_3\cos\theta_0 - p/2)}\right], \\ -\frac{2\beta(u)\beta(-u)}{(1 + k')\sqrt{k'}} &= \frac{\tau + \mathrm{i}(\kappa w_3 + \beta\cos\theta_0/2)}{w_1\sqrt{1+\kappa^2}} = \frac{w_1\sqrt{1+\kappa^2}}{\tau - \mathrm{i}(\kappa w_3 + \beta\cos\theta_0/2)}, \\ \frac{\beta(u)}{\beta(-u)} &= \frac{\tau - (w_3\cos\theta_0 - p/2)}{w_1\sin\theta_0} = \frac{w_1\sin\theta_0}{\tau + w_3\cos\theta_0 - p/2}. \end{aligned} \tag{3.7.33}$$

In the end, we have found a matrix function that satisfies the algebraic constraints (3.7.29) and (3.7.31):

$$\begin{aligned}\Psi(u) &= \frac{1}{\sqrt{\gamma}}\begin{pmatrix}\beta(-u)a_1 - \beta(u)b(u)a_2 & -\beta(u)b(u)a_1 + \beta(-u)a_2\\ \beta(u)a_1^* - \beta(-u)b(u)a_2^* & \beta(-u)b(u)a_1^* - \beta(u)a_2^*\end{pmatrix},\\ \gamma &= |a_1|^2 + |a_2|^2 - 2\theta\sqrt{k'}(1+k')^{-1}(a_1a_2^* + a_1^*a_2),\end{aligned} \tag{3.7.34}$$

from which we have

$$\Psi^{-1}(u) = \Psi^\dagger(u^*)r(u), \tag{3.7.35}$$

where $r(u) = [\beta(u)\beta^*(u^*) + \beta(-u)\beta^*(-u^*)]^{-1}$.

In the domain Γ_-, additional differential constraints on choosing the matrix $\Psi(u,x,t)$ consist in reducing the functions $\partial_x\chi_-\chi_-^{-1}$ and $\partial_t\chi_-\chi_-^{-1}$ to the expressions

$$\begin{aligned}\partial_x\chi_-\chi_-^{-1} &= U = -\mathrm{i}\{w_1(S_1\sigma_1 + S_2\sigma_2) + w_3S_3\sigma_3\},\\ \partial_t\chi_-\chi_-^{-1} &= V = -\mathrm{i}\{w_1([\mathbf{S}\times\partial_x\mathbf{S}]_1\sigma_1 + [\mathbf{S}\times\partial_x\mathbf{S}]_2\sigma_2)\\ &\quad + w_3[\mathbf{S}\times\partial_x\mathbf{S}]_3\sigma_3 - 2w_1^2S_3\sigma_3 - 2w_1w_3(S_1\sigma_1 + S_2\sigma_2)\},\end{aligned} \tag{3.7.36}$$

where $\mathbf{S}^2 = 1$.

Relations (3.7.28), (A.17), and (3.7.33)–(3.7.35) enable us to conclude that the left parts of formulas (3.7.36) include the desired algebraic-structure terms only when the following equations are satisfied:

$$\begin{aligned}a_1\partial_xa_2 - a_2\partial_xa_1 &= a_1a_2(\mathrm{i}p - \beta\cos\theta_0),\\ a_1^*\partial_xa_1 + a_2^*\partial_xa_2 - a_1\partial_xa_1^* - a_2\partial_xa_2^* &= \mathrm{i}p(|a_2|^2 - |a_1|^2),\\ a_2^*\partial_xa_1 + a_1^*\partial_xa_2 - a_1\partial_xa_2^* - a_2\partial_xa_1^* &= \beta\cos\theta_0(a_1a_2^* - a_2a_1^*),\\ a_1\partial_ta_2 - a_2\partial_ta_1 &= -a_1a_2(\beta p\sin^2\theta_0 + \mathrm{i}\omega),\\ a_1^*\partial_ta_1 + a_2^*\partial_ta_2 - a_1\partial_ta_1^* - a_2\partial_ta_2^* &= \mathrm{i}\omega(|a_1|^2 - |a_2|^2),\\ a_2^*\partial_ta_1 + a_1^*\partial_ta_2 - a_1\partial_ta_2^* - a_2\partial_ta_1^* &= \beta p\sin^2\theta_0(a_1a_2^* - a_2a_1^*).\end{aligned}$$

This system is boiled down to linear equations having a simple solution

$$a_{1,2} = \exp\left(\pm\frac{\beta}{2}(x\cos\theta_0 + pt\sin^2\theta_0 + \Delta_\pm) \mp \frac{\mathrm{i}\varphi_\pm}{2}\right), \tag{3.7.37}$$

where $\varphi_\pm = px - \omega t + \delta_\pm$, and $\delta_\pm$ and $\Delta_\pm$ are real constants of integration.

If the functions $a_{1,2}(x,t)$ are chosen in line with formulas (3.7.37), the Jost solution has the correct asymptotics (3.7.32). In this context, the first equation of (3.7.36) makes it possible to find the components of the vector $\mathbf{S}$:

$$S_1 - iS_2 = \frac{1}{\gamma}\left[(a_1^2 + a_2^2)\sin\theta_0 - \frac{2a_1 a_2 \theta}{\sqrt{1+\kappa^2}}(1 - i\kappa\cos\theta_0)\right],$$
$$S_3 = \frac{1}{\gamma}\left[(|a_1|^2 - |a_2|^2)\cos\theta_0 + \frac{i\kappa\theta\sin\theta_0}{\sqrt{1+\kappa^2}}(a_1 a_2^* - a_1^* a_2)\right].$$

We have exactly solved the Landau–Lifshitz equations for describing the breather-free domain wall-spin wave interaction. In more detailed notation, their solution has the form

$$S_1 - iS_2 = e^{-i\delta_2}\,\frac{\sin\theta_0\cosh(y + i\Phi) - \theta(1+\kappa^2)^{-1/2}(1 - i\kappa\cos\theta_0)}{\cosh y - \theta(1+\kappa^2)^{-1/2}\sin\theta_0\cos\Phi},$$
$$S_3 = \frac{\sinh y\cos\theta_0 - \theta\kappa(1+\kappa^2)^{-1/2}\sin\theta_0\sin\Phi}{\cosh y - \theta(1+\kappa^2)^{-1/2}\sin\theta_0\cos\Phi}, \tag{3.7.38}$$

where

$$y = \beta(\cos\theta_0 x + p\sin^2\theta_0 t + \Delta_1), \quad \Delta_1 = (\Delta_+ + \Delta_-)/2,$$
$$\Phi = \omega t - px - \delta_1, \quad \omega = \cos\theta_0(p^2 + \beta^2), \quad \theta = \pm 1,$$
$$\delta_1 = \frac{\delta_+ + \delta_-}{2}, \quad \delta_2 = \frac{\delta_+ - \delta_-}{2}.$$

In dimensionless variables, the width of the domain wall interacting with a spin wave is of order $(\beta\cos\theta_0)^{-1}$, and its velocity is opposite to the direction of the spin wave propagation and is determined by the relation

$$V = -p\tan\theta_0\sin\theta_0.$$

For $p = \theta_0 = 0$, solution (3.7.38) handles an isolated, motionless domain wall (3.5.7) not exposed to a magnetization wave, with the parameter δ_2 governing its topology.

For $p = 0$ and $0 < \cos\theta_0 < 1$, at spatial infinity, we have the uniform precession of vector **S** with frequency $\omega = \beta^2\cos\theta_0$; the precession directions, both left and right from the center of the domain wall, are opposite. Although the domain wall does not move as a whole, it pulsates. The component S_3 oscillates between the extreme positions shown in Figure 3.10. At times when $\theta\cos\Phi = 1$, the function $S_3(x,t)$ reaches the values of ± 1 at points with coordinates $\bar{x}_\pm = (\beta\cos\theta_0)^{-1}(-\beta\Delta_1 \mp \ln\tan(\theta_0/2))$. At other times, the extreme values satisfy the inequality $\cos\theta_0 \le |S_3|_{max} < 1$ and are reached at points that are periodically shifted to spatial infinity where $|S_3|_{max} \to \cos\theta_0$. At the center of the domain wall (for $y = 0$), the magnetization is time-independent and remains the same as in the lack of a precession wave: $\mathbf{S} = (-\theta\cos\delta_2, -\theta\sin\delta_2, 0)$.

In the general case, for convenience, we rewrite solution (3.7.38) in the form

$$\mathbf{S} = \frac{1}{A}(\mathbf{m} + \mathbf{n}), \quad A \equiv \cosh y - \frac{\theta}{\sqrt{1+\kappa^2}}\sin\theta_0\cos\Phi.$$

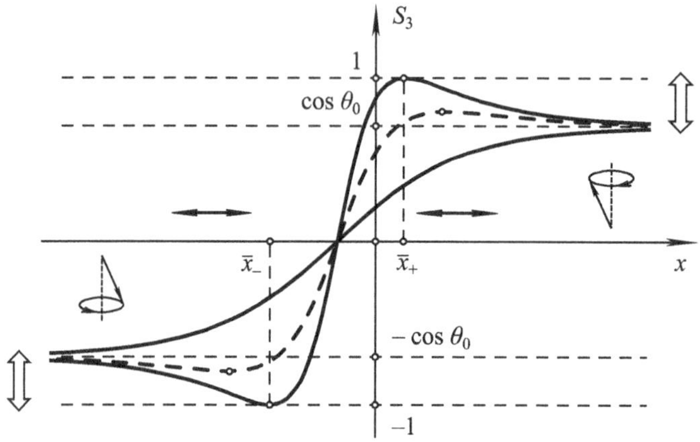

Figure 3.10: S_3-projection pulsations between two extreme positions depicted by solid lines.

Here **n** is the vector with components

$$n_1 = \theta(1+\kappa^2)^{-1/2}[\kappa\cos\theta_0\sin\delta_2 - \cos\delta_2],$$
$$n_2 = -\theta(1+\kappa^2)^{-1/2}[\kappa\cos\theta_0\cos\delta_2 + \sin\delta_2],$$
$$n_3 = \sinh y\cos\theta_0.$$

The vector **m** admits the representation

$$\mathbf{m} = \mathbf{e_1}\cos\Phi + \mathbf{e_2}\sin\Phi; \quad \mathbf{e_1} = \cosh y\sin\theta_0(\cos\delta_2, \sin\delta_2, 0),$$
$$\mathbf{e_2} = \sinh y\sin\theta_0(\sin\delta_2, -\cos\delta_2, 0) - \theta\kappa(1+\kappa^2)^{-1/2}\sin\theta_0(0,0,1).$$

The vectors **n**, $\mathbf{e_1}$, and $\mathbf{e_2}$ are time-unchanged at each point along the Ox-axis (for each constant value of y) in the frame of reference connected to the domain wall. Since phase Φ is linearly time-dependent,

$$\Phi = \frac{(p^2+\beta^2\cos^2\theta_0)t}{\cos\theta_0} + \left(\Delta_1 - \frac{y}{\beta}\right)\frac{p}{\cos\theta_0} - \delta_1,$$

the end of the vector **m** traces an elliptic trajectory in the plane defined by the vectors $\mathbf{e_1}$ and $\mathbf{e_2}$. Accordingly, the vector $\mathbf{m}+\mathbf{n}$ flows around the surface of the cone with axis coinciding with the direction of the vector **n**.

The vector **S** is codirectional with the vector $\mathbf{m}+\mathbf{n}$ and differs from it by the factor $1/A > 0$. Therefore the passage from the vector $\mathbf{m}+\mathbf{n}$ to the vector **S** instigates a scaling transformation for the admissible values of the former to fall on the surface of the sphere $\mathbf{S}^2 = 1$ (Figure 3.11).

Thus, in the frame of reference connected to the domain wall, the vector **S** precesses with frequency $(p^2+\cos^2\theta_0\beta^2)/\cos\theta_0$ around local axes parallel to the vector **n**.

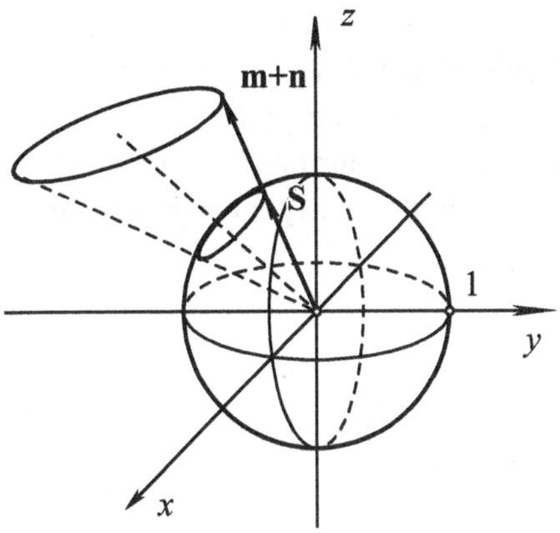

Figure 3.11: The projection of a vector **m** + **n** onto the surface of a unit-radius sphere is performed along the generators of the cone.

The position of the vector **n** at the domain wall center (for $y = 0$) characterizes its internal structure through the parameter

$$\xi = \frac{n_2}{n_1} = \frac{\kappa \cos\theta_0 \cos\delta_2 + \sin\delta_2}{\cos\delta_2 - \kappa\cos\theta_0 \sin\delta_2}, \qquad \kappa = \frac{p}{\beta}.$$

The values of $\kappa\cos\theta_0\cos\delta_2 + \sin\delta_2 = 0$ and $\cos\delta_2 - \kappa\cos\theta_0\sin\delta_2 = 0$ correspond to walls of the "Neel" and "Bloch" types, respectively. We have used quotation marks because the neglected magnetostatics and energies of both walls in our model are the same.

Such a situation suggests that the domain wall structure varies depending on the ratio between the momentum density $p\cos\theta_0$ of a wave incident on it and the anisotropy parameter β. Under the influence of a magnetization wave, the Bloch wall becomes intermediate between the Bloch and Neel walls. On the other hand, even when exposed to a wave, walls with predominant "Bloch"-type magnetization distribution can appear. For such walls, magneto-static interactions are not so important.

Note also that for quasi-one-dimensional ferromagnetic "easy-axis" anisotropy chains, the theoretical results may be collocated directly with experimental data. In quasi-one-dimensional crystals, magneto-static interactions and dissipative losses are small. They can be taken into account within the soliton perturbation theory [13, 24, 19].

From a mathematical point of view, a distinguishing feature of the problem at hand is that the matrix function $\Psi(u)$ (3.7.34) of the Riemann problem has double poles and is not expressed in terms of simple projection matrices. It is worth underscoring that a normalization condition for the Riemann problem (3.7.31) to be completely determined was not used. Only the asymptotic behavior of function (3.7.32) was stated for $u \in \Gamma$ as $x \to \pm\infty$.

Certainly, the Riemann problem and its solution can be reformulated in terms of the original spectral parameter λ. For this, it suffices to use formulas (3.7.33). The two-sheeted Riemann surface of the function $\tau = \tau(\lambda)$ and the contour $\Gamma = \{\lambda : \operatorname{Im} \tau(\lambda) = 0\}$ are depicted in Figure 3.12. The above relations once again indicate the complexity of the "dressing" procedure in terms of the spectral parameter λ or the parameter introduced in [30]. Perhaps, this explains the fact that the expressions for magnetization given in [30] have an incorrect denominator and therefore do not satisfy the Landau–Lifshitz equations.

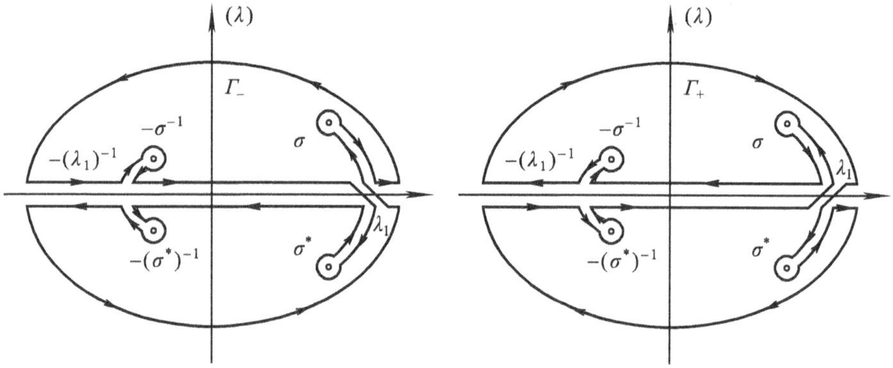

Figure 3.12: Sheets $\Gamma_\pm$ of the Riemann surface of the function $\tau = \tau(\lambda)$; $\sigma = (\sqrt{1+\kappa^2} + \kappa)\exp i\theta_0$, $\lambda_1 = \kappa \cos\theta_0 + \sqrt{\kappa^2 \cos^2\theta_0 + 1}$.

3.8 Interaction of a nonlinear wave with breathers and nuclei of magnetization reversal

Get out on a flat board, six feet long, two feet wide, and roughly oval in shape. Lie down upon it like a small boy on a coaster and paddle with your hands out to deep water, where the waves begin to crest.

J. London "The cruise of the Snark"

This section deals with analytical solving of the Landau–Lifshitz model describing the interaction between a breather and a high-amplitude precession wave in an easy-axis ferromagnet [15–17, 14]. At certain parameter values, a breather is a solitary domain, i. e., a wide area bounded by two domain walls in an overmagnetized material. It is established that the solitary domain moves toward the wave with the same speed as a separate domain wall. Having passed through the domain, the wave acquires a specific phase shift. This circumstance can be used to detect and diagnose magnetization reversal nuclei in ferromagnets. Generally, a breather has internal oscillations and, in contrast to a domain wall, can move along the sample even under no-wave conditions. This fact affects its notation and complicates the analysis. The features of the breather's oscillatory regimes near the boundaries of the region of its existence are revealed. It is shown that the speed of the breather can be controlled by changing the wavenumber and amplitude of a pumping wave.

3.8.1 Constructing breather solutions

The problem posed here differs from that in Section 3.7 only in the shape of the asymptotics $\mathbf{S}_0^{(\pm)}$ as $x \to \pm\infty$. Now, at both infinities, we have

$$\mathbf{S} \to \mathbf{S}_0^{(\pm)} = (\sin\theta_0 \cos(\varphi_\pm),\ \sin\theta_0 \sin(\varphi_\pm),\ \cos\theta_0). \tag{3.8.1}$$

The angle θ_0 characterizes the wave amplitude, $\varphi_\pm = px - \omega t + \delta_\pm$, $\omega = (p^2+\beta^2)\cos\theta_0$ is the precession frequency, p is the wavenumber, and $\delta_\pm = \text{const}$. Further, we demonstrate that the phase shifts $\delta_\pm$ cannot be arbitrary; they carry information about the breather that modulates the wave (3.8.1).

As solutions of system (3.7.6), we define the Jost functions χ_1 and χ_2 with the asymptotic behavior $\chi_{1,2} \to \chi_{1,2}^{(0)}$ as $x \to \mp\infty$. Here $\chi_{1,2}^{(0)}$ are the exact solutions of system (3.7.6):

$$\chi_{1,2}^{(0)}(u) = \exp\left(-\frac{\mathrm{i}\sigma_3}{2}\varphi_\mp\right)\begin{pmatrix}\beta(-u) & -\beta(u)\\ \beta(u) & \beta(-u)\end{pmatrix}\exp\left(\frac{\tau}{\mathrm{i}}\sigma_3\eta\right),$$
$$\eta = x - (2w_3 + p\cos\theta_0)t.$$

They correspond to the magnetization distributions (3.8.1).

As before, the set $\Gamma = \{u : \operatorname{Im}\tau(u) = 0\}$ meets the continuous spectrum of the scattering problem. On the contour Γ, the Jost solutions are interconnected by the former relation (3.7.14). However, the new Jost functions adhere to constraints that are marginally different from the previous ones:

$$\begin{aligned}\chi_{1,2}(-u+K+2\mathrm{i}K') &= -\sigma_3\chi_{1,2}(u)\sigma_3,\\ \chi_{1,2}^*[(-u+2\mathrm{i}K')^*] &= -\mathrm{i}\sigma_2\chi_{1,2}(u),\\ \chi_{1,2}^*(u^*) &= \frac{(1+k')\sqrt{k'}}{2\beta(u)\beta(-u)}\sigma_2\chi_{1,2}(u)\sigma_2 \quad (u\in\Gamma).\end{aligned} \tag{3.8.2}$$

Therefore, although the transition matrix has the same structure (3.7.15), its elements satisfy other constraints:

$$\begin{aligned}a(-u+K+2\mathrm{i}K') &= a(u), & b(-u+K+2\mathrm{i}K') &= -b(u),\\ a^*[(-u+2\mathrm{i}K')^*] &= a(u), & b^*[(-u+2\mathrm{i}K')^*] &= b(u).\end{aligned}$$

Let us introduce the functions $\chi_+ = (\chi_1^{(1)}, \chi_2^{(2)})$ and $\chi_- = (\chi_2^{(1)}, \chi_1^{(2)})$, which admit analytic continuation from the contour Γ to the domains Γ_+ and Γ_-, respectively.

Relation (3.7.14) yields representations for the elements of the transition matrix:

$$a(u) = \frac{\det\chi_+(u)}{\det\chi_2^{(0)}(u)}, \quad a^*(u^*) = \frac{\det\chi_-(u)}{\det\chi_2^{(0)}(u)}.$$

Hence the functions $a(u)$ and $a^*(u^*)$ can be analytically continued from the contour Γ to the domains Γ_+ and Γ_-, respectively, with the exception perhaps of the points where $\det\chi_{1,2}^{(0)}(u) = \beta^2(-u) + \beta^2(u) = 0$.

The zeros of the coefficients $a(u)$ and $a^*(u^*)$ in the domains Γ_- and Γ_+ are responsible for a discrete spectrum of the scattering problem. In view of reductions (3.8.2), the function has two special zeros (3.7.23) and groups of zeros (3.7.24). The former induce a domain wall to emerge, and the latter (each of them) take into account the presence of one breather against the background of the wave (3.8.1).

Under the boundary conditions (3.8.1), the problem at hand either is domain wall-free or involves pairs of domain walls with opposite topological charges. In other words, this problem includes either no special zeros (3.7.23) or multiple ones, which give rise to weakly coupled exponentially polynomial solitons. For definiteness, the zeros of the functions $a(u)$ and $a^*(u^*)$ are assumed to be simple, and the special zeros (3.7.23) can therefore be ignored.

For further analysis, it is convenient to extract the seeding solution $\chi_2^{(0)}$ from the functions $\chi_\pm$ and introduce the matrix functions $\Psi_\pm$:

$$\chi_-(u) = \Psi_-(u)\chi_2^{(0)}(u), \quad \frac{\chi_+(u)}{a(u)} = \Psi_+(u)\chi_2^{(0)}(u).$$

Then relation (3.7.14) reduces to the conjugation condition for the analytic functions $\Psi_\pm$ on the contour Γ:

$$\Psi_-(x,t,u) = \Psi_+(x,t,u)G(x,t,u) \quad (u \in \Gamma), \tag{3.8.3}$$

where

$$G(x,t,u) = \chi_2^{(0)}(x,t,u)\begin{pmatrix} 1 & -b^*(u^*) \\ -b(u) & 1 \end{pmatrix}(\chi_2^{(0)})^{-1}(x,t,u).$$

Thus we again come to the Riemann problem. Its distinguishing feature is that reductions (3.8.2) rather “rigidly” specify the structure of matrix functions $\Psi_\pm$. In the domains of their analyticity, these functions satisfy the following constraints:

$$\begin{aligned} \Psi_\pm(-u + K + 2iK') &= \sigma_3\Psi_\pm(u)\sigma_3, \\ \Psi_\pm^*[(-u + 2iK')^*] &= \sigma_2\Psi_\pm(u)\sigma_2. \end{aligned} \tag{3.8.4}$$

On the contour Γ, one of them can be expressed through the other:

$$\Psi_+^{-1}(u) = \Psi_-^\dagger(u^*), \quad u \in \Gamma.$$

Finally, the condition for the contour Γ can be rewritten in terms of only one function Ψ_-:

$$\Psi_-^\dagger(u^*)\Psi_-(u) = G(u). \tag{3.8.5}$$

Let $0 < u_0 < K$ be the real root of the equation

$$\mathrm{i}\,\mathrm{dn}(u_0 + \mathrm{i}K') = \tan\frac{\theta_0 + \theta_4}{2}.$$

At the point $u = u_0$, the function $\tau(u)$ has a simple pole (see (3.7.10)), and the right-hand side of formula (3.8.5) has an essential singularity. On the other hand, the left-hand side of eq. (3.8.5) in a small neighborhood of the point $u - u_0$ is a function differentiable over the parameter u. The contradiction can be avoided only under the constraint $b(u_0) = 0$ or, equivalently, under the unitarity condition for the matrix function $\Psi_-(u)$ at the point $u = u_0$:

$$\Psi_-^\dagger(u_0)\Psi_-(u_0) = I. \tag{3.8.6}$$

Next, we get a formula for calculating the magnetization. The function Ψ_- satisfies the differential equations

$$\partial_x\Psi_- = U\Psi_- - \Psi_- U_0, \quad \partial_t\Psi_- = V\Psi_- - \Psi_- V_0, \tag{3.8.7}$$

where the matrix functions U and V are defined by formulas (3.7.6), and U_0 and V_0 are the values of U and V at $\mathbf{S} = \mathbf{S}_0^{(+)}$. From this, in particular, we obtain the equality $U = \partial_x\Psi_-\Psi_-^{-1} + \Psi_- U_0\Psi_-^{-1}$. Its left- and right-hand sides have simple poles at $u = u_0$. Equating the residues at this point and taking into account eq. (3.8.6), we deduce a formula for calculating the magnetization:

$$S_k\sigma_k = \Psi_-(u_0)(\mathbf{S}_0^{(+)})_k\sigma_k\Psi_-^\dagger(u_0). \tag{3.8.8}$$

This formula is used after knowing the solutions of the Riemann problem on a torus (3.8.3).

Soliton states correspond to solutions to the Riemann problem with zeros of the coefficients $a(u)$ and $a^*(u^*)$ in the domains Γ_+ and Γ_- when $b = b^* \equiv 0$. In this case the functions $\Psi_-(u)$ and $\Psi_-^\dagger(u^*)$ are meromorphic in the domain $\Gamma_- \cup \Gamma_+$. The condition of their conjugation on the contour Γ takes a simple form:

$$\Psi_-^\dagger(u^*)\Psi_-(u) = \Psi_-(u)\Psi_-^\dagger(u^*) = I \quad (u \in \Gamma).$$

Let us construct the simplest soliton function $\Psi_-^\dagger(u^*)$ for describing a breather against the background of the wave (3.8.1). Due to (3.7.24), such a function has four simple poles. Therefore it can be expanded in terms of the Weierstrass zeta functions as follows (see Appendix):

$$\Psi_-^\dagger(u^*) = A_0 + A\zeta_1 + B\zeta_2 + C\zeta_3 + D\zeta_4, \tag{3.8.9}$$

where $\zeta_i = \zeta(u-\mu_i)$, $\mu_1 = \mu$, $\mu_2 = -\mu-K-2iK'$, $\mu_3 = -\mu^*+2iK'$, and $\mu_4 = \mu^*+K$. The periods of the Weierstrass functions are $2K$, $4iK'$, $\mu \in \Gamma_-$. Bearing in mind the transformation properties of the Weierstrass functions [3, 1] and the symmetry conditions (3.8.4), we express the matrices B, C, and D in terms of A:

$$B = -\boldsymbol{\sigma}_3 A \boldsymbol{\sigma}_3, \quad C = -\boldsymbol{\sigma}_2 A^* \boldsymbol{\sigma}_2, \quad D = \boldsymbol{\sigma}_1 A^* \boldsymbol{\sigma}_1.$$

In doing so, the matrix A_0 appears as

$$A_0 \equiv \begin{pmatrix} r & 2\eta_2 A_{12} \\ 2\eta_2 A_{21} & r^* \end{pmatrix},$$

where $\eta_2 = \zeta(-K - 2iK'|K, 2iK')$.

The matrix function (3.8.9) has periods $2K$ and $4iK'$ only if the sum of the residues of the function $\Psi_-^\dagger(u^*)$ is equal to zero at its poles (see [1] and Appendix). This requirement will be satisfied if the off-diagonal elements of the matrix A satisfy the constraint

$$A_{12} = -A_{21}^*. \tag{3.8.10}$$

Further calculations should be conducted under assumption (3.8.10). The final result will confirm its correctness.

The reductions specify the shape of the matrix function (3.8.9). Let it be $A = \tilde{A}R$, where $R = \operatorname{diag}(r, r^*)$, $r \neq 0$. Then we can present the outcome in a form convenient for further analysis:

$$\Psi_-^\dagger(u^*) = \Psi^\dagger(u^*)R,$$
$$\Psi^\dagger(u^*) = I + \begin{pmatrix} \tilde{A}_{11}\zeta_{12}(u) - \tilde{A}_{22}^*\zeta_{34}(u) & \tilde{A}_{12}\varphi(u) \\ \tilde{A}_{21}\varphi(u) & \tilde{A}_{22}\zeta_{12}(u) - \tilde{A}_{11}^*\zeta_{34}(u) \end{pmatrix},$$
$$\zeta_{12}(u) = \zeta_1 - \zeta_2, \quad \zeta_{34}(u) = \zeta_3 - \zeta_4, \quad \varphi(u) = \zeta_1 + \zeta_2 - \zeta_3 - \zeta_4 + 2\eta_2.$$

Conditions for the lack of poles in the product of the matrix functions

$$\Psi_-(u)\Psi_-^\dagger(u^*) = R^\dagger \Psi(u)\Psi^\dagger(u^*)R = I \tag{3.8.11}$$

are equivalent to one independent matrix equation

$$\tilde{A}^\dagger \Psi^\dagger(u^*)|_{u=\mu^*} = 0.$$

The latter has a nontrivial solution only under the condition (see [27] and Section 2.9.2):

$$\tilde{A}_{\alpha\beta} = m_\alpha X_\beta, \quad \alpha, \beta = 1, 2.$$

Given this condition and the identity

$$\varphi(\mu^*) + \zeta_{34}(\mu^*) = -\zeta_{12}^*(\mu^*),$$

we have derived two equations:

$$\begin{aligned} X_1 &= -\frac{m_1^*}{|m_1|^2\zeta_{12}(\mu^*) - |m_2|^2\zeta_{12}^*(\mu^*)}, \\ X_2 &= \frac{m_2^*}{|m_1|^2\zeta_{12}^*(\mu^*) - |m_2|^2\zeta_{12}(\mu^*)}. \end{aligned} \tag{3.8.12}$$

The constructed matrix $A_{\alpha\beta}$ holds for constraint (3.8.10).

The denominators appearing in formulas (3.8.12) can be eliminated by redefining the parameters r and r^*. Then the matrix function $\Psi^\dagger(u^*)$ takes the form

$$\Psi^\dagger(u^*) = \begin{pmatrix} -|m_1|^2\varphi_1(u) - |m_2|^2\varphi_2(u) & m_1 m_2^*\varphi(u) \\ -m_2 m_1^*\varphi(u) & |m_1|^2\varphi_2(u) + |m_2|^2\varphi_1(u) \end{pmatrix}, \tag{3.8.13}$$

where

$$\varphi_1(u) = \zeta_{12}(u) - \zeta_{12}(\mu^*), \quad \varphi_2(u) = \zeta_{34}(u) + \zeta_{12}^*(\mu^*).$$

Note that the functions $\varphi(u)$ and $\varphi_{1,2}(u)$ constitute a representation of the reduction group:

$$\begin{aligned} \varphi_{1,2}(-u + K + 2\mathrm{i}K') &= \varphi_{1,2}(u), \quad \varphi(-u + K + 2\mathrm{i}K') = -\varphi(u); \\ \varphi_{2,1}^*[(-u + 2\mathrm{i}K')^*] &= -\varphi_{1,2}(u), \quad \varphi^*[(-u + 2\mathrm{i}K')^*] = \varphi(u). \end{aligned}$$

Now we express combinations of the Weierstrass zeta functions $\varphi_{1,2}(u)$ and $\varphi(u)$ in terms of the Jacobi functions. To shorten the notation, we introduce the notations $\tilde{\mu} = \mu + \mathrm{i}K' + K/2$; $\tilde{s} = \mathrm{sn}(\tilde{\mu}, k)$, etc.; $s = \mathrm{sn}(u + \mathrm{i}K' + K/2, k)$, etc. Then we have

$$\begin{aligned} \varphi(u) &= -\frac{k^2(k' + |\tilde{d}|^2)sc}{(d - \tilde{d})(k' + \tilde{d}^* d)}, \\ \varphi_1(u) &= -\frac{k^2\tilde{s}\tilde{c}}{\tilde{d} + \tilde{d}^*}\left(\frac{d + \tilde{d}^*}{d - \tilde{d}}\right), \\ \varphi_2(u) &= -\frac{k^2\tilde{s}^*\tilde{c}^*}{\tilde{d} + \tilde{d}^*}\left(\frac{d\tilde{d} - k'}{d\tilde{d}^* + k'}\right). \end{aligned} \tag{3.8.14}$$

These and other identities, as always, are easiest to check by expanding their right- and left-hand sides either into "simple fractions" or into "factors" in terms of Weierstrass' functions (see Appendix).

For specific calculations, we use the following identities:

$$\begin{aligned}
&\varphi_1^*(u^*)\varphi(u) + \varphi^*(u^*)\varphi_2(u) = 0,\\
&\varphi_1(u)\varphi_2^*(u^*) + \varphi_2(u)\varphi_1^*(u^*) + \varphi(u)\varphi^*(u^*) = c_2,\\
&\varphi_1^*(u^*)\varphi_1(u) = \varphi_2^*(u^*)\varphi_2(u) = c_1,\\
&\varphi^2(u) - \varphi_1^2(u) - \varphi_2^2(u) = -\frac{c_2}{c_1}\varphi_1(u)\varphi_2(u),\\
&\frac{\beta(u)}{\beta(-u)} = \frac{1}{\sqrt{k'}}\mathrm{dn}\left(u + \frac{K}{2}\right),
\end{aligned} \tag{3.8.15}$$

where the constants c_1 and c_2 have the form

$$c_1 = \frac{k^4|\tilde{s}\tilde{c}|^2}{(\tilde{d} + \tilde{d}^*)^2}, \quad c_2 = \frac{2}{(\tilde{d} + \tilde{d}^*)^2}\left[(1 + k')|\tilde{d}|^2 + \left|k' + \tilde{d}\right|^2\right].$$

From relation (3.8.11) we find the modulus of the number r:

$$\left(RR^\dagger\right)^{-1} \equiv |r|^{-2}I = \Psi(u)\Psi^\dagger(u^*).$$

When constructed, the right-hand side of the equality is u-independent and is a positive-definite diagonal matrix. Direct calculations involving identities (3.8.15) yield

$$|r|^{-2} = \left(|m_1|^4 + |m_2|^4\right)c_1 + |m_1 m_2|^2 c_2. \tag{3.8.16}$$

Solving eqs. (3.7.1), we prefer another representation of $|r|^{-2}$ from (3.8.16), replacing the constants $c_{1,2}$ by combinations of the numbers $\varphi_{1,2}(u_0)$ and $\varphi(u_0)$ (see (3.8.15)).

Since eqs. (3.8.7) for U and V provide the lack of extra poles, the dependence of the vector $\mathbf{m} = (m_1, m_2)$ and phase $\varphi_0 = \arg(r)$ on the coordinates x and t is observed. The solutions to these equations are

$$\mathbf{m} = \chi_2^{(0)}(\mu)\mathbf{e}, \quad \varphi_0 = \text{const}, \tag{3.8.17}$$

where $\mathbf{e} = (e_1, e_2)$ is an arbitrary constant complex vector.

By direct calculation of the coefficient

$$a(u) = \det\Psi_-^\dagger(u^*) = -\frac{\varphi_1(u)\varphi_2(u)}{c_1} = \frac{\varphi(u)}{\varphi^*(u^*)} \tag{3.8.18}$$

we make sure that together with the prescribed analytical properties, the function $a(u)$ has poles at the points μ_i, zeros at the complex conjugate points, and satisfies the condition $a^*(u^*)a(u) = 1$.

To find the constant φ_0, we use the connection formula (3.7.14) for the Jost solutions on the contour Γ:

$$\Psi_-(u)\chi_2^{(0)}(u) = \chi_2(u)\begin{pmatrix} 1 & 0 \\ 0 & a^*(u^*) \end{pmatrix}.$$

The requirement of coinciding asymptotics of the right- and left-hand parts of the above equality leads to

$$\varphi_0 = \mathrm{i}\ln\left(-\frac{\tilde{s}|\tilde{c}|}{\tilde{c}|\tilde{s}|}\right) = \pi - \arg\frac{\tilde{s}}{\tilde{c}}. \tag{3.8.19}$$

Using formulas (3.8.8), (3.8.13), (3.8.16), (3.8.17), and (3.8.19), we have solved the Landau–Lifshitz equations for describing a breather against the background of the nonlinear, arbitrary-amplitude spin-wave (3.8.1). The outcome can be written as

$$\begin{aligned} S_3 &= \frac{(|\alpha|^2 - |\beta|^2)\cos\theta_0 - (\alpha^*\beta + \alpha\beta^*)\sin\theta_0}{|\alpha|^2 + |\beta|^2}, \\ S_1 - \mathrm{i}S_2 &= \frac{\exp(-\mathrm{i}\varphi_+ - 2\mathrm{i}\varphi_0)[2\alpha\beta\cos\theta_0 + (\alpha^2 - \beta^2)\sin\theta_0]}{|\alpha|^2 + |\beta|^2}. \end{aligned} \tag{3.8.20}$$

Here

$$\begin{aligned} &\varphi_0 = \pi - \arg(\tilde{s}/\tilde{c}), \quad \varphi_+ = px - \omega t + \delta_+, \\ &\alpha = |n_2|^2\frac{\varphi_1(u_0)}{\varphi(u_0)} - |n_1|^2\frac{\varphi_1^*(u_0)}{\varphi^*(u_0)}, \quad \beta = n_1 n_2^*, \\ &n_1 = -\left(\mathrm{i}\sqrt{k'}\frac{\tilde{s}}{\tilde{c}}\exp\gamma + \exp(-\gamma)\right), \quad n_2 = \exp\gamma - \mathrm{i}\sqrt{k'}\frac{\tilde{s}}{\tilde{c}}\exp(-\gamma), \\ &\gamma = -\mathrm{i}\tau(\mu)[x - (2w_3(\mu) + p\cos\theta_0)t] + \frac{1}{2}\ln\frac{e_1}{e_2}. \end{aligned}$$

According to (3.8.14), the coefficients $\varphi(u_0)$ and $\varphi_1(u_0)$ are expressed in terms of Jacobi elliptic functions of the arguments $\tilde{u}_0 = u_0 + \mathrm{i}K' + K/2$ and $\tilde{\mu} = \mu + \mathrm{i}K' + K/2$. In the same terms, it is convenient to write the coefficients $\tau(\mu)$ and $w_{1,3}(\mu)$:

$$\begin{aligned} \tau(\mu) &= \frac{\mathrm{i}\beta k^2(1-k')(k'-\tilde{d}^2)s_0^2c_0^2}{2(k'-d_0^2)(\tilde{d}-d_0)(k'-\tilde{d}d_0)}, \quad w_1 = \frac{\mathrm{i}\beta k^2(1-k')\tilde{s}\tilde{c}s_0c_0}{2(\tilde{d}-d_0)(k'-\tilde{d}d_0)}, \\ w_3(\mu) &= \frac{\mathrm{i}\beta[\tilde{d}d_0(1+k')^2 - (k'+\tilde{d}^2)(k'+d_0^2)]}{2(1+k')(\tilde{d}-d_0)(k'-\tilde{d}d_0)}. \end{aligned} \tag{3.8.21}$$

Here $\mu \in \Gamma_-$ is a complex parameter, and for simplicity, we have introduced the notations $s_0 = \mathrm{sn}(\tilde{u}_0)$, $\tilde{s} = \mathrm{sn}(\tilde{\mu})$, etc.

According to the found solution (3.8.20), the magnetization wave gains an additional phase shift after passing through the breather. Therefore the constants $\delta_\pm$ in formulas (3.8.1) turn out to be interconnected:

$$\delta_+ - \delta_- = 4\arg(\tilde{s}/\tilde{c}) \pmod{2\pi}.$$

Note also that due to the identities

$$\kappa = \frac{p}{\beta} = \frac{\mathrm{i}(1-k')d_0}{k'-d_0^2}, \quad \sin\theta_0 = \frac{2\mathrm{i}(1-k')\sqrt{k'}s_0c_0}{k'-d_0^2},$$
$$\cos\theta_0 = \left(\frac{d_0^2+k'}{k'-d_0^2}\right)\left(\frac{1-k'}{1+k'}\right),$$

we may use k' and u_0 instead of the wave parameters p and θ_0. It is also worth emphasizing that there is a nontrivial algebraic connection between the quantities k', p, and $\sin\theta_0$:

$$\frac{4k'}{(1+k')^2} = \frac{\sin^2\theta_0}{\kappa^2+1}.$$

Let us briefly characterize the solution obtained. The breather width (3.8.20) is of order $-[\operatorname{Im}\tau(\mu)]^{-1}$, and its speed is equal to

$$V = \frac{2\operatorname{Im}[\tau(\mu)w_3(\mu)]}{\operatorname{Im}\tau(\mu)} + p\cos\theta_0.$$

The breather, as a localized modulation of the wave (3.8.1), travels like a particle against the background of the wave. In the soliton localization region, additional magnetization pulsations occur due to the wave (3.8.1). In a laboratory coordinate system, their frequency is

$$\Omega = 2\operatorname{Re}[\tau(\mu)(2w_3(\mu) + p\cos\theta_0)].$$

In the reference frame connected to the breather, their frequency is equal to

$$\bar{\omega} = \Omega - 2V\operatorname{Re}\tau(\mu).$$

Over the breather localization region, these pulsations propagate with phase velocity V_{ph}, emerging at one of its edges and disappearing at another edge:

$$V_{\mathrm{ph}} = \frac{\Omega}{2\operatorname{Re}\tau(\mu)}.$$

The quantity $[2\operatorname{Re}\tau(\mu)]^{-1}$ gives the characteristic scale of the breather's additional spatial modulations associated with the pulsations propagating.

3.8.2 "Destruction" scenarios for a breather by a wave

... what should be actively destroyed, one first needs to grab it hard.

F. Kafka. Aphorisms

A breather exists only in the range of physical parameter values of the problem, which satisfy the inequality $\mathrm{Im}\,\tau(\mu) < 0$. As $\mathrm{Im}\,\tau(\mu) \to -0$, the breather localization region extends unlimitedly, with the magnetization oscillations in the breather behaving in various manners.

For values of $\mu \in \Gamma_-$ near the horizontal sections of the contour $\Gamma = \{u : \mathrm{Im}\,\tau(u) = 0\}$, the magnetic soliton (3.8.20) is a superposition of linear modes under the real dispersion law (3.7.5) (where $\tilde{p} = 2\tau(\mu)$).

Near the inclined boundaries of the region Γ_- (see Figure 3.13), soliton velocity is sufficiently great everywhere, except for the neighborhoods of the points $\mu = \mathrm{i}K'$ (mod$(K, 2\mathrm{i}K')$). By emitting breathers the system, as it were, tends to lower its energy.

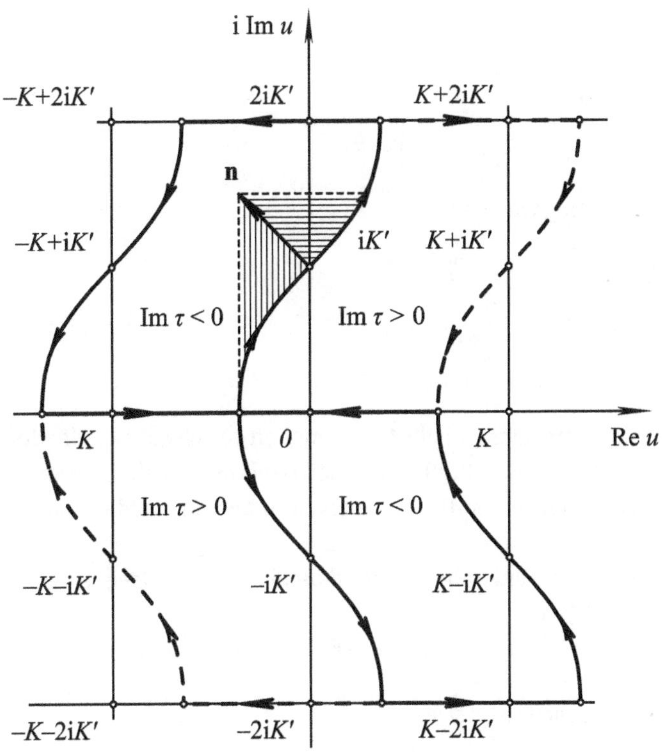

Figure 3.13: A normal **n** to the contour Γ at a branch point; in the horizontal hatched region: $V_{\mathrm{eff}} > V_g$; in the vertical hatched region: $V_{\mathrm{eff}} < V_g$.

When μ lies near a branch point of the function $\tau(\mu)$, $\mu = \mathrm{i}K' + \varepsilon$, where ε is a small complex number, the scenario of the magnetization oscillations appears to be easy to observe experimentally. In this case the magnetization distribution is given by the approximate formulas

$$S_3 \approx \frac{(|\tilde{\alpha}|^2 - |\tilde{\beta}|^2)\cos\theta_0 - (\tilde{\alpha}^*\tilde{\beta} + \tilde{\alpha}\tilde{\beta}^*)\sin\theta_0}{|\tilde{\alpha}|^2 + |\tilde{\beta}|^2},$$
$$S_1 - \mathrm{i}S_2 \approx \frac{\exp(-\mathrm{i}\varphi_+)[2\tilde{\alpha}\tilde{\beta}\cos\theta_0 + (\tilde{\alpha}^2 - \tilde{\beta}^2)\sin\theta_0]}{|\tilde{\alpha}|^2 + |\tilde{\beta}|^2} \tag{3.8.22}$$

with

$$\tilde{\alpha} = \frac{2\kappa\,\mathrm{Re}\,\tilde{\gamma}}{\sqrt{1+\kappa^2}} + \mathrm{i}(|\tilde{\gamma}|^2 + 1)\frac{\cos\theta_0}{\sin\theta_0}, \quad \tilde{\beta} = -2\,\mathrm{Im}\,\tilde{\gamma} - \mathrm{i}(|\tilde{\gamma}|^2 - 1),$$
$$\tilde{\gamma} = \frac{1}{\tilde{\varepsilon}}\sinh\big[\tilde{\varepsilon}(\cos\theta_0\sqrt{1+\kappa^2}(x - (V_g + V)t)$$
$$+ \mathrm{i}\kappa\sin\theta_0(x - (V_g + V_{ph})t))\beta\sin\theta_0\big],$$
$$\tilde{\varepsilon} = \frac{\varepsilon}{2}(1 + k'),$$

where

$$V_{ph} = \frac{p^2 + \beta^2}{p}\cos\theta_0$$

is the phase velocity of a pumping wave, and

$$V_g = 2p\cos\theta_0$$

is the group wave velocity. Here

$$V = -p\tan\theta_0\sin\theta_0$$

is the velocity, which formally coincides with the forced motion velocity of an individual domain wall. When we pass from eqs. (3.8.20) and (3.8.21) to eq. (3.8.22), the integration constant entering into γ is chosen so that the formulas for α and β should contain no zeroth-order-in-ε terms.

The solution of (3.8.22) controls emerging localized oscillations with frequency

$$\Omega = 2\,\mathrm{Re}\,\tilde{\varepsilon}\big[\tan\varphi_\varepsilon\sqrt{1+\kappa^2}(V_g + V)\cos\theta_0 + \kappa(V_g + V_{ph})\sin\theta_0\big]\beta\sin\theta_0$$

in an extended region with characteristic size

$$d \backsim \left(\mathrm{Re}\,\tilde{\varepsilon}(1-\eta)\beta\sin\theta_0\cos\theta_0\sqrt{1+\kappa^2}\right)^{-1},$$

where

$$\eta = \sqrt{-\frac{V}{V_{ph}}}\tan\varphi_\varepsilon, \quad \varphi_\varepsilon = \arg(\varepsilon).$$

This region moves as a whole against the background of the wave (3.8.1) with speed

$$V_{\text{eff}} = V_g + \frac{V - \eta V_{\text{ph}}}{1 - \eta}.$$

The latter essentially depends on the value of $\arg(\varepsilon)$. Since $\mu \in \Gamma_-$, the range of the values of $\arg(\varepsilon)$ is limited by the condition $\eta < 1$. For $\eta = V/V_{\text{ph}} < 0$, the point $\mu = \mathrm{i}K' + \varepsilon$ resides on the normal to the contour Γ. In this case, $V_{\text{eff}} = V_g$. The normal serves as a boundary between regions where $V_{\text{eff}} > V_g$ and $V_{\text{eff}} < V_g$ (Figure 3.13).

When the condition

$$\eta = \frac{V_g + V}{V_g + V_{\text{ph}}}$$

is fulfilled, the oscillation localization region is motionless. In particular, if ε is real, then the soliton velocity vanishes either when pumped by a uniform magnetization precession $p - 0$ or under the conditions $p \neq 0$ and $\theta_0 = \arccos(\sqrt{3}/3) \approx 54.7°$. At the angles $\arccos(\sqrt{3}/3) < \theta_0 < \pi/2$, the breather moves toward the wave (3.8.1), and at the precession angles $0 < \theta_0 < \arccos(\sqrt{3}/3)$, the breather travels in the same direction as the wave propagates.

The frequency Ω of the internal breather oscillations in the region under consideration (before the wave may destroy the soliton) is much lower than the frequency ω of a pumping wave. So, in the case of uniform magnetization precession (at $p = 0$), we have

$$\Omega = \beta^2 \operatorname{Re} \tilde{\varepsilon} \sin^2 \theta_0 \cos \theta_0 \ll \omega = \beta^2 \cos \theta_0.$$

When we resort to dimensional variables, the above expressions for the soliton width d, its velocity V_{eff}, and frequencies ω and Ω gain additional multipliers $\sqrt{a}$, $\gamma M_0 \sqrt{a}$, and γM_0, accordingly, where γ is the gyromagnetic ratio, M_0, is the nominal magnetization, and a is the exchange interaction constant.

Both the nature of the oscillations of the S_3-component (3.8.22) and the shape of their envelope dramatically depend on the magnitude of $|\varepsilon|$. Approximation (3.8.22) holds only for small values of $|\varepsilon|$. At $|\varepsilon| \simeq 1$, the envelope shape is weakly time-dependent (Figure 3.14). Then formula (3.8.22) has a low accuracy; however, it allows us to get an idea of the breather's structure in the general case, i. e., far away from the point $\mu = \mathrm{i}K'$.

For $|\varepsilon| \ll 1$, the pulsations of magnetization are smoothed out. The function $S_3(x,t)$ has only two maximums and one minimum at each time t. At some moments of time, the maxima of S_3 become equal to +1. Its global minimum is equal to $S_3(0,0) = \cos\theta_0(1 - 4\sin^2\theta_0) \geq -1$. In this context, when $|\varepsilon| \ll 1$, solution (3.8.22) defines the magnetization oscillations of the S_3-component having a large amplitude in an extensive region. Figure 3.15 depicts a special case of such oscillations (when ε is real). Generally, the extremum points of the function $S_3(x,t)$ reach their limit positions at different moments of time.

In the limit as $\varepsilon \to 0$, the oscillation period grows unlimitedly, and expression (3.8.22) describes an aperiodic generation process of localized modulations of the magnetization field of the S_3-component, followed by their disappearance, whereas the

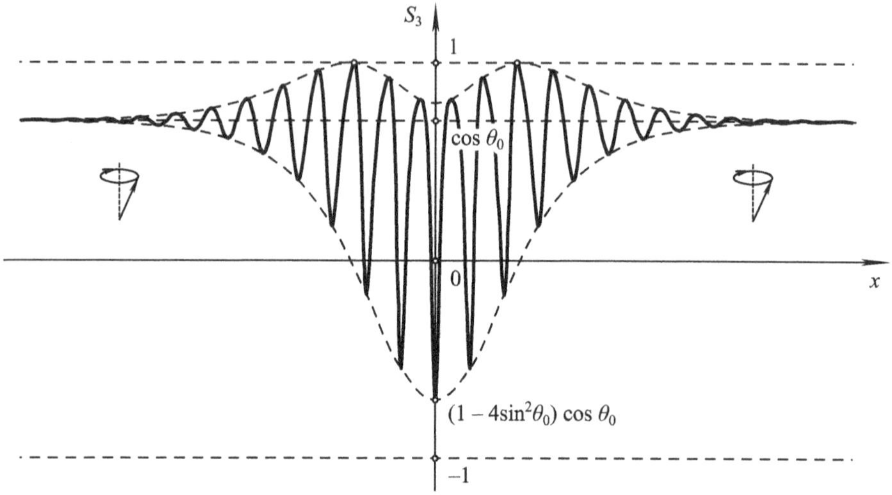

Figure 3.14: Magnetization oscillations of the S_3-component at $t = 0$ and $|\varepsilon| \simeq 1$.

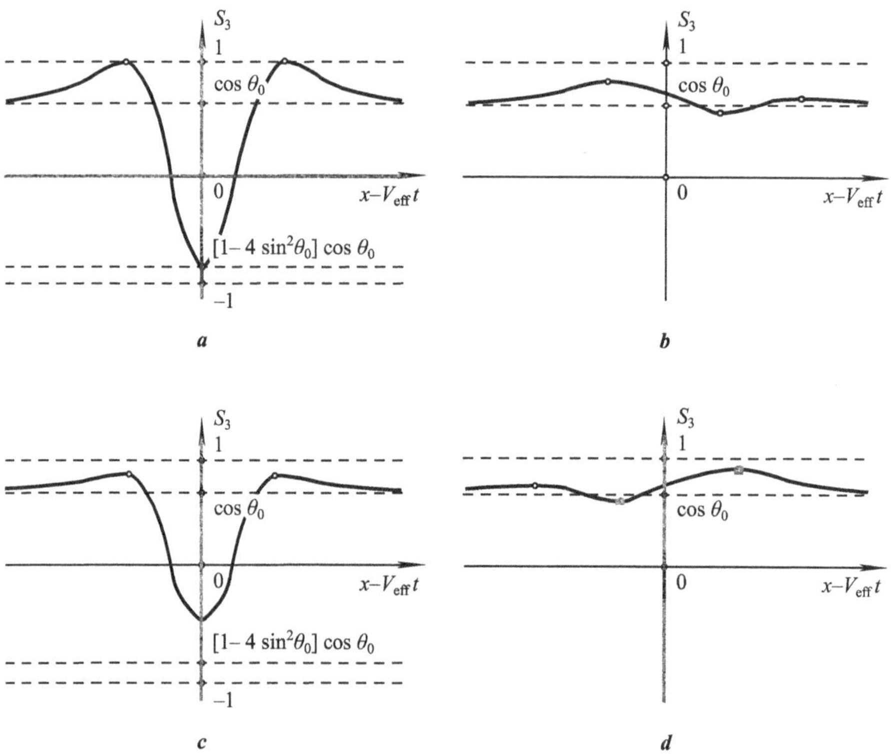

Figure 3.15: Magnetization oscillations of the S_3-component at time moments $t = 0$ (*a*), $t = T/4$ (*b*), $t = T/2$ (*c*), and $t = 3T/4$ (*d*), where T is the period of oscillations, and ε is chosen to be small and real.

exponential-trigonometric dependence of $\tilde{y}$ on x and t is replaced by a linear one. When exposed to such short-term but strong "bursts" of pumping wave modulations, various instruments and devices may be malfunctioning. By their nature their occurrence resembles the formation of "killer waves" in oceanology.

3.8.3 A solitary domain against the background of a wave

Let us discuss the most interesting particular case of solution (3.8.20), which corresponds to the interaction between a solitary domain ("magnetization reversal nuclei") and a nonlinear spin wave. From a mathematical point of view, in this case the zeros (3.7.24) of the coefficient $a^*(u^*)$ approach the special values (3.7.23), so that $\tilde{\mu} = 2iK' + \varepsilon$. The complex parameter ε determines the internal structure and extent of the solitary domain. In particular, the larger the domain size, the smaller the $|\varepsilon|$. For sufficiently small values of $|\varepsilon|$, the magnetization distribution (3.8.20) is well approximated by the expression

$$
\begin{aligned}
S_3 \approx{}& \tau\Bigg(\cos\theta_0\left[\frac{|\varepsilon_0|^2}{2}\cosh 2y - 1\right] \\
&+ \frac{\kappa\sin\theta_0}{\sqrt{1+\kappa^2}}[\varepsilon_0\sinh(y+i\Phi) + \varepsilon_0^*\sinh(y-i\Phi)]\Bigg), \\
S_1 + iS_2 \approx{}& \tau\exp(i\Phi_0)\Bigg(\frac{1}{4}\sin\theta_0\exp(-i\Phi) \\
&\times[\varepsilon_0^2\exp(-2y) + \varepsilon_0^{*2}\exp(2y)] - \frac{i}{\sqrt{1+\kappa^2}}[\varepsilon_0\exp(-y) \\
&\times(1+i\kappa\cos\theta_0) + \varepsilon_0^*\exp(y)(1-i\kappa\cos\theta_0)] \\
&- \sin\theta_0\exp(i\Phi)\Bigg),
\end{aligned}
\tag{3.8.23}
$$

where

$$
\begin{aligned}
&\varepsilon_0 = (1+k')\varepsilon, \quad \kappa = \frac{p}{\beta}, \\
&y = -\beta x\cos\theta_0 - p\beta t\sin^2\theta_0 + \ln\left|\frac{e_1}{e_2}\right|, \\
&\tau^{-1} = \frac{|\varepsilon_0|^2}{2}\cosh 2y + 1 - \frac{i\sin\theta_0}{\sqrt{1+\kappa^2}}[\varepsilon_0\sinh(y+i\Phi) - \varepsilon_0^*\sinh(y-i\Phi)]; \\
&\omega = (p^2+\beta^2)\cos\theta_0, \\
&\Phi \approx \omega t - px + \arg\left(\frac{e_1}{e_2}\right); \quad \Phi_0 = \arg\left(\frac{e_1}{e_2}\right) + \delta_+.
\end{aligned}
$$

The approximate formula (3.8.23) is derived by matching the asymptotics of the exact solution (3.8.20) for $|\varepsilon| \ll 1$ in the domains $|y| < 1$ and $|y| > 1$. Near the walls bounding the domain, we have the following estimate: $|\varepsilon| \exp\lfloor y| = O(1)$ ($\lfloor y| > 1$, $|\varepsilon| < 1$). Therefore expansion (3.8.23) retains the summands containing small coefficients $O(\varepsilon^2)$ before $\exp(2\lfloor y|)$.

Small wave numbers ($p \approx 0$) and large precession angles $\theta_0 \approx \pi/2$ of the wave (3.8.1) break down the applicability conditions $|\partial_x \Phi + p|/|p| \ll 1$ and $|\partial_x y + \beta \cos\theta_0|/\beta\cos\theta_0 \ll 1$ for the approximate solution. A reference frame connected to one of the boundaries of the solitary domain confirms the justification of approximation (3.8.23) as follows: In the limit as $\varepsilon \to 0$ and under the condition $|\varepsilon| \exp(|y|) = \text{const}$, expression (3.8.23) reduces to the exact solution (3.7.38) for describing an individual domain wall against the background of a pumping wave.

According to (3.8.23), the magnetization distribution centered in a solitary domain has the form

$$S_1 + \mathrm{i}S_2 \approx -\sin\theta_0 \exp\left[-\mathrm{i}\left(px - \omega t - 2\arg\left(\frac{e_1}{e_2}\right) - \delta_+\right)\right],$$
$$S_3 \approx -\cos\theta_0.$$

As $x \to \pm\infty$, the magnetization performs an inhomogeneous circular precession around the Ox_3-axis tied with the propagation of the wave (3.8.1):

$$S_3 \approx \cos\theta_0, \quad S_1 + \mathrm{i}S_2 \approx \sin\theta_0 \exp(\mathrm{i}\varphi_\pm), \quad x \to \pm\infty;$$
$$\delta_+ - \delta_- = 4\arg\varepsilon_0, \quad \varphi_\pm = px - \omega t + \delta_\pm.$$

The positions of the domain walls bounding the domain are defined by the condition $S_3 \approx 0$. Hence we find the domain size

$$l \sim \frac{2}{\beta\cos\theta_0} \ln\frac{2}{|\varepsilon_0|}.$$

Because the spin wave propagates, the magnetization precesses within the domain walls in an inhomogeneous elliptical manner with nutational oscillations of the precession axis. The domain wall width is $d \sim 1/\beta\cos\theta_0 \ll l$. Inside each wall, the projection of S_3 changes sharply from one of its quasiequilibrium values, $S_3 \approx \cos\theta_0$, to another, $S_3 \approx -\cos\theta_0$. The centers of gravity of the walls fluctuate around the middle positions (Figure 3.16). The hatched and solid arrows indicate that the left- and right-hand walls of the solitary domain oscillate in turn.

Recall that both a solitary domain and a domain wall are always immobile in the absence of a wave. It is both interesting and significant that a domain moves as a whole toward a nonlinear wave with the same velocity as a separate domain wall as a result of their interaction:

$$V = -p\tan\theta_0 \sin\theta_0.$$

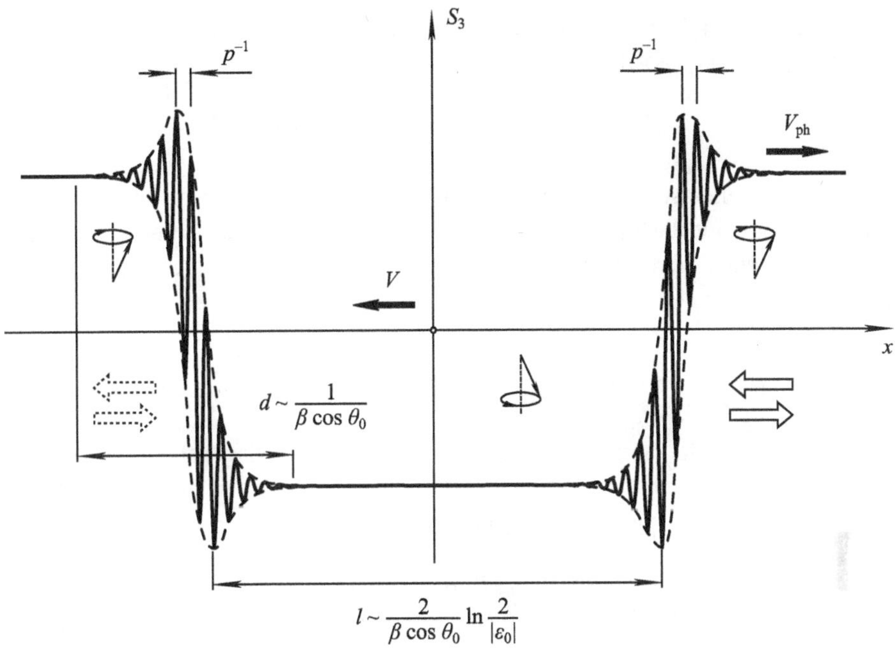

Figure 3.16: Forced motion of a solitary domain in the field of an arbitrary-amplitude spin wave.

This evinces that the velocity of the forced motion of domain walls does not depend on their chirality. It is crucial that the velocity of forced motion of domain walls and solitary domains can be controlled by varying the wavenumber and amplitude of a pumping wave.

3.9 Integrals of motion for solitons against the background of a wave

If there is no law on this matter, then it will appear.

B. Gates

A nontrivial set of conservation principles, including analogs to the conventional laws of conservation of energy and momentum, ensures the structural stability of solitons against the background of a wave. Let us give an algorithm for constructing integrals of motion for solitons against the background of precession wave.

To start with, we consider a problem of the interaction between a magnetization wave and a breather. In this case, the analytic element $a^*(u^*)$ of the transition matrix $T(u)$ in the region Γ_- on is time-independent. Therefore it can be used as a generating function for integrals of motion. For further analysis, it is convenient to present the function $a^*(u^*)$ in the form

$$a^*(u^*) = \lim_{x\to-\infty} [(\chi_1^{(0)}(x,t,u))^{-1}\chi_2(x,t,u)]_{11}. \tag{3.9.1}$$

Next, we demonstrate that physically meaningful integrals of motion can be obtained by expanding the function $\ln a^*(u^*)$ into an asymptotic series in powers of $(u-u_0)$. For this goal, we need to expand the solution χ_2 in powers of $(u-u_0)$. Let us represent the function χ_2 as follows:

$$\chi_2 = \exp\left[-\frac{\mathrm{i}\sigma_3}{2}\varphi_+\right]N[I+\Phi]\exp\left(\frac{\tau}{\mathrm{i}}\sigma_3\eta + Z\right), \tag{3.9.2}$$

where I is the identity matrix, Φ and Z are antidiagonal and diagonal matrix functions of the variables x, t, and u, respectively. As $x\to+\infty$, they are such that $\Phi, Z\to 0$, and the condition $\chi_2\to\chi_2^{(0)}$ yields the form of the matrix N:

$$N = \begin{pmatrix} \beta(-u) & -\beta(u) \\ \beta(u) & \beta(-u) \end{pmatrix}.$$

The matrix N has the properties

$$\begin{aligned} N^{-1}(u) &= N^\dagger(u^*)[\beta(u)\beta^*(u^*) + \beta(-u)\beta^*(-u^*)]^{-1}, \\ N^{-1}\sigma_1 N &= w_1\sin\theta_0\tau^{-1}\sigma_3 + (w_3\cos\theta_0 - p/2)\tau^{-1}\sigma_1, \\ N^{-1}\sigma_2 N &= \sigma_2, \\ N^{-1}\sigma_3 N &= (w_3\cos\theta_0 - p/2)\tau^{-1}\sigma_3 - w_1\sin\theta_0\tau^{-1}\sigma_1. \end{aligned} \tag{3.9.3}$$

Using formulas (3.9.3) and representation (3.9.2) plugged into the second equation of (3.7.6), we arrive at the system

$$\begin{aligned} -\mathrm{i}\tau\sigma_3 + \partial_x Z &= -\mathrm{i}A_3\sigma_3 - \mathrm{i}(A_1\sigma_1 + A_2\sigma_2)\Phi, \\ -\mathrm{i}\tau\Phi\sigma_3 + \Phi\partial_x Z + \partial_x\Phi &= -\mathrm{i}(A_1\sigma_1 + A_2\sigma_2) - \mathrm{i}A_3\sigma_3\Phi. \end{aligned} \tag{3.9.4}$$

The functions $A_i(x,t,u)$ parameterized as $\mathbf{S} = (\sin\theta\cos\varphi, \sin\theta\sin\varphi, \cos\theta)$ have the form

$$\begin{aligned} A_1 &= -\tau^{-1}(w_3\cos\theta - p/2)w_1\sin\theta_0 + \\ &\quad + w_1\tau^{-1}\sin\theta\cos(\varphi-\varphi_+)(w_3\cos\theta_0 - p/2), \\ A_2 &= w_1\sin\theta\sin(\varphi-\varphi_+), \\ A_3 &= \tau^{-1}(w_3\cos\theta - p/2)(w_3\cos\theta_0 - p/2) \\ &\quad + w_1^2\tau^{-1}\sin\theta\cos(\varphi-\varphi_+)\sin\theta_0. \end{aligned}$$

We represent the solution to eqs. (3.9.4) and the coefficients $A_i(x,t,u)$ and $\tau(u)$ as a power series in $(u-u_0)$:

$$\Phi=\sum_{n=0}^{\infty}\Phi_n(u-u_0)^n,\quad Z=\sum_{n=0}^{\infty}Z_n(u-u_0)^n,$$
$$A_i=\sum_{n=-1}^{\infty}A_i^{(n)}(u-u_0)^n,\quad \tau=\sum_{n=-1}^{\infty}\tau^{(n)}(u-u_0)^n.$$

The coefficients Φ_n and Z_n are determined by the chain of recurrent relations

$$\begin{aligned}&\partial_x\Phi_n-\mathrm{i}\sum_{m=0}^{n+1}A_3^{(n-m)}[\Phi_m,\sigma_3]\\&\quad-\mathrm{i}\sum_{m=0}^{n}\sum_{k=0}^{n+1}\Phi_{n-m}(A_1^{(m-k)}\sigma_1+A_2^{(m-k)}\sigma_2)\Phi_k\\&\quad+\mathrm{i}(A_1^{(n)}\sigma_1+A_2^{(n)}\sigma_2)=0,\\&Z_n=\mathrm{i}\sigma_3\int_x^{+\infty}[A_3^{(n)}-\tau^{(n)}]\mathrm{d}x'\\&\quad+\mathrm{i}\int_x^{+\infty}\sum_{m=0}^{n+1}[A_1^{(n-m)}\sigma_1+A_2^{(n-m)}\sigma_2]\Phi_m\mathrm{d}x',\end{aligned}$$

where $n=0,1,2,\ldots$, $\Phi_0=(1+f_3)^{-1}(f_1\sigma_1+f_2\sigma_2)\sigma_3$,

$$\begin{pmatrix}f_1\\f_2\\f_3\end{pmatrix}=\begin{pmatrix}\cos\theta_0&0&-\sin\theta_0\\0&1&0\\\sin\theta_0&0&\cos\theta_0\end{pmatrix}\begin{pmatrix}\tilde{S}_1\\\tilde{S}_2\\\tilde{S}_3\end{pmatrix},$$
$$\tilde{\mathbf{S}}=(\sin\theta\cos(\varphi-\varphi_+),\ \sin\theta\sin(\varphi-\varphi_+),\ \cos\theta).$$

Bearing in mind the series for the function χ_2 and using representation (3.9.1), we can expand the function $\ln a^*(u^*)$ into a power series in $(u-u_0)$:

$$\ln a^*(u^*)=\frac{\mathrm{i}}{2}P-\frac{\mathrm{i}\sqrt{k'}}{\beta\sin\theta_0}H(u-u_0)+\cdots.\qquad(3.9.5)$$

The first summands are responsible for the breather's momentum P and energy H. The breather's momentum is understood as the difference between the total momentum of the system and the momentum of the magnetization wave (3.8.1):

$$P=\int_{-\infty}^{+\infty}[p\cos\theta_0-\partial_x\varphi\cos\theta]\mathrm{d}x.\qquad(3.9.6)$$

Similarly, we evaluate the breather's energy:

$$H = \frac{1}{2} \int\limits_{-\infty}^{+\infty} [(\partial_x \theta)^2 + (\partial_x \varphi)^2 \sin^2 \theta + \beta^2 \sin^2 \theta - (p^2 + \beta^2) \sin^2 \theta_0] \mathrm{d}x. \tag{3.9.7}$$

Using relations (3.8.14) and (3.8.21) and the explicit expression (3.8.18) for the coefficient $a^*(u^*)$, we find

$$\begin{aligned} P &= -2\mathrm{i} \ln a^*(u_0) = -2\mathrm{i} \ln \frac{\varphi^*(u_0)}{\varphi(u_0)} = -4 \arg \varphi(u_0) + \mathrm{mod}(2\pi), \\ H &= \mathrm{i}\beta(k')^{-1/2} \sin \theta_0 \frac{\mathrm{d} \ln a^*(u^*)}{\mathrm{d}u}\bigg|_{u=u_0} = -8 \mathrm{Im}\, \tau(\mu). \end{aligned} \tag{3.9.8}$$

We can see from formulas (3.9.8) that the parameters u_0 and k' of the magnetization wave (3.8.1) essentially affect the breather's momentum and energy. This circumstance should be taken into account when theoretically describing the physical properties of a system formed by a wave and solitons.

In addition, another integral of motion is also directly deduced from the equations of motion (3.7.1) as follows:

$$N_0 = \int\limits_{-\infty}^{+\infty} (S_3^{(0)} - S_3) \mathrm{d}x = \int\limits_{-\infty}^{+\infty} (\cos \theta_0 - \cos \theta) \mathrm{d}x.$$

In the semiclassical quantization of the energy of solitons, the integral of motion N_0 implies the so-called number of spin deviations [12]. However, the family of conserved quantities (3.9.5) does not contain N_0.

It should be pointed out that the equalities

$$\begin{aligned} &\varphi(-\mathrm{i}K' + K/2) = w_1(-\mathrm{i}K' + K/2) = 0, \\ &w_3(-\mathrm{i}K' + K/2) = -\mathrm{i}\frac{\beta}{2}, \quad \tau(-\mathrm{i}K' + K/2) = -\frac{\mathrm{i}\beta \cos \theta_0 - p}{2} \end{aligned}$$

are valid at the point $u = -\mathrm{i}K' + K/2$ and that the matrix functions $\chi_{1,2}^{(0)}$ and $\Psi_\pm$ become diagonal at this point. In view of these remarks, from the second equation of (3.7.6) we get

$$N_0 = \frac{2}{\beta} \ln \left[\Psi_-\left(x, t, u = -\mathrm{i}K' + \frac{K}{2} \right) \right]_{11} \Bigg|_{x=-\infty}^{x=+\infty} = \frac{4}{\beta} \ln \left| \frac{k^2 \tilde{s} \tilde{c}}{(1 + \tilde{d})(\tilde{d} - k')} \right|.$$

A similar algorithm enables us to determine the momentum and energy of the domain wall against the background of the magnetization wave. Here, as a generating function of the integrals of motion, the function $b(u)$ should be resorted to admit the representation

$$b(u) = \lim_{x\to-\infty}[(\chi_1^{(0)}(x,t,u))^{-1}\chi_2(x,t,u)]_{21}.$$

Using the expansions already found for χ_2, we obtain

$$\ln b(u) = \frac{\mathrm{i}}{2}P - \frac{\mathrm{i}\sqrt{k'}}{\beta\sin\theta_0}H(u-u_0) + \cdots.$$

The field momentum P and energy H of the domain wall are given by the same formulas (3.9.6) and (3.9.7) as for the breather.

On the other hand, the function $b(u)$ has already been defined explicitly (see (3.7.26) and (A.17)). Therefore

$$P = -2\mathrm{i}\ln\left(\frac{(\mathrm{i}+1)}{(\mathrm{i}-1)}\frac{(\mathrm{dn}u_0 - \mathrm{i}\sqrt{k'})}{(\mathrm{dn}u_0 + \mathrm{i}\sqrt{k'})}\right) + \mathrm{mod}(2\pi), \quad H = 2\beta\cos\theta_0. \tag{3.9.9}$$

As already mentioned, a particular case of the interaction between a solitary domain and a magnetization wave flows out of the general solution (3.8.20), which describes the breather in the limit $\tilde{\mu} = 2\mathrm{i}K' + \varepsilon$, $|\varepsilon| \ll 1$, where the parameter $\varepsilon|^{-1}$ replies for the length of the domain (see (3.8.23)). It is easy to check that in the limit as $\varepsilon \to 0$, the energy of the solitary domain (3.9.8) reduces to the sum of the domain walls' energies (3.9.9).

4 Dynamics of a quasi-one-dimensional “easy-plane” anisotropy ferromagnet

4.1 Solitons against the background of a magnetization wave and the “dressing” method

Analogy and experience are the two sticks on which we trudge along the road of reasoning.

W. Friedrich

Constructing soliton solutions of integrable models under different asymptotic conditions at spatial infinity requires a deep insight into the analyticity properties of the matrix functions of the Riemann problem. Meanwhile, soliton solutions as behaving asymptotically alike at infinity are easier to find. The point is that there is some degree of arbitrariness in choosing a contour and meromorphic functions of the Riemann problem. Ultimately, the “dressing” procedure to be implemented contains two most important aspects: Additional poles in the “dressing” equations should be eliminated, and reductions that guarantee proper transformation properties of newly generated solutions should be taken into account. This claim proves to be constructive since it makes it relatively simple to build exact solutions to integrable models. Let us use an example of deriving soliton-like solutions to the quasi-one-dimensional model of an easy-plane anisotropy ferromagnet (3.1.19) to illustrate the opening potential.

It is easy to check that the equations of an easy-plane ferromagnet (3.1.19) have a particular solution such as a precession wave of considerable amplitude (Figure 4.1):

$$\mathbf{S}^{(0)} = (\cos\theta_0 \cos\varphi_0,\ \cos\theta_0 \sin\varphi_0,\ \sin\theta_0), \tag{4.1.1}$$

where the angle $\theta_0 = \text{const}$ determines the amplitude of the wave, the phase is equal to $\varphi_0 = pz - \omega t$, p is the wave number, and the precession frequency is $\omega = (p^2 - \rho^2)\sin\theta_0$.

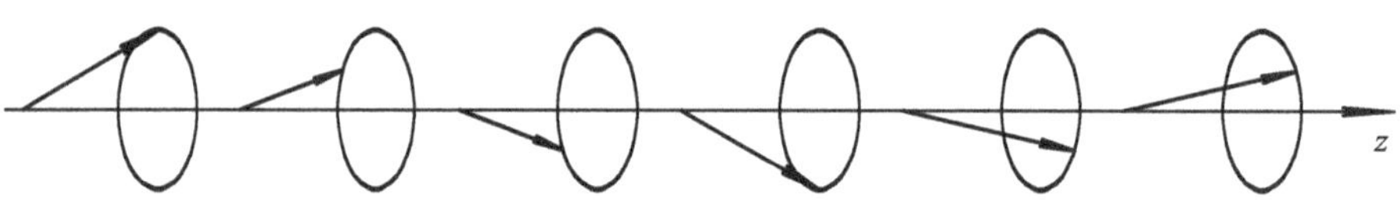

Figure 4.1: A spatially nonlocalized precession wave.

Let us apply the “dressing” method to build new soliton-like solutions to the model (3.1.19) through solution (4.1.1) as a “seeding” one. As a result, we get analytical solutions for describing the interaction between a nonlocalized wave (4.1.1) and solitary waves [5, 6]. Moreover, these are precisely those solitary waves that correspond to typical magnetic solitons of an easy-plane ferromagnet not subjected to a pumping wave (4.1.1). The

https://doi.org/10.1515/9783112235973-004

pumping wave-soliton interaction changes their physical characteristics: width, velocity, energy, and momentum. At certain values of the parameters, the wave destroys magnetic solitons.

A linear analysis of the stability of the nonlinear wave (4.1.1) shows that it can be stable only for wave numbers that satisfy the constraint

$$0 < p^2 < \rho^2. \tag{4.1.2}$$

Therefore, as $z \to \pm\infty$, solutions to model (3.1.19) with asymptotics of type (4.1.1) should be considered only under condition (4.1.2)

In addition, this section covers a complete classification of soliton-like excitations induced by an external magnetic field directed along an easy-plane ferromagnet anisotropy axis. For studying such states, the paper [14] uses the Hirota method.

Note that it is possible to directly compare the predicted results and experimental data for quasi-one-dimensional magnetic crystals with easy-plane anisotropy [22, 12, 16, 20, 21, 11]. A typical representative of such crystals is a Cs Ni F_3 compound. Inelastic neutron scattering experiments indicate the existence of soliton-like excitations in Cs Ni F_3.

Further analysis exploits the Riemann problem with canonical normalization to the identity matrix at the point $\lambda = \infty$. To match the canonical normalization with the $U - V$ pair (3.1.15), (3.1.20), we pass from the auxiliary linear system

$$\partial_z \varphi = U\varphi, \quad \partial_t \varphi = V\varphi \tag{4.1.3}$$

to the gauge-equivalent system

$$\partial_z \chi = \tilde{U}(\lambda)\chi, \quad \partial_t \chi = \tilde{V}(\lambda)\chi, \tag{4.1.4}$$

where $\chi = g^\dagger \varphi$,

$$\tilde{U}(\lambda) = g^\dagger (U(\lambda) - U(\lambda = \infty))\, g,$$
$$\tilde{V}(\lambda) = g^\dagger (V(\lambda) - V(\lambda = \infty))\, g,$$
$$g = \exp\left[-\frac{i\sigma_3 \rho}{2}\left(\int_{z_0}^{z} S_3(z', t)\, dz' + \int_{t_0}^{t} [\mathbf{S} \times \partial_z \mathbf{S}]_3\, dt'|_{z=z_0}\right)\right].$$

A precession wave evokes the asymptotic behavior of solutions to model (3.1.19) in the form

$$S_3 \to \tilde{S}_3 = \text{const}, \quad [\mathbf{S} \times \partial_z \mathbf{S}]_3 \to [\tilde{\mathbf{S}} \times \partial_z \tilde{\mathbf{S}}]_3 = \text{const}$$

as $z \to -\infty$ Therefore the formula for g can be written more simply:

$$g = \exp\left[-\frac{\mathrm{i}\sigma_3 p}{2}\left(\int_{-\infty}^{z}(S_3(z',t)-\tilde{S}_3)\mathrm{d}z' + \tilde{S}_3 z + [\tilde{\mathbf{S}}\times\partial_z\tilde{\mathbf{S}}]_3\, t\right)\right].$$

Here we have accounted for the fact that as $z \to \pm\infty$, the asymptotics of solutions produced by the procedure proposed, although they belong to the class of nonlinear-in-shape waves (4.1.1), do not coincide with the "seeding" solution (4.1.1).

For a given solution (4.1.1) to model (3.1.19), the auxiliary linear system (4.1.3) has the solution

$$\begin{aligned}
f_0(\lambda) &= \exp\left(-\frac{\mathrm{i}\sigma_3}{2}\varphi_0\right)A(\lambda)\exp(\sigma_3\eta(\lambda)),\\
A(\lambda) &= \begin{pmatrix} a(\lambda)+\kappa(\lambda) & -b(\lambda)\\ b(\lambda) & a(\lambda)+\kappa(\lambda)\end{pmatrix},\\
a(\lambda) &= w_3(\lambda)\sin\theta_0 - p/2, \quad b(\lambda) = w_1(\lambda)\cos\theta_0,\\
\eta(\lambda) &= \mathrm{i}\,\kappa(\lambda)[(2w_3(\lambda)+p\sin\theta_0)t - z], \quad \kappa^2(\lambda) = a^2(\lambda)+b^2(\lambda).
\end{aligned}$$

Hence the solution of the gauge-equivalent system (4.1.4) is

$$\chi_0(\lambda) = g_0^{\dagger} f_0(\lambda), \quad g_0 = \exp\left[-\frac{\mathrm{i}p\sigma_3}{2}(z\sin\theta_0 + t\,p\cos^2\theta_0)\right]. \tag{4.1.5}$$

A new solution for the auxiliary linear system (4.1.4) should be sought as follows:

$$\chi(\lambda) = \Psi(\lambda)\chi_0(\lambda).$$

Representation (4.1.4) implies that the matrix function $\Psi(\lambda)$ can be chosen normalized to the identity matrix at the point $\lambda = \infty$ ($\Psi(\lambda = \infty) = I$) and satisfying the constraints

$$\Psi(\lambda) = \sigma_1\Psi^*(-\lambda^*)\sigma_1, \quad \Psi^{\dagger}(\lambda^*)\Psi(\lambda) = I. \tag{4.1.6}$$

New matrix functions $\tilde{U}(z,t,\lambda)$ and $\tilde{V}(z,t,\lambda)$ are generated by the formulas

$$\begin{aligned}
\tilde{U} &= \partial_z\Psi(\lambda)\,\Psi^{-1}(\lambda) + \Psi(\lambda)\,\tilde{U}_0(\lambda)\,\Psi^{-1}(\lambda),\\
\tilde{V} &= \partial_t\Psi(\lambda)\,\Psi^{-1}(\lambda) + \Psi(\lambda)\,\tilde{V}_0(\lambda)\,\Psi^{-1}(\lambda),
\end{aligned} \tag{4.1.7}$$

where $\tilde{U}_0$ and $\tilde{V}_0$ are the values of the matrix functions $\tilde{U}$ and $\tilde{V}$ for $\mathbf{S} = \mathbf{S}^{(0)}$.

Having taken a residue at the point $\lambda = 1$ of the first equality of (4.1.7), we obtain the important relation

$$S_k\sigma_k = g\,\Psi(\lambda=1)\,g_0^{\dagger}\,S_k^{(0)}\sigma_k\,g_0\Psi^{\dagger}(\lambda=1)\,g^{\dagger}, \tag{4.1.8}$$

which determines both the magnetization vector $\mathbf{S}(z,t)$ through both elements of the matrix $\Psi(z,t,\lambda)$ and the "seeding" solution $\mathbf{S}^{(0)}$.

The assigned reduction (4.1.6) and the requirement that expressions (4.1.7) lack additional poles allow us to explicitly express the matrix functions $\Psi(z,t,\lambda)$ for soliton-like states. It is simple to confirm directly that the formulated criteria cause the matrix functions to be meromorphic in the λ-plane:

$$\Psi_{\alpha\beta}(\lambda) = \delta_{\alpha\beta} + \sum_{i,j=1}^{n} \frac{(m^i)_\alpha (m^j)^*_\beta}{\lambda - \lambda_j} \frac{\partial \ln \det D}{\partial d_{ji}}, \tag{4.1.9}$$

where D is the $n \times n$ matrix with elements

$$d_{ij} = \frac{(\mathbf{m}^{i*} \cdot \mathbf{m}^j)}{(\lambda_i - \lambda_j^*)}.$$

Two-dimensional complex vectors $\mathbf{m}^i$ and complex numbers λ_j are related to each other in one of two possible ways:

$$\begin{aligned} &n = 2N, \quad \lambda_{k+N} = -\lambda_k^* \quad (\lambda_k \neq -\lambda_k^*), \quad \mathbf{m}^{k+N} = \sigma_1 (\mathbf{m}^k)^*, \\ &k = 1, 2, \ldots, N, \quad N = 1, 2, \ldots; \\ &n = 1, 2, \ldots, N, \quad \lambda_i^* = -\lambda_i, \quad (m^k)_1^* = (m^k)_2 \exp(\mathrm{i}\varphi_k). \end{aligned} \tag{4.1.10}$$

Formula (4.1.10) contains the quantities φ_k as arbitrary real parameters.

For $\lambda_k \neq -\lambda_k^*$, the vectors $\mathbf{m}^k$ appear as

$$\mathbf{m}^k = \chi_0(z, t, \lambda_k^*) \mathbf{c}^k, \tag{4.1.11}$$

where $\mathbf{c}^k$ are arbitrary constant complex vectors. Using solution (4.1.5), we arrive at the explicit form of the vectors $\mathbf{m}^k$:

$$\begin{aligned} (m^k)_1 &= \exp(-\mathrm{i}y)[(a(\lambda_k^*) + \kappa(\lambda_k^*)) \exp \gamma_k - b(\lambda_k^*) \exp(-\gamma_k)], \\ (m^k)_2 &= \exp(\mathrm{i}y)[b(\lambda_k^*) \exp(\gamma_k) + (a(\lambda_k^*) + \kappa(\lambda_k^*)) \exp(-\gamma_k)], \\ y &= \frac{1}{2}(\rho \sin\theta_0 - p)[(p \sin\theta_0 + \rho)t - z], \\ \gamma_k &= \eta(\lambda_k^*) + \frac{1}{2} \ln \frac{(c^k)_1}{(c^k)_2}. \end{aligned} \tag{4.1.12}$$

For purely imaginary poles $\lambda_k = -\lambda_k^*$, formulas (4.1.11) still define the vectors $\mathbf{m}^k$; however, the constant vectors $\mathbf{c}^k$ are no longer arbitrary. They must be chosen in such a way that the constraints

$$(m^k)_1^* = (m^k)_2 \exp(\mathrm{i}\varphi_k)$$

should be fulfilled. Here the phase φ_k is real. It is easy to check that such a condition holds for formulas (4.1.12) if we admit that

$$\frac{1}{2}\ln\frac{(c^k)_1}{(c^k)_2} = \mathrm{i}\,\kappa(\lambda_k^*)\delta_k + \frac{\pi\,\mathrm{i}}{4}, \qquad \delta_k = \delta_k^* = \text{const.}$$

In doing so, the parameters $\kappa(\lambda_k^*)$ are either real or purely imaginary numbers.

Relation (4.1.8) generates a new class N-soliton solutions to model (3.1.19). We first find the component $S_3(z,t)$ of the magnetization field from the “dressing” equation (4.1.8). For this, it suffices to know the matrix function $\Psi(z,t,\lambda)$ (4.1.9). A simple integration of the expression for $S_3(z,t)$ and algebraic operations are required for calculating the remaining magnetization components.

Solutions for the easy-plane ferromagnet model (3.1.9) ($\xi = z$, $\tilde{\beta} = -\rho^2$) exposed to a time-dependent external magnetic field directed along the anisotropy axis can be easily obtained from the solutions to the field-free model (3.1.19). For this, we merely resort to transformation (3.1.10), which leads to an additional precession of the magnetization vector around the field direction.

Note that the magnetic solitons obtained by the “dressing” of the precession wave (4.1.1) are not localized in space. Moreover, it turns out that even a one-soliton solution of the equations for an easy-plane ferromagnet describes several types of magnetic excitations. The following sections are devoted to the study of these inhomogeneous states.

4.2 Scattering of a precession wave of arbitrary amplitude by algebraic solitons and magnetization rotation waves. Destruction of solitons

Complicated systems lead to unexpected implications.
One of the formulations of Murphy’s generalized uncertainty principle

The present section examines magnetic excitations in an easy-plane anisotropy ferromagnet through the matrix function (4.1.9) with one purely imaginary pole $\lambda_1 = -\lambda_1^*$.

Let us first discuss a one-soliton solution when there is no external magnetic field:

$$\begin{aligned} S_3 &= -\frac{p}{\rho} - \frac{2\,\mathrm{i}\,\kappa^2}{\rho\,(\mathrm{i}\,a - b\,\cosh 2\gamma)}, \quad \gamma = \mathrm{i}\kappa\big[(2w_3 + p\sin\theta_0)t - z + \delta_1\big], \\ S_1 - \mathrm{i}\,S_2 &= -\exp\big[\mathrm{i}(\rho\,z\sin\theta_0 + \rho\,p\,t\cos^2\theta_0)\big] \\ &\quad\times\left[\frac{2w_1}{\rho}\sin\theta_0 + \frac{\kappa\rho\,\sinh 2\gamma + 2w_3(a\,\cosh 2\gamma + \mathrm{i}\,b)}{\rho(\mathrm{i}\,a - b\,\cosh 2\gamma)}\cos\theta_0\right]. \end{aligned} \tag{4.2.1}$$

Here and below, for brevity, we omit the argument λ_1^* of the functions $a(\lambda_1^*)$, $b(\lambda_1^*)$, $\kappa(\lambda_1^*)$, and $w_{1,3}(\lambda_1^*)$.

Suppose that the quantity $\mathrm{i}\kappa(\lambda_1^*)$ is real (and negative for definiteness) in formulas (4.2.1). Then, outside the region of a linear size $d = |2\kappa|^{-1}$, the solution asymptotically approaches the precession wave coinciding in shape with the “seeding” solution (4.1.1):

$$\begin{aligned}
&S_3 \to \sin\tilde{\theta}, \\
&S_1 - \mathrm{i}S_2 \to \exp[\mathrm{i}(\tilde{\omega}t - \tilde{p}z \pm \Delta\varphi/2)]\cos\tilde{\theta}, \\
&\tilde{\omega} = \sin\tilde{\theta}(\tilde{p}^2 - \rho^2),
\end{aligned} \tag{4.2.2}$$

as $z \to \pm\infty$.

The amplitude $\sin\tilde{\theta}$, the wavenumber $\tilde{p}$, and the phase shift $\Delta\varphi$ of the precession wave (4.2.2) are expressed in terms of the parameters of the "seeding" solution:

$$\begin{aligned}
\sin\tilde{\theta} &= -\frac{p}{\rho}, \quad \tilde{p} = -\rho\sin\theta_0, \\
\frac{\Delta\varphi}{2} &= -\arctan\left[\frac{2b}{\mathrm{i}\kappa\rho\cos\theta_0}(w_1\sin\theta_0 - w_3 a b^{-1}\cos\theta_0)\right].
\end{aligned} \tag{4.2.3}$$

Consequently, solution (4.2.1) can be interpreted as the scattering of the magnetization wave (4.2.2) by some localized magnetic excitation. A feature of the scattering is its non-reflective nature (we have an incident wave and a transmitted one as $z \to -\infty$ and $z \to +\infty$, respectively). After passing through the localized excitation, the wave (4.2.2) gains a phase factor to fix a turn of the vector **S** in the plane (x_1, x_2) through the angle $\Delta\varphi = \text{const}$.

We express the width d of the localized excitation and the phase shift $\Delta\varphi$ in terms of the physical parameters of the problem:

$$\begin{aligned}
d &= \left[\rho^2\left(1 - \frac{\tilde{p}^2}{\rho^2}\right)\cos^2\tilde{\theta} - (V - V_0)^2\right]^{-1/2}, \\
\Delta\varphi &= -2\arctan(d\,[V\sin\tilde{\theta} - \tilde{p}(1 + \sin^2\tilde{\theta})]).
\end{aligned} \tag{4.2.4}$$

Here V and V_0 are the group velocities of the localized excitation and magnetization waves (4.2.2), respectively:

$$V = 2w_3 + \tilde{p}\sin\tilde{\theta}, \quad V_0 = \frac{\partial\tilde{\omega}}{\partial\tilde{p}} = 2\tilde{p}\sin\tilde{\theta}. \tag{4.2.5}$$

In formula (4.2.5) the value of $2w_3$ is equal to the velocity of self-localized excitation under no impact of the wave (4.2.2) occurred (when $\tilde{p} = \tilde{\omega} = 0$). The addition is numerically equal to the wave momentum density:

$$p_0 = \tilde{p}\sin\tilde{\theta}. \tag{4.2.6}$$

From eqs. (4.2.4) we can infer that a localized magnetic excitation exists only when $0 \le \tilde{p}^2 < \rho^2$. The energy density of the wave (4.2.2) has the form

$$w = \frac{1}{2}\left[\cos^2\tilde{\theta}\left(\frac{p_0^2}{\sin^2\tilde{\theta}} - \rho^2\right) + \rho^2\right].$$

Therefore the condition $\tilde{p}^2 < \rho^2$ stands for the smallness of the contribution of the exchange energy as compared to that of the anisotropy energy:

$$\frac{p_0^2}{\sin^2\tilde{\theta}} < \rho^2. \tag{4.2.7}$$

When the velocity V of the localized excitation belongs to the interval

$$\min\{V_1,\ V_2\} < V < \max\{V_1,\ V_2\}, \quad V_{1,2} = V_0 \pm \rho\sqrt{1+\tilde{p}^2/\rho^2}\cos\tilde{\theta},$$

condition (4.2.7) is fulfilled. According to (4.2.4), the phase shift $\Delta\varphi$ conveys information about the velocity magnitude of the localized excitation and its width.

Let us first look at the structure of the localized excitation without the magnetization wave (4.2.2). Suppose that $\tilde{p} = \tilde{\omega} = 0$. Then we have

$$\begin{aligned} &S_3 = -\frac{2\,\mathrm{i}\,w_1}{\rho\cosh 2\gamma}, \quad S_2 = \frac{2\,w_3}{\rho\cosh 2\gamma}, \quad S_1 = \tanh 2\gamma, \\ &\gamma = \mathrm{i}\,w_1(2\,w_3\,t - z + \delta_1). \end{aligned} \tag{4.2.8}$$

In [1], direct integration of equations (3.1.19) yields solution (4.2.8) for describing a magnetization turning wave moving at a speed of $2w_3$. The wave asymptotics is uniform as $z \to \pm\infty$: $S_{2,3} \to 0$, and $S_1 \to \pm 1$ (Figure 4.2). As shown in [23], when exposed to any slight perturbations, the shape of solution (4.2.8) for the one-dimensional model at hand is stable. It becomes unstable when changing the equations of motion. For example, it can be seen upon taking a ferromagnet's three-dimensionality into account. As a result, turning waves are assumed to be observed in quasi-one-dimensional magnetic crystals.

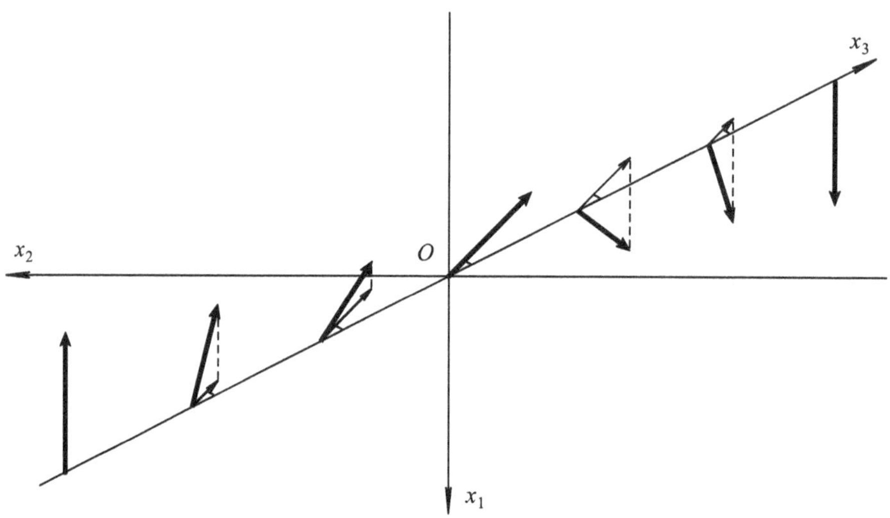

Figure 4.2: Distribution of magnetization in the turning wave. The projection of the vector **S** onto a plane forms an angle r with the axis Ox_3: $\cos r = -2\mathrm{i}w_1/\rho$, $\sin r = 2w_3/\rho$.

When $\tilde{p} \neq 0$ and $\tilde{\omega} \neq 0$, the phase of the quantity $S_1 - iS_2$ alters and receives an additional term $\tilde{\omega}t - \tilde{p}z$, which leads to phase modulation of the turning wave. In the limit as $z \to \pm\infty$, none of the components of the vector **S** reaches a maximum value equal to unity.

The energy (momentum) density that meets solution (4.2.1) is the sum of two terms [5, 6]. The former coincides with the energy (momentum) density of the wave (4.2.2), and the latter can be treated as the energy (momentum) density of a localized magnetic excitation. It should be stressed that despite the above additivity, only the "weakened principle of asymptotic superposition" holds because the parameters of the precession wave (4.2.2) still affect the energy (momentum). This circumstance must be kept in mind, for example, to describe the thermodynamic properties of the collective excitations considered.

The total energy of the localized excitation has the form

$$E = \frac{2}{d}. \tag{4.2.9}$$

According to eqs. (4.2.4) and (4.2.9), as the wave number $|\tilde{p}|$ ($\tilde{p}^2 < \rho^2$) increases, the width d of the localized excitation enlarges indefinitely, and its energy drops. In this case, the localized excitation spreads out as the precession wave field impulse (4.2.6) increases. For given parameters $\tilde{p}$ and $\sin\tilde{\theta}$ of the pumping wave (4.2.2), the maximum energy is concentrated in the localized excitation, provided that its velocity coincides with the group velocity of the wave.

Note that the spatial- and time-exponential dependence typical of a localized magnetic excitation for $\kappa^2 = a^2 + b^2 < 0$ degenerates into a rational one in the limit as $\mathrm{i}\,a - b \to 0$ or at

$$\mathrm{i}\lambda_1 \to \frac{\cos\theta_0 \pm \sqrt{1-(p/\rho)^2}}{\sin\theta_0 - p/\rho}$$

In this limit, solution (4.2.1) appears as

$$\begin{aligned} S_3 &= -\frac{p}{\rho} + \frac{4\mathrm{i}\,b}{\rho\,(1-4b^2\sigma^2)}, \qquad \sigma = (2w_3 + p\sin\theta_0)t - z + \delta_1, \\ S_1 - \mathrm{i}\,S_2 &= \exp[\rho\,\mathrm{i}\,(z\sin\theta_0 + p\,t\cos^2\theta_0)] \\ &\quad \times \left[-\frac{2w_1}{\rho}\sin\theta_0 + \frac{4\,\mathrm{i}\,b\,\sigma\rho + 2\,\mathrm{i}\,w_3(1+4\,b^2\sigma^2)}{(1-4\,b^2\sigma^2)\,\rho}\cos\theta_0\right] \end{aligned} \tag{4.2.10}$$

and describes scattering without reflection of the magnetization wave (4.2.2) by an algebraic soliton.

For $p_0^2/\sin^2\tilde{\theta} > \rho^2$, there exists no localized excitation ($\kappa^2 = a^2 + b^2 > 0$). This is because the kinetic energy of the precession wave (4.2.2) is greater than the anisotropy energy responsible for this excitation to arise. In this case, solution (4.2.1) can be represented as a “superposition” of two nonlocalized magnetization waves with characteristics

$$(2w_3 + p\sin\theta_0)t - z = \text{const}, \quad \rho z \sin\theta_0 + \rho\, p\, t \cos^2\theta_0 = \text{const}.$$

Propagating, the first of them changes all components of the vector **S**, and the second impacts only the S_1- and S_2-components. When $\tilde{p}^2 > \rho^2$, this statement agrees with condition (4.1.2), under which the precession wave is unstable regarding long-wavelength modulations.

Let us dwell on nonlinear excitations of a ferromagnet subjected to a magnetic field perpendicular to its “easy plane”: $\mathbf{H} = (0, 0, H)$. Then the Landau–Lifshitz equations (3.1.9) can be easily solved: $\xi = z$, $\tilde{\beta} = -\rho^2$ using transformation (3.1.10) and solutions (4.2.1) and (4.2.10) considered above. In particular, it is possible to reproduce the results of [9, 17, 13, 2], where solutions found with homogeneous asymptotics as $z \to \pm\infty$ describe turning waves and algebraic solitons in a constant magnetic field ($H = \text{const}$) without a running pumping wave. Suppose that $\tilde{p} = 0$ in relations (4.2.1) and (4.2.10). Besides, the equilibrium values of the magnetization as $z \to \pm\infty$ in the magnetic field are determined by the relations

$$\sin\tilde{\theta} = \frac{H}{\rho^2} \quad (|H| < \rho^2), \quad \sin\tilde{\theta} = \text{sign} H \quad (|H| > \rho^2). \tag{4.2.11}$$

To avoid misunderstandings, we recall that in this case the wave-number $\tilde{p}$ and amplitude $\sin\tilde{\theta}$ of the pumping wave (4.2.2) differ from those of the “seeding” solution (see formulas (4.1.1), (4.2.2), and (4.2.3)).

It is convenient to express the final result in terms of the physical parameters of the problem, such as the soliton velocity V and the magnitude of the magnetic field H. The condition $\text{Re}\,(\mathrm{i}\kappa) \neq 0$, which controls the region of existence of self-localized excitations, is reduced to the constraint

$$|H| < H_c = \rho\sqrt{\rho^2 - V^2} \leq \rho^2. \tag{4.2.12}$$

Employing eqs. (4.2.11) and (4.2.12), from (3.1.10) and (4.2.1) we find

$$\begin{aligned}
S_3 &= \frac{H}{\rho^2} - \frac{A^2 - (H/\rho^2)^2}{H/\rho^2 + A\cosh 2\gamma}, \\
S_1 - \mathrm{i}\, S_2 &= \frac{1}{H/\rho^2 + A\cosh 2\gamma}\left[\frac{\sinh 2\gamma}{\rho\, d} - \frac{\mathrm{i}\, V}{\rho}\left(\frac{H}{\rho^2}\cosh 2\gamma + A\right)\right], \\
d^{-1} &= \sqrt{\rho^2 - V^2 - H^2/\rho^2}, \quad 2\gamma = d^{-1}(z - Vt + \delta_1), \\
V &= 2w_3, \quad A = -\text{sign}(\mathrm{i}\, w_1)\sqrt{1 - V^2/\rho^2}.
\end{aligned} \tag{4.2.13}$$

In formulas (4.2.13) the magnitude of A is equal to that of S_3 at the center of the soliton. When $H = 0$, eq. (4.2.13) coincides with the turning wave (4.2.8). In addition, the sign of the magnitude of A plays no role. Under a magnetic field, the nature of the solution essentially depends on the direction of the projection of A: $AH > 0$ along the field or $AH < 0$ against the field (see Figure 4.3). If $AH > 0$, then as H grows ($H \to H_c$, $H < H_c$), the width of the turning wave increases indefinitely, and its amplitude tends to zero. If $AH < 0$, then as H grows, the amplitude and width of the soliton increase, and in the limit $H \to H_c$, solution (4.2.13) degenerates into an algebraic soliton [9]:

$$
\begin{aligned}
S_3 &= \frac{H}{\rho^2}\left(1 - \frac{2\rho^2}{\rho^2 + H^2\sigma^2}\right), \quad \sigma = z - Vt + \delta_1, \\
S_1 - \mathrm{i}\, S_2 &= -\frac{\rho^2}{\rho^2 + H^2\sigma^2}\left[\frac{2H\sigma}{\rho} + \frac{\mathrm{i}V}{\rho}\left(1 - \frac{H^2\sigma^2}{\rho^2}\right)\right], \\
V &= \operatorname{sign} w_3 \sqrt{\rho^2 - H^2/\rho^2}.
\end{aligned}
\tag{4.2.14}
$$

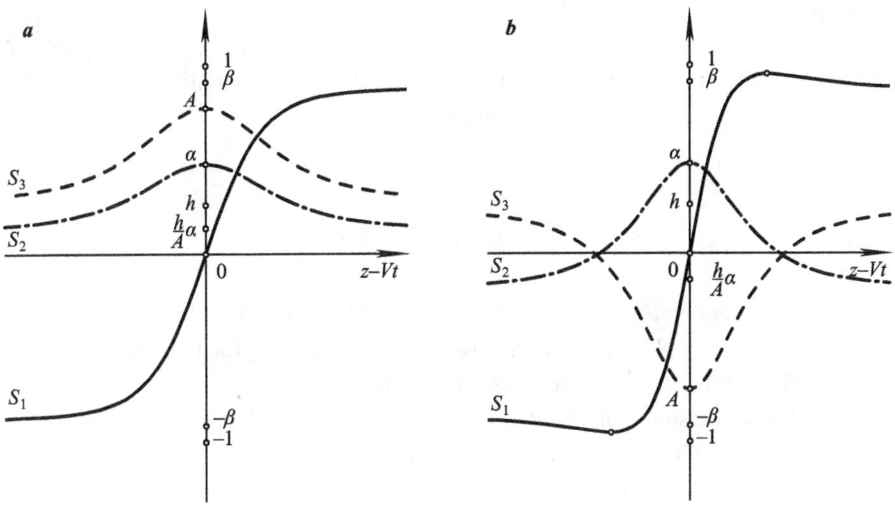

Figure 4.3: Magnetization projections as a function of coordinates for the turning wave (4.2.13) in a magnetic field: *a*) for $AH > 0$, *b*) for $AH < 0$; $\alpha = \sqrt{1 - A^2}$, $\beta = \sqrt{1 - h^2/A^2}$, $h = H/\rho^2$ $(H > 0)$.

Based on the results of [23], we believe that solution (4.2.13) is stable within the one-dimensional model, at least in the region of low magnetic fields.

Let us give the arguments of [17, 13, 8] to understand the reason for the existence of turning waves (4.2.13) only in a certain range of magnetic fields when $|H| < \rho^2$. This range of values of H produces the following spectrum of linear spin-wave modes (magnons) [24, 25]:

$$\varepsilon = |p|\sqrt{p^2 + \rho^2 - H^2/\rho^2}, \tag{4.2.15}$$

where p is the momentum. When $|H| \ll \rho^2$, the dispersion law (4.2.15) is linear in momentum. The energy E of soliton (4.2.8) is less than that of a magnon with the same momentum, as shown by a comparison of eq. (4.2.15) with the dispersion law for nonlinear excitations (4.2.13). In particular, when $H = 0$, we have [13]

$$E = 2\rho\left|\sin\frac{p}{2}\right| < \varepsilon \cong \rho|p|. \tag{4.2.16}$$

However, as the field strength grows ($|H| \to \rho^2$, $|H| < \rho^2$), magnons obeying a linear dispersion law continuously turn into magnons behaving along a quadratic dispersion law, with soliton (4.2.13) remaining stable until after the field $|H|$ overcomes a certain threshold.

Let a constant magnetic field $|H| \geq \rho^2$ oriented along the anisotropy axis $\mathbf{n} = (0, 0, 1)$ fall into a given region of inhomogeneity. Then, when imposing on it, a uniform alternating magnetic field perpendicular to the vector $\mathbf{n}$ effectively generates running spin waves [18, 19]. We believe that this method can excite not only a low-amplitude spin wave but also a high-amplitude precession wave (4.2.2) when $|H| < \rho^2$. If the pumping field is small, then the value of the angle $\tilde{\theta}$ is determined by the relation $\sin\tilde{\theta} \cong H/\rho^2$.

For explaining how the running precession wave

$$\begin{gathered}\mathbf{S} = (\cos\tilde{\theta}\cos\tilde{\varphi},\ \cos\tilde{\theta}\sin\tilde{\varphi},\ \sin\theta),\\ \sin\tilde{\theta} = \frac{H}{\rho^2},\quad \tilde{\varphi} = \tilde{p}z - \tilde{p}^2 t\sin\tilde{\theta},\end{gathered} \tag{4.2.17}$$

interacts with the turning waves (4.2.13) and (4.2.14) in a constant magnetic field, the equations for an easy-plane ferromagnet can be solved applying formulas (3.1.10), (4.2.1), and (4.2.10). In doing so, we should take into account that $\tilde{p} \neq 0$ and $\sin\tilde{\theta} = H/\rho^2$.

Note that the existence of solitons (4.2.8) and (4.2.13) in the field of the magnetization wave (4.2.2) or (4.2.17) confirms the stability of these excitations.

We presume that solitons in an "easy-plane" anisotropy ferromagnet can be detected and studied via the measurement of the phase shift gained by the precession wave after interaction with them.

4.3 Breathers against the background of a precession wave

Every solution breeds new problems.
Murphy's fifth corollary

Let us focus on an easy-plane ferromagnet's magnetic excitations that can be obtained using the matrix function (4.1.9) with two complex poles λ_1 and $-\lambda_1^*$ ($\lambda_1 \neq -\lambda_1^*$).

In the absence of an external magnetic field, we obtain a one-soliton solution for model (3.1.19) of the following form:

$$\begin{aligned} S_3 &= \sin\theta_0(1-2|\beta|^2) - \cos\theta_0(\alpha\beta + \alpha^*\beta^*), \\ S_1 - \mathrm{i}S_2 &= \exp[\mathrm{i}(-\varphi_0 + \Phi(z,t))] \\ &\quad \times [2\alpha^*\beta\sin\theta_0 - \cos\theta_0(\beta^2 - \alpha^{*2})], \\ \Phi(z,t) &= 2\arctan\left[\frac{(\lambda_1-\lambda_1^*)(|m_1|^2-|m_2|^2)}{\mathrm{i}(\lambda_1+\lambda_1^*)(|m_1|^2+|m_2|^2)}\right], \\ \alpha &= \frac{|m_1|^2\lambda_1^*(1-\lambda_1^2) + |m_2|^2\lambda_1(1-\lambda_1^{*2})}{|1-\lambda_1^2|\,(|m_1|^2\lambda_1^* + |m_2|^2\lambda_1)}, \\ \beta &= -\frac{(\lambda_1^2-\lambda_1^{*2})\,m_1m_2^*\exp(2\mathrm{i}y)}{|1-\lambda_1^2|\,(|m_1|^2\lambda_1 + |m_2|^2\lambda_1^*)}. \end{aligned} \tag{4.3.1}$$

Here the vector $\mathbf{m} = (m_1, m_2)$ is determined by relations (4.1.12).

Suppose that $\mathrm{Re}\,(\mathrm{i}\kappa) \neq 0$ (for definiteness, $\mathrm{Re}\,(\mathrm{i}\kappa) < 0$). Then, as $z \to \pm\infty$, solution (4.3.1) outside the domain with a characteristic size of order $|2\mathrm{Re}\,(\mathrm{i}\kappa)|^{-1}$ approaches the asymptotic values:

$$\begin{aligned} &S_3 \to \sin\theta_0, \quad S_1 - \mathrm{i}\,S_2 \to \exp\left[\mathrm{i}\left(-\varphi_0 \pm \frac{\Delta\varphi}{2}\right)\right]\cos\theta_0, \\ &\frac{\Delta\varphi}{2} = \frac{1}{\mathrm{i}}\ln\left[\frac{(\kappa+a)b^*}{(\kappa^*+a^*)b}\right]. \end{aligned} \tag{4.3.2}$$

The present problem, against that considered previously, involves the wave-number p and amplitude $\sin\theta_0$ of the pumping wave (4.3.2) coinciding with those for the "seeding" solution (4.1.1). Solution (4.3.1) describes the interaction between a pumping wave and a pulsating, spatially localized excitation, a breather. As the past case, passaging the wave (4.3.2) through a breather is analogous to passaging a beam of light through a transparent plate with certain refractive index. After passing through the breather, the magnetization wave (4.3.2) gains a phase factor that determines a turn of the S_1- and S_2-components in the (x_1, x_2)-plane by an angle $\Delta\varphi$.

To reveal the structure of breather (4.3.1), we should examine it in the absence of a precession wave (for $\sin\theta_0 = p = 0$). Then we have

$$\begin{aligned} S_3 &= 2\sin 2\delta(\cosh 2\varepsilon - \cos 2\delta)^{-1}\tau \\ &\quad \times [\sin s\,\cosh y\,\cos\delta\,\sinh\varepsilon + \sinh y\,\cos s\,\sin\delta\,\cosh\varepsilon], \\ S_2 &= \tau(\cosh 2\varepsilon - \cos 2\delta)^{-1} \\ &\quad \times \left[\sin^2 2\delta\,\sin s\,\sinh y - \cosh y\,\cos s\,\sin 2\delta\,\sinh 2\varepsilon\right], \\ S_1 &= -\tau\left(\sin^2\delta\,\cos^2 s - \cos^2\delta\,\cosh^2 y + \frac{2\cos^2\delta\,\sin^2\delta}{\sinh^2\varepsilon + \sin^2\delta}\right), \end{aligned} \tag{4.3.3}$$

$$y = \frac{2\rho \cosh\varepsilon \sin\delta}{\cosh 2\varepsilon - \cos 2\delta}(z - Vt) + y_0, \quad V = \frac{2\rho \tanh\varepsilon\,(\cosh^2\varepsilon + \cos^2\delta)}{\cosh 2\varepsilon - \cos 2\delta},$$
$$s = \frac{2\rho \cos\delta \sinh\varepsilon}{\cosh 2\varepsilon - \cos 2\delta}\left(z - \frac{2\rho \coth\varepsilon\,(\sinh^2\varepsilon - \sin^2\delta)\,t}{\cosh 2\varepsilon - \cos 2\delta}\right) + s_0,$$
$$\tau^{-1} = \cosh^2 y \cos^2\delta + \cos^2 s \sin^2\delta.$$

Here $\lambda_1 = \exp(\varepsilon + i\delta)$ $(\lambda_1 \neq -\lambda_1^*)$, and y_0, s_0, ε, δ are real parameters. The quantity V has the sense of the speed of the breather as a whole.

At the center of the soliton (for $y = 0$), the projection S_1 varies within the limits

$$\frac{\sin^2\delta - \sinh^2\varepsilon}{\sin^2\delta + \sinh^2\varepsilon} \leq S_1(y=0) \leq \frac{\cosh^2\varepsilon \sin^2\delta - \sinh^2\varepsilon \cos^2\delta}{\sin^2\delta + \sinh^2\varepsilon}.$$

For $\varepsilon = 0$, the magnetization at the center of the soliton is time-independent and is equal to its equilibrium value $\mathbf{S} = (-1, 0, 0)$ (see Figure 4.4). In this case the breather speed is equal to zero, and solution (4.3.3) takes a simple form:

$$\begin{aligned} S_3 &= 2\tau \cos\delta \sinh y \cos s, \quad S_2 = 2\tau \cos^2\delta \sinh y \sin s,\\ S_1 &= \tau[\cos^2\delta\,(\sinh^2 y - 1) - \sin^2\delta \cos^2 s],\\ \tau^{-1} &= \cos^2\delta \cosh^2 y + \sin^2\delta \cos^2 s,\\ y &= \frac{\rho z}{\sin\delta} + y_0, \quad s = \frac{\rho^2 t \cos\delta}{\sin^2\delta} + s_0. \end{aligned} \tag{4.3.4}$$

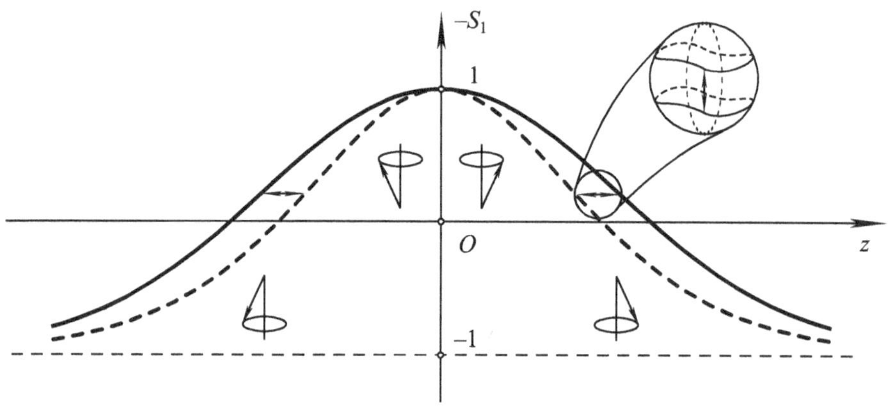

Figure 4.4: Pattern of magnetization fluctuations in breather (4.3.4).

The projection of the S_1-component vanishes at points with coordinates

$$z = \pm z_0(t) - y_0 \rho^{-1} \sin\delta, \tag{4.3.5}$$

where

$$z_0(t) = \rho^{-1} \sin\delta \operatorname{arsinh}[1 + \tan^2\delta \cos^2 s]^{1/2}. \tag{4.3.6}$$

The distance $d = 2|z_0(t)|$ between the points (4.3.5) is taken as the characteristic size of the breather. According to eq. (4.3.6), its size meets a periodic function of time and varies within the limits

$$2\rho^{-1}|\sin\delta| \operatorname{arsinh} 1 \le d \le 2\rho^{-1}|\sin\delta| \operatorname{arsinh}(|\cos\delta|^{-1}).$$

The S_3- and S_2-components are described by odd functions of the parameter y. They reach their extreme values at the points with coordinates (4.3.5). Hence this means that at a given moment of time t, the function $S_\alpha(z,t)$ $(\alpha = 2,3)$ takes simultaneously maximal and minimal values at points with coordinates $z = z_0(t) - y_0\rho^{-1}\sin\delta$ and $z = -z_0(t) - y_0\rho^{-1}\sin\delta$, respectively.

With the passage of time, the size of the breather periodically changes, and, simultaneously, the magnetization vector in the soliton performs an elliptical precession around the axis Ox_1. The precession phases in the regions $y > 0$ and $y < 0$ differ by π (Figure 4.4).

The pulsating soliton (4.3.4) is the simplest representative of a whole class of multi-soliton solutions to the equations for an "easy-plane" anisotropy ferromagnet. The energy of latter,

$$E = \frac{4\rho}{\sin\delta}, \tag{4.3.7}$$

is greater than the sum of the energy of two turning waves (compare formulas (4.2.16) and (4.3.7)). Therefore, under certain conditions, the breather can break up into two turning waves. Let us delve into this claim.

For $\cos\delta \ll 1$, the breather's magnetization frequency $\omega = \rho^2\cos\delta/\sin^2\delta$ is close to zero. At a moment of time t, when $\cos s = 1$, expression (4.3.4) represents a superposition of two turning waves that differ in the sign of the S_3-component projection. The distance between the waves amounts to the order of $2\rho^{-1}\sin\delta\ln(2\tan\delta) \gg 1$. In the limit as $\delta \to \pi/2$ for $s_0 = \pi/2$, solution (4.3.4) changes to

$$\begin{aligned} S_3 &\approx -\frac{2\rho^2 t \sinh y}{\cosh^2 y + (\rho^2 t)^2}, \quad S_2 \approx \frac{2\sinh y}{\cosh^2 y + (\rho^2 t)^2}, \\ S_1 &= -\frac{1 + (\rho^2 t)^2 - \sinh^2 y}{\cosh^2 y + (\rho^2 t)^2}, \end{aligned} \tag{4.3.8}$$

where $y = \rho z + y_0$. At $t = 0$, under the approximations made above, eq. (4.3.8) represents a localized excitation coinciding with breather (4.3.4). At large t $(t \to +\infty)$, solution (4.3.8) describes a superposition of two turning waves separated by a distance of order $2\rho^{-1}\ln(2\rho^2 t)$. Thus, at a low frequency ω (for $\delta \approx \pi/2$), the pulsating soliton is not stable. Expression (4.3.8) (at $t > 0$) governs the process of its decay into two turning waves.

For $\varepsilon \neq 0$, the breather's magnetization behaves otherwise than that described above. In the interval $0 < \sinh^4 \varepsilon \leq \sin^2 \delta$, the $-S_1$-component has the shape as depicted in Figure 4.5*a*. The "top" of the soliton performs a circular motion along the curve $1-2-3-4$. The value of $|\varepsilon| = \operatorname{arsinh}(\sqrt{|\sin \delta|})$ is sensitive. When $\sinh^4 \varepsilon > \sin^2 \delta$, the $-S_1$-component achieves a second maximum (see Figure 4.5*b*). As $z \to \pm\infty$, the magnetization tends to the equilibrium value $\mathbf{S} = (1, 0, 0)$

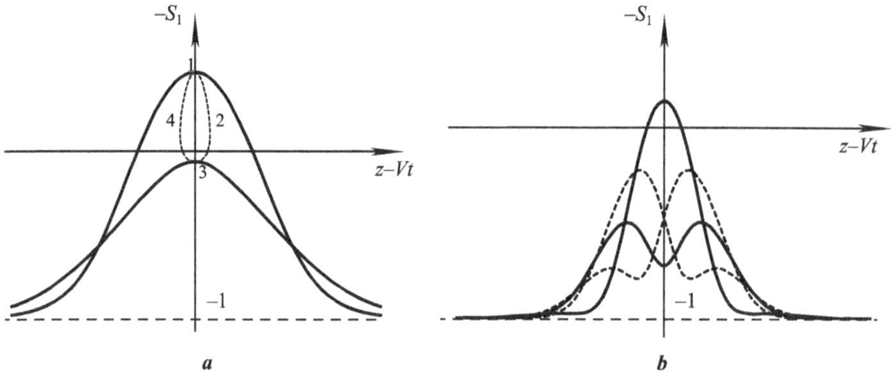

Figure 4.5: The breather's $-S_1$-component for magnetization: *a*) for $0 < |\varepsilon| \leq \operatorname{arsinh}(\sqrt{|\sin\delta|})$, *b*) for $|\varepsilon| > \operatorname{arsinh}(\sqrt{|\sin\delta|})$. The extremum moves along the curve $1-2-3-4$ clockwise when $\tanh\varepsilon/\tan\delta < 0$ and counterclockwise when $\tanh\varepsilon/\tan\delta > 0$. In the case a), the abscissas of the points 2 and 4 are $(z-Vt)_{2,4} \approx \pm(\tanh^2\varepsilon\,(\sin^2\delta + \sinh^2\varepsilon))(p|\cos\delta|(\cosh^2\varepsilon + \cos^2\delta))^{-1}$.

For an "easy-plane" ferromagnetic under a magnetic field, its breather equations can be easily solved using eqs. (4.3.1) and transformation (3.1.10). For a constant magnetic field without a running wave of magnetization ($p = 0$), we should calculate the equilibrium value of the magnetization as $z \to \pm\infty$ through relations (4.2.11) (in this case, $\tilde{\theta} = \theta_0$).

As the field grows, the **S**-vector precession axis takes a turn, and the breather magnetization nutation amplitude diminishes. At low H $(|H| < p^2)$, an analytical expression for a breather whose center oscillates periodically but does not shift on average was obtained in [2] by direct integration of the Landau–Lifshitz equations. For $p = 0$ and particularly choosing the parameter $\lambda = \lambda_1 = \exp i\delta$, formulas (3.1.10), (4.2.11), and (4.3.1) provide this solution.

In a sufficiently strong magnetic field $|H| > p^2$, solutions (3.1.10), (4.2.11), and (4.3.1) are transformed into a precessional soliton found in [9, 17]:

$$\arctan\frac{S_2}{S_1} = \frac{\sin 2\varepsilon}{\sin 2\delta}y - t\left(H - \frac{2p^2\cosh 2\varepsilon}{\cosh 2\varepsilon - \cos 2\delta}\right)$$
$$+ \arctan(\tanh y\,\tan\delta\,\cot\varepsilon) + \arg\left(\frac{c_1}{c_2}\right), \tag{4.3.9}$$

$$S_3 = 1 - \frac{\sin^2 2\delta}{(\cosh 2\varepsilon - \cos 2\delta)(\sinh^2 y + \cos^2 \delta)},$$

$$y = \frac{\rho \sin 2\delta}{\cosh 2\varepsilon - \cos 2\delta}(z - Vt) + \ln\left|\frac{c_1}{c_2}\right|, \quad V = \frac{2\rho \sinh 2\varepsilon}{\cosh 2\varepsilon - \cos 2\delta}.$$

In formulas (4.3.9) the real parameters ε and δ are connected to the pole $\lambda_1 = \exp(\varepsilon + i\delta)$ of the matrix function (4.1.9). According to [17, 2], the precessional soliton (4.3.9) can be interpreted as a bound state of magnons with a quadratic dispersion law [24]:

$$\varepsilon = p^2 - \rho^2 + H, \quad H \geq \rho^2.$$

In the limit as $\delta \to \pi/2$, $\varepsilon = 0$, and $H = \rho^2$, the precessional soliton (4.3.9) degenerates into an algebraic soliton (4.2.14) with zeroth velocity.

For $H > \rho^2$ and $p \neq 0$, the solutions (3.1.10), (4.2.11), and (4.3.1) describe the scattering of a small-amplitude spin wave

$$\mathbf{S} = (\cos\theta_0 \cos\varphi_0, \ \cos\theta_0 \sin\varphi_0, \ \sin\theta_0),$$

$$\varphi_0 \approx pz - t(p^2 - \rho^2 + H), \quad \sin\theta_0 \cong 1$$

by the precessional soliton (4.3.9)

It should be emphasized that the degeneracy of the ground state of an easy-plane ferromagnet when $|H| < \rho^2$ initiates several types of magnetic solitons at low magnetic fields. A strong magnetic field (when $|H| > \rho^2$) removes the degeneracy of the ground state and simplifies soliton-like excitations.

4.4 Application of the easy-plane ferromagnet model for analyzing solitons in spiral structures

As far as I know, all physical results a priori have their source in symmetry.

G. Weil

A distinct subgroup of magnetic crystals are those with a helicoidal magnetic structure as the ground state. There are various mechanisms for such structures to emerge. Inhomogeneous relativistic interactions that are theoretically described by Lifshitz invariants in the expansion of the free energy of a magnet may cause helicoidal structures to form in inversion-center-free magnetic crystals [7, 3, 15]. Resorting to [4, 10], we show that the integrable model of an "easy-plane" anisotropy ferromagnet is isomorphic to one of the models of magnets with helicoidal structure.

Let us consider a quasi-one-dimensional inversion-center-free ferromagnetic crystal with energy density

$$w = \frac{\alpha}{2}(\partial_{z'}\mathbf{M})^2 + \frac{\tilde{\beta}}{2}M_3^2 - \kappa(M_1\partial_{z'}M_2 - M_2\partial_{z'}M_1). \tag{4.4.1}$$

Here $\mathbf{M}(z', t')$ is magnetization per unit length along the Ox_3-axis, z' and t' are the space-time coordinates, $\mathbf{M}^2 = M_0^2 = \text{const}$, $\alpha > 0$ is the exchange constant, $\tilde{\beta} = \rho^2 > 0$ is the anisotropy constant of the "easy-plane" type, and κ is the constant under the Lifshitz invariant, $|\kappa| \ll \alpha/a$, where a is the lattice constant.

The conditions $\alpha > 0$ and $\tilde{\beta} > 0$ now do not provide a stable, homogeneous ground state. Only a magnetization distribution of a helicoidal magnetic structure can stand behind the stable ground state of a magnetic crystal:

$$\mathbf{M} = M_0(\cos pz', \ \sin pz', \ 0), \tag{4.4.2}$$

where $p = \kappa/\alpha$. The period of the helicoidal structure is much longer than the crystallographic periods $2\pi/|p| = 2\pi\alpha/|\kappa| \gg a$ and is usually incommensurable with them.

As before, we turn to the dimensionless variables

$$\mathbf{M} = -M_0\mathbf{S}, \quad z = \alpha^{-1/2}z', \quad t = \gamma M_0 t', \quad g = \kappa\alpha^{-1/2}.$$

Then the Landau–Lifshitz equations for a helicoidal ferromagnet take the form

$$\partial_t \mathbf{S} = [\mathbf{S} \times \partial_z^2 \mathbf{S}] - \rho^2(\mathbf{n} \cdot \mathbf{S})[\mathbf{S} \times \mathbf{n}] + 2g\,\partial_z \mathbf{S}(\mathbf{n} \cdot \mathbf{S}), \quad \mathbf{S}^2 = 1, \tag{4.4.3}$$

where $\mathbf{n} = (0, 0, 1)$.

Further, it is useful to rewrite eqs. (4.4.3) in the parameterization of $\mathbf{S}$ by the angles θ and φ:

$$\mathbf{S} = (\cos\theta\cos\varphi, \ \cos\theta\sin\varphi, \ \sin\theta). \tag{4.4.4}$$

Then they appear as

$$\begin{aligned} \cos\theta\,\partial_t\varphi &= -\partial_z^2\theta - \sin\theta\cos\theta(\partial_z\varphi - g)^2 \\ &\quad + (\rho^2 + g^2)\sin\theta\cos\theta, \\ \partial_t \sin\theta &= \partial_z[\cos^2\theta(\partial_z\varphi - g)]. \end{aligned} \tag{4.4.5}$$

Remarkably, the conversion

$$\tilde{\varphi} = \varphi - gz, \quad \tilde{\rho}^2 = \rho^2 + g^2$$

transfers eqs. (4.4.5) into the Landau–Lifshitz equations for an "easy-plane" anisotropy ferromagnet. Here the anisotropy constant of an easy-plane ferromagnet is equal to $\tilde{\rho}^2 = \rho^2 + g^2$, and the angles θ and $\tilde{\varphi}$ ascertain the magnetization distribution:

$$\tilde{\mathbf{S}} = (\cos\theta\cos\tilde{\varphi}, \ \cos\theta\sin\tilde{\varphi}, \ \sin\theta). \tag{4.4.6}$$

Thus soliton solutions to the ferromagnet model (4.4.1) with the helicoidal structure (4.4.2) can be found by choosing appropriate solutions to the "easy-plane" anisotropy ferromagnet model. Namely, it is necessary to apply solutions (4.4.6) for an "easy-plane" ferromagnet with constant anisotropy $\tilde{\rho}^2 = \rho^2 + g^2$ and asymptotics

$$\tilde{\mathbf{S}} \to (\cos\tilde{\varphi}_\pm,\ \sin\tilde{\varphi}_\pm,\ 0), \quad \tilde{\varphi}_\pm = \text{const},$$

as $z \to \pm\infty$. Then, after the replacement $\tilde{\varphi} \to \varphi = \tilde{\varphi} + gz$, solutions (4.4.6) pass into solutions (4.4.4) to model (4.4.3) to describe nonlinear excitations of a ferromagnet with helicoidal structure.

So the solution of the turning wave type (4.2.8) is transformed into a solution that describes a soliton moving with velocity $V = \tilde{\rho}\cos r$ ($|V| < \tilde{\rho}$ in a spiral structure:

$$S_3 = \frac{\sin r}{\cosh\eta}, \quad S_1 + \mathrm{i}\,S_2 = \left(\tanh\eta + \frac{\mathrm{i}\cos r}{\cosh\eta}\right)\exp(\mathrm{i}gz), \qquad (4.4.7)$$
$$\eta = \tilde{\rho}\sin r(z - t\tilde{\rho}\cos r).$$

Here r is a real parameter. Soliton (4.4.7) is associated with a spatially localized change in the magnetic spiral phase and with the escape of magnetic moments from the basal plane (x_1, x_2) (see Figure 4.6). For the same momentum, the soliton energy (4.4.7) lies below the energy of a linear spin wave propagating in a helicoidal structure along the anisotropy axis [4].

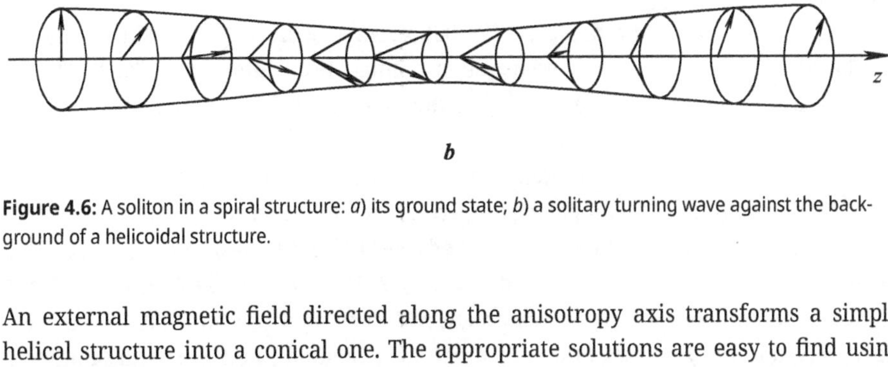

Figure 4.6: A soliton in a spiral structure: *a*) its ground state; *b*) a solitary turning wave against the background of a helicoidal structure.

An external magnetic field directed along the anisotropy axis transforms a simple helical structure into a conical one. The appropriate solutions are easy to find using formulas (3.1.10) and (4.2.11). According to the findings secured previously, two types of turning waves are possible to form against the background of a conical helical structure, with their energy less than that of a linear spin-wave mode with the same momentum.

5 Integration of the equations for a ferromagnet with biaxial magnetic anisotropy

5.1 Direct and inverse problems for the Landau–Lifshitz equations for an anisotropic ferromagnet

A method is necessary to find the truth.

R. Descartes

The present section deals with an algorithm for integrating eq. (3.1.14) by the inverse scattering method for two types of boundary conditions:

$$\mathbf{S} \to (0,0,1) \quad \text{as} \quad x \to +\infty, \quad \mathbf{S} \to (0,0,\varepsilon) \quad \text{as} \quad x \to -\infty. \tag{5.1.1}$$

Since the case $\varepsilon = 1$ contains no solutions of the type of domain walls, below it is referred to as nontopological (NT); the topological (T) case corresponds to $\varepsilon = -1$. Numerical calculations, particular solutions, the existence of a $U-V$ pair, and a proof of complete integrability (see references in Section 2.1) raised the question of solving the Cauchy problem for this equation. Despite the existence of the $U-V$ pair approach, the main challenge faced in solving the Cauchy problem is that representations (3.1.15) and (3.1.21) generate a novel class of spectral problems (an elliptic $U-V$ pair) previously unheard of. The authors of [14, 15] were the first to design a procedure for such problems to be exactly solved within an SU(2)-asymmetric chiral model. As for the biaxial ferromagnetic model (3.1.14), the "dressing" procedure for the NT-case $\mathbf{S} \to (0,0,1)$ as $|x| \to \infty$ and for an arbitrary "seeding" solution were independently advanced in [26, 27, 29] and [11], respectively.

It is worth pointing out that [26, 27, 29, 30] offer a specific procedure for solving the regular Riemann problem based on a system of singular equations with Cauchy kernel on a torus and therefore give a solution to the inverse scattering problem for the NT-case.

Somewhat later, another "dressing" procedure was proposed [6], reflected further in [24, 23]. Without going into comparing these two "dressing" procedures, we note that the progress in modification of the second method (see [4, 5]) led to the finite-gap integration of eq. (3.1.14) for $J_1 = J_2 \neq J_3$ [2, 3] and $J_1 \neq J_2 \neq J_3$ [1, 7] involving the data of [10, 16]. However, not only was the Cauchy problem with boundary conditions (5.1.1) far from being solved in the T-case, but the direct scattering method could afford no solution in the NT-case either.

The present section illustrates the results obtained in [12, 13] concerning the solution to the direct scattering problem for $\varepsilon = \pm 1$. Using an analog of the Fourier transform on a torus, we succeed in constructing a closed integration scheme based on the Gel'fand–

https://doi.org/10.1515/9783112235973-005

Levitan–Marchenko equations. Such an algorithm makes it possible to solve the Cauchy problem for eq. (3.1.14) by means of the inverse scattering method.

5.1.1 Transformations of the direct scattering problem

Consider first the direct scattering problem to define two fundamental matrices of solutions to the first equation of (2.1.2), (3.1.15), and (3.1.21) (Jost matrices), having the following asymptotic behavior:

$$\Phi \to \exp[-\mathrm{i}\, w_3(\lambda)\, x\, \varepsilon\, \sigma_3] \quad \text{as} \quad x \to -\infty,$$
$$\Psi \to \exp[-\mathrm{i}\, w_3(\lambda)\, x\, \sigma_3] \quad \text{as} \quad x \to +\infty.$$

It is convenient to combine the description of NT- and T-cases. Then the reductions for $\Psi(\lambda)$ and $\Phi(\lambda)$,

$$\begin{aligned} \Phi^*(\lambda) &= \sigma_2 \Phi(\lambda^*)\sigma_2, \quad \Phi(\lambda + 2K) = \sigma_3 \Phi(\lambda)\sigma_3, \\ \Phi(\lambda + 2\mathrm{i}K') &= \sigma_3 \Phi^*(\lambda^*)\sigma_3, \\ \Psi^*(\lambda) &= \sigma_2 \Psi(\lambda^*)\sigma_2, \quad \Psi(\lambda + 2K) = \sigma_3 \Psi(\lambda)\sigma_3, \\ \Psi(\lambda + 2\mathrm{i}K') &= \sigma_3 \Psi^*(\lambda^*)\sigma_3, \end{aligned} \tag{5.1.2}$$

imply that the transition matrix $T(\lambda)$, introduced in the usual way

$$\begin{aligned} &\Phi(\lambda, x) = \Psi(\lambda, x)\, T(\lambda) \\ &(\operatorname{Im}\lambda = 0, 2K', \ 0 \le \operatorname{Re}\lambda \le 4K, \ \operatorname{mod}(4K, 4iK')), \end{aligned} \tag{5.1.3}$$

has the following shape:

$$\begin{aligned} &T(\lambda) = \begin{pmatrix} a(\lambda) & -b^*(\lambda^*) \\ b(\lambda) & a^*(\lambda^*) \end{pmatrix}, \quad \det T(\lambda) = 1; \\ &a(\lambda + 2K) = a(\lambda), \quad b(\lambda + 2K) = -b(\lambda), \\ &a(\lambda + 2\mathrm{i}K') = a^*(\lambda^*), \quad b(\lambda + 2\mathrm{i}K') = -b^*(\lambda^*). \end{aligned} \tag{5.1.4}$$

To solve the direct scattering problem, we need to find the dependence on the spectral parameter λ in the kernels of the triangular representation of Jost matrices. We can guess this dependence along the solution to the first equation of (2.1.2) according to the perturbation theory (near the solution $\mathbf{S} = (0, 0, 1)$ up to the third order inclusive). Appropriate speculations and calculations are similar to those used in the third chapter and are not given here.

Let us go straight to the rigorous proofs. We show that the matrix solutions Φ and Ψ to the spectral problems (2.1.2), (3.1.15), and (3.1.21) have the following triangular representations:

$$\Phi(x,\lambda) = k_3(x)\exp[-\mathrm{i}\,\varepsilon\, w_3(\lambda)\, x\, \sigma_3] + \int_{-\infty}^{x} \mathrm{d}y \left(k_5(x,y) + \sum_{a=1}^{2} k_a(x,y)\, w_a \sigma_a + k_4(x,y)\, w_1 w_2\, \sigma_3 \right) \exp(-\mathrm{i}\,\varepsilon\, w_3\, y\, \sigma_3); \tag{5.1.5}$$

$$\Psi(x,\lambda) = l_3(x)\exp[-\mathrm{i}\, w_3(\lambda)\, x\, \sigma_3] + \int_{x}^{+\infty} \mathrm{d}y \left(l_5(x,y) + \sum_{a=1}^{2} l_a(x,y)\, w_a \sigma_a + l_4(x,y)\, w_1 w_2\, \sigma_3 \right) \exp(-\mathrm{i}\, w_3\, y\, \sigma_3), \tag{5.1.6}$$

where

$$k_3(x) = \sqrt{\frac{1+\varepsilon S_3(x)}{2}} \cos\left(\frac{1}{2} \int_{-\infty}^{x} \mathrm{d}y [1 - \varepsilon S_3(y)] \partial_y \arctan\left[\frac{S_1(y)}{S_2(y)} \right] \right),$$
$$\frac{k_1(x,x) + \mathrm{i}\, k_2(x,x)}{k_3(x) + k_{4,y}(x,x)} = -\mathrm{i}\, \frac{S_1(x) + \mathrm{i}\, S_2(x)}{1 + \varepsilon\, S_3(x)}; \tag{5.1.7}$$

$$l_3(x) = \sqrt{\frac{1+S_3(x)}{2}} \cos\left(\frac{1}{2} \int_{x}^{+\infty} \mathrm{d}y [1 - S_3(y)] \partial_y \arctan\left[\frac{S_1(y)}{S_2(y)} \right] \right),$$
$$\frac{l_1(x,x) + \mathrm{i}\, l_2(x,x)}{l_3(x) - l_{4,y}(x,x)} = \mathrm{i}\, \frac{S_1(x) + \mathrm{i}\, S_2(x)}{1 + S_3(x)}. \tag{5.1.8}$$

Here we have introduced the following notations through $k_{4,x}(x,y)$, $k_{4,y}(x,y)$, for $\partial_x k_4(x,y)$, $\partial_y k_4(x,y)$, etc. For simplicity, we omit the t-dependence of the functions.

Due to reductions (5.1.2), the kernels $k_5(x,y)$ and $l_5(x,y)$ are real; the kernels $k_a(x,y)$ and $l_a(x,y)$ ($a = 1, 2, 4$) are purely imaginary. The nonsingularity of functions (5.1.5) and (5.1.6) at the poles of the coefficients $w_j(\lambda)$ ($j = 1, 2, 3$) requires the fulfillment of zero conditions on the diagonal $x = y$:

$$k_4(x,x) = l_4(x,x) = 0.$$

Here an important circumstance should be noted. The λ-dependence in the kernels of the triangular representations (5.1.5) and (5.1.6) is formed through the functions 1, w_1, w_2, $w_1 w_2$. The latter are not accidental; they constitute a basis in the space of all polynomials of w_1, w_2, and w_3, where connections (3.1.16) are taken into account, and two elements are equivalent if they differ by the factor w_3. The third chapter has already dealt with a similar kernel constructions made up of the coefficients of a $U - V$ pair. Besides, the solutions of earlier known, simpler direct scattering methods implicitly contain such kernels. It is clear that choosing their λ-dependence always leads to certain differential equations for the matrix coefficients. Indeed, plugging eq. (5.1.5) into the first equation of (2.1.2), we arrive at the system of linear equations

$$\begin{aligned}
&k_{1,x}(x,y) + S_2(x)k_{4,yy}(x,y) - \varepsilon S_3(x)k_{1,y}(x,y)\\
&\quad = -\mathrm{i}\, S_1(x)k_5(x,y) + \rho^2(k')^2 S_2(x)k_4(x,y),\\
&k_{2,x}(x,y) - S_1(x)k_{4,yy}(x,y) - \varepsilon S_3(x)k_{2,y}(x,y)\\
&\quad = -\mathrm{i}\, S_2(x)k_5(x,y) - \rho^2 S_1(x)k_4(x,y),\\
&k_{4,x}(x,y) + \varepsilon S_3(x)k_{4,y}(x,y) = S_1(x)k_2(x,y) - S_2(x)k_1(x,y),\\
&k_{5,x}(x,y) + \varepsilon S_3(x)k_{5,y}(x,y) - \mathrm{i}\sum_{a=1}^{2} S_a(x)k_{a,yy}(x,y)\\
&\quad = -\mathrm{i}\, S_1(x)\rho^2 k_1(x,y) - \mathrm{i}\,\rho^2(k')^2 S_2(x)k_2(x,y)
\end{aligned} \tag{5.1.9}$$

and the conditions on the diagonal:

$$\begin{aligned}
&k_{3,x}(x) + k_5(x,x)(1-\varepsilon S_3(x)) + \mathrm{i}\sum_{a=1}^{2} S_a(x)k_{a,y}(x,y)|_{y=x} = 0,\\
&\mathrm{i}\sum_{a=1}^{2} k_a(x,x)S_a(x) = (1-\varepsilon S_3(x))k_3(x), \quad k_4(x,x) = 0,\\
&k_1(x,x)S_2(x) - S_1(x)k_2(x,x) = (1-\varepsilon S_3(x))k_{4,y}(x,y)|_{y=x}.
\end{aligned} \tag{5.1.10}$$

To investigate the unique solvability of this system, it is convenient to introduce the auxiliary variables

$$q_1 = S_1k_2 - S_2k_1 \quad \text{and} \quad q_2 = S_1k_1 + S_2k_2.$$

Then, due to the first three equations of (5.1.9), the variables q_1 and k_5 are expressed through q_2 and k_4 and their derivatives:

$$\begin{aligned}
k_5(x,y) = \frac{\mathrm{i}}{S_\perp^2}\Big(&q_{2,x}(x,y) - \varepsilon S_3(x)q_{2,y}(x,y)\\
&+ \rho^2k^2 S_1(x)S_2(x)k_4(x,y) + q_1(x,y)\,\partial_x \arctan\left[\frac{S_1(x)}{S_2(x)}\right]\\
&- q_2(x,y)\,\frac{1}{2}\,\partial_x \ln S_\perp^2(x)\Big), \quad S_\perp^2 = S_1^2 + S_2^2;
\end{aligned} \tag{5.1.11}$$

$$q_1(x,y) = k_{4,x}(x,y) + \varepsilon S_3(x)k_{4,y}(x,y). \tag{5.1.12}$$

Substitution of eqs. (5.1.11) and (5.1.12) into system (5.1.9)–(5.1.10) leads the latter to the following system of two second-order linear differential equations:

$$\begin{aligned} &q_{2,xx} - q_{2,yy} - \varepsilon S_{3,x} q_{2,y} + (\partial_x + \varepsilon S_3 \partial_y)\Big[\rho^2 k^2 S_1 S_2 k_4 + \\ &\qquad + q_1 \partial_x \arctan\Big(\frac{S_1}{S_2}\Big) - \frac{1}{2} q_2 \partial_x \ln S_\perp^2\Big] - \mathrm{i} S_\perp^4 k_5 \partial_x S_\perp^2 \\ &\quad = \rho^2 [q_1 k^2 S_1 S_2 - q_2 (S_1^2 + (k')^2 S_2^2)], \\ &k_{4,xx} - k_{4,yy} = -\varepsilon S_{3,x} k_{4,y} + \frac{1}{2} q_1 \partial_x \ln S_\perp^2 \\ &\qquad + q_2 \partial_x \arctan\Big(\frac{S_1}{S_2}\Big) - (S_1^2 + (k')^2 S_2^2) k_4 \rho^2 \end{aligned} \tag{5.1.13}$$

and the ordinary differential equation

$$2\mathrm{i} k_{3,x} + \mathrm{i} k_3 \frac{\varepsilon S_{3,x}}{1 + \varepsilon S_3} + \Big(\partial_x \arctan \frac{S_1}{S_2}\Big)(\partial_x + \varepsilon S_3 \partial_y) k_4(x, y)|_{y=x} = 0 \tag{5.1.14}$$

with boundary conditions

$$\begin{aligned} &k_4(x, x) = 0, \quad k_4(x, -\infty) = q_2(x, -\infty) = 0, \\ &q_2(x, x) = \mathrm{i}(S_3 - 1) k_3(x). \end{aligned}$$

The condition $\det \Phi(\lambda) = 1$, together with representation (5.1.5), yields the important relation

$$\det \Phi(\lambda = 0) = k_3^2(x) - k_{4,y}^2(x, x) - k_1^2(x, x) - k_2^2(x, x) = 1.$$

Combining the latter and eqs. (5.1.10), we can deduce a simple algebraic relation between $k_3(x)$ and $k_{4,y}(x, x)$ to solve eq. (5.1.14) explicitly. We get

$$\begin{aligned} &k_3(x) = \sqrt{\frac{1 + \varepsilon S_3}{2}} \cos\varphi, \quad \mathrm{i} k_{4,y}(x, x) = \sqrt{\frac{1 + \varepsilon S_3}{2}} \sin\varphi, \\ &\partial_x \varphi(x) = -\frac{1}{2}(1 - \varepsilon S_3) \partial_x \arctan \frac{S_1}{S_2}. \end{aligned}$$

To prove the existence of solutions to eqs. (5.1.13) and (5.1.14), we introduce new coordinates

$$\xi = \frac{x + y}{2}, \quad \eta = \frac{x - y}{2}$$

and variables

$$v_k = k_{4,\xi}, \quad v_q = q_{2,\xi}, \quad u_k = k_{4,\eta}, \quad u_q = q_{2,\eta}.$$

Then the solution to eqs. (5.1.13) and (5.1.14) is equivalent to the Cauchy–Goursat problem for a hyperbolic system of six equations with variables k_4, v_k, u_k, q_2, u_q, and v_q. The boundary conditions are set on the characteristic $\eta = 0$ and on the straight line $\xi + \eta = -\infty$. From the theory of characteristics [28, 17] it follows that the solution exists and is unique.

The matrix Ψ can be examined in a similar manner. In this instance, the formal replacement

$$k_\mu(x,y) \to -l_\mu(x,y), \quad k_3(x) \to l_3(x)$$

of relations (5.1.9) and (5.1.10) with $\varepsilon = 1$ brings about a system of equations for the kernels $l_\mu(x,y)$ ($\mu = 1,2,4,5$) and $l_3(x)$, as well as the conditions on the diagonal. Representation (5.1.5)–(5.1.8) and eqs. (5.1.11)–(5.1.14) give a complete solution to the direct scattering problem.

Representations (5.1.5) and (5.1.6) are useful for proving the analytic properties of Jost solutions. Suppose the distribution of magnetization rather quickly tends to its limiting values as $|x| \to \infty$. Then the analyticity domains for the columns of the Jost matrices and for the elements of the transition matrix are determined by the analytic continuation with respect to the parameter λ and the condition that the operators in representations (5.1.5) and (5.1.6) are contractive.

For $\varepsilon = 1$, we find that the functions

$$\Psi_{\alpha 1} \exp(\mathrm{i}\, w_3\, x), \quad \Phi_{\alpha 2} \exp(-\mathrm{i}\, w_3\, x) \quad (\alpha = 1,2)$$

admit an analytic continuation to the domain

$$D_- = \{\lambda \mid 0 < \operatorname{Im}\lambda < 2K',\ 0 < \operatorname{Re}\lambda < 4K\},$$

whereas the functions

$$\Psi_{\alpha 2} \exp(-\mathrm{i}\, w_3\, x), \quad \Phi_{\alpha 1} \exp(\mathrm{i}\, w_3\, x), \quad a(\lambda)$$

can be analytically continued to the domain

$$D_+ = \{\lambda \mid -2K' < \operatorname{Im}\lambda < 0,\ 0 < \operatorname{Re}\lambda < 4K\}.$$

In view of reductions (5.1.4), the zeros $\lambda_n^{(1)}$ of the function $a(\lambda)$ are split into two groups:

$$\begin{aligned} &\lambda = u_j - \mathrm{i}K', \quad u_j - \mathrm{i}K' + 2K, \quad j = 1,2,\dots,n, \\ &\lambda = u_s - \mathrm{i}K' \pm \mathrm{i}\theta_s, \quad u_s - \mathrm{i}K' \pm \mathrm{i}\theta_s + 2K, \quad s = 1,2,\dots,m, \end{aligned}$$

where $0 < u_j < 2K$ and $0 < \theta_s < K'$.

Equation (5.1.3) means that when $\lambda = \lambda_n^{(1)}$, the functions $\Phi_{\alpha 1}$ and $\Psi_{\alpha 2}$ are proportional:

$$\Phi_{\alpha 1}(\lambda_n^{(1)}) = b_n^{(1)} \Psi_{\alpha 2}(\lambda_n^{(1)}).$$

For $\varepsilon = 1$, the set

$$r^{(1)} = \frac{b(\lambda)}{a(\lambda)} \quad (\operatorname{Im}\lambda = 0, 2K',\ 0 < \operatorname{Re}\lambda < 4K), \quad \{\lambda_n^{(1)}\}, \quad \{b_n^{(1)}\}$$

involves scatter data. This set evolves in accordance with the equations

$$\partial_t a = 0, \quad \partial_t b = -4\mathrm{i}\, w_1(\lambda) w_2(\lambda)\, b, \quad \partial_t b_n^{(1)} = -4\mathrm{i}\, w_1(\lambda_n^{(1)}) w_2(\lambda_n^{(1)}) b_n^{(1)}.$$

For $\varepsilon = -1$, the functions $\Phi_{\alpha 1} \exp(-\mathrm{i}\, w_3\, x)$, $\Psi_{\alpha 1} \exp(-\mathrm{i}\, w_3\, x)$, and $b(\lambda)$ admit an analytic continuation to the domain D_-. In the domain D_- the zeros $\lambda = \lambda_n^{(-1)}$ of the function $b(\lambda)$, due to reductions (5.1.4), are divided into two groups:

$$\begin{aligned} &u_j + \mathrm{i}K', \quad u_j + \mathrm{i}K' + 2K, \quad j = 1, 2, \ldots, n; \\ &u_s + \mathrm{i}K' \pm \mathrm{i}\theta_s, \quad u_s + \mathrm{i}K' \pm \mathrm{i}\theta_s + 2K, \quad s = 1, 2, \ldots, m, \end{aligned} \tag{5.1.15}$$

where $0 < u_j < 2K$ and $0 < \theta_s < K'$. The functions $\Phi_{\alpha 2} \exp(\mathrm{i}\, w_3\, x)$ and $\Psi_{\alpha 2}(-\mathrm{i}\, w_3\, x)$ can be analytically extended to the domain D_+.

The scatter data time-dependence

$$r^{(-1)} = \frac{a(\lambda)}{b(\lambda)} \quad (\operatorname{Im}\lambda = 0, 2K',\ 0 < \operatorname{Re}\lambda < 4K), \quad \{\lambda_n^{(-1)}\}, \quad \{b_n^{(-1)}\}$$

obeys the equations

$$\begin{aligned} \partial_t b = 0, \quad \partial_t a &= -4\mathrm{i}\, w_1(\lambda) w_2(\lambda)\, a, \\ \partial_t b_n^{(-1)} &= -4 w_1(\lambda_n^{(-1)}) w_2(\lambda_n^{(-1)}) b_n^{(-1)}, \end{aligned}$$

where $b_n^{(-1)}$ is the coefficient of proportionality in the ratio

$$\Phi_{\alpha 1}(\lambda_n^{(-1)}) = b_n^{(-1)} \Psi_{\alpha 1}(\lambda_n^{(-1)}).$$

5.1.2 Inverse spectral transform

Let us move on to solving the inverse problem. For rationally λ-dependent U–V pairs, the inverse problem always exploits the Fourier transform if the Jost functions in the asymptotics are irreducible representations of the shift group with respect to the variable x. In our case the Fourier transform described below is the key to solving the problem.

Suppose Γ_ε is a closed contour as depicted in Figure 5.1:

$$\begin{aligned} \Gamma_\varepsilon = \{\lambda \mid 0 < \operatorname{Re}\lambda < 2K,\ \operatorname{Im}\lambda = 0; \quad 0 < -\varepsilon \operatorname{Im}\lambda < 2K',\ \operatorname{Re}\lambda = 4K; \\ 4K > \operatorname{Re}\lambda > 0,\ \operatorname{Im}\lambda = -2\varepsilon K'; \quad 2K' > -\varepsilon \operatorname{Im}\lambda > 0,\ \operatorname{Re}\lambda = 0\}. \end{aligned}$$

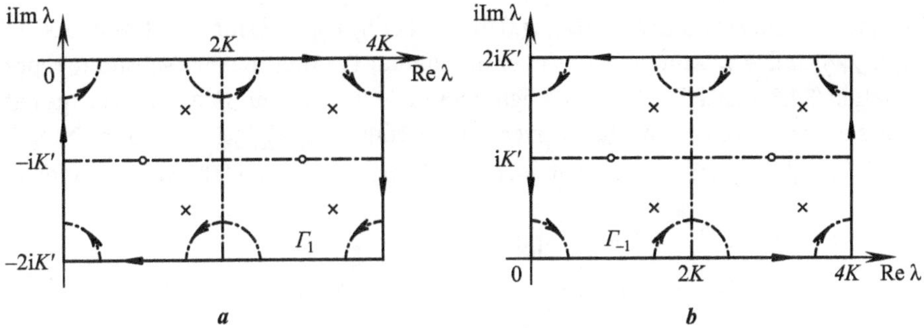

Figure 5.1: Integration contours Γ_1 ($\varepsilon = 1$) (*a*) and Γ_{-1} ($\varepsilon = -1$) (*b*) (a solid line) and a system of contours $C_i^{(\varepsilon)}$ ($i = 1, 2, 3, 4$) (a dashed line) used in deriving the equations of the inverse scattering problem. Location of "domain" (∘) and "breather" (×) zeros of the functions $a(\lambda)$ ($\varepsilon = 1$) (*a*) and $b(\lambda)$ ($\varepsilon = -1$) (*b*).

Then an integral over the contour Γ_ε of doubly periodic functions $q(\lambda)$ with periods $4K$ and $4iK'$ is reduced to an integral over the interval $0 < \lambda < 2K$ as follows:

$$\int_{\Gamma_\varepsilon} d\lambda\, q(\lambda) \exp(i\, w_3\, x) = \int_0^{2K} d\lambda\, \bar{q}(\lambda) \exp(i\, w_3\, x),$$

where

$$\bar{q}(\lambda) = q(\lambda) + q(\lambda + 2K) - q(-\lambda + 2iK') - q(-\lambda + 2K + 2iK').$$

Note that the functions w_1, w_2, and w_3 possess the following properties:

$$w_s(\lambda + 2K) = -w_s(\lambda) \quad (s = 1, 2), \quad w_3(\lambda + 2K) = w_3(\lambda),$$
$$w_1(\lambda + 2iK') = w_1(\lambda), \quad w_j(\lambda + 2iK') = -w_j(\lambda) \quad (j = 2, 3).$$

Therefore we get

$$\begin{aligned} &\overline{w_s} = \overline{w_s w_3} = \overline{w_s^2 w_j} = 0 \quad (s, j = 1, 2, 3); \\ &\overline{w_1 w_2} = 4\, w_1 w_2 = -4\rho \frac{d\, w_3}{d\lambda}; \\ &\int_{\Gamma_\varepsilon} d\lambda\, w_1 w_2 \exp[i\, w_3(\lambda)\, x] = 8\pi\rho\, \delta(x). \end{aligned} \tag{5.1.16}$$

To derive the equations of the inverse scattering problem, we consider one of relations (5.1.3) in the form

$$\Phi_{\alpha 1}(x, \lambda) \exp(i\, w_3\, \varepsilon y) = [\Psi_{\alpha 1}(x, \lambda) a(\lambda) + b(\lambda) \Phi_{\alpha 2}(x, \lambda)] \exp(i\, w_3\, \varepsilon y), \quad y \ge x. \tag{5.1.17}$$

Next, we multiply eq. (5.1.17) by $\tau(\lambda)/a(\lambda)$ for $\varepsilon = 1$ or by $\tau(\lambda)/b(\lambda)$ for $\varepsilon = -1$, where $\tau(\lambda) = w_3, w_1, w_2$, and $w_1 w_2$, and integrate over the contour Γ_ε. After that, we substitute representation (5.1.6) instead of Ψ into the right-hand side. The contour integration on the left-hand side is replaced by an integral over a closed contour $\tilde{\Gamma}_\varepsilon + \sum_{i=1}^4 C_i^{(\varepsilon)}$ minus an integral over $\sum_{i=1}^4 C_i^{(\varepsilon)}$. A system of the contours $C_i^{(\varepsilon)}$ is determined by the following equations:

$$\begin{aligned} C_1^{(\varepsilon)} &= \{\lambda \mid \lambda = r\exp(\mathrm{i}\,\varepsilon\,\varphi),\ \pi \le \varphi \le 2\pi\}, \\ C_2^{(\varepsilon)} &= \{\lambda \mid \lambda = 2K + r\exp(\mathrm{i}\,\varepsilon\,\varphi),\ \pi \le \varphi \le 2\pi\}, \\ C_3^{(\varepsilon)} &= \{\lambda \mid \lambda = r\exp(\mathrm{i}\,\varepsilon\,\varphi) + 2K - 2\,\mathrm{i}\,\varepsilon K',\ 0 \le \varphi \le \pi\}, \\ C_4^{(\varepsilon)} &= \{\lambda \mid \lambda = -2\,\mathrm{i}K'\varepsilon + r\exp(\mathrm{i}\,\varepsilon\,\varphi),\ 0 \le \varphi \le \pi\}. \end{aligned}$$

Due to the analyticity in the domain D_ε, the integrals over the contour

$$\tilde{\Gamma}_\varepsilon + \sum_{i=1}^{4} C_i^{(\varepsilon)}$$

of the functions

$$\begin{aligned} &\tau(\lambda)a^{-1}(\lambda)\,\Phi_{\alpha 1}\exp(\mathrm{i}\,w_3 y) \quad (\text{at } \varepsilon = 1), \\ &\tau(\lambda)b^{-1}(\lambda)\,\Phi_{\alpha 1}\exp(-\mathrm{i}\,w_3 y) \quad (\text{at } \varepsilon = -1) \end{aligned}$$

are equal to the residues of these functions at the zeros of the functions $a(\lambda)$ and $b(\lambda)$, respectively.

The integrals over a system of contours $\sum_{i=1}^4 C_i^{(\varepsilon)}$ cancel together with singular contributions of the right-hand side. Let us see how this happens.

Firstly, representations (5.1.5) and (5.1.6), together with the Riemann–Lebesgue lemma, imply

$$\begin{aligned} \Phi\exp(\mathrm{i}\,\varepsilon\,w_3\,x\,\sigma_3)|_{\lambda=0} &= k_3(x) + k_{4,y}(x,x)\sigma_3 - \varepsilon\,k_2(x,x)\sigma_1 + k_1(x,x)\sigma_2\,\varepsilon, \\ \Psi\exp(\mathrm{i}\,w_3\,x\,\sigma_3)|_{\lambda=0} &= l_3(x) - l_{4,y}(x,x)\sigma_3 - l_1(x,x)\sigma_2 - l_2(x,x)\sigma_1. \end{aligned} \tag{5.1.18}$$

Then, eqs. (5.1.7), (5.1.8), and (5.1.18) produce an important relation for the functions $a(\lambda) = \Phi_{11}\Psi_{22} - \Phi_{21}\Psi_{12}$ and $b(\lambda) = \Psi_{11}\Phi_{21} - \Phi_{11}\Psi_{21}$ at $\lambda = 0$:

$$b(0) = 0 \quad (\varepsilon = 1); \quad a(0) = 0 \quad (\varepsilon = -1). \tag{5.1.19}$$

Suppose that $\varepsilon = 1$. We multiply the eq. (5.1.17) by $w_1 w_2\, a^{-1}(\lambda)$ and integrate over the contour Γ_1. The singular term on the right-hand side is equal to

$$8\pi p\,\delta_{\alpha 1}\big[l_3(x) - l_{4,y}(x,x)\big]\delta(x-y). \tag{5.1.20}$$

Formulas (5.1.2), (5.1.4), and (5.1.17)–(5.1.19) transform the singular term on the left-hand side so that it coincides with expression (5.1.20) as follows:

$$- \int_{\Sigma C_i^{(1)}} \mathrm{d}\lambda \frac{w_1 w_2}{a(\lambda)} \Phi_{\alpha 1}(x,\lambda) \exp(\mathrm{i}\, w_3 y)$$

$$= \sum_{\lambda=0,\, 2K,\, -2\mathrm{i}K',\, 2K-2\mathrm{i}K'} \frac{\Phi_{\alpha 1}(x,\lambda)}{a(\lambda)} \exp(\mathrm{i}\, w_3 x) \lim_{r\to 0} \int_{-r}^{r} \mathrm{d}\lambda \left(-\frac{\rho^2}{\lambda^2}\right) \exp\left[\frac{\mathrm{i}\rho}{\lambda}(y-x)\right]$$

$$= 8\pi\rho\, \delta(y-x)\delta_{\alpha 1} \mathrm{Re}[\Psi_{11}(x,\lambda) \exp(\mathrm{i}\, w_3 x)]|_{\lambda=0}.$$

The other singular terms cancel similarly. Nonsingular terms give equations of the Gel'fand–Levitan–Marchenko type. Bearing in mind the symmetry properties of the coefficients w_i (5.1.16) and reductions (5.1.4), we boil down the final result to four equations (integrating by parts the terms $l_4 w_1 w_2$ in eq. (5.1.6) allows for avoiding the last "trap"):

$$\begin{aligned}
&\int_x^{+\infty} \mathrm{d}s\, [l_1(x,s)F^{\varepsilon}_{1,y}(s+y) - \mathrm{i}\,\varepsilon\, l_2(x,s)F^{\varepsilon}_{2,y}(s+y)] = \varepsilon\, l_{4,y}(x,y),\\
&\int_x^{+\infty} \mathrm{d}s\, [l_1(x,s)\Pi^{\varepsilon}_{1,y}(s+y) - \mathrm{i}\varepsilon l_2(x,s)\Pi^{\varepsilon}_{2,y}(s+y)] = \mathrm{i}\varepsilon l_5(x,y),\\
l_3(x)F^{\varepsilon}_1(s+y) + &\int_x^{+\infty} \mathrm{d}s\, [l_5(x,s)F^{\varepsilon}_1(x+y) - \mathrm{i}\, l_{4,s}(x,s)\Pi^{\varepsilon}_1(s+y)] = \mathrm{i}\,\varepsilon\, l_2(x,y),\\
l_3(x)F^{\varepsilon}_2(x+y) + &\int_x^{+\infty} \mathrm{d}s\, [l_5(x,s)F^{\varepsilon}_2(s+y) - \mathrm{i}\, l_{4,s}(x,s)\Pi^{\varepsilon}_2(s+y)] = l_1(x,y).
\end{aligned} \tag{5.1.21}$$

System (5.1.21) ties the kernels l_1/l_3, l_2/l_3, $l_{4,s}/l_3$, and l_5/l_3 with the kernels $F^{\varepsilon}_i(x)$ and Π^{ε}_i $(i = 1,2)$ to be determined through scatter data:

$$\begin{aligned}
F^{\varepsilon}_i(x) &= -\frac{\mathrm{i}\,\varepsilon}{4\rho} \sum_n r^{(\varepsilon)}_n w_i(\lambda^{(\varepsilon)}_n) \exp(\mathrm{i}\,\varepsilon\, w_3(\lambda^{(\varepsilon)}_n)\, x)\\
&\quad - \frac{1}{8\pi\rho} \int_{\Gamma_\varepsilon} \mathrm{d}\lambda\, w_i(\lambda) r^{(\varepsilon)}(\lambda) \exp(\mathrm{i}\, w_3(\lambda)\, \varepsilon\, x),\\
\Pi^{\varepsilon}_i(x) &= -\frac{\mathrm{i}\,\varepsilon}{4\rho} \sum_n r^{(\varepsilon)}_n w_i(\lambda^{(\varepsilon)}_n) w_1(\lambda^{(\varepsilon)}_n) w_2(\lambda^{(\varepsilon)}_n)\\
&\quad \times w_3^{-1}(\lambda^{(\varepsilon)}_n) \exp(\mathrm{i}\, w_3(\lambda^{(\varepsilon)}_n) x) - \frac{1}{8\pi\rho} \int_{\Gamma_\varepsilon} \mathrm{d}\lambda\, r^{(\varepsilon)}(\lambda) w_3^{-1}(\lambda)\\
&\quad \times w_i(\lambda) w_1(\lambda) w_2(\lambda) \exp(\mathrm{i}\,\varepsilon\, w_3\, x).
\end{aligned}$$

Here

$$r^{(1)}(\lambda) = \frac{b(\lambda)}{a(\lambda)}, \quad r^{(1)}_n = \frac{b^{(1)}_n}{a_n}; \quad r^{(-1)}(\lambda) = \frac{a(\lambda)}{b(\lambda)}, \quad r^{(-1)}_n = \frac{b^{(-1)}_n}{b_n},$$

where a_n^{-1} and b_n^{-1} are the residues of the functions $a^{-1}(\lambda)$ and $b^{-1}(\lambda)$, respectively, at zeros $\lambda_n^{(\varepsilon)}$.

According to eqs. (5.1.8), the solution to eqs. (5.1.21) with given kernels generates the complex function $S_+/(1+S_3)$ and, consequently, the function **S**, as well as the Jost functions along formulas (5.1.5)–(5.1.8). It should be stressed that the solutions to the Landau–Lifshitz equations are most conveniently represented as $S_+/(1+S_3)$, which was previously used in the Hirota method [8, 9, 21]. Here and further, $S_+ = S_1 + \mathrm{i}\,S_2$.

Soliton solutions of system (5.1.21) correspond to reflectorless potentials:

$$r^{(\varepsilon)}(\lambda) \equiv 0.$$

Let us look at the simplest case where $\varepsilon = -1$ and the coefficient $b(\lambda)$ has zeros (5.1.15) ($n = 1$, $m = 0$). Then from eqs. (5.1.21) we get

$$\begin{aligned}
l_1(x,y) &= \frac{\mathrm{i}}{2\rho} w_2(\lambda_1) r_1^{(-1)} l_3(x) \exp(-\mathrm{i}\, w_3(\lambda_1)(x+y)),\\
l_2(x,y) &= -\frac{1}{2\rho} w_1(\lambda_1) r_1^{(-1)} l_3(x) \exp(-\mathrm{i}\, w_3(\lambda_1)(x+y)),\\
l_5(x,y) &= l_4(x,y) = 0 \quad (\lambda_1 = u_1 + \mathrm{i}K'),
\end{aligned}$$

where, due to reductions (5.1.2), the parameter $r_1^{(-1)}$ is purely imaginary. The solution follows from eq. (5.1.8) and describes a domain boundary:

$$\frac{S_+}{1+S_3} = \frac{1}{2\rho} r_1^{(-1)} (w_2(\lambda_1) - w_2(\lambda_1)) \exp[-2\mathrm{i}\, w_3(\lambda_1)x + 4\mathrm{i}\, w_1(\lambda_1) w_2(\lambda_1) t].$$

We will analyze soliton solutions in more detail in the next section when discussing the “dressing” procedure.

We prove the existence and uniqueness of solutions to system (5.1.21) only for $\varepsilon = 1$ and without a soliton sector. Suppose the inhomogeneous part in eqs. (5.1.21) is equal to zero. Next, we multiply each equation sequentially by $\varepsilon\, l_{4,y}$, $\mathrm{i}\varepsilon l_5$, $-\mathrm{i}\varepsilon l_2$, and l_1, sum up the outcomes, and integrate them over y. In the long run, we arrive at the following equality:

$$\begin{aligned}
&\int_x^{+\infty} \mathrm{d}y\, [l_1^2(x,y) + l_2^2(x,y) + l_{4,y}^2(x,y) - l_5^2(x,y)]\\
&= \int_x^{+\infty} \mathrm{d}s \int_x^{+\infty} \mathrm{d}y\, [l_5(x,s)\, l_1(x,y)(F_2^{\varepsilon}(s+y) + \mathrm{i}\,\varepsilon\, \Pi_{1,y}^{\varepsilon}(s+y))\\
&\quad + l_5(x,s)\, l_2(x,y)(-\mathrm{i}\,\varepsilon\, F_1^{\varepsilon}(s+y) + \Pi_{2,y}^{\varepsilon}(s+y))\\
&\quad + l_2(x,y)\, l_{4,s}(x,s)(-\varepsilon\, \Pi_1^{\varepsilon}(s+y) - \mathrm{i}\, F_{2,y}^{\varepsilon}(s+y))\\
&\quad + l_1(x,s)\, l_{4,y}(x,y)(\varepsilon\, F_{1,y}^{\varepsilon}(s+y) - \mathrm{i}\,\Pi_2^{\varepsilon}(s+y))].
\end{aligned} \tag{5.1.22}$$

By Parseval’s identity for expansion into the Fourier integral, we have

$$\int_X^{+\infty} \mathrm{d}y\, l_a^2(x,y) = \frac{\rho}{2\pi}\int_0^{2K} \mathrm{d}\lambda\, w_1(\lambda)w_2(\lambda)|l_a(x,\lambda)|^2,$$

$$l_a(x,\lambda) = \int_X^{+\infty} \mathrm{d}y\, l_a(x,y)\exp(\mathrm{i}\, w_3 y), \quad a = 1,2,(4,y),5 \quad (0 < \lambda < 2K).$$

The real part of relation (5.1.22) can be written as

$$\begin{aligned}\int_0^{2K} \mathrm{d}\lambda \Big[& w_1w_2(|l_1(x,\lambda)|^2 + |l_2(x,\lambda)|^2 + |l_{4,y}(x,\lambda)|^2 + |l_5(x,\lambda)|^2) \\ & -((1-w_1^2)w_1^{-1}|l_5(x,\lambda)||l_1(x,\lambda)|\cos(\varphi_{l_5} + \varphi_{l_1} + \varphi_r - 3\pi/2) \\ & + |l_5(x,\lambda)||l_2(x,\lambda)|\cos(\varphi_{l_5} + \varphi_{l_2} + \varphi_r)(1-w_2^2)w_2^{-1} \\ & + |l_2(x,\lambda)||l_{4,y}(x,\lambda)|\cos(\varphi_{l_2} + \varphi_{4,y} + \varphi_r)(w_3^2 - w_1^2)w_3^{-1}w_1^{-1} \\ & + |l_1(x,\lambda)||l_{4,y}(x,\lambda)|\cos(\varphi_{l_1} + \varphi_{l_{4,y}} + \varphi_r + \pi/2) \\ & \times (w_3^2 - w_2^2)w_2^{-1}w_3^{-1})|r^{(1)}|^2 \Big] = 0,\end{aligned}$$

$\varphi_r = \arg r^{(1)}$, and $\varphi_{l_a} = \arg l_a$ $(a = 1,2,(4,y),5)$. Hence we conclude that when

$$|r^{(1)}(\lambda)| < \min\left\{\left|\frac{w_\alpha}{1-w_\alpha^2}\right|, \left|\frac{w_\beta\, w_3}{w_\beta^2 - w_3^2}\right|\right\}, \quad \alpha,\beta = 1,2, \tag{5.1.23}$$

the homogeneous system matched to system (5.1.21) has no nonzero solutions in $L_2(x,\infty)$. As Fredholm’s alternative theorem claims [25], a solution to the linear system (5.1.21) exists and is unique under condition (5.1.23).

Equations (5.1.7), (5.1.8), and (5.1.21) are the key ones in the inverse scattering method and allow us to trace how the initial distribution of magnetization evolves.

5.2 The method of “dressing” on a torus. Soliton solutions with asymptotics S → (0, 0, 1) as *x* → ±∞

To complicate is easy. To simplify is difficult.

Bruno Munari

This section covers a method for integrating the Landau–Lifshitz equations

$$\partial_t \mathbf{S} = [\mathbf{S} \times \partial_x^2 \mathbf{S}] + [\mathbf{S} \times J\mathbf{S}], \quad \mathbf{S}^2 = 1;$$

$$J = \operatorname{diag}(J_1, J_2, J_3), \quad J_3 > J_2 > J_1, \tag{5.2.1}$$

based on the Riemann problem for doubly periodic functions of a spectral parameter.

As already noted, the "dressing" procedure for a biaxial ferromagnet was proposed in [11, 26, 27, 29, 30, 6]. However, these works construct no solution for an isolated domain wall. Moreover, they do not analyze soliton solutions that describe a breather and the scattering of two opposite-chirality domain walls. Direct integration of eqs. (5.2.1) has given many exact solutions [22]. However, only the simplest cases tie these solutions to the inverse scattering method [6]. Later in this chapter, we modify the technique of [26, 27, 29, 30] and discuss the above issues. For completeness, some formulas in this section repeat the formulas in the previous one.

For definiteness, we assume that the distribution of magnetization as $|x| \to \infty$ tends rather quickly to equilibrium values:

$$\mathbf{S} \to (0,0,1) \quad \text{as} \quad x \to +\infty, \qquad \mathbf{S} \to (0,0,\varepsilon) \quad \text{as} \quad x \to -\infty, \tag{5.2.2}$$

where $\varepsilon = \pm 1$.

The ferromagnet equations (5.2.1) are equivalent to the compatibility condition (3.1.15) of the system of linear differential equations for the matrix function φ:

$$\partial_x \varphi = U\varphi, \quad \partial_t \varphi = V\varphi. \tag{5.2.3}$$

According to formulas (3.1.15) and (3.1.21), the matrices $U(\lambda, x, t)$ and $V(\lambda, x, t)$ are expressed in terms of $\mathbf{S}(x,t)$-vector components with coefficients $w_\alpha(\lambda)$ $(\alpha = 1, 2, 3)$ to be parameterized by Jacobi elliptic functions with modulus

$$k = \sqrt{\frac{J_2 - J_1}{J_3 - J_1}}.$$

In this context, the matrices $\varphi(\lambda)$, $U(\lambda)$, and $V(\lambda)$ turn out to be doubly periodic functions with periods $4K(k)$ and $4\mathrm{i}K'(k)$, where K are K' are the first-kind complete elliptic integrals.

The fundamental solutions $\varphi_\pm(\lambda)$ to the auxiliary linear system (5.2.3) are given by the asymptotic conditions

$$\varphi_+ \to \chi_2^{(0)} \equiv \exp(-\mathrm{i}\,\sigma_3[w_3(\lambda)\,x - 2w_1(\lambda)w_2(\lambda)\,t]), \quad x \to +\infty;$$
$$\varphi_- \to \chi_1^{(0)} \equiv \exp(-\mathrm{i}\,\varepsilon\sigma_3[w_3(\lambda)\,x - 2w_1(\lambda)w_2(\lambda)\,t]), \quad x \to -\infty.$$

Here the spectral parameter λ belongs to the contour

$$\Gamma = \{\lambda \mid |\operatorname{Re}\lambda| \le 2K, \ \operatorname{Im}\lambda = 0, 2K'\}.$$

Due to the double periodicity of the functions $\varphi_\pm(\lambda)$, it is sufficient to analyze their analyticity properties within the fundamental rectangle as in Figure 5.2.

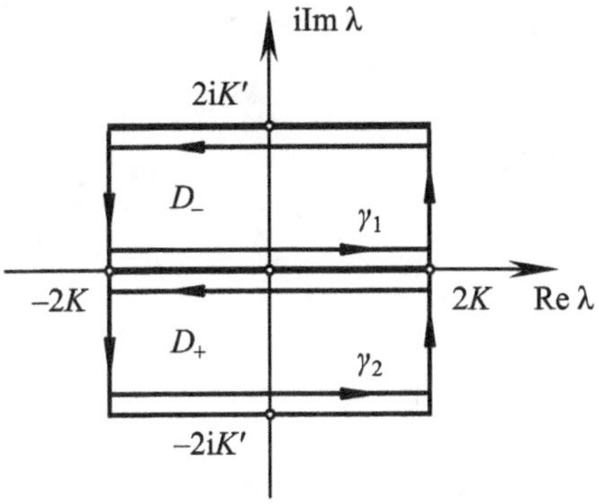

Figure 5.2: A fundamental rectangle in the λ-plane. Bold lines designate the contour Γ. The rectangle sides $|\mathrm{Re}\,\lambda| \le 2K$ and $\mathrm{Im}\,\lambda = \pm 2K'$ are treated as equivalent.

On the contour Γ, the Jost solutions $\varphi_{\pm}(\lambda)$ exhibit oscillatory behavior as $|x| \to \infty$, so the manifold Γ corresponds to a continuous spectrum of the scattering problem.

For $\lambda \in \Gamma$, the basic solutions are related to each other:

$$\varphi_{-}(\lambda) = \varphi_{+}(\lambda)\,T(\lambda). \tag{5.2.4}$$

The transition matrix $T(\lambda)$ depends only on the spectral parameter λ.

The Jost solutions satisfy the constraints

$$\begin{aligned} \varphi_{\pm}(\lambda + 2K) &= \sigma_3\,\varphi_{\pm}(\lambda)\,\sigma_3, \\ \varphi_{\pm}^{*}[(\lambda - 2\mathrm{i}K')^{*}] &= \sigma_3\,\varphi_{\pm}(\lambda)\,\sigma_3, \\ \varphi_{\pm}^{*}(\lambda^{*}) &= \sigma_2\,\varphi_{\pm}(\lambda)\,\sigma_2. \end{aligned} \tag{5.2.5}$$

The idea of proving relations (5.2.5) should be explained using the example of the first of them. The functions $\varphi_{\pm}(\lambda + 2K)$ and $\sigma_3\varphi_{\pm}(\lambda)\sigma_3$ meet the same differential equations because

$$U(\lambda + 2K) = \sigma_3\,U(\lambda)\,\sigma_3, \quad V(\lambda + 2K) = \sigma_3\,V(\lambda)\,\sigma_3.$$

In addition, their asymptotics are identical because

$$w_3(\lambda + 2K) = w_3(\lambda), \quad w_1(\lambda + 2K)\,w_2(\lambda + 2K) = w_1(\lambda)\,w_2(\lambda).$$

The uniqueness of solutions to linear differential equations under identical asymptotic conditions binds the equality

$$\varphi_{\pm}(\lambda + 2K) = \sigma_3\,\varphi_{\pm}(\lambda)\,\sigma_3$$

for all x and t.

Reductions (5.2.5) make it possible to specify the shape of the matrix $T(\lambda)$:

$$T(\lambda) = \begin{pmatrix} a(\lambda) & -b^*(\lambda^*) \\ b(\lambda) & a^*(\lambda^*) \end{pmatrix},$$

where

$$\begin{aligned} a(\lambda + 2K) &= a(\lambda), \quad & b(\lambda + 2K) &= -b(\lambda), \\ a^*[(\lambda - 2\mathrm{i}K')^*] &= a(\lambda), \quad & b^*[(\lambda - 2\mathrm{i}K')^*] &= -b(\lambda). \end{aligned} \tag{5.2.6}$$

Since $\operatorname{Sp} U(\lambda) = 0$, $\det\varphi_\pm(\lambda)$ does not depend on x. Hence we obtain $\det\varphi_\pm(\lambda) = \det\chi_{2,1}^{(0)} = 1$ and, as a consequence,

$$\det T = a(\lambda)a^*(\lambda^*) + b(\lambda)b^*(\lambda^*) = 1. \tag{5.2.7}$$

Further analysis depends on the sign of the parameter ε.

Consider first the case $\varepsilon = 1$ when the magnetization distribution has the same asymptotic behavior at infinity:

$$\mathbf{S} \to (0,0,1) \quad \text{as} \quad |x| \to \infty.$$

Invoking standard arguments, it is easy to show that the columns $\varphi_+^{(1)}(\lambda)$ and $\varphi_-^{(2)}(\lambda)$ of Jost solutions continue analytically from the contour Γ to the domain

$$D_- = \{\lambda \mid \operatorname{Im} w_3 < 0\} \equiv \{\lambda \mid |\operatorname{Re}\lambda| \le 2K,\ 0 < \operatorname{Im}\lambda < 2K'\},$$

whereas the columns $\varphi_-^{(1)}(\lambda)$ and $\varphi_+^{(2)}(\lambda)$ continue to the domain

$$D_+ = \{\lambda \mid \operatorname{Im} w_3 > 0\} \equiv \{\lambda \mid |\operatorname{Re}\lambda| \le 2K,\ -2K' < \operatorname{Im}\lambda < 0\}.$$

It is useful to introduce new fundamental solutions $F_+(\lambda)$ and $F_-(\lambda)$ to system (5.2.3) so that they are analytic functions in the domains D_+ and D_-, respectively:

$$F_+ = (\varphi_-^{(1)}, \varphi_+^{(2)}), \quad F_- = (\varphi_+^{(1)}, \varphi_-^{(2)}).$$

On the contour Γ the basic solutions F_+, F_-, φ_+, and φ_- are interconnected by the same relations (3.6.34) as for an "easy-axis" anisotropy ferromagnet. Their consequences, in particular, are representations of the coefficients $a(\lambda)$ and $a^*(\lambda^*)$:

$$a(\lambda) = \det F_+(\lambda), \quad a^*(\lambda^*) = \det F_-(\lambda).$$

From this we can infer that the coefficient $a(\lambda)$ ($a^*(\lambda^*)$) continues analytically from the contour Γ to the domain D_+ (D_-).

In the domains of their analyticity, the functions $a(\lambda)$ and $a^*(\lambda^*)$ can have a finite number of zeros that correspond to a discrete spectrum of the scattering problem. Suppose, for definiteness, that all zeros of the function $a(\lambda)$ are simple. Then due to reductions (5.2.6), they must have the form (mod $(4K, 4iK')$):

$$\begin{aligned} &\lambda = u_j - iK', \ u_j - iK' + 2K, \quad j = 1, 2, \ldots, n; \\ &\lambda = u_s - iK' \pm i\theta_s, \ u_s - iK' \pm i\theta_s + 2K, \quad s = 1, 2, \ldots, m, \end{aligned} \tag{5.2.8}$$

where $0 < u_j < 2K$ and $0 < \theta_s < K'$.

Further, we will see that each pair of zeros of the form (5.2.8) describes domain boundary. The domain wall connects different equilibrium values of the magnetization. In our problem the magnetization distributions are identical as $|x| \to \infty$. Therefore the domain walls are either absent or appear in pairs with opposite chiralities so that the uniform distribution of magnetization at infinity should not change: $n = 2r$, $r = 1, 2, \ldots$.

Next, it is convenient to extract the "seeding" solution $\chi_2^{(0)}(\lambda)$ from the functions $F_\mp(\lambda)$ and enter new matrix functions $\Psi_\mp(\lambda)$:

$$F_-(\lambda) = \Psi_-(\lambda)\chi_2^{(0)}(\lambda), \quad \frac{F_+(\lambda)}{a(\lambda)} = \Psi_+(\lambda)\chi_2^{(0)}(\lambda).$$

Equalities (3.6.34) yield the connection condition for the functions $\Psi_\pm(\lambda, x, t)$ on the contour Γ:

$$\Psi_-(\lambda, x, t) = \Psi_+(\lambda, x, t)G(\lambda, x, t), \tag{5.2.9}$$

where

$$\begin{aligned} G(\lambda, x, t) &= \chi_2^{(0)}(\lambda, x, t)G_0(\lambda)(\chi_2^{(0)}(\lambda, x, t))^{-1} \\ &= \begin{pmatrix} 1 & -b^*(\lambda^*, x, t) \\ -b(\lambda, x, t) & 1 \end{pmatrix}, \\ b(\lambda, x, t) &= b(\lambda)\exp[2i(w_3(\lambda)x - 2w_1(\lambda)w_2(\lambda)t)]. \end{aligned}$$

We have arrived at the classical Riemann problem (5.2.9) on a Riemann surface as a torus, since all functions in formula (5.2.9) are doubly periodic ones in the spectral parameter λ.

Constraints (5.2.5) can be easily transferred to new matrix functions:

$$\begin{aligned} \Psi_\pm(\lambda + 2K) &= \sigma_3\Psi_\pm(\lambda)\sigma_3, \\ \Psi_\pm^*[(\lambda - 2iK')^*] &= \sigma_3\Psi_\pm(\lambda)\sigma_3, \\ \Psi_-^*(\lambda^*) &= \sigma_2\Psi_+(\lambda)\sigma_2\, a(\lambda). \end{aligned} \tag{5.2.10}$$

The first two relations in (5.2.10) can be continued from the contour Γ to the analyticity domains for the functions $\Psi_{\pm}(\lambda)$. The last equality is valid only on the contour Γ, where it can be rewritten in a form more convenient for further analysis:

$$\Psi_{+}^{-1}(\lambda) = \Psi_{-}^{\dagger}(\lambda^{*}). \tag{5.2.11}$$

Then relation (5.2.9) can be represented in terms of only one function Ψ_{-}:

$$\Psi_{-}^{\dagger}(\lambda^{*})\Psi_{-}(\lambda) = G(\lambda), \quad \lambda \in \Gamma. \tag{5.2.12}$$

The left-hand side of formula (5.2.12) is a function infinitely differentiable in a neighborhood of the point $\lambda = 0$. For $x, t \neq 0$, the right-hand side of the equality contains the off-diagonal elements of the matrix $G(\lambda)$ that are not differentiable at the point $\lambda = 0$ since the coefficients $w_i(\lambda)$ $(i = 1, 2, 3)$ have a singularity at this point. The contradiction can be avoided only under the constraint $b(\lambda = 0) = 0$ or, equivalently, under the condition that the matrix function $\Psi_{-}(\lambda)$ is unitary at the point $\lambda = 0$:

$$\Psi_{-}^{\dagger}(\lambda = 0)\,\Psi_{-}(\lambda = 0) = I. \tag{5.2.13}$$

Due to connection (5.2.11), the matrix function $\Psi_{+}(\lambda)$ must have the same property:

$$\Psi_{+}^{\dagger}(\lambda = 0)\,\Psi_{+}(\lambda = 0) = I. \tag{5.2.14}$$

The function $\Psi_{-}(\lambda, x, t)$ satisfies the differential equation

$$\partial_x \Psi_{-} = U\Psi_{-} - \mathrm{i}\, w_3 \Psi_{-}\sigma_3.$$

Hence we get the equality

$$U(\lambda) = \partial_x \Psi_{-}(\lambda)\Psi_{-}^{-1}(\lambda) - \mathrm{i}\, w_3(\lambda)\Psi_{-}(\lambda)\sigma_3\Psi_{-}^{-1}(\lambda).$$

The left- and right-hand sides of this equality have simple poles at the point $\lambda = 0$ (see formulas (3.1.15) and (3.1.21)). Equating the residues at these poles, we deduce a working formula for computing the magnetization in a ferromagnet with biaxial anisotropy:

$$S_k\sigma_k = \Psi_{-}(\lambda = 0)\,\sigma_3\Psi_{-}^{\dagger}(\lambda = 0). \tag{5.2.15}$$

Here we have taken relation (5.2.13) into account. Formula (5.2.15) becomes constructive if the Riemann problem on a torus (5.2.9) has a solution as $\Psi_{-}(\lambda, x, t)$.

A distinctive feature of the Riemann problem on a torus is that the symmetry properties (5.2.10) largely fix the structure of the matrix function $\Psi_{-}(\lambda, x, t)$.

5.2.1 Breathers of a biaxial ferromagnet

Let us focus on the model soliton solutions under the condition

$$b(\lambda) = b^*(\lambda) \equiv 0.$$

Then we have

$$\Psi_-^\dagger(\lambda^*)\Psi_-(\lambda) = \Psi_-(\lambda)\Psi_-^\dagger(\lambda^*) = I.$$

By the relation $\det \Psi_-(\lambda) = a^*(\lambda^*)$ the zeros of the function $\Psi_-(\lambda)$ coincide with the zeros of the coefficient $a^*(\lambda^*)$ and are located at the points

$$u_j + \mathrm{i}K', \; u_j + \mathrm{i}K' + 2K, \quad j = 1, 2, \dots, 2r; \tag{5.2.16}$$
$$u_s + \mathrm{i}K' \pm \mathrm{i}\theta_s, \; u_s + \mathrm{i}K' \pm \mathrm{i}\theta_s + 2K, \quad s = 1, 2, \dots, m, \tag{5.2.17}$$

where $0 < u_j < 2K$ and $0 < \theta_s < K'$. The matrices of $\Psi_-(\lambda)$ and $\Psi_-^\dagger(\lambda^*)$ being mutually inverse, the function $\Psi_-^\dagger(\lambda^*)$ must have poles at the points (5.2.16) and (5.2.17).

The next step is to construct the simplest meromorphic function $\Psi_-^\dagger(\lambda^*)$ with four poles (5.2.17). It can be expanded in terms of the Weierstrass zeta-functions as follows (see Appendix):

$$\Psi_-^\dagger(\lambda^*) = A_0 + A\zeta_1 + B\zeta_2 + C\zeta_3 + D\zeta_4, \tag{5.2.18}$$

where

$$\zeta_i = \zeta(\lambda - \mu_i), \quad \mu_1 = u_1 + \mathrm{i}K' + \mathrm{i}\theta, \quad \mu_2 = u_1 + \mathrm{i}K' - \mathrm{i}\theta,$$
$$\mu_3 = \mu_1 + 2K, \quad \mu_4 = \mu_2 + 2K.$$

Here the periods of the Weierstrass functions are $4K$ and $4\mathrm{i}K'$.

The operation $f(\lambda) \to f(\lambda + 2K)$ transforms the Weierstrass zeta-functions into each other with accuracy up to an additive term:

$$\zeta_1 \to \zeta_3 + 2\eta_1, \quad \zeta_2 \to \zeta_4 + 2\eta_1, \quad \zeta_3 \to \zeta_1, \quad \zeta_4 \to \zeta_2. \tag{5.2.19}$$

After the transformation

$$f(\lambda) \to f^*[(\lambda - 2\mathrm{i}K')^*],$$

they pass into each other:

$$\zeta_1 \to \zeta_2, \quad \zeta_2 \to \zeta_1, \quad \zeta_3 \to \zeta_4, \quad \zeta_4 \to \zeta_3. \tag{5.2.20}$$

Based on formulas (5.2.19) and (5.2.20) and the first two symmetry conditions in (5.2.10), we express the matrices B, B, C, and D through A:

$$B = \sigma_3 A^* \sigma_3, \quad C = \sigma_3 A \sigma_3, \quad D = A^*.$$

For this, we need the matrix A_0 to have the following structure:

$$A_0 = \begin{pmatrix} r_1 & -\eta_1 A_{12} + \eta_1 A_{12}^* \\ -\eta_1 A_{21} + \eta_1 A_{21}^* & r_2 \end{pmatrix},$$

where r_1 and r_2 are real parameters.

The matrix function (5.2.18) have periods $4K$ and $4iK'$ if the sum of the residues of the function $\Psi_-^\dagger(\lambda^*)$ at its poles μ_i is equal to zero (see Appendix):

$$A + \sigma_3 A^* \sigma_3 + \sigma_3 A \sigma_3 + A^* = 0. \tag{5.2.21}$$

Condition (5.2.21) is fulfilled if the diagonal elements of the matrix A are imaginary numbers:

$$A_{11} = -A_{11}^*, \quad A_{22} = -A_{22}^*. \tag{5.2.22}$$

Further calculations confirm the correctness of eqs. (5.2.22).

The reductions have specified the algebraic structure of the matrix function (5.2.18). Suppose that $A = \tilde{A}R$, where $R = \text{diag}(r_1, r_2)$ with $r_1, r_2 \neq 0$. Then we get an ultimately convenient outcome to proceed:

$$\Psi_-^\dagger(\lambda^*) = \tilde{\Psi}^\dagger(\lambda^*)R,$$

where

$$\tilde{\Psi}^\dagger = I + \begin{pmatrix} \tilde{A}_{11}\varphi(\lambda) & \tilde{A}_{12}\alpha(\lambda) + \tilde{A}_{12}^*\beta(\lambda) \\ \tilde{A}_{21}\alpha(\lambda) + \tilde{A}_{21}^*\beta(\lambda) & \tilde{A}_{22}\varphi(\lambda) \end{pmatrix},$$

$$\varphi(\lambda) = \zeta_1 - \zeta_2 + \zeta_3 - \zeta_4, \quad \alpha(\lambda) = \zeta_1 - \zeta_3 - \eta_1, \quad \beta(\lambda) = \zeta_4 - \zeta_2 + \eta_1.$$

The no-poles conditions in the product of the matrix functions,

$$\Psi_-(\lambda)\Psi_+^\dagger(\lambda^*) \equiv R\tilde{\Psi}(\lambda)\tilde{\Psi}^\dagger(\lambda^*)R = I, \tag{5.2.23}$$

are equivalent to one independent matrix equation

$$\tilde{A}^\dagger \tilde{\Psi}^\dagger(\lambda^*)|_{\lambda=\mu_1^*} = 0 \tag{5.2.24}$$

to be solved nontrivially, conditional upon only the degenerate matrix $\tilde{A}$:

$$\tilde{A}_{\alpha\beta} = m_\alpha X_\beta. \tag{5.2.25}$$

Knowing that $\beta(\mu_1^*) = 0$, from eqs. (5.2.24) and (5.2.25) we find

$$X_1 = -\frac{m_1^*}{|m_2|^2\alpha(\mu_1^*) + |m_1|^2\varphi(\mu_1^*)}, \quad X_2 = -\frac{m_2^*}{|m_1|^2\alpha(\mu_1^*) + |m_2|^2\varphi(\mu_1^*)}.$$

The quantities $\alpha(\mu_1^*)$ and $\varphi(\mu_1^*)$ turn out to be purely imaginary numbers, so condition (5.2.22) holds.

The matrix $R^2 = \operatorname{diag}(r_1^2, r_2^2)$ can be expressed through relation (5.2.23):

$$R^{-2} = \tilde{\Psi}(\lambda)\tilde{\Psi}^\dagger(\lambda). \tag{5.2.26}$$

By construction the right-hand side of eq. (5.2.26) is a positive-definite λ-independent diagonal matrix. At $\lambda = 0$, eq. (5.2.26) defines the matrix R with accuracy up to a factor $N = \operatorname{diag}(\varepsilon_1, \varepsilon_2)$, where $\varepsilon_1^2 = \varepsilon_2^2 = 1$. The ambiguity noted is not essential for constructing solutions of the ferromagnet model.

Using the identities

$$\alpha(\lambda) = \zeta_1 - \zeta_3 - \eta_1 = \frac{1}{2}\frac{\operatorname{dn} v}{\operatorname{sn} v \operatorname{cn} v}\bigg|_{v=(\lambda-\mu_1)/2} = \frac{1+\operatorname{dn} v}{2\operatorname{sn} v}\bigg|_{v=\lambda-\mu_1},$$

$$\beta(\lambda) = \zeta_4 - \zeta_2 + \eta_1 = -\frac{k^2}{4\alpha^*(\lambda^*)}, \quad \varphi(\lambda) - \varphi(\mu_1^*) = -\frac{\alpha(\mu_1^*)\alpha(\lambda)}{\alpha^*(\lambda^*)},$$

we represent the matrix $\Psi_-^\dagger(\lambda)N$ in terms of the Jacobi functions of the modulo $k = \sqrt{(J_2 - J_3)/(J_3 - J_1)}$:

$$\Psi_-^\dagger(\lambda^*)N = r\begin{pmatrix} \mathrm{i}\,\alpha(\mu_1^*)[|m_2|^2 + |m_1|^2\frac{\alpha(\lambda)}{\alpha^*(\lambda^*)}] & -\mathrm{i}[m_1 m_2^*\alpha(\lambda) + \frac{k^2 m_2 m_1^*}{4\alpha^*(\lambda^*)}] \\ -\mathrm{i}[m_1^* m_2\alpha(\lambda) + \frac{k^2 m_2^* m_1}{4\alpha^*(\lambda^*)}] & \mathrm{i}\,\alpha(\mu_1^*)[|m_1|^2 + |m_2|^2\frac{\alpha(\lambda)}{\alpha^*(\lambda^*)}] \end{pmatrix}, \tag{5.2.27}$$

where

$$r^{-2} = -\alpha^2(\mu_1^*)(|m_1|^4 + |m_2|^4) + \delta|m_1 m_2|^2 + \frac{k^2}{4}[m_1^2(m_2^*)^2 + m_2^2(m_1^*)^2].$$

When calculating the matrix $R^{-2} = r^{-2}I$, a combination of elliptic functions appears:

$$\delta = \alpha(\lambda)\alpha^*(\lambda^*) - \frac{\alpha^2(\mu_1^*)[\alpha^2(\lambda) + (\alpha^*(\lambda^*))^2] - k^4/16}{\alpha(\lambda)\alpha^*(\lambda^*)}.$$

By expanding in terms of zeta functions (see Appendix) it is easy to verify that this expression is λ-independent. Assuming, for definiteness, that $\lambda = 0$, we obtain the following representation for the parameter δ:

$$\delta = |\alpha(0)|^2 - \frac{\alpha^2(\mu_1^*)[\alpha^2(0) + (\alpha^*(0))^2] - k^4/16}{|\alpha(0)|^2}.$$

Assuming that $\lambda = \mu^*$, we obtain the equivalent expression

$$\delta = -\frac{1 + \operatorname{dn}\vartheta}{2\operatorname{sn}^2\vartheta}\operatorname{cn}\vartheta|_{\vartheta=\mu-\mu^*}.$$

Note that

$$\begin{aligned} a(\lambda) &= \det \Psi_-^\dagger(\lambda^*) = \varepsilon_1\varepsilon_2 \frac{\alpha(\lambda)}{\alpha^*(\lambda^*)} \\ &\equiv -\varepsilon_1\varepsilon_2 \frac{\operatorname{cn}(\lambda - \nu) - \mathrm{i}\operatorname{sn}(\lambda - \nu)}{\operatorname{cn}(\lambda - \nu^*) + \mathrm{i}\operatorname{sn}(\lambda - \nu^*)}\bigg|_{\nu=u_1+\mathrm{i}\theta_1}. \end{aligned} \tag{5.2.28}$$

The algebraic structure of the coefficient $a(\lambda)$ is more illustrative if we expand the function $a(\lambda)$ in terms of Weierstrass' sigma-functions with periods $4K$ and $4\mathrm{i}K'$:

$$a(\lambda) \sim \frac{\sigma(\lambda - \mu_2^*)\sigma(\lambda - \mu_4^*)}{\sigma(\lambda - \mu_1)\sigma(\lambda - \mu_3)}\exp(-2\eta_3\lambda).$$

As before, the evolution of the vectors $\mathbf{m} = (m_1, m_2)$ obey the equations that ensure no extra poles in the expressions:

$$\begin{aligned} U &= \Psi_-(\lambda)[-\partial_x \Psi_-^\dagger(\lambda^*) - \mathrm{i}\, w_3(\lambda)\sigma_3\Psi_-^\dagger(\lambda^*)], \\ V &= \Psi_-(\lambda)[-\partial_t \Psi_-^\dagger(\lambda^*) + 2\mathrm{i}\, w_1(\lambda)w_2(\lambda)\sigma_3\Psi_-^\dagger(\lambda^*)]. \end{aligned} \tag{5.2.29}$$

The solution to these equations is

$$\mathbf{m} = \chi_2^{(0)}(\mu_1, x, t)\mathbf{c}, \tag{5.2.30}$$

where $\mathbf{c} = (c_1, c_2)$ is an arbitrary constant complex vector.

Employing formulas (5.2.15), (5.2.27), and (5.2.30), we find the breather solution to the equations for a biaxial ferromagnet:

$$\begin{aligned} S_3 &= \frac{A^2(\cosh^2 y - \sin^2\varepsilon) - B_-^2 - k^2\cos^2\phi}{A^2(\cosh^2 y - \sin^2\varepsilon) + B_-^2 + k^2\cos^2\phi}, \\ S_1 + \mathrm{i}S_2 &= \pm\frac{2\mathrm{i}A\cosh(y + \mathrm{i}\varepsilon)[B_+\cos\phi + \mathrm{i}B_-\sin\phi]}{A^2(\cosh^2 y - \sin^2\varepsilon) + B_-^2 + k^2\cos^2\phi}, \end{aligned} \tag{5.2.31}$$

where

$$\begin{aligned} &A = 2\mathrm{i}\,\alpha(\mu_1^*), \quad B_\pm = |\alpha(0)| \pm \frac{k^2}{4|\alpha(0)|}, \quad \varepsilon = \frac{1}{2\mathrm{i}}\ln\frac{\alpha^*(0)}{\alpha(0)}, \\ &y = \ln\left|\frac{m_2}{m_1}\right| = \ln\left|\frac{c_2}{c_1}\right| + \mathrm{i}(w_3 - w_3^*)x - 2\mathrm{i}(w_1 w_2 - w_1^* w_2^*)t, \\ &\phi = \frac{1}{\mathrm{i}}\ln\frac{m_1^* m_2}{|m_1 m_2|} = \arg\frac{c_2}{c_1} + (w_3 + w_3^*)x - 2(w_1 w_2 + w_1^* w_2^*)t. \end{aligned}$$

Here and below, we omit the argument μ_1 for the functions $w_a(\mu_1)$ $(a = 1, 2, 3)$. We have the following relations:

$$2a^2(\mu_1^*) + \delta - \frac{k^2}{2} = \frac{(k')^2}{\mathrm{cn}^2[(\mu_1 - \mu_1^*)/2,\, k]} = -A^2 \sin^2 \varepsilon + B_-^2.$$

In the limit as $k \to 0$, the model of a biaxial ferromagnet turns into the model of an "easy-axis" anisotropy ferromagnet. Namely, in the limit as $k \to 0$, $\mu_1 = \text{const}$, solution (5.2.31) becomes solution (3.4.20) (3.4.22) to describe a precessing soliton in an "easy-axis" ferromagnet. Recall that the internal structure of a soliton in an "easy-axis" ferromagnet depends significantly on the sign of the precession frequency. The sign points to the direction for the vector **S** to precess around the anisotropy axis. For a biaxial ferromagnet, solution (5.2.31) describes not only the precession of magnetization in a soliton but also its pulsations. To avoid confusion, let us determine the positive frequency of oscillations of the magnetization in the breather (5.2.31):

$$\omega = 2|w_1 w_2 + w_1^* w_2^*|.$$

In the general case, solution (5.2.31) handles a spatially localized excitation that moves as a whole with speed

$$V = \frac{2(w_1 w_2 - w_1^* w_2^*)}{w_3 - w_3^*}. \tag{5.2.32}$$

The breather magnetization changes abruptly in regions with a typical size of order κ^{-1}:

$$\kappa = |w_3 - w_3^*|.$$

At small values of κ, we can observe a low-amplitude soliton. As in an easy-axis ferromagnet, a low-amplitude soliton in a biaxial ferromagnet can be represented as a linear spin wave localized in space. However, in an easy-axis ferromagnet, the latter is a precession wave with $S_3 \approx \text{const}$, whereas in a biaxial magnet, not only the spin wave phase $\varphi = \arctan(S_2/S_1)$ changes but also the magnitude of S_3. The order parameter projection of S_3 is additionally periodically modulated with a characteristic distance of order p^{-1} (see Figure 5.3):

$$p = |w_3 + w_3^*|.$$

The periodic relief moves along the crest of the soliton with speed

$$V_{\mathrm{ph}} = \frac{\omega}{p}.$$

In Figure 5.3 the distance between the envelopes is proportional to κ.

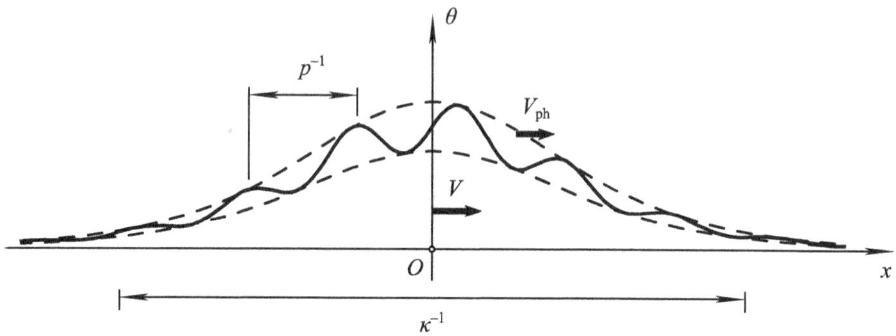

Figure 5.3: Distribution of magnetization in a low-amplitude soliton, $S_3 = \cos\theta$.

Generally, in the reference frame connected to the soliton, the projection of S_3 oscillates as a whole with amplitude $2\,a\,[b\,(1+b)]^{-1}$, where

$$a = \frac{2A^2}{k^2}(\cosh^2 y - \sin^2\varepsilon), \quad b = \frac{1}{k^2}[A^2(\cosh^2 y - \sin^2\varepsilon) + B_-^2].$$

Moreover, with increasing κ, the soliton envelope gradients become large, and the periodic relief is almost indiscernible. It is worth emphasizing that the moving relief speed V_{ph} is equal to the soliton speed V if $u_1 = K/2$ or $u_1 = 3K/2$.

To understand the structure of solution (5.2.31), we should dwell on immobile magnetic solitons [22]. The soliton speed (5.2.32) vanishes at $u_1 = K$ or at $u_1 = 0$. Let us examine the first case.

When

$$u_1 = K \quad \text{and} \quad K' < \tilde{\theta} \equiv \theta_1 + K' < 2K',$$

the distribution of magnetization has the form

$$\begin{aligned} S_3 &= \frac{A^2\cosh^2 y - B^2 - k^2\cos^2\phi}{A^2\cosh^2 y + B^2 + k^2\cos^2\phi}, \\ S_1 + \mathrm{i}S_2 &= \frac{2\,\mathrm{i}\,\varepsilon A\,\cosh y\,[\mathrm{i}k'\mathrm{cn}(\tilde{\theta},k')\sin\phi + \mathrm{dn}(\tilde{\theta},k')\cos\phi]}{A^2\cosh^2 y + B^2 + k^2\cos^2\phi}. \end{aligned} \tag{5.2.33}$$

Here

$$A = \frac{\mathrm{dn}(\tilde{\theta},k')\,\mathrm{cn}(\tilde{\theta},k')}{\mathrm{sn}(\tilde{\theta},k')}, \quad B = k'\mathrm{cn}(\tilde{\theta},k'),$$
$$y = y_0 + 2pk'\mathrm{sn}(\tilde{\theta},k')x, \quad \phi = \phi_0 - 4p^2k'\mathrm{cn}(\tilde{\theta},k')\,\mathrm{dn}(\tilde{\theta},k')t,$$

y_0 and ϕ_0 are real parameters, and $\varepsilon^2 = 1$.

Far away from the boundary values of the parameter $\tilde{\theta} = K', 2K'$, the distribution of S_3 oscillates with large amplitude (Figure 5.4*a*). The vector **S** performs a precession around the Ox_3-axis with frequency

$$\omega = -4\rho^2 k' \operatorname{cn}(\tilde{\theta}, k') \operatorname{dn}(\tilde{\theta}, k').$$

It varies within the range of $0 \le \omega \le 4\rho^2 k'$. When

$$\omega \to 4\rho^2 k' = \sqrt{(J_3 - J_1)(J_3 - J_2)} \quad \text{as} \quad \tilde{\theta} \to 2K',$$

the parameter $\kappa = 2\rho k' \operatorname{sn}(\tilde{\theta}, k')$ tends to zero, and solution (5.2.33) converts into the expression for linear spin waves with wavenumber $p = 0$. At small κ, the projection of S_3 has a usual soliton shape, but its value at the center of the soliton (at $y = 0$) changes periodically with time (Figure 5.4*b*).

Soliton (5.2.33) behaves rather interestingly at low frequencies ($\tilde{\theta} \cong K'$). Here the ferromagnet clearly manifests its biaxiality. Although, as in a uniaxial magnet at a low precession frequency, breather (5.2.33) can be treated as a bound state of two domain boundaries; now the boundaries oscillate relative to each other (Figure 5.4*c*), the soliton seems to breathe. Suppose the characteristic breather size is the distance between the points of vanishing the projection of S_3. These points are the roots of the equation

$$\cosh^2 y = \frac{1}{A^2}(B^2 + k^2 \cos^2 \phi) \ge 1. \tag{5.2.34}$$

It is useful to express the characteristic size of the breather in terms of the problem's physical parameters κ and ω [22]:

$$\Delta = \frac{2}{\kappa} \operatorname{arcosh}\left[\frac{\kappa^2}{\omega}\left(\sqrt{1 + \left(\frac{J_2 - J_1}{2\omega}\right)^2} + \frac{J_2 - J_1}{2\omega}\cos 2\phi\right)\right]^{\frac{1}{2}}. \tag{5.2.35}$$

In terms of the same parameters, we express the angle θ that determines the projection of $S_3 = \cos\theta$:

$$\operatorname{tg}^2\frac{\theta}{2} = \frac{\kappa^2}{\omega \cosh^2 y}\left(\sqrt{1 + \left(\frac{J_2 - J_1}{2\omega}\right)^2} + \frac{J_2 - J_1}{2\omega}\cos 2\phi\right). \tag{5.2.36}$$

Formulas (5.2.35) and (5.2.36) stand for a significant dependency of both the magnetization distribution and the pattern of domain boundary oscillations in the breather on the ratio between the parameters κ, $J_2 - J_1$, and ω. The angle θ at the center of the soliton (at $y = 0$) does not reach the limiting value $\theta = \pi$. The greater the divergence of domain boundaries during oscillations, the greater the change in the projection of S_3 at the center of the breather.

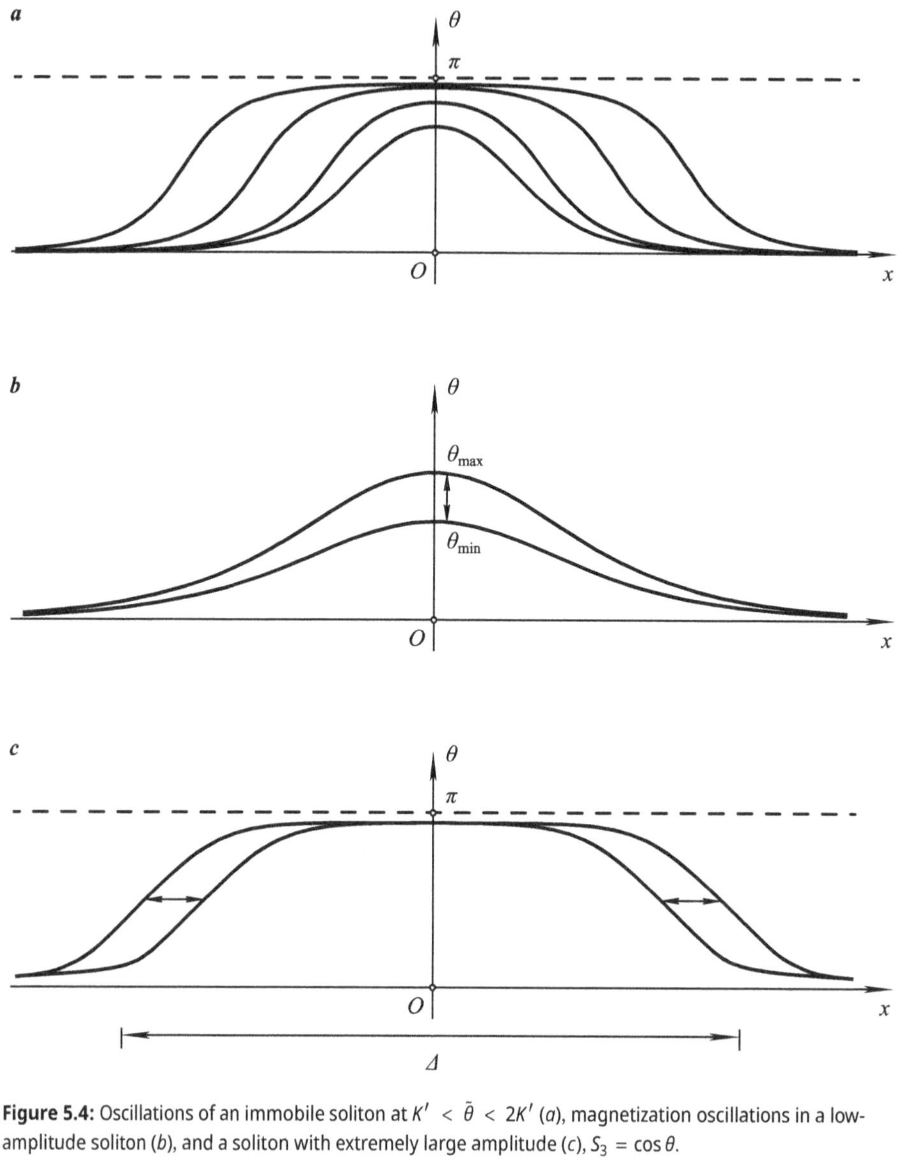

Figure 5.4: Oscillations of an immobile soliton at $K' < \tilde{\theta} < 2K'$ (*a*), magnetization oscillations in a low-amplitude soliton (*b*), and a soliton with extremely large amplitude (*c*), $S_3 = \cos\theta$.

For $J_2 - J_1 \ll \omega \ll \kappa^2$, the amplitude of the breather pulsations is small compared to the width ($\sim \kappa^{-1}$) of the domain boundaries enclosing the breather:

$$\Delta_{\max} - \Delta_{\min} \cong \frac{J_2 - J_1}{\kappa\omega} \ll \frac{1}{\kappa}.$$

In this case the average size of the remagnetized region is

$$\frac{\Delta_{\max} + \Delta_{\min}}{2} \cong \frac{1}{\kappa} \ln \frac{4\kappa^2}{\omega} \gg \frac{1}{\kappa} \approx \frac{1}{\sqrt{J_3 - J_2}}.$$

It is much larger than the width of the domain boundaries and coincides with the size of a soliton in an easy-axis ferromagnet.

For $(J_2 - J_1) \gg 2\omega$, the amplitude of relative oscillations of the domain boundaries grows. The minimum soliton size is frequency-independent:

$$\Delta_{\min} \approx \frac{2}{\kappa}\operatorname{arcosh}\frac{\kappa}{\sqrt{J_2 - J_1}}.$$

However, its maximum size increases as the frequency goes down:

$$\Delta_{\max} \cong \frac{2}{\kappa}\operatorname{arcosh}\left[\frac{\kappa^2}{J_2 - J_1} + \frac{\kappa^2(J_2 - J_1)}{\omega^2}\right]^{\frac{1}{2}} \approx \frac{1}{\kappa}\ln\frac{4\kappa^2(J_2 - J_1)}{\omega^2}.$$

With such parameters, breather (5.2.33) differs significantly from a precessional soliton in an "easy-axis" ferromagnet: the amplitude of its pulsations becomes huge.

When the average soliton size is large, the magnetization distribution (5.2.33) near the points $y_{1,2}$ ($y_1 > 0, y_2 = -y_1$) that are the roots of eq. (5.2.34) has the shape typical of isolated domain boundaries:

$$S_3 \cong \pm\tanh\xi, \quad S_1 + \mathrm{i}S_2 = \mathrm{i}\,\varepsilon\,\frac{\mathrm{i}k'\operatorname{cn}(\tilde\theta, k')\sin\phi + \operatorname{dn}(\tilde\theta, k')\cos\phi}{\cosh\xi\sqrt{B^2 + k^2\cos^2\phi}}, \tag{5.2.37}$$

where $\xi = y - y_{1,2}$, $\varepsilon^2 = 1$. Formulas (5.2.37) control a periodical transition of the domain boundaries enclosing the soliton from Bloch to Neel ones and vice versa. This occurs during the process of oscillations relative to the common "center of gravity". The walls become purely Bloch ones at a maximum distance ($\phi = \pi n$, n is an integer) and purely Néel ones at a maximum approach ($\phi = \pi/2 + \pi n$).

In the limit $\phi_0 = -\pi/2$, $\tilde\theta \to K'$, solution (5.2.33) takes the form

$$S_3 = \frac{k^2\cosh^2 y - k'^2 - \tilde\phi^2}{k^2\cosh^2 y + k'^2 + \tilde\phi^2}, \quad S_1 + \mathrm{i}S_2 = \frac{2\mathrm{i}\,\varepsilon\,\cosh y(\mathrm{i}k' + \tilde\phi)}{k^2\cosh^2 y + k'^2 + \tilde\phi^2}, \tag{5.2.38}$$

where $y = 2\rho k'x + y_0$ and $\tilde\phi = 4\rho^2 k^2 k' t$. Expression (5.2.38) describes the decay of the breather into two domain boundaries (see Figure 5.5). As $t \to +\infty$, the domain boundaries enclosing the breather mutually move away along the logarithmic law:

$$\Delta \sim \frac{2}{\kappa}\ln t.$$

Their relative velocity $2V = 2\ln t/(\kappa t)$ tends to zero, and excitation (5.2.38) degenerates into two immobile Bloch boundaries. We can say that solution (5.2.38) governs the scattering of two Bloch domain boundaries until their velocity vanishes at infinity. With such a process, the direction of magnetization in the domain boundaries changes by π.

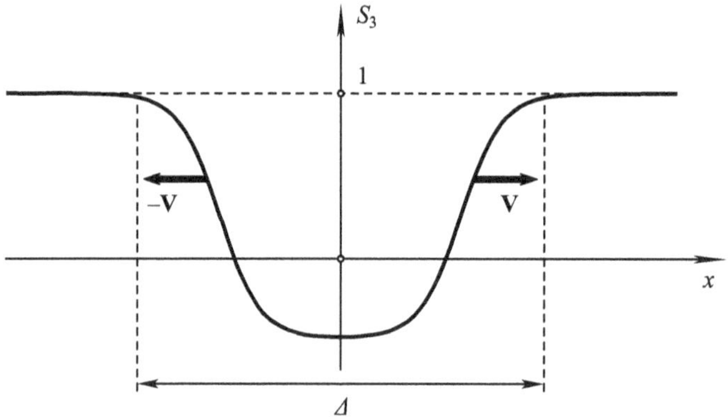

Figure 5.5: Decay of a breather into two domain boundaries traveling away from each other ($t \gg 1$).

Earlier, by directly integrating the Landau–Lifshitz equations, another explicit solution was found [22], also corresponding to a fixed breather:

$$\begin{aligned} S_3 &= \frac{A^2 \sinh^2 y - B^2 - k^2 \mathrm{sn}^2(\tilde{\theta}, k') \cos^2 \phi}{A^2 \sinh^2 y + B^2 + k^2 \mathrm{sn}^2(\tilde{\theta}, k') \cos^2 \phi}, \\ S_1 + \mathrm{i} S_2 &= \pm \frac{2A \sinh y \, [\mathrm{i} B \sin \phi + \mathrm{dn}(\tilde{\theta}, k') \cos \phi]}{A^2 \sinh^2 y + B^2 + k^2 \mathrm{sn}^2(\tilde{\theta}, k') \cos^2 \phi}, \end{aligned} \tag{5.2.39}$$

where

$$\begin{gathered} y = y_0 + \frac{2\rho x}{\mathrm{sn}(\tilde{\theta}, k')}, \quad \phi = \phi_0 + \frac{4\rho A t}{\mathrm{sn}^2(\tilde{\theta}, k')}, \\ A = \mathrm{dn}(\tilde{\theta}, k') \, \mathrm{cn}(\tilde{\theta}, k'), \quad B = \mathrm{cn}(\tilde{\theta}, k'), \end{gathered}$$

and y_0 and ϕ_0 are real parameters. This solution can be derived from the general formula (5.2.31) for the parameter values

$$u_1 = 0, \quad K' < \tilde{\theta} \equiv K' + \theta_1 < 2K'.$$

Solution (5.2.39) explains how the vector **S** precesses around the Ox_3-axis with frequency

$$\omega = \frac{4\rho^2 \mathrm{dn}(\tilde{\theta}, k') |\mathrm{cn}(\tilde{\theta}, k')|}{\mathrm{sn}^2(\tilde{\theta}, k')} > 4\rho^2 = J_3 - J_1.$$

Unlike the previous case, the precession is clockwise (recall that $\mathrm{cn}(\tilde{\theta}, k') < 0$).

Solution (5.2.39) is a generalization of solution (3.4.33) for an "easy-axis" anisotropy ferromagnet. Breather (5.2.39) has approximately the same shape as the soliton outlined in Figure 5.4*c*. However, for $y = 0$, the projection of S_3 at the center of breather (5.2.39) reaches the value $S_3 = -1$ ($\theta = \pi$). As in a uniaxial ferromagnet, the precession phases in breather (5.2.39) differ by π when $y > 0$ and $y < 0$. The nature of the relative oscillations of tied boundaries in breather (5.2.39) differs from that in breather (5.2.33). At the maximum distance, the domain boundaries are purely Nèel, and at the maximum approach, they are Bloch.

Within the limit $\phi_0 = -\pi/2$, $\omega \to 0$ ($\tilde{\theta} \to K'$), solution (5.2.48) takes the form

$$S_3 = \frac{k^2 \sinh^2 y - 1 - \tilde{\phi}^2}{k^2 \sinh^2 y + 1 + \tilde{\phi}^2}, \quad S_1 + \mathrm{i}S_2 = \mp \frac{2k \sinh y\,(\mathrm{i} + \tilde{\phi})}{k^2 \sinh^2 y + 1 + \tilde{\phi}^2}, \tag{5.2.40}$$

where $y = y_0 + 2px$ and $\tilde{\phi} = 4p^2k^2t$. Expression (5.2.40) describes the decay of the breather into two purely Neel domain boundaries until their velocity zeros at infinity.

5.2.2 Interaction of two domain walls

Let us examine a soliton solution describing the collision of two domain walls. For this purpose, we should construct a meromorphic doubly periodic matrix function $\Psi_-^\dagger(\lambda^*)$ with poles of the type (5.2.16) at four points of the fundamental rectangle:

$$\mu_1 = u_1 + \mathrm{i}K', \quad \mu_2 = \mu_1 + 2K, \quad \nu_1 = u_2 + \mathrm{i}K',$$
$$\nu_2 = \nu_1 + 2K \quad (\mathrm{mod}(4K, 4\mathrm{i}K')).$$

We expand the function $\Psi_-^\dagger(\lambda^*)$ in terms of the Weierstrass zeta-functions:

$$\Psi_-^\dagger(\lambda^*) = A_0 + A\zeta_1 + A_1\zeta_2 + B\zeta_3 + B_1\zeta_4, \tag{5.2.41}$$

where

$$\zeta_1 = \zeta(\lambda - \mu_1), \quad \zeta_2 = \zeta(\lambda - \mu_2), \quad \zeta_3 = \zeta(\lambda - \nu_1), \quad \zeta_4 = \zeta(\lambda - \nu_2).$$

The periods of the zeta-functions are $4K$ and $4\mathrm{i}K'$.

Reductions (5.2.10) specify the shape of the matrices in representation (5.2.41):

$$A_1 = \sigma_3 A \sigma_3, \quad A^* = \sigma_3 A \sigma_3, \quad B_1 = \sigma_3 B \sigma_3, \quad B^* = \sigma_3 B \sigma_3,$$
$$A_0 = \begin{pmatrix} r_1 & -\eta_1(A_{12} + B_{12}) \\ -\eta_1(A_{21} + B_{21}) & r_2 \end{pmatrix}, \tag{5.2.42}$$

where r_1 are r_2 are real parameters.

The double-periodicity condition for function (5.2.41) is equivalent to the equality

$$\mathrm{diag}(A + B) = 0. \tag{5.2.43}$$

As before, we carry out further calculations under assumption (5.2.43). The final result will confirm its correctness.

Using formulas (5.2.42) and (5.2.43), we write the matrix function (5.2.41) as

$$\Psi_-^\dagger(\lambda^*) = \tilde{\Psi}^\dagger(\lambda^*)R, \quad R = \mathrm{diag}(r_1, r_2) \quad (r_1, r_2 \neq 0),$$
$$\tilde{\Psi}^\dagger(\lambda^*) = \left[I + \begin{pmatrix} \tilde{A}_{11}\chi(\lambda) & \tilde{A}_{12}\alpha_1(\lambda) + \tilde{B}_{21}\alpha_2(\lambda) \\ \tilde{A}_{21}\alpha_1(\lambda) + \tilde{B}_{21}\alpha_2(\lambda) & \tilde{A}_{22}\chi(\lambda) \end{pmatrix}\right] R,$$

where

$$\chi(\lambda) = \zeta_1 + \zeta_2 - \zeta_3 - \zeta_4, \quad \alpha_1(\lambda) = \zeta_1 - \zeta_2 - \eta_1, \quad \alpha_2(\lambda) = \zeta_3 - \zeta_4 - \eta_1.$$

The problem at hand has a nontrivial solution only when the matrices $\tilde{A}$ and $\tilde{B}$ are degenerate. Suppose that

$$\tilde{A}_{\alpha\beta} = m_\alpha X_\beta, \quad \tilde{B}_{\alpha\beta} = n_\alpha Y_\beta.$$

Next, we will see that requirements (5.2.42) and (5.2.43) are fulfilled if

$$\begin{gathered} \mathbf{m} = (\tilde{m}_1, \mathrm{i}\,\tilde{m}_2), \quad \mathbf{n} = (\tilde{n}_1, \mathrm{i}\,\tilde{n}_2), \\ \tilde{m}_1 X_1 = -\tilde{n}_1 Y_1, \quad \tilde{m}_2 X_2 = -\tilde{n}_2 Y_2, \end{gathered} \tag{5.2.44}$$

where $\tilde{m}_i$ and $\tilde{n}_i$ $(i = 1, 2)$ are real parameters.

The no-poles condition in the product of the matrix functions (5.2.23) enables finding the vectors **X** and **Y** by reduction to two independent matrix equations

$$\tilde{A}^\dagger \tilde{\Psi}^\dagger(\lambda^*)|_{\lambda=\mu_1^*} = 0, \quad \tilde{B}^\dagger \tilde{\Psi}^\dagger(\lambda^*)|_{\lambda=\nu_1^*} = 0.$$

Their solutions are

$$\begin{gathered} X_1 = -\tilde{n}_1 \gamma^{-1}, \quad Y_1 = \tilde{m}_1 \gamma^{-1}, \\ X_2 = \mathrm{i}\,\tilde{n}_2 \delta^{-1}, \quad Y_2 = -\mathrm{i}\,\tilde{m}_2 \delta^{-1}, \\ \gamma = \tilde{n}_1 \tilde{m}_1 \chi(\nu_1^*) + \tilde{n}_2 \tilde{m}_2\, \alpha_1(\nu_1^*), \\ \delta = \tilde{n}_1 \tilde{m}_1 \alpha_1(\nu_1^*) + \tilde{n}_2 \tilde{m}_2 \chi(\nu_1^*). \end{gathered} \tag{5.2.45}$$

When composing formula (5.2.45), we took into account that

$$\chi(\nu_1^*) = \chi(\mu_1^*), \quad \alpha_1(\nu_1^*) = -\alpha_2(\mu_1^*).$$

The quantities $\chi(\nu_1^*)$ and $\alpha_1(\nu_1^*)$ being real, condition (5.2.42) is satisfied. Constraints (5.2.43) also turn out to be satisfied.

The no-extra-poles requirement in relations (5.2.29) specifies the x- and t-dependence of the vectors **m** and **n**:

$$\mathbf{m} = \chi_2^{(0)}(\mu_1, x, t)\mathbf{c}, \quad \mathbf{n} = \chi_2^{(0)}(\nu_1, x, t)\mathbf{d},$$

where **c** and **d** are constant complex vectors. Since $\chi_2^{(0)}(\mu_1)$ and $\chi_2^{(0)}(\nu_1)$ are real diagonal matrices, eqs. (5.2.44) hold under choosing

$$\mathbf{c} = (c_1, \mathrm{i}\, c_2), \quad \mathbf{d} = (d_1, \mathrm{i}\, d_2),$$

where c_i and d_i $(i = 1, 2)$ are real numbers. Then we have

$$\begin{aligned} \tilde{m}_1 &= \exp[-\varphi(u_1)]\, c_1, \quad \tilde{m}_2 = \exp\varphi(u_1)\, c_2, \\ \tilde{n}_1 &= \exp[-\varphi(u_2)]\, d_1, \quad \tilde{n}_2 = \exp\varphi(u_2)\, c_2, \\ \varphi(u) &= \rho\, x\, \mathrm{dn}\, u - 2(\rho k)^2 t\, \mathrm{sn}\, u\, \mathrm{cn}\, u. \end{aligned} \tag{5.2.46}$$

The elements of the diagonal matrix R (with accuracy up to sign) are determined by eq. (5.2.26): its right-hand side includes a λ-independent combination of elliptic functions. Suppose that $\lambda = 0$ in the right-hand side of eq. (5.2.26). Then for further analysis, it is convenient to parameterize the matrix R.

We have the identities

$$\begin{aligned} \chi(\lambda) - \chi(\nu_1^*) &= -\frac{a_1(\lambda)a_2(\lambda)}{a_1(\nu_1)}, \\ a_1(\lambda) &= f(\lambda - \mu_1), \quad a_2(\lambda) = f(\lambda - \nu_1), \\ f(\lambda - \mu) &= \frac{\mathrm{dn}\, v}{2\, \mathrm{sn}\, v\, \mathrm{cn}\, v}\Big|_{v=(\lambda-\mu)/2} = \frac{1 + \mathrm{dn}\, v}{2\, \mathrm{sn}\, v}\Big|_{v=\lambda-\mu}. \end{aligned}$$

In this problem, the parameter $\mu_1(\nu_1)$ is specific: $\mu_1 = u_1 + \mathrm{i}K'$. Therefore the functions $a_{1,2}$ have additional transformation properties:

$$f(\lambda - \mu_1) = \frac{k}{2}[\mathrm{sn}(\lambda - u_1) + \mathrm{i}\, \mathrm{cn}(\lambda - u_1)], \quad a_{1,2}^*(\lambda^*)a_{1,2}(\lambda) = \frac{k^2}{4}.$$

Using these identities, we represent the matrix function

$$\Psi_-^\dagger(\lambda^*)N, \quad N = \mathrm{diag}(\varepsilon_1, \varepsilon_2) \quad (\varepsilon_1^2 = \varepsilon_2^2 = 1),$$

in terms of the Jacobi functions of modulo $k = \sqrt{(J_2 - J_1)/(J_3 - J_1)}$:

$$\Psi_1^\dagger(\lambda^*)N = r\begin{pmatrix} a_1(\lambda)a_2(\lambda)\tilde{n}_1\tilde{m}_1 + k^2\tilde{n}_2\tilde{m}_2/4 & \mathrm{i}a_1(\nu_1)[\tilde{m}_1\tilde{n}_2 a_1(\lambda) - \tilde{n}_1\tilde{m}_2 a_2(\lambda)] \\ \mathrm{i}a_1(\nu_1)[-\tilde{m}_2\tilde{n}_1 a_1(\lambda) + \tilde{n}_2\tilde{m}_1 a_2(\lambda)] & a_1(\lambda)a_2(\lambda)\tilde{n}_2\tilde{m}_2 + k^2\tilde{n}_1\tilde{m}_1/4 \end{pmatrix}, \tag{5.2.47}$$

where

$$r^{-2} = \frac{k^2}{4}\left[\frac{k^2}{4}(\tilde{n}_1^2\tilde{m}_1^2 + \tilde{n}_2^2\tilde{m}_2^2) + a_1^2(\nu_1)(\tilde{m}_2^2\tilde{n}_1^2 + \tilde{n}_2^2\tilde{m}_1^2) + \frac{\delta}{2}\,\tilde{n}_1\tilde{n}_2\tilde{m}_1\tilde{m}_2\right].$$

When calculating r, a λ-independent combination of elliptic functions appears:

$$\frac{\delta}{2} = \alpha_1(\lambda)\alpha_2(\lambda) + \alpha_1^*(\lambda^*)\alpha_2^*(\lambda^*) - \alpha_1^2(v_1)\left(\frac{\alpha_1(\lambda)}{\alpha_2(\lambda)} + \frac{\alpha_2(\lambda)}{\alpha_1(\lambda)}\right).$$

Hence, putting $\lambda = 0$, we find

$$\delta = k^2(\operatorname{sn} u_1 \operatorname{sn} u_2 - \operatorname{cn} u_1 \operatorname{cn} u_2) - 4\alpha_1^2(v_1)(\operatorname{sn} u_1 \operatorname{sn} u_2 + \operatorname{cn} u_1 \operatorname{cn} u_2),$$

and we obtain the following relations:

$$\begin{aligned} a(\lambda) &= \det \Psi_-^\dagger(\lambda^*) = 4\,\varepsilon_1\varepsilon_2 k^{-2}\alpha_1(\lambda)\alpha_2(\lambda) \\ &\equiv -\varepsilon_1\varepsilon_2[\operatorname{cn}(\lambda - u_1) - \mathrm{i}\operatorname{sn}(\lambda - u_1)][\operatorname{cn}(\lambda - u_2) - \mathrm{i}\operatorname{sn}(\lambda - u_2)]. \end{aligned} \tag{5.2.48}$$

The soliton solution constructed using formulas (5.2.15), (5.2.46), and (5.2.47) has the form

$$\begin{aligned} S_3 &= D^{-1}[k^2\cosh(y_1 + y_2) - 4\alpha_1^2(v_1)\cosh(y_2 - y_1) + \Delta], \\ S_1 - \mathrm{i}S_2 &= 4\,\mathrm{i}\,\varepsilon D^{-1}\alpha_1(v_1)k\,[\operatorname{sn} u_2 \sinh y_1 - \operatorname{sn} u_1 \sinh y_2 \\ &\quad + \mathrm{i}(\operatorname{cn} u_1 \cosh y_2 - \operatorname{cn} u_2 \cosh y_1)]. \end{aligned} \tag{5.2.49}$$

Here $\varepsilon = \pm 1$,

$$\begin{aligned} D &= k^2\cosh(y_1 + y_2) + 4\alpha_1^2(v_1)\cosh(y_2 - y_1) + \delta, \\ \Delta &= \delta + 4\alpha_1^2(v_1)(\operatorname{sn} u_1 \operatorname{sn} u_2 + \operatorname{cn} u_1 \operatorname{cn} u_2), \\ y_{1,2} &= 2\rho\, x \operatorname{dn} u_{1,2} - (2\rho k)^2 t \operatorname{sn} u_{1,2} \operatorname{cn} u_{1,2} + y_{1,2}^{(0)}, \\ y_1^{(0)} &= \ln\frac{c_2}{c_1}, \quad y_2^{(0)} = \ln\frac{d_2}{d_1}. \end{aligned}$$

Expression (5.2.49) describes the scattering of two domain boundaries (Figure 5.6). The width of the sth domain boundary is of order κ_s^{-1}, where $\kappa_s = 2\rho \operatorname{dn} u_s$ $(s = 1, 2)$. Its velocity v_s can be calculated by the relation

$$v_s = 2\rho k^2 \frac{\operatorname{sn} u_s \operatorname{cn} u_s}{\operatorname{dn} u_s} \quad (0 < u_s < 2K).$$

To clarify the nature of the interaction of domain walls, it is convenient to “sit down” on one of them and observe how the second one changes. Mathematically, this conforms with the transition to a reference frame moving at the first domain boundary speed v_1. Introducing the new variable $\xi = x - v_1 t$, we get

$$y_1 = \kappa_1\xi + y_1^{(0)}, \quad y_2 = \kappa_2[\xi - (v_2 - v_1)t] + y_2^{(0)}.$$

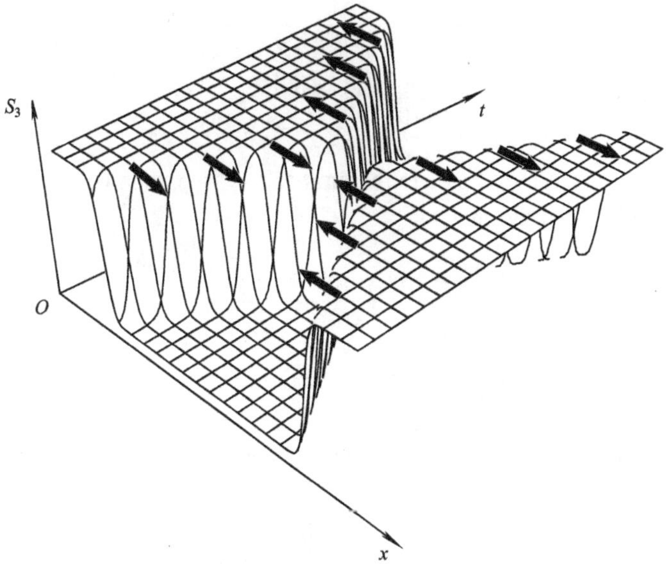

Figure 5.6: Scattering of two domain boundaries (5.2.49). The arrows indicate the motion directions of the boundaries.

Let $0 < v_1 < v_2$ for definiteness. Then in the limit as $\xi = \text{const}$, $t \to -\infty$, solution (5.2.49) turns into an expression describing the first domain boundary:

$$S_3 \cong \tanh \tilde{y}_1, \quad S_1 - \mathrm{i}S_2 \approx -\frac{\varepsilon(\operatorname{cn} u_1 + \mathrm{i} \operatorname{sn} u_1)}{\cosh \tilde{y}_1},$$
$$\tilde{y}_1 = \kappa_1\left(x - v_1 t + \frac{1}{\kappa_1} \ln \frac{k}{2a_1(v_1)}\right) + y_1^{(0)}.$$

On the other hand, in the limit as $t \to -\infty$, for $\xi = (v_2 - v_1)t$, we have

$$y_2 = \text{const}, \quad y_1 \to -\infty,$$

from which we find

$$S_3 \cong -\tanh \tilde{y}_2, \quad S_1 - \mathrm{i}S_2 = \frac{\varepsilon(\operatorname{cn} u_2 - \mathrm{i} \operatorname{sn} u_2)}{\cosh \tilde{y}_2},$$
$$\tilde{y}_2 = \kappa_2\left(x - v_2 t - \frac{1}{\kappa_2} \ln \frac{k}{2a_1(v_1)}\right) + y_2^{(0)}.$$

Thus in the limit as $t \to -\infty$, solution (5.2.49) contains two well-separated domain boundaries. Since $v_2 > v_1$, the second domain boundary is to the left of the first one.

Consider now the limit as ξ is finite and $t \to +\infty$. At the same time, $y_1 = \text{const}$, $y_2 \to -\infty$, and

$$S_3 \cong -\tanh \bar{y}_1, \quad S_1 - \mathrm{i}S_2 = -\frac{\varepsilon(\operatorname{cn} u_1 - \mathrm{i}\operatorname{sn} u_1)}{\cosh \bar{y}_1},$$

$$\bar{y}_1 = \kappa_1\left(x - v_1 t - \frac{1}{\kappa_1}\ln\frac{k}{2a_1(v_1)}\right) + y_1^{(0)}.$$

If we assume that $\xi = (v_2 - v_1)t$ and pass to the limit as $t \to +\infty$, then $y_2 = \text{const}$, and $y_1 \to +\infty$. Consequently, we arrive at

$$S_3 \cong \tanh \bar{y}_2, \quad S_1 - \mathrm{i}S_2 \cong \frac{\varepsilon(\operatorname{cn} u_2 + \mathrm{i}\operatorname{sn} u_2)}{\cosh \bar{y}_2},$$

$$\bar{y}_2 = \kappa_2\left(x - v_2 t + \frac{1}{\kappa_2}\ln\frac{k}{2a_1(v_1)}\right) + y_2^{(0)}.$$

In other words, as $t \to +\infty$, the second domain boundary appears to the right of the first one.

We have convinced ourselves that solution (5.2.49) demonstrates structural stability under mutually interacting the domain walls. Tracing only the asymptotics of the solution and leaving aside the moment of their collision itself, we can believe that the domain walls pass through each other. In this case, each domain wall preserves its velocity and width. Only the signs of the azimuthal angles of the magnetization vector, and the phase of motion of each domain wall change. Such a process can be interpreted as an elastic interaction.

When $u_1 = s$ and $u_2 = 2K - s$, the velocities of the domain walls are equal in magnitude and oppositely directed. With a certain choice of the parameters $y_{1,2}^{(0)}$, in the limit as $s \to K$ and $s \to 0$, solution (5.2.49) reduces to expressions (5.2.38) and (5.2.40) to describe the scattering (or decay) of two weakly coupled domain walls.

There are two ways of constructing N-soliton solutions to the model of a biaxial anisotropy ferromagnet. The former is to generalize the above procedure and is based on the expansion of the matrix function of the Riemann problem in $4N$ linearly independent Weierstrass zeta-functions. This method will be discussed in detail in Chapter 7. The latter consists in recurrently constructing multisolitons; the “dressing” method builds an N-soliton solution through an existing $N - 1$ soliton solution using the functions of the Riemann problem with zeros, given in the present section.

5.3 Soliton solutions to the model of a biaxial ferromagnet with asymptotics $\mathbf{S} \to (0, 0, \pm 1)$ as $x \to \pm\infty$

When we get an issue right, we need to refine it from superfluous ideas and reduce it to its simplest elements.

R. Descartes

Let us look into the procedure of integrating the equations for a biaxial ferromagnet with asymptotic conditions (5.2.2), where $\varepsilon = -1$.

In this case the fundamental solutions $\Phi_{\pm}(\lambda)$ to system (5.2.3), which are analytic in the domains $D_{\pm}$, have a different form

$$\Phi_{-}(\lambda) = (\varphi_{+}^{(1)}(\lambda), \varphi_{-}^{(1)}(\lambda)), \quad \Phi_{+}(\lambda) = (\varphi_{-}^{(2)}(\lambda), \varphi_{+}^{(2)}(\lambda)).$$

In the domains of their analyticity, the matrix functions $\Phi_{\pm}(\lambda)$ satisfy the following constraints:

$$\Phi_{\mp}(\lambda + 2K) = \pm\sigma_3\Phi_{\mp}(\lambda), \quad \Phi_{\mp}^{*}[(\lambda - 2\mathrm{i}K')^{*}] = \pm\sigma_3\Phi_{\mp}(\lambda).$$

Simultaneously, on the contour Γ, we have

$$\Phi_{\mp}^{*}(\lambda^{*}) = \pm\sigma_2\Phi_{\pm}(\lambda)\sigma_1. \tag{5.3.1}$$

It is worth emphasizing that many algebraic relations of this problem, with accuracy up to designations, formally coincide with those for an “easy-axis” anisotropy ferromagnet. So, on the contour Γ, formulas (3.6.7) interconnect all fundamental solutions. Consequently, we can infer the following equalities:

$$b(\lambda) = \det\Phi_{-}(\lambda), \quad b^{*}(\lambda^{*}) = -\det\Phi_{+}(\lambda).$$

These indicate that the coefficients $b(\lambda)$ and $b^{*}(\lambda^{*})$ can be analytically continued from the contour Γ to the domains D_{-} and D_{+}, respectively.

In the domains of their analyticity, the functions $b(\lambda)$ and $b^{*}(\lambda^{*})$ can have a finite number of zeros. The zeros of the coefficient $b^{*}(\lambda^{*})$ are complex conjugate regarding the zeros of the function $b(\lambda)$. In view of reductions (5.2.6), the zeros of the function $b(\lambda)$ constitute two sets

$$u_j + \mathrm{i}K', \quad u_j + \mathrm{i}K' + 2K, \quad j = 1, 2, \ldots, n; \tag{5.3.2}$$

$$u_s + \mathrm{i}K' \pm \mathrm{i}\theta_s, \quad u_s + \mathrm{i}K' \pm \mathrm{i}\theta_s + 2K, \quad s = 1, 2, \ldots, m, \tag{5.3.3}$$

where $0 < u_j < 2K$ and $0 < \theta_s < K'$.

As noted earlier, each group of two zeros of type (5.3.2) defines one domain boundary. The problem under consideration always contains domain boundaries, and their total number must be odd: $n = 2r - 1$, $r = 1, 2, \ldots$. Otherwise, the asymptotic conditions $\mathbf{S} \to (0, 0, \pm 1)$ fail as $x \to \pm\infty$.

Each group of four zeros of type (5.3.3) is responsible for a single breather. As we have seen, the breathers are localized and do not affect the magnetization distribution at infinity. Therefore the present problem may exclude the zeros (5.3.3) parameterizing breathers. Below, dispersion relations will substantiate these qualitative considerations.

The connection between the Jost functions (5.2.4) is equivalent to the condition of conjugation of the analytic functions $\Phi_\pm(\lambda)$ on the contour Γ:

$$\Phi_-(\lambda) = -\frac{1}{b^*(\lambda^*)}\,\Phi_+(\lambda)\sigma_3\begin{pmatrix} 1 & a(\lambda) \\ a^*(\lambda^*) & 1 \end{pmatrix}. \tag{5.3.4}$$

Soliton solutions to the model are now determined by another condition

$$a(\lambda) = a^*(\lambda^*) \equiv 0.$$

For the soliton sector, using reduction (5.3.1), we rewrite eq. (5.3.4) in terms of the matrix function $\Phi_-(\lambda)$ only:

$$\Phi_-^\dagger(\lambda^*)\Phi_-(\lambda) = I.$$

Let us extract the "seeding" solution $\chi_2^{(0)}(\lambda)$ from this function (this operation does not contradict relations (3.6.7)):

$$\Phi_-(\lambda) = \Psi_-(\lambda)\chi_2^{(0)}(\lambda).$$

Here the function $\Psi_-(\lambda)$ is a doubly periodic function and meromorphic in the fundamental rectangle $D_- \cup D_+$. It has zeros at the points (5.3.2) and (5.3.3) and poles at complex conjugate points.

Next, we rewrite the Riemann problem with zeros in terms of the function $\Psi_-(\lambda)$:

$$\Psi_-^\dagger(\lambda^*)\Psi_-(\lambda) = \Psi_-(\lambda)\Psi_-^\dagger(\lambda^*) = I, \quad \lambda \in \Gamma; \tag{5.3.5}$$

$$\Psi_-(\lambda + 2K) = \sigma_3\Psi_-(\lambda), \quad \Psi_-^*[(\lambda - 2\mathrm{i}K')^*] = \sigma_3\Psi_-(\lambda). \tag{5.3.6}$$

A no-breather solution to the domain boundary problem (5.3.5) is of particular interest. The matrix function $\Psi_-^\dagger(\lambda^*)$ should be sought for such a solution as an expansion in terms of the Weierstrass zeta functions:

$$\Psi_-^\dagger(\lambda^*) = A_0 + A\zeta(\lambda - \mu_1) + B\zeta(\lambda - \mu_2),$$

where

$$\mu_1 = u_1 + \mathrm{i}K', \quad \mu_2 = \mu_1 + 2K$$

and $4K$ and $4\mathrm{i}K'$ are the periods of the Weierstrass functions.

As before, reductions (5.3.6) rather "rigidly" fix the algebraic structure of the matrix function $\Psi_-^\dagger(\lambda^*)$:

$$\Psi_-^\dagger(\lambda^*) = \begin{pmatrix} s_1 & \mathrm{i}\,c_1 a(\lambda) \\ s_2 & \mathrm{i}\,c_2 a(\lambda) \end{pmatrix},$$

where

$$\alpha(\lambda) = \zeta(\lambda - \mu_1) - \zeta(\lambda - \mu_2) - \eta_1 = \frac{k}{2}[\mathrm{sn}(\lambda - u_1) + \mathrm{i}\,\mathrm{cn}(\lambda - u_1)],$$

and $s_{1,2}$ and $c_{1,2}$ are real parameters.

Applying the identity

$$\alpha(\lambda)\alpha^*(\lambda^*) = \frac{k^2}{4},$$

we reduce condition (5.3.5) to the equations

$$s_1^2 + s_2^2 = 1, \quad c_1^2 + c_2^2 = 4/k^2, \quad s_1c_1 + s_2c_2 = 0,$$

which allow for the parameterization

$$s_1 = \cos\varphi, \quad s_2 = \sin\varphi, \quad c_1 = \frac{2}{k}\sin\psi, \quad c_2 = \frac{2}{k}\cos\psi, \quad \psi = -\varphi + \pi n,$$

where n is an integer.

Thus the desired solution to the Riemann problems (5.3.5) and (5.3.6) appears as

$$\Psi_-^\dagger(\lambda^*) = \begin{pmatrix} \cos\varphi & -2ik^{-1}\alpha(\lambda)\sin\varphi \\ \sin\varphi & 2ik^{-1}\alpha(\lambda)\cos\varphi \end{pmatrix}\begin{pmatrix} 1 & 0 \\ 0 & \varepsilon \end{pmatrix}, \tag{5.3.7}$$

where $\varepsilon = \pm 1$. Hence we can deduce the expression for the coefficient $b(\lambda)$:

$$\begin{aligned} b(\lambda) &= \det\Psi_-(\lambda) = [\det\Psi_-^\dagger(\lambda^*)]^{-1} = \frac{k\varepsilon}{2i\alpha(\lambda)} \\ &\equiv -\varepsilon[\mathrm{cn}(u - u_1) + \mathrm{i}\,\mathrm{sn}(u - u_1)]. \end{aligned}$$

The no-extra poles requirement for the "dressing" formulas (5.2.29) at the points $\lambda = \mu_{1,2}, \mu_{1,2}^*$ leads to the following differential equations for φ:

$$\partial_x\varphi = -\mathrm{i}\,w_3(\mu_1)\sin 2\varphi, \quad \partial_t\varphi = 2\mathrm{i}\,w_1(\mu_1)w_2(\mu_1)\sin 2\varphi.$$

They have the trivial solution

$$\tan\varphi = \exp(-y), \quad y = 2\rho\, x\,\mathrm{dn}\,u_1 - (2\rho k)^2 t\,\mathrm{sn}\,u_1\,\mathrm{cn}\,u_1 + y_0, \tag{5.3.8}$$

where y_0 is a real constant.

The soliton solution obtained using formulas (5.2.15), (5.3.7), and (5.3.8), as expected, describes the domain boundary:

$$S_3 = \tanh y, \quad S_1 - \mathrm{i}S_2 = \varepsilon\,\frac{\mathrm{cn}\,u_1 + \mathrm{i}\,\mathrm{sn}\,u_1}{\cosh y}. \tag{5.3.9}$$

The width of the domain boundary is of order κ^{-1}, where $\kappa = 2\rho \,\mathrm{dn}\, u_1$, and its velocity v is

$$v = 2\rho k^2 \frac{\mathrm{sn}\, u_1 \;\mathrm{cn}\, u_1}{\mathrm{dn}\, u_1} \qquad (0 < u_1 < 2K).$$

Indicatively, the problem illustrated above requires constructing soliton solutions to integrable models with different asymptotics at spatial infinity through the matrix functions of the Riemann problem that cannot always be stated in terms of projection matrices. For specifying the algebraic structure of these matrix functions, the decisive role belongs to their analytic properties and reductions. Each model reduction is unique.

5.4 The regular Riemann problem on a torus

Gentlemen, for Gaussian rigor we have today no time.

C. Jacobi

For a biaxial ferromagnet, weakly nonlinear magnetization waves with asymptotic behavior at infinity

$$\mathbf{S} \to (0,0,1) \quad \text{as} \quad x \to \pm\infty$$

can be theoretically described using the regular Riemann problem on a torus. Formulas (5.2.9) and (5.2.10), where the coefficients $a(\lambda)$ and $a^*(\lambda^*)$ have no zeros in their analyticity domains, are necessary for this problem to be solved.

Let us connect the solution of the regular Riemann problem with a system of singular integral equations. For this purpose, we invoke the Cauchy integral on a torus whose kernel has the shape [26, 27, 29]

$$M(\mu,\lambda) = \zeta(\mu-\lambda) - \zeta(\mu-a) + \zeta(\lambda-b) - \zeta(a-b),$$

where a and b are arbitrary points on the torus that do not lie on the contour Γ, the Weierstrass zeta functions have periods $4K$ and $4\mathrm{i}K'$. The differential form $M(\mu,\lambda)\,\mathrm{d}\mu$ has first-order poles at points $\mu = \lambda$ and $\mu = a$ with residues ± 1. The kernel $M(\mu,\lambda)$ represents a doubly periodic function of both variables μ and λ with periods $4K$ and $4\mathrm{i}K'$. This function has first-order poles at the points $\lambda = \mu$ and $\lambda = b$ with residues ∓ 1 and is equal to zero for any μ at the point $\lambda = a$.

The Cauchy-type integral

$$F(\lambda) = \frac{1}{2\pi \mathrm{i}} \oint\limits_C \varphi(\mu) M(\mu,\lambda)\mathrm{d}\mu,$$

where C is a closed contour on the torus, is a doubly periodic piecewise analytic function that vanishes at the point $\lambda = a$. The function $F(\lambda)$ has a first-order pole at the point $\lambda = b$ with residue

$$\frac{1}{2\pi i} \oint_C \varphi(\mu) d\mu. \tag{5.4.1}$$

If $\varphi(\mu)$ is the boundary value of a function that is analytical in a domain bounded by the contour C, then there is no pole at the point $\lambda = b$ because the integral (5.4.1) vanishes.

In our case, the function

$$\Phi(\lambda) = \begin{cases} \Psi_-(\lambda), & \lambda \in D_-, \\ \Psi_+(\lambda), & \lambda \in D_+, \end{cases}$$

is piecewise analytic on the torus. Let Φ_0 be the value of this function at the point a. We choose the contour $\gamma = \gamma_1 \cup \gamma_2$, where γ_1 and γ_2 are the contours bounding the domains D_- and D_+, respectively (Figure 5.2). We use positive traversal directions so that when traversing the contour $\gamma_1(\gamma_2)$, the domain $D_-(D_+)$ remains to the left of the contour. Then by the Cauchy theorem we get

$$\Phi(\lambda) = \Phi_0 + \frac{1}{2\pi i} \int_\gamma \Phi(\mu) M(\mu, \lambda) d\mu. \tag{5.4.2}$$

The integration performs a traversal of segments on the contour γ, on which $\operatorname{Re}\mu = \pm 2K$, in opposite directions, and the integrand takes the same values on them. Therefore formula (5.4.4) saves an integration only along the contour Γ. We express the integral (5.4.2) in terms of jumps of the function $\Phi(\lambda)$ on the contour Γ:

$$\Phi(\lambda) = \Phi_0 + \frac{1}{2\pi i} \int_\Gamma [\Psi_-(\mu) - \Psi_+(\mu)] M(\mu, \lambda) d\mu.$$

Using eq. (5.2.9), we modify the outcome:

$$\Phi(\lambda) = \Phi_0 + \frac{1}{2\pi i} \int_\Gamma \Psi_+(\mu)[G(\mu) - I] M(\mu, \lambda) d\mu. \tag{5.4.3}$$

Suppose that λ tends to a point $\nu \in \Gamma$ from the domain D_+. Then by the Sokhotski–Plemelj formulas relation (5.4.3) yields a singular integral equation for calculating $\Psi_+(\nu)$:

$$\frac{1}{2}\Psi_+(\nu)[G(\nu) + I] = \Phi_0 + \frac{1}{2\pi i} \mathrm{V.p.} \int_\Gamma \Psi_+(\mu)[G(\mu) - I] M(\mu, \lambda) d\mu. \tag{5.4.4}$$

The letters V. ;p. before the integral stand for calculating in the sense of the Cauchy principal value. Reductions (5.2.6) and (5.2.10) allow for reducing the integration over the contour Γ in eq. (5.4.4) to an integration over the segment $[0, 2K]$:

$$\frac{1}{2}\Psi_+(\nu)[G(\nu)+I] = \Phi_0 + \frac{1}{2\pi \mathrm{i}}\mathrm{V.p.}\int_0^{2K}[\rho(\mu)M(\mu,\nu) + \sigma_3\rho(\mu)\sigma_3 M(\mu-2K,\nu) - \sigma_3\rho^*(\mu)\sigma_3 M(\mu+2\mathrm{i}K',\nu) - \rho^*(\mu)M(\mu-2K+2\mathrm{i}K',\nu)]\,\mathrm{d}\mu, \tag{5.4.5}$$

where $\rho(\mu) = \Psi_+(\mu)[G(\mu) - I]$. It suffices to find the solution of eqs. (5.4.5) within the interval $0 < \nu < 2K$. By applying reductions (5.2.10) to the remaining segments of the contour Γ this solution is recovered. Based on the known values of the function $\Psi_+(\nu)$ on the contour Γ, the Cauchy formula (5.4.3) provides the values of the functions $\Psi_\pm(\nu)$ in the domains $D_\pm$.

Next, we demonstrate that this problem can avoid an arduous choice of the normalization point $\lambda = a$ and calculating the magnitude of $\Phi(a) = \Phi_0$. To this end, note that $\Phi(\lambda)$ is an analytic function in the domain $D_+ \cup D_-$ and is continuous there up to the boundary of γ. Therefore by the Cauchy theorem we have

$$\oint_\gamma \Phi(\mu)\mathrm{d}\mu = \int_0^{2K}[\rho(\mu) + \sigma_3\rho(\mu)\sigma_3 - \sigma_3\rho^*(\mu)\sigma_3 - \rho^*(\mu)]\,\mathrm{d}\mu = 0.$$

Here we have modified the contour integral, bearing in mind the periodicity properties of $\Phi(\lambda)$ and reductions (5.2.10), to express it through the jumps $\rho(\mu) = \Psi_+(\mu)[G(\mu) - I]$ of the function $\Phi(\lambda)$ on the contour Γ. This implies two simple identities

$$\int_0^{2K}[\rho_{ss}(\mu) - \rho_{ss}^*(\mu)]\,\mathrm{d}\mu = 0 \quad (s = 1, 2), \tag{5.4.6}$$

where

$$\rho_{11}(\mu) = -[\Psi_+(\mu,x,t)]_{12}b(\mu,x,t),$$
$$\rho_{22}(\mu) = -[\Psi_+(\mu,x,t)]_{21}b^*(\mu^*,x,t).$$

On the right-hand side of representation (5.4.3), we group up the terms and rewrite it using reductions (5.2.10) as

$$\Phi(\lambda) = F_0 + \frac{1}{2\pi\mathrm{i}}\int_0^{2K}\begin{pmatrix}\rho_{11}(\mu)\phi(\mu-\lambda) - \rho_{11}^*(\mu)\phi(\mu-\lambda+2\mathrm{i}K') & \rho_{12}(\mu)a(\mu-\lambda) + \rho_{12}^*(\mu)a(\mu-\lambda+2\mathrm{i}K') \\ \rho_{21}(\mu)a(\mu-\lambda) + \rho_{21}^*(\mu)a(\mu-\lambda+2\mathrm{i}K') & \rho_{22}(\mu)\phi(\mu-\lambda) - \rho_{22}^*(\mu)\phi(\mu-\lambda+2\mathrm{i}K')\end{pmatrix}\mathrm{d}\mu, \tag{5.4.7}$$

where the matrix F_0 does not depend on the spectral parameter λ, and

$$\rho_{21}(\mu) = -[\Psi_+(\mu,x,t)]_{22} b(\mu,x,t),$$
$$\rho_{12}(\mu) = -[\Psi_+(\mu,x,t)]_{11} b^*(\mu^*,x,t).$$

The functions

$$\alpha(\lambda) = \zeta(\lambda) - \zeta(\lambda - 2K) - \eta_1 = \frac{1 + \operatorname{dn}\lambda}{2\operatorname{sn}\lambda} \quad \text{and} \quad \phi(\lambda) = \zeta(\lambda) + \zeta(\lambda - 2K)$$

have the transformational properties

$$\begin{aligned} \alpha(\lambda \pm 2K) &= -\alpha(\lambda), \quad \alpha^*[(\lambda \pm 2\mathrm{i}K')^*] = \alpha(\lambda + 2\mathrm{i}K'), \\ \alpha^*[(\lambda \pm 2\mathrm{i}K')^* + 2\mathrm{i}K'] &= \alpha(\lambda); \\ \phi[(\lambda \pm 2K)] &= \phi(\lambda) \pm 2\eta_1, \\ \phi^*[(\lambda \pm 2\mathrm{i}K')^*] &= \phi(\lambda + 2\mathrm{i}K') - 2\eta_3 \pm 2\eta_3, \\ \phi^*[(\lambda \pm 2\mathrm{i}K')^* + 2\mathrm{i}K'] &= \phi(\lambda) - 2\eta_3 \pm 2\eta_3. \end{aligned} \tag{5.4.8}$$

Now we can infer that the first two reductions (5.2.10) and formulas (5.4.6) and (5.4.8) instigate the matrix F_0 to be real and diagonal in representation (5.4.7):

$$F_0 = \operatorname{diag}(f_1, f_2).$$

This simplifies the calculations.

It is convenient to introduce new matrix functions

$$\Psi_\pm(\mu,x,t) = F_0(x,t)\tilde{\Psi}_\pm(\mu,x,t).$$

Then from eq. (5.4.7) we deduce a closed equation for calculating $\tilde{\Psi}_\pm$:

$$\frac{1}{2}\tilde{\Psi}_+(\nu)[G(\nu) + I] = I + \frac{1}{2\pi\mathrm{i}} \mathrm{V.p.} \int_0^{2K} \begin{pmatrix} \tilde{\rho}_{11}(\mu)\phi(\mu-\nu) - \tilde{\rho}_{11}^*(\mu)\phi(\mu-\nu+2\mathrm{i}K') & \tilde{\rho}_{12}(\mu)\alpha(\mu-\nu) + \tilde{\rho}_{12}^*(\mu)\alpha(\mu-\nu+2\mathrm{i}K') \\ \tilde{\rho}_{21}(\mu)\alpha(\mu-\nu) + \tilde{\rho}_{21}^*(\mu)\alpha(\mu-\nu+2\mathrm{i}K') & \tilde{\rho}_{22}(\mu)\phi(\mu-\nu) - \tilde{\rho}_{22}^*(\mu)\phi(\mu-\nu+2\mathrm{i}K') \end{pmatrix} \mathrm{d}\mu,$$

where

$$\tilde{\rho}(\mu) = \tilde{\Psi}_+(\mu)[G(\mu) - I].$$

Using relation (5.2.14), we can express the quantity $F_0^2 = \operatorname{diag}(f_1^2, f_2^2)$ in terms of $\tilde{\Psi}_+$:

$$F_0^{-2} = \tilde{\Psi}_+(\lambda = 0)\tilde{\Psi}_+^\dagger(\lambda = 0). \tag{5.4.9}$$

Then from eq. (5.2.15) we restore the solution to the Landau–Lifshitz model:

$$S_k\sigma_k = F_0\tilde{\Psi}_+(\lambda = 0)\sigma_3\tilde{\Psi}_+^\dagger(\lambda = 0)F_0.$$

It describes spin waves with a predominance of dispersion effects over nonlinearity effects. Here we have taken into account the fact that according to formulas (5.2.11) and (5.2.13), $\Psi_-(\lambda = 0) = \Psi_+(\lambda = 0)$.

Equality (5.4.9) defines the matrix F_0 with accuracy up to a factor $N = \mathrm{diag}(\varepsilon_1, \varepsilon_2)$, where $\varepsilon_1^2 = \varepsilon_2^2 = 1$. However, this is not essential, because after the transformations

$$S_3 \to S_3, \quad S_1 \to -S_1, \quad S_2 \to -S_2,$$

the solutions to the Landau–Lifshitz model (3.1.14) turn into solutions with the same boundary conditions (5.2.2) at infinity.

It is important to note that any solution to the model for a biaxial ferromagnet with asymptotics (5.2.2) can be constructed by the method of recurrent "dressing" of elementary solutions resorting to the Riemann problems with zeros and the regular Riemann problem on a torus.

5.5 Integrals of motion and the spectrum of nonlinear excitations

Without the useful, there are no pleasant things.

German proverb

Recall that the model of a biaxial ferromagnet with the asymptotic conditions (5.2.2) for $\varepsilon = 1$ includes both the time-independent coefficient of the transition matrix

$$a(\lambda) = \det(\varphi_-^{(1)}(\lambda),\ \varphi_+^{(2)}(\lambda)) \tag{5.5.1}$$

and the time-independent coefficient

$$b(\lambda) = \det(\varphi_+^{(1)}(\lambda),\ \varphi_-^{(1)}(\lambda))$$

under conditions (5.2.2) for $\varepsilon = -1$. Hence it follows that $a(\lambda)$ and $b(\lambda)$ can be used as generating functions of integrals of motion.

Let us show that by expanding the functions $\ln a(\lambda)$ or $\ln b(\lambda)$ into a series in powers of λ we ascertain physically meaningful conservation laws. To do this, it is necessary to find expansions of the Jost solutions $\varphi_\pm(\lambda)$ in powers of λ.

We represent the Jost function $\varphi_-(\lambda)$ in the form

$$\varphi_-(\lambda) = (I + \Phi)\exp\left[-\mathrm{i}\,\sigma_3\,\varepsilon\,(w_3\,x - 2\,w_1 w_2\,t) + Z\right], \quad \varepsilon = \pm 1, \tag{5.5.2}$$

where Φ and Z are antidiagonal and diagonal matrix functions of the variables λ, x, and t, respectively, such that

$$\Phi, Z \to 0 \quad \text{as} \quad x \to -\infty.$$

After plugging representation (5.5.2) into the first equation of (5.2.3) and separating the diagonal part from the antidiagonal one, we get

$$\begin{aligned} &\partial_x \Phi - 2\mathrm{i} w_3 \Phi \sigma_3 S_3 - \mathrm{i}\Phi(w_1 S_1 \sigma_1 + w_2 S_2 \sigma_2)\Phi + \mathrm{i}(w_1 S_1 \sigma_1 + w_2 S_2 \sigma_2) = 0, \\ &\partial_x Z = -\mathrm{i} w_3 (S_3 - \varepsilon)\sigma_3 - \mathrm{i}(w_1 S_1 \sigma_1 + w_2 S_2 \sigma_2)\Phi. \end{aligned} \tag{5.5.3}$$

Next, we expand the functions $w_\alpha(\lambda)$ $(\alpha = 1, 2, 3)$ as power series in λ:

$$w_\alpha(\lambda) = \sum_{k=-1}^{\infty} w_\alpha^{(k)} \lambda^k.$$

We have the following relations: $w_\alpha^{(-1)} = \rho$, $w_\alpha^{(2k)} = 0$, $k = 0, 1, 2, \ldots$. The solution to eqs. (5.5.3) should be sought as power series in λ:

$$\Phi = \sum_{n=0}^{\infty} \lambda^n \begin{pmatrix} 0 & -\omega_n^*(x) \\ \omega_n(x) & 0 \end{pmatrix}, \quad Z = \sum_{n=0}^{\infty} \lambda^n \begin{pmatrix} \zeta_n(x) & 0 \\ 0 & \zeta_n^*(x) \end{pmatrix}.$$

Then we arrive at a chain of recurrence relations for calculating the functions $\omega_n(x)$ and $\zeta_n(x)$:

$$\begin{aligned} &\partial_x \omega_n - 2\mathrm{i} \sum_{m=0}^{n+1} w_3^{(n-m)} S_3\, \omega_m \\ &\quad - \mathrm{i} \sum_{\{p+k+q=n\}} \omega_p (w_1^{(k)} S_1 - \mathrm{i} w_2^{(k)} S_2)\omega_q + \mathrm{i}(w_1^{(n)} S_1 + \mathrm{i} w_2^{(n)} S_2) = 0, \\ &\zeta_n = -\mathrm{i} w_3^{(n)} \int_{-\infty}^{x} \mathrm{d}x' [S_3(x') - \varepsilon] \\ &\quad - \mathrm{i} \int_{-\infty}^{x} \mathrm{d}x' \sum_{m=0}^{n+1} [w_1^{(n-m)} S_1(x') - \mathrm{i} w_2^{(n-m)} S_2(x')]\omega_m(x'), \end{aligned}$$

where $n = 0, 1, 2, \ldots, \sum_{\{p+k+q=n\}}$ stands for the sum of the products of three multipliers whose indices satisfy the constraints $p + k + q = n$, and

$$\omega_0 = \frac{S_1 + \mathrm{i} S_2}{\varepsilon + S_3}.$$

The formal replacement

$$\varepsilon \to 1, \quad \int_{-\infty}^{x} \to \int_{+\infty}^{x}$$

yields an asymptotic expansion of the Jost function $\varphi_+(\lambda)$ in powers of λ from the expansion of the function $\varphi_-(\lambda)$.

Using the series for the Jost solutions and relation (5.5.1), we find the expansion of the function $\ln a(\lambda)$ into a power series in λ:

$$\ln a(\lambda) = -\frac{\mathrm{i}}{2}P_1 - \frac{\mathrm{i}}{4\rho}H\lambda + \cdots. \tag{5.5.4}$$

The first two terms of eq. (5.5.4) involves the field momentum P_1 and energy H of the ferromagnet under the boundary conditions (5.2.2) for $\varepsilon = 1$:

$$\begin{aligned} P_1 &= \int_{-\infty}^{+\infty} \frac{S_1\partial_x S_2 - S_2\partial_x S_1}{1+S_3}\,\mathrm{d}x, \\ H &= \int_{-\infty}^{+\infty} \left[\frac{1}{2}(\partial_x \mathbf{S})^2 + 2\rho^2(1 - S_3^2 - k^2 S_2^2)\right]\mathrm{d}x. \end{aligned} \tag{5.5.5}$$

A similar calculation under the boundary conditions (5.2.2) for $\varepsilon = -1$ gives the formula

$$\ln b(\lambda) = \frac{\mathrm{i}}{2}P_2 + \frac{\mathrm{i}}{4\rho}H\lambda + \cdots. \tag{5.5.6}$$

Formulas (5.5.4) and (5.5.6) differ not only in the signs standing before the terms but also in another appearance of the system momentum:

$$P_2 = \int_{-\infty}^{+\infty} \partial_x S_3 \arctan\frac{S_2}{S_1}\,\mathrm{d}x. \tag{5.5.7}$$

The scientific literature [22, 20, 32] treats the total momentum of a ferromagnet through formula (5.5.5). It is easy to see that the integrands in relations (5.5.5) and (5.5.7) differ by a total derivative. Therefore, under the boundary conditions (5.2.2) and for $\varepsilon = 1$, the states of a ferromagnet can be calculated by any of formulas (5.5.5) and (5.5.7). However, under the boundary conditions of the domain wall type and for $\varepsilon = -1$, expressions (5.5.5) and (5.5.7) differ by a nonzero term. To choose the correct one, we calculate the momentum of the moving domain wall (5.3.9). Formula (5.5.5) gives the value $P_1 = 0$, and formula (5.5.7) gives $P_2 = -2\arctan(\operatorname{sn} u_1/\operatorname{cn} u_1)$. We can see that formula (5.5.7) is more preferable to compute the total momentum of a ferromagnet with different equilibrium magnetization values at spatial infinity [31].

Let us take a look at the study of the spectrum of nonlinear excitations of a biaxial ferromagnet. To find the spectrum of excitations, it is necessary to deduce dispersion relations on a torus. For states of a ferromagnet with homogeneous asymptotics of magnetization at infinity ($\varepsilon = 1$), this problem was solved in [31] without a detailed analysis of the scattering method. Using symmetry properties, the function $a(\lambda)$ was constructed, and a connection between the expansion coefficients of the function $\ln a(\lambda)$ and the integrals of motion is found. We show that for the cases $\varepsilon = 1$ or $\varepsilon = -1$, dispersion relations for the coefficients $a(\lambda)$ or $b(\lambda)$ can be directly derived not resorting to the Cauchy kernel on a torus but rather by applying the usual Cauchy theorem to the domains D_+ or D_- involving the function $w_3(\lambda)\rho^{-1} = \mathrm{cs}(\lambda, k)$, which determines the essential features of the "seeding" solutions $\chi_{1,2}^{(0)}$. Further analysis implies the expansion of the function $\mathrm{cs}(\lambda) \equiv \mathrm{cn}\,\lambda/\mathrm{sn}\,\lambda$ in terms of Weierstrass zeta functions:

$$\frac{\mathrm{cn}\,\lambda}{\mathrm{sn}\,\lambda} = \zeta(\lambda) + \zeta(\lambda - 2K) - \zeta(\lambda - 2\mathrm{i}K') \\ - \zeta(\lambda - 2K - 2\mathrm{i}K') - 2\eta_3, \tag{5.5.8}$$

where $4K$ and $4\mathrm{i}K$ are the periods of the Weierstrass functions.

First, we discuss states of a ferromagnetic with the same equilibrium values of magnetization at infinity $\varepsilon = 1$. In this case the function $a(\lambda)$ is analytic in the domain D_+, whereas, generally speaking, $b(\lambda)$ has no properties of analyticity. We address the results obtained earlier for typical solitons (see formulas (5.2.28) and (5.2.48)) and extract from the function $a(\lambda)$ its soliton part, which is related to the zeros of the coefficient $a(\lambda)$:

$$a(\lambda) = \sigma \prod_{j=1}^{m} \frac{\mathrm{cn}(\lambda - v) - \mathrm{i}\,\mathrm{sn}(\lambda - v)}{\mathrm{cn}(\lambda - v^*) + \mathrm{i}\,\mathrm{sn}(\lambda - v^*)}\bigg|_{v=u_j+\mathrm{i}\theta_j} \\ \times \prod_{s=1}^{n} [\mathrm{cn}(\lambda - u_s) - \mathrm{i}\,\mathrm{sn}(\lambda - u_s)]\tilde{a}(\lambda). \tag{5.5.9}$$

Here $\sigma = \pm 1$, $m = 0, 1, 2, \ldots$. The multiplier $\tilde{a}(\lambda)$ has no zeros in the domain D_+. Therefore the function $\ln \tilde{a}(\lambda)$ is analytic in D_+. If n is an even number in formula (5.5.9), then the coefficient $\tilde{a}(\lambda)$ satisfies the same reductions as $a(\lambda)$:

$$\tilde{a}(\lambda + 2K) = \tilde{a}(\lambda), \quad \tilde{a}(\lambda - 2\mathrm{i}K') = \tilde{a}^*(\lambda^*). \tag{5.5.10}$$

It is worth emphasizing that only for the values $n = 2r$, $r = 0, 1, 2, \ldots$, factorization (5.5.9) leads to correct dispersion relations, consistent with the analyticity and symmetry properties of the function $a(\lambda)$.

Consider the integral

$$\frac{1}{4\pi\mathrm{i}} \int_{\gamma_2} \ln \tilde{a}(\lambda') \frac{\mathrm{cn}(\lambda' - \lambda)}{\mathrm{sn}(\lambda' - \lambda)}\, \mathrm{d}\lambda', \tag{5.5.11}$$

where the contour γ_2 bounds the domain D_+. As before, we take a positive traversal along the contour γ_2 (see Figure 5.2). For $\lambda \in D_+$, the integrand in (5.5.11) represents the boundary value of a function that is analytic inside the contour γ_2 everywhere except for simple poles at the points λ and $\lambda + 2K$ (see expansion (5.5.8)). According to reductions

(5.5.10), the residues of $\ln \tilde{a}(\lambda)$ and $\ln \tilde{a}(\lambda+2K)$ coincide at these poles. From here, by the Cauchy theorem we conclude that

$$\frac{1}{4\pi \mathrm{i}} \int_{\gamma_2} \ln \tilde{a}(\lambda') \frac{\mathrm{cn}(\lambda'-\lambda)}{\mathrm{sn}(\lambda'-\lambda)} \,\mathrm{d}\lambda' = \ln \tilde{a}(\lambda).$$

The symmetry properties (5.5.10) allow us to reduce the integration over the contour γ_2 to an integration over the segment $[0, 2K]$:

$$\ln \tilde{a}(\lambda) = -\frac{1}{2\pi \mathrm{i}} \int_0^{2K} \ln|\tilde{a}(\lambda')|^2 \frac{\mathrm{cn}(\lambda'-\lambda)}{\mathrm{sn}(\lambda'-\lambda)} \,\mathrm{d}\lambda'. \tag{5.5.12}$$

Let $\lambda = \nu - \mathrm{i}\varepsilon$ in formula (5.5.15), where $0 < \nu < 2K$, $\varepsilon > 0$. We pass to the limit as $\varepsilon \to +0$. By the Sokhotski–Plemelj formula,

$$\begin{aligned}
&\int_0^{2K} \ln|\tilde{a}(\lambda')|^2 \frac{\mathrm{cn}(\lambda'-\nu+\mathrm{i}0)}{\mathrm{sn}(\lambda'-\nu+\mathrm{i}0)} \,\mathrm{d}\lambda' \\
&\qquad = -\pi \mathrm{i} \ln|\tilde{a}(\lambda)|^2 + \mathrm{V.p.} \int_0^{2K} \ln|\tilde{a}(\lambda')|^2 \frac{\mathrm{cn}(\lambda'-\nu)}{\mathrm{sn}(\lambda'-\nu)} \,\mathrm{d}\lambda',
\end{aligned}$$

and we find the dispersion relation

$$\mathrm{i}\arg[\ln \tilde{a}(\lambda)] = -\frac{1}{2\pi \mathrm{i}} \mathrm{V.p.} \int_0^{2K} \ln|\tilde{a}(\lambda')|^2 \frac{\mathrm{cn}(\lambda'-\nu)}{\mathrm{sn}(\lambda'-\nu)} \,\mathrm{d}\lambda'.$$

Reusing the Sokhotski–Plemelj formula restores the values of the function $\ln \tilde{a}(\lambda)$ in its modulus:

$$\begin{aligned}
\ln \tilde{a}(\lambda - \mathrm{i}0) &\equiv \ln|\tilde{a}(\lambda)| + \mathrm{i}\arg \ln \tilde{a}(\lambda) \\
&= -\frac{1}{2\pi \mathrm{i}} \int_0^{2K} \ln|\tilde{a}(\lambda')|^2 \frac{\mathrm{cn}(\lambda'-\lambda+\mathrm{i}0)}{\mathrm{sn}(\lambda'-\lambda+\mathrm{i}0)} \,\mathrm{d}\lambda',
\end{aligned}$$

where $\lambda \in [0, 2K]$.

Within the range $0 < \nu < 2K$, the modulus of the function $a(\lambda)$ can be expressed in terms of the coefficient $b(\lambda)$ through the unimodularity property for the transition matrix (5.2.7):

$$|a(\lambda)|^2 = 1 - |b(\lambda)|^2.$$

On the other hand, representation (5.5.9) yields

$$|a(\lambda)|^2 = |\tilde{a}(\lambda)|^2.$$

Finally, we obtain the following expression for the nonsoliton factor $\tilde{a}(\lambda)$:

$$\tilde{a}(\lambda) = \exp\left(\frac{i}{2\pi} \int_0^{2K} \ln(1 - |b(\lambda')|^2) \frac{\operatorname{cn}(\lambda' - \lambda + i0)}{\operatorname{sn}(\lambda' - \lambda + i0)} \, d\lambda' \right).$$

Taking the logarithm of eq. (5.5.9) and expanding both parts in powers of λ, we get spectral representations for the integrals of motion. For example, for the momentum P, we find

$$P = \operatorname{mod}(2\pi) + \int_0^{2K} 2\rho \frac{\operatorname{cn}(\lambda, k)}{\operatorname{sn}(\lambda, k)} n(\lambda) d\lambda + 2i \sum_{j=1}^{m} \ln\left(\frac{\operatorname{cn} v + i \operatorname{sn} v}{\operatorname{cn} v^* - i \operatorname{sn} v^*} \right)\Bigg|_{v=u_j+i\theta_j} - 2 \sum_{s=1}^{n} \arctan \frac{\operatorname{sn} u_s}{\operatorname{cn} u_s}, \tag{5.5.13}$$

where

$$n(\lambda) = -\frac{1}{2\pi\rho} \ln(1 - |b(\lambda)|^2) > 0.$$

The symbol mod(2π) denotes the constant term $2i \ln \sigma = 0, \pm 2\pi$, which can be neglected. The integral in eq. (5.5.13) converges despite the pole singularity of the factor $\operatorname{cs}(\lambda, k)$ at the point $\lambda = 0$. This is because for the magnetization distributions $\mathbf{S}(x)$ tending to their asymptotic values as $x \to \pm\infty$ faster than any power of x, the function $n(\lambda)$ tends to zero faster than any power of λ.

The spectral expansion of the energy of the system has the form

$$H = \int_0^{2K} 4\rho^2 \frac{\operatorname{dn}(\lambda, k)}{\operatorname{sn}^2(\lambda, k)} n(\lambda) d\lambda + 8\rho \sum_{j=1}^{m} \operatorname{Re} \operatorname{dn}(u_j + i\theta_j) + 4\rho \sum_{s=1}^{n} \operatorname{dn} u_s. \tag{5.5.14}$$

We recall once again that in formulas (5.5.13) and (5.5.14), the number n is even.

Let us interpret the results. The first term in formulas (5.5.13) and (5.5.14) is continuous and meets low-amplitude spin-wave oscillations near the ground state with frequency

$$\omega = 4\rho^2 \frac{\operatorname{dn} \lambda}{\operatorname{sn}^2 \lambda} = 4 w_1 w_2$$

and wave number

$$p = 2\rho \frac{\operatorname{cn} \lambda}{\operatorname{sn} \lambda} = 2 w_3,$$

which satisfy the dispersion law for linear spin waves:

$$\omega^2 = (p^2 + J_3 - J_1)(p^2 + J_3 - J_2).$$

The quantity $n(\lambda)\mathrm{d}\lambda$ gives the number of spin-wave modes ranged from λ to $\lambda + \mathrm{d}\lambda$. The second term can be treated as a contribution of breathers. The third term makes a contribution of domain walls. Remarkably, in the spectral representations for the integrals of motion, we have succeeded in completely separating the contributions of all degrees of freedom of the ferromagnet.

Now we briefly discuss the states of a ferromagnet with different equilibrium magnetization values at infinity ($\varepsilon = -1$). For $\varepsilon = -1$, the function $b(\lambda)$ turns out to be analytic in the domain D_-, and the coefficient $a(\lambda)$ has no certain analytic properties. In the same way as before, we extract its soliton part from the function $b(\lambda)$:

$$b(\lambda) = \sigma \prod_{j=1}^{m} \frac{\mathrm{cn}(\lambda - \nu^*) + \mathrm{i}\,\mathrm{sn}(\lambda - \nu^*)}{\mathrm{cn}(\lambda - \nu) - \mathrm{i}\,\mathrm{sn}(\lambda - \nu)}\Bigg|_{\nu = u_j + \mathrm{i}\theta_j} \times \prod_{s=1}^{n} [\mathrm{cn}(\lambda - u_s) + \mathrm{i}\,\mathrm{sn}(\lambda - u_s)]\tilde{b}(\lambda), \tag{5.5.15}$$

where $\sigma = \pm 1$, $m = 0, 1, 2, \ldots$. The desired symmetry properties of the function (5.2.6) can be derived only by having chosen the number n odd in formula (5.5.15). This results in the fact that the function $\tilde{b}(\lambda)$ must have the symmetry properties

$$\tilde{b}(\lambda + 2K) = \tilde{b}(\lambda), \quad \tilde{b}(\lambda + 2\mathrm{i}K') = \tilde{b}(\lambda^*). \tag{5.5.16}$$

These properties differ from those of the function $b(\lambda)$. At the same time, reductions (5.5.16) coincide up to notations with those for $\varepsilon = 1$. Only the values $n = 2r-1$, $r = 1, 2, \ldots$, ensure a proper dispersion relation for the multiplier $\tilde{b}(\lambda)$:

$$\tilde{b}(\lambda) = \exp\left(\frac{1}{2\pi \mathrm{i}} \int_0^{2K} \ln(1 - |a(\lambda')|^2) \frac{\mathrm{cn}(\lambda' - \lambda - \mathrm{i}0)}{\mathrm{sn}(\lambda' - \lambda - \mathrm{i}0)}\, \mathrm{d}\lambda' \right).$$

For states of a ferromagnet with different equilibrium magnetization values at infinity, the spectral expansions of the total system momentum and energy are determined by formulas (5.5.13) and (5.5.14), where we should set $n = 2r - 1$, $r = 1, 2, \ldots$, and

$$n(\lambda) = -\frac{1}{2\pi\rho} \ln(1 - |a(\lambda)|^2).$$

We conclude by noting that although solitary wave-type solutions for magnets with a complicated form of anisotropy energy exist [19, 18, 33], model (5.2.1) is the most general integrable equation for ferromagnets. This statement is proved as follows: Suppose that the matrices $U(\lambda)$ and $V(\lambda)$ are arbitrarily (but locally) $\mathbf{S}$- and $\partial_x \mathbf{S}$-dependent. Then we require that the compatibility condition (2.1.1) should be equivalent to the Landau–Lifshitz equation with anisotropy energy $w^{(a)}$. It turns out that this requirement is fulfilled only when $w^{(a)}$ is a quadratic function of magnetization.

6 Nonlinear waves and solitons in a system of two crystallo-graphically nonequivalent magnetic sublattices

The present chapter focuses on the results of studies of nonlinear waves and solitons in ferrimagnets. The primary objective of our works [7, 4] is to test the existence and possibility of describing the properties of bound states of a huge number of magnons – solitons in are widely spread and convenient media with two magnetic sublattices for experimental research. Section 6.1 contains the effective equations for a ferrimagnet and their $U - V$ pair. Section 6.2 covers the inverse scattering method to comprehensively explain the nonlinear dynamics of a ferrimagnet. Section 6.3 utilizes the matrix Riemann problem to formulate the method of integrating the equations for a ferrimagnet and analyzes soliton solutions. Section 6.4 shows that noninteracting nonlinear waves and solitons can properly describe any state of a ferrimagnet. Their dispersion laws are established. Section 6.5 explores two-dimensional solitons in an isotropic ferromagnet [6], which are closely related to the integrable model of a ferrimagnet.

6.1 Effective equations for a ferrimagnet and their $U - V$ pair

Set problems if their solution only you have.

Consequence of Burke's rule

We restrict ourselves to studying a simple ferrimagnet in the exchange approximation [22] to elucidate the main features of nonlinear waves and solitons in magnetic media with nonequivalent magnetic sublattices. Behind the description of the magnetic properties of such a system, the magnetic moment $\mathbf{M}_\nu(\mathbf{r},\ t)$ ($\nu = 1, 2$) density of each of its two sublattices lies. The dependence of a ferrimagnet energy on the distribution of magnetization has the form

$$W = \int d^3\mathbf{r} \left(\frac{1}{2} \sum_{i=1}^{3} [\alpha_1(\partial_i \mathbf{M}_1 \cdot \partial_i \mathbf{M}_1) + \alpha_2(\partial_i \mathbf{M}_2 \cdot \partial_i \mathbf{M}_2) + 2\alpha_3(\partial_i \mathbf{M}_1 \cdot \partial_i \mathbf{M}_2)] + \eta\, (\mathbf{M}_1 \cdot \mathbf{M}_2) \right),$$

where $\eta > 0$ is the constant of homogeneous exchange between sublattices, α_1, α_2, and α_3 are the constants of inhomogeneous exchange, and

$$|\mathbf{M}_\nu(\mathbf{r}, t)| = M_{0\nu} = \text{const}, \quad M_{01} \neq M_{02}.$$

https://doi.org/10.1515/9783112235973-006

In order of magnitude, $\alpha_s \sim \eta\, a^2$, where a is the lattice constant. It is convenient to pass from the magnetizations $\mathbf{M}_1$ and $\mathbf{M}_2$ to the normalized ferro- and antiferromagnetism vectors

$$\mathbf{m} = \frac{\mathbf{M}_1 + \mathbf{M}_2}{\sqrt{2(M_{01}^2 + M_{02}^2)}} \quad \text{and} \quad \mathbf{l} = \frac{\mathbf{M}_1 - \mathbf{M}_2}{\sqrt{2(M_{01}^2 + M_{02}^2)}};$$
$$\mathbf{m}^2 + \mathbf{l}^2 = 1, \quad (\mathbf{m} \cdot \mathbf{l}) = \frac{M_{01}^2 - M_{02}^2}{2(M_{01}^2 + M_{02}^2)}. \tag{6.1.1}$$

Magnetic structure ferrimagnet substances are well studied [11, 8]. The rare earth ferrite garnets (so named because their lattice structure resembles that of the mineral garnet) are a striking illustration of those. Their general formula is $R_3Fe_5O_{12}$, where R is a rare earth ion. There are two magnetic sublattices that can be identified in ferrites. Iron atoms constitute the former, and the latter consists of rare earth ions. The exchange interaction between iron atoms is the strongest one. Therefore, at room temperature, only iron atoms are ordered, and ferrite garnet exhibits a low magnetic anisotropy. Such a feature, apart from its high electrical resistance, makes ferrite a perfect soft magnetic material. In the ground state the vectors $\mathbf{M}_1$ and $\mathbf{M}_2$ are antiparallel, and $\mathbf{m}^2 \ll \mathbf{l}^2$. For iron–yttrium garnet and magnetite, the ratios $\mathbf{m}^2/\mathbf{l}^2$ are 0.04 and 0.1, respectively.

Modern technologies make it possible to grow thin films of ferrite-garnet single crystals with almost any predictable properties on nonmagnetic substrates. Because of this, ferrites are highly sought-after for their application in sophisticated devices.

The equation of motion for the vectors $\mathbf{m}$ and $\mathbf{l}$ follows from the Landau–Lifshitz equations for the sublattice magnetizations:

$$\partial_t \mathbf{M}_\nu = \gamma_\nu \left[\mathbf{M}_\nu \times \frac{\delta W}{\delta \mathbf{M}_\nu}\right], \quad \nu = 1, 2, \tag{6.1.2}$$

where $\gamma_\nu = \text{const}$ is the magneto-mechanical ratio for the νth sublattice. These equations are currently too arduous to solve completely.

Equations (6.1.2) can be substantially simplified using physically justified approximations. The latter were first proposed in [1, 2] for analyzing the nonlinear dynamics of a two-sublattice antiferromagnet. Suppose that the characteristic size d of magnetic inhomogeneities is much larger than the lattice constant a. Then the terms containing α can be neglected in comparison with the terms containing η:

$$\frac{|\alpha_s[\mathbf{m} \times \Delta\mathbf{l}]|}{|\eta[\mathbf{m} \times \mathbf{l}]|} \sim \frac{|\alpha_s[\Delta\mathbf{m} \times \mathbf{l}]|}{|\eta[\mathbf{m} \times \mathbf{l}]|} \sim \left|\frac{\alpha_s}{\eta d^2}\right| \sim \frac{a^2}{d^2} \ll 1.$$

In addition, if $\mathbf{m}^2 \ll \mathbf{l}^2$, then we believe that the length of the vector $\mathbf{l}$ is unchangeable, that is, $\mathbf{l}^2 = 1$. Then system (6.1.2) appears as

$$
\begin{aligned}
\partial_t \mathbf{l} &= \sqrt{\frac{M_{01}^2 + M_{02}^2}{2}} \Big(\eta(\gamma_1 + \gamma_2)[\mathbf{l} \times \mathbf{m}] \\
&\quad - \frac{1}{2}[\gamma_1(\alpha_1 - \alpha_3) - \gamma_2(\alpha_2 - \alpha_3)][\mathbf{l} \times \Delta \mathbf{l}] \Big), \\
\partial_t \mathbf{m} &= \sqrt{\frac{M_{01}^2 + M_{02}^2}{2}} \Big(\eta(\gamma_1 - \gamma_2)[\mathbf{l} \times \mathbf{m}] \\
&\quad - \frac{1}{2}[\gamma_1(\alpha_1 - \alpha_3) + \gamma_2(\alpha_2 - \alpha_3)][\mathbf{l} \times \Delta \mathbf{l}] \Big).
\end{aligned}
\tag{6.1.3}
$$

Exploiting (6.1.1) and the first equation of (6.1.3), we express $\mathbf{m}$ in terms of the vector $\mathbf{l}$ and its derivatives:

$$
\begin{aligned}
\mathbf{m} &= -\sqrt{\frac{2}{M_{01}^2 + M_{02}^2}} \frac{[\mathbf{l} \times \partial_t \mathbf{l}]}{\eta(\gamma_1 + \gamma_2)} + \frac{(M_{01}^2 - M_{02}^2)}{2(M_{01}^2 + M_{02}^2)} \mathbf{l} \\
&\quad + \frac{\gamma_1(\alpha_3 - \alpha_1) + \gamma_2(\alpha_2 - \alpha_3)}{2\eta(\gamma_1 + \gamma_2)} [\mathbf{l} \times [\mathbf{l} \times \Delta \mathbf{l}]].
\end{aligned}
\tag{6.1.4}
$$

Note that when substituting the expression for $\mathbf{m}$ into the second equation of (6.1.3), the last summand in (6.1.4) brings to terms containing third derivatives of the vector $\mathbf{l}$. They are small in the long-wavelength limit compared to those containing the second derivatives of $\mathbf{l}$. Consequently, the last term in the expression for $\mathbf{m}$ can be neglected. Plugging eq. (6.1.4) into the first equation of (6.1.3) yields a closed equation for the vector $\mathbf{l}$:

$$
[\mathbf{l} \times (c^{-2}\, \partial_t^2 \mathbf{l} - \Delta \mathbf{l})] + \beta\, c^{-1}\, \partial_t \mathbf{l} = 0,
\tag{6.1.5}
$$

where

$$
\begin{aligned}
c^2 &= \frac{1}{2}(M_{01}^2 + M_{02}^2)\eta\, \gamma_1 \gamma_2 (\alpha_1 + \alpha_2 - 2\alpha_3), \\
\beta &= \sqrt{\frac{\eta}{\gamma_1 \gamma_2 (\alpha_1 + \alpha_2 - 2\alpha_3)}} \left(\frac{M_{02}^2 - M_{01}^2}{M_{01}^2 + M_{02}^2} \cdot \frac{\gamma_1 + \gamma_2}{2} + \gamma_1 - \gamma_2 \right).
\end{aligned}
\tag{6.1.6}
$$

The formal limit as $c \to \infty$ and $\beta c^{-1} = \text{const}$ corresponds to a rigid connection between the sublattices ($\eta \to +\infty$). In this context, the effective equation (6.1.5) coincides with the integrable Landau–Lifshitz equation for a Heisenberg ferromagnet [13, 21]. For $M_{01} = M_{02}$ and $\gamma_1 = \gamma_2$, system (6.1.4)–(6.1.5) describes waves in an isotropic two-sublattice antiferromagnet [1, 2] and is equivalent to an integrable model of an $\mathbf{n}$-field on a sphere S^2 [19, 16].

The scaling transformations

$$
x \to \frac{x}{\beta}, \quad t \to \frac{t}{c\beta},
$$

turn the effective equation (6.1.5) into an equation convenient for further analysis:

$$[\mathbf{l} \times (\partial_t^2 \mathbf{l} - \Delta \mathbf{l})] + \partial_t \mathbf{l} = 0. \tag{6.1.7}$$

This chapter addresses magnetization waves propagating along the Ox-axis. In this case, $\mathbf{l} = \mathbf{l}(x, t)$, and therefore eq. (6.1.7) takes the form

$$[\mathbf{l} \times (\partial_t^2 \mathbf{l} - \partial_x^2 \mathbf{l})] + \partial_t \mathbf{l} = 0. \tag{6.1.8}$$

To explore the nonlinear dynamics of a ferrimagnet in the quasi-one- dimensional case, we should resort to the inverse scattering method. The key point in applying the method is the possibility of representing eq. (6.1.8) through the compatibility condition

$$\partial_x U(\lambda) - \partial_t V(\lambda) + [U(\lambda), V(\lambda)] = 0 \tag{6.1.9}$$

for the system of linear differential equations:

$$\partial_x T = V(\lambda)\, T, \quad \partial_t T = U(\lambda)\, T \tag{6.1.10}$$

for auxiliary fields $T(x, t, \lambda)$. Here the matrices $U(\lambda)$ and $V(\lambda)$ depend on the vector $\mathbf{l}$ and its derivatives and on the spectral parameter λ.

Generally, there is no regular way of constructing a $U - V$ pair. The well-known Estabrook–Wahlquist method [23] resolves this issue through studying the algebra of differential operators. We apply the simpler technique presented in Chapter 2.

It should be pointed out that the invariance of eqs. (6.1.7) with respect to a group of spin rotations, its linear realization on the fields l_i $(i = 1, 2, 3)$, and the form of eqs. (6.1.8) and (6.1.9) make searching of $U(\lambda)$ and $V(\lambda)$ in the matrix representation of the algebra of the group SU(2) advisable. Assuming a local dependence, the matrices $U(\lambda)$ and $V(\lambda)$ must be expanded in the linearly independent vectors $\mathbf{l}$, $\partial_t \mathbf{l}$, $\partial_x \mathbf{l}$, $[\mathbf{l} \times \partial_x \mathbf{l}]$, $[\mathbf{l} \times \partial_t \mathbf{l}]$. The expansion coefficients are uniquely determined by the equivalence of eqs. (6.1.8) and (6.1.9). As a result, we arrive at

$$V(\lambda) = -\left(\gamma A + \delta B + \frac{\mathrm{i}\Lambda \hat{l}}{2}\right), \quad U(\lambda) = -\left(\delta A + \gamma B - \frac{\mathrm{i}\gamma^2 \hat{l}}{2}\right), \tag{6.1.11}$$

where the two-dimensional representation of the algebra of the group SU(2) in terms of the Pauli matrices σ_i is chosen:

$$\sigma_1 = \begin{pmatrix} 0 & 1 \\ 1 & 0 \end{pmatrix}, \quad \sigma_2 = \begin{pmatrix} 0 & -\mathrm{i} \\ \mathrm{i} & 0 \end{pmatrix}, \quad \sigma_3 = \begin{pmatrix} 1 & 0 \\ 0 & -1 \end{pmatrix}, \quad I = \begin{pmatrix} 1 & 0 \\ 0 & 1 \end{pmatrix},$$

$$\sigma_i \sigma_j = I \delta_{ij} + \mathrm{i}\, \varepsilon_{ijk} \sigma_k,$$

and

$$A = \frac{i}{2}[\mathbf{l} \times \partial_t \mathbf{l}]_a \sigma_a, \quad B = \frac{i}{2}[\mathbf{l} \times \partial_x \mathbf{l}]_a \sigma_a, \quad \hat{l} = l_a \sigma_a \quad (a = 1, 2, 3). \tag{6.1.12}$$

The coefficients γ, δ, and Λ are related by the algebraic equations

$$\delta^2 - 2\delta - \gamma^2 = 0, \quad \Lambda = \gamma(1 - \delta), \tag{6.1.13}$$

which allow for uniformization by rational functions of the parameter λ:

$$\delta(\lambda) = \frac{2}{1 - \lambda^2}, \quad \gamma(\lambda) = \frac{2\lambda}{1 - \lambda^2}, \quad \Lambda(\lambda) = -\frac{2\lambda\,(1 + \lambda^2)}{(1 - \lambda^2)^2}. \tag{6.1.14}$$

Direct verification allows us to convince ourselves that the compatibility condition (6.1.9) exactly coincides with eq. (6.1.8). The found $U - V$ pair has, as the Thirring model [12], two simple and two double poles. However, it differs significantly from the Thirring model in the matrix structure.

Thus the present section derives the effective equations (6.1.8) of the dynamics of a ferrimagnet and shows their integrability. The appropriate $U - V$ pair is given by formulas (6.1.11)–(6.1.14).

6.2 Integration of dynamic equations by the inverse scattering method

If during the construction or in the proof we admit something that has not been previously proven, but requires argumentation, then we consider this assumption to be dubious in itself and worthy of investigation and call it a lemma.

Proclus' Commentary on Euclid

The present section solves the Cauchy problem for eqs. (6.1.8), i. e., we offer a constructive method for finding the sublattice magnetizations $\mathbf{M}_1$ and $\mathbf{M}_2$ at any moment of time for preset values of these quantities at the initial moment $t = 0$.

To start with, we consider the direct scattering method for the system

$$\partial_x T = V(\lambda)\, T \tag{6.2.1}$$

and determine the scattering coefficients for given values of $\mathbf{l}(x, t = 0)$ and $\partial_t \mathbf{l}(x, t = 0)$ and the boundary conditions

$$\mathbf{l}(x, t) \to (0, 0, 1), \quad \partial_t \mathbf{l} \to 0 \quad \text{as} \quad |x| \to \infty.$$

For this purpose, we introduce the Jost functions $\varphi(x, \lambda)$, $\bar{\varphi}(x, \lambda)$, $\Psi(x, \lambda)$, and $\bar{\Psi}(x, \lambda)$ as solutions to eq. (6.2.1) with asymptotics

$$\varphi \to \begin{pmatrix}1\\0\end{pmatrix}\exp\left(-\frac{\mathrm{i}\Lambda x}{2}\right), \quad \bar{\varphi} \to \begin{pmatrix}0\\1\end{pmatrix}\exp\left(\frac{\mathrm{i}\Lambda x}{2}\right)$$
$$\text{as} \quad x \to -\infty, \tag{6.2.2}$$
$$\Psi \to \begin{pmatrix}1\\0\end{pmatrix}\exp\left(-\frac{\mathrm{i}\Lambda x}{2}\right); \quad \bar{\Psi} \to \begin{pmatrix}0\\1\end{pmatrix}\exp\left(\frac{\mathrm{i}\Lambda x}{2}\right)$$
$$\text{as} \quad x \to +\infty.$$

In view of the anti-Hermitian matrix V ($V^\dagger = -V$) and the above boundary conditions, the Jost functions are related as follows:

$$\bar{\varphi}(\lambda) = -\mathrm{i}\sigma_2\,\varphi^*(\lambda^*), \quad \bar{\Psi}(\lambda) = -\mathrm{i}\sigma_2\Psi^*(\lambda^*). \tag{6.2.3}$$

For real values of the parameter λ, each of the two sets of fundamental solutions for $\varphi(x,\lambda)$, $\bar{\varphi}(x,\lambda)$ and $\Psi(x,\lambda)$, $\bar{\Psi}(x,\lambda)$ can be treated as a basic one, so we have the representation

$$\begin{aligned}\varphi(x,\lambda) &= a(\lambda)\Psi(x,\lambda) + b(\lambda)\bar{\Psi}(x,\lambda),\\ \bar{\varphi}(x,\lambda) &= \bar{a}(\lambda)\bar{\Psi}(x,\lambda) + \bar{b}(\lambda)\Psi(x,\lambda),\end{aligned} \tag{6.2.4}$$

where $a(\lambda)$, $\bar{b}(\lambda)$, $\bar{a}(\lambda)$, and $b(\lambda)$ are the elements of the transition matrix.

For any pair of α, $\tilde{\beta}$ of solutions to system (6.2.1), the Wronskian

$$W(\alpha,\tilde{\beta}) = \det(\alpha,\tilde{\beta})$$

does not depend on the coordinate x. Hence

$$W(\varphi,\bar{\varphi}) = W(\Psi,\bar{\Psi}) = 1.$$

The symmetry properties of system (6.2.1) implies that the Jost solutions satisfy the reduction

$$\begin{aligned}\bar{\varphi}(x,\lambda^{-1}) &= -\hat{l}\,\bar{\varphi}(x,\lambda), \quad \bar{\Psi}(x,\lambda^{-1}) = -\hat{l}\,\bar{\Psi}(x,\lambda),\\ \varphi(x,\lambda^{-1}) &= \hat{l}\,\varphi(x,\lambda), \quad \Psi(x,\lambda^{-1}) = \hat{l}\,\Psi(x,\lambda).\end{aligned} \tag{6.2.5}$$

The listed properties of the Jost functions lead to unitarity and symmetry relations for the elements of the transition matrix:

$$\bar{a}(\lambda) = a^*(\lambda^*), \quad \bar{b}(\lambda) = -b^*(\lambda^*),$$
$$a(\lambda)\,a^*(\lambda^*) + b(\lambda)\,b^*(\lambda^*) = 1, \tag{6.2.6}$$
$$a(\lambda^{-1}) = a(\lambda), \quad b(\lambda^{-1}) = -b(\lambda). \tag{6.2.7}$$

To determine the analytic properties of the Jost functions, we should search for representations in terms of transformation operators for them to map solutions of eq. (6.2.1) with "potential" $\mathbf{l}_0 = (0,0,1)$ into solutions with "potential" $\mathbf{l} \neq \mathbf{l}_0$. Let us show that for the functions φ and Ψ, such representations have the form

$$\begin{aligned}\varphi(x,\lambda) &= \begin{pmatrix}1\\0\end{pmatrix}\exp\left(-\frac{\mathrm{i}\Lambda x}{2}\right)\\ &+ \int_{-\infty}^{x} \mathrm{d}y\,[\gamma(\lambda)K_1(x,y) + \delta(\lambda)K_2(x,y)\\ &+ \gamma^2(\lambda)K_3(x,y) + \Lambda(\lambda)K_4(x,y)]\exp\left(-\frac{\mathrm{i}}{2}\Lambda(\lambda)y\right);\end{aligned} \tag{6.2.8}$$

$$\begin{aligned}\Psi(x,\lambda) &= \begin{pmatrix}1\\0\end{pmatrix}\exp\left(-\frac{\mathrm{i}\Lambda x}{2}\right)\\ &+ \int_{x}^{+\infty} \mathrm{d}y[\gamma(\lambda)L_1(x,y) + \delta(\lambda)L_2(x,y)\\ &+ \gamma^2(\lambda)L_3(x,y) + \Lambda(\lambda)L_4(x,y)]\exp\left(-\frac{\mathrm{i}}{2}\Lambda(\lambda)y\right).\end{aligned} \tag{6.2.9}$$

As in Chapter 3, it would be instructive to first use perturbation theory to reveal a λ-dependence of the transformation operator kernels and to consider a physically justified case where the fields $\mathbf{l}(x,t)$ are close to the ground state:

$$\mathbf{l} = \mathbf{l}^{(0)} + \varepsilon\mathbf{l}^{(1)} + \varepsilon^2\mathbf{l}^{(2)} + \cdots, \quad \mathbf{l}^{(0)} = (0,0,1).$$

Here we have entered a formal parameter ε for a visual grouping of the summands of the same order of smallness.

Using the method of variation of arbitrary constants, we create the fundamental matrix of solutions Φ ($\Phi_{\alpha 1} = \varphi_\alpha, \Phi_{\alpha 2} = \bar{\varphi}_\alpha, \alpha = 1,2$) in the form of a series $\Phi = \sum_{m=0}^{\infty}\varepsilon^m\Phi^{(m)}$. To calculate matrix functions, we have a chain of recurrent relations:

$$\begin{aligned}\Phi^{(0)} &= \exp\left(-\mathrm{i}\frac{\Lambda\sigma_3}{2}x\right),\\ \Phi^{(m)}(x,\lambda) &= \int_{-\infty}^{x}\sum_{k=1}^{m}\Phi^{(0)}(x-y,\lambda)\,V^{(k)}(y,\lambda)\,\Phi^{(m-k)}(y,\lambda)\,\mathrm{d}y;\\ V^{(0)} &= -\mathrm{i}\frac{\Lambda\sigma_3}{2}, \quad V^{(m)} = \sum_{i=1}^{3}V_i^{(m)}\sigma_i, \quad V = \sum_{m=0}^{\infty}V^{(m)}\varepsilon^m.\end{aligned}$$

Let us discuss the first two terms of the expansion and transform them into a triangular form by changing the integration variables. We get

$$\Phi^{(1)} = \int_{-\infty}^{x} K^{(1)}(x,y)\Phi^{(0)}(y)dy, \quad K^{(1)}(x,y) = \frac{1}{2}V^{(1)}\left(\frac{x+y}{2}\right),$$

$$\Phi^{(2)} = \int_{-\infty}^{x} K^{(2)}(x,y)\Phi^{(0)}(y)dy + \int_{-\infty}^{x} V_3^{(2)}(y)\sigma_3\Phi^{(0)}(x)dy,$$

where

$$K^{(2)}(x,y) = \frac{1}{2}\sum_{a=1}^{2} V_a^{(2)}\left(\frac{y+x}{2}\right)\sigma_a + \frac{1}{4}\int_{-\infty}^{y} V^{(1)}\left(\frac{s-y+2x}{2}\right)V^{(1)}\left(\frac{s+x}{2}\right)dx.$$

Consequently, the kernel $K^{(1)}(x,y)$ depends on the parameter λ only through the functions $\gamma(\lambda)$, $\delta(\lambda)$, $\Lambda(\lambda)$, and the kernel $K^{(2)}(x,y)$ contains two groups of summands. The former includes summands with the coefficients γ, δ, Λ, γ^2, $\delta\gamma$, and δ^2. The latter involves summands with the factors $\Lambda\gamma$, $\Lambda\delta$, Λ^2. Integration by parts reduced them to the first group. In addition, we can show by direct calculations that the above transformations result in disappearing the “dangerous” summand

$$\int_{-\infty}^{x} V_3^{(2)}(y)\sigma_3\Phi^{(0)}(x)dy$$

of the matrix function $\Phi^{(2)}(x,\lambda)$. In the process the λ-dependence of the kernel of the triangular matrix representation becomes

$$K^{(2)}(x,y) = \sum_{i=1}^{4} K_i^{(2)}(x,y)f_i(\lambda), \quad f_i(\lambda) = (\gamma^2(\lambda),\ \gamma(\lambda),\ \delta(\lambda),\ \Lambda(\lambda)).$$

Let us discuss the following terms of the Born approximation. The relations

$$\gamma^3 = -\gamma + \Lambda - \delta\Lambda, \quad \delta\gamma^2 = \gamma^2 - \gamma\Lambda \tag{6.2.10}$$

and formulas (6.1.13) disclose the fact that the correction kernels $\Phi^{(3)}$ incorporate the λ-dependence only through summands with the factors $f_i(\lambda)$ and $f_j(\lambda)\Lambda(\lambda)$. Integration by parts can reduce the last group of summands to summands with the factors $f_i(\lambda)$. It is clear that such a procedure can be carried out in any order within the perturbation theory. Here we leave aside the proof of the convergence of the perturbation series and the lack of “dangerous” terms. The analysis conducted above indicates the expediency of studying the representation of φ in the form (6.2.8).

Relations (6.2.5) and (6.2.8) yield the following constraints:

$$\begin{aligned} &\rho(x) = \hat{l}\rho(x), \quad K_a(x,y) = -\hat{l}K_a(x,y), \quad a = 1,2; \\ &K_b(x,y) = \hat{l}K_b(x,y), \quad b = 3,4; \end{aligned} \tag{6.2.11}$$

$$\hat{\imath}\begin{pmatrix}1\\0\end{pmatrix} = \begin{pmatrix}1\\0\end{pmatrix} + 2\int_{-\infty}^{x} \mathrm{d}y\, \exp\left(-\frac{\mathrm{i}\Lambda y}{2}\right) K_2(x,y).$$

For a rigorous proof of representation (6.2.8), we insert it into eq. (6.2.1). Bearing in mind connections (6.1.13) and (6.2.10), we can boil down all the integral expressions by integrating by parts to summands with the λ-independent factors and to terms with the coefficients γ, δ, and γ^2. Next, we multiply the resulting equation successively by $\gamma\,\delta$, γ^2, δ, and $\gamma^2\delta$ and use the Fourier transform defined below (see (6.2.21) and (6.2.22)). Finally, we compose the following system of linear differential equations for the transformation kernels:

$$\partial_x K_1 - \partial_y K_1 + 2\mathrm{i}B\,\partial_y K_3 = A(2\mathrm{i}\,\partial_y K_4 - K_2 + K_3) - BK_1, \tag{6.2.12}$$

$$\partial_x K_3 + \partial_y K_3 = -AK_1 - B(K_2 + K_3), \tag{6.2.13}$$

$$\partial_x K_2 - \partial_y K_2 = 2B(\mathrm{i}\,\partial_y K_4 - K_2) - 2\mathrm{i}A\,\partial_y K_3, \tag{6.2.14}$$

$$\partial_x\partial_y K_4 + \partial_y^2 K_4 = A(\partial_y K_2 - \partial_y K_3) + B\partial_y K_1 \tag{6.2.15}$$

and deduce the additional conditions

$$K_2(x,x) = \mathrm{i}\,AK_3(x,x) - \frac{B}{2}\rho(x),$$

$$K_1(x,x) = \mathrm{i}\,BK_3(x,x) - \frac{A}{2}\rho(x), \tag{6.2.16}$$

$$\partial_x\rho(x) = -\mathrm{i}\,(AB + BA)\,\rho(x) - 2(A^2 + B^2 + \mathrm{i}\,A)\,K_3(x,x), \tag{6.2.17}$$

where

$$\rho(x) = \begin{pmatrix}1\\0\end{pmatrix} + 2\mathrm{i}K_4(x,x). \tag{6.2.18}$$

The quantities A and B depend only on the variable x (t is the representation parameter).

Let us show that solving eqs. (6.2.12)–(6.2.18) is the Goursat problem (the characteristic Cauchy problem [18, 9]).

Note first that eq. (6.2.13) for $y = x$ and eq. (6.2.17) are reduced to a system of ordinary differential equations for computing the functions $\rho(x)$ and $K_3(x,x)$. Its solution with the initial conditions

$$\rho(-\infty) = \begin{pmatrix}1\\0\end{pmatrix}, \quad K_3(-\infty,\ -\infty) = 0$$

exists and, according to (6.2.16), uniquely defines $K_1(x,x)$ and $K_2(x,x)$.

It is easy to ascertain that eqs. (6.2.12)–(6.2.18) are a system of hyperbolic type equations and straight lines

$$\zeta \equiv x + y = \text{const}, \quad \eta \equiv y - x = \text{const}$$

are its double characteristics. The values of $K_1(\zeta, 0)$ and $K_2(\zeta, 0)$ are given on the characteristic $\eta = 0$, and the values of K_2 and K_4 are set for $y = -\infty$. Therefore the proof of the existence and uniqueness of representation (6.2.8) can be based on iterations of integral Volterra-type equations that are equivalent to system (6.2.12)–(6.2.18). This makes it possible to outline constraints on the functions A, B, and $\mathbf{l}$, under which the kernels are unique and continuous.

Ultimately, the number of equations for the direct scattering problem shrinks to five equations for five scalar functions n_i to be conveniently defined by the following formulas:

$$K_a = \begin{pmatrix} -1 \\ (l_3 - 1)l_-^{-1} \end{pmatrix} n_a, \quad a = 1, 2;$$

$$K_b = \begin{pmatrix} 1 \\ (1 - l_3)l_-^{-1} \end{pmatrix} n_b, \quad b = 3, 4;$$

$$\rho = \begin{pmatrix} 1 \\ (1 - l_3)l_-^{-1} \end{pmatrix} n_5,$$

where $l_- = l_1 - \mathrm{i}\, l_2$. The integral representation for the solution of $\Psi(x, \lambda)$ is found in a similar way.

Let us look into the analytic properties of Jost functions. The structure of representations (6.2.8) and (6.2.9) suggests that the functions $\varphi(x, \lambda) \exp(\mathrm{i}\Lambda x/2)$ and $\Psi(x, \lambda) \exp(\mathrm{i}\Lambda x/2)$ can be analytically continued with respect to the spectral parameter λ in the domain where the integral operators are contractive. In fact, the domain of analyticity is determined by the behavior of the function $\exp[\mathrm{i}\Lambda(x-y)/2]$ as $|x-y| \to \infty$. Since

$$\mathrm{Im}\Lambda(\lambda) = \frac{\cosh^2 \rho + \cos^2 \varphi}{(\cosh^2 \rho - \cos^2 \varphi)^2} \sinh \rho \, \sin \varphi \quad (\lambda = \exp(\rho + \mathrm{i}\varphi), \quad \rho, \varphi \in R),$$

the functions $\Psi(x, \lambda) \exp(\mathrm{i}\Lambda x/2)$ and $\bar{\varphi}(x, \lambda) \exp(-\mathrm{i}\Lambda x/2)$ admit an analytic continuation from the contour

$$\Gamma = \{\lambda \mid \varphi = 0, \pi \ (-\infty < \rho < \infty); \ \rho = 0 \ (0 < \varphi < 2\pi)\}$$

in the domain

$$D_2 = \{\lambda \mid \rho < 0, \ 0 < \varphi < \pi; \ \rho > 0, \ \pi < \varphi < 2\pi\}$$

(see Figure 6.1), whereas the functions $\bar{\Psi}(x, \lambda) \exp(-\mathrm{i}\Lambda x/2)$ and $\varphi(x, \lambda) \exp(\mathrm{i}\Lambda x/2)$ can be continued in the domain

$$D_1 = \{\lambda \mid \rho > 0, \ 0 < \varphi < \pi; \ \rho < 0, \ \pi < \varphi < 2\pi\}.$$

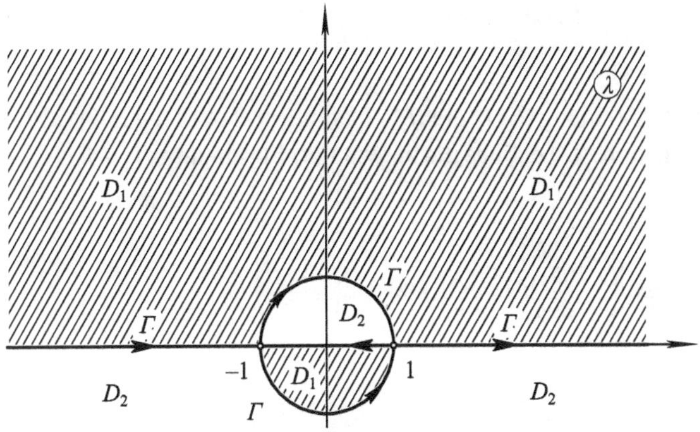

Figure 6.1: Domains of analyticity: D_1 for the functions $\varphi(\lambda)$, $\bar{\Psi}(\lambda)$, and $a(\lambda)$ and D_2 for the functions $\bar{\varphi}(\lambda)$, $\Psi(\lambda)$, and $\bar{a}(\lambda)$ inside the complex λ-plane.

From the connection between the elements of the transition matrix and the Jost solutions

$$a(\lambda) = W(\varphi, \bar{\Psi}), \quad b(\lambda) = -W(\varphi, \Psi) \tag{6.2.19}$$

it follows that the function $a(\lambda)$ is analytic in the domain D_1. Its zeros constitute a set $\{\lambda_n\}$ $(n = 1, 2, \dots, 2N)$. The latter is invariant to inversion: $\lambda_k = \lambda_{k+N}^{-1}$, $k = 1, 2, \dots, N$. The zeros $\{\lambda_n\}$ coincide with the discrete spectra of system (6.2.1), and

$$\bar{\Psi}(\lambda_n) = \bar{b}_n \varphi(\lambda_n) \quad (\bar{b}_{k+N} = -\bar{b}_k). \tag{6.2.20}$$

This concludes our discussion of the direct scattering method. Formulas (6.2.8), (6.2.9), (6.2.12)–(6.2.18), (6.2.19), and (6.2.20) give a constructive way to determine the scattering data

$$r(\lambda) = \frac{b(\lambda)}{a(\lambda)} \quad (\lambda \in \Gamma), \quad \lambda_n, \ \bar{b}_n$$

using the preassigned initial fields $l_i(x, t = 0)$ and $\partial_t l_i(x, t = 0)$.

Next, we discuss the inverse scattering problem, i. e., recovering the matrix $V(x, t, \lambda)$ from scattering data. Solving the inverse problem relies on a system of linear integral equations. It can be greatly simplified if an inverse transformation to representations (6.2.8) and (6.2.9) is found or, in other words, if we succeed in expressing the kernels of transformation operators in terms of Jost functions. Let us solve this problem first.

The function $\Lambda(\lambda)$ performs a four-sheeted covering $\Gamma \to R$. It is invariant under inversion of $q_1: \lambda \to 1/\lambda$ and fractional-linear transformation of $q_2: \lambda \to \mathrm{i}(\lambda + \mathrm{i})/(\lambda - \mathrm{i})$ to map the circle $|\lambda| = 1$ to the real R-axis. This fundamental property reduces the integral

$$\int_\Gamma f(\lambda)\exp(\pm i\Lambda x/2)d\lambda$$

with an arbitrary function $f(\lambda)$ along the contour Γ to an integral over the interval $I = (-1 < \lambda < 1)$, where the function $\Lambda(\lambda)$ defines the homeomorphic mapping $I \to R$ $(-\infty < \Lambda < \infty, d\Lambda/d\lambda < 0, \lambda \in I)$:

$$\int_\Gamma f(\lambda)\exp\left(\pm\frac{i\Lambda x}{2}\right)d\lambda = -\int_{-1}^{1}\bar{f}(\lambda)\exp\left(\pm\frac{i\Lambda x}{2}\right)d\lambda,$$

where

$$\bar{f}(\lambda) = f(\lambda) + \frac{2}{(\lambda+i)^2}f\left(\frac{i-\lambda}{i+\lambda}i\right) - \frac{1}{\lambda^2}f\left(\frac{1}{\lambda}\right) + \frac{2}{(\lambda-i)^2}f\left(\frac{\lambda+i}{\lambda-i}i\right).$$

An analog of the inverse Fourier transform can be constructed after having found an analog of the delta function. Solving the appropriate functional equation and using expressions for γ, δ, and Λ in terms of the parameter λ, we obtain useful relations

$$\int_\Gamma q(\lambda)\exp\left(\pm\frac{i\Lambda x}{2}\right)d\lambda = 0,$$

$$q(\lambda) = (\gamma\delta^2,\ \gamma^3\delta,\ \gamma\delta,\ \gamma^3,\ \gamma^2,\ \delta^2,\ \delta); \tag{6.2.21}$$

$$\int_\Gamma q(\lambda)\exp\left(\pm\frac{i\Lambda x}{2}\right)d\lambda = 8\pi\delta(x),$$

$$q(\lambda) = (-\gamma^2\delta,\ \gamma^4,\ -\gamma^2\delta^2,\ \gamma^4\delta,\ -\delta^3,\ -\delta^3\gamma^2,\ \gamma^4\delta^2). \tag{6.2.22}$$

To derive the equations of the inverse scattering problem, we express the field $\bar{\Psi}(x,\lambda)$ from eq. (6.2.4) through $\varphi(x,\lambda)$ and $\bar{\varphi}(x,\lambda)$ and divide the resulting equation by the coefficient $a(\lambda)$. As a result, we get

$$\frac{\bar{\Psi}(x,\lambda)}{a(\lambda)} = -\frac{\bar{b}(\lambda)}{a(\lambda)}\varphi(x,\lambda) + \bar{\varphi}(x,\lambda).$$

Subtract from the left- and right-hand sides of this equation the expression

$$\begin{pmatrix}1\\0\end{pmatrix}\exp\left(-\frac{i\Lambda}{2}x\right),$$

multiply the result by

$$\exp\left(-\frac{i\Lambda}{2}z\right) \quad (z < x),$$

and then sequentially by $\gamma\delta$, γ^2, δ, and $\gamma^2\delta$. Next, we integrate the equalities obtained over a contour Γ_1 that encloses the domain D_1 and consists of the contour Γ (with the exception of a quarter of circles of infinitely small radius around the singular points $\lambda = \pm 1$) and an infinitely distant contour, as in Figure 6.1. The analyticity of the functions $\bar{\Psi}(x,\lambda)\exp(-\mathrm{i}\Lambda x/2)$ and $a(\lambda)$ inside the domain D_2 and the relation

$$\frac{\bar{\Psi}(x,\lambda)}{a(\lambda)} \to \begin{pmatrix}1\\0\end{pmatrix}\exp\left(-\frac{\mathrm{i}\Lambda x}{2}\right) \quad \text{as} \quad |\lambda| \to \infty$$

allow us to apply the Jordan lemma and the Cauchy residue theorem to integrate the left-hand side of each of the equalities. Each result is equal to the product of $2\pi\mathrm{i}$ and the sum of the residues of the integrand at the zeros of the function $a(\lambda)$. Integrating the right-hand sides of the equalities exploits formulas (6.2.21) and (6.2.22) and reductions (6.2.7) and (6.2.20). Finally, we can compose a system of integral equations of the inverse scattering problem:

$$\int_{-\infty}^{x} \mathrm{d}s\,[K_2(x,s)F(\gamma\delta^2, s+y) + K_3(x,s)F(\gamma^3\delta, s+y)$$
$$+ K_4(x,s)F(\Lambda\gamma\delta, s+y)] + \begin{pmatrix}1\\0\end{pmatrix}F(\gamma\delta, x+y) + \bar{K}_1(x,y) = 0; \tag{6.2.23}$$

$$\int_{-\infty}^{x} \mathrm{d}s[K_1(x,s)F(\gamma^3, s+y) + K_3(x,s)F(\gamma^4, s+y)$$
$$+ K_4(x,s)F(\Lambda\gamma^2, s+y)]$$
$$+ \begin{pmatrix}1\\0\end{pmatrix}F(\gamma^2, x+y) + \bar{K}_2(x,y) - \bar{K}_3(x,y) = 0; \tag{6.2.24}$$

$$\int_{-\infty}^{x} \mathrm{d}s\,[K_1(x,s)F(\gamma\delta, s+y) + K_2(x,s)F(\delta^2, s+y)]$$
$$+ \bar{K}_3(x,y) = 0; \tag{6.2.25}$$

$$\int_{-\infty}^{x} \mathrm{d}s[K_1(x,s)F(\gamma^3\delta, s+y) + K_2(x,s)F(\gamma^2\delta^2, s+y)]$$
$$+ \bar{K}_2(x,y) - \bar{K}_3(x,y) + 2\mathrm{i}\,\partial_y \bar{K}_4(x,y) = 0. \tag{6.2.26}$$

They tie the kernels

$$K_i(x,y), \quad \bar{K}_i(x,y) = -\mathrm{i}\sigma_2 K_i^*(x,y), \quad i = 1,2,3,4,$$

with the kernels $F(\xi(\lambda), x)$:

$$F(\xi(\lambda),x)=\frac{\mathrm{i}}{4}\sum_n\frac{\bar b_n\xi_n}{a_n}\exp\left(-\frac{\mathrm{i}\Lambda(\lambda_n)}{2}x\right)$$
$$+\frac{1}{8\pi}\int_\Gamma\frac{\bar b(\lambda)}{a(\lambda)}\xi(\lambda)\exp\left(-\frac{\mathrm{i}\Lambda(\lambda)}{2}x\right)\mathrm{d}\lambda,\qquad(6.2.27)$$

where $a_n=\partial_\lambda a(\lambda)|_{\lambda\to\lambda_n}$. The analysis shows that the conditions $\bar b(\lambda=1)=\bar b(\lambda=-1)=0$ (see (6.2.6) and (6.2.7)) prevent eqs. (6.2.23)–(6.2.27) from containing singular terms.

It should be emphasized that the simplicity of deriving the equations lies in the existence of the domain I as a coset-space: $I=\Gamma/G$ is the main bundle of the manifold Γ regarding the discrete transformation group G with generators q_1 and q_2. This circumstance makes it possible to apply the Fourier transform to R and brings to integral equations of the inverse scattering problem.

The evolutionary dependence of the scattering data is given by simple equations

$$\partial_t a=\partial_t\lambda_n=0,\quad \partial_t\bar b(\lambda)=\mathrm{i}y^2(\lambda)\bar b(\lambda),\quad \partial_t\bar b_n=\mathrm{i}y^2(\lambda_n)\bar b_n.\qquad(6.2.28)$$

The initial condition

$$\mathbf{M}_1(x,t=0),\quad \mathbf{M}_2(x,t=0),\quad \partial_t\mathbf{M}_1(x,t)|_{t=0},\quad \partial_t\mathbf{M}_2(x,t)|_{t=0}$$

sets the scattering data at the initial time $t=0$ as a result of solving the direct scattering problem. Equations (6.2.28) guarantee the time-dependence of these data, whereas formulas (6.2.27) provide for finding the kernels $F(\xi(\lambda),x,t)$. The kernels $K_i(x,y,t)$ are the solution of the Gel'fand–Levitan–Marchenko-type equations (6.2.23)–(6.2.26), and formula (6.2.11) determines the explicit form of $\mathbf{l}(x,t)$.

Multisoliton solutions to the ferrimagnet equations match the case $\bar b(\lambda)=b(\lambda)\equiv 0$ and will be discussed in the next section.

6.3 Solitons in a ferrimagnet

What you are looking for, you will find only by searching everything.

Booba's law

The inverse scattering method makes it possible to completely integrate a ferrimagnet model with equilibrium distributions of magnetization at infinity. However, it is not the only way to obtain exact solutions. For physical applications, solutions with different asymptotics are often of interest. Although the above method is formally applicable in this case, solving the direct and inverse scattering problems is cumbersome due to the lack of an analog of the Fourier transform and triangular representation. Moreover, difficulties met in determining the λ-dependencies of the transformation kernels also make themselves felt. As a modified scheme, the "dressing" method using the Riemann problem has the necessary versatility and efficiency. To apply this scheme to our case, we introduce new fundamental solution matrices Ψ_1 and Ψ_2:

$$(\Psi_1^{-1})_{\alpha 1} = a^{-1}(\lambda)\varphi_\alpha(\lambda)\exp\left(\frac{i\Lambda}{2}x\right), \quad (\Psi_1^{-1})_{\alpha 2} = a^{-1}(\lambda)\bar{\Psi}_\alpha \exp\left(-\frac{i\Lambda}{2}x\right);$$
$$(\Psi_2)_{\alpha 1} = \Psi_\alpha(\lambda)\exp\left(\frac{i\Lambda}{2}x\right), \quad (\Psi_2)_{\alpha 2} = \bar{\varphi}_\alpha(\lambda)\exp\left(-\frac{i\Lambda}{2}x\right).$$

From eqs. (6.2.3) and (6.2.4) it follows that on the contour Γ, they satisfy the conditions

$$\Psi_1\Psi_2 = T_0(\lambda)G_0(\lambda)T_0^{-1}(\lambda) \equiv G(\lambda); \quad G_0(\lambda) = \begin{pmatrix} 1 & \bar{b}_0(\lambda) \\ -b_0(\lambda) & 1 \end{pmatrix}; \tag{6.3.1}$$
$$\Psi_2^\dagger(\lambda^*) = \Psi_1^{-1}(\lambda) \quad (\lambda \in \Gamma),$$

where $b_0(\lambda) = b(\lambda, t = 0)$, $\bar{b}_0(\lambda) = \bar{b}(\lambda, t = 0)$, and

$$T_0(\lambda) = \exp\left[\frac{i}{2}(-\Lambda x + \gamma^2 t)\sigma_3\right]$$

is a particular solution of eqs. (6.1.10) for $\mathbf{l}_0 = (0, 0, 1)$. Then solving the inverse scattering problem is reduced to solving the Riemann problem, i. e., to finding the matrices Ψ_1 and Ψ_2. The latter are analytic in the domains D_1 and D_2 and satisfy the conjugation condition (6.3.1) on the contour Γ. This scenario allows for generalizations and leads to a method for constructing new solutions to eqs. (6.1.9) and (6.1.10) from an arbitrary particular solution T_0, U_0, and V_0 to these equations.

The matrices T_1, T_2, U, and V of the new solution are determined using the gauge transformation (Darboux transformation) carried out by the matrices Ψ_1 and Ψ_2:

$$T_1 = \Psi_1^{-1}(\lambda)\, T_0(\lambda), \quad T_2 = \Psi_2(\lambda)\, T_0(\lambda);$$
$$U = (\partial_t \Psi_2 + \Psi_2 U_0)\Psi_2^{-1} = (\partial_t \Psi_1^{-1} + \Psi_1^{-1} U_0)\Psi_1; \tag{6.3.2}$$
$$V = (\partial_x \Psi_2 + \Psi_2 V_0)\Psi_2^{-1} = (\partial_x \Psi_1^{-1} + \Psi_1^{-1} V_0)\Psi_1. \tag{6.3.3}$$

The matrix functions Ψ_1 and Ψ_2 are solutions to the Riemann problem (6.3.1) on the contour Γ with the preshaped matrix $G_0(\lambda)$ and the matrix T_0 corresponding to a particular solution $\mathbf{l}_0(x, t)$.

The algebraic structure of a $U - V$ pair makes it possible to canonically normalize the functions of the Riemann problem:

$$\Psi_a(\lambda = \infty) = I \quad (a = 1, 2).$$

Reduction (6.2.5) can be written as

$$\Psi_2(\lambda^{-1}) = \hat{l}\,\Psi_2(\lambda)\hat{l}_0 \equiv \Psi_2(\lambda = 0)\hat{l}_0\Psi_2(\lambda)\hat{l}_0.$$

The choice of $G_0(\lambda) = I$ is responsible for soliton solutions. Then $|a(\lambda)| = 1$ on the contour Γ, and the function $a(\lambda) \equiv a_s(\lambda)$ is defined by its zeros and poles:

$$a_s(\lambda) = \prod_{i=1}^{N}\left(\frac{\lambda - \lambda_i}{\lambda - \lambda_i^*}\right)\left(\frac{\lambda - \lambda_i^{-1}}{\lambda - (\lambda_i^*)^{-1}}\right). \tag{6.3.4}$$

Let us construct the matrix functions $\Psi_1(\lambda)$ and $\Psi_2(\lambda)$ that are meromorphic in the complex λ-plane and degenerate at zeros and poles of the coefficient $a(\lambda)$, respectively:

$$\det\Psi_1(\lambda) = a(\lambda), \quad \det\Psi_2(\lambda) = a^*(\lambda^*).$$

For the "seeding" solution $\mathbf{l}_0 = (0, 0, 1)$ ($\hat{l}_0 = \sigma_3$), we perform a procedure similar to that described in Chapter 3 to find

$$\begin{aligned}(\Psi_1)_{\alpha\beta} &= \delta_{\alpha\beta} - \sum_{i,j=1}^{2N}\frac{(m_i)_\alpha(m_j^*)_\beta}{\lambda - \lambda_i^*}\frac{\partial\ln\det\|a\|}{\partial a_{ji}},\\ \Psi_2(\lambda) &= \Psi_1^\dagger(\lambda^*),\end{aligned} \tag{6.3.5}$$

where $\|a\|$ is the matrix with elements $a_{ij} = (\mathbf{m}_i^* \cdot \mathbf{m}_j)(\lambda_i - \lambda_j^*)^{-1}$,

$$\mathbf{m}_k = T_0(\lambda_k^*)\mathbf{c}_k; \tag{6.3.6}$$

$$\lambda_{k+N} = \lambda_k^{-1}, \quad \mathbf{m}_{k+N} = \hat{l}_0\,\mathbf{m}_k \quad (k = 1, 2, \ldots, N), \tag{6.3.7}$$

where $\mathbf{c}_k$ are arbitrary constant two-dimensional vectors. The dependence (6.3.6) on the "seeding" solution to system (6.2.1),

$$T_0(\lambda) = \exp\left[\frac{i\lambda\sigma_3}{(1-\lambda^2)^2}(2\lambda t + (1+\lambda^2)x)\right],$$

ensures the absence of "extra" poles in formulas (6.3.2) and (6.3.3) at the points $\lambda = \lambda_i$, λ_i^*.

In the general case, where $G_0(\lambda) \neq I$, the matrix functions $\Psi_1(\lambda)$ and $\Psi_2(\lambda)$ are factorized as follows:

$$\Psi_1(\lambda) = \Psi_1^{(0)}(\lambda)\varphi_1(\lambda), \quad \Psi_2(\lambda) = \varphi_2(\lambda)\Psi_2^{(0)}(\lambda).$$

Here formulas (6.3.5)–(6.3.7) compute $\Psi_{1,2}^{(0)}(\lambda)$, and $\varphi_1(\lambda)$ and $\varphi_2(\lambda)$ are analytic matrix functions in the domains D_1 and D_2. Solving the Riemann problem is simplified by the reduction

$$\Psi_2^\dagger(\lambda^*) = \Psi_1(\lambda) \quad (\lambda \in \Gamma)$$

and the relation

$$\Psi_1^{-1}(\lambda) = \sigma_2 \Psi_1^{\mathrm{T}}(\lambda)\sigma_2 [\det \Psi_1(\lambda)]^{-1}.$$

As in Section 2.9.1, we apply the Cauchy theorem to derive a closed integral equation of the Fredholm type ($\lambda \in D_2$) for $\varphi_2(\lambda)$:

$$\varphi_2(\lambda) = I + \frac{1}{2\pi \mathrm{i}} \int_\Gamma \frac{\sigma_2 \varphi_2^*(\lambda')\sigma_2 [I - G(\lambda')]}{(\lambda' - \lambda) \det \varphi_1(\lambda')} \mathrm{d}\lambda', \tag{6.3.8}$$

where

$$G(\lambda) = [\Psi_1^{(0)}(\lambda)]^{-1} T_0(\lambda) G_0(\lambda) T_0^{-1}(\lambda) [\Psi_2^{(0)}(\lambda)]^{-1}.$$

The function $\det \varphi_1(\lambda)$ is expressed in terms of $\det G_0(\lambda)$. Indeed, from eq. (6.3.1) it follows that $|\det \varphi_1(\lambda)| = \sqrt{\det G_0(\lambda)}$. The phase of the function $\det \varphi_1(\lambda)$ is determined through its modulus using a dispersion relation for $\ln \det \varphi_1(\lambda)$. Formulas (6.3.5) and (6.3.8) allow us to construct $\Psi_a(x, t)$ lying on the particular solution $\mathbf{l}_0(x, t)$ to gain two-sublattice soliton $\mathbf{l}(x, t)$:

$$\hat{l} = \Psi_2(\lambda = 0)\, \hat{l}_0. \tag{6.3.9}$$

The most difficult part of this procedure is to construct the matrix function $T_0(x, t, \lambda)$. Nevertheless, it is simply solved when $\mathbf{l}_0(x, t)$ depends on only one variable t (or x). Then eqs. (6.1.10) can be merely integrated using a gauge transformation that diagonalizes the matrix V (or U). Note that the "dressing" procedure outlined here classifies many exact solutions but does not exhaust the variety of all solutions. The reason is that for some "seeding" solutions, the analytical properties of the matrix functions $\Psi_1(\lambda)$ and $\Psi_2(\lambda)$ are defined on the Riemann surfaces of genus $g \geq 1$. In this context, the matrix Riemann problem requires a separate discussion. One of these problems is discussed in Chapter 3, and the other in Chapter 9.

Let us present a two-sublattice soliton solution to model (6.1.8), obtained by "dressing" a particular solution $\mathbf{l}_0 = (0,\ 0,\ 1)$:

$$\begin{gathered}
l_1 + \mathrm{i}\, l_2 = -\frac{e^{-\mathrm{i}S}}{\tau \sin\varepsilon\ \sinh\rho} [\sin(\varepsilon + \mathrm{i}\rho)\, e^{y} - \sin(\varepsilon - \mathrm{i}\rho)\, e^{-y}], \\
l_3 = 1 - \frac{2}{\tau}, \quad \tau = \frac{\cosh^2 y}{\sin^2\varepsilon} + \frac{\sinh^2 y}{\sinh^2\rho}, \\
S + \mathrm{i}y = \frac{\beta}{\sinh^2(\rho - \mathrm{i}\varepsilon)} [x \cosh(\rho - \mathrm{i}\varepsilon) + c\, t],
\end{gathered} \tag{6.3.10}$$

where $\rho \neq 0, \varepsilon \neq 0$, and π are arbitrary real constants ($\lambda_1 = \exp(\rho + i\varepsilon)$). Here we have returned to the original dimensional variables and omitted the constants associated with the invariance of eq. (6.1.8) regarding space-time translations.

Solution (6.3.10) describes a precessional soliton. In a coordinate system moving at the soliton speed

$$V = -\frac{2c\,\cos\varepsilon\cosh\rho}{\cos^2\varepsilon + \cosh^2\rho},$$

the vectors $\mathbf{l}$ and $\mathbf{m}$ perform a nonuniform procession around the axis $\mathbf{n} = (0, 0, 1)$ with frequency $\omega = \partial_t S + V\partial_x S$. The projection l_3 takes the value $\cos 2\varepsilon$ at the center of the soliton and tends to the equilibrium value $l_3 = 1$ at infinity. The behavior of the combination $l_\perp = \sqrt{l_1^2 + l_2^2}$ is affected by the magnitude of ε (see Figure 6.2).

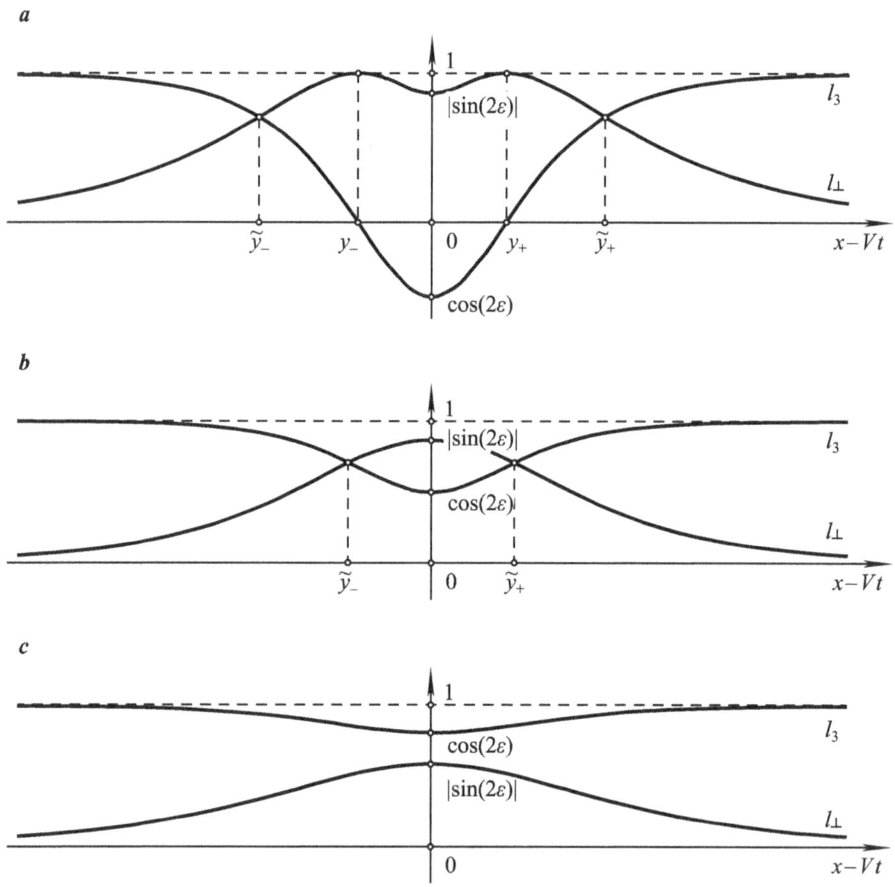

Figure 6.2: Components of the antiferromagnetism vector: *a*) for $\sin^2\varepsilon > 1/2$, *b*) for $(2-\sqrt{2})/4 < \sin^2\varepsilon \leq 1/2$, *c*) for $\sin^2\varepsilon \leq (2-\sqrt{2})/4$.

In Figure 6.2*a*,*b*, we see that

$$y_\pm = \pm\text{arsinh}\sqrt{\frac{-\sinh^2\rho\,\cos 2\varepsilon}{\sinh^2\rho + \sin^2\varepsilon}}, \quad \tilde{y}_\pm = \pm\text{arsinh}\sqrt{\frac{\sinh^2\rho\,[(4+2\sqrt{2})\sin^2\varepsilon - 1]}{\sinh^2\rho + \sin^2\varepsilon}},$$

where the values of ε such that the subduplicate expressions are nonnegative.

Let $d = (2\,\partial_x y)^{-1}$ be the width of a soliton. Then we can express it through its physical parameters, the precession's speed and frequency:

$$d = \frac{1 - V^2c^{-2}}{\beta\sqrt{1 - V^2c^{-2} - (1 + 2\omega\beta^{-1}c^{-1})^2}}. \tag{6.3.11}$$

From eq. (6.3.11) it follows that the localized excitation (6.3.9) exists only under the condition

$$V^2 < c^2\left[1 - \left(1 + \frac{2\omega}{\beta c}\right)^2\right].$$

It is easy to show that the region boundary of existence of the soliton obeys the dispersion law of linear quasiacoustic magnons in a frame of reference moving with the group velocity.

With an immobile soliton (for $\varepsilon = \pi/2$), we have

$$\tau = \cosh^2 y + \frac{\sinh^2 y}{\sinh^2\rho}, \quad y = \frac{2\beta x \sinh\rho}{\cosh^2\rho}, \quad S = -\frac{\beta c\, t}{\cosh^2\rho}.$$

The sublattice magnetization projections $\mathbf{M}_1$ and $\mathbf{M}_2$ onto the Ox_3-axis are shown in Figure 6.3. The behavior of the magnetization of each of the sublattices qualitatively coincides with the solution for an isotropic ferromagnet. Interestingly, the nonequivalence of the sublattices of the ferrimagnet provokes a magnetization reversal region at the center of the soliton.

In the limit as $\rho \to 0$, solution (6.3.10) degenerates into a precessing algebraic soliton that collates to a power-behaved x-dependence rather than an exponential one:

$$l_3 = 1 - \frac{2}{\tau}, \quad l_1 + \mathrm{i}\,l_2 = -\frac{2}{\tau}\cot\varepsilon\, e^{-\mathrm{i}S}[y\sin\varepsilon + \mathrm{i}\cos\varepsilon],$$
$$\tau = y^2 + \frac{1}{\sin^2\varepsilon}, \quad S = -\frac{\beta}{\sin^2\varepsilon}[x\cos\varepsilon + c\,t],$$
$$y = \frac{\beta}{\sin^4\varepsilon}[x\sin\varepsilon(1 + \cos^2\varepsilon) + c\,t\sin 2\varepsilon].$$

In this case, the plot shapes for l_3 and $l_\perp$ in Figure 6.2 do not change qualitatively. The general N-soliton solution of model (6.1.8) is determined by setting $2N$ zeros of the func-

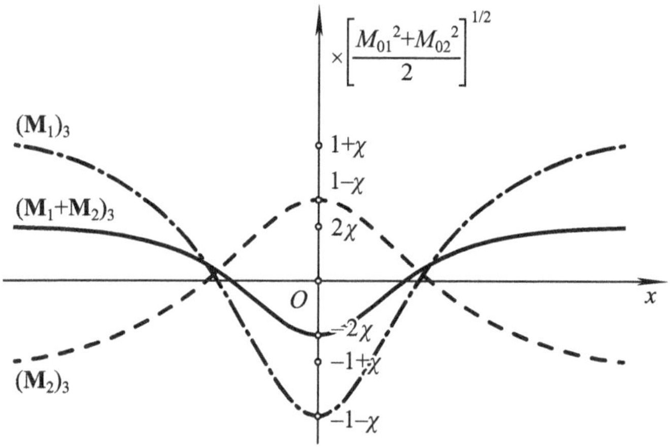

Figure 6.3: Projections onto the Ox_3-axis of sublattice magnetizations; $\sinh\rho > 1$, $\varepsilon = \pi/2$, $\chi = (M_{01}^2 - M_{02}^2)[2(M_{01}^2 + M_{02}^2)]^{-1} > 0$.

tion $a(\lambda)$ and constant vectors $\mathbf{c}_k$. This solution describes a sequence of pair collisions of N localized waves of the type (6.3.10).

Further, we discuss the validity of the approximations used in passing from eq. (6.1.2) to eqs. (6.1.4) and (6.1.5). The long-wavelength approximation limits the width d of the soliton:

$$\frac{a}{d} = a|\beta\ \sin\varepsilon\ \sinh\rho|\left(\frac{\cosh^2\rho + \cos^2\varepsilon}{\cosh^2\rho - \cos^2\varepsilon}\right) \ll 1.$$

Direct calculations show that the condition $\mathbf{m}^2 \ll \mathbf{l}^2$ is equivalent to the inequality

$$\frac{|M_{01}^2 - M_{02}^2|}{2\,(M_{01}^2 + M_{02}^2)}\left[1 + \left(\frac{\cosh^2\rho\ \sin^2\varepsilon - \sinh^2\rho\ \cos^2\varepsilon}{\cosh^2\rho - \cos^2\varepsilon}\right)^2\right] \ll 1.$$

Both inequalities are valid, for example, for $\sinh\rho > 1$.

6.4 Elementary excitations in a ferrimagnet

The joys of a naturalist: lifting nature's skirts.

Jean Rostand

The dynamics of a ferrimagnet can be completely described based on the fact that system (6.1.8) complies with an infinite set of conservation laws. This section discloses a relationship between the most important of them, the energy, momentum, and total magnetic moment and scattering data. This makes it possible to fully analyze the spectrum and dispersion laws of all nonlinear excitations in a ferrimagnet.

For deriving the conservation laws, we exploit asymptotic trace formulas. We express the first component of the Jost function φ as

$$\varphi_1 = \exp\left(-\frac{\mathrm{i}\Lambda}{2} x + \int_{-\infty}^{x} \chi(y,\lambda)\,\mathrm{d}y \right).$$

As a consequence of eq. (6.2.1), the function $\chi(x,\lambda)$ satisfies the Riccati-type equation

$$-\frac{\Lambda^2}{4} - \mathrm{i}\chi\Lambda + \chi^2 + \det V + \left(\frac{\mathrm{i}\Lambda}{2} - \chi\right)\frac{\partial_x V_{12}}{V_{12}} - \partial_x\left(\frac{V_{11}}{V_{12}}\right)V_{12} + \partial_x\chi = 0. \tag{6.4.1}$$

Since $\varphi_1 \exp(\mathrm{i}\Lambda x/2)$ tends to the time-independent function $a(\lambda)$ as $x \to +\infty$ (see (6.2.2) and (6.2.4)), the quantity $\chi(\lambda, x)$ is the density of the integrals of motion:

$$\ln a(\lambda) = \int_{-\infty}^{+\infty} \chi(x,\lambda)\,\mathrm{d}x. \tag{6.4.2}$$

Expanding $\chi(x,\lambda)$ into asymptotic power series λ^{-1}, $\lambda \pm 1$, and using eq. (6.4.1), we obtain three discrete series of integrals of motion. The parameterization $\mathbf{l} = (\sin\theta\cos\Psi, \sin\theta\sin\Psi, \cos\theta)$ yields some of them:

$$\lambda\chi(\lambda)|_{\lambda\to\infty} = \mathrm{i}(\partial_t\Psi\sin^2\theta + \cos\theta - 1); \tag{6.4.3}$$

$$\chi(\lambda = 1) = \frac{\mathrm{i}}{4}\left[(\partial_t\theta + \partial_x\theta)^2 + \sin^2\theta(\partial_t\Psi + \partial_x\Psi)^2 + 2(\cos\theta - 1)\partial_x\Psi\right]; \tag{6.4.4}$$

$$\chi(\lambda = -1) = -\frac{\mathrm{i}}{4}\left[(\partial_t\theta - \partial_x\theta)^2 + \sin^2\theta(\partial_t\Psi - \partial_x\Psi)^2 - 2(\cos\theta - 1)\partial_x\Psi\right]. \tag{6.4.5}$$

The equations of motion (6.1.8) correspond to the density of the Lagrange function

$$L = (\cos\theta - 1)\,\partial_t\Psi + \frac{1}{2}[(\partial_t\theta)^2 - (\partial_x\theta)^2 + (\partial_t\Psi)^2\sin^2\theta - (\partial_x\Psi)^2\sin^2\theta]. \tag{6.4.6}$$

The cyclic coordinate Ψ matches the canonical momentum density

$$p_\Psi = \frac{\partial L}{\partial(\partial_t\Psi)} = \partial_t\Psi\sin^2\theta + \cos\theta - 1$$

and the integral of motion

$$N = \int_{-\infty}^{+\infty} p_\Psi(x)\,\mathrm{d}x. \tag{6.4.7}$$

Implementing the standard procedure [14, 3], we find expressions for the energy E and field momentum P of a ferrimagnet in terms of the Lagrange function:

$$E = \int \mathrm{d}x \left[\frac{\partial L}{\partial(\partial_t \theta)} \partial_t \theta + \frac{\partial L}{\partial(\partial_t \Psi)} \partial_t \Psi - L \right]$$
$$= \frac{1}{2} \int \mathrm{d}x \, [(\partial_t \theta)^2 + (\partial_x \theta)^2 + ((\partial_t \Psi)^2 + (\partial_x \Psi)^2) \sin^2 \theta];$$
$$P = \int \mathrm{d}x \left[\frac{\partial L}{\partial(\partial_t \theta)} \partial_x \theta + \frac{\partial L}{\partial(\partial_t \Psi)} \partial_x \Psi \right]$$
$$= \int \mathrm{d}x [\partial_t \theta \, \partial_x \theta + \partial_t \Psi \, \partial_x \Psi + (\cos\theta - 1)\partial_x \Psi].$$

Let us express the energy E of a ferromagnetic, field momentum P, and the quantity N in terms of the scattering data. For this purpose, we introduce the function $\ln a_1(\lambda) \equiv \ln[a(\lambda)/a_s(\lambda)]$, which is analytic in the domain D_1, with the meromorphic (soliton) function $a_s(\lambda)$ defined by formula (6.3.4). Using the Cauchy theorem and dispersion relations (see Section 2.5), we get ($\lambda \in D_1$)

$$\ln a_1(\lambda) = \frac{1}{2\pi \mathrm{i}} \int_\Gamma \frac{\ln a_1(\lambda')}{\lambda' - \lambda} \mathrm{d}\lambda' = \frac{1}{\pi \mathrm{i}} \int_\Gamma \frac{\ln |a_1(\lambda')|}{\lambda' - \lambda} \mathrm{d}\lambda';$$
$$\ln a(\lambda) = \ln \prod_{i=1}^{N} \left(\frac{\lambda - \lambda_i}{\lambda - \lambda_i^*} \right) \left(\frac{\lambda - \lambda_i^{-1}}{\lambda - \lambda_i^{*-1}} \right)$$
$$- \frac{2}{\mathrm{i}\pi} \int_0^\pi \frac{\lambda \ln |a\,(\exp(\mathrm{i}\varphi))| \sin\varphi}{(\exp(-\mathrm{i}\varphi) - \lambda)(\exp(\mathrm{i}\varphi) - \lambda)} \mathrm{d}\varphi \tag{6.4.8}$$
$$- \frac{1}{\mathrm{i}\pi} \int_{-1}^{1} \ln|a(\lambda')| \left(\frac{1}{\lambda' - \lambda} - \frac{1}{\lambda'(1 - \lambda\lambda')} \right) \mathrm{d}\lambda'.$$

Then using formulas (6.4.2)–(6.4.8), we arrive at

$$N = -\mathrm{i}\lambda \ln a(\lambda)|_{\lambda \to \infty}$$
$$= - \int_{-1}^{1} N(\lambda)\, \mathrm{d}\lambda + \int_0^\pi N(\varphi)\, \mathrm{d}\varphi - 4 \sum_{i=1}^{N} \sin\varepsilon_i \, \sinh\rho_i; \tag{6.4.9}$$
$$E = \mathrm{i} \ln \frac{a(\lambda = -1)}{a(\lambda = 1)}$$
$$= \int_{-1}^{1} E_{\mathrm{a}}(\lambda) N(\lambda)\, \mathrm{d}\lambda + \int_0^\pi N(\varphi) E_{\mathrm{op}}(\varphi)\, \mathrm{d}\varphi + 4 \sum_{i=1}^{N} \arctan\left(\frac{\sin\varepsilon_i}{\sinh\rho_i} \right); \tag{6.4.10}$$
$$P = -\mathrm{i} \ln[a(\lambda = 1) a(\lambda = -1)]$$
$$= - \int_{-1}^{1} P_{\mathrm{a}}(\lambda) N(\lambda)\, \mathrm{d}\lambda + \int_0^\pi N(\varphi) P_{\mathrm{op}}(\varphi)\, \mathrm{d}\varphi + 4 \sum_{i=1}^{N} \arctan(\coth\rho_i \, \tan\varepsilon_i). \tag{6.4.11}$$

Here we have applied the parameterization $\lambda_k = \exp(\rho_k + i\varepsilon_k) \in D_1$ and introduced the notations

$$N(\lambda) = \frac{1}{\pi\lambda^2}(\lambda^2 - 1)\ln|a(\lambda)|, \quad N(\varphi) = \frac{2}{\pi}\sin\varphi \ln|a(\lambda = \exp(i\varphi))|,$$
$$E_a(\lambda) = \gamma^2(\lambda), \quad E_{op}(\varphi) = -\gamma^2(\lambda = \exp(i\varphi)) = \frac{1}{\sin^2\varphi},$$
$$P_a(\lambda) = \Lambda(\lambda), \quad P_{op}(\varphi) = \Lambda(\lambda = \exp(i\varphi)) = \frac{\cos\varphi}{\sin^2\varphi}.$$

Note that eqs. (6.2.6) and (6.2.7) lead to the relations

$$|a(\lambda)|^2 + |b(\lambda)|^2 = 1 \quad (\lambda \in R); \quad |a(\lambda)|^2 - |b(\lambda)|^2 = 1 \quad (\lambda = \exp(i\varphi)),$$

guaranteeing the positivity of $N(\lambda)$ and $N(\varphi)$.

To interpret the conservation laws (6.4.9)–(6.4.11), we look into weakly excited states of the model (6.1.8):

$$l_1 + i\, l_2 \approx \int_{-\infty}^{+\infty} \sum_{p=1,2} \xi_p(k)\exp(i\omega_p t + ikx)dk, \quad |\xi_p(k)| \ll 1; \qquad l_3 \approx 1 \tag{6.4.12}$$

with the dispersion equation $(\omega^2 - k^2)^2 - \omega^2 = 0$. The latter relies on the quasiacoustic (Goldstone) and quasioptical oscillation modes:

$$\omega_1(k) = \frac{1}{2}(\sqrt{1+4k^2} - 1); \quad \omega_2(k) = \frac{1}{2}(1 + \sqrt{1+4k^2}). \tag{6.4.13}$$

For the linear modes (6.4.12), the direct scattering problem involves the formula

$$b(\lambda) = \varphi_2 \exp\left(-\frac{i\Lambda(\lambda)}{2}x\right)\Big|_{x\to\infty}$$

to bring about to

$$b(\lambda) \approx \pi i \sum_{p=1,2} \xi_p(k)\exp(it\,\omega_p(k))[\gamma(\lambda)\,\omega_p(k) + (\delta(\lambda) - 1)\,k],$$

where $k = \Lambda(\lambda)$. Expressing λ through k, it is easy to verify that the integrals of motion (6.4.9)–(6.4.11) are contributed by only the acoustic oscillation mode for $\lambda \in [-1, 1]$ and by only the optical mode for $\lambda = \exp(i\varphi)$, $\varphi \in [0, \pi]$, respectively. Therefore the first two terms in these integrals are responsible for the additive contribution of the linear magnetization modes, the optical branch (with the magnon density $N(\varphi)$) and the acoustic branch (with the magnon density $N(\lambda)$), which satisfy the dispersion equations

$$E_a^2 + E_a - P_a^2 = 0; \quad E_{op}^2 - E_{op} - P_{op}^2 = 0.$$

The remaining summands in formulas (6.4.9)–(6.4.11) correspond to the contributions of solitons.

Next, we express a single-soliton energy E_s through its momentum P_s and the integral of motion

$$N_s = 4\sin\varepsilon\,\sinh\rho.$$

The physical meaning of this integral will become clear in the following discussion. As a result, we have

$$E_s = 4\arctan\left[\frac{N_s}{8\cos^2(P_s/4)}\left(\sqrt{1+\frac{16}{N_s^2}\sin^2(P_s/2)}-1\right)\right]. \tag{6.4.14}$$

In the limit as $P_s \to 0$, expression (6.4.14) modifies into the dispersion law for N_s linear Goldstone modes with a quasimomentum P_s/N_s (see (6.4.13)):

$$E_s \approx \frac{N_s}{2}\left(\sqrt{1+4(P_s/N_s)^2}-1\right).$$

Therefore a soliton of a ferrimagnet can be interpreted as a bound state of a large number of N_s quasiacoustic magnons. Simple calculations show that at finite values of P_s and N_s, the energy E_s/N_s per bound magnon is less than that of a free magnon at the same fixed momentum P_s/N_s per quasiparticle.

It should be emphasized that excitation of solitons may be caused by thermal fluctuations or a high-frequency magnetic field due to their "gapless" nature.

The periodic dependence of the soliton energy on its momentum values ranged in the interval of $0 \le P \le 2\pi$ encloses the soliton's physically different states. In the initial dimensional variables, the energy and momentum of the soliton contain the factors

$$E_0 = \frac{\beta}{2}(M_{01}^2 + M_{02}^2)(\alpha_1 + \alpha_2 - 2\alpha_3), \quad P_0 = \frac{E_0}{c}.$$

Therefore they are proportional to the difference in the magnetizations of the sublattices of a ferrimagnet (see definition (6.1.6)).

Further studies [10] demonstrated that the account for the energy of uniaxial anisotropy separates new types of solutions, domain walls. In addition, such a system offers two- and three-dimensional dynamic solitons with finite energy.

Thus the present section underlines the fact that the integrals of motion of strongly excited states of a ferrimagnet can be represented as additive contributions from three independent types of excitations, such as quasioptical and quasiacoustic magnons and solitons. Solitons can be interpreted as bound states of a large number of magnons.

We would like to highlight a certain universality of the integrable model (6.1.8). It describes not only the dynamics of a ferrimagnet but also, for example, nonlinear spin waves in liquid helium [20]. In the next section, we will apply this model to analyze two-dimensional magnetization waves in a Heisenberg ferromagnet.

6.5 Two-dimensional precession waves in an isotropic ferromagnet

I know what I'm running from, but I don't know what I'm looking for.

M. Montaigne

It is interesting and important that complex-valued scaling transformations of space-time coordinates save the integrability property of a nonlinear equation but change the type of the equation and the nature of its solutions. This section based on the results of our papers [6, 5] dives into how, by means of complex changes of variables, the equations of one-dimensional dynamics of a ferrimagnet can be facilitated to a model suitable for analyzing two-dimensional magnetization distributions in an isotropic ferromagnet.

Consider the structure and dynamics of magnetization distributions described by the Landau–Lifshitz equation

$$\partial_t \mathbf{S} = [\mathbf{S} \times (\partial_x^2 \mathbf{S} + \partial_y^2 \mathbf{S})], \quad \mathbf{S}^2 = 1. \tag{6.5.1}$$

Although the complete integrability of eq. (6.5.1) has not yet been established to date, the inverse scattering method has been successfully applied for describing the dynamics of axially symmetric distributions $\mathbf{S}(\mathbf{r}, t)$ [15] (see Chapter 2). The absence of magnetization waves localized in space with radial symmetry is proven. Any magnetization distribution localized at the initial moment of time and depending only on the distance from the symmetry axis spreads out over time.

In this section, we show that eq. (6.5.1) admits a wide class of exact solutions of the form

$$\mathbf{S} = \mathbf{S}(x, \tau), \quad \tau = y - \alpha t, \tag{6.5.2}$$

where α is a constant parameter. For $\alpha = 0$, distributions (6.5.2) describe two-dimensional instantons and nonlinear topological defects such as disclinations in isotropic ferro- and antiferromagnets [6, 5, 17]. In what follows, we confine ourselves to discussing two-dimensional precessional waves that retain their shape with time, in contrast to the axially symmetric waves, as in [15]. When precessional waves interact with each other, an analog of the asymptotic superposition principle is fulfilled.

The nonlinear dynamics of two-dimensional magnetization waves of the kind (6.5.2) is determined by the equation

$$\alpha \, \partial_\tau \mathbf{S} + [\mathbf{S} \times (\partial_x^2 \mathbf{S} + \partial_\tau^2 \mathbf{S})], \quad \mathbf{S}^2 = 1. \tag{6.5.3}$$

After the formal replacement

$$\tau = \frac{t}{\alpha}, \quad \tilde{x} = \frac{x}{\mathrm{i}\,\alpha}, \quad \mathbf{l} = \mathbf{S}, \quad \lambda = -\mathrm{i}\,\tilde{\lambda},$$

the two-dimensional dynamic equation (6.5.3) and its $U-V$ pair

$$\begin{aligned}&\partial_x\chi = V(\lambda)\chi, \quad \partial_\tau\chi = U(\lambda)\chi,\\&\partial_x U(\lambda) - \partial_\tau V(\lambda) + [U(\lambda), V(\lambda)] = 0\end{aligned} \tag{6.5.4}$$

can be deduced from eq. (6.1.8) for the one-dimensional dynamics of a ferrimagnet and the corresponding $U-V$ pair (6.1.11)–(6.1.14). Next, we omit the tilde icon above the new variables. The matrices $U(\lambda)$ and $V(\lambda)$ have the shape

$$\begin{aligned}V(\lambda) &= -\left(\gamma A + \delta B - \frac{\mathrm{i}\alpha}{2}\Lambda\hat{S}\right),\\U(\lambda) &= -\left(\delta A - \gamma B + \frac{\mathrm{i}\alpha\gamma^2}{2}\hat{S}\right),\end{aligned} \tag{6.5.5}$$

where

$$\begin{aligned}&A = \frac{\mathrm{i}}{2}[\mathbf{S}\times\partial_\tau\mathbf{S}]_a\sigma_a, \quad B = \frac{\mathrm{i}}{2}[\mathbf{S}\times\partial_x\mathbf{S}]_a\sigma_a,\\&\hat{S} = S_a\sigma_a \quad (a = 1, 2, 3).\end{aligned} \tag{6.5.6}$$

The coefficients γ, δ, and λ are functions of the spectral parameter λ:

$$\delta(\lambda) = \frac{2}{1+\lambda^2}, \quad \gamma(\lambda) = \frac{2\lambda}{1+\lambda^2}, \quad \Lambda(\lambda) = \frac{2\lambda(1-\lambda^2)}{(1+\lambda^2)^2}. \tag{6.5.7}$$

They satisfy the relations

$$\delta^2 - 2\delta + \gamma^2 = 0, \quad \Lambda = \gamma(\delta - 1).$$

The next step is to use the "dressing method" to integrate model (6.5.3). Representations (6.5.4) and (6.5.5)–(6.5.7) imply that the matrix $S_k\sigma_k$ can be expressed in terms of the matrix function $\chi(\lambda = 0)$:

$$\hat{S} \equiv S_k\sigma_k = \chi(\lambda = 0). \tag{6.5.8}$$

The fulfillment of the conditions $S_k^* = S_k$ and $\mathbf{S}^2 = 1$ guarantees additional restrictions on the choice of solutions $\chi(\lambda)$:

$$\begin{aligned}&\chi^\dagger(\lambda^*) = \chi^{-1}(\lambda), \quad \chi(\lambda = \infty) = I;\\&\chi(-\lambda^{-1}) = \hat{S}\chi(\lambda).\end{aligned} \tag{6.5.9}$$

It should be pointed out that the latter do not contradict systems (6.5.4) and (6.5.5)–(6.5.7).

Let us set forth the procedure for finding soliton solutions to model (6.5.3). Let $\mathbf{S}^{(0)}$ be a particular solution to model (6.5.3), and let $\chi_0(\lambda)$ be the corresponding solution to the auxiliary system (6.5.4). Now we represent a new solution to model (6.5.3) as follows:

$$\chi(\lambda) = \Psi_1^{-1}(\lambda)\chi_0(\lambda), \quad \chi(\lambda) = \Psi_2(\lambda)\chi_0(\lambda),$$

where the matrix functions $\Psi_{1,2}(\lambda)$ are meromorphic in the complex λ-plane. On the real λ-axis, they are connected by the conjugation condition

$$\Psi_1(\lambda)\Psi_2(\lambda) = I, \quad \Psi_2(\lambda) = \Psi_1^\dagger(\lambda).$$

In addition, from expressions (6.5.4) and (6.5.9) it follows that these functions satisfy the canonical normalization

$$\Psi_{1,2}(\lambda = \infty) = I$$

and the nonlinear reduction

$$\begin{aligned}\Psi_2(-\lambda^{-1}) &= \Psi_2(\lambda = 0)\hat{S}^{(0)}\Psi_2(\lambda)\hat{S}^{(0)},\\ \Psi_1(-\lambda^{-1}) &= \hat{S}^{(0)}\Psi_2(\lambda)\hat{S}^{(0)}\Psi_1(\lambda = 0).\end{aligned}$$

The same technique as in Chapter 3 makes it not difficult to find a general solution to the formulated Riemann problem:

$$\begin{aligned}(\Psi_1(\lambda))_{\alpha\beta} &= \delta_{\alpha\beta} - \sum_{i,j=1}^{2N} \frac{(m_i)_\alpha (m_j)^*_\beta}{\lambda - \lambda_i^*} \frac{\partial \ln \det \|D\|}{\partial D_{ji}}, \quad \alpha, \beta = 1, 2;\\ \Psi_2(\lambda) &= \Psi_1^\dagger(\lambda^*).\end{aligned} \tag{6.5.10}$$

Here $\|D\|$ is the matrix with elements

$$D_{ij} = (\mathbf{m}_i^* \cdot \mathbf{m}_j)(\lambda_i - \lambda_j^*)^{-1}, \quad \mathbf{m}_{k+N} = \hat{S}^{(0)}\mathbf{m}_k, \quad \mathbf{m}_k = \chi_0(\lambda_k^*)\mathbf{c}_k,$$

where $\lambda_{k+N} = -\lambda_k^{-1}$, with λ_k being complex numbers, and $\mathbf{c}_k$ are arbitrary constant complex vectors, $k = 1, 2, \ldots, N$.

According to identity (6.5.8), new exact solutions to model (6.5.3) are given by the relation

$$S_k\sigma_k = \Psi_1^\dagger(\lambda = 0)\hat{S}^{(0)}. \tag{6.5.11}$$

We analyze the one-soliton solution to the Landau–Lifshitz equations (6.5.3), obtained by "dressing" the trivial solution to these equations, $\mathbf{S}^{(0)} = (0, 0, 1)$ ($\hat{S}^{(0)} = \sigma_3$). In this case, we have

$$\chi_0(\lambda) = \exp\left[-\frac{\mathrm{i}\alpha\lambda\sigma_3}{(1+\lambda^2)^2}(2\lambda t - (1-\lambda^2)x)\right]\left(\frac{\lambda + \mathrm{i}\sigma_3}{\lambda + \mathrm{i}}\right).$$

Here we have introduced the λ-dependent multiplier to perform reductions (6.5.9). It is not essential for the calculation of the vectors $\mathbf{m}_k$ because it leads only to a change in arbitrary constants $\mathbf{c}_k$. Applying formulas (6.5.10) and (6.5.11), we find

$$\begin{aligned} S_1 + iS_2 &= -2\mu^{-1} \exp[i(k_1\xi + k_2\eta - \Omega t)] \\ &\quad \times (\tanh\rho \sinh n + i\cot\delta \cosh n), \\ S_3 &= 1 - \frac{2}{\mu}; \quad \mu = \frac{\cosh^2 n}{\sin^2\delta} - \frac{\sinh^2 n}{\cosh^2\rho}, \quad n = \frac{1}{2d}[\xi - Vt], \\ d &= \frac{\cos^2\delta + \sinh^2\rho}{2a\sin\delta\cosh\rho}, \quad (k_1, k_2) = \frac{a(-\cos\delta\sinh\rho, 1)}{\sinh^2\rho + \cos^2\delta}, \\ \Omega &= \frac{a^2(\cosh^2\rho + \sin^2\delta - 2\cosh^2\rho\sin^2\delta)}{(\sinh^2\rho + \cos^2\delta)^2}. \end{aligned} \tag{6.5.12}$$

Here ρ and δ are real parameters ($\lambda_1 = \exp(\rho + i\delta) \neq i$), $\xi = (\mathbf{q}\cdot\mathbf{r})$, $\eta = (\mathbf{p}\cdot\mathbf{r})$ are new coordinates in the xOy-plane, and $\mathbf{q}$ and $\mathbf{p}$ are two mutually perpendicular unit vectors:

$$\mathbf{q} = (\sinh^2\rho + \cos^2\delta)^{-1}(\sinh^2\rho - \cos^2\delta, -2\cos\delta\sinh\rho), \quad \mathbf{p} = (q_2, -q_1).$$

Let d be the width of a strip enclosed by two straight lines perpendicular to the vector $\mathbf{q}$ in the xOy-plane. Suppose that this strip moves at a constant speed $V = aq_2$ and, beyond it, the magnetization distribution predicted from (6.5.12) approaches the equilibrium value $\mathbf{S} \approx (0, 0, 1)$. Although soliton (6.5.12) in the direction of $\mathbf{q}$ is unlimited and, consequently, its total energy must be infinite, we are interested in exploring similar solutions since the properties of physically realizable solitons of sufficiently large length can be revealed.

When moving along any straight line in the xOy-plane, parallel to the vector $\mathbf{q}$ ($\eta =$ const), the two-dimensional soliton (6.5.12) qualitatively resembles a one-dimensional soliton, as in Chapter 2. In this cross-section, in a coordinate system moving together with the wave, the vector $\mathbf{S}$ performs precession around the Oz-axis with frequency $\omega = \Omega - k_1 V$. Moreover, the component S_3 and the initial phase of the precession are functions of the variable $\xi - Vt$. The initial precession phase φ_0 is the sum of the linear function of $\xi - Vt$ and an inflection-type function:

$$\varphi_0 = k_2\eta + k_1(\xi - Vt) + \arctan(\cot\delta\coth\rho\coth n).$$

The edges of the pit-shaped plot of dependence of S_3 on $\xi - Vt$ tend to the value $S_3 \to 1$ as $|\xi - Vt| \to \infty$. At the center of the soliton, S_3 reaches its minimum value, $S_3 = \cos 2\delta$. For $\delta = \pi/2$ ($\rho \neq 0$), the soliton velocity vanishes, and the magnetization distribution at the center of the soliton is close to another equilibrium value, $\mathbf{S} \approx (0, 0, -1)$.

Unlike a one-dimensional soliton, the wave (6.5.12) is not uniform in the direction perpendicular to $\mathbf{q}$. In the cross-section perpendicular to $\mathbf{q}$, solution (6.5.12) describes a nonlocalized precession wave that runs along the strip indicated above in the xOy-plane, as along a peculiar waveguide (Figure 6.4). In a coordinate system moving at a speed of $V\mathbf{q}$, the frequencies of the traveling wave and its wave vector are equal to ω and $\mathbf{k}$, respectively.

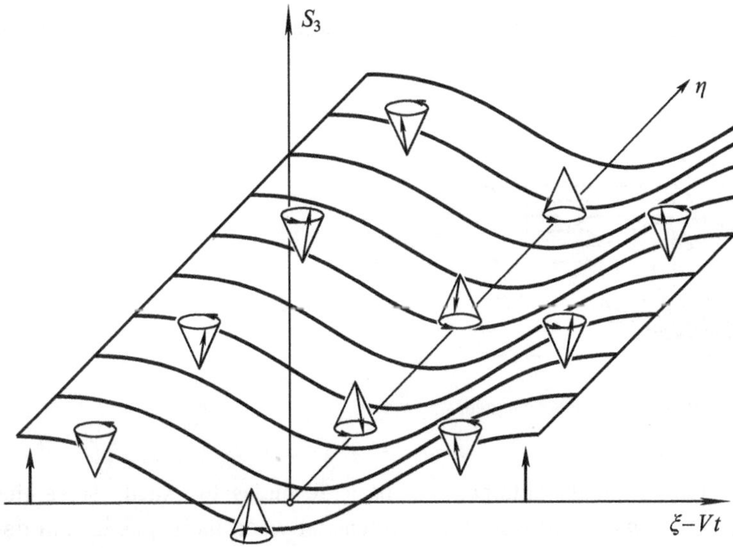

Figure 6.4: Distribution of magnetization in a two-dimensional precessional soliton.

As the parameter δ diminishes, the soliton width d increases indefinitely. Its velocity V and wave vector tend to finite values, and the projection S_3 approaches the value $S_3 \approx 1$, so that for any finite x and t, solution (6.5.12) coincides with a small-amplitude spin wave and the dispersion law $\Omega = \mathbf{k}^2$.

For a two-dimensional soliton, the energy (momentum) density and energy flux depend only on the variable $\xi - Vt$, with the momentum density and energy flux divided into two summands. The former is responsible for transferring the energy (momentum) during the propagation of a nonlinear wave as a whole in the direction of $\mathbf{q}$. The latter pertains to the energy (momentum) transfer in the transverse direction of $\mathbf{p}$. The total energy E and the total momentum $\mathbf{P}$ (both per unit of length) oriented along $\mathbf{p}$ are given by the formulas

$$E = \frac{4|\alpha \sin\delta| \cosh\rho}{\sinh^2\rho + \cos^2\delta}, \quad \mathbf{P} = 2\mathbf{p} \ln \frac{\cosh\rho - \sin\delta}{\cosh\rho + \sin\delta} + 4\mathbf{q} \arctan(\tanh\rho \tan\delta).$$

Multisoliton solutions describe the scattering of the precessional solitons considered above. Since each wave is unlimited in one direction, any two waves always intersect in some region of the xOy-plane (Figure 6.5). Near the intersection, the interaction of the waves is intense, and the magnetization distribution has a rather complicated shape. However, far away enough from the region of interaction, each of the nonlinear waves restores its original shape and velocity, as upon exciting a single soliton. To clarify, we inspect a segment on one of the nonlinear waves and observe how its magnetization distribution behaves. We can show that after each passage of the remaining system waves through the observation segment, the selected wave restores its shape and velocity and changes in its initial precession phase, and the coordinates of the center are additively summed.

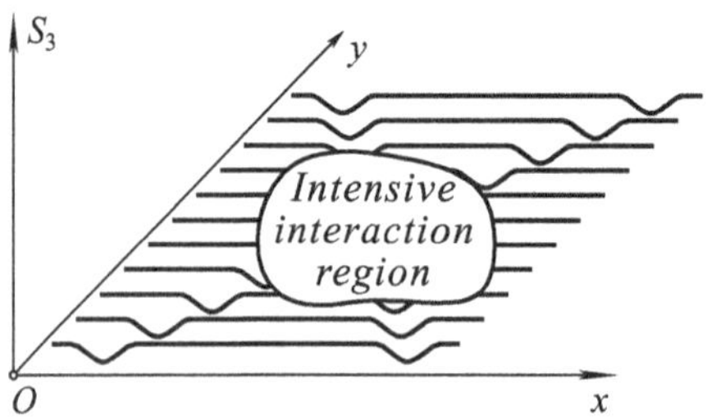

Figure 6.5: Interaction of two precessional magnetization waves.

Thus in this section, we have obtained the basic formulas to exactly solve the Landau–Lifshitz equations for an isotropic ferromagnet when its magnetization depends only on the variables x and $y - at$. In addition, we have analyzed the main features of the dynamics of solitons localized in only one direction.

7 Application of the matrix Riemann problem to the macroscopic dynamics of Goldstone excitations in ordered magnets

7.1 Basic integrable models of the method of phenomenological Lagrangians

> Plato's doctrine that God is a geometer seems truer today than ever. We all see more and more clearly that the most general mathematical formulation is the most physically fruitful as well... Mathematical formulas effectively control physical phenomena and can even lead to their discovery.
>
> A. Sommerfeld. Ways of knowledge in physics

The present chapter deals with the scheme proposed by the authors of [5, 7–9, 12, 4] for integrating the Andreev–Volkov–Marchenko–Zheltukhin equations (see (1.6.19))

$$
\begin{aligned}
&a_{lj}\,\partial_t\omega_{j,0}(x,t) - c_{lj}\,\partial_x\omega_{j,1}(x,t) - b_n\,\varepsilon_{lnm}\,\omega_{m,0}(x,t)\\
&\quad - \varepsilon_{lnm}a_{nk}\,\omega_{k,0}(x,t)\,\omega_{m,0}(x,t) + \varepsilon_{lnm}c_{nk}\,\omega_{k,1}(x,t)\,\omega_{m,1}(x,t) = 0.
\end{aligned}
\tag{7.1.1}
$$

To write down them simpler, we have entered the following notations for Cartan's forms:

$$
\omega_m(A,\partial_t A) \equiv \omega_{m,0}(x,t), \quad \omega_m(A,\partial_x A) \equiv \omega_{m,1}(x,t).
$$

At the Euler angles $(A_1, A_2, A_3) \equiv (\theta, \varphi, \psi)$, a matrix $g(x,t) \in \mathrm{SU}(2)$, matrix elements $D(x,t) \in \mathrm{SO}(3)$, and Cartan's forms are

$$
\begin{aligned}
g &= \begin{pmatrix} \exp[\frac{\mathrm{i}}{2}(\varphi+\psi)]\cos(\theta/2) & \mathrm{i}\exp[\frac{\mathrm{i}}{2}(\psi-\varphi)]\sin(\theta/2)\\ \mathrm{i}\exp[-\frac{\mathrm{i}}{2}(\psi-\varphi)]\sin(\theta/2) & \exp[-\frac{\mathrm{i}}{2}(\varphi+\psi)]\cos(\theta/2)\end{pmatrix};\\
\partial_\mu g\, g^\dagger &= \frac{\mathrm{i}}{2}\sigma_k\,\omega_{k,\mu}, \quad \mu = 0,1; \quad g^\dagger\sigma_p g = \sigma_k D_{kp}
\end{aligned}
\tag{7.1.2}
$$

or, more explicitly,

$$
\begin{aligned}
\omega_{1,\mu} &= \partial_\mu\varphi\ \sin\theta\sin\psi + \partial_\mu\theta\ \cos\psi,\\
\omega_{2,\mu} &= \partial_\mu\varphi\ \sin\theta\cos\psi - \partial_\mu\theta\ \sin\psi,\\
\omega_{3,\mu} &= \partial_\mu\varphi\ \cos\theta + \partial_\mu\psi.
\end{aligned}
\tag{7.1.3}
$$

As noted in Chapter 1, eqs. (7.1.1) describe, in the exchange approximation, the nonlinear dynamics of long-wavelength spin Goldstone excitations. The right shifts $g(x,t) \to g(x,t)h_0$, $h_0 \in \mathrm{SU}(2)$, on the group SU(2) implement the exchange symmetry related to the conserved current with the charge density:

https://doi.org/10.1515/9783112235973-007

$$S_{k0} = D_{kn}[b_n + a_{ni}\,\omega_{i,0}].$$

Moreover, the vector $\mathbf{S}_0$ determines the magnetization of the medium through the magneto-mechanical ratio γ:

$$M_k = -\hbar\gamma S_{k0} = -\hbar\gamma D_{kn}[b_n + a_{ni}\,\omega_{i,0}]. \tag{7.1.4}$$

With a certain choice of the parameters a_{ik}, c_{ik}, and b_i for spontaneous symmetry breaking, equations (7.1.1) are behind the nonlinear dynamics of a number of magnets. In the degenerate case ($\det a = \det b = 0$), only equations for two independent Goldstone fields are left in eqs. (7.1.1) due to the projection matrices. For example, as shown in Chapter 1, under the conditions $a_{ij} = 0$, $c_{ij} = b_i b_j - \delta_{ij}\mathbf{b}^2$, and $a_{ij}, c_{ij} \propto (b_i b_j - \delta_{ij}\mathbf{b}^2)$, eqs. (7.1.1) govern the dynamics of a Heisenberg ferromagnet and a two-sublattice ferrimagnet, respectively.

Further, we ascertain that eqs. (7.1.1) under the constraints

$$a_{ij} \propto c_{ij}, \quad b_i = 0 \tag{7.1.5}$$

or

$$a_{ij} \propto c_{ij} = \mathrm{diag}(c_1, c_1, c_3), \quad \mathbf{b} = (0, 0, b_3) \tag{7.1.6}$$

are integrable by the inverse scattering method. For $\mathbf{b} = 0$, approximation (7.1.6) holds for a noncollinear antiferromagnet $YMnO_3$. Let us dwell on this in more detail.

A six-sublattice antiferromagnet $YMnO_3$ is classified as a substance with the chemical formula $AMnO_3$ (A = Y, Sc, Mo – Lu), manganite [16, 10]. Manganites belong to the hexagonal syngony (space group C_{6v}^3). Only manganese atoms have magnetic moments arranged three in a row in two alternating basal planes shifted relative to each other by half a period along the hexagonal axis. When ordered, the atoms of each triple create an equilateral triangle. The planes of both triangles coincide and include an angle with the hexagonal axis (for $YMnO_3$, the angle is equal to $\gamma = 90°$). Figure 7.1 schematically depicts the arrangement of atoms Mn and the $YMnO_3$ compound's configuration pattern. The arrows indicate the directions of the magnetic moments. The angles between the magnetic moments amount to 120°. Atoms marked with crosses are shifted relative to other atoms by half a period along an axis perpendicular to the plane of the figure (Ox_3 is the sixth-order hexagonal axis). A model of the $YMnO_3$ magnetic structure in the projection along the Ox_3-axis can be scrutinized in Figure 7.2. Manganese atoms are marked with black circles, and oxygen atoms are labeled with light circles.

For the $YMnO_3$ compound, the explicit form of the parameters in the general expression for the phenomenological Lagrangian (1.6.13) is given in [2]:

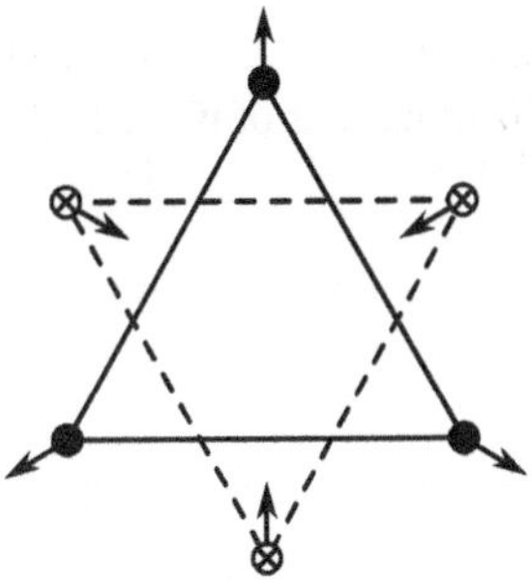

Figure 7.1: Directions of magnetic moments of manganese atoms in the compound $YMnO_3$. (See explanations in the text.)

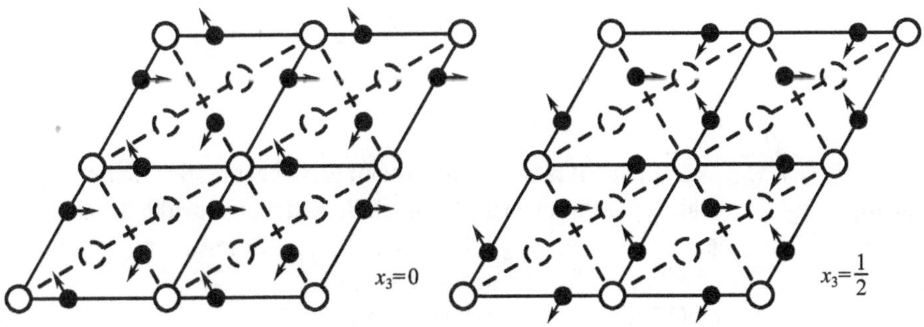

Figure 7.2: Model of the $YMnO_3$ magnetic structure in the projection along the Ox_3-axis. (See explanations in the text.)

$$\begin{aligned} b_i = 0, \quad d_{i,l} = 0, \quad a_{ij} &= A\delta_{ij} + B\delta_{i3}\delta_{3j}, \\ c_{ik,lm} &= \Lambda_1\delta_{ik}\delta_{l3}\delta_{m3} + \Lambda_2\delta_{i3}\delta_{k3}\delta_{l3}\delta_{m3} \\ &\quad + (\Lambda_3\delta_{ik} + \Lambda_4\delta_{i3}\delta_{k3})(\delta_{l1}\delta_{m1} + \delta_{l2}\delta_{m2}) \\ &\quad + \Lambda_5(\delta_{i2}\delta_{k2}\delta_{l1}\delta_{m1} + \delta_{i1}\delta_{k1}\delta_{l2}\delta_{m2} - 2\,\delta_{i2}\delta_{k1}\delta_{l1}\delta_{m2}) \\ &\quad + \Lambda_6(\delta_{i1}\delta_{k1}\delta_{l1}\delta_{m1} + \delta_{i2}\delta_{k2}\delta_{l2}\delta_{m2} + 2\,\delta_{i1}\delta_{k2}\delta_{l1}\delta_{m2}). \end{aligned} \tag{7.1.7}$$

Here A, B, and Λ are the phenomenological constants.

Having compared the coefficients in formulas (7.1.1), (7.1.6), and (1.6.19), (7.1.7) we can infer that under condition (7.1.6), for $\mathbf{b} = 0$, the model equations considered in this section cover one-dimensional magnetic excitations propagating along the sixth-order axis in the $YMnO_3$ compound.

Also note that the equations of the kind (7.1.1) and (7.1.6) find applications in optics to explain Raman scattering [11, 18].

In accordance with the general theory set forth in Chapter 1, eqs. (7.1.1), (7.1.5), and (7.1.6) describe the quasi-one-dimensional nonlinear dynamics of a certain class of noncollinear ferri- and antiferromagnets. A symmetric special case of eqs. (7.1.1) and (7.1.6), where $a_{ij} \propto c_{ij} \propto \delta_{ij}$, manages for $\mathbf{b} = (0, 0, b_3)$ the nonlinear dynamics of disordered no-collinear ferrimagnets, and for $\mathbf{b} = 0$, it responsible for the dynamics of nondissipative spin glass [2].

7.2 $U - V$ pair building. Basic ratios

If you view a problem closely enough, you will recognize yourself as part of the problem.

Ducharm's Axiom

Under the constraints $a_{ij} \propto c_{ij}$, the spin dynamics equations (7.1.1) and the Maurer–Cartan structural equations (1.6.10) can be written in a compact form

$$\begin{aligned} a_{lj}\partial_\mu \omega_{j,}{}^{\mu} - b_n\, \varepsilon_{lnm}\, \omega_{m,0} - \varepsilon_{lnm} a_{nk}\, \omega_{k,\mu}\, \omega_{m,}{}^{\mu} &= 0, \\ \varepsilon^{\mu\nu}\Big(\partial_\mu \omega_{s,\nu} + \frac{1}{2}\, \varepsilon_{snp}\, \omega_{n,\mu}\, \omega_{p,\nu} \Big) &= 0, \end{aligned} \tag{7.2.1}$$

where $x^\mu = (t, x)$ ($\mu = 0, 1$), and it is accepted, without loss of generality, that $a_{ij} = c_{ij} = \mathrm{diag}(a_1, a_2, a_3)$. The “raising” and “lowering” of the indices denoted by the Greek letters are carried out using the metric tensor $g_{\mu\nu} = \mathrm{diag}(1, -1)$, $\varepsilon_{\mu\nu}$ is the antisymmetric tensor ($\varepsilon_{01} = 1$, $\varepsilon^{01} = -1$), and $\varepsilon^{\mu\nu}\varepsilon_{\nu\rho} = \delta^\mu_\rho$ is the Kronecker symbol.

We represent eqs. (7.2.1) as the compatibility condition

$$\varepsilon^{\mu\nu}\partial_\mu\partial_\nu\Phi = 0 \tag{7.2.2}$$

for a system of linear differential equations for the complex matrices $\Phi(x, t, \lambda)$:

$$\partial_\nu\Phi = V_\nu\Phi. \tag{7.2.3}$$

The invariance of eqs. (7.2.2) with respect to the Lorentz group (for $b_i = 0$) and right shifts on the group SU(2) entrain a general form of the matrix functions $V_\nu(x, t, \lambda)$. For the latter, the forms $\omega_{i,\mu}$ constitute a complete set of invariants, and therefore in linear- and local-in-$\omega_{i,\mu}$ approximations, we have

$$V_\mu = \frac{\mathrm{i}}{2}(c_{lk}\, \omega_{k,\mu} + \varepsilon_{\mu\nu} d_{lk}\, \omega_{k,}{}^{\nu} + f_{l\mu})\sigma_l. \tag{7.2.4}$$

Here c_{ik}, d_{ik}, and $f_{i\mu}$ are as yet undefined functions of the spectral parameter λ, and σ_i are the Pauli matrices.

Plugging (7.2.3) into (7.2.2), we obtain that representations (7.2.2)–(7.2.4) are equivalent to eqs. (7.2.1) if the nonlinear algebraic equations for the coefficients c_{ik} are valid, which, in particular, lead to the ratio

$$(c^{\mathrm{T}}c)_{lp} = 0 \quad (l \neq p).$$

Hence, without loss of generality, we can assume that only the following components are nonzero:

$$c_{12} = c_2, \quad c_{21} = c_1, \quad c_{33} = c_3, \quad d_{12} = d_2, \quad d_{21} = d_1, \quad d_{33} = d_3.$$

Then the coefficients d_i are expressed in a simple way through c_i:

$$d_1 = d^{-1}(c_1c_3 + c_2)(c_1c_2 + c_3) \tag{7.2.5}$$

(and cyclic permutation of indices 1, 2, 3). Here we have introduced the following designations:

$$d = d_1\, d_2\, d_3, \quad c = c_1\, c_2\, c_3.$$

In addition, the algebraic equations tie the coefficients c_{ik} with the parameters a_{ik} = diag(a_1, a_2, a_3) as follows:

$$\frac{a_1 - a_2}{a_3} = \frac{(c_1^2 - c_2^2)\,(c_1c_2 + c_3)}{(c_1^2 + c_2^2 + c)\,c_3 + c_1c_2} \tag{7.2.6}$$

(and cyclic permutation of indices 1, 2, 3), producing the constraints

$$b_i = 0 \quad (i = 1, 2, 3) \quad \text{for } a_1 \neq a_2 \neq a_3 \tag{7.2.7}$$

or

$$b_1 = b_2 = 0, \quad b_3 \neq 0 \quad \text{for } a_1 = a_2 \neq a_3. \tag{7.2.8}$$

The coefficients $f_{l\mu}$ are nonzero only in the second case. As will be clear from what follows, their calculation meets no difficulties. The above constraints impose strict requirements on the structure of the integrable equation.

Equations (7.2.6) are not independent. It is easy to check that one of them is a consequence of the other two. Since the number of independent equations (two) is less than the number of the unknowns c_1, c_2, and c_3, system (7.2.6) admits λ-dependent solutions. In the process, the nonzero components c_{ik} and d_{ik} can be uniformized as meromorphic functions of the parameter λ.

Let us first discuss the axisymmetric case (7.2.8). In the parameterization of the SU(2) group elements by Euler angles (see formulas (1.6.12) and (7.1.3)), under constraints (7.2.8), the Lagrangian density for eqs. (7.2.1) has the form

$$\begin{aligned} L &= \frac{1}{2}(a_{ik}\, \omega_{i,}{}^{\mu} \omega_{k,\mu} + 2\, b_3\, \omega_{3,0}) \\ &\equiv \frac{1}{2}\Big(a_1[(\partial_t\varphi)^2 \sin^2\theta + (\partial_t\theta)^2 - (\partial_x\varphi)^2 \sin^2\theta - (\partial_x\theta)^2] \end{aligned}$$

$$+ a_3[(\partial_t\varphi \ \cos\theta + \partial_t\psi)^2 - (\partial_x\varphi \ \cos\theta + \partial_x\psi)^2]$$
$$+ 2 b_3(\partial_t\varphi \ \cos\theta + \partial_t\psi)\Big). \tag{7.2.9}$$

Hence the formal change

$$\tilde{\psi} = \psi + \frac{b_3}{a_3}\, t$$

modifies the density of the Lagrangian and, consequently, the first equation in (7.2.1) to the case $b_3 = 0$:

$$a_{lj}\,\partial^{\mu}\omega_{j,\mu}(x,t) - \varepsilon_{lnm}\, a_{n\,k}\,\omega_{k,}{}^{\mu}(x,t)\,\omega_{m,\mu}(x,t) = 0. \tag{7.2.10}$$

It is the case that we will consider further. Against models (7.2.1) and (7.2.8), where $b_3 \neq 0$, which cover media with spontaneous magnetization in the ground state, i. e., ferrimagnets, model (7.2.10) offers the dynamics of noncollinear antiferromagnets.

Recall that long-period helicoidal magnetic inversion-center-free structures are theoretically described by Lifshitz invariants of the kind $\Lambda(l_1\partial_x l_2 - l_2\partial_x l_1)$ in the expansion of the magnetic energy (**l** is the vector of ferro- or antiferromagnetism, and Λ is the phenomenological constant). Some antiferromagnets with a magnetic symmetry group of the crystalline class C_{2h}^2 admit an invariant $\Lambda[(\mathbf{l}\cdot\partial_x\mathbf{m}) - (\mathbf{m}\cdot\partial_x\mathbf{l})]$, where **m** and **l** are the vectors of ferro- and antiferromagnetism. Then their helicoidal structure is of exchange origin [17]. In the Andreev–Volkov–Marchenko–Zheltukhin approach, such invariants correspond to the summands $d_{i,l}\,\omega_i(A, \partial_l A)$ in expression (1.6.13) for the Goldstone mode Lagrangian. As noted in [6], the method of phenomenological Lagrangians can be used, for example, in analyzing the nonlinear dynamics of Goldstone modes in helicoidal structures of compounds of the type MnOOH. It is interesting and important that the transformation

$$\tilde{\psi} = \psi + \frac{b_3}{a_3}t + \frac{d_3}{a_3}x$$

eliminates such an addition as $-d_3\omega_{3,1}$ to the Lagrangian density (7.2.9) and leads to eqs. (7.2.10). A similar situation takes place in Chapter 4 when discussing nonlinear excitations against the background of the helicoidal structure of a ferromagnet. Thus we can easily come up with exact solutions describing the quasi-one-dimensional dynamics of a wide class of noncollinear ferri- and antiferromagnets with a helicoidal structure by choosing suitable solutions to the integrable axisymmetric model (7.2.10).

The final formulas are simplified by passing from the variables x and t to the new variables

$$\xi = \frac{t + x}{2}, \quad \eta = \frac{t - x}{2}.$$

In the new variables the equations of dynamics of the axisymmetric chiral model (7.2.10) and the Maurer–Cartan equations appear as

$$\partial_\eta \omega_{l,\xi} + \partial_\xi \omega_{l,\eta} + h\,\varepsilon_{ln3}(\omega_{n,\eta}\,\omega_{3,\xi} + \omega_{n,\xi}\,\omega_{3,\eta}) = 0, \tag{7.2.11}$$

$$\partial_\eta \omega_{l,\xi} - \partial_\xi \omega_{l,\eta} + \varepsilon_{lnm}\,\omega_{n,\eta}\,\omega_{m,\xi} = 0, \tag{7.2.12}$$

where $h = (a_3 - a_1)/a_1$. The Cartan forms $\omega_{k,\xi}$ and $\omega_{k,\eta}$ are given by the previous formulas (7.1.2) and (7.1.3); only differentiations are made with respect to the variables ξ and η, rather than with respect to the coordinates $x^\mu = (t, x)$.

The constraints $a_{ij} = \mathrm{diag}(a_1, a_1, a_3)$ raise the symmetry of eqs. (7.2.11) and (7.2.12) up to the group SU(2) × SO(2) to include also left shifts

$$g \to h_0 g, \quad h_0 = \exp\left(\frac{\mathrm{i}}{2}\alpha\sigma_3\right) \in \mathrm{SO}(2),$$

(α is a real parameter) apart from the right shifts ($g \to gh$, $h \in \mathrm{SU}(2)$) on the group SU(2), which are because of the isotropic nature of the exchange forces. According to eq. (7.1.2), the left shifts modify the Cartan forms along the formulas

$$\omega'_{1,\mu} = \omega_{1,\mu}\cos\alpha + \omega_{2,\mu}\sin\alpha, \quad \omega'_{2,\mu} = \omega_{2,\mu}\cos\alpha - \omega_{1,\mu}\sin\alpha, \quad \omega'_{3,\mu} = \omega_{3,\mu}.$$

The Lagrangian of the axisymmetric model is invariant under such transformations, and this results in the additional conservation law (see (7.2.11))

$$\partial_\eta \omega_{3,\xi} + \partial_\xi \omega_{3,\eta} = 0. \tag{7.2.13}$$

Equations (7.2.11) and (7.2.12) can be pictured as the compatibility conditions

$$\partial_\eta U - \partial_\xi V + [U, V] = 0 \tag{7.2.14}$$

for the auxiliary linear system

$$\partial_\xi \Phi = U\Phi, \quad \partial_\eta \Phi = V\Phi. \tag{7.2.15}$$

Using the uniformization of eqs. (7.2.5), (7.2.6), and (7.2.8) by rational functions of the spectral parameter λ, we find a representation for the complex matrices $U(\lambda)$ and $V(\lambda)$:

$$\begin{aligned} U &\equiv V_0 + V_1 = \frac{\mathrm{i}}{2}\left[l_2(\sigma_1\,\omega_{2,\xi} - \sigma_2\,\omega_{1,\xi}) + l_1\sigma_3\,\omega_{3,\xi}\right], \\ V &\equiv V_0 - V_1 = \frac{\mathrm{i}}{2}\left[l_4(\sigma_1\,\omega_{2,\eta} - \sigma_2\,\omega_{1,\eta}) + l_3\sigma_3\,\omega_{3,\eta}\right]. \end{aligned} \tag{7.2.16}$$

The explicit form of the coefficients l_r ($r = 1, 2, 3, 4$) is affected by the range of the parameter h. In what follows, we will treat eqs. (7.2.11) and (7.2.12) only when $h > -1$. Local deviations of spins from equilibrium lower the total energy of a magnet, and therefore the stability criterion is violated for magnetic structures corresponding to the values of $h < -1$. If $h = -1$, then, as shown in Chapter 1, eqs. (7.2.11) and (7.2.12) reduce to the integrable equations discussed in Chapter 6.

For $-1 < h < 0$:

$$l_1 = \frac{\sqrt{-h}\,(p^2+\lambda^2)}{p^2-\lambda^2}, \quad l_2 = \frac{\lambda\,(p^2-1)}{p^2-\lambda^2},$$
$$l_3 = \frac{\sqrt{-h}\,(p^2\lambda^2+1)}{p^2\lambda^2-1}, \quad l_4 = \frac{\lambda\,(p^2-1)}{p^2\lambda^2-1}, \quad p^2 = \frac{1+\sqrt{-h}}{1-\sqrt{-h}}. \tag{7.2.17}$$

For $h > 0$:

$$l_1 = \frac{\lambda^2 - 2\lambda\sqrt{h} - 1}{(\lambda-q)(\lambda+q^{-1})}, \quad l_2 = -\frac{\lambda^2+1}{(\lambda-q)(\lambda+q^{-1})},$$
$$l_3 = \frac{\lambda^2 + 2\lambda\sqrt{h} - 1}{(\lambda+q)(\lambda-q^{-1})}, \quad l_4 = -\frac{\lambda^2+1}{(\lambda+q)(\lambda-q^{-1})}, \tag{7.2.18}$$
$$q = \frac{\sqrt{1+h}-1}{\sqrt{h}} = \frac{\sqrt{h}}{\sqrt{1+h}+1}.$$

When $h = 0$, eqs. (7.2.11) and (7.2.12) become simpler and bring about the equations of the main chiral field, which are integrated along the procedure proposed in [15, 19, 14]. A set of interacting spherical tops with the inertia moment tensor $a_{ik} = a_1\delta_{ik}$ serves as a mechanical analog of the main chiral field model. In this case the matrices U and V change their shape:

$$U(\lambda) = \frac{\mathrm{i}\,\omega_{k,\xi}\sigma_k}{2(1+\lambda)}, \quad V(\lambda) = \frac{\mathrm{i}\,\omega_{k,\eta}\sigma_k}{2(1-\lambda)}. \tag{7.2.19}$$

Here it is worth pointing out a number of important formulas related to representations (7.2.14) and (7.2.15).

The relations

$$\mathrm{Sp}\,U = \mathrm{Sp}\,V = 0$$

and eqs. (7.2.15), as usual, imply that $\det\Phi$ does not depend on the variables ξ and η. At the same time, under the conditions

$$U^\dagger(\lambda^*) = -U(\lambda), \quad V^\dagger(\lambda^*) = -V(\lambda),$$

the solutions $\Phi^{-1}(\lambda)$ and $\Phi^\dagger(\lambda^*)$ of systems (7.2.15) may be chosen as coinciding.

Besides, for $h \neq 0$, representations (7.2.16)–(7.2.18) generate the involution relations

$$\Phi(-\lambda) = \sigma_1\Phi^*(\lambda^*)\sigma_1, \quad -1 < h < 0,$$
$$\Phi(-\lambda^{-1}) = \sigma_3\Phi(\lambda)\sigma_3, \quad h > 0, \tag{7.2.20}$$

for solutions to system (7.2.15). When $h = 0$, there are no such constraints.

In the long run, comparing the definition of Cartan's forms (7.1.2) with eqs. (7.2.15)–(7.2.19) and taking into account the identities

$$\sigma_1 = \exp\left(\frac{i}{4}\pi\sigma_3\right)\sigma_2 \exp\left(-\frac{i}{4}\pi\sigma_3\right), \quad \sigma_2 = -\exp\left(\frac{i}{4}\pi\sigma_3\right)\sigma_1 \exp\left(-\frac{i}{4}\pi\sigma_3\right),$$

we deduce an explicit expression for the matrices $g(\xi, \eta)$ in terms of certain values of the matrix function $\Phi(\xi, \eta, \lambda)$:

$$\begin{aligned}
g &= \frac{1}{\sqrt{\det\Phi(\lambda = 1)}} \exp\left(-\frac{i}{4}\pi\sigma_3\right)\Phi(\lambda = 1) \quad \text{for } -1 < h < 0;\\
g &= \frac{1}{\sqrt{\det\Phi(\lambda = 0)}} \Phi(\lambda = 0) \quad \text{for } h = 0;\\
g &= \frac{1}{\sqrt{\det\Phi(\lambda = 0)}} \exp\left(-\frac{1}{4}\pi\sigma_3\right)\Phi(\lambda = 0) \quad \text{for } h > 0.
\end{aligned} \tag{7.2.21}$$

Here we have incorporated the ξ- and η-variable-independent factors $[\det\Phi]^{-1/2}$ into relations (7.2.21) to satisfy the unimodularity condition for the matrix $g(\xi, \eta)$.

Since the main formulas given in this section depend on the range of the parameter h, the algorithms for integrating eqs. (7.2.11) and (7.2.12) turn out to be different for different values.

7.3 Nonlinear dynamics of nondissipative spin glass. Model of the main chiral field

As far as the propositions of mathematics refer to reality, they are not certain; and as far as they are certain, they do not refer to reality.

A. Einstein

In [2], it is assumed that spin glass is a nondissipative system at frequencies exceeding the characteristic time of elastic stress relaxation. Therefore the spin glass dynamics can be described through integrating the equations for the principal chiral field:

$$\partial_\eta \omega_{l,\xi} + \partial_\xi \omega_{l,\eta} = 0, \quad \partial_\eta \omega_{l,\xi} - \partial_\xi \omega_{l,\eta} + \varepsilon_{lnm}\, \omega_{n,\eta}\, \omega_{m,\xi} = 0. \tag{7.3.1}$$

The Lagrangian density of the model can be written in several ways:

$$\begin{aligned}
L &= \frac{1}{2}\, \omega_{s,\xi}(\xi, \eta)\omega_{s,\eta}(\xi, \eta) = \frac{1}{2}[\omega_{s,0}^2(x, t) - \omega_{s,1}^2(x, t)]\\
&= \frac{1}{4}[(\partial_t D_{kp})^2 - (\partial_x D_{kp})^2] = \mathrm{Sp}(\partial_t g\, \partial_t g^\dagger - \partial_x g\, \partial_x g^\dagger).
\end{aligned} \tag{7.3.2}$$

Here we have introduced dimensionless variables and applied scaling to eliminate the coefficients in these equations. To write down the chain of eqs. (7.3.2), we have used the definitions of Cartan's forms in terms of the matrices $D \in \mathrm{SO}(3)$ and $g \in \mathrm{SU}(2)$:

$$\omega_{n,\mu} = -\frac{1}{2}\,\varepsilon_{nsp}\,D_{ks}\,\partial_\mu D_{kp}; \quad \partial_\mu g\, g^\dagger = \frac{\mathrm{i}}{2}\,\sigma_k\,\omega_{k,\mu}. \tag{7.3.3}$$

As shown in Chapter 1, by virtue of the identity $g^\dagger \sigma_p g = \sigma_k D_{kp}$, definitions (7.3.3) are equivalent.

Remarkably, the Lagrangian density (7.3.2) is an invariant form under both left and right shifts on the group SO(3) (or SU(2)). The first equation of (7.3.1) can be interpreted as the law of conservation of currents generated by such a symmetry. Moreover, the equations and the Lagrangian of the model are invariant under Lorentz transformations.

Next, we introduce the vector fields

$$\mathbf{A} = \frac{1}{2}(\omega_{1,\xi},\, \omega_{2,\xi},\, \omega_{3,\xi}), \quad \mathbf{B} = \frac{1}{2}(\omega_{1,\eta},\, \omega_{2,\eta},\, \omega_{3,\eta})$$

and rewrite eqs. (7.3.1) in a form convenient for further analysis:

$$\partial_\eta \mathbf{A} = [\mathbf{A} \times \mathbf{B}], \quad \partial_\xi \mathbf{B} = [\mathbf{B} \times \mathbf{A}]. \tag{7.3.4}$$

As we can from eqs. (7.3.4), the length of the vector **A** does not depend on η, and the length of the vector **B** does not depend on ξ:

$$[\mathbf{A}(\xi,\eta)]^2 = A_0^2(\xi), \quad [\mathbf{B}(\xi,\eta)]^2 = B_0^2(\eta). \tag{7.3.5}$$

Here the high spatio-temporal symmetry of the spin glass manifests itself as an addition to the symmetry of the exchange interaction. Hence, in turn, the previous definition of the system energy (1.6.15) needs to be modified. Indeed, using formula (1.6.15), we obtain an expression for the system energy:

$$\begin{aligned} E &= \frac{1}{2}\int [\omega_{s,0}^2(x,t) + \omega_{s,1}^2(x,t)]\mathrm{d}x \\ &= \int [\mathbf{A}^2(\xi,\eta) + \mathbf{B}^2(\xi,\eta)]\mathrm{d}x = \int [A_0^2(\xi) + B_0^2(\eta)]\mathrm{d}x, \end{aligned}$$

which does not depend on the kind of the fields $\mathbf{A}(\xi,\eta)$ and $\mathbf{B}(\xi,\eta)$.

The proper definition of energy requires resolving constraints (7.3.5) and introducing new independent fields. In this case, a difficulty arises in disclosing the Lagrangian nature of the equations for new fields. In other words, they must obey some principle of least action. Similar challenges are met in the discussion of all integrable models in this chapter. Using the main chiral field model as an example, following [15, 14], we show how to overcome them.

Suppose, for simplicity, that the vectors **A** and **B** have unit length:

$$\mathbf{A}^2 = 1, \quad \mathbf{B}^2 = 1. \tag{7.3.6}$$

In the three-dimensional space, the linearly independent vectors **A**, **B**, and $[\mathbf{A}\times\mathbf{B}]$ create a local basis. The dynamic equations (7.3.4) can be interpreted as the result of expansion of the vectors $\partial_\eta \mathbf{A}$ and $\partial_\xi \mathbf{B}$ in terms of the local-basis vectors. The expansions of the vectors $\partial_\xi \mathbf{A}$ and $\partial_\eta \mathbf{B}$ in the same vectors must appear as

$$\begin{aligned} \partial_\xi \mathbf{A} &= [\mathbf{A} \times (a_1 \mathbf{B} + a_2 [\mathbf{A} \times \mathbf{B}])], \\ \partial_\eta \mathbf{B} &= [\mathbf{B} \times (b_1 \mathbf{A} + b_2 [\mathbf{A} \times \mathbf{B}])], \end{aligned} \tag{7.3.7}$$

where a_1, a_2, b_1, and b_2 are numeric functions. When writing down representation (7.3.7), we have taken into account the identities as consequences of the normalization (7.3.6):

$$(\partial_\xi \mathbf{A} \cdot \mathbf{A}) = (\partial_\eta \mathbf{B} \cdot \mathbf{B}) = 0.$$

Calculating the derivatives $\partial_\xi(\mathbf{A} \cdot \mathbf{B})$ and $\partial_\eta(\mathbf{A} \cdot \mathbf{B})$ using formulas (7.3.4) and (7.3.7), we can find the coefficients a_2 and b_2. Assuming that $\cos\alpha = (\mathbf{A} \cdot \mathbf{B})$, we get

$$a_2 = \frac{\partial_\xi \alpha}{\sin\alpha}, \quad b_2 = -\frac{\partial_\eta \alpha}{\sin\alpha}.$$

Calculating the derivatives $(\partial_\xi \mathbf{A} \cdot \partial_\xi \mathbf{B})$ and $(\partial_\eta \mathbf{A} \cdot \partial_\eta \mathbf{B})$, we find representations for the coefficients a_1 and b_1:

$$a_1 = \frac{M}{\sin\alpha}, \quad b_1 = \frac{N}{\sin\alpha}, \quad M = -\frac{(\partial_\xi \mathbf{A} \cdot \partial_\xi \mathbf{B})}{\sin\alpha}, \quad N = -\frac{(\partial_\eta \mathbf{A} \cdot \partial_\eta \mathbf{B})}{\sin\alpha}.$$

The compatibility conditions

$$\partial_\eta \partial_\xi \mathbf{A} = \partial_\xi \partial_\eta \mathbf{A}, \quad \partial_\eta \partial_\xi \mathbf{B} = \partial_\xi \partial_\eta \mathbf{B}$$

give rise to a system of three equations

$$\partial_\eta M = -\frac{\partial_\xi \alpha\, N}{\sin\alpha}, \quad \partial_\xi N = -\frac{\partial_\eta \alpha\, M}{\sin\alpha}, \tag{7.3.8}$$

$$\partial_\xi \partial_\eta \alpha + \sin\alpha - \frac{MN}{\sin\alpha} = 0. \tag{7.3.9}$$

From eqs. (7.3.8) we arrive at

$$\partial_\eta \left(M \tan\frac{\alpha}{2} \right) + \partial_\xi \left(N \tan\frac{\alpha}{2} \right) = 0,$$

so that

$$\partial_\xi \beta = M \tan\frac{\alpha}{2}, \quad \partial_\eta \beta = -N \tan\frac{\alpha}{2}.$$

As a result, we reduce eqs. (7.3.8) and (7.3.9) to a closed system for calculating the fields α and β:

$$\begin{aligned} \partial_\xi\partial_\eta\alpha + \sin\alpha + \frac{\cos(\alpha/2)}{2\sin^3(\alpha/2)}\,\partial_\xi\beta\,\partial_\eta\beta &= 0, \\ \partial_\xi\partial_\eta\beta - \frac{\partial_\xi\alpha\,\partial_\eta\beta + \partial_\eta\alpha\,\partial_\xi\beta}{\sin\alpha} &= 0. \end{aligned} \tag{7.3.10}$$

It is easy to persuade ourselves that the density of the Lagrange function for eqs. (7.3.10) has the form

$$\begin{aligned} L &= \frac{1}{2}\,\partial_\xi\alpha\,\partial_\eta\alpha + \frac{1}{2}\,\partial_\xi\beta\,\partial_\eta\beta\,\cot^2\frac{\alpha}{2} + \cos\alpha - 1 \\ &= \frac{1}{2}[(\partial_t\alpha)^2 - (\partial_x\alpha)^2] + \frac{1}{2}[(\partial_t\beta)^2 - (\partial_x\beta)^2]\,\cot^2\frac{\alpha}{2} + \cos\alpha - 1. \end{aligned}$$

Then the standard procedure produces the generalized impulses

$$p_\alpha = \partial_t\alpha, \quad p_\beta = \partial_t\beta\cot^2\frac{\alpha}{2}$$

and energy density

$$\begin{aligned} p_\alpha\partial_t\alpha + p_\beta\partial_t\beta - L &= \frac{1}{2}[(\partial_t\alpha)^2 + (\partial_x\alpha)^2] \\ &\quad + \frac{1}{2}[(\partial_t\beta)^2 + (\partial_x\beta)^2]\,\cot^2\frac{\alpha}{2} + 1 - \cos\alpha. \end{aligned}$$

Returning to the fields **A** and **B**, we derive the following expression for the system energy:

$$\begin{aligned} E = \int\Big[\frac{1}{4}\Big([\partial_\xi(\mathbf{A}\cdot\mathbf{B})]^2 + [\partial_\eta(\mathbf{A}\cdot\mathbf{B})]^2 + (\partial_\xi\mathbf{A}\cdot\partial_\xi\mathbf{B})^2 \\ +(\partial_\eta\mathbf{A}\cdot\partial_\eta\mathbf{B})^2\Big)\,[1-(\mathbf{A}\cdot\mathbf{B})^2]^{-1} + 1 - (\mathbf{A}\cdot\mathbf{B})\Big]\,dx. \end{aligned}$$

A simpler expression for the energy can be derived by resorting to the following parameterization of the unit vectors **A** and **B**:

$$\mathbf{A} = (\sin\theta_1\cos\varphi_1, \sin\theta_1\cos\varphi_1, \cos\theta_1), \quad \mathbf{B} = (\sin\theta_2\cos\varphi_2, \sin\theta_2\cos\varphi_2, \cos\theta_2)$$

In the independent angular variables θ_i and φ_i, eqs. (7.3.4) correspond to the density of the Lagrange function

$$\begin{aligned} L &= (1-\cos\theta_1)(\partial_t - \partial_x)\varphi_1 \\ &\quad + (1-\cos\theta_2)(\partial_t + \partial_x)\varphi_2 - \sin\theta_1\sin\theta_2\cos(\varphi_1-\varphi_2) - \cos\theta_1\cos\theta_2. \end{aligned}$$

This function corresponds to the density of the Hamiltonian function

$$(1-\cos\theta_1)\partial_x\varphi_1 - (1-\cos\theta_2)\partial_x\varphi_2 + \sin\theta_1\sin\theta_2\cos(\varphi_1-\varphi_2) + \cos\theta_1\cos\theta_2$$

and an alternative expression for the energy of the system in terms of the fields **A** and **B**:

$$E = \int\left[\frac{A_1\partial_x A_2 - A_2\partial_x A_1}{1-A_3} - \frac{B_1\partial_x B_2 - B_2\partial_x B_1}{1-B_3} + (\mathbf{A}\cdot\mathbf{B}) - 1\right]\mathrm{d}x.$$

Here, as before, we assume that the fields behave asymptotically as $\mathbf{A},\mathbf{B}\to(0,0,1)$ as $|x|\to\infty$.

Generally, the field normalization (7.3.5) differs from (7.3.6). Suppose that

$$\mathbf{A}(\xi,\eta) = A_0(\xi)\,\tilde{\mathbf{A}}(\tilde{\xi},\tilde{\eta}),\quad \mathbf{B}(\xi,\eta) = B_0(\eta)\,\tilde{\mathbf{B}}(\tilde{\xi},\tilde{\eta}),$$

where

$$\tilde{\xi} = \int A_0(\xi)\,\mathrm{d}\xi,\quad \tilde{\eta} = \int B_0(\eta)\,\mathrm{d}\eta.$$

Then we can consider the previous case in the variables $\tilde{\mathbf{A}}$, $\tilde{\mathbf{B}}$, $\tilde{\xi}$, and $\tilde{\eta}$.

To construct soliton solutions to model (7.3.1), we use the "dressing" procedure for partial solutions using the matrix Riemann problem with zeros in the complex λ-plane. First, we choose the solution to eqs. (7.3.1) as a "seeding" one:

$$\begin{aligned} g_0(\xi,\eta) &= \exp\left(\frac{i\sigma_3}{2}[A(\xi)+B(\eta)]\right),\\ \omega^{(0)}_{3,\xi} &= \partial_\xi A(\xi),\quad \omega^{(0)}_{3,\eta} = \partial_\eta B(\eta),\\ \omega^{(0)}_{a,\xi} &= \omega^{(0)}_{a,\eta} = 0,\quad a = 1,2, \end{aligned} \tag{7.3.11}$$

where $A(\xi)$ and $B(\eta)$ are arbitrary functions of the variables ξ and η, respectively. In this case, nonzero elements of the matrix $D^{(0)}\in \mathrm{SO}(3)$,

$$D^{(0)}_{11} = D^{(0)}_{22} = \cos(A+B),\quad D^{(0)}_{21} = -D^{(0)}_{12} = \sin(A+B),\quad D^{(0)}_{33} = 1,$$

describe local rotations of the spin moments of the medium around the Ox_3-axis, with the values of the antiferromagnetic vectors of the equilibrium exchange structure changing (see Chapter 1).

The solution to the auxiliary linear systems (7.2.15) and (7.2.19) corresponding to such a choice has the form

$$\Phi_0(\xi,\eta,\lambda) = \exp\left[\frac{i\sigma_3}{2}\left(\frac{A(\xi)}{1+\lambda} + \frac{B(\eta)}{1-\lambda}\right)\right].$$

It is worth stressing that the general "dressing" algorithm imposes the same restrictions as in Chapter 2 on the solutions to the linear systems (7.2.15) and (7.2.19). Therefore we immediately give an explicit form of a new solution to the auxiliary system, which corresponds to N solitons:

$$\begin{aligned}\Phi(\xi,\eta,\lambda) &= \Psi(\xi,\eta,\lambda)\,\Phi_0(\xi,\eta,\lambda),\\ \det\Psi &= \prod_{k=1}^{N}\frac{(\lambda-\mu_k)}{(\lambda-\mu_k^*)},\quad \Psi^{-1}(\lambda)=\Psi^\dagger(\lambda^*),\\ [\Psi(\lambda)]_{\sigma\sigma'} &= \delta_{\sigma\sigma'}+\sum_{k,n=0}^{N}\frac{(m_k)_\sigma(m_n)^*_{\sigma'}}{\lambda-\mu_n^*}\frac{\partial\ln\det A}{\partial A_{nk}},\end{aligned}\tag{7.3.12}$$

where $\sigma,\sigma'=1,2$; A is an $N\times N$-matrix with the elements

$$A_{ij}=\frac{(\mathbf{m}_i^*\cdot\mathbf{m}_j)}{\mu_i^*-\mu_j}=-A_{ji}^*,\quad \mathbf{m}_j=\Phi_0(\xi,\eta,\mu_j)\mathbf{c}_j,$$

μ_k are independent simple complex zeros of the Riemann problem, and $\mathbf{c}_j=(c_j^{(1)},c_j^{(2)})$ are arbitrary complex vectors.

Next, we find multisoliton solutions to eqs. (7.3.1) for the main chiral field using formulas (7.2.21) and (7.3.12):

$$g(\xi,\eta)=\Psi(\xi,\eta,\lambda=0)\,g_0(\xi,\eta)\left[\prod_{k=1}^{N}\frac{\mu_k^*}{\mu_k}\right]^{1/2}.$$

Let us show that calculating Cartan's forms reduces to algebraic operations. In terms of Ψ, eqs. (7.2.15) can be written as

$$\begin{aligned}U(\lambda)&=[\partial_\xi\Psi(\lambda)+\Psi(\lambda)U_0(\lambda)]\Psi^\dagger(\lambda),\\ V(\lambda)&=[\partial_\eta\Psi(\lambda)+\Psi(\lambda)\,V_0(\lambda)]\Psi^\dagger(\lambda),\end{aligned}\tag{7.3.13}$$

where U_0 and V_0 are the values of the matrix functions U and V (7.2.19) on the "seeding" solution (7.3.11). Equating the residues at the poles of eqs. (7.3.13), we obtain the Cartan forms for multisoliton states:

$$\begin{aligned}\omega_{k,\xi}\sigma_k &= \Psi(\lambda=-1)\,\sigma_3\Psi^\dagger(\lambda=-1)\,\partial_\xi A(\xi),\\ \omega_{k,\eta}\sigma_k &= \Psi(\lambda=1)\,\sigma_3\Psi^\dagger(\lambda=1)\,\partial_\eta B(\eta).\end{aligned}$$

A one-soliton solution to the model in terms of the matrix g has the form

$$\begin{aligned}g(\xi,\eta)&=\left(1+\frac{\mu-\mu^*}{\mu^*}P\right)g_0(\xi,\eta)\sqrt{\frac{\mu^*}{\mu}},\\ P&=\frac{1}{2\cosh y}\begin{pmatrix}\exp(y) & \exp(iS)\\ \exp(-iS) & \exp(-y)\end{pmatrix},\end{aligned}\tag{7.3.14}$$

$$y = \mu''\left(\frac{A(\xi)}{|1+\mu|^2} - \frac{B(\eta)}{|1-\mu|^2}\right) + \ln\left|\frac{c^{(1)}}{c^{(2)}}\right|,$$

$$S = \frac{1+\mu'}{|1+\mu|^2}A(\xi) + \frac{1-\mu'}{|1-\mu|^2}B(\eta) + \arg\frac{c^{(1)}}{c^{(2)}}, \quad \mu \equiv \mu' + \mathrm{i}\mu''.$$

When

$$A(\xi) = (t+x)a, \quad B(\eta) = (t-x)b \quad (a, b = \text{const}),$$

solution (7.3.14) offers a solitary wave with constant velocity V and localization region d:

$$V = \frac{b|1+\mu|^2 - a|1-\mu|^2}{b|1+\mu|^2 + a|1-\mu|^2}, \quad d \sim \frac{|1-\mu^2|^2}{|\mu''\,(b|1+\mu|^2 + a|1-\mu|^2)|}.$$

When $ab > 0$, the soliton velocity does not exceed in magnitude unity. The magnetization distribution matching such a solution is uniformly asymptotical at spatial infinity: $\mathbf{M} = -\hbar\gamma(0, 0, a+b)$. Hence it follows that the linear functions make it possible to describe solitons in disordered ferrimagnets. Sections 7.1 and 7.2 have already addressed the relationship between the models of spin glass and a disordered noncollinear ferrimagnet.

7.4 Dynamics of a noncollinear antiferromagnet with a structure of the $YMnO_3$-type. Axisymmetric chiral model

Until now, I have been engaged in solving problems because, in studying the sciences, examples are more useful than rules.

Isaac Newton

7.4.1 Model integration procedure

Let us dive into the procedure for integrating eqs. (7.2.11) and (7.2.12) for $-1 < h < 0$. Further analysis uses a normalization of the Riemann problem matrix functions to the identity matrix at a point $\lambda = \infty$ (the canonical normalization). To harmonize such a normalization with the form of $U-V$–pairs, we introduce another matrix function $\chi(\lambda)$:

$$\Phi(\lambda) = \Phi(\lambda = \infty)\,\chi(\lambda),$$

$$\Phi(\lambda = \infty) = \exp\left[\frac{\mathrm{i}\sigma_3}{2}\sqrt{-h}\left(-\int_{\xi_0}^{\xi}\omega_{3,\xi'}(\xi',\eta)\,\mathrm{d}\xi' + \int_{\eta_0}^{\eta}\omega_{3,\eta'}(\xi_0,\eta')\,\mathrm{d}\eta'\right)\right]. \tag{7.4.1}$$

Here we have written the solution $\Phi(\lambda = \infty)$ to system (7.2.15)–(7.2.17) for $\lambda = \infty$, bearing in mind the conservation law (7.2.13). The system of equations for χ,

$$\partial_\xi \chi = \Phi^\dagger(\lambda = \infty)\big[U(\lambda) - U(\lambda = \infty)\big]\,\Phi(\lambda = \infty)\chi \equiv \tilde{U}(\lambda)\chi,$$
$$\partial_\eta \chi = \Phi^\dagger(\lambda = \infty)\big[V(\lambda) - V(\lambda = \infty)\big]\,\Phi(\lambda = \infty)\chi \equiv \tilde{V}(\lambda)\chi, \tag{7.4.2}$$

is gauge equivalent to system (7.2.15)–(7.2.17). We write its solution as

$$\chi(\lambda) = \Psi(\lambda)\chi_0(\lambda), \tag{7.4.3}$$

where $\chi_0(\lambda)$ is the solution of the linear equations (7.4.2) as a particular solution to the original axisymmetric chiral models (7.2.11) and (7.2.12). Remarkably, apart from the main chiral field model (7.3.1), the axisymmetric and even asymmetric chiral models (7.2.7) and (7.2.10) have the particular solution (7.3.11). Further, for definiteness, we choose the latter as a "seeding" solution to the nonlinear equations (7.2.11) and (7.2.12). Then the corresponding solution to the linear system (7.4.2) is

$$\chi_0(\lambda) = \exp\left[\mathrm{i}\sigma_3\left(\frac{p^2(p^2-1)A(\xi)}{(p^2+1)(p^2-\lambda^2)} + \frac{p^{-2}(p^{-2}-1)B(\eta)}{(p^{-2}+1)(p^{-2}-\lambda^2)}\right)\right]. \tag{7.4.4}$$

The Riemann problem with canonical normalization shapes the matrix $\Psi(\lambda)$. When $\lambda \in R$, it is unitary and meets the following constraint because of the involution (7.2.20):

$$\Psi(-\lambda) = \sigma_1 \Psi^*(\lambda^*)\sigma_1. \tag{7.4.5}$$

We find such a matrix function for soliton solutions to models (7.2.11) and (7.2.12) by applying the conventional techniques covered in Chapters 2 and 3:

$$[\Psi(\lambda)]_{\sigma\sigma'} = \delta_{\sigma\sigma'} + \sum_{i,j=0}^{2N} \frac{(m_i)_\sigma (m_j)^*_{\sigma'}}{\lambda - \mu_j^*}\frac{\partial \ln \det A}{\partial A_{ji}},$$
$$\det\Psi = \prod_{j=1}^{2N}\frac{(\lambda-\mu_j)}{(\lambda-\mu_j^*)}, \quad \Psi^{-1}(\lambda) = \Psi^\dagger(\lambda^*), \tag{7.4.6}$$

where $\sigma, \sigma' = 1, 2$, and A is the $2N \times 2N$ matrix with elements

$$A_{ij} = \frac{(\mathbf{m}_i^* \cdot \mathbf{m}_j)}{\mu_i^* - \mu_j} = -A_{ji}^*.$$

The complex numbers μ_j and $\mathbf{m}_j$ vectors should be subject to the constraints

$$\mu_{k+N} = -\mu_k^*, \quad \mathbf{m}_{k+N} = \sigma_1 \mathbf{m}_k^*, \quad \mathbf{m}_k = \chi_0(\xi, \eta, \mu_k)\mathbf{c}_k, \quad k = 1, 2, \ldots, N,$$

to satisfy both reduction (7.4.5) and the condition of lacking extra poles in the "dressing relations". Here, $\mathbf{c}_k$ are arbitrary constant complex vectors.

The "seeding" solutions $\tilde{U}_0(\lambda)$ and $\tilde{V}_0(\lambda)$ from the "dressing" relations

$$\begin{aligned}
&\Phi^\dagger(\lambda=\infty)\big[U(\lambda)-U(\lambda=\infty)\big]\,\Phi(\lambda=\infty)\\
&\quad=\big[\partial_\xi\Psi(\lambda)+\Psi(\lambda)\tilde{U}_0(\lambda)\big]\,\Psi^\dagger(\lambda),\\
&\Phi^\dagger(\lambda=\infty)\big[V(\lambda)-V(\lambda=\infty)\big]\,\Phi(\lambda=\infty)\\
&\quad=\big[\partial_\eta\Psi(\lambda)+\Psi(\lambda)\,\tilde{V}_0(\lambda)\big]\,\Psi^\dagger(\lambda)
\end{aligned} \tag{7.4.7}$$

produce the Cartan forms for the multisoliton states of the (7.2.11) and (7.2.12).

To do this, we compute the residues of eqs. (7.4.7) at the poles $\lambda = p$ and $\lambda = p^{-1}$, respectively. We get the equations

$$\begin{aligned}
&\Phi^\dagger(\lambda=\infty)\Big(\frac{p^2+1}{2p}(\omega_{2,\xi}\sigma_1-\omega_{1,\xi}\sigma_2)+\omega_{3,\xi}\sigma_3\Big)\Phi(\lambda=\infty)\\
&\quad=\Psi(\lambda=p)\{\partial_\xi A(\xi)\sigma_3\}\Psi^\dagger(\lambda=p),\\
&\Phi^\dagger(\lambda=\infty)\Big(\frac{p^2+1}{2p}(\omega_{2,\eta}\sigma_1-\omega_{1,\eta}\sigma_2)+\omega_{3,\eta}\sigma_3\Big)\Phi(\lambda=\infty)\\
&\quad=\Psi(\lambda=p)\{\partial_\eta B(\eta)\sigma_3\}\Psi^\dagger(\lambda=p).
\end{aligned} \tag{7.4.8}$$

For definiteness, we have written formulas (7.4.8) with (7.3.11) as a particular solution of the model. Generally, the right-hand side of each of the formulas in the curly bracket contains a value that corresponds to the "seeding" solution to the same expression in the bracket of the left-hand side of the equality.

Further, we carry out calculations in the following sequence. First, from (7.4.8) we find $\omega_{3,\xi}$ and $\omega_{3,\eta}$. Then from eq. (7.4.1) we obtain $\Phi(\lambda=\infty)$. Finally, using the expression for $\Phi(\lambda=\infty)$, we solve eqs. (7.4.8) for $\omega_{1,\xi}$, $\omega_{1,\eta}$, $\omega_{2,\xi}$, and $\omega_{2,\eta}$.

The equations of motion (7.2.11) and (7.2.12) can be explicitly solved in terms of the fields $g(\xi,\eta)\in SU(2)$ by means of formulas (7.2.21), (7.4.1), (7.4.3), (7.4.4), and (7.4.6):

$$g=\exp\Big(-\frac{i}{4}\pi\sigma_3\Big)\frac{\Phi(\lambda=\infty)\Psi(\lambda=1)\chi_0(\lambda=1)}{\sqrt{\det[\Psi(\lambda=1)\chi_0(\lambda=1)]}}.$$

This concludes the discussion of the procedure for integrating the equations of the axisymmetric models (7.2.11) and (7.2.12) for the range of $-1<h<0$.

When $h>0$, the gauge transformation

$$\Phi(\lambda)=\Phi(\lambda=\infty)\chi(\lambda) \tag{7.4.9}$$

maintains an agreement between the canonical normalization of the Riemann problem and representations (7.2.14)–(7.2.16) and (7.2.18). In this case, we can shape the matrices $\Phi(\lambda=\infty)$ explicitly in terms of g:

$$\Phi(\lambda=\infty)=\sigma_3\exp\Big(\frac{i}{4}\pi\sigma_3\Big)g\,\sigma_3. \tag{7.4.10}$$

Using the identity

$$g^{\dagger}\sigma_p g = \sigma_k D_{kp} \quad (D_{ki} \in \mathrm{SO}(3),\ D_{ki}D_{kj} = \delta_{ij},\ \det D = 1),$$

we can compose a system of linear differential equations for determining $\chi(\xi, \eta, \lambda)$ as follows:

$$\begin{aligned}
&\partial_\xi \chi = \tilde{U}(\lambda)\chi, \quad \partial_\eta \chi = \tilde{V}(\lambda)\chi, \\
&\tilde{U}(\lambda) = \frac{\mathrm{i}}{2}\sigma_3([l_1(\lambda) - 1]\sigma_k D_{k3}\,\omega_{3,\xi} - [l_2(\lambda) + 1]\sigma_k D_{ka}\,\omega_{a,\xi})\sigma_3, \\
&\tilde{V}(\lambda) = \frac{\mathrm{i}}{2}\sigma_3([l_3(\lambda) - 1]\sigma_k D_{k3}\,\omega_{3,\eta} - [l_4(\lambda) + 1]\sigma_k D_{ka}\,\omega_{a,\eta})\sigma_3.
\end{aligned} \tag{7.4.11}$$

Here summation over repeated indices is implied, and $a = 1, 2$; $k = 1, 2, 3$. The compatibility condition for system (7.4.11), as well as the original representations (7.2.14)–(7.2.16) and (7.2.18), are equivalent to eqs. (7.2.11) and (7.2.12) for all values of the parameter λ.

As before, we choose (7.3.11) as a particular solution to the nonlinear equations (7.2.11)–(7.2.12). Then the “seeding” solution to the linear system (7.4.11) is

$$\chi_0(\lambda) = \exp\left(\frac{\mathrm{i}\sigma_3}{2}([l_1(\lambda) - 1]A(\xi) + [l_3(\lambda) - 1]B(\eta))\right). \tag{7.4.12}$$

Soliton states correspond to “dressed” solutions to the linear system (7.4.11) of the following form:

$$\chi(\lambda) = \Psi(\lambda)\chi_0(\lambda), \tag{7.4.13}$$

where $\Psi(\lambda)$ is the solution to the Riemann problem with zeros. The constraints formulated in the previous section for the functions $\Psi(\lambda)$ are easily transferred to the matrices $\Psi(\lambda)$:

$$\begin{aligned}
&\Psi^{-1}(\lambda) = \Psi^{\dagger}(\lambda^*), \quad \Psi(\lambda = \infty) = I, \\
&\Psi(-\lambda^{-1}) = \Psi(\lambda = 0)\, Z\, \Psi(\lambda) Z^{\dagger}, \quad Z = \chi_0(\lambda = 0)\,\sigma_3.
\end{aligned} \tag{7.4.14}$$

For the “seeding” solution (7.4.12), we have $\chi_0(\lambda = 0) = I$.

As in Chapter 3, constructing the soliton solutions using the nonlinear reduction (7.4.14) causes no difficulties. The matrix function Ψ is given by

$$[\Psi(\lambda)]_{\sigma\sigma'} = \delta_{\sigma\sigma'} + \sum_{i,j=0}^{2N} \frac{(m_i)_\sigma (m_j)^*_{\sigma'}}{\lambda - \mu_j^*} \frac{\partial \ln \det A}{\partial A_{ji}}, \qquad \det \Psi = \prod_{j=1}^{2N} \frac{\lambda - \mu_j}{\lambda - \mu_j^*},$$

where $\sigma, \sigma' = 1, 2$, and A is the $2N \times 2N$ matrix with elements

$$A_{ij} = \frac{(\mathbf{m}_i^* \cdot \mathbf{m}_j)}{\mu_i^* - \mu_j}.$$

Constraint (7.4.14) and the condition of lacking extra poles in the "dressing" relations hold if

$$\mu_{k+N} = -\mu_k^{-1}, \quad \mathbf{m}_{k+N} = \sigma_3 \mathbf{m}_k, \quad \mathbf{m}_k = \chi_0(\xi, \eta, \mu_k)\mathbf{c}_k, \quad k = 1, 2, \ldots, N,$$

with arbitrary constant complex vectors $\mathbf{c}_k$.

The "dressing" relations (7.4.7) give rise to the products $D_{ka}\,\omega_{a,\xi}$, $D_{ka}\,\omega_{a,\eta}$, $D_{k3}\,\omega_{3,\xi}$, and $D_{k3}\,\omega_{3,\eta}$ ($a = 1, 2; k = 1, 2, 3$) to define the density of the magnetic moment of a medium according to eq. (7.1.4). However, their calculations are too cumbersome (compared with the case $-1 < h < 0$). Let us discuss the key points.

First, from eqs. (7.2.21), (7.4.9), (7.4.10), and (7.4.13) we find the matrix elements of D_{k3}:

$$g^\dagger \sigma_3 g \equiv \sigma_k D_{k3} = \sigma_3 \Psi(\lambda = 0). \tag{7.4.15}$$

Now formula (7.4.15) does not contain the multiplier $\det^{-1/2}[\Psi(\lambda = 0)]$ since it is equal to unity due to reductions. Let us find $\omega_{3,\xi}$ and $\omega_{3,\eta}$ from the equations

$$\begin{aligned}
&\sigma_3(\sigma_k D_{ka}\,\omega_{a,\xi} - \sqrt{h+1}\,\sigma_k D_{k3}\,\omega_{3,\xi})\sigma_3 \\
&\quad = \Psi(\lambda = q)\,\sigma_3\left(\sigma_k D_{ka}^{(0)}\omega_{a,\xi}^{(0)}\right. \\
&\qquad \left. - \sqrt{h+1}\,\sigma_k D_{k3}^{(0)}\omega_{3,\xi}^{(0)}\right)\sigma_3 \Psi^\dagger(\lambda = q), \\
&\sigma_3(\sigma_k D_{ka}\,\omega_{a,\eta} - \sqrt{h+1}\,\sigma_k D_{k3}\,\omega_{3,\eta})\sigma_3 \\
&\quad = \Psi(\lambda = -q)\,\sigma_3\left(\sigma_k D_{ka}^{(0)}\omega_{a,\eta}^{(0)}\right. \\
&\qquad \left. - \sqrt{h+1}\,\sigma_k D_{k3}^{(0)}\omega_{3,\eta}^{(0)}\right)\sigma_3 \Psi^\dagger(\lambda = -q).
\end{aligned} \tag{7.4.16}$$

The latter are the residues of the "dressing" relations (7.4.7) at the poles $\lambda = q$ and $\lambda = -q$ (the index "zero" marks the quantities corresponding to the "seeding" solution). For this purpose, we multiply eqs. (7.4.16) by σ_3 on the left and by $\sigma_3 D_{p3}\sigma_p$ on the right, calculate a spur of each of the expressions obtained, and take into account the orthogonality condition for the matrices D. As a result, the formulas that determine $\omega_{3,\xi}$ and $\omega_{3,\eta}$ can be a little simplified by taking into account reduction (7.4.14):

$$\begin{aligned}
\sqrt{h+1}\,\omega_{3,\xi} &= -\frac{1}{2}\,\mathrm{Sp}\left[\Psi(\lambda = q)\,\sigma_3\left(\sigma_k D_{ka}^{(0)}\omega_{a,\xi}^{(0)}\right.\right. \\
&\qquad \left.\left. - \sqrt{h+1}\,\sigma_k D_{k3}^{(0)}\omega_{3,\xi}^{(0)}\right)\Psi^\dagger(\lambda = -q^{-1})\right], \\
\sqrt{h+1}\,\omega_{3,\eta} &= -\frac{1}{2}\,\mathrm{Sp}\left[\Psi(\lambda = -q)\,\sigma_3\left(\sigma_k D_{ka}^{(0)}\omega_{a,\eta}^{(0)}\right.\right. \\
&\qquad \left.\left. - \sqrt{h+1}\,\sigma_k D_{k3}^{(0)}\omega_{3,\eta}^{(0)}\right)\Psi^\dagger(\lambda = q^{-1})\right].
\end{aligned}$$

The explicit form of the quantities D_{k3}, $\omega_{3,\xi}$, and $\omega_{3,\eta}$ makes it possible to find the products $D_{ik}\,\omega_{k,\xi}$ and $D_{ik}\,\omega_{k,\eta}$ using eqs. (7.4.16).

In addition, the approach suggested provides an explicit solution to the equations of motion (7.2.11) and (7.2.12) in terms of the fields $g(\xi,\eta) \in \mathrm{SU}(2)$. The matrix $g(\xi,\eta)$ parameterized by Euler angles as in eq. (7.1.2) is convenient to write the final outcome. Relation (7.4.15) determines two Euler angles, θ and φ:

$$\sigma_k D_{k3} \equiv g^\dagger \sigma_3 g = \begin{pmatrix} \cos\theta & \mathrm{i}\,\sin\theta\exp(-\mathrm{i}\varphi) \\ -\mathrm{i}\,\sin\theta\exp(\mathrm{i}\varphi) & -\cos\theta \end{pmatrix} = \sigma_3\Psi(\lambda = 0).$$

Using the explicit expressions for the Cartan forms in terms of $\omega_{3,\xi}$ and $\omega_{3,\eta}$ through the Euler angles (7.1.3), we get differential equations for the parameter ψ:

$$\partial_\xi\psi = \omega_{3,\xi} - \partial_\xi\varphi\,\cos\theta, \quad \partial_\eta\psi = \omega_{3,\eta} - \partial_\eta\varphi\,\cos\theta.$$

After integrating them, we have

$$\psi = \int_{\xi_0}^{\xi}[\omega_{3,\xi'}(\xi',\eta) - \partial_{\xi'}\varphi(\xi',\eta)\,\cos(\theta(\xi',\eta))]\mathrm{d}\xi' - \int_{\eta_0}^{\eta}[\omega_{3,\eta'}(\xi_0,\eta') - \partial_{\eta'}\varphi(\xi_0,\eta')\,\cos(\theta(\xi_0,\eta'))]\mathrm{d}\eta'. \tag{7.4.17}$$

Here eq. (7.4.17) takes into account the conservation law (7.2.13). Since the Lagrangian (7.2.9) of the axisymmetric model is ψ-field-independent, the additive constant can be chosen arbitrarily when determining the angle ψ.

So, in this section, we have developed an algorithm for exact integration of eqs. (7.2.10) in the axisymmetric case ($a_1 = a_2 \neq a_3$). In the asymmetric case ($a_1 \neq a_2 \neq a_3$), the technique of integrating by the Riemann matrix problem needs to be generalized and will be set forth separately.

In conclusion, we make a general remark concerning the solutions of the model considered. In terms of the vector fields

$$\mathbf{A}(\xi,\eta) = \frac{\sqrt{1+h}}{2}(\omega_{1,\xi}, \omega_{2,\xi}, \sqrt{1+h}\,\omega_{3,\xi}),$$
$$\mathbf{B}(\xi,\eta) = \frac{\sqrt{1+h}}{2}(\omega_{1,\eta}, \omega_{2,\eta}, \sqrt{1+h}\,\omega_{3,\eta}),$$

the equations for the axisymmetric models (7.2.11) and (7.2.12) take the form

$$\begin{aligned} &\partial_\eta\mathbf{A} + [\mathbf{B}\times\mathbf{A}] + \frac{2h}{1+h}(\mathbf{n}\cdot\mathbf{B})[\mathbf{A}\times\mathbf{n}] = 0, \\ &\partial_\xi\mathbf{B} + [\mathbf{A}\times\mathbf{B}] + \frac{2h}{1+h}(\mathbf{n}\cdot\mathbf{A})[\mathbf{B}\times\mathbf{n}] = 0, \end{aligned} \tag{7.4.18}$$

where $\mathbf{n} = (0,0,1)$. It follows that the length of the vector $\mathbf{A}$ does not depend on the variable η and the length of the vector $\mathbf{B}$ does not depend on ξ: $\mathbf{A}^2 = A_0^2(\xi)$, $\mathbf{B}^2 = B_0^2(\eta)$. The “dressing” procedure makes the functions $A_0^2(\xi)$ and $B_0^2(\eta)$ equal to the squares of the lengths of the vectors $\mathbf{A}_0$ and $\mathbf{B}_0$ for the “seeding” solution:

$$[\mathbf{A}(\xi,\eta)]^2 = \mathbf{A}_0^2 = A_0^2(\xi), \quad [\mathbf{B}(\xi,\eta)]^2 = \mathbf{B}_0^2 = B_0^2(\eta). \tag{7.4.19}$$

To prove this statement, it suffices to square eqs. (7.4.8) for $-1 < h < 0$ or eqs. (7.4.16) for $h > 0$ and calculate a spur from the resulting expressions. Ultimately, both cases lead to formulas (7.4.19).

For the fields $\mathbf{A}$ and $\mathbf{B}$ with normalization $\mathbf{A}^2 = \mathbf{B}^2 = 1$ and asymptotic behavior $\mathbf{A}, \mathbf{B} \to (0,0,1)$ as $|x| \to \infty$, the expression for the system energy has the form

$$E = \int \Bigg[\frac{A_1\partial_x A_2 - A_2\partial_x A_1}{1-A_3} - \frac{B_1\partial_x B_2 - B_2\partial_x B_1}{1-B_3} + (\mathbf{A}\cdot\mathbf{B}) - \frac{2h}{1+h}(\mathbf{n}\cdot\mathbf{A})(\mathbf{n}\cdot\mathbf{B}) - \frac{1-h}{1+h} \Bigg] \mathrm{d}x.$$

Connections (7.4.19) should be taken into account to compute the energy of the system. In addition, due to the physical conditions of the problem, relations (7.4.19) are suitable for only solutions to eqs. (7.4.18) with the quantities $\mathbf{A}^2$ and $\mathbf{B}^2$ as preset functions. Such a scenario takes place, for example, when considering applications of model (7.4.18) to the Raman scattering problem in optics [11, 18]. When implementing the “dressing” procedure, the necessary class of exact solutions can be easily distinguished by an appropriate choice of a particular solution.

7.4.2 Multisublattice magnetic solitons

The method outlined in the preceding section allows us to run a thorough analysis of the nonlinear dynamics of multisublattice magnets described by the phenomenological equations (7.2.10) (for ($a_1 = a_2 \neq a_3$)). This section covers soliton-like excitations obtained by “dressing” the particular solution (7.3.11). With a linear dependence of the functions $A(\xi)$ and $B(\eta)$ on the appropriate arguments, the “seeding” solution (7.3.11) governs the static magnetization distribution. In the general case the distribution of magnetization $\mathbf{M}^{(0)}$ for solution (7.3.11) consists of two spin waves of arbitrary shape propagating in opposite directions:

$$\mathbf{M}^{(0)} = -\gamma\hbar a_3 \mathbf{n}\,\partial_t \left[A\left(\frac{t+x}{2}\right) + B\left(\frac{t-x}{2}\right) \right], \tag{7.4.20}$$

where $\mathbf{n} = (0,0,1)$.

For $-1 < h < 0$, the final formulas for the one-soliton solution of the model in terms of Euler angles are

$$\begin{aligned}\cos\theta &= 1+\frac{(\mu^*+\mu)^2}{2\tau|\mu|^2|1-\mu^2|^2}, \quad \tau=\frac{\cosh^2 y}{(\mu^*-\mu)^2}+\frac{1}{4|\mu|^2},\\ \varphi &= -S+\frac{2p^2A}{p^2+1}+\frac{2p^{-2}B}{p^{-2}+1}-C,\\ \psi &= -\frac{\pi}{2}+S-\sqrt{-h}\,A+\sqrt{-h}\,B-C,\end{aligned} \tag{7.4.21}$$

where

$$\begin{aligned}C &\equiv \arctan\left[\frac{\mathrm{i}(\mu^*-\mu)(1+|\mu|^2)}{(\mu^*+\mu)(1-|\mu|^2)}\tanh y\right],\\ y &= \ln\left|\frac{m^{(1)}}{m^{(2)}}\right| = \ln\left|\frac{c^{(1)}}{c^{(2)}}\right| + \mathrm{i}(\mu^2-\mu^{*2})\sqrt{-h}\left[\frac{p^2A}{|p^2-\mu^2|^2}-\frac{p^{-2}B}{|p^{-2}-\mu^2|^2}\right],\\ S &= \arg\frac{m^{(1)}}{m^{(2)}} = \arg\frac{c^{(1)}}{c^{(2)}} + \frac{(2p^2-\mu^{*2}-\mu^2)}{|p^2-\mu^2|^2}\,p^2\sqrt{-h}\,A\\ &\quad -\frac{(2p^{-2}-\mu^{*2}-\mu^2)}{|p^{-2}-\mu^2|^2}\,p^{-2}\sqrt{-h}\,B.\end{aligned}$$

We express the distribution of the magnetization **M** for the one-soliton solution (7.4.21) through the matrix $g(\xi,\eta)$ and Cartan's forms. The procedure for calculating the latter is mentioned in the previous section. After simple algebraic transformations, we find

$$\begin{aligned}\frac{M_3}{\gamma\hbar a_1} &= -\frac{1}{2}\sum_{\lambda=p,p^{-1}}\left[\alpha^*\delta\, c_-(\lambda)+\alpha\,\delta^*c_+(\lambda)\right.\\ &\qquad \left.+(1+h)\,c_3(\lambda)(1-2|\delta|^2)\right];\\ \frac{M_1-\mathrm{i}M_2}{\gamma\hbar a_1} &= -\frac{1}{2}\exp(-\mathrm{i}\chi)\sum_{\lambda=p,p^{-1}}\left[\alpha^2c_+(\lambda)\right.\\ &\qquad \left.-\delta^2c_-(\lambda)-2\alpha\,\delta\,(1+h)\,c_3(\lambda)\right],\end{aligned} \tag{7.4.22}$$

where the quantities α, δ, γ, and c are determined by the formulas

$$\begin{aligned}c_+(\lambda) &= c_-^*(\lambda) = \frac{4(\mu^2-\mu^{*2})\,\lambda^2c(\lambda)}{[\mu^*\exp(y)+\mu\exp(-y)]^2(1+\lambda^2)}\left(\frac{\mu^*\exp(y)}{\lambda^2-\mu^{*2}}+\frac{\mu\exp(-y)}{\lambda^2-\mu^2}\right),\\ c_3(\lambda) &= c(\lambda)\left(1+\frac{\lambda^2(\mu^*+\mu)^2}{2\tau|\mu|^2|\lambda^2-\mu^2|^2}\right), \quad c(p)=\partial_\xi A, \quad c(p^{-1})=\partial_\eta B,\\ \alpha &= \frac{\mu(1-\mu^{*2})\exp(y)+\mu^*(1-\mu^2)\exp(-y)}{[\mu\exp(y)+\mu^*\exp(-y)]|1-\mu^2|},\\ \delta &= \frac{\mu^2-\mu^{*2}}{[\mu^*\exp(y)+\mu\exp(-y)]|1-\mu^2|}, \quad \chi=-S+\frac{2p^2A}{p^2+1}+\frac{2p^{-2}B}{p^{-2}+1}.\end{aligned}$$

It is important to note that in the case $-1 < h < 0$, the calculation of the medium's magnetization caused by solitons requires no integration. To shape the matrix $\Phi(\lambda = \infty)$, integration is needed. However, along with the matrix $\Phi^{-1}(\lambda = \infty)$, it should be canceled out in the full expression for the medium's magnetic moment density **M**. Thus **M** is an algebraic function of the elements of the matrix $\Psi(\lambda)$. This circumstance facilitates the analysis of the asymptotic values of the magnetization field.

For $h > 0$, the one-soliton solution deduced by the algorithm in the previous section has the form

$$\cos\theta = 1 + \frac{1}{2\tau}, \quad \tau = |\mu|^2\left[\left(\frac{\cosh y}{\mu^* - \mu}\right)^2 + \left(\frac{\sinh y}{1+|\mu|^2}\right)^2\right], \qquad (7.4.23)$$
$$\varphi = -S + C, \quad \psi = A + B + S + C,$$

where

$$C = \arctan\left[\frac{\mathrm{i}(\mu^* - \mu)(1-|\mu|^2)}{(\mu^* + \mu)(1+|\mu|^2)}\tanh y\right],$$
$$y = \ln\left|\frac{m^{(1)}}{m^{(2)}}\right| = \ln\left|\frac{c^{(1)}}{c^{(2)}}\right| + \frac{q}{2\mathrm{i}}(1+q^2)\sqrt{h+1}(\mu^* - \mu)(1+|\mu|^2)(\tilde{A} - \tilde{B}),$$
$$S = \arg\frac{m^{(1)}}{m^{(2)}} = \arg\frac{c^{(1)}}{c^{(2)}} - \frac{1}{2}(1+q^2)\sqrt{h+1}([2(1-q^2)|\mu|^2$$
$$+ q(|\mu|^2 - 1)(\mu^* + \mu)]\tilde{A} + [2(1-q^2)|\mu|^2 - q(|\mu|^2-1)(\mu^*+\mu)]\tilde{B}),$$
$$\tilde{A} \equiv \frac{A}{|(\mu - q)(\mu q + 1)|^2}, \quad \tilde{B} \equiv \frac{B}{|(\mu + q)(\mu q - 1)|^2}.$$

For $h > 0$, the medium's magnetization **M** tied with solitons is expressed through combinations of $D_{ik}\,\omega_{k,\xi}$ and $D_{ik}\,\omega_{k,\eta}$. As before, **M** is an algebraic function of the elements of the matrix $\Psi(\lambda)$.

The one-soliton solution (7.4.23) corresponds to the following magnetization of the medium:

$$\frac{M_3}{\gamma\hbar a_1} = -\frac{\sqrt{1+h}}{2}\left(A_0 + \frac{A_1}{\tau} + \frac{A_2}{\tau^2}\right),$$
$$A_0 = \sqrt{h+1}(\partial_\xi A + \partial_\eta B), \; A_2 = \frac{|\mu|^2}{2}q^2\sqrt{h+1}(\partial_\xi\tilde{A} + \partial_\eta\tilde{B}),$$
$$A_1 = \frac{\sqrt{h}}{2q}(\partial_\xi A + \partial_\eta B) + q(\mu + \mu^*)(1-|\mu|^2)(\partial_\xi\tilde{A} - \partial_\eta\tilde{B})$$
$$+ \left(\sqrt{h+1}\,|\mu|^2 - 2\,|\mu|^2 - \frac{q^2}{2}\,|1-\mu^2|^2\right)(\partial_\xi\tilde{A} + \partial_\eta\tilde{B});$$
$$\frac{M_1 - \mathrm{i}M_2}{\gamma\hbar a_1} = \frac{\sqrt{1+h}}{2}\exp(\mathrm{i}S)\sum_{\lambda=q,-q} c(\lambda)\left(\frac{\sqrt{h}}{2q\tau}\left[\frac{\mu^*+\mu}{\mu^*-\mu}\cosh y + \frac{1-|\mu|^2}{1+|\mu|^2}\sinh y\right]\right. \qquad (7.4.24)$$

$$
\begin{aligned}
&\times\left[1+\frac{|\mu|^2(1+\lambda^2)}{2\tau|(\lambda-\mu)(\lambda\mu+1)|^2}\right]-\frac{\mu}{\tau(1+\lambda\mu)(\lambda-\mu)}\\
&\times\left[\frac{\lambda(1-|\mu|^2)-\mu^*-\mu}{\mu^*-\mu}\cosh y-\frac{1-|\mu|^2+\lambda(\mu^*+\mu)}{1+|\mu|^2}\sinh y\right]\\
&\times\left(1+\frac{\exp y}{2\tau(1+\lambda\mu^*)(\lambda-\mu^*)}\left[\frac{\lambda(1+|\mu|^2)+\mu-\mu^*}{\mu^*-\mu}\cosh y\right.\right.\\
&\left.\left.+\frac{1+|\mu|^2+\lambda(\mu^*+\mu)}{1+|\mu|^2}\sinh y\right]\right)\Bigg);\\
c(q)&\equiv\partial_\xi A,\quad c(-q)\equiv\partial_\eta B.
\end{aligned}
$$

Further, we proceed to an analysis of the obtained results. A linear dependence of the functions $A(\xi)$ and $B(\eta)$ on the appropriate arguments

$$A(\xi)=2u_1\xi,\quad B(\eta)=2u_2\eta \tag{7.4.25}$$

brings about the simplest form of one-soliton solutions. Here u_1 and u_2 are real constants. In this case, the variables y and S take the form

$$y=y_0+\kappa_1(x-Vt),\quad S=S_0+\omega t+\kappa_2(x-Vt), \tag{7.4.26}$$

where y_0, S_0, $\kappa_{1,2}$, V, and ω are real parameters.

The above choice of the "seeding" solution suggests the presence of spontaneous magnetization in the ground state of the medium. According to eqs. (7.4.22) and (7.4.24), one-soliton states match spatially localized magnetization waves with the asymptotics:

$$\mathbf{M}\to-\gamma\hbar a_1(1+h)(u_1+u_2)\mathbf{n}\quad\text{as}\quad x\to\pm\infty,$$

where $\mathbf{n}=(0,0,1)$. Returning to the relationship between the models of noncollinear antiferro- and ferrimagnets in Section 7.2, we can infer that the solutions obtained are responsible for nonlinear excitations in noncollinear ferrimagnets. The localization area of each magnetic excitation moves at a constant speed V. In the soliton-fixed coordinate system, the magnetization precesses with constant frequency

$$
\begin{aligned}
\Omega&=|\partial_t\chi+V\partial_x\chi|\quad(-1<h<0),\\
\Omega&=|\partial_t S+V\partial_x S|\quad(h>0)
\end{aligned}
$$

around the Ox_3-axis, with $M_1^2+M_2^2$, M_3^2, and the initial precession phase depending on $x-Vt$ (see Figure 7.3). In this context the length of the density vector of the medium's magnetic moment $\mathbf{M}^2$ changes in space and with time. Additional degrees of freedom in multisublattice solitons expand the range of admissible velocities of their motion.

The plot of the function $\cos\theta$ in eqs. (7.4.21) and (7.4.23) depicts a solitary pit-shaped wave that moves with speed $V=-\partial_t y/\partial_x y$ and has the asymptotics $\cos\theta\to 1$ as $|x|\to\infty$

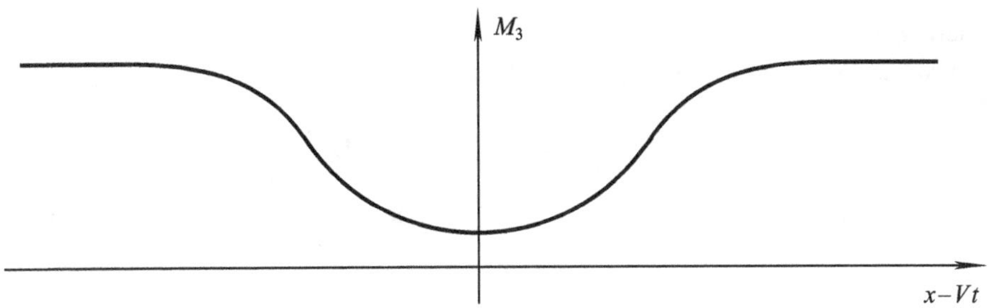

Figure 7.3: A precessional soliton in a noncollinear ferrimagnet.

(see Figure 7.4). The characteristic pit size is $\sim 2(\partial_x y)^{-1}$. For $|c^{(1)}| = |c^{(2)}|$, in the limit as $\mu \to \pm\mathrm{i}$, solution (7.4.21) degenerates into a weakly localized algebraic soliton centered at $\cos\theta = -1$. Its velocity of motion is equal to $V = (u_2 - u_1)/(u_1 + u_2)$. It is curious that only the exchange forces cause the formation of the remagnetized region. Chapter 6 explains this circumstance by the nonequivalence of the magnetic sublattices of a ferrimagnet.

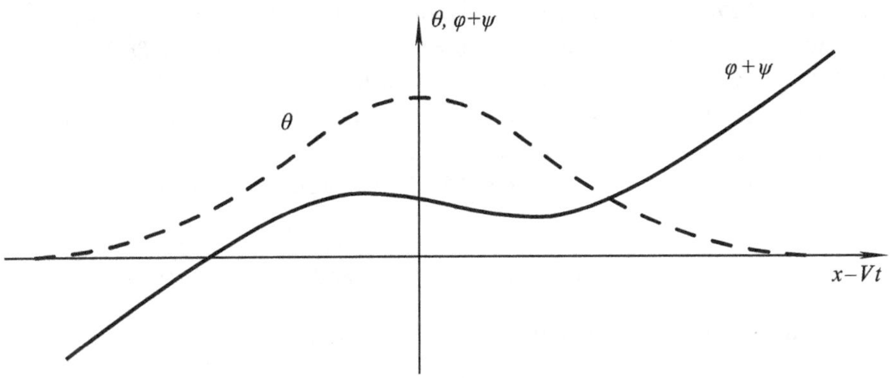

Figure 7.4: Semiclassical magnon fields for the simplest soliton in a noncollinear ferrimagnet. The shift of the plot $\varphi + \psi$ for along the y-axis is proportional to time.

The field $\varphi - \psi$ is a linear function of x and t, whereas $\varphi + \psi$ is the sum of a linear function of x and t and an inflection type function that depends on $x - Vt$.

In the general case, formulas (7.4.22) and (7.4.24) control localized excitation against the "background" of nonlocalized magnetization waves (7.4.20). The nature of the solution becomes clearer if we consider a special case where the functions $A(\xi)$ and $B(\eta)$ have the form

$$A(\xi) = 2u_1\xi + u_3\sin(2k\xi), \quad B(\eta) = 2u_2\eta, \tag{7.4.27}$$

where u_i are real constants ($i = 1, 2, 3$). For $u_3 = 0$, the obtained solution coincides with that considered earlier. When $u_3 \neq 0$, the solution has the asymptotics

$$\mathbf{M} \to -\gamma\hbar a_1(1+h)\,\mathbf{n}\left[u_1 + u_2 + k\,u_3 \cos(k(x+t))\right] \quad \text{as} \quad x \to \pm\infty.$$

Consequently, it can be interpreted (at least for small u_3) as the interaction between the spin wave

$$\mathbf{M}_0 = -\gamma\hbar a_1(1+h)\,\mathbf{n}\,k\,u_3 \cos(k(x+t)) \tag{7.4.28}$$

and the localized magnetic excitation discussed above. Such an interaction leads, in particular, to oscillations of the "center of gravity" of the excitation.

As the parameter

$$r = (\partial_x y)^2|_{u_3=0}\left[\frac{\partial y}{\partial(\sin 2k\xi)}\right]^{-2}$$

increases, both the width of the localized excitation and the level of its amplitude modulation decrease. Under the condition $r \gg 1$, the localized excitation is well noticeable against the background of the spin wave. A small-amplitude spin wave runs across the region of localization of the magnetic excitation moving as a whole with average velocity V. All three components of the medium's magnetization vector change as the wave travels forward. If $|dk| \gg 1$, then the length of the localized excitation includes several periods of the traveling wave (Figure 7.5). At a sufficiently large amplitude spin-wave ($r \ll 1$), the latter destroys the localized excitation.

To illustrate solutions (7.4.22) and (7.4.24) in the localized magnetic excitation-free region, we choose

$$A(\xi) = \sin 2\xi, \quad B(\eta) = \sin 2\eta.$$

Then the magnetization distribution obeys the following dependence:

$$M_k = \sum_{\alpha=1,2} m_k^{(\alpha)}(\sin 2\xi, \cos 2\eta)\cos[t - (-1)^\alpha x]. \tag{7.4.29}$$

The vector functions $\mathbf{m}^{(\alpha)}$ in eq. (7.4.29) are defined by formulas (7.4.22) and (7.4.24) and have a rather complicated form. This solution describes the magnetization wave as a result of the amplitude and phase modulation of the spin wave (7.4.28). It has interesting features. According to eq. (7.4.29), there is no uniform distribution of energy over all wave numbers over time. The decay of the wave (7.4.29) into a stable set of solitons is also not observed. In fact, the energy is time-periodically redistributed between a certain number of interacting modes.

Soliton distributions of magnetization are rational functions of the matrix elements of the Riemann problem. Therefore, as in Chapter 2, the scattering process in a system of N magnetic excitations localized against the background of the wave (7.4.20) can be

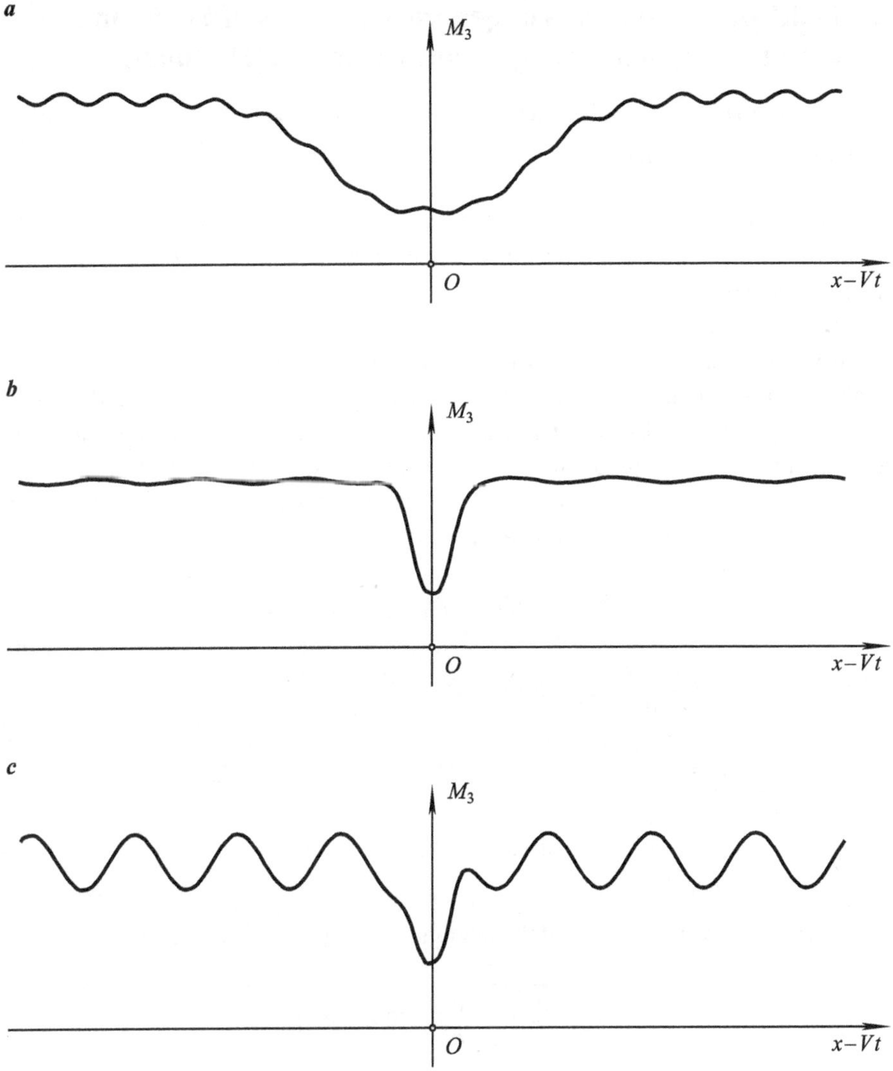

Figure 7.5: A precessional soliton against the background of a traveling spin wave: *a*) for $r \gg 1$, $|dk| \gg 1$, *b*) for $r \gg 1$, $|dk| \sim 1$, *c*) for $r \ll 1$.

comprehensively explored. For this purpose, an N-soliton matrix should be factorized into the product of one-soliton matrices. In particular, writing the functions $A(\xi)$ and $B(\eta)$ as in (7.4.27), a two-soliton solution describes the elastic collision of two isolated solitons. The only result of the collision is changes in the coordinates of the solitons' centers and a phase shift of the field $M_1 + iM_2$ for each soliton. These changes are additively summed in successive collisions of magnetic solitons localized against the background of a spin wave.

7.5 Multisoliton solutions of general equations of Goldstone excitation dynamics (asymmetric chiral SU(2)-model)

Inside every large problem is a small problem struggling to get out.

Hoare's Law of Large Problems

This section deals with a direct method for constructing soliton solutions of eqs. (7.2.1) for the cases $a_1 \neq a_2 \neq a_3$ and $\mathbf{b} = 0$ (an asymmetric chiral SU(2)-model). The matrix Riemann problem with zeros on a torus underlies this technique. The calculation algorithm is similar to that used in Chapter 5. Let us dwell on the key points.

In Section 7.2, we showed that eqs. (7.2.1) admit a representation in the form of a compatibility condition for the auxiliary linear systems (7.2.3) and (7.2.4). The coefficients c_{ik} and d_{ik} obey the algebraic equations (7.2.5)–(7.2.7). Their solution depends on the values of the parameters a_{ik}. Without loss of generality, we accept that $a_{ik} = \operatorname{diag}(a_1, a_2, a_3)$, $0 < a_1 < a_2 < a_3$. For further analysis, it is useful to uniformize eqs. (7.2.5)–(7.2.7) in terms of a complex parameter u:

$$\begin{aligned} c_{21} &= -\frac{(a_3 - a_1)}{a_3}\,\mathrm{cn}u\,\mathrm{dn}u \cdot a(u), \quad c_{12} = \frac{a_3 - a_1}{a_3}\,\mathrm{cn}u \cdot a(u), \\ c_{33} &= \frac{a_3 - a_1}{a_3}\,\mathrm{dn}u \cdot a(u), \quad d_{21} = \sqrt{\frac{a_3 - a_1}{a_2}\frac{a_1}{a_3}}\,\mathrm{sn}u \cdot a(u), \\ d_{12} &= -\frac{\sqrt{a_2(a_3 - a_1)}}{a_3}\,\mathrm{sn}u\,\mathrm{dn}u \cdot a(u), \\ d_{33} &= -\sqrt{\frac{a_3 - a_1}{a_2}}\,\mathrm{sn}u\,\mathrm{cn}u \cdot a(u), \end{aligned} \tag{7.5.1}$$

where $a(u) = 1/(\mathrm{cn}^2 u - a_1/a_3)$, and the modulus of Jacobi elliptic functions is

$$k = \sqrt{\frac{a_3(a_2 - a_1)}{a_2(a_3 - a_1)}} \quad (0 < k < 1).$$

In this case an additional module is equal to

$$k' = \sqrt{1 - k^2} = \sqrt{\frac{a_1(a_3 - a_2)}{a_2(a_3 - a_1)}}.$$

The coefficients c_{ik} and d_{ik} are meromorphic doubly periodic functions of the parameter u with periods $4K$ and $4iK'$, where $K = K(k)$ and $K' = K(k')$ are elliptic integrals of the first kind. In the complex u-plane, in the fundamental rectangle with sides $4K$ and $4iK'$, the simple poles are the only singularities of the functions $c_{ik}(u)$ and $d_{ik}(u)$.

As in the previous sections of this chapter, all formulas become easier if we passes from variables x and t to new variables $\xi = (t + x)/2$ and $\eta = (t - x)/2$. In the variables ξ

and η, the equations of motion of the model at hand and the Maurer–Cartan structural equations appear as

$$\begin{aligned} a_1(\partial_\eta \omega_{1,\xi} + \partial_\xi \omega_{1,\eta}) - (a_2 - a_3)(\omega_{2,\xi}\,\omega_{3,\eta} + \omega_{2,\eta}\,\omega_{3,\xi}) &= 0, \\ \partial_\eta \omega_{1,\xi} - \partial_\xi \omega_{1,\eta} + \omega_{2,\eta}\,\omega_{3,\xi} - \omega_{2,\xi}\,\omega_{3,\eta} &= 0. \end{aligned} \tag{7.5.2}$$

The cyclic permutation of indices 1, 2, and 3 adds four more equations.

Equations (7.5.2) are equivalent to the compatibility condition of the linear system

$$\partial_\xi \Phi = U\Phi, \quad \partial_\eta \Phi = V\Phi, \tag{7.5.3}$$

where the matrices U and V are defined by the formulas

$$\begin{aligned} U \equiv V_0 + V_1 &= \frac{\mathrm{i}}{2}\,[\sigma_1(c_{12} - d_{12})\omega_{2,\xi} \\ &\quad + \sigma_2(c_{21} - d_{21})\omega_{1,\xi} + \sigma_3(c_{33} - d_{33})\omega_{3,\xi}]\,, \\ V \equiv V_0 - V_1 &= \frac{\mathrm{i}}{2}\,[\sigma_1(c_{12} + d_{12})\omega_{2,\eta} \\ &\quad + \sigma_2(c_{21} + d_{21})\omega_{1,\eta} + \sigma_3(c_{33} + d_{33})\omega_{3,\eta}]\,. \end{aligned} \tag{7.5.4}$$

Since the coefficients $c_{ik}(u)$ and $d_{ik}(u)$ are doubly periodic functions of the spectral parameter u, it suffices to analyze the properties of solutions to system (7.5.3) within an elementary cell with sides $4K$ and $4\mathrm{i}K'$ in the complex u-plane.

As before, the relations $\mathrm{Sp}\,U = \mathrm{Sp}\,V = 0$ and eqs. (7.5.3) yield ξ- and η-independent $\det \Phi$. The conditions

$$U^\dagger(u^*) = -U(u), \quad V^\dagger(u^*) = -V(u)$$

enable us to infer that the solutions $\Phi^{-1}(u)$ and $\Phi^\dagger(u^*)$ to systems (7.5.3) may be chosen coinciding.

The transformation properties of coefficients (7.5.1) allow us to easily check that the solutions to systems (7.5.3) and (7.5.4) hold for the constraints

$$\Phi(u) = \sigma_3 \Phi(u + 2K)\sigma_3, \quad \Phi(u) = \Phi^*(u^* + 2\mathrm{i}K').$$

Representations (7.5.1), (7.5.3), and (7.5.4), as in Section 7.2, imply an explicit expression for the matrices $g(\xi, \eta) \in \mathrm{SU}(2)$ through the value of the matrix function $\Phi(\xi, \eta, u)$ at the point $u = 0$:

$$g = \exp\left(-\frac{\mathrm{i}}{4}\pi\sigma_3\right)\Phi(u = 0)[\det \Phi(u = 0)]^{-1/2}. \tag{7.5.5}$$

For definiteness, we choose eq. (7.3.11) as a particular solution to the nonlinear equations (7.5.2) of the asymmetric SU(2)-model. The solution to the linear equations (7.5.3) corresponding to such a choice has the form

$$\Phi_0(\xi,\eta,u) = \exp\left(\frac{\mathrm{i}\sigma_3}{2}\left[(c_{33}(u) - d_{33}(u))A(\xi)\right.\right.$$
$$\left.\left.+(c_{33}(u) + d_{33}(u))B(\eta)\right]\right). \tag{7.5.6}$$

In the context of the "dressing" method, we represent new solutions to the linear system (7.5.3) as follows:

$$\Phi(u) = \Psi(u)\Phi_0(u). \tag{7.5.7}$$

For soliton states, the new matrix function $\Psi(u)$ is a meromorphic doubly periodic function of the parameter u with periods $4K$ and $4\mathrm{i}K'$ and satisfies the reductions

$$\begin{aligned}\Psi^{-1}(u) &= \Psi^{\dagger}(u^*), \quad \Psi(u) = \sigma_3\Psi(u+2K)\sigma_3,\\ \Psi(u) &= \Psi^*(u^* + 2\mathrm{i}K').\end{aligned} \tag{7.5.8}$$

According to the first of eqs. (7.5.8), the zeros of the function $\Psi(u)$ coincide with the poles of the function $\Psi^{\dagger}(u^*)$. From the last two equalities it follows that the zeros of the function $\Psi(u)$ are located symmetrically with respect to the straight line $\operatorname{Im} u = K'$ and split into two groups

$$\mu_a^{(1)} = u_a + \mathrm{i}K', \quad \mu_a^{(2)} = u_a + \mathrm{i}K' + 2K \tag{7.5.9}$$

and

$$\begin{aligned}&\mu_a^{(1)} = u_a + \mathrm{i}K' + \mathrm{i}\,\theta_a, \quad \mu_a^{(2)} = \mu_a^{(1)*} + 2\mathrm{i}K',\\ &\mu_a^{(3)} = \mu_a^{(1)} + 2K, \quad \mu_a^{(4)} = \mu_a^{(1)*} + 2K + 2\mathrm{i}K',\\ &(a = 1, 2, \ldots, N),\end{aligned} \tag{7.5.10}$$

where $-2K \le u_a \le 0$ and $0 < \theta_a < K'$.

First, we look at the second case, where the matrix function $\Psi(u)$ has simple zeros at the points (7.5.10). As in Chapter 5, the matrix function $\Psi^{\dagger}(u^*)$ needs to be sought as an expansion in Weierstrass' zeta functions $\zeta(u) \equiv \zeta(u|2K, 2\mathrm{i}K')$ with periods $4K$ and $4\mathrm{i}K'$ or, rather, in terms of functions

$$f_a^{(i)}(u) = \zeta(u - \mu_a^{(i)}) + c^{(i)}, \quad c^{(1)} = -\eta_3 = -c^{(2)}, \quad c^{(3)*} = -\eta_2 = c^{(4)}.$$

The latter, as follows from the relations

$$\begin{aligned}f_a^{(s)}(u - 2K) &= f_a^{(s+2)}(u) - \eta_1,\\ f_a^{(2s-1)}(u + 2\mathrm{i}K') &= [f_a^{(2s)}(u^*)]^* \quad (s = 1, 2),\end{aligned}$$

constitute a representation of the reduction group (7.5.8) with accuracy up to an arbitrary constant term. Here and below, we use the conventional abbreviated notations [3, 1]

$$\zeta(2K|2K,2\mathrm{i}K') = \eta_1, \quad \zeta(-2K-2\mathrm{i}K'|2K,2\mathrm{i}K') = \eta_2 \equiv -\eta_1-\eta_3,$$
$$\zeta(2\mathrm{i}K'|2K,2\mathrm{i}K') = \eta_3.$$

Then we have

$$\Psi^\dagger(u^*) = \left(I + \sum_{\alpha=1}^{N}\sum_{i=1}^{4} R_i^\alpha f_\alpha^{(i)}(u)\right)C, \tag{7.5.11}$$

where the matrices R_i^α and C are u-independent. Since the functions $\zeta(u)$ are quasiperiodic (see [3, 1] and Appendix), representation (7.5.11) has periods $4K$ and $4\mathrm{i}K'$ only if

$$\sum_{\alpha=1}^{N}\sum_{i=1}^{4} R_i^\alpha = 0. \tag{7.5.12}$$

As in Chapter 5, we carry out further calculations under assumption (7.5.12). An ultimate outcome confirms the correctness of the constructions.

It is easy to show that the last two of constraints (7.5.8) lead to the relations

$$R_1^\alpha = R_2^{\alpha*} = \sigma_3 R_3^\alpha \sigma_3 = \sigma_3 R_4^{\alpha*}\sigma_3, \quad C = C^* = \sigma_3 C^* \sigma_3. \tag{7.5.13}$$

It follows that C is a diagonal matrix with real elements.

We define the explicit form of the matrices R_i^α as follows. Under the condition that the residues at the poles of the product $\Psi(u)\Psi^\dagger(u^*)$ vanish, due to relations (7.5.13), we obtain N independent matrix equations

$$[R_1^\alpha]^\dagger \Psi^\dagger(u^*)|_{u=\lambda_\alpha} = 0 \quad (\alpha = 1,2,\dots,N), \tag{7.5.14}$$

which are simplified by the notation $\lambda_\alpha = \mu_\alpha^{(1)*}$. The elements of the degenerate matrices R_1^α can be written as

$$(R_1^\alpha)_{ab} = (m^\alpha)_a (X^\alpha)_b \quad (a,b = 1,2;\ \alpha = 1,2,\dots,N). \tag{7.5.15}$$

The substitution of (7.5.15) into eqs. (7.5.14) determines the vectors $\mathbf{X}^\alpha$ from the system of linear equations

$$\begin{aligned}\mathbf{m}^{\beta*} + \sum_{\alpha=1}^{N} \big[&(\mathbf{m}^{\beta*}\cdot\mathbf{m}^\alpha)\mathbf{X}^\alpha f_\alpha^{(1)}(\lambda_\beta) + (\mathbf{m}^{\beta*}\cdot\mathbf{m}^{\alpha*})\mathbf{X}^{\alpha*} f_\alpha^{(2)}(\lambda_\beta)\\ &+ (\mathbf{m}^{\beta*}\cdot\sigma_3\mathbf{m}^\alpha)\mathbf{X}^\alpha\sigma_3 f_\alpha^{(3)}(\lambda_\beta)\\ &+ (\mathbf{m}^{\beta*}\cdot\sigma_3\mathbf{m}^{\alpha*})\mathbf{X}^{\alpha*}\sigma_3 f_\alpha^{(4)}(\lambda_\beta)\big] = 0.\end{aligned} \tag{7.5.16}$$

Here we have

$$\mathbf{m}^\alpha = (m_1^\alpha, m_2^\alpha), \quad \mathbf{X}^\alpha = (X_1^\alpha, X_2^\alpha),$$

$$(\mathbf{m}^{\alpha} \cdot \mathbf{m}^{\beta}) = m_1^{\alpha *} m_1^{\beta} + m_2^{\alpha *} m_2^{\beta}.$$

The result of solving eqs. (7.5.16) is easier to present in terms of N-dimensional matrices F_a and Q_a and vectors $\mathbf{p}_a = (m_a^1, m_a^2, \dots, m_a^N)$ and $\mathbf{X}_a = (X_a^1, X_a^2, \dots, X_a^N)$:

$$\begin{aligned}
\mathbf{X}_a &= -\left[F_a - Q_a F_a^{*-1} Q_a^*\right]^{-1} \left(\mathbf{p}_a^* - Q_a F_a^{*-1} \mathbf{p}_a\right); \\
(F_a)_{\beta\alpha} &= (\mathbf{m}^{\beta *} \cdot \mathbf{m}^{\alpha}) f_a^{(1)}(\lambda_\beta) \\
&\quad - (-1)^a (\mathbf{m}^{*\beta} \cdot \sigma_3 \mathbf{m}^{\alpha}) f_a^{(3)}(\lambda_\beta), \\
(Q_a)_{\beta\alpha} &= (\mathbf{m}^{\beta *} \cdot \mathbf{m}^{\alpha *}) f_a^{(2)}(\lambda_\beta) \\
&\quad - (-1)^a (\mathbf{m}^{\beta *} \cdot \sigma_3 \mathbf{m}^{*\alpha}) f_a^{(4)}(\lambda_\beta).
\end{aligned} \tag{7.5.17}$$

The constants in the definition of the functions $f_a^{(s)}(u)$ are chosen so that the following identities hold:

$$F_a^* = -F_a^{\mathrm{T}}, \quad Q_a^{\mathrm{T}} = -Q_a. \tag{7.5.18}$$

The symmetry properties (7.5.18) greatly facilitate the calculations. For example, for a one-soliton solution ($N = 1$), the numbers Q_a are equal to zero, and F_a are purely imaginary quantities. For a two-soliton solution, the matrices in formula (7.5.17) have a simple form

$$\begin{aligned}
\left[F_a - Q_a F_a^{*-1} Q_a^*\right]^{-1} &= \frac{1}{\det F_a - |(Q_a)_{12}|^2} \begin{pmatrix} (F_a)_{22} & -(F_a)_{12} \\ -(F_a)_{21} & (F_a)_{11} \end{pmatrix}, \\
\left[F_a - Q_a F_a^{*-1} Q_a^*\right]^{-1} Q_a F_a^{*-1} &= -\frac{Q_a}{\det F_a - |(Q_a)_{12}|^2}.
\end{aligned}$$

In addition, formulas (7.5.13) and (7.5.15)–(7.5.18) imply relation (7.5.12).

Using the first equality of reductions (7.5.8), we express the matrix C in terms of R_i^{α}:

$$\left(C^{-1}\right)^2 = \left(I + \sum_{i=1}^{4} \sum_{\alpha=1}^{N} R_i^{\alpha \dagger} f_\alpha^{(i)*}(u^*)\right) \left(I + \sum_{j=1}^{4} \sum_{\beta=1}^{N} R_j^{\beta} f_\beta^{(j)}(u)\right). \tag{7.5.19}$$

The previous analysis states that the right-hand side of eq. (7.5.19) is poles-free and is a positive-definite u-independent diagonal matrix. If we put $u = 0$ in eq. (7.5.19), we can obtain a convenient representation for the matrix C.

Note that the matrix C is defined up to multiplication by a matrix $W = \mathrm{diag}(\varepsilon_1, \varepsilon_2)$ with $\varepsilon_1^2 = \varepsilon_2^2 = 1$. It is further shown that this ambiguity is not essential for constructing new solutions of the model by the "dressing" method. This is because the model equations (7.5.2) are invariant regarding the replacement $\omega_1 \to -\omega_1$, $\omega_2 \to -\omega_2$, and $\omega_3 \to \omega_3$.

The vectors $\mathbf{m}^\alpha$ evolve as additional pole-free equations dictate within the Riemann problem in the "dressing" relations (7.3.13). The solutions to these equations are expressed through $\Phi_0(\lambda, \xi, \eta)$:

$$\mathbf{m}^\alpha = \Phi_0(\lambda_\alpha^*, \xi, \eta)\mathbf{c}^\alpha,$$

where $\mathbf{c}^\alpha$ are arbitrary constant complex vectors ($\alpha = 1, 2, \dots, N$).

The final formulas for the Riemann problem's matrices $\Psi(u)$ and $\Psi^\dagger(u^*)$ contain combinations of the Weierstrass zeta functions, which can be expressed in terms of the more popular Jacobi elliptic functions:

$$\begin{aligned}
&\alpha_\beta(u) \equiv f_\beta^{(1)}(u) - f_\beta^{(3)}(u) = \frac{\mathrm{dn} v}{2\,\mathrm{sn} v\,\mathrm{cn} v} = \frac{1 + \mathrm{dn} 2v}{2\,\mathrm{sn} 2v},\\
&v = \frac{u - \mu_\beta^{(1)}}{2}; \quad \varphi_\beta(u) - \varphi_\beta(\lambda_\beta) = \alpha_\beta(\lambda_\beta)\frac{\alpha_\beta(u)}{\alpha_\beta^*(u^*)},\\
&\varphi_\beta(u) = f_\beta^{(2)}(u) + f_\beta^{(4)}(u) - f_\beta^{(1)}(u) - f_\beta^{(3)}(u);\\
&f_\beta^{(2)}(u) - f_\beta^{(4)}(u) = \frac{k^2}{4\,\alpha_\beta^*(u^*)}.
\end{aligned} \tag{7.5.20}$$

Identities (7.5.20) have already been used before. Chapter 3 and Appendix carry their proofs.

In particular, using formulas (7.5.20), we can write the elements of the one-soliton matrix $\Psi^\dagger(u^*)$ as

$$\begin{aligned}
&[\Psi^\dagger(u^*)]_{11} = \mathrm{i}\, r\, \alpha(\lambda_1)\left(|m_1|^2\frac{\alpha(u)}{\alpha^*(u^*)} + |m_2|^2\right),\\
&[\Psi^\dagger(u^*)]_{12} = \mathrm{i}\, r\left(-m_1 m_2^*\,\alpha(u) + m_1^* m_2\,\frac{k^2}{4\,\alpha^*(u^*)}\right),\\
&[\Psi^\dagger(u^*)]_{21} = \mathrm{i}\, r\left(-m_1^* m_2\,\alpha(u) + m_1 m_2^*\,\frac{k^2}{4\,\alpha^*(u^*)}\right),\\
&[\Psi^\dagger(u^*)]_{22} = \mathrm{i}\, r\, \alpha(\lambda_1)\left(|m_2|^2\frac{\alpha(u)}{\alpha^*(u^*)} + |m_1|^2\right).
\end{aligned} \tag{7.5.21}$$

In relations (7.5.21), we have omitted the matrix $W = \mathrm{diag}(\varepsilon_1, \varepsilon_2)$ with $\varepsilon_1^2 = \varepsilon_2^2 = 1$. The quantities $r > 0$, $\alpha(u)$, and $\det\Psi^\dagger(u^*)$ are expressed in terms of Jacobi elliptic functions with modulus k:

$$\begin{aligned}
&r^{-2} = -\alpha^2(\lambda_1)(|m_1|^4 + |m_2|^4) + \delta|m_1 m_2|^2 - (m_1^2 m_2^{*2} + m_1^{*2} m_2^2)k^2/4,\\
&\delta = \frac{k^4}{16|\alpha(0)|^2} + |\alpha(0)|^2 - \alpha^2(\lambda_1)\left(\frac{\alpha(0)}{\alpha^*(0)} + \frac{\alpha^*(0)}{\alpha(0)}\right);\\
&\det\Psi^\dagger(u^*) = \frac{\alpha(u)}{\alpha^*(u^*)}; \quad \alpha(u) = \frac{\mathrm{dn}\bar{u}}{2\,\mathrm{sn}\bar{u}\,\mathrm{cn}\bar{u}}, \quad \bar{u} = \frac{u - \lambda_1^*}{2}.
\end{aligned}$$

The next step is to pass to constructing the functions $\Psi(u)$ with specific zeros (7.5.9). The examples given in Chapters 3–5 show that accounting for the specific zeros in the "dressing" procedure may change the ground state of the system at infinity and lead to solutions describing uncompensated domains. This does not happen if the function $\Psi(u)$ contains an even number of different groups of simple zeros (7.5.9). Keeping in mind the formulated criterion and reductions (7.5.8), we search for $\Psi^{\dagger}(u^*)$ in the form of the following expansion in terms of the Weierstrass zeta functions:

$$\Psi^{\dagger}(u^*) = \left(I + \sum_{\alpha=1}^{2N} [R^{\alpha} g_{\alpha}^{(1)}(u) + \sigma_3 R^{\alpha} \sigma_3 g_{\alpha}^{(2)}(u)] \right) C, \quad R^{\alpha} = R^{\alpha *},$$
$$g_{\alpha}^{(1)}(u) = \zeta(u - \mu_{\alpha}^{(1)}) + \eta_3, \quad g_{\alpha}^{(2)}(u) = \zeta(u - \mu_{\alpha}^{(2)}) - \eta_2.$$

Here C is a diagonal real matrix, and

$$\mu_{\alpha}^{(1)} = u_{\alpha} + \mathrm{i}K', \quad \mu_{\alpha}^{(2)} = \mu_{\alpha}^{(1)} + 2K, \quad -2K \le u_{\alpha} \le 0.$$

As before, the elements of the degenerate matrices R^{α} should be represented as

$$(R^{\alpha})_{ab} = (m^{\alpha})_a (X^{\alpha})_b \quad (a, b = 1, 2;\ \alpha = 1, 2, \dots, 2N).$$

The matrix R^{α} is real, for example, if the vectors $\mathbf{m}^{\alpha}$ and $\mathbf{X}^{\alpha}$ are real. Then the function $\Psi^{\dagger}(u^*)$ doubly periodic under the constraint

$$\sum_{\alpha=1}^{2N} \mathrm{diag}[m_1^{\alpha} X_1^{\alpha}, m_2^{\alpha} X_2^{\alpha}] = 0. \tag{7.5.22}$$

Further, we assume that the vectors $\mathbf{m}^{\alpha}$ are real. The reality of the vectors $\mathbf{X}^{\alpha}$ and relation (7.5.22) are consequences of the reality of $\mathbf{m}^{\alpha}$.

The condition for the residues to vanish at the poles of the product $\Psi(u)\Psi^{\dagger}(u^*)$ produces algebraic equations for calculating the vectors $\mathbf{X}^{\alpha} = (X_1^{\alpha}, X_2^{\alpha})$ $(\alpha, \beta = 1, 2, \dots, 2N)$. As before, it is convenient to write their solution in terms of $2N$-dimensional matrices G_a and vectors $\mathbf{p}_a = (m_a^1, m_a^2, \dots, m_a^{2N})$ and $\mathbf{X}_a = (X_a^1, X_a^2, \dots, X_a^{2N})$:

$$\mathbf{X}_a = -G_a^{-1} \mathbf{p}_a, \quad [G_a^{-1}]_{\alpha\beta} = \frac{\partial \ln \det G_a}{\partial [G_a]_{\beta\alpha}},$$
$$(G_a)_{\beta\alpha} = (\mathbf{m}^{\beta} \cdot \mathbf{m}^{\alpha}) g_{\alpha}^{(1)}(\lambda_{\beta}) - (-1)^a (\mathbf{m}^{\beta} \cdot \sigma_3 \mathbf{m}^{\alpha}) g_{\alpha}^{(2)}(\lambda_{\beta}),$$

where $a, b = 1, 2$, and $\lambda_{\beta} = \mu_{\beta}^{(1)*}$. The matrices G_a have the symmetry properties

$$G_a^{\mathrm{T}} = -G_a, \quad G_a^* = G_a.$$

It is easy to verify that the consequences of the reality of $\mathbf{m}^{\alpha}$ are the reality of the vectors $\mathbf{X}^{\alpha}$ and identity (7.5.22).

As before, the evolution of the vectors $\mathbf{m}^a$ is determined by the formula

$$\mathbf{m}^a = \Phi_0(\lambda_a^*, \xi, \eta)\mathbf{c}^a, \quad \lambda_a^* \equiv \mu_a^{(1)} = u_a + \mathrm{i}K'.$$

For the "seeding" solution (7.5.6), the matrix $\Phi_0(\lambda_a^*, \xi, \eta)$ is real. Hence the vectors $\mathbf{m}^a$ are real only when arbitrary constant vectors $\mathbf{c}^a = (c_1^a, c_2^a)$ are real. Without loss of generality, the numbers $c_{1,2}^a$ can be chosen so that the condition $c_1^a c_2^a > 0$ should be fulfilled (there is no summation over a).

We give an expression for the elements of a one-soliton matrix (up to the insignificant matrix $W = \mathrm{diag}(\varepsilon_1, \varepsilon_2)$):

$$\begin{aligned}
[\Psi^\dagger(u^*)]_{11} &= r(\alpha_1(u)\alpha_2(u)\, m_1^1 m_1^2 + k^2 m_2^1 m_2^2/4),\\
[\Psi^\dagger(u^*)]_{12} &= r\,\alpha_1(\lambda_2^*)\,[-m_1^1 m_2^2\, \alpha_1(u) + m_2^1 m_1^2\, \alpha_2(u)],\\
[\Psi^\dagger(u^*)]_{21} &= r\,\alpha_1(\lambda_2^*)[-m_1^2 m_2^1\, \alpha_1(u) + m_2^2 m_1^1\, \alpha_2(u)],\\
[\Psi^\dagger(u^*)]_{22} &= r(\alpha_1(u)\alpha_2(u)\, m_2^1 m_2^2 + k^2 m_1^1 m_1^2/4),
\end{aligned} \tag{7.5.23}$$

where

$$\alpha_\beta(u) = \left.\frac{1 + \mathrm{dn}v}{2\,\mathrm{sn}v}\right|_{v=u-\mu_\beta^{(1)}} = \frac{k}{2}[\mathrm{sn}(u - u_\beta) + \mathrm{i}\,\mathrm{cn}(u - u_\beta)], \quad \beta = 1, 2;$$

$$\begin{aligned}
r^{-2} &= \left[\frac{k}{2}\right]^4 \Big((m_1^1 m_1^2)^2 + (m_2^1 m_2^2)^2\\
&\quad + [2k^{-1}\alpha_1(\lambda_2^*)]^2\,[(m_1^1 m_2^2)^2 + (m_2^1 m_1^2)^2] + 2\delta\, m_1^1 m_2^1\, m_1^2 m_2^2\Big),\\
\delta &= \mathrm{sn}u_1\mathrm{sn}u_2 - \mathrm{cn}u_1\mathrm{cn}u_2 - [2k^{-1}\alpha_1(\lambda_2^*)]^2(\mathrm{sn}u_1\mathrm{sn}u_2 + \mathrm{cn}u_1\mathrm{cn}u_2).
\end{aligned}$$

It is easy to check that

$$\begin{aligned}
\det\Psi^\dagger(u^*) &= 4k^{-2}\alpha_1(u)\alpha_2(u)\\
&= -[\mathrm{cn}(u - u_1) - \mathrm{i}\,\mathrm{sn}(u - u_1)][\mathrm{cn}(u - u_2) - \mathrm{i}\,\mathrm{sn}(u - u_2)].
\end{aligned}$$

This completes constructing the Riemann problem's multisoliton matrices for an asymmetric chiral model. It is worth stressing that the reductions in the Riemann problems for the biaxial ferromagnet's Landau–Lifshitz equations and for an asymmetric chiral model do not literally coincide but are close. Therefore formulas (5.2.47) and (7.5.23) for the matrices of the Riemann problems (5.2.27) and (7.5.21) differ little from each other. Consequently, from a qualitative point of view, the solutions to the equations for the asymmetric chiral model should be close to those for the model of a biaxial ferromagnet.

According to formulas (7.5.5) and (7.5.7), using the chosen "seeding" solution Φ_0 and the multisoliton matrices Ψ, we can find new solutions to the asymmetric chiral model in terms of the fields $g(\xi, \eta) \in \mathrm{SU}(2)$:

$$g = \exp\left(-\frac{i}{4}\pi\sigma_3\right)\frac{\Psi(u=0)\Phi_0(u=0)}{\sqrt{\det[\Psi(u=0)\Phi_0(u=0)]}}.$$

Next, we draw working formulas for Cartan's "dressed" forms. It should be emphasized that the matrix functions $U(\lambda)$ and $V(\lambda)$ (7.5.4) have simple poles at the points that are the roots of the equation $\operatorname{cn}^2 u = a_1/a_3$. Suppose that u_0 is the root of the equation $\operatorname{cn} u = \sqrt{a_1/a_3}$ in the interval $0 \le \operatorname{Re} u_0 \le 2K$, $\operatorname{Im} u_0 = 0$. Then we have

$$\operatorname{sn} u_0 = \sqrt{\frac{a_3 - a_1}{a_3}}, \quad \operatorname{cn} u_0 = \sqrt{\frac{a_1}{a_3}}, \quad \operatorname{dn} u_0 = \sqrt{\frac{a_1}{a_2}}.$$

Equating the residues at the poles $u = \pm u_0$ of the "dressing" relations (7.3.13), we derive simple formulas expressing the new Cartan forms in terms of the "seeding" ones:

$$\begin{aligned}
&\sigma_1\sqrt{a_2}\,\omega_{2,\xi} - \sigma_2\sqrt{a_1}\,\omega_{1,\xi} + \sigma_3\sqrt{a_3}\,\omega_{3,\xi} \\
&\quad = \Psi(u=u_0)\big[\sigma_1\sqrt{a_2}\,\omega_{2,\xi}^{(0)} - \sigma_2\sqrt{a_1}\,\omega_{1,\xi}^{(0)} + \sigma_3\sqrt{a_3}\,\omega_{3,\xi}^{(0)}\big]\Psi^\dagger(u=u_0); \\
&\sigma_1\sqrt{a_2}\,\omega_{2,\eta} - \sigma_2\sqrt{a_1}\,\omega_{1,\eta} + \sigma_3\sqrt{a_3}\,\omega_{3,\eta} \\
&\quad = \Psi(u=-u_0)\big[\sigma_1\sqrt{a_2}\,\omega_{2,\eta}^{(0)} - \sigma_2\sqrt{a_1}\,\omega_{1,\eta}^{(0)} + \sigma_3\sqrt{a_3}\,\omega_{3,\eta}^{(0)}\big]\Psi^\dagger(u=-u_0).
\end{aligned} \tag{7.5.24}$$

We define the vector fields

$$\begin{aligned}
\mathbf{A}(\xi,\eta) &= \frac{(a_1+a_2+a_3)}{2\sqrt{a_1 a_2 a_3}}(\sqrt{a_1}\omega_{1,\xi},\ \sqrt{a_2}\omega_{2,\xi},\ \sqrt{a_3}\omega_{3,\xi}), \\
\mathbf{B}(\xi,\eta) &= \frac{(a_1+a_2+a_3)}{2\sqrt{a_1 a_2 a_3}}(\sqrt{a_1}\omega_{1,\eta},\ \sqrt{a_2}\omega_{2,\eta},\ \sqrt{a_3}\omega_{3,\eta}).
\end{aligned}$$

As in the previous sections, relations (7.5.24) imply that the lengths of the vectors **A** and **B** constructed by the "dressing procedure" coincide with those for the "seeding" solutions: $\mathbf{A}^2 = \mathbf{A}_0^2$, $\mathbf{B}^2 = \mathbf{B}_0^2$. For the particular solution (7.3.11), the length of the vector **A** does not depend on η, and the length of the vector **B** does not depend on ξ. This is not accidental because the equations for the asymmetric model (7.5.2) in terms of the fields **A** and **B** have the form

$$\begin{aligned}
\partial_\eta\mathbf{A} + [\mathbf{B}\times\mathbf{A}] + 2(a_1+a_2+a_3)^{-1}[\mathbf{A}\times a\mathbf{B}] &= 0, \\
\partial_\xi\mathbf{B} + [\mathbf{A}\times\mathbf{B}] + 2(a_1+a_2+a_3)^{-1}[\mathbf{B}\times a\mathbf{A}] &= 0,
\end{aligned} \tag{7.5.25}$$

where $a = \operatorname{diag}(a_1, a_2, a_3)$.

For the fields **A** and **B** with normalization $\mathbf{A}^2 = \mathbf{B}^2 = 1$ and asymptotic behavior $\mathbf{A}, \mathbf{B} \to (0,0,1)$ as $|x| \to \infty$, the system energy can be expressed as

$$\begin{aligned}
E = \int\bigg[&\frac{A_1\partial_x A_2 - A_2\partial_x A_1}{1 - A_3} - \frac{B_1\partial_x B_2 - B_2\partial_x B_1}{1 - B_3} \\
&+ (\mathbf{A}\cdot\mathbf{B}) - \frac{2}{a_1+a_2+a_3}\sum_{i=1,2} a_i A_i B_i + \frac{a_3 - a_1 - a_2}{a_1+a_2+a_3}\bigg]dx.
\end{aligned}$$

Let us first look into the one-soliton solution to the asymmetric model, obtained by "dressing" the particular solution (7.3.11) using the Riemann problem with zeros (7.5.10). It is convenient to write the final result in terms of the Euler angles θ, φ, and ψ, which define, in particular, the orientation of three mutually perpendicular antiferromagnetic vectors of an equilibrium exchange structure. After simple transformations, we get

$$\cos\theta = 1 - \frac{2(\cosh 2\sigma - \cos 2S)}{\cosh 2\sigma - \cos 2S + [2ik^{-1}\alpha(\lambda_1)]^2(\cosh 2y + \cos 2\gamma)}, \tag{7.5.26}$$

$$\begin{aligned} \varphi + \psi &= -\frac{\pi}{2} + A(\xi) + B(\eta) + \frac{1}{\mathrm{i}}\ln\frac{\cosh(y - \mathrm{i}\gamma)}{\cosh(y + \mathrm{i}\gamma)}, \\ \psi - \varphi &= -\frac{\pi}{2} - A(\xi) - B(\eta) + \frac{1}{\mathrm{i}}\ln\frac{\sinh(\sigma + \mathrm{i}S)}{\sinh(\sigma - \mathrm{i}S)}, \end{aligned} \tag{7.5.27}$$

where the real parameters y, S, σ, and γ are given by the relations

$$y = \ln\left|\frac{m_1}{m_2}\right|, \quad S = \arg\frac{m_1}{m_2}, \quad \alpha(0) = \frac{k}{2}\exp(\sigma + \mathrm{i}\gamma).$$

Let us discuss the simplest solution. We choose the functions $A(\xi)$ and $B(\eta)$ as in (7.4.25). Then the variables y and S take the form (7.4.26). Expression (7.5.26) for $\cos\theta$ is a spatially localized pit-shaped wave, which moves as a whole with speed $V = -\partial_t y/\partial_x y$. However, against solution (7.4.23) to axisymmetric model, the profile of the pit undergoes additional pulsations with frequency $\Omega = 2\partial_t S$ (Figure 7.6). It is interesting that up to designations, expression (7.5.26) for $\cos\theta$ coincides with formula (5.2.31) for the third component of the magnetization vector in a biaxial ferromagnet. Therefore, according to the results of Chapter 5, the pit in Figure 7.6 becomes the most extended and deepest when the complex parameter of the breather $\mu_1^{(1)} = \lambda_1^*$ approaches specific values of (7.5.9). In this case the pit is limited by walls with a width of order $(\partial_x y)^{-1}$, between which $\cos\theta$ rapidly alters from one limiting value $\cos\theta \approx 1$ to another $\cos\theta \approx -1$. The condition $S_3 = \cos\theta \approx 0$ controls the distance between the walls (the length of the pit); it turns out to be much greater than their width. The walls bounding the pit perform oscillations about the breather's "center of gravity". In the limit as $\mu_1^{(1)} \to u_1 + \mathrm{i}K'$, the amplitude of oscillations increases so much that the breather breaks up into two kinks moving away from each other (see Figure 5.5).

Formula (7.5.27) for the field $(\varphi + \psi)$ resembles solution (7.4.23) to an axisymmetric model. At the same time, the field $(\psi - \varphi)$ is described not just by a linear function of x and t but by the sum of linear and periodic functions of these variables.

The magnetization distribution corresponding to a multisublattice soliton has a homogeneous asymptotics

$$\mathbf{M} \to -\gamma\hbar a_3(u_1 + u_2)\mathbf{n} \quad \text{as} \quad x \to \pm\infty,$$

where $\mathbf{n} = (0, 0, 1)$, and involves an extremely complicated, spatially localized pulsating formation moving at a constant speed. In the region of soliton localization, the axes of

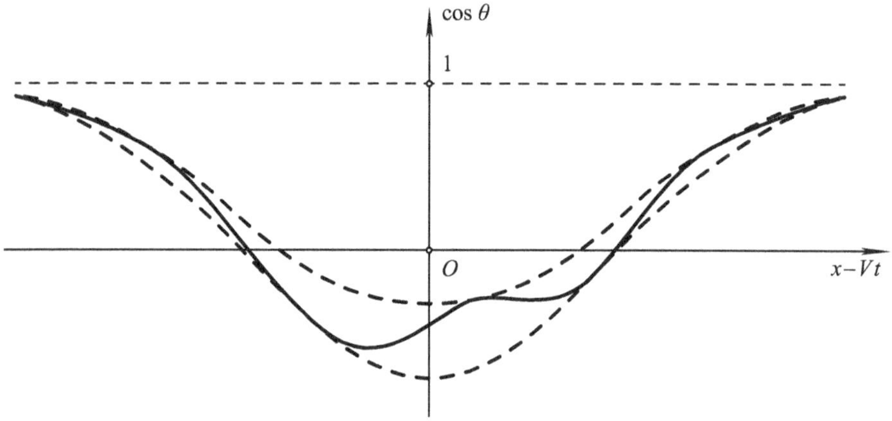

Figure 7.6: Field distribution for a pulsating soliton.

local rotations of the magnetization vector and the positions of the antiferromagnetic vectors of the exchange structure oscillate in time and change in space with increasing distance from the soliton center.

Next, we focus on the simplest solution to an asymmetric model, obtained by "dressing" the particular solution (7.3.11) using the matrices $\Psi(u)$ (7.5.23) with specific zeros (7.5.9), resulting in terms of Euler angles:

$$\begin{aligned}\cos\theta &= 1 - \frac{2D^2[\cosh(y_2 - y_1) - \cos(\gamma_2 - \gamma_1)]}{\cosh(y_2 + y_1) - \cos(\gamma_2 + \gamma_1) + D^2[\cosh(y_2 - y_1) - \cos(\gamma_2 - \gamma_1)]};\\ \varphi + \psi &= \frac{\pi}{2} + A(\xi) + B(\eta) + \frac{1}{\mathrm{i}}\ln\frac{\sinh[(y_2 + y_1 + \mathrm{i}(\gamma_2 + \gamma_1))/2]}{\sinh[(y_2 + y_1 - \mathrm{i}(\gamma_2 + \gamma_1))/2]},\\ \psi - \varphi &= \frac{\pi}{2} - A(\xi) - B(\eta) + \frac{1}{\mathrm{i}}\ln\frac{\sinh[(y_2 - y_1 - \mathrm{i}(\gamma_2 - \gamma_1))/2]}{\sinh[(y_2 - y_1 + \mathrm{i}(\gamma_2 - \gamma_1))/2]}.\end{aligned} \tag{7.5.28}$$

The real parameters D, y_β, and γ_β are defined by the formulas

$$D = 2k^{-1}\alpha_1(\lambda_2^*), \quad y_\beta = \ln\frac{m_2^\beta}{m_1^\beta}, \quad \gamma_\beta = \mathrm{am}(u_\beta, k), \quad \beta = 1, 2,$$

where $\mathrm{am}(u, k)$ is the amplitude of u:

$$\exp[\mathrm{i}\,\mathrm{am}(u, k)] = \mathrm{cn}(u, k) + \mathrm{i}\,\mathrm{sn}(u, k).$$

When choosing linear functions $A(\xi)$ and $B(\eta)$ (7.4.25), the variables y_β take the form

$$y_\beta = y_\beta^{(0)} + \kappa_\beta\,(x - V_\beta t),$$

where $y_\beta^{(0)}$, κ_β, and V_β are real parameters. Then expression (7.5.28) for the field $\cos\theta$ differs only by the constant $D^2\cos(\gamma_2-\gamma_1)$ from the projection S_3 of the magnetization vector **S** of a biaxial ferromagnet in the solution that describes elastic collisions of two domain walls (5.2.49). Hence the field $\cos\theta$ governs elastic collisions of two kinks, as shown in Figure 5.6. In the frame of reference connected with one of them when the kinks diverge over long distances (in the limit $y_{1,2}=\text{const}, y_{2,1}\to\pm\infty$), solution (7.5.28) matches a separate kink. As a result, the asymptotics of the functions $\varphi\pm\Psi$ differ from the "seeding" values only in constant summands.

In the general case, choosing the different functions $A(\xi)$ and $B(\eta)$, we obtain pulsating solitons or solitary kink-type waves against the background of nonlocalized magnetization waves and can thoroughly describe their interactions.

The Gel'fand–Levitan–Marchenko formalism-based integration algorithm for eqs. (7.5.25) with homogeneous asymptotic behavior of the fields **A** and **B** at infinity is discussed in [13]. It is shown that in this case the elementary excitations of the system create an ideal gas consisting of solitons and quasiparticles with continuous spectrum.

8 Weakly nonlinear magneto-static and magneto-elastic solitons

When significantly exposed, a system cannot be properly described in terms of Fourier harmonics (or linear quasiparticles). This is because the interaction between harmonics intensifies. However, if the wave amplitudes are not too large, then the harmonic interaction is weak. In this case the nonlinear dynamics of a real physical system can be correctly approximated within the nonlinear reductive perturbation theory (RPT) [22]. To date, RPT is the only bridge between the linear theory and the theory of solitons. When constructing RPT, as the physics of a problem dictates, a real system must contain spatio-temporal regions to incorporate slow spatial and temporal variables through scaling transformations. The ultimate goal is to describe the system in terms of the theory of solitons. Since a soliton is known to be the result of a balance between dispersion and nonlinearity, the scaling transformations should be chosen so as to match the spatio-temporal response of the system to a perturbation and a linear dispersion law and to balance the effects of dispersion and nonlinearity. With an appropriate choice of scaling transformations, the lower orders of RPT yield a closed effective equation for the weakly nonlinear dynamics of a real system. In practice, it is important that RPT disclosing hidden algebraic symmetries, as a rule, leads to equations that have amazing universality and turn out to be integrable or close to them. This achieves a detailed description of the real system within the inverse scattering method or soliton perturbation theory.

To predict new types of solitons and nonlinear defects and to change the understanding about a medium's microstructure and properties, building nonlinear models is necessary. On this path, the theory can claim to be an analytical description of the thermodynamic, kinetic, mechanical, electrical, and other properties of essentially nonlinear systems.

There are various versions of nonlinear perturbation theory to simplify models for weakly nonlinear dynamics of a number of magnetic, elastic, and magneto-elastic systems with an unusual ground state or nontrivial nature of complicated-in-description interactions caused by superstructures, spontaneous deformations, and manifestations of magnetostatics. The publications [56, 47, 50, 51, 48, 46, 49, 52] focus on the features of the weakly nonlinear dynamics of exchange-dipole spin waves in ferro- and antiferromagnetic plates due to the competition between exchange and nonlocal magnetostatic dispersions and the influence of the sample surface. It has been established in [56, 47, 50, 51, 48] that the modulations of the activation spin wave envelope cannot be always described by such a traditionally used model as the nonlinear Schrödinger equation [64, 94, 37–39, 36, 13, 81, 16, 15]. Alternative models, including nonlocal models, are proposed, and new types of nonlinear excitations are predicted. The nonlinear dynamics of gapless exchange-dipole waves in antiferromagnetic films has been studied in [46, 49, 52]. The works [14, 44, 53, 45] have described the transfer of energy in the region

https://doi.org/10.1515/9783112235973-008

of intense resonant interaction between activation and gapless modes in antiferromagnets with helicoidal structure and in magnetoelastic media and have explored various soliton regimes in the vicinity of a nonlinear resonance. Far from resonance, the weakly nonlinear dynamics of magnetoelastic waves in ferro- and antiferromagnets has been considered in [89, 72, 33, 86]. Three-wave interactions in antiferromagnets, satisfying the synchronism conditions, are studied in [68]. Near orientational phase transitions, two-sublattice magnetoelastic solitons, including those associated with different resonantly coupled branches of the linear spectrum (see [80] and references therein), have been investigated under driven electric and magnetic fields and external mechanical stress.

This chapter illustrates the effectiveness of the RPT methods by using the theoretical description of low-amplitude exchange-dipole solitons in ferro- and antiferromagnetic plates and magneto-elastic solitons in ferromagnets. The media listed are not equivalent in terms of the intensity of nonlinear interactions. The most significant nonlinearity addresses magnetic media. In magnetoelastic media the elastic anharmonicity is mainly induced by the magnetic subsystem.

8.1 Nonlinear dynamics of exchange-dipole spin waves in a finite-thickness ferromagnetic plate

Don't throw anything away.

The first rule of smart repair

Research on the features of excitation and propagation of nonlinear spin waves in magnetic plates (films) is aided by the range of nonlinear phenomena observed in films, the basic physical properties of these materials, and their potential applications in computer technology and microelectronics (different modulators and multipliers, components of logic devices), diagnostics, and manufacturing control. Interesting features of film materials are either absent or weakly expressed in bulk samples. Ferromagnetic films of yttrium iron garnets are thought to be thin if their thickness and length do not exceed tens of microns and tens of centimeters, respectively. They can exhibit ferromagnetic properties within the range of 1 GHz to 20 GHz. This frequency interval is thoroughly explored due to the possibility of excitation of exchange-dipole spin waves propagating along the films. One of the main results of this study is to discover solitons of a spin wave envelope in ferromagnetic films [37–39, 36].

The weakly nonlinear dynamics of spin waves in magnetic films can be theoretically described by the local nonlinear Schrödinger equation (NLSE) [64, 94, 37–39, 36, 13, 81, 16, 15]. Its solutions under various initial and boundary conditions have been studied by the inverse scattering method [73, 84, 10, 25, 27, 11, 12] and numerical methods [81, 16, 15]. Usually, for the long-wavelength region, a simplified NLSE derivation involving the nonlinear dispersion relation $\omega = \omega(k, |\varphi|)$ for spin waves is used. It implies that

$\omega = \omega(k, |\varphi|)$ is differentiable over the wavenumber k. However, the nonlinear dispersion relation $\omega = \omega(k, |\varphi|)$ is a nondifferentiable function in the region of small wave numbers ($|kd| \ll 1$, where d is the film thickness) [19]. Consequently, the NLSE derivation is not suitable. When it comes to a consistent consideration of the effects triggered by the exchange and dipole–dipole interactions, issues about the structure of soliton-like states and about the data interpretation in terms of the shape of the "envelope" and its coordinate and functional dependencies require a more comprehensive examination. In the event of the comparability-in-magnitude of the exchange and dipole energies in linear problems posed for these effects, the specifics of the nonlinear problems is largely determined not by the ratio of energies but rather by the competition between two types of spatial dispersion, exchange and magneto-static ones. In addition, the problem becomes not only non-one-dimensional but also nonlocal when the magnetostatics is accounted for, necessitating a careful study of the boundary conditions on the sample surface.

When studying the processes of propagation of activation waves with wave vector $\mathbf{k}$ in a ferromagnetic plate, it is necessary to accept the following hierarchy of scales: Firstly, the continuum approximation holds, i. e., $k^{-1} \gg a$, where a is the lattice constant, and $k \equiv |\mathbf{k}|$. Secondly, the magneto-static approximation is perceived to be sufficient to describe magnetization oscillations, i. e., $k \gg \omega/c$, where ω is the oscillation frequency, and c is the speed of light [4]. For real values of the appropriate parameters, we have $\omega/c \approx 1 \div 10\ \text{cm}^{-1}$. Thirdly, the resulting equations should govern the spatial modulations of the magnetization with a characteristic scale of λ. The latter must be larger than k^{-1} but smaller than the sample's linear size (length L). The current section investigates the weakly nonlinear dynamics of exchange-dipole spin excitations in a ferromagnetic plate across the following range of scales:

$$L \gg \lambda \gg k^{-1} \gg a, \quad \lambda \ll c/\omega.$$

In this interval a correct description of the interaction of spin waves is possible within local equations of dynamics. The present section constructs effective local equations using a version of reductive perturbation theory (RPT) proposed in [48]. The approach mentioned above is not suitable for the region of too small k. Nonetheless, it offers notable advantages for the region of moderate and large values of wave numbers, enabling us to overcome the issues connected to the nonlinearity of the problem as well as the nonlocality and non-one-dimensionality of magneto-static interactions.

It is known (see [19, 4, 28, 69, 70, 23, 24, 35]) that the spatial quantization of the wave vector components of a spin wave along the normal to the plate surface affects the exchange-dipole spectrum. In this context the latter can be described by separate curves related to a set of waves running along the plate. The approach in [48] makes it possible to construct equations for envelopes of any spin-wave mode to evolve from this spectrum for plates of arbitrary thickness.

When there are no peculiarities in the dispersion law for spin waves, the effective equation reduces to the NLSE. It should be especially emphasized that when deriving the NLSE, the wave interaction constant accounts for the inhomogeneities of the magnetization distribution along the normal to the plate and nontrivially depends both on the wave vector of the fundamental harmonics and on the plate thickness. Therefore the NLSE derivation proposed in [94] and afterward having become a conventional one due to expanding the nonlinear dispersion relation $\omega = \omega(k, |\varphi|)$ in powers of the spin wave amplitude $|\varphi|$ leads to wrong results in the general case. This approach argues that an amplitude-frequency dependency should be the same as in the theory of uniform ferromagnetic resonance: $\omega = \omega(k = 0, |\varphi|)$. As a result, the features of the interaction between the inhomogeneous-over-thickness plate and the spin waves with close values of nonzero wave vectors, propagating along the plate, have been ignored.

Recall that in the lower-energy branch of the spectrum of spin waves, two types of dispersion, exchange and magneto-static ones, compete with each other. Consequently, the dependence $\omega = \omega(k)$ has an inflection point (a point of zero dispersion) in the region of moderate values of the wave numbers. In the vicinity of the inflection point, a change in the mode of wave propagation takes place. In this region, although the effective equation remains local, it does not boil down to the NLSE. The features of the nonlinear dynamics of magneto-static waves in the vicinity of the zero dispersion point will be discussed separately.

8.1.1 Reductive perturbation theory

Let d be the thickness of a ferromagnetic plate in the Oz-direction, unbounded in the xOy-plane, and pierced by a uniform external constant magnetic field $\mathbf{H}_0$ oriented along the normal to its surface (along the Oz-axis). For simplicity, we take into account only the isotropic exchange interaction. Suppose that the plate is uniformly magnetized to saturation in the ground state and spin waves propagate along the Ox-axis. The dynamics of the magnetization vector is governed by the Landau–Lifshitz equation. Since we consider low-amplitude waves, in the initial equations, we restrict ourselves to cubic terms in the amplitude of deviations of the magnetization $M_0\mathbf{n}$ ($\mathbf{n} = (0, 0, 1)$) from the ground state:

$$\mathbf{M} = M_0\mathbf{n} + \mathbf{m}; \quad m_z \cong -\frac{m_x^2 + m_y^2}{2M_0}. \tag{8.1.1}$$

In this approximation, the Landau–Lifshitz equations take the following form:

$$\begin{aligned} \partial_t m_x &= -\gamma[m_y(H_0 + H_z^{(\mathrm{m})} + \alpha\Delta m_z) - (M_0 + m_z)\alpha\Delta m_y], \\ \partial_t m_y &= -\gamma\left[(M_0 + m_z)(H_x^{(\mathrm{m})} + \alpha\Delta m_x)\right. \\ &\qquad \left. - m_x(H_0 + H_z^{(\mathrm{m})} + \alpha\Delta m_z)\right]. \end{aligned} \tag{8.1.2}$$

We conduct an analysis for the case of free surface spins:

$$\partial_z m_i|_{z=\pm d/2} = 0, \quad i = x, y. \tag{8.1.3}$$

This experiment in such a situation (a normally magnetized plate in the absence of spin pinning) is apparently difficult to implement. Nevertheless, it is useful to mentally design it to reveal a number of features that may occur in more complicated cases.

A demagnetizing field $\mathbf{H}^{(\mathrm{m})}$ satisfies the equations of magnetostatics

$$\Delta\varphi = 4\pi \operatorname{div} \mathbf{m}; \quad \mathbf{H}^{(\mathrm{m})} = -\nabla\varphi; \quad |z| < d/2;$$
$$\Delta\tilde{\varphi} = 0; \quad \mathbf{H}^{(\mathrm{m})} = -\nabla\tilde{\varphi}; \quad |z| > d/2.$$

Here φ and $\tilde{\varphi}$ are the scalar magneto-static potentials inside and outside the plate, respectively. The conditions for their sewing at the plate boundary are

$$\varphi|_{z=\pm d/2} = \tilde{\varphi}|_{z=\pm d/2}; \quad -\partial_z\tilde{\varphi}|_{z=\pm d/2} = (-\partial_z\varphi + 4\pi(M_0 + m_z))|_{z=\pm d/2}.$$

For what follows, it is important that the boundary value problem can be reformulated in such a way as to completely eliminate the use of the potentia $\tilde{\varphi}$ outside the plate. For this purpose, in the potentials φ and $\tilde{\varphi}$, we should select the terms corresponding to the equilibrium state:

$$\varphi = 4\pi M_0 z + f; \quad \tilde{\varphi} = \begin{cases} 2\pi M_0 d + \tilde{f}, & z \geq d/2, \\ -2\pi M_0 d + \tilde{f}, & z \leq -d/2. \end{cases}$$

Then the equations of magnetostatics and the boundary conditions take the form

$$\Delta f = 4\pi \operatorname{div} \mathbf{m}, \quad -\partial_z\tilde{f}|_{z=\pm d/2} = (-\partial_z f + 4\pi m_z)|_{z=\pm d/2}; \tag{8.1.4}$$
$$\Delta\tilde{f} = 0, \quad \tilde{f}(z = \pm d/2) = f(z = \pm d/2); \quad \tilde{f} \to 0 \quad (|z| \to \infty). \tag{8.1.5}$$

The solution to the Dirichlet boundary value problem (8.1.5) by determining a potential $\tilde{f}$ outside the plate can be written as

$$\begin{aligned} \tilde{f}(x,z) &= \frac{1}{\pi}\int_{-\infty}^{+\infty} \frac{f(x', z = d/2)(z - d/2)}{(x' - x)^2 + (z - d/2)^2}\,\mathrm{d}x', \quad z \geq \frac{d}{2}; \\ \tilde{f}(x,z) &= -\frac{1}{\pi}\int_{-\infty}^{+\infty} \frac{f(x', z = -d/2)(z + d/2)}{(x' - x)^2 + (z + d/2)^2}\,\mathrm{d}x', \quad z \leq -\frac{d}{2}. \end{aligned} \tag{8.1.6}$$

Using integrals (8.1.6), it is easy to express the limiting values of the derivative $\partial_z\tilde{f}\,|_{z=\pm d/2}$ in terms of f:

$$\partial_z\tilde{f}|_{z=\pm d/2} = \pm\partial_x \hat{\mathrm{H}}\, f(x', z = \pm d/2) = \pm\hat{\mathrm{H}}\,\partial_{x'} f(x', z = \pm d/2). \tag{8.1.7}$$

Here Ĥ is the Hilbert transform operator:

$$\hat{\mathrm{H}}\,u = \frac{1}{\pi}\mathrm{V.p.}\int_{-\infty}^{+\infty} \frac{u(\xi')}{\xi' - \xi}\,\mathrm{d}\xi'.$$

After substituting eq. (8.1.7) into eq. (8.1.4), we arrive at an effective boundary-value problem for determining the potential f:

$$\begin{aligned} \Delta f &= 4\pi \operatorname{div}\mathbf{m}; \\ \mp\hat{\mathrm{H}}\,\partial_{x'} f\left(z = \pm\frac{d}{2}\right) &= [-\partial_z f + 4\pi m_z]|_{z=\pm d/2}. \end{aligned} \tag{8.1.8}$$

Thus we can restrict ourselves to solving the magneto-static equations inside the plate. However, in this case the conditions on the plate surface become integro-differential. Physically, the appearance of such boundary conditions is a consequence of long-range dipole–dipole forces. The last circumstance is overlooked in [64, 94] when deriving the NLSE for magneto-static spin waves.

The linearized problems (8.1.2), (8.1.3), and (8.1.8) have solutions in the form of plane waves $\propto \exp(ikx + i\,\omega(k)\,t)$. Our goal is to describe the modulations of these linear modes caused by harmonic interaction. For low-amplitude waves, the interactions are weak, so it is natural that the modulations of rapidly oscillating waves are assumed to be also weak. The solution to eqs. (8.1.2), (8.1.3), and (8.1.8) should be sought as follows [22, 82, 83]:

$$\begin{aligned} m_j &= \sum_{n=1}^{+\infty} \varepsilon^n \sum_{l=-\infty}^{+\infty} m_j^{(n,l)}(z, X_1, \tau)\exp i\theta_l, \quad j = x, y, \\ f &= \sum_{n=1}^{+\infty} \varepsilon^n \sum_{l=-\infty}^{+\infty} f^{(n,l)}(z, X_1, \tau)\exp i\theta_l, \quad \theta_l = l(kx + \omega t), \end{aligned} \tag{8.1.9}$$

where ε is a small parameter characterizing the deviation of the system from the linear regime, $X_1 = \epsilon(x + c_g\,t)$, c_g is the group velocity, and X_1 and $\tau = \varepsilon^2 t$ are slow variables. To balance the effects of dispersion and nonlinearity in the RPT equations, an appropriate choice of scaling transformations of spatial and temporal variables is needed. Ultimately, the self-consistency of the results obtained justifies this choice. The parameter c_g and the dispersion law $\omega = \omega(k)$ are determined in the process of solving the problem. The reality of the fields f and m_i is guaranteed by the following conditions:

$$m_i^{(n,l)*} = m_i^{(n,-l)}; \quad f^{(n,l)*} = f^{(n,-l)}.$$

Note that here and below, ε acts as a formal expansion parameter to visually group up terms of the same order of magnitude. At the end of the calculations, we put $\varepsilon = 1$. In the meantime, having deduced the series (8.1.9), by comparing the orders of magnitude of its terms, we can easily identify a physical small parameter and thereby establish the applicability span for the asymptotic expansion (8.1.9). In this problem the physical small expansion parameter is $(k\lambda)^{-1} \ll 1$.

The potential expansion (8.1.9) in powers of the parameter ε leads to the expansion of the field $\mathbf{H}^{(\mathrm{m})}$ of the following form:

$$\begin{aligned}
\mathbf{H}^{(\mathrm{m})} &= -\nabla\varphi = -4\pi M_0\mathbf{n} \\
&\quad + \sum_{n=1}^{+\infty} \varepsilon^n \sum_{l=-\infty}^{+\infty} (\mathbf{H}^{(\mathrm{m})})^{(n,l)} \exp[\mathrm{i}l(kx+\omega t)]; \\
(H_x^{(\mathrm{m})})^{(1,l)} &= -\mathrm{i}\,k\,lf^{(1,l)}; \\
(H_x^{(\mathrm{m})})^{(n,l)} &= -\mathrm{i}\,k\,lf^{(n,l)} - \partial_{X_1} f^{(n-1,l)}, \quad n \ge 2; \\
(H_z^{(\mathrm{m})})^{(n,l)} &= -\partial_z f^{(n,l)}, \quad n \ge 1.
\end{aligned} \tag{8.1.10}$$

After plugging (8.1.9) and (8.1.10) into eqs. (8.1.2) and (8.1.8) and equating the coefficients at the same powers of ε, we arrive at a system of linked equations for the coefficients $m_i^{(n,l)}$ and $f^{(n,l)}$.

We get the boundary conditions for the perturbation theory equations by expanding in powers of the parameter ε the initial boundary conditions (8.1.3) and (8.1.8). In particular, requirement (8.1.3) gives

$$\partial_z m_i^{(n,l)}|_{z=\pm d/2} = 0.$$

The specificity of the problem at hand is that the boundary condition in (8.1.8) bears an integral nature. As a result, connections emerge between the coefficients $f^{(n,l)}$ with different n. After substituting (8.1.9) into eqs. (8.1.8) for the potential f, the left-hand side of the second formula in (8.1.8) contains two groups of terms, with $l = 0$ and $l \neq 0$. Since the kernel of the Hilbert operator is a homogeneous function, the relations of the type

$$\hat{\mathrm{H}}\,\partial_{x'} f^{(n,0)}(X_1') = \varepsilon\,\hat{\mathrm{H}}\,\partial_{X_1'} f^{(n,0)}(X_1')$$

are satisfied for the first group of summands. They indicate that the terms with index n constitute the boundary conditions in the following order of the RPT. For the coefficients $f^{(n,0)}$, we have the boundary conditions

$$\begin{aligned}
&[-\partial_z f^{(1,0)} + 4\pi m_z^{(1,0)}]|_{z=\pm d/2} = 0; \\
&[-\partial_z f^{(n,0)} + 4\pi m_z^{(n,0)}]|_{z=\pm d/2} = \mp\hat{\mathrm{H}}\,\partial_{X_1'} f^{(n-1,0)}\left(X_1', z = \pm\frac{d}{2}\right), \quad n \ge 2.
\end{aligned}$$

The second group of summands (with $l \neq 0$) contains integrals of the form

$$\begin{aligned}
&\hat{\mathrm{H}}\,\partial_{x'}\,[f^{(n,l)}(X_1')\exp(\mathrm{i}\,k\,l\,x')] \\
&= \hat{\mathrm{H}}[(\varepsilon\,\partial_{X_1'} f^{(n,l)}(X_1') + \mathrm{i}\,k\,l f^{(n,l)}(X_1'))\exp(\mathrm{i}\,k\,l\,x')].
\end{aligned} \tag{8.1.11}$$

An analysis shows that when calculating the integral (8.1.11), the dependence of the integrand on the slow variable X_1 can be neglected in the first orders of the perturbation theory ($n = 1, 2, 3$). Consequently, we can exploit the relation

$$\hat{\mathrm{H}} \exp(\mathrm{i}\, k\, l\, x') = \mathrm{i}\, \mathrm{sign}(k\, l)\, \exp(\mathrm{i}\, k\, l\, x).$$

It should be emphasized that the following orders of the RPT require a more careful expansion of the integral (8.1.11) in powers of ε. However, they can be left aside when deriving effective dynamic equations. As we will see from further analysis, equations for the weakly nonlinear dynamics of spin waves are obtained in the third order of RPT ($n = 3$). In this case the boundary conditions for the coefficients $f^{(n,l)}$ with $l \neq 0$ ($n = 1, 2, 3$) take a simple local form

$$[-\partial_z f^{(1,l)} + 4\pi m_z^{(1,l)}]\big|_{z=\pm d/2} = \pm|kl| f^{(1,l)}\left(X_1, z = \pm\frac{d}{2}\right);$$

$$[-\partial_z f^{(n,l)} + 4\pi m_z^{(n,l)}]\big|_{z=\pm d/2} = \pm|kl| f^{(n,l)}\left(X_1, z = \pm\frac{d}{2}\right)$$
$$\mp \mathrm{i}\, \mathrm{sign}\,(kl)\, \partial_{X_1} f^{(n-1,l)}\left(X_1, z = \pm\frac{d}{2}\right), \quad n = 2, 3.$$

It is remarkable that all magneto-static boundary value problems admit an explicit solution. The results of calculating the potentials $f^{(n,l)}$ with $l, k \neq 0$ ($n = 1, 2, 3$) can be combined by the formula

$$\sum_{n=1}^{3} \varepsilon^n \sum_{l=-\infty, l\neq 0}^{l=+\infty} f^{(n,l)}(z, X_1, \tau) \exp \mathrm{i}\theta_l = \sum_{n=1}^{3} \varepsilon^n \sum_{l=-\infty, l\neq 0}^{l=+\infty} \exp \mathrm{i}\theta_l$$
$$\times \sum_{m=0}^{2} \frac{\varepsilon^m}{m!} (-\mathrm{i}\,\partial_{lk})^m \left(\hat{G}_l \partial_{X_1}^m m_x^{(n,l)} + \left(\frac{\partial_z \hat{G}_l}{\mathrm{i}\, k\, l}\right) \partial_{X_1}^m m_z^{(n,l)} \right). \tag{8.1.12}$$

Here $\hat{G}_l$ is the integral operator

$$\hat{G}_l\, \varphi = -2\pi\, \mathrm{i}\, \mathrm{sign}(kl) \int_{-d/2}^{d/2} \exp(-|lk(z - z')|) \varphi(z')\, \mathrm{d}z'.$$

To obtain an effective dynamic equation, we also need $f^{(2,0)}$. Next, we find

$$f^{(2,0)} = 4\pi \int_{-d/2}^{z} m_z^{(2,0)}(z', X_1, \tau) \mathrm{d}z'. \tag{8.1.13}$$

Here and below, eq. (8.1.1) implies that $m_z^{(n,l)}$ is expressed through $m_j^{(n,l)}$ ($j = x, y$).

Using relations (8.1.12) and (8.1.13) in the equations of magnetization dynamics (8.1.2), we express the demagnetizing fields $\mathbf{H}^{(m)}$ in terms of $m_j^{(n,l)}$. As a result, the dynamic equations become integro-differential. It is important that integrating is carried out only over the variable z, which characterizes the magnetization inhomogeneity along the normal to the plate. The method is successful enough due to the possibility of separating the variables z, X_1, and τ in each RPT's order. Carrying out this procedure successively, we deduce an effective local equation describing the propagation of spin waves along a plate.

8.1.2 First-order perturbation theory. Spectrum of spin excitations

The first order in ε of perturbation theory gives a system of homogeneous linear equations for identifying the spectrum of spin waves and eigenfunctions that characterize the inhomogeneity of the magnetization distribution along the normal to a plate for $l = \pm 1$ (harmonics are $\propto \exp(\pm \mathrm{i}(\omega t + kx))$). For $l \neq \pm 1$, an operator of a homogeneous linear system is not degenerate, and therefore the system has only trivial solutions: $m_i^{(1,l)} = f^{(1,l)} = 0$. For $l = 1$, the solution for $m_i^{(1,1)}$ can be represented as

$$m_i^{(1,1)}(X_1, \tau, z) = S_i(z)\,\Psi(X_1, \tau), \quad i = x, y,$$

where $\Psi(X_1, \tau)$ is an arbitrary function of slow variables to be defined later. The vector function $\mathbf{S}(z) = (S_x, S_y)$ is a solution to the system of integro-differential equations

$$\hat{\mathrm{L}}_1 \mathbf{S} = 0; \quad \partial_z \mathbf{S}|_{z=\pm d/2} = 0. \tag{8.1.14}$$

They correspond to the eigenvalue of $\omega = \omega(k)$. Here we have introduced the operator

$$\hat{\mathrm{L}}_1 = \begin{pmatrix} \mathrm{i}\,\omega & \omega_H - \gamma M_0 \alpha(\partial_z^2 - k^2) \\ \omega_H - \gamma M_0 \alpha(\partial_z^2 - k^2) + \mathrm{i}k\gamma M_0 \hat{G}_1 & -\mathrm{i}\,\omega \end{pmatrix},$$

where $\omega_H = \gamma H_0$. Different eigenvalues of $\omega(k)$ match different branches of the spin wave spectrum.

The developed perturbation theory is equally suitable for studying the interaction of spin waves belonging to any of the branches of the spectrum of linear modes. Next, we describe the interaction of waves belonging to any one branch of the spectrum, for example, the lowest in energy. The specificity of a particular branch of the spectrum in the final formulas makes itself felt only in the form of the dispersion law and the magnitude of the wave interaction constant.

We show that problem (8.1.14) for eigenvalues and eigenfunctions admits an explicit solution. From the first equation of system (8.1.14) we express S_x through S_y,

$$S_x = -\frac{1}{\mathrm{i}\,\omega}[\omega_H - \gamma M_0 \alpha(\partial_z^2 - k^2)]S_y,$$

and substitute it into the second equation:

$$\begin{aligned}&\left([\omega_H - \gamma M_0\alpha(\partial_z^2 - k^2) + \mathrm{i}\gamma M_0 k\hat{G}_1]\right.\\&\left.\times [\omega_H - \gamma M_0\alpha(\partial_z^2 - k^2)] - \omega^2\right) S_y = 0.\end{aligned} \tag{8.1.15}$$

Let us act by the operator $\partial_z^2 - k^2$ on eq. (8.1.15). In view of the identity

$$(\partial_z^2 - k^2)\hat{G}_1\varphi = 4\pi\,\mathrm{i}\,k\,\varphi,$$

which is valid for an arbitrary function φ, we get the outcome

$$\begin{aligned}&\left([\omega_H(\partial_z^2 - k^2) - \gamma M_0\alpha(\partial_z^2 - k^2)^2 - \omega_M k^2]\right.\\&\left.\times [\omega_H - \gamma M_0\alpha(\partial_z^2 - k^2)] - \omega^2\right) S_y = 0.\end{aligned} \tag{8.1.16}$$

Here we have accepted the notation $\omega_M = 4\pi\gamma M_0$. The solution to eq. (8.1.16) has the form

$$S_y = \sum_{j=1}^{6} C_j \exp(\mathrm{i}f_j z), \tag{8.1.17}$$

where C_j are integration constants, and $f_i = f_i(k, \omega)$ are the roots of the equation

$$\begin{aligned}&k^2[\Omega^2 + \omega_M\Omega - \omega^2] + f^2[\Omega^2 - \omega^2] = 0;\\&\Omega = \omega_H + \gamma M_0\alpha(f^2 + k^2).\end{aligned} \tag{8.1.18}$$

Since eq. (8.1.18) is cubic with respect to f^2, the roots are $\pm f_1$, $\pm f_2$, and $\pm f_3$. Relation (8.1.18) for the parameters f_j is equivalent to that given in [19], where a different approach was used to calculate the spectrum of spin waves.

Equation (8.1.18) has been obtained as a result of the action of the operator $(\partial_z^2 - k^2)$ on eq. (8.1.15). Consequently, generally speaking, solution (8.1.17) does not satisfy eq. (8.1.15). Namely, after substituting eq. (8.1.17) into relation (8.1.15), the left-hand side of eq. (8.1.15) does not vanish but gives some solution to the homogeneous equation

$$(\partial_z^2 - k^2)\,\varphi = 0.$$

If in this solution, we equate the coefficients of linearly independent functions of the kind $\exp(\pm kz)$ to zero, we obtain conditions on the parameters C_j, under which the function (8.1.17) holds for the original integro-differential equation (8.1.15):

$$\sum_{j=1}^{6} \frac{C_j}{|k| \pm \mathrm{i}f_j} \exp\left(\mp \mathrm{i}f_j \frac{d}{2}\right) = 0. \tag{8.1.19}$$

The boundary conditions $\partial_z \mathbf{S}|_{z=\pm d/2} = 0$ generate four more relations:

$$\sum_{j=1}^{6} f_j C_j \exp\left(\pm i f_j \frac{d}{2}\right) = 0;$$
$$\sum_{j=1}^{6} C_j (f_j^2 + k^2) f_j \exp\left(\pm i f_j \frac{d}{2}\right) = 0. \tag{8.1.20}$$

The system of homogeneous linear equations for the constants C_j has a nontrivial solution only if its determinant vanishes. This requirement gives a transcendental equation for determining permissible values of f_j and therefore the spectrum of spin waves $\omega = \omega(k)$ (see (8.1.18)). A simple analysis of relations (8.1.18)–(8.1.20) leads to an important conclusion: When an external field is perpendicular to the plate surface, the spectrum of linear modes depends only on $|\mathbf{k}|$ (see [19]). This circumstance means that the dispersion law for spin waves becomes nondifferentiable at the point $\mathbf{k} = 0$. Therefore, in the region of small wave numbers, the traditional algorithm for deriving the NLSE by differentiating the dispersion law of spin waves over the components of the vector $\mathbf{k}$ is not applicable. The weakly nonlinear dynamics of long-wavelength magneto-static excitations turns out to be nonlocal and is described by another model [56, 47, 50, 51].

8.1.3 Approximate solution to the boundary value problem for eigenvalues

Suppose that a plate is of finite thickness. Then, to gain approximate (but sufficiently accurate) solutions to the eigenvalue problem (8.1.14), an effective approach [35] is based on the expansion of $\mathbf{S}(z)$ over a complete system of functions $\mathbf{S}_n^p(z)$, which is orthogonal on a segment $|z| \le d/2$:

$$\mathbf{S}(z) = \sum_{p=1}^{2} \sum_{n=0}^{+\infty} C_n^p \, \mathbf{S}_n^p(z);$$
$$\mathbf{S}_n^p(z) = \frac{[i(-1)^p, 1]}{\sqrt{1+\delta_{n0}}} \cos\left[\kappa_n\left(z + \frac{d}{2}\right)\right], \quad \kappa_n = \frac{\pi n}{d}. \tag{8.1.21}$$

The functions $\mathbf{S}_n^p(z)$ correspond to spin-wave resonant oscillations in the standing wave regime. The coefficients C_n^p control the amplitudes of various modes; n is the number of nodes in the magnetization distribution along the normal to the plate. As a result, the boundary eigenvalue problem (8.1.14) is reduced to an algebraic infinite homogeneous system defining the constants C_n^p [35]:

$$[\omega + \Omega_n(k)] C_n^1 + \frac{\omega_M}{2} \sum_{n'} P_{nn'} [C_{n'}^1 - C_{n'}^2] = 0,$$
$$[\omega - \Omega_n(k)] C_n^2 + \frac{\omega_M}{2} \sum_{n'} P_{nn'} [C_{n'}^1 - C_{n'}^2] = 0, \tag{8.1.22}$$

where

$$
\begin{aligned}
P_{nn'} &= \frac{k^2}{k_{n'}^2}\left(\delta_{nn'} - \frac{k^2}{k_n^2}F_n\left[\frac{1+(-1)^{n+n'}}{2}\right]\left[(1+\delta_{n0})(1+\delta_{n'0})\right]^{-\frac{1}{2}}\right),\\
F_n &= \frac{2}{|kd|}\left[1-(-1)^n\exp(-|kd|)\right], \quad \omega_M = 4\pi\gamma M_0,\\
\Omega_n(k) &= \omega_H + \gamma M_0\,\alpha\,k_n^2, \quad k_n^2 = k^2 + \kappa_n^2, \quad \omega_H = \gamma H.
\end{aligned}
$$

Equations (8.1.22) contain the summands with $n \neq n'$, which characterize the mutual influences of different types of modes. In the region of wave numbers, where different branches of the spectrum are quite far from each other, the off-diagonal part of the linear system operator (8.1.22) can be taken into account within the perturbation theory [40]. In the main approximation (for $n = n'$), from eqs. (8.1.22) we find

$$
\begin{aligned}
\omega^2 &= \Omega_n(k)\left[\Omega_n(k) + \omega_M P_{nn}\right],\\
C_n^2 &= -\frac{\omega_M}{2}P_{nn}\left[\omega - \Omega_n(k) - \frac{\omega_M}{2}P_{nn}\right]^{-1} C_n^1.
\end{aligned}
\tag{8.1.23}
$$

It is worth underscoring that expansion (8.1.21) allows us to correctly approximate only the dynamics of bulk spin waves propagating along the plate. The basis functions used are unsuitable for describing surface waves.

8.1.4 Nonlinear evolution equation for a spin wave envelope at $\partial_k^2\omega \neq 0$

In the second order of $O(\varepsilon^2)$ within the RPT, the functions $m_i^{(2,l)}$ $(i = x, y)$ are nonzero only when $l = 0, \pm1$, and ±2. We have found [48]

$$
\begin{aligned}
m_i^{(2,0)} &= A_i(z)\,|\Psi|^2, \quad \mathbf{m}^{(2,1)} = -\mathrm{i}\,\partial_k \mathbf{S}\,\partial_{X_1}\Psi + \mathbf{S}\,\Psi_1,\\
m_i^{(2,2)} &= B_i(z)\,\Psi^2; \quad i = x, y.
\end{aligned}
$$

Here $\Psi_1(X_1, \tau)$ is the second arbitrary function of slow variables. The functions $A_j(z)$ and $B_j(z)$ are responsible for the inhomogeneity of the magnetization distribution along the normal to the plate. They can be computed explicitly and constitute sets of exponentials [48]. Note that the vector function $\mathbf{m}^{(2,1)} = (m_x^{(2,1)}, m_y^{(2,1)})$ meets an inhomogeneous linear system of the type

$$
\hat{L}_1\mathbf{m}^{(2,1)} = \mathbf{q}.
$$

The operator $\hat{L}_1$ being degenerate, the system is solvable only under the orthogonality condition [55]

$$\int_{-d/2}^{+d/2} (\mathbf{P} \cdot \mathbf{q})\, \mathrm{d}z = 0,$$

where $\mathbf{P} = (S_y, -S_x)$ is the solution to the adjoint homogeneous equation $\mathbf{P}\hat{L}_1 = 0$. The orthogonality condition is fulfilled when the parameter c_g and the group velocity of spin waves coincide: $c_g = \partial_k \omega$.

An effective equation for the spin wave envelope can be derived in the third order of $O(\epsilon^3)$ within the perturbation theory when considering an inhomogeneous linear system for the function $\mathbf{m}^{(3,1)} = (m_x^{(3,1)}, m_y^{(3,1)})$. The linear operator of this system again coincides with $\hat{L}_1$. The solvability condition for an inhomogeneous system leads to a closed equation for the function $\Psi(X_1, \tau)$ of slow variables; it is an arbitrary function in the lower orders of perturbation theory. This equation is effective for the dynamics of spin waves. As a result, we find that the quasi-one-dimensional dynamics of spin waves in a finite-thickness plate is described by the NLSE

$$\mathrm{i}\,\partial_\tau \Psi - \frac{1}{2}\partial_k^2 \omega(k)\, \partial_{X_1}^2 \Psi + g|\Psi|^2 \Psi = 0, \tag{8.1.24}$$

where the interaction constant of spin waves bears an integral nature, depends on k, and is given by the expression

$$g = \int_{-d/2}^{d/2} (\mathbf{P} \cdot \mathbf{f})\, \mathrm{d}z \left[\int_{-d/2}^{d/2} (\mathbf{P} \cdot \sigma_3 \mathbf{S}) \mathrm{d}z \right]^{-1}, \tag{8.1.25}$$

where $\mathbf{f} = (f_x, f_y)$ with

$$f_y(\mathbf{S}) = c_x(\mathbf{S}) - \mathrm{i}\gamma \Big((\partial_z \hat{G}_1)\, [S_x A_x + S_x^* B_x + S_y A_y + S_y^* B_y] + \frac{1}{M_0}\Big[|\mathbf{S}|^2 (\mathrm{i}\,k\, \hat{G}_1 S_x) + \frac{1}{2}\, (\mathrm{i}\,k\, \hat{G}_1 S_x)^* S^+ S^- \Big] \Big),$$

$$f_x(\mathbf{S}) = c_y(\mathbf{S}), \quad S^{\pm} = S_x \pm \mathrm{i}\, S_y, \quad |\mathbf{S}|^2 = |S_x|^2 + |S_y|^2, \quad \sigma_3 = \begin{pmatrix} 1 & 0 \\ 0 & -1 \end{pmatrix};$$

$$c_j = -\mathrm{i}\gamma \Big(\frac{\alpha}{M_0} \Big[\frac{S_j^*}{2} (\partial_z^2 - 4k^2) S^+ S^- - \frac{S^+ S^-}{2} (\partial_z^2 - k^2) S_j^* + S_j \partial_z^2 |\mathbf{S}|^2 - |\mathbf{S}|^2 (\partial_z^2 - k^2) S_j \Big] - \frac{4\pi}{M_0} S_j |\mathbf{S}|^2 + A_j (\partial_z \hat{G}_1) S_x + B_j (\partial_z \hat{G}_1)^* S_x^* + S_j^* \Big[(\partial_z \hat{G}_2)\, B_x - \frac{1}{4\,\mathrm{i}\,k\, M_0}\, (\partial_z^2 \hat{G}_2)\, S^+ S^- \Big] \Big), \quad j = x, y.$$

The functions $\mathbf{S}$ and Ψ describe the slow evolution of the fundamental harmonic envelope:

$$m_x^{(1,1)} + \mathrm{i}\, m_y^{(1,1)} = S^+ \Psi.$$

For the lower-energy branch of spin waves in a thin plate, the magnetization distribution along the normal to the plate is close to a uniform one [78]. Therefore we can neglect the dependence of the functions S_i, A_i, and B_i $(i = x, y)$ on the z-coordinate. The problem can be significantly facilitated by defining the specified functions from the RPT equations averaged over the plate thickness. This yields [48]

$$\begin{aligned} \omega^2 &= \Omega(k)[\Omega(k) + \omega_M \sigma(k)], \quad \Omega(k) = \omega_H + \gamma M_0\, \alpha\, k^2, \\ \sigma(k) &= 1 - \frac{1 - \exp(-|kd|)}{|kd|}, \quad S_y = -\mathrm{i}\frac{\omega}{\Omega} S_x, \quad A_i = B_i = 0. \end{aligned} \tag{8.1.26}$$

Without loss of generality, we further set $S_x = 1$. The average description (8.1.26) corresponds exactly to the branch of the spectrum with $n = 0$ in eqs. (8.1.22). In this case the magnetization distribution has no nodes along the normal to the plate. According to [35], only this branch is excited by an alternating magnetic field in a perpendicularly magnetized film with free surface spins. As a result, the wave interaction constant takes the form

$$\begin{aligned} g = \frac{\gamma\Omega}{2\omega M_0} &\left[(4\pi - ak^2)\left(1 + \frac{\omega^2}{\Omega^2}\right)^2 \right. \\ &\left. + \left(\frac{3}{2}\alpha k^2 + 2\pi\mu(k)\right)\left(1 - \frac{\omega^2}{\Omega^2}\right)^2 - 2\pi\sigma(k)\left(3 + \frac{\omega^2}{\Omega^2}\right)\right], \end{aligned} \tag{8.1.27}$$

where $\mu(k) = [1 - \exp(-|2kd|)]\, |2kd|^{-1}$.

Equation (8.1.24) is known to admit exponential solitons if its parameters satisfy the Lighthill criterion [91]

$$g\, \partial_k^2 \omega < 0. \tag{8.1.28}$$

Let the magnetization distribution along the normal to the surface be close to uniform due to the plate parameters. We use the results of the averaged description in eqs. (8.1.26) and (8.1.27) to test criterion (8.1.28). As numerical analysis claims, in the case of sufficiently thick films (Figure 8.1) for the accepted values of real parameters, requirement (8.1.28) holds for a fairly wide range of changes in $|kd|$ up to $|kd| \approx 10$.

As for thinner films (Figure 8.2), requirement (8.1.28) for the same values of real parameters is already violated when $|kd| \approx 1$; near this point, the quantity $\partial_k^2 \omega(k)$ vanishes and then changes its sign, becoming positive. In this range of wave numbers, the dispersion summand in eq. (8.1.24) is close to zero.

When $k = 0$, expression (8.1.27) for the wave interaction constant coincides with that obtained in [94] from the expansion in frequency amplitude of a uniform ferromagnetic resonance. At the same time the dependencies shown in Figures 8.1 and 8.2 indicate the sensitivity of the specified constant to the plate parameters and to the values

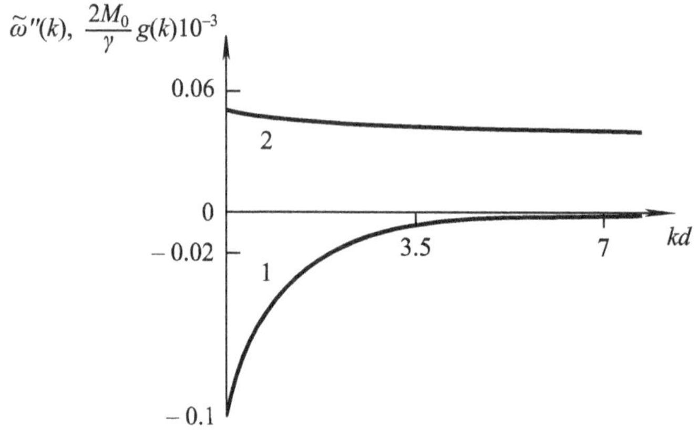

Figure 8.1: Dependencies of the second derivative of the dispersion law on a wave vector k, $\tilde{\omega}''(k) = \omega_H^{-1}\partial^2\omega/\partial(kd)^2$ (curve 1), and the wave interaction parameter g (curve 2) for a perpendicularly magnetized $d = 1$-μm-thick film with free surface spins. It is thought that $\alpha = 5 \cdot 10^{-11}$ cm^2 and $\omega_M/\omega_H = 1/2$.

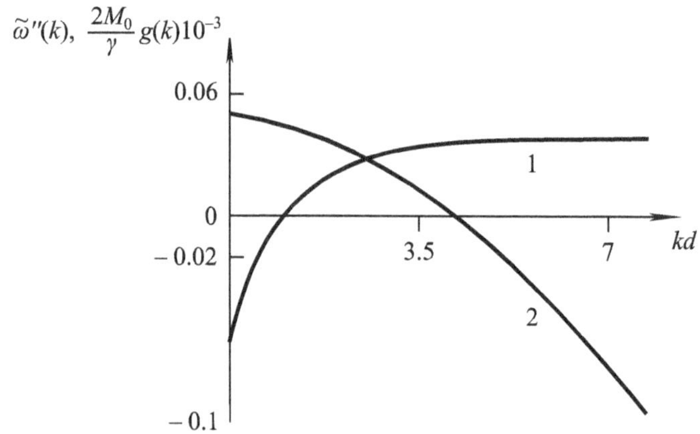

Figure 8.2: Dependencies of the second derivative of the dispersion law on a wave vector k, $\tilde{\omega}''(k) = \omega_H^{-1}\partial^2\omega/\partial(kd)^2$ (curve 1), and the wave interaction parameter g (curve 2) for a perpendicularly magnetized $d = 0.1$-μm-thick film with free surface spins. It is thought that $\alpha = 5 \cdot 10^{-11}$ cm^2 and $\omega_M/\omega_H = 1/2$.

of the fundamental harmonic wave vector. Therefore the interaction constant cannot be properly predicted by expanding the uniform ferromagnetic resonance frequency in terms of magnetization.

One of the important issues of the problem at hand is to create conditions for generating solitons in the region (8.1.28). Let λ, T, and A be the characteristic spatial and temporal scales of modulation of spin waves and their amplitude, respectively. Since the effects of nonlinearity and dispersion are balanced during the formation of solitons, we may get the following estimates from inequality (8.1.24):

$$\lambda^2 \backsim \frac{|\partial_k^2 \omega(k)|}{gA^2}, \quad T \backsim \frac{1}{gA^2}.$$

Suppose that pulses of duration τ_0 give rise to the creation of solitons. Then $\lambda \backsim c_g \tau_0$. Next, an estimate for the spin wave threshold amplitude at which solitons start to form is [94]

$$gA^2 \backsim \frac{|\partial_k^2 \omega|}{(c_g \tau_0)^2}.$$

In this case the inequality $\lambda \gg k^{-1}$ must be fulfilled, i. e.,

$$\tau_0\, d\, \omega_H\, k \frac{\partial \tilde{\omega}(kd)}{\partial (kd)} \gg 1; \quad \tilde{\omega}(k) = \frac{\omega(k)}{\omega_H}.$$

Assuming that $\omega_H \approx 10^{10}$ Hz and $\tau_0 \approx 10 : 20$ ns and taking into account the numerical values of the derivative of the dispersion law (see Figs. 8.1 and 8.2), we can see that $k \gg 10^3 \div 10^4\ \text{cm}^{-1}$ in the region of existence of soliton states. In this context the inequality $L \gg k^{-1}$ is always true in experiment because the linear dimensions of the sample L, as a rule, are of the order of 2–5 mm [16].

8.2 Quasisolitons in the vicinity of the point of zero dispersion

> If you do not understand a particular word in a piece of technical writing, ignore it. The piece will make perfect sense without it.
>
> Cooper's law

An analysis of the spectrum of spin waves in ferromagnetic films has led to an interesting conclusion. At least, the branch of the spectrum with $n = 0$ (see eqs. (8.1.23)) matches an interval of wave numbers where the second derivative of the dispersion law is close to zero. The existence of such a region is the result of competition between exchange and dipole–dipole interactions. The description of weakly nonlinear dynamics within the NLSE (8.1.24) in this range of wave numbers becomes incorrect. This is because the role of dispersion described by higher derivatives regarding a slow coordinate rises. To obtain an appropriate evolutionary equation, the above algorithm of RPT should be modified.

We have found that when $\partial_k^2 \omega = 0$ and $\partial_k^3 \omega \neq 0$, the balance of the effects between dispersion and nonlinearity reflects the following form of scaling transformations:

$$m_j = \sum_{n=3}^{\infty} \varepsilon^n \sum_{l=-\infty}^{+\infty} m_j^{(n,l)}(z, X, \tau) \exp i\theta_l, \quad \theta_l = \omega t + kx, \quad j = x, y;$$

$$f = \sum_{n=3}^{\infty} \varepsilon^n \sum_{l=-\infty}^{+\infty} f^{(n,l)}(z, X, \tau) \exp i\theta_l, \quad X = \varepsilon^2 (x + c_g t),$$

$$\tau = \varepsilon^6 t, \quad c_g = \partial_k \omega,$$

where ε is the expansion parameter within the RPT.

An effective equation for the spin wave envelope can be composed in the ninth order in the parameter ε as a condition for the solvability of the inhomogeneous linear system that determines $m_j^{(9,1)}$. Since the calculations are similar to those already considered, we present the final result [54]:

$$\begin{aligned} &\mathrm{i}\,\partial_\tau \varphi + \frac{\mathrm{i}}{2}\,\partial_k^3 \omega\, \partial_X^3 \varphi + g|\varphi|^2 \varphi = 0; \\ &m_x^{(3,1)} + \mathrm{i}\, m_y^{(3,1)} = S^+(z)\varphi(X,\tau). \end{aligned} \tag{8.2.1}$$

Here the value of $\partial_k^3 \omega$ is calculated at the point where $\partial_k^2 \omega(k) = 0$. Despite the differences in intermediate calculations, the expressions for the functions $S_j(z)$ $(j = x, y)$ and the wave interaction constant g coincide with the previous ones (see eqs. (8.1.14) and (8.1.25)). Further, for definiteness, we suppose that $g > 0$ and $\partial_k^3 \omega > 0$.

Scenarios for the evolution of various initial distributions of magnetization have been analytically and numerically analyzed in [54].

It has been found that in the anomalous region, an initial spatially localized distribution of magnetization first “dumps” part of the energy, radiating a nonlinear wave train. It is interesting that when $\partial_k^2 \omega = 0$, the field momentum of the distribution of magnetization depends on the field phase gradient φ:

$$P = \frac{4}{\gamma M_0} \int_{-\infty}^{+\infty} \partial_X \theta\, R^2 \mathrm{d}X.$$

Here we have introduced the amplitude R and phase θ of the field φ:

$$\varphi = R \exp \mathrm{i}\theta.$$

The phase gradients of the emitted wave train and the remaining part of the magnetization distribution differ in sign. As a result, the “dumped” radiation and the remaining localized part of the magnetization distribution move in the same direction, and this does not contradict the momentum conservation law.

Equation (8.2.1) is similar to the Korteweg–de Vries (KdV) model in terms of the type of dispersion and to the NLSE in terms of the nature of the nonlinearity. By analogy with the data known for these models [3, 71, 2] we suggest that in the spatio-temporal interval $|X| \le (\partial_k^3 \omega\, \tau)^{1/3}$, the evolution of the nonlinear wave train is determined by self-similar solutions of the type

$$\varphi = \tau^{-1/2} \chi(\eta), \quad \eta = \tau^{-1/3} X. \tag{8.2.2}$$

The self-similar wave regime (8.2.2) develops once the initial magnetization distribution acquires on the left a power-law asymptotics of the type $|\varphi|^2 \propto (-X)^{-3}$ as $X \to -\infty$. Interestingly, eq. (8.2.1) has the following exact solution in the region $X < 0$:

$$\varphi = \sqrt{\frac{8}{3}\left|\frac{\partial_k^3\omega}{g}\right|}\frac{\sqrt[4]{7}}{\sqrt{2(a-X)^3}}\exp\left(-\mathrm{i}\,\mathrm{sign}(g\,\partial_k^3\omega)\sqrt{\frac{7}{4}}\ln(a-X)+\mathrm{i}\,\delta\right).$$

The oscillations on the left-hand side of the initial magnetization distribution in Figure 8.3 can probably be interpreted as the result of the development of instabilities in the above solution.

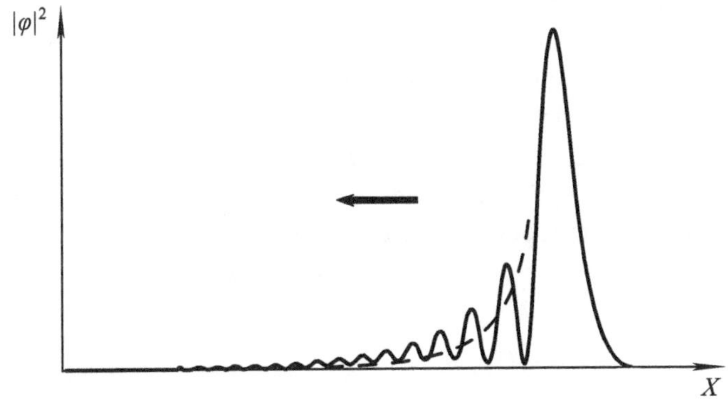

Figure 8.3: "Cleavage" of a nonlinear wave train from an initial localized distribution of magnetization. The dashed line indicates the power dependence $|\varphi|^2 \propto (-X)^{-3}$.

The function $\chi(\eta)$ in eq. (8.2.2) can never be calculated analytically; however, its asymptotic behavior can be explored:

$$\begin{aligned}
\chi(\eta) &= R_1 \exp \mathrm{i}\theta + R_2 \exp(-\mathrm{i}\theta) + O[(-\eta)^{-3/2}],\\
\theta &= \frac{2}{3}\sqrt{\frac{2}{\partial_k^3\omega}}(-\eta)^{-3/2},\\
R_1 &= R_1^{(0)} \exp\left[\frac{3}{2}\,\mathrm{i}\,g\ln(-\eta)\,(2|R_2^{(0)}|^2 + |R_1^{(0)}|^2)\right],\\
R_2 &= R_2^{(0)} \exp\left[\frac{3}{2}\,\mathrm{i}\,g\ln(-\eta)\,(2|R_1^{(0)}|^2 + |R_2^{(0)}|^2)\right], \quad \eta \to -\infty;\\
\chi(\eta) &= O\left(\exp\left(-\frac{2}{3}\sqrt{\frac{2}{\partial_k^3\omega}}\eta^{3/2}\right)\right), \quad \eta \to +\infty.
\end{aligned} \tag{8.2.3}$$

At the boundary of the region of applicability of the self-similar solution, the complex constants $R_i^{(0)}$ $(i = 1, 2)$ in eqs. (8.2.3) become dependent on the slow variable $\xi = X/\tau$. As

$\tau \to \infty$ and $X/\tau = \text{const}$, the self-similar solutions (8.2.3) are stitched with the following asymptotic solution to eq. (8.2.1):

$$\varphi = \sum_{m=1}^{\infty} \tau^{-m+1/2} \sum_{l=-\infty}^{+\infty} f_{m,l}(\xi,\tau) \exp[\mathrm{i}(2l-1)\theta],$$
$$\theta = \frac{\tau}{3}\left(\partial_k^3\omega\right)^{-1/2}(-2\xi)^{3/2},$$
$$f_{m,l}(\xi,\tau) = \exp[\mathrm{i}\tau\chi_{m,l}(\xi,\tau)]\left(R_{m,l}^{(0,0)}(\xi) + \sum_{n=1}^{\infty}\sum_{k=0}^{n} \frac{(\ln\tau)^k}{\tau^n} R_{m,l}^{(n,k)}(\xi) \right),$$
$$\chi_{m,l}(\xi,\tau) = \sum_{n=1}^{\infty}\sum_{k=0}^{n} \frac{(\ln\tau)^k}{\tau^n}\chi_{m,l}^{(n,k)}(\xi), \quad \xi = \frac{X}{\tau}, \quad \xi < 0.$$

Having “dumped” a part of the energy, the initial distribution of magnetization is transformed into a long-lived soliton-like formation that travels at an almost constant speed from right to left. Changes in the shape of such a “bright” quasisoliton occur for two reasons: firstly, this object continues to radiate forward (to the left) low-amplitude waves; secondly, the quasisoliton leaves behind a quasistatic low-amplitude tail. At large times the latter involves an almost static periodic wave with small oscillations $|\varphi|$ around a certain constant value (Figure 8.4).

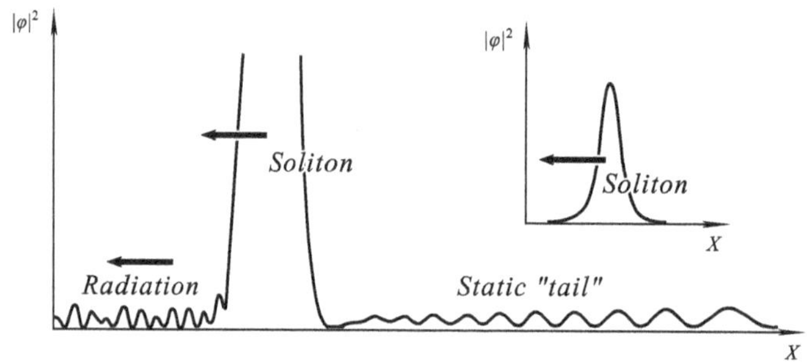

Figure 8.4: A traveling “bright” quasisoliton. At the bottom of the soliton (when the figure is zoomed), a quasistatic “tail” and low-amplitude forward radiation are noticeable.

According to [30, 29], we compute the coordinate of the “center of gravity” $\langle X\rangle$ of the quasisoliton and its velocity v:

$$\langle X\rangle = \frac{1}{N}\int_{-\infty}^{+\infty} XR^2\,\mathrm{d}X, \quad N = \int_{-\infty}^{+\infty} R^2\,\mathrm{d}X,$$
$$v = \frac{\mathrm{d}\langle X\rangle}{\mathrm{d}\tau} = -\frac{\partial_k^3\omega}{N}\int_{-\infty}^{+\infty}[(\partial_X R)^2 + (\partial_X\theta)^2 R^2]\,\mathrm{d}X.$$

The quantity N is an integral of motion. Thus the quasisoliton velocity depends on the amplitude and phase gradients of the field φ. Its movement direction is determined only by the sign of the third derivative of the dispersion law (in our case, the sign is "+", so the quasisoliton moves to the left).

The findings of numerical simulation state that in the localization region of a bright quasisoliton, the phase gradient of the field φ is $\partial_X \arg\varphi = k_0 \approx \text{const} < 0$. This gradient conforms to a quasisoliton size Δ greater than the wavelength modulating its profile ($\Delta > |k_0|^{-1}$).

The next step is to utilize the above information for constructing an approximate solution of eq. (8.2.1), followed by a description of the internal structure of the quasisoliton. To construct an asymptotic solution in powers of a parameter $|k_0|^{-1}\Delta^{-1}$, the following version of the nonlinear perturbation theory should be employed:

$$\varphi = R(\varsigma)\exp\left(\mathrm{i}\left[k_0 X + \frac{\partial_k^3\omega}{6}k_0^3\tau + \tau\sum_{n=1}^{\infty}\varepsilon^n\Omega^{(n)} + \chi(\varsigma)\right]\right), \tag{8.2.4}$$

where ε is a small parameter characterizing the slow evolution of the waveshape, $\varsigma = \varepsilon(X + v\tau)$,

$$v = v^{(0)} + \sum_{n=1}^{\infty}\varepsilon^n v^{(n)}, \quad R(\varsigma) = \sum_{n=1}^{\infty}\varepsilon^n R^{(n)}(\varsigma), \quad \chi = \sum_{n=1}^{\infty}\varepsilon^n\chi^{(n)}(\varsigma).$$

The equations appearing in each order in the parameter ε are solved sequentially. The functions $\chi^{(n)}(\varsigma)$ are expressed through the coefficients $R^{(n)}(\varsigma)$, and for the correction $R^{(1)}(\varsigma)$, we have the closed equation

$$\Omega^{(2)}R^{(1)} + \frac{1}{2}\,\partial_k^3\omega\, k_0\,\partial_k^3 R^{(1)} - g\left(R^{(1)}\right)^3 = 0.$$

Among its bounded solutions, there are spatially localized and periodic ones. The integration constants and the parameters $\Omega^{(n)}$ and $v^{(n)}$ for both types of solutions can be chosen so that no secular terms should arise due to phase shifts of the wave and its translation, which are proportional to $R^{(1)}$ and $\partial_\varsigma R^{(1)}$. For a localized wave, after simple calculations, we find

$$v^{(0)} = \frac{k_0^2}{2}\partial_k^3\omega, \quad v^{(2)} = -\frac{\kappa^2}{6}\partial_k^3\omega, \quad \Omega^{(2)} = \frac{\kappa^2}{2}\left|\partial_k^3\omega\right|,$$

$$v^{(1)} = \Omega^{(1)} = R^{(2)} = \Omega^{(3)} = \Omega^{(4)} = 0, \quad R^{(1)} = \kappa\sqrt{\left|\frac{\partial_k^3\omega\, k_0}{g}\right|}\operatorname{sech}(\kappa\varsigma),$$

$$\chi^{(1)} = -\frac{\kappa}{2k_0}\tanh(\kappa\varsigma), \quad \frac{\Omega^{(2)}}{g} > 0, \quad g\,k_0\,\partial_k^3\omega < 0.$$

Here κ is the parameter that determines the size of the quasisoliton ($\Delta \sim \kappa^{-1} < |k_0|^{-1}$). Such a soliton cannot be stationary because it resonantly interacts with a radiation wave whose phase velocity coincides with the soliton velocity $v \approx v^{(0)} + v^{(2)}$.

It is interesting and important that low-amplitude radiation cannot be described in powers of a physically small parameter $\delta = O(|k_0|^{-1}\Delta^{-1})$ within the traditional perturbation theory (8.2.4), because its amplitude C is nonanalytically δ-dependent [54, 90, 57, 41–43]:

$$C \sim \delta^{-1} \exp\left(-\frac{\pi}{2\delta}\right).$$

Here $\delta = |\kappa/r_0| < 1$, and $r_0 = -3k_0 > 0$ is the wave number that obeys the "Cherenkov resonance" criterion [54]. To describe the radiation connected with a soliton, the papers [90, 57, 41–43] propose special versions of the perturbation theory, close to the Wentzel–Kramers–Brillouin method. These papers treat a third-order dispersion summand as a small perturbation for the NLSE with quadratic dispersion. In [54], we have analyzed an opposite limiting case conditional upon dominating higher-order dispersion effects. We have found explicit expressions for describing the parameters of a "bright" quasisoliton, altering in time under radiation.

Let us discuss the possibility of excitation near the point of zero dispersion of spatially nonlocalized monochromatic nonlinear waves:

$$\varphi_0 = \sqrt{w_0}\, \exp[i(\Omega_0\tau + k_0X)], \quad \Omega_0 = \frac{1}{6}\partial_k^3\omega\, k_0^3 + g\, w_0. \tag{8.2.5}$$

Here $w_0 = \text{const} > 0$, and k_0 is a change in the wave number when shifting from the point of zero dispersion, where $\partial_k^2\omega = 0$. A linear stability analysis conducted states that the wave (8.2.5) is stable with respect to long-wave perturbations only under the condition $gk_0\partial_k^3\omega > 0$. In our case, $g\partial_k^3\omega > 0$, and therefore the wave (8.2.5) is stable when $k_0 > 0$. When $k_0 < 0$, the modulation instability of the wave (8.2.5) continues to develop with the formation of "bright" quasisolitons discussed previously.

Investigating long-wavelength low-amplitude modulations of the wave (8.2.5) requires searching for approximate solutions to eq. (8.2.1) in the following way:

$$\begin{aligned} \varphi_0 &= \sqrt{w(X,\tau)}\, \exp[i(\Omega_0\tau + k_0X + \chi(X,\tau))], \\ w(X,\tau) &= w_0(1 + v(X,\tau)), \end{aligned} \tag{8.2.6}$$

where $|v| \ll 1$, $|\partial_X\chi| \ll |k_0|$, and $|\partial_\tau\chi| \ll |\Omega_0|$. In the long-wavelength limit, the spatial fourth-order derivatives in the linear terms of the equations for the functions v and χ can be neglected. At the same time, only the nonlinear terms quadratic over the fields v and χ, containing the minimum number of derivatives, should be left. In the above approximation the parameter v is expressed in terms of the field χ:

$$w_0\, g\, v = \hat{L}\chi, \quad \hat{L} = \partial_\tau - \frac{k_0^2}{2}\,\partial_k^3\omega\,\partial_X.$$

To calculate χ, we can resort to the closed equation [54]

$$\left[-\hat{L}^2 - \frac{1}{3}\partial_k^3\omega\,\hat{L}\,\partial_X^3 + \partial_k^3\omega\,k_0\,g\,\partial_X^2 - \left(\frac{\partial_k^3\omega}{2}k_0\right)^2\partial_X^4\right]\chi$$
$$+\frac{\partial_k^3\omega}{2}\partial_X\left[gw_0(\partial_X\chi)^2 + 2k_0\,\partial_X\chi\,\hat{L}\chi\right] + \frac{\partial_k^3\omega}{2}k_0\,\hat{L}(\partial_X\chi)^2 = 0. \tag{8.2.7}$$

Equation (8.2.7) admits bounded stationary-profile wave-like solutions. In particular, under certain conditions, there are self-localized waves of the following form:

$$\partial_\xi\chi = -\frac{\gamma}{\delta}\,\mathrm{sech}^2\left(\frac{\sqrt{\gamma}\xi}{2}\right),\quad \xi = X + u\tau, \tag{8.2.8}$$

where the constant parameters $\gamma > 0$ and δ are expressed in terms of the coefficients of eq. (8.2.7), and $u = \text{const}$.

Figure 8.5 displays the result of the numerical simulation of the solutions to eq. (8.2.1) after pulsed excitation of the inhomogeneity against the background of the harmonic wave (8.2.5) ($k_0 > 0$). Once begun evolving, the inhomogeneity against the background of the wave (8.2.6) provokes a long-lived localized state, a "dark" quasisoliton, whose profile is in good agreement with solution (8.2.8).

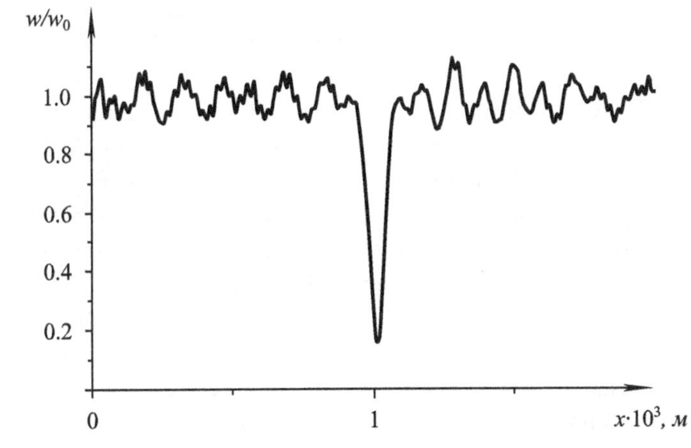

Figure 8.5: A "dark" quasisoliton against the background of a nonlinear monochromatic wave: $k_0 d \approx 0.1$, $\partial_k^3\tilde{\omega} \approx 0.04$, $\tilde{g} \approx 0.9$, and $d \approx 1\,\mu$m. The plate length in the wave propagation direction is ≈ 0.5 cm.

Solution (8.2.8) resembles a soliton of the KdV equation. This is not a coincidence. Quasisolitons (8.2.8) can be interpreted as bound states of low-amplitude Goldstone modes propagating against the background of the wave (8.2.5) with an increment of the wavenumber $k_0 > 0$. Goldstone modes have dispersion laws $\lambda_i(p)$ ($i = 1, 2$):

$$\lambda_i(p) = \left(\frac{k_0^2}{2}\,\partial_k^3\omega + \varepsilon_i\, m\right)p + \left(\frac{1}{6}\,\partial_k^3\omega + \varepsilon_i\, n\right)p^3,$$

where

$$\varepsilon_1 = 1, \quad \varepsilon_2 = -1, \quad m^2 = g\, k_0 w_0\, \partial_k^3\omega > 0, \quad n = \left(k_0\partial_k^3\omega\right)^2/(8m).$$

We suggest that

$$k_0 p^2 \ll 4g\, w_0/\partial_k^3\omega.$$

An analysis shows that the mutual interaction of the modes with $\varepsilon_1 = 1$ is the most intensive against the background of the wave (8.2.5). When the wave modulations (8.2.5) are generated by the excitation and interaction of only such modes, it is convenient to switch to a reference frame moving with the phase velocity of these modes, $v = m + k_0^2\,\partial_k^3\omega/2$, and to use new variables $\xi = X + v\tau$ and $T = \tau$. At that point the dependence of the field $\chi(\xi, T)$ on the variable T will be weak. Consequently, in eq. (8.2.8), both terms with $\partial_T^2\chi$ and $\partial_T\partial_X^3\chi$ and the nonlinear terms containing the derivatives $\partial_T\chi$ can be ignored. Finally, the closed equation (8.2.7) reduces to the KdV equation integrable by the inverse scattering method:

$$\partial_T f + \left(\frac{1}{6}\,\partial_k^3\omega + n\right)\partial_\xi^3 f - \frac{\partial_k^3\omega}{4}\left(3k_0 + \frac{g\, w_0}{m}\right)\partial_\xi f^2 = 0, \tag{8.2.9}$$

where $f = \partial_\xi\chi$. It is easy to check that solution (8.2.8) generalizes the one-soliton solution to the KdV model (8.2.9).

Interestingly, this problem is connected to the previous one due to the possibility of theoretically describing the modulations of the "tail" of a "bright" quasisoliton. In the main approximation the "tail" appertains to the static wave (8.2.5), where $\Omega_0 = 0$, and $k_0 = -(6gw_0/\partial_k^3\omega)^{1/3} < 0$. The linear analysis claims that such a wave is unstable when $k_0 < 0$. However, computer simulation gives an unexpected result. The low-amplitude wave behind the "bright" quasisoliton remains unchanged for a long time, suffering only small spatial periodic modulations. The time-independent "modulated structure" (8.2.6) ($\Omega_0 = 0$) exists because its energy per spatial period is less than that of the state with $\varphi = 0$. Nonlinear interactions under certain conditions can stabilize the static wave.

In [54] an approximate solution to eq. (8.2.7) is found to describe the modulations of the tail of a "bright" quasisoliton:

$$\varphi \approx \sqrt{w_0(1 + v(X))}\,\exp\left[-\mathrm{i}\left(\frac{6g\, w_0}{\partial_k^3\omega}\right)^{1/3} X + \mathrm{i}\chi(X)\right],$$

$$v = b\,\mathrm{dn}^2\eta - a, \quad \eta = mX, \quad \partial_X\chi = -c\, v.$$

The positive constant parameters $b > a$, c, and m and the Jacobi function modulus are expressed in terms of the parameters of eq. (8.2.1) and w_0.

The listed tendencies for quasisolitons to emerge are quite natural. This is because the emission of a wave train or the excitation of nonlocalized waves causes them to shift from the point of zero dispersion to the region of wave numbers, where the typical quadratic-dispersion NLSE admits the generation of "bright" or "dark" solitons.

As mentioned earlier, the nonlocal nature of the connection between the magnetization and the magnetic field induced by the long-range dipole–dipole interaction cannot be neglected in the region of too small wave numbers. In this region, the effective equations for thin ferromagnetic films turn out to be integro-differential [56, 47, 50, 51]. The nonlocal part of the magneto-static dispersion smooths out the inhomogeneities of the magnetization distribution in the films. Consequently, purely exponential magnetic solitons described by local equations of the NLSE type cannot be inherent to thin ferromagnetic films in the long-wavelength limit. At the same time, for such films, new weakly localized exchange-dipole states of the type of algebraic solitons have been predicted and analytically described [56, 47, 50, 51]. The next section discusses related issues when analyzing magneto-static waves in an antiferromagnetic film.

8.3 Nonlocal dynamics of Goldstone excitations in an antiferromagnetic film

If I found any new truths in the sciences, I can say that they follow from, or depend on, five or six principal problems which I succeeded in solving and which I regard as so many battles where the fortunes of war were on my side.

R. Descartes

Antiferromagnetic films have relatively recently become the object of investigation due not only to their unique properties but also to new technologies in manufacturing them as high-quality samples by molecular beam epitaxy [62, 63]. In [13] the weakly nonlinear dynamics of activation exchange-dipole spin waves in antiferromagnetic films has been studied within the NLSE. However, there is a wide class, for example, of two-sublattice easy-plane anisotropy antiferromagnets, where gapless (Goldstone) modes are more important because less energy is required to excite them. For such materials, an external magnetic field perpendicular to the film plane changes not only the dispersion law of spin waves but also the intensity of their interaction.

Magneto-static spin waves can exist in antiferromagnets even under the magnetization-free condition for their ground state [31, 63]. The previously noted features of the interaction of exchange-dipole spin waves in ferromagnetic films are inherent to antiferromagnetic films as well. In addition, when exposed to an external magnetic field, several magnetic sublattices provoke the formation of new multisoliton states in antiferromagnetic films with a different behavior compared to ferromagnets.

This section discusses the results concerning integro-differential equations describing the interaction of Goldstone exchange-dipole waves in thin antiferromagnetic films

[46, 49, 52]. Algebraic solitons and multisoliton regimes are predicted. The soliton relaxation process is analytically covered.

8.3.1 Problem formulation

Suppose we have an antiferromagnetic film with "easy-plane" anisotropy. A constant magnetic field $\mathbf{H}$ and the anisotropy axis are directed perpendicular to the plate surface along the Oz-axis: $\mathbf{H} = H\mathbf{n}$, $\mathbf{n} = (0, 0, 1)$ (see Figure 8.6).

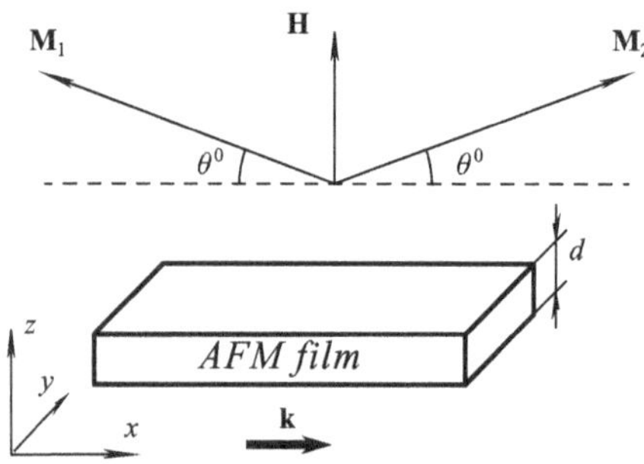

Figure 8.6: Geometry of a problem in describing nonlinear excitations in an antiferromagnet film with "easy-plane" anisotropy.

Within a simple antiferromagnet two-sublattice model, the energy density has the form [4]

$$\begin{aligned} w &= w_1 + w_2; \quad w_2 = -\frac{1}{2}((\mathbf{M}_1 + \mathbf{M}_2)\cdot \mathbf{H}^{(\mathrm{m})}); \\ w_1 &= \frac{\alpha_1}{2}[(\partial_i \mathbf{M}_1)^2 + (\partial_i \mathbf{M}_2)^2] \\ &\quad + \alpha_2(\partial_i \mathbf{M}_1 \cdot \partial_i \mathbf{M}_2) + \delta(\mathbf{M}_1 \cdot \mathbf{M}_2) - \frac{1}{2}\beta[(\mathbf{M}_1\cdot \mathbf{n})^2 + (\mathbf{M}_2 \cdot \mathbf{n})^2] \\ &\quad - \beta'(\mathbf{M}_1 \cdot \mathbf{n})(\mathbf{M}_2 \cdot \mathbf{n}) - H((\mathbf{M}_1 + \mathbf{M}_2)\cdot \mathbf{n}), \end{aligned}$$

where $\mathbf{M}_i$ $(i = 1, 2)$ are sublattice magnetization vectors, $\mathbf{M}_1^2 = \mathbf{M}_2^2 = M_0^2 = \mathrm{const}$, α_1, α_2, and δ are the inhomogeneous and homogeneous exchange constants, respectively, and β and β' are the magnetic anisotropy constants. It is known that $\beta' > \beta$ for an easy-plane anisotropy antiferromagnet. A demagnetizing field $\mathbf{H}^{(m)}$ is determined by the equations of magnetostatics

$$\begin{aligned} -\Delta\varphi + 4\pi \operatorname{div}(\mathbf{M}_1 + \mathbf{M}_2) = 0; \quad \mathbf{H}^{(m)} &= -\nabla\varphi; \\ \Delta\tilde{\varphi} = 0; \quad \mathbf{H}^{(m)} &= -\nabla\tilde{\varphi}, \end{aligned} \tag{8.3.1}$$

where φ and $\tilde{\varphi}$ are magnetic scalar potentials inside and outside the plate, respectively.

For further analysis, it is convenient to use the following parameterization of the sublattice magnetization vectors:

$$\mathbf{M}_i = M_0(\cos\theta_i \cos\varphi_i,\ \cos\theta_i \sin\varphi_i,\ \sin\theta_i), \quad i = 1, 2.$$

The minimum energy condition dictates the main uniform state of the system:

$$\int \frac{\delta w}{\delta \theta_i}\, \mathrm{d}^3\mathbf{r} = \int \frac{\delta w}{\delta \varphi_i}\, \mathrm{d}^3\mathbf{r} = 0.$$

It implies the equilibrium values of the angles:

$$\varphi_1^{(0)} - \varphi_2^{(0)} = \pi, \quad \theta_1^{(0)} = \theta_2^{(0)} = \theta^0,$$
$$\sin\theta^0 = \frac{H}{2M_0(\delta + 4\pi - \beta_2)}, \quad \beta_2 = \frac{\beta + \beta'}{2}.$$

Here the equations describing the dynamics of nonlinear excitations can be written using the Lagrange function with density:

$$L = \frac{M_0}{g}(\sin\theta_1\, \partial_t\varphi_1 + \sin\theta_2\, \partial_t\varphi_2) - w, \tag{8.3.2}$$

where g is the gyromagnetic ratio. The system of these equations appears as follows:

$$\begin{aligned} \frac{\partial L}{\partial \theta_i} - \partial_j\left[\frac{\partial L}{\partial(\partial_j\theta_i)}\right] - \frac{\partial w_2}{\partial \theta_i} &= 0, \\ \frac{\partial L}{\partial \varphi_i} - \partial_t\left[\frac{\partial L}{\partial(\partial_t\varphi_i)}\right] - \partial_j\left[\frac{\partial L}{\partial(\partial_j\varphi_i)}\right] - \frac{\partial w_2}{\partial \varphi_i} &= 0, \quad i = 1, 2. \end{aligned} \tag{8.3.3}$$

The additional summands $\partial w_2/\partial\theta_i$ and $\partial w_2/\partial\varphi_i$ in eqs. (8.3.3) explain the fact that the field $\mathbf{H}^{(m)}$ is $\mathbf{M} = \mathbf{M}_1 + \mathbf{M}_2$-resulting-magnetization-dependent. Therefore, when varied over the fields θ_i and φ_i, the Lagrangian accounts for the variations of the field $\mathbf{H}^{(m)}$ [4].

The boundary conditions for the problem to be free from spin pinning on the plate surface are given by the relations

$$\begin{aligned} &\partial_z\varphi_i|_{\pm d/2} = \partial_z\theta_i|_{\pm d/2} = 0, \quad \varphi|_{\pm d/2} = \tilde{\varphi}|_{\pm d/2}, \\ &-\partial_z\varphi|_{\pm d/2} = [-\partial_z\varphi + 4\pi M_0(\sin\theta_1 + \sin\theta_2)]|_{\pm d/2}, \end{aligned} \tag{8.3.4}$$

where d is the thickness of the plate.

8.3.2 An effective equation for a weakly nonlinear dynamics of Goldstone modes

For easy-plane anisotropy antiferromagnetic films, the exchange-dipole spectrum of linear spin waves, apart from activation branches, also has a gapless (Goldstone) branch. The present section focuses on the interaction of low-amplitude Goldstone modes in a region where the energy of an external action is insufficient to excite activation waves. The problem has two characteristic spatial scales: Δ is the size of the magnetic inhomogeneity (wavelength or soliton size), and d is the film thickness.

We dwell on spin-wave excitations under the following conditions:

$$\max\left[\left(\alpha + \alpha''\frac{\tan^2\theta^0}{2\pi d}\right), d\right] \ll \Delta,$$
$$\frac{d}{\delta\Delta} \ll \left[\frac{H}{2M_0(\delta + 4\pi - \beta_2)}\right]^2 \ll 1. \tag{8.3.5}$$

Here and further, we have

$$\alpha = \frac{\alpha_1 + \alpha_2}{2}, \quad \alpha'' = \frac{\alpha_1 - \alpha_2}{2}.$$

The dispersion law for Goldstone modes in this range of fields and spatial scales is exchange-magneto-static and has the form

$$\omega^2 = (2gM_0)^2\alpha'' k_x^2 \cos^2\theta^0(\delta + 4\pi - \beta_2 - 2\pi d|k_x|).$$

It should be stressed that the method proposed in [47, 50, 5] is the simplest one to produce demagnetizing fields responsible for the magneto-static part of the spin-wave spectrum. Further, for definiteness, the waves are thought to propagate along the Ox-axis (for the problem's chosen geometry, the dispersion law is independent of the direction of wave propagation in the plate plane).

To derive effective equations describing the interaction of low-amplitude Goldstone waves, we use the reductive perturbation theory based on the coordinate stretching method [22]. The scale transformations of the RPT are different outside and inside the plate.

Inside the plate, the solutions to problems (8.3.1),(8.3.3), and (8.3.4) should be sought as follows:

$$\varphi_i = \varphi_i^{(0)}(\xi,\tau,z) + \sum_{n=1}^{\infty}\varepsilon^n\varphi_i^{(n)}(\xi,\tau,z), \quad \theta_i = \theta^0 + \sum_{n=1}^{\infty}\varepsilon^n\theta_i^n(\xi,\tau,z),$$
$$\varphi = \varphi^{(0)}(\xi,\tau,z) + \sum_{n=1}^{\infty}\varphi^{(n)}(\xi,\tau,z)\,\varepsilon^n;$$
$$\xi = \varepsilon(x + s\,t), \quad s = 2M_0 g\cos\theta^0\sqrt{\alpha''(\delta + 4\pi - \beta_2)}, \quad \tau = \varepsilon^2 t.$$

Here ε is a small parameter characterizing the deviation of the system from the ground state:

$$\frac{d}{\delta\Delta}\cot\theta^0 \sim \theta_i - \theta^0 = O(\varepsilon) \ll 1.$$

Outside the film, the magneto-static potential is

$$\tilde{\varphi} = \tilde{\varphi}^{(0)}(\eta, z, \tau) + \sum_{n=1}^{\infty} \varepsilon^n \tilde{\varphi}^{(n)}(\eta, z, \tau), \quad \eta = x + st, \quad \tau = \varepsilon^2 t.$$

These two versions of the perturbation theory provide effective equations for Goldstone excitations by sewing. A similar approach was used earlier in [74] to describe waves in a stratified fluid. As in Section 8.1, the advantage of this method relies on the possibility of separating the variables ξ and z in perturbation theory equations as well. The chain of perturbation theory equations needs to be solved sequentially. The expansion in the parameter ε of requirements (8.3.4) yields necessary boundary conditions. It is important that the conditions

$$\partial_z \theta_i^{(n)}|_{\pm d/2} = \partial_z \varphi_i^{(n)}|_{\pm d/2} = 0 \quad (i = 1, 2)$$

in the lower orders of the perturbation theory lead to the fact that the functions $\varphi_i^{(0)}$, $\varphi_i^{(1)}, \theta_i^{(1)}$ $(i = 1, 2)$, $\varphi_1^{(n)} + \varphi_2^{(n)}$ $(n = 2, 3)$, and $\theta_1^{(2)} + \theta_2^{(2)}$ become z-independent. This circumstance greatly simplifies the problem but is not decisive. In principle, the method makes it possible to take into account both the inhomogeneity of spin waves and real parameters along the normal to the film and more complicated exchange boundary conditions on its surface. From the equations of the perturbation theory in the orders of ε^0 and ε we find

$$\begin{aligned}
&\varphi_1^{(0)}(\xi, \tau) = \pi + \varphi_2^{(0)}(\xi, \tau), \quad \theta_1^{(1)} = \theta_2^{(1)} = \gamma\, \partial_\xi \varphi_1^{(0)}, \\
&\gamma = \sqrt{\alpha''/(\delta + 4\pi - \beta_2)}; \\
&\varphi^{(0)} = 8\pi M_0 \sin\theta^0 z, \quad |z| \le d/2; \\
&\tilde{\varphi}^{(0)} = \begin{cases} 4\pi M_0\, d \sin\theta^0, & z \ge d/2, \\ -4\pi M_0\, d\sin\theta^0, & z \ge -d/2. \end{cases}
\end{aligned} \tag{8.3.6}$$

Outside the plate, to determine the magneto-static potential $\tilde{\varphi}^{(1)}$ (the order of $O(\varepsilon)$), we should solve the Dirichlet problem. In particular, as in Section 8.1 (see (8.1.17)), it is easy to express the limiting values of the derivatives $\partial_z \tilde{\varphi}^{(1)}(z, \eta)|_{z=\pm d/2}$ through $\partial_\xi \varphi^{(1)}(z, \xi)|_{z=\pm d/2}$:

$$\partial_z \tilde{\varphi}^{(1)}|_{z=\pm d/2} = \varepsilon\, \hat{\mathrm{H}}\, \partial_{\xi'} \varphi^{(1)}(\xi', z = d/2) = O(\varepsilon). \tag{8.3.7}$$

Formulas (8.3.7) state that the limiting values of the derivative $\partial_z\tilde{\varphi}^{(1)}|_{\pm d/2}$, indeed, belong to the next order of $O(\varepsilon^2)$ in the reductive perturbation theory. In the first order with respect to ε, the following simpler problem governs the magneto-static potential inside the plate:

$$\begin{aligned}&\partial_z^2\varphi^{(1)} = 0, \quad |z| \le d/2;\\&[-\partial_z\varphi^{(1)} + 8\pi M_0\cos\theta^0\theta^{(1)}(\varepsilon,\tau)]|_{z=\pm d/2} = 0.\end{aligned} \tag{8.3.8}$$

Its solution has the form

$$\varphi^{(1)} = 8\pi M_0\cos\theta^0\,\theta_1^{(1)}(\xi,\tau)z.$$

An analysis of the equations for the next orders $O(\varepsilon^2)$ and $O(\varepsilon^3)$ is conducted in a similar way. To compute the magneto-static potential $\varphi^{(2)}$ (the order of $O(\varepsilon^2)$) inside the plate, we should resort to solving the boundary value problem

$$\begin{aligned}&\partial_z^2\varphi^{(2)} = 0, \quad |z| \le d/2;\\&-\partial_z\tilde{\varphi}^{(1)}|_{z=\pm d/2} = \Big[-\partial_z\varphi^{(2)} + 4\pi M_0\big(\cos\theta^0(\theta_1^{(2)}+\theta_2^{(2)}) - \sin\theta^0(\theta^{(1)})^2\big)\Big]\Big|_{z=\pm d/2}.\end{aligned} \tag{8.3.9}$$

Since the functions $\theta^{(1)}$ and $\theta_2^{(2)} + \theta_1^{(2)}$ are z-independent, the solution to problem (8.3.9) is trivial. As a result, the second and third order of the perturbation theory with respect to the parameter ε give a closed effective integro-differential equation for determining the function $\theta_1^{(1)}(\xi,\tau)$:

$$\begin{aligned}&\partial_\tau\theta_1^{(1)} - \nu\,\hat{\mathrm{H}}(\partial_{\xi}^2\theta_1^{(1)}) + q\,\partial_\xi(\theta_1^{(1)})^2 = 0,\\&\nu = 2\pi d\,M_0\gamma\cos\theta^0, \quad q = \frac{3}{2}\,s\tan\theta^0.\end{aligned} \tag{8.3.10}$$

It coincides with the Benjamin–Ono equation [74, 6]. The Benjamin–Ono model admits multisoliton excitations [2, 67] and can be thoroughly explored by the inverse scattering method [1, 26].

8.3.3 Analysis of soliton solutions. Exchange relaxation of solitons

An exchange-magneto-static soliton is less localized than typical magnetic solitons of exchange and exchange-relativistic origin. A one-soliton solution to eq. (8.3.10) is described by an “algebraic” wave rather than an exponential one:

$$\begin{aligned}&\theta_1^{(1)} = -\frac{\sigma\Delta}{(\xi+\vartheta\tau)^2+\Delta^2}, \quad \sigma = \frac{4\pi d\cot\theta^0}{3[\delta+4\pi-\beta_2]},\\&\Delta = 2\pi\,gM_0 d\,\gamma\,\vartheta^{-1}\cos\theta^0 > 0,\end{aligned} \tag{8.3.11}$$

where ϑ is a positive real parameter. In the initial variables x and t, soliton (8.3.11) corresponds to a localized perturbation propagating with velocity $\vartheta + s$. Note that the soliton's amplitude is proportional to the excess ϑ of its velocity over the phase velocity s of spin waves. Apparently, solitons appear in a threshold manner, as the amplitude of the exchange-magneto-static waves grows, and make themselves felt only in dynamics. The radiation of the solitons lowers the system energy. The soliton size 2Δ must be greater than the thickness of the plate (see (8.3.5)). This requirement limits the range of values for the parameter ϑ:

$$\frac{\vartheta}{s} \sim \frac{d}{\Delta\delta} \ll 1.$$

Equation (8.3.10) controls the dynamics of a Lagrangian system. Formulas (8.3.6)–(8.3.9) can be used to show that the approximation at hand reduces the complete expression (8.3.2) for the density of the Lagrange function to the effective one:

$$L_{\text{eff}} = c\left[\frac{1}{2}\,\partial_\xi\chi\,\partial_\tau\chi - \frac{1}{2}\,\nu\,\partial_\xi\chi\,\partial_\xi\hat{\mathrm{H}}\,\partial_{\xi'}\chi + \frac{1}{3}\,q(\partial_\xi\chi)^3\right]. \tag{8.3.12}$$

Here $c = 4M_0a^2(g\gamma)^{-1}\cos\theta^0$, and a is a lattice constant. We have introduced a potential $\partial_\xi\chi = \theta_1^{(1)}$ instead of a field $\theta_1^{(1)}$. Over-the-field-χ variation of an action corresponding to L_{eff} gives eq. (8.3.10). Using relation (8.3.12), it is easy to obtain expressions for the energy E and field momentum P:

$$\begin{aligned} E &= \int\left[\frac{\partial L_{\text{eff}}}{\partial(\partial_\tau\chi)}\partial_\tau\chi - L_{\text{eff}}\right]\mathrm{d}\xi \\ &= c\int\left[\frac{1}{2}\,\nu\,\theta_1^{(1)}\partial_\xi\hat{\mathrm{H}}\,\theta_1^{(1)} - \frac{1}{3}\,q(\theta_1^{(1)})^3\right]\mathrm{d}\xi, \\ P &= \int \mathrm{d}\xi\,\frac{\partial L_{\text{eff}}}{\partial(\partial_\tau\chi)}\,\partial_\xi\chi = \frac{c}{2}\int(\theta_1^{(1)})^2\mathrm{d}\xi. \end{aligned} \tag{8.3.13}$$

Plugging eqs. (8.3.11) into eqs. (8.3.13), we come up with

$$E = d\pi\sigma a\,\Delta^{-1}M_0\cos\theta^0, \quad P = \pi(\sigma\,a)^2M_0\cos\theta^0(g\gamma\,\Delta)^{-1}. \tag{8.3.14}$$

Note that the relations obtained refer to a system moving at a speed $-s$. The dispersion law for spin waves in this coordinate system has the form

$$\Omega(k) = k\,s - \omega(k) = 2\pi d\gamma\,M_0g\cos\theta^0|k|k. \tag{8.3.15}$$

Formulas (8.3.14) and (8.3.15) allow us to interpret soliton (8.3.11) as a complex of N non-interacting magnons with wave vectors $k = (2\Delta)^{-1}$, where 2Δ is a parameter featuring the soliton width. The number N is given by the formula

$$N = \pi(\sigma\,a)^2M_0\cos\theta^0(\gamma\mu_B)^{-1} \sim \pi[8\pi dM_0(3aH)^{-1}]^2 \gg 1,$$

where μ_B is the Bohr magneton. Given that $g = 2\mu_B/\hbar$, formulas (8.3.14) can be rewritten as

$$P = N\hbar k, \quad E = N\hbar\,\Omega(k), \quad k = (2\Delta)^{-1},$$

where $\Omega(k)$ coincides with the dispersion law (8.3.15) for spin waves, $k > 0$.

The soliton energy can also be presented as

$$E = \frac{P^2}{2M_{\text{eff}}}.$$

This can treat the soliton as a particle-like formation with effective mass

$$M_{\text{eff}} = Nm_0,$$

where $m_0 \approx \hbar(4\pi d\gamma M_0 \cos\theta^0)^{-1}$ is the effective mass of a magnon obeying the dispersion law (8.3.15). We have arrived at the outcome that is characteristic of an "algebraic" soliton; namely, its energy, momentum, and mass are the sums of N appropriate quantities for an individual magnon. It should be underscored, however, that the number N here is not a free physical parameter (like velocity ϑ or momentum P) but is dictated by external conditions, such as the plate thickness d and the magnitude of a magnetic field. From this point of view, the energy of a soliton is P-dependent and turns out to be N times less than that of a magnon with the same momentum. The integrability of the model is said to guarantee the stability of soliton (8.3.11). Multisoliton solutions to eq. (8.3.10) govern elastic collisions of such "algebraic" solitons.

The vectors

$$\mathbf{M} = \mathbf{M}_1 + \mathbf{M}_2 \quad \text{and} \quad \mathbf{L} = \mathbf{M}_1 - \mathbf{M}_2$$

of ferro- and antiferromagnetism, respectively, regulate the magnetic state of an antiferromagnet. In terms of $\theta_1^{(1)}$ and $\varphi_1^{(0)}$, they come out as

$$\begin{aligned} \mathbf{M} &= 2M_0(0,\ 0,\ \sin\theta^0 + \cos\theta^0\,\theta_1^{(1)}) + O(\varepsilon^2), \\ \mathbf{L} &= 2M_0(\cos\theta^0\cos\varphi_1^{(0)},\ \cos\theta^0\sin\varphi_1^{(0)},\ 0) + O(\varepsilon). \end{aligned} \tag{8.3.16}$$

In the soliton localization region, the component M_z is less than at infinity; therefore it describes a "dark" soliton. The depth of the minimum is proportional to the parameter ϑ. According to (8.3.6) and (8.3.11), we get

$$\varphi_1^{(0)} = -\frac{\sigma}{\gamma}\arctan\left(\frac{\xi + \vartheta\tau}{\Delta}\right). \tag{8.3.17}$$

In view of eqs. (8.3.16) and (8.3.17), the vector $\mathbf{L}$ lies in the film plane and makes many turns in the soliton localization region, turning through the angle

$$\Delta\varphi_1^{(0)} = \frac{\pi\sigma}{\gamma} \sim \frac{8\pi^2 dM_0}{3aH} \gg \pi.$$

Interestingly, for an unbounded antiferromagnetic sample, the dynamics of low-amplitude waves is described by the KdV-type local equations. The latter have different forms depending on the direction of wave propagation and the magnitude of an external magnetic field. In this case, ordinary “exponential” solitons emerge rather than “algebraic” solitons, as in a thin plate. This because the contribution of magnetostatics is less important for unbounded samples, and wave propagation is determined mainly by short-range exchange interactions. For a thin plate, the picture is opposite: long-range dipole–dipole forces predominate, and “algebraic” solitons arise in the main approximation due to surface magnetic charges. Other types of magneto-static charges, such as bulk ones, only refine the internal structure of solitary waves.

The model we have designed corresponds to real antiferromagnets. For specific assessments, we can take material parameters of $MnCO_3$ with the Neel point $T_N = 32.5$ K, of hematite ($\alpha - Fe_2O_3$) with $T_N = 950$ K, which is “easy-planar” above the Morin point $T_M = 260$ K, and of iron borate $FeBO_3$ with $T_N = 348$ K. Conditions (8.3.5) can be easily created in such systems if $d \approx 5$–10 μm. Let us evaluate the parameter ϑ for $FeBO_3$. Given that $\delta \approx 10^3$ and $M_0 \approx 1040$ G, we obtain that $\vartheta \leq 10^2$ cm/sec under conditions (8.3.5). The effective soliton velocity $s + \vartheta$ turns out to be quite large, $\sim 10^4 - 10^5$ cm/sec, and close to the speed of sound in these materials. Apparently, magneto-static solitons can be detected by their resonant interaction with an elastic subsystem of an antiferromagnet.

Accounting for energy dissipation is an important problem in the physics of solitons. The calculation of a decrement of spin waves due to exchange interactions leads to the expression $\Gamma = \mu k^2$ ($\mu = \text{const} > 0$) [32]. We take relaxation into account phenomenologically by adding the term $-\mu\,\partial_\xi^2\theta_1^{(1)}$ to the left-hand side of eq. (8.3.10):

$$\partial_\tau\theta_1^{(1)} - \nu\,\hat{H}(\partial_{\xi'}^2\theta_1^{(1)}) - \mu\,\partial_\xi^2\theta_1^{(1)} + q\,\partial_\xi(\theta_1^{(1)})^2 = 0. \tag{8.3.18}$$

It is remarkable that the built-up model (8.3.18) admits a wide class of exact solutions [79, 60]. The simplest of them looks like

$$\begin{gathered} \theta_1^{(1)} = -\sigma\,\frac{\Delta(\tau) + \mu\nu^{-1}[\xi + \xi_0(\tau)]}{[\xi + \xi_0(\tau)]^2 + \Delta^2(\tau)}, \\ \xi_0(\tau) = \frac{\nu}{\mu}[\Delta(\tau) - \delta_0], \quad \Delta(\tau) = \sqrt{2\mu\tau + \delta_0^2}, \; \delta_0 = \text{const} > 0. \end{gathered} \tag{8.3.19}$$

In the limit as $\mu \to 0$, solution (8.3.19) coincides with (8.3.11). From (8.3.19) it follows that the localized excitation spreads out over time: its size increases, whereas the speed and amplitude decrease.

8.4 Nonlinear magneto-elastic resonance of long and short waves in magnets

Observations are an abundant source of discoveries both in a world of subjective phenomena and in a world of real phenomena perceived by our senses.

Ch. Hermite. Correspondence with Stieltjes

Numerous noteworthy peculiarities of magneto-acoustic phenomena in crystals are the object of sought-after investigation. The point is that a magnetic subsystem imposes a significant nonlinearity on an elastic one by means of magnetostriction. The ground state of the medium is nontrivial because spontaneous deformations are associated with an equilibrium distribution of magnetization. This leads to a renormalization of elastic moduli and magnetic anisotropy constants. Amazing effects such as a strong decrease in the velocity of a quasiacoustic mode, the presence of an additional contribution to the gap in the spectrum of activation modes, birefringence, and rotation of the polarization plane of quasiacoustic waves can be observed [87, 21, 88]. Similar effects make themselves felt mainly in "easy-plane" anisotropy magnets when the crystallographic magnetic anisotropy in the basal plane is small and the magneto-elastic coupling constants are large. Rare earth metals as dysprosium and terbium at temperatures below a certain critical temperature, equal to 87 K and 219 K, respectively, belong to such ferromagnets. The noted features can be significant in antiferromagnets due to their enhancements by uniform exchange interaction and also near the points of orientational phase transitions.

In the same systems, magneto-elastic interactions furnish the specifics of excitation, interaction, and propagation of low-amplitude waves [75, 59, 76, 9]. Note that the dynamics of weakly nonlinear perturbations in miscellaneous magneto-elastic media can be studied within universal integrable models that allow for the formation of solitons. In the event of a low-energy external impact on the system and thereby of no activation spin waves, gapless magneto-elastic modes evolve along the Korteweg–de Vries-type equation [86]. Far from a resonance of activation and gapless modes, the dynamics of activation spin waves can be thoroughly described using the nonlinear Schrödinger equation [89, 72]. If the interacting waves gather for a "triad" that satisfies the conditions of "synchronism" ($\mathbf{k}_1+\mathbf{k}_2 = \mathbf{k}_3, \omega_1+\omega_2 = \omega_3$; $k_i, \omega_i, i = 1, 2, 3$, are the wave vectors and frequencies of the interacting modes, respectively), then three-wave resonance equations handle the energy exchange between the modes [68].

The wave resonance-interaction region includes the spectrum's different mutually intensively interacting branches. Therefore, to distinguish one of them is wrong. It is necessary to reconstruct the reductive perturbation theory so that all resonant modes should have the same rights. This requires some ingenuity [22, 93]. This section outlines a technique suitable for finding evolutionary equations in the case of the resonant interaction of waves, looking at different branches of the spectrum of linear modes. Its essence is to transition from the original field variables to normal modes. Moreover, when describing the interaction of normal modes, we apply approximations that are self-consistent with the results of RPT for the limiting cases where one of the branches of the linear mode spectrum dominates. As a result, we obtain effective equations of weakly nonlinear dynamics, which have a wider range of applicability than the usual variants of RPT.

It is worth emphasizing that media with dispersion relations that allow for the existence of activation and gapless (Goldstone) waves sometimes produce a special type of energy exchange between the gapless and activation branches of the spectrum of linear modes. Due to beats, activation modes with close wave vectors can transfer energy and excite Goldstone waves. As Benny showed [22, 7, 8], this process has a resonant nature and is most effective when the group velocity of the activation wave c_g is close to the phase velocity of the gapless wave: $c_g \approx c_p$. The condition $c_g = c_p$ can be achieved purely geometrically by exploring the shape of the dispersion curves (Figure 8.7).

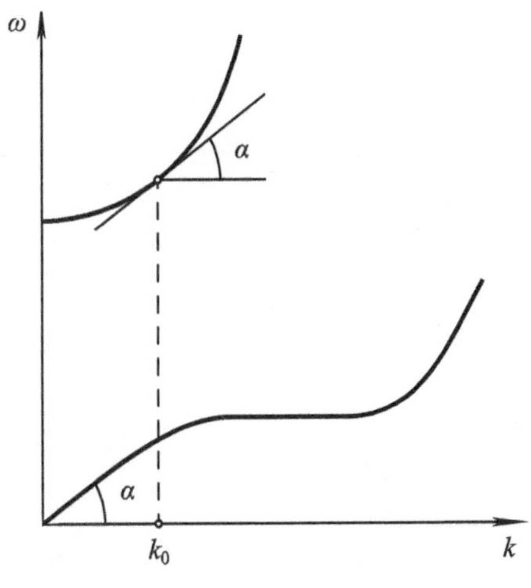

Figure 8.7: Dispersion laws under which nonlinear resonance of long- and short-waves is possible.

At some point $k = k_0$, the slope of the tangent line to the activation wave's dispersion law curve $\omega = \omega_a(k)$ coincides with the group velocity of the activation mode. At the point $k = 0$ the phase velocity c_p coincides with the slope of the tangent line to the gapless branch of the spectrum. Recall that the Benney resonance is the exchange of energy between waves with different characteristic spatial scales. Short-wave activation modes actuate long-wave gapless ones. It turns out that the two spectrum branches that are thought to be independent in the region of distinct wavelengths are connected. The Benny mechanism is not always implemented but implies a certain type of nonlinear interactions between the activation and gapless modes. A resonantly excited gapless mode evolves according to the quadratic-in-amplitude interaction of activation waves.

The interaction between a spin wave and branches of elastic oscillations is known to give rise to the phenomenon of magneto-acoustic resonance. The resonance between elastic and spin subsystems of a crystal arises conditional upon coinciding with their own unperturbed frequencies and can already be described within the linear theory [4]. This section, following the work [45], draws attention to the fact that in magnets, a special type of resonance between the gapless and activation branches of the spectrum of magneto-elastic waves is always possible, the nonlinear Benney resonance. Using a specific example of a tetragonal "easy-plane" anisotropy ferromagnet, we have analyzed general regularities governing the interaction of low-amplitude magneto-elastic waves in magnets. In particular, we have shown that in the vicinity of the resonance of long and short waves, the Zakharov–Benney integrable model [22] effectively handles the weakly nonlinear dynamics of magneto-elastic excitations.

An expression for the free energy density F of an "easy-plane" anisotropy tetragonal ferromagnet can be written as [21]

$$\begin{aligned}
F(\mathbf{r}) &= F_{\mathrm{m}}(\mathbf{r}) + F_{\mathrm{e}}(\mathbf{r}) + F_{\mathrm{me}}(\mathbf{r}),\\
F_{\mathrm{m}}(\mathbf{r}) &= \frac{1}{2}E\,a^2(\partial_i \mathbf{m})^2 + \frac{1}{2}K_1^0 m_z^2 + K_2^0 m_z^4\\
&\quad + \frac{1}{2}K_\Box^0 m_x^2 m_y^2 - (\mathbf{M}\cdot\mathbf{H}) + \frac{\mathbf{H}_M^2}{8\pi},\\
F_{\mathrm{e}}(\mathbf{r}) &= \frac{1}{2}C_{11}(\varepsilon_{xx}^2 + \varepsilon_{yy}^2) + C_{12}\,\varepsilon_{xx}\varepsilon_{yy} + C_{13}(\varepsilon_{xx} + \varepsilon_{yy})\varepsilon_{zz}\\
&\quad + \frac{1}{2}C_{33}\,\varepsilon_{zz}^2 + 2C_{44}(\varepsilon_{yz}^2 + \varepsilon_{zx}^2) + 2C_{66}\,\varepsilon_{xy}^2,
\end{aligned} \tag{8.4.1}$$

$$\begin{aligned}
F_{\mathrm{me}}(\mathbf{r}) &= [B_1(\varepsilon_{xx} + \varepsilon_{yy}) + B_3\,\varepsilon_{zz}]\mathbf{m}^2\\
&\quad + (B_{11}\,m_x^2 + B_{12}\,m_y^2)\varepsilon_{xx} + (B_{12}\,m_x^2 + B_{11}\,m_y^2)\varepsilon_{yy}\\
&\quad + B_{33} m_z^2\,\varepsilon_{zz} + 2B_{44}(m_y m_z \varepsilon_{yz} + m_x m_z \varepsilon_{xz}) + 2B_{66}\,m_x m_y \varepsilon_{xy}.
\end{aligned} \tag{8.4.2}$$

Here $\mathbf{M} = M_0\mathbf{m}$ is the magnetization vector of the medium, $\mathbf{M}^2 = M_0^2 = \text{const}$, E is the exchange interaction constant (a is the lattice parameter), K_1^0, K_2^0, and $K_\Box^0$ are constants of magnetic anisotropy ($K_1^0 > 0$, $K_1^0 \gg K_2^0, K_\Box^0$), $\mathbf{H}$ is the external magnetic field, and the field $\mathbf{H}_M$ is determined by the equations of magneto-statics. The constants C_{sp} determine the elastic energy, B_s and B_{sp} manage the magneto-elastic energy, $\varepsilon_{ij} = (\partial_j u_i + \partial_i u_j)/2$ is the strain tensor ($i, j = x, y, z$), and $\mathbf{u}$ is the displacement vector of points of the medium. Equation (8.4.1) leaves aside the elastic anharmonicity itself because the latter is usually small compared to that induced by the magnetic subsystem on the elastic one. For the same reason, we neglect the nonlinear dependence of the strain tensor ε_{ij} on the derivatives of the displacement vector $\mathbf{u}$ and consider the interaction between the elastic subsystem and the magnetic one (8.4.2) in the linear-in-strain approximation.

As follows from eq. (8.4.2), the magneto-elastic energy $F_{\mathrm{me}}(\mathbf{r})$ contains terms that describe the magneto-strictive effects due to the uniform rotation of the magnetization vector. These are not the only linear-in-strain terms. We account for the dependence of the exchange integrals on the deformations of the medium by expanding the isotropic exchange in a series in ε_{ij}. Then the expression for the free energy includes terms related to the magnetization inhomogeneity [34]:

$$F^{(1)}_{\mathrm{me}} = G_{ijkl}(\partial_{x_i} m_n)(\partial_{x_j} m_n)\, \varepsilon_{kl}. \tag{8.4.3}$$

The tensor ε_{kl} can be regarded as the sum of two terms describing spontaneous and dynamic deformations (see below). The energy (8.4.3) also contains two contributions. The former (from spontaneous deformations) controls only a renormalization of the exchange interaction constant and can be taken into account trivially. The contribution of dynamic deformations in the long-wave approximation to (8.4.3) leads, in comparison with that of (8.4.2), to corrections of the next order of smallness. In this monograph, we have ignored the energy (8.4.3). On the other hand, interaction (8.4.3) may turn out to be important, for example, for a chain of spins (an "easy-plane" anisotropy one-dimensional ferromagnet $CsNiF_3$ is a typical example). As [77] illustrates, the system of equations for atomic displacements and spin deflections can be boiled down to the Davydov equation [18]. References [72, 33] offer refinements to this model.

Our approach is not precise. The magnetic and elastic subsystems influence each other softly. This is evident from the fact that not only the symmetric strain tensor ε_{ij} but also the antisymmetric combinations $\omega_{ij} = (\partial_j u_i - \partial_i u_j)/2$ define the expression for free energy in the general case. Neglecting the rotations of the magneto-elastic medium's volume elements is equivalent to excluding ω_{ij}. For quasi-one-dimensional movements, the antisymmetric part of the tensor $\partial_j u_i$ does not significantly change outcomes and only slightly renormalizes the theoretical constants. When describing strongly nonuniform states, the expression for the elastic energy should include invariants dependent on the second spatial derivatives of the displacement field [85]. We also note that beyond the linear-in-strain approximation, unit mass-related quantities, rather than those per unit volume of the ferromagnet, must be applied. Consequently, we should distinguish the spatial (Euler) coordinates from the material (Lagrangian) ones associated with particles of the body [4].

It is known that the approximate theory of magneto-elastic interaction properly describes experimental data secured during research on magneto-elastic effects connected to the resonant interaction between elastic and spin waves. We will apparently not make a big mistake by using the approximate version [61].

In what follows, it is convenient to use the following parameterization for the magnetization vector:

$$\mathbf{M} = M_0\mathbf{m} = M_0(\sin\theta\ \sin\varphi,\ \sin\theta\ \cos\varphi,\ \cos\theta).$$

The ground state of the system is immediate from the minimum energy condition

$$\frac{\partial F}{\partial \varepsilon_{ij}^0} = 0; \qquad \frac{\partial F}{\partial \varphi_0} = \frac{\partial F}{\partial \theta_0} = 0,$$

where ε_{ij}^0 are spontaneous deformations, and the angles θ_0 and φ_0 stand behind the equilibrium orientation of $\mathbf{m}$. For simplicity, we restrict ourselves to the case $\mathbf{m}\|\mathbf{H}\|\mathbf{x}$. Then, with $\theta_0 = \varphi_0 = \pi/2$, the spontaneous deformations ε_{ij}^0 are given by the formulas [21]

$$\begin{aligned}
\varepsilon_{xx}^0 &= \Delta^{-1}\,[(C_{11} - C_{12})(C_{13}B_3 - C_{33}B_1) \\
&\quad -B_{11}(C_{11}C_{33} - C_{13}^2) + B_{12}(C_{12}C_{33} - C_{13}^2)]\,, \\
\varepsilon_{yy}^0 &= \Delta^{-1}\,[(C_{11} - C_{12})(C_{13}B_3 - C_{33}B_1) \\
&\quad +B_{11}(C_{12}C_{33} - C_{13}^2) - B_{12}(C_{11}C_{33} - C_{13}^2)]\,, \\
\varepsilon_{zz}^0 &= \Delta^{-1}[C_{13}(2B_1 + B_{11} + B_{12}) - B_3(C_{11} + C_{12})](C_{11} - C_{12}),
\end{aligned}$$

where

$$\Delta = (C_{11} - C_{12})[C_{33}(C_{11} + C_{12}) - 2\,C_{13}^2].$$

The stability conditions for this state have been studied in [21] and are reduced to

$$H_A + H > 0 \quad \text{and} \quad H_\square + H > 0,$$

where

$$\begin{aligned}
H_A &= K_1^*/M_0, \quad H_\square = K_\square^*/M_0, \\
K_1^* &= K_1^0 + 2[B_{33}\,\varepsilon_{zz}^0 - B_{11}\,\varepsilon_{xx}^0 - B_{12}\,\varepsilon_{yy}^0] - B_{44}^2/C_{44}, \\
K_\square^* &= K_\square^0 - \frac{B_{66}^2}{C_{66}} + 2\,\frac{(B_{11} - B_{12})^2}{C_{11} - C_{12}}.
\end{aligned} \tag{8.4.4}$$

Here K_1^* and $K_\square^*$ are magnetostriction-renormalized anisotropy constants.

The present section deals with weakly nonlinear quasi-one-dimensional excitations propagating along the Oy-axis. In the linear approximation, the above geometry of the problem ties only the branch of transverse sound waves with spin waves.

8.4.1 Solitons caused by the interaction of quasiacoustic modes

Further analysis requires to separate spontaneous deformations from total deformations, $\varepsilon_{ij} = \tilde{\varepsilon}_{ij} + \varepsilon_{ij}^0$ ($\tilde{\varepsilon}_{ij}$ is responsible for vibrations near the ground state), and to count out angles from their equilibrium values $\theta_0 = \varphi_0 = \pi/2$, corresponding to the position of the magnetization vector $\mathbf{m} = (1, 0, 0)$: $\chi = \varphi - \pi/2$, $\eta = \theta - \pi/2$.

Restricting ourselves to considering small-amplitude perturbations, we write the equations of dynamics up to cubic terms in $\chi, \eta, \partial_y \tilde{u}_i \equiv \tilde{u}_{yi}$ ($i = x, y, z$). The problem has three characteristic spatial scales: the lattice constant a, the wavelength of activation modes, $\lambda = k^{-1}$, and the size of magnetic inhomogeneities, d. In this context, $\lambda \gg d \gg a$. In the long-wave cubic-term approximation, the dependence on the derivatives of the field χ and the distortion field $\tilde{u}_{yi}$ can be ignored. The analysis shows that here we omit small-order corrections

$$O\left(\frac{E}{K_1}(ak)^2\right),\quad O\left(\frac{E}{K_1}\frac{a^2k}{d}\right),\quad O\left(\frac{E}{K_1}\left(\frac{a}{d}\right)^2\right)$$

in the effective interaction constants. In addition, the angle η describing the magnetization exit from the basis plane is small due to the strong "easy-plane" anisotropy, and therefore the nonlinear-in-term η-dependence can be neglected. The smallness criterion for η is the magnitude of the ratio

$$\frac{\eta}{\chi} \approx \frac{\sqrt{H_\square + H_{me6} + 4\pi M_0 + H}}{\sqrt{H_A + H_{me4} + H}} \ll 1,$$

where

$$H_{me4} = \frac{B_{44}^2}{M_0 C_{44}},\quad H_{me6} = \frac{B_{66}^2}{M_0 C_{66}},$$

and relations (8.4.4) yield $H_\square$ and H_A. Then we succeed in eliminating η from the equations to get the following system:

$$\begin{aligned}
&\left(\frac{M_0}{\gamma}\right)^2 \partial_t^2\chi + [Ea^2\partial_y^2 - K_1 - M_0H][(Ea^2\partial_y^2 - K_\square^0 - M_0H\\
&\quad - K_3)\chi + B_{66}\tilde{u}_{yx}] - (K_1 + M_0H)\left(\frac{2}{3}(K_3 + 4K_\square^0)\chi^3\right.\\
&\quad \left. + \frac{H}{6}M_0\chi^3 - 2(B_{11} - B_{12})\tilde{u}_{yy}\chi - 2B_{66}\tilde{u}_{yx}\chi^2\right) = 0;\\
&\partial_t^2\tilde{u}_{yx} - S_{66}^2\partial_y^2\tilde{u}_{yx} + B_{66}\,\rho^{-1}\partial_y^2\chi - \frac{2}{3}B_{66}\,\rho^{-1}\partial_y^2\chi^3 = 0;\\
&\partial_t^2\tilde{u}_{yy} - S_{11}^2\partial_y^2\tilde{u}_{yy} - (B_{11} - B_{12})\rho^{-1}\partial_y^2\chi^2 = 0.
\end{aligned} \tag{8.4.5}$$

Here

$$S_{ij}^2 = \frac{C_{ij}}{\rho},\quad K_1 = K_1^* + \frac{B_{44}^2}{C_{44}},\quad K_3 = 4\pi M_0^2 + 2\frac{(B_{11} - B_{12})^2}{C_{11} - C_{12}},$$

γ is the gyromagnetic ratio, and ρ is the density of matter.

As follows from eqs. (8.4.5), the linear approximation connects magnons only with transverse sound waves. The spectrum of linear excitations has branches such as [21]

a) activation one with the dispersion law

$$\omega_a^2 = \omega_0^2 + fk^2, \tag{8.4.6}$$

where

$$\omega_0^2 = \gamma^2[H + H_\square + 4\pi M_0 + H_{me6}][H_A + H_{me4} + H],$$
$$f = E\,a^2\gamma^2 M_0^{-1}[H_A + H_\square + H_{me4} + H_{me6} + 4\pi M_0 + 2H]$$
$$+ H_{me6}\,S_{66}^2\,[H_\square + H_{me6} + 4\pi M_0 + H]^{-1};$$

b) dispersionless one for longitudinal sound waves:

$$\omega = S_{11}k;$$

c) Goldstone one associated with transverse sound waves:

$$\omega_G^2(k) = S^2k^2 + r\,k^4, \tag{8.4.7}$$

where

$$S^2 = S_{66}^2(1 - H_{me6}[H_\square + H_{me4} + 4\pi M_0 + H]^{-1}),$$
$$r = -\frac{H_{me6}S_{66}^2}{\omega_0^2}\left(\frac{S^2 - Ea^2\gamma^2M_0^{-1}(H_A + H_{me4} + H)}{H_\square + H_{me6} + 4\pi M_0 + H}\right).$$

In the problem geometry at hand, the "softening" of the transverse sound velocity S due to magnetostriction is insignificant because usually $4\pi M_0 \gg H_{me}$. The dependence of the energy gap ω_0 on the external magnetic field is such that $\min \omega_0 = \omega_0(H = 0)$.

In this section, we have studied the interaction of quasiacoustic (Goldstone) modes (8.4.7), provided that the energy of the external perturbation is small and no activation waves are excited. To derive an effective equation for the evolution of Goldstone modes, we use a coordinate stretching-based version of PRT. A solution to the equation system (8.4.5) should be sought in the form

$$\tilde{u}_{yx} = \sum_{n=1}^{\infty} \varepsilon^n \tilde{u}_{yx}^{(n)}(\xi, \tau), \quad \tilde{u}_{yy} = \sum_{n=2}^{\infty} \varepsilon^n \tilde{u}_{yy}^{(n)}(\xi, \tau),$$
$$\chi = \sum_{n=1}^{\infty} \varepsilon^n \chi^{(n)}(\xi, \tau), \tag{8.4.8}$$

where $\xi = \varepsilon(y + S\,t)$ and $\tau = \varepsilon^3 t$. The parameter ε characterizes the deviation of the system from equilibrium. Slow variables τ and ξ are chosen so as to agree the space-time response of the medium to the disturbance with the dispersion law (8.4.7) and to balance the effects of dispersion and nonlinearity [22]. After plugging eqs. (8.4.8) into system (8.4.5), we get a chain of equations that expresses the "highest" coefficients (over the index n) through "lower" ones. The idea of the method is that under appropriately chosen scaling transformations, a certain order of the perturbation theory (8.4.8) gives a closed equation for the coefficients. This is the shortened equation of nonlinear dynamics.

In this case the coefficients $\chi^{(1)}$ and $\tilde{u}_{yy}^{(2)}$ are expressed through the field $\tilde{u}_{yx}^{(1)} \equiv e$:

$$\chi^{(1)} = B_{66}\left[M_0(H_\square + H_{me6} + H + 4\pi M_0)\right]^{-1} e;$$
$$\tilde{u}_{yy}^{(2)} = (B_{11} - B_{12})\rho^{-1}\left(S^2 - S_{11}^2\right)^{-1}\left(\chi^{(1)}\right)^2.$$

This field evolves in accordance with the modified Korteweg–de Vries equation (mKdV)

$$2S\,\partial_\tau e + r\,\partial_\xi^3 e - q\,\partial_\xi e^3 = 0, \tag{8.4.9}$$

where

$$q = 2H_{me6}\left(\frac{B_{66}S_{66}}{M_0(H_\square + H_{me6} + H + 4\pi M_0)^2}\right)^2 \times \left(4\pi M_0 + 2H_{me3} + \frac{5H}{4} + H_{me3}\frac{S_{11}^2 - S_{12}^2}{S^2 - S_{11}^2}\right),$$

$$H_{me3} = \frac{(B_{11} - B_{12})^2}{M_0(C_{11} - C_{12})}.$$

It is known [58] that eq. (8.4.9) has soliton solutions only if $rq < 0$. Since $4\pi M_0 \gg H_{me}$ and the anisotropy in the basal plane is small, we can observe solitons only in substances with $r < 0$, that is, under the condition

$$S^2 > \frac{Ea^2\gamma^2}{M_0}(H_A + H_{me4} + H).$$

The bisoliton solution of eq. (8.4.9) has the form

$$e = \sqrt{\frac{4S\lambda}{q}}\,\mathrm{sech}\left(\sqrt{\frac{S\lambda}{|r|}}\zeta\right), \quad \tilde{u}_x = 4\sqrt{\frac{|r|}{q}}\,\arctan\exp\left(\sqrt{\frac{S\lambda}{|r|}}\zeta\right), \tag{8.4.10}$$

where $\zeta = y + (S + \lambda)t + y_0$ with real parameters y_0 and $\lambda > 0$. As eqs. (8.4.10) suggest, the soliton moves at a speed $V = S + \lambda$ greater than that of transverse sound. When the soliton travels, the magnetization oscillations are adjusted in a quasiequilibrium way to the transverse elastic deformations: $m_y \propto \tilde{u}_{yx} = e$. The displacement of the medium's points generates the component m_y of the ferromagnetism vector, as depicted in Figure 8.8*b*.

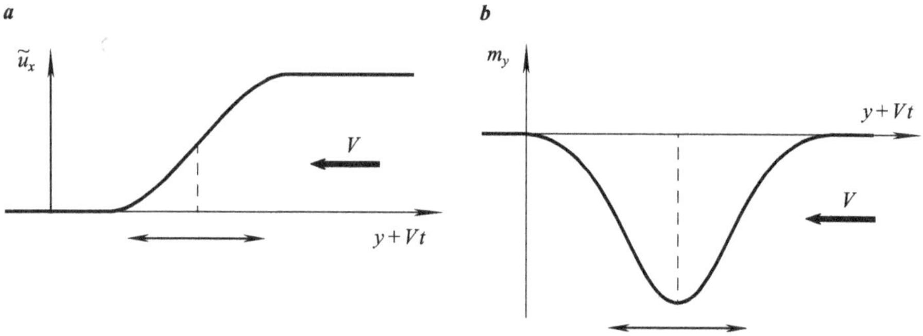

Figure 8.8: Quasi-equilibrium connection between strain a) and magnetization b) in a magneto-elastic soliton (8.4.10).

The magnetization distribution tied with a soliton represents a stationary wave profile $\mathbf{m} \approx (1 - \chi^2/2, -\chi, 0)$, $\chi \approx \chi^{(1)}$ with asymptotics $\mathbf{m} \to (1, 0, 0)$ as $|y| \to \infty$ (see Figures 8.8b and 8.9). In turn, due to magnetostriction, the magnetization wave induces weaker elastic longitudinal strains ($\tilde{u}_{yy} \propto (B_{11} - B_{12})m_y^2 \propto e^2$).

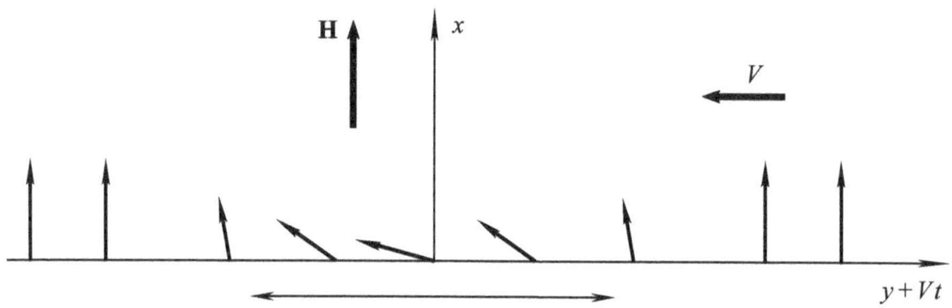

Figure 8.9: Magnetization distribution in the basal plane for soliton (8.4.10) emerged as a result of the interaction of gapless modes.

Besides kinks, the mKdV equation (8.4.10) admits one more type of localized states, breathers [58]. Each breather involves a localized impulse along which the oscillation runs, originating at one of its edges and disappearing at another:

$$\tilde{u}_x = -2\sqrt{\frac{2|r|}{q}} \arctan\left(\frac{\beta}{\alpha} \sin(\alpha\,\tilde{\varsigma} + \delta\,\tilde{\tau} - \varphi)\operatorname{sech}(2\beta\,\tilde{\varsigma} + \gamma\,\tilde{\tau} + \psi)\right),$$

where

$$\tilde{\varsigma} = \sqrt{\frac{q}{2|r|}}\xi, \quad \tilde{\tau} = -\frac{q}{4S}\sqrt{\frac{q}{2|r|}}\tau;$$
$$\gamma = 8\beta(3\alpha^2 - \beta^2), \quad \delta = 8\alpha(\alpha^2 - 3\beta^2).$$

The stability of mKdV solitons has been studied in [86]. Of the instabilities, soliton corrugation is possible, which has been experimentally observed in [17].

8.4.2 Interaction of activation waves

The nonlinear Schrödinger equation (NLSE) is known to regulate the weakly linear dynamics of spin waves far from a resonance of activation and Goldstone modes [89, 72]. We employ the multiscale method reported in [22] to derive the NLSE. The equation system (8.4.5) should be solved as follows:

$$V = \sum_{l=1}^{+\infty} \varepsilon^l \sum_{n=-\infty}^{+\infty} V_n^{(l)}(\xi, \tau) \exp[\mathrm{i}\, n(\omega_a(k)t + ky)].$$

Here $V \equiv (\chi, \tilde{u}_{yx}, \tilde{u}_{yy})$, ε is a small parameter characterizing the deviation of the system from an equilibrium state $V = 0$, $\xi = \varepsilon(y + c_g t)$ and $\tau = \varepsilon^2 t$ are slow variables, and $c_g = \partial_k \omega_a$ is the group velocity of activation waves. We execute scale transformations that comply with the activation mode dispersion law (8.4.6). Reality of V is guaranteed by the condition $V_n^{(l)} = V_{-n}^{(l)^*}$. Our goal is to compose an equation responsible for a slow evolution of the fundamental harmonic $V_{\pm 1}^{(1)}$ after its interaction with other harmonics. The procedure for sequential determination of the coefficients $V_n^{(l)}$ is consistent if there are no secular terms in the equations of the reductive perturbation theory. This requirement establishes a closed effective equation for the envelope of the fundamental harmonic. Since the NLSE derivation procedure was offered in Section 8.1, we note only an essential nuance. In magneto-elastic media the kernel of an operator $\hat{L}$ of the linearized problem $\hat{L}V = 0$, as a rule, includes not only the fundamental harmonic ($V_{\pm 1}^{(1)} \neq 0$) but also a slowly changing wave $V_0^{(2)}(\xi, \tau)$. In this case, the average fluxes of type $\partial_\xi^2 N : V_1^{(1)} V_{-1}^{(1)}$ (N is a tensor of constants) also turn out to be nonzero due to the slow ξ-dependence of the envelope. Such fluxes can cause, if not eliminated, a secular response in $V_0^{(2)}$. Having got rid of the secular terms, we arrive at a system of equations for the envelope and the slowly changing wave $V_0^{(2)}$. Usually, the average fluxes of type $\partial_\xi^2 N : V_1^{(1)} V_{-1}^{(1)}$ and the slow field $V_0^{(2)}$ are equal to zero. The physics-based mathematical procedure explains the aforementioned specificity as follows: Beats of close wavelength activation modes can pump up and excite that of the sound branches, the evolution of which can be traced by a quadratic-in-amplitude term of the activation modes. Ultimately, the field $V_0^{(2)}$ provides a renormalization of the activation wave self-action with a “dangerous” resonant multiplier $\propto (S_{11}^2 - c_g^2)^{-1}$. References [44, 53] offer a detailed discussion of the same problem.

The problem at hand regards the field $V_0^{(2)}$ as a dispersion-free sound branch $\tilde{u}_{yy}$ whose evolution is defined by the square of the activation wave amplitude. The cubic-in-amplitude approximation links the quasiacoustic mode (8.4.7) with an activation mode, and therefore the former has no resonant impact on the envelope evolution.

Far from resonance, when c_g differs dramatically from S_{11}, the field $\tilde{u}_{yy}$ is not free and follows the mean square of the magnetization oscillation:

$$\chi = \Psi(\xi,\tau)\exp\left[\mathrm{i}(\omega_a t + ky)\right] + \mathrm{c.\,c.},$$
$$\tilde{u}_{yy} = 2(B_{11} - B_{12})\rho^{-1}\left(c_g^2 - S_{11}^2\right)^{-1}|\Psi|^2.$$

Here and below, the abbreviation "c. c." stands for a complex conjugate (term). Then the effective evolution equation has the form

$$\mathrm{i}\,\partial_t\Psi - \frac{1}{2}\,\partial_k^2\omega_a\partial_\xi^2\Psi - \left(g + \frac{2g_2(B_{11} - B_{12})}{\rho(S_{11}^2 - c_g^2)}\right)|\Psi|^2\Psi = 0, \tag{8.4.11}$$

$$\begin{aligned} g &= \frac{\gamma^2}{\omega_0}(H_A + H_{me4} + H) \\ &\quad\times(4\pi M_0 + 4(H_\square + H_{me6}) - 6H_{me3} + H/4); \\ g_2 &= \frac{\gamma^2}{M_0\omega_0}(B_{11} - B_{12})(H_A + H_{me4} + H). \end{aligned} \tag{8.4.12}$$

The k-dependence of the constant g should be neglected due to its yielding small corrections of the order $O(S_{66}\,k\,H_{m_e6}/(4\pi\omega_0 M_0))$. Once again, we draw attention on the appearance of the "dangerous" multiplier $\propto (S_{11}^2 - c_g^2)^{-1}$ in eq. (8.4.11).

8.4.3 Magneto-elastic Benney resonance. Connection with integrable models

To theoretically explain the dynamics of low-amplitude waves in the vicinity of a resonance, where $c_g \approx S_{11}$, it is necessary to consider activation and sound waves on an equal footing. The RPT versions used above, where one of the modes is in the priority, turn out to be unsuitable. Although challenging, there is a straightforward way to overcome this obstacle and generate efficient equations for weakly nonlinear dynamics that are suitable even in close proximity to the resonance.

Let us pass from the fields $\tilde{u}_{yx}$ and χ to new dynamic variables (e, v), which correspond to normal modes of a linearized problem, i. e.,

$$\partial_t^2 e + \omega_{\mathrm{G}}^2(-\mathrm{i}\,\partial_y)e \cong 0; \quad \partial_t^2 v + \omega_a^2(-\mathrm{i}\,\partial_y)v \cong 0. \tag{8.4.13}$$

The explicit form of the operators $\omega_a(-\mathrm{i}\,\partial_\xi)$ and $\omega_{\mathrm{G}}(-\mathrm{i}\,\partial_\xi)$ obey the dispersion laws of $\omega_a(k)$ (8.4.6) and $\omega_{\mathrm{G}}(k)$ (8.4.7). Let $\hat{\mathrm{S}}(\nabla)$ be a differential operator defining the transition to the new variables:

$$\begin{pmatrix} e \\ v \end{pmatrix} = \hat{\mathrm{S}}(\nabla)\begin{pmatrix} \tilde{u}_{yx} \\ \chi \end{pmatrix}, \quad \begin{pmatrix} \tilde{u}_{yx} \\ \chi \end{pmatrix} = \hat{\mathrm{S}}^{-1}(\nabla)\begin{pmatrix} e \\ v \end{pmatrix}. \tag{8.4.14}$$

Next, we act with the operator $\hat{S}(\nabla)$ on the first two equations of system (8.4.5). Then by appropriately choosing $\hat{S}(\nabla)$ the linear part of the system acquires the form (8.4.13). In the nonlinear terms of the equations deduced, we retain only the terms that lead to the NLSE and express them through the fields e and v using transformation (8.4.14). In the long-wave approximation, it suffices to calculate $\hat{S}(\nabla)$ and $\hat{S}^{-1}(\nabla)$ up to terms containing the second derivatives over the coordinate:

$$\hat{S}(\nabla) = \begin{pmatrix} 1 & -a\,\omega_0^{-2}\partial_y^2 \\ -b\,\omega_0^{-2} & 1 \end{pmatrix}, \quad \hat{S}^{-1}(\nabla) = \begin{pmatrix} 1 & a\,\omega_0^{-2}\partial_y^2 \\ b\,\omega_0^{-2} & 1 \end{pmatrix}, \tag{8.4.15}$$

where

$$a = \frac{B_{66}}{\rho}; \quad b = \frac{\gamma^2}{M_0} B_{66}(H_A + H_{me4} + H).$$

It can be shown that other terms disappear after averaging over fast time oscillations of the activation modes. Approximation (8.4.15) turns out to be sufficient to reproduce the findings of the previous sections. The formulas

$$\chi = b\,\omega_0^{-2} e + v; \quad \tilde{u}_{yx} = e + a\,\omega_0^{-2}\partial_y^2 v$$

found in approximation (8.4.15) are compatible with both versions of the reductive perturbation theory, even if the slow variable choice in different versions is not identical. In the cubic terms of the equations, we can neglect the spatial derivatives of v due to their providing small corrections of the order $O((kS_{66})^2 H_{me6}/(4\pi M_0 \omega_0^2))$.

Suppose that the modulation of activation waves is slow. Then we can also exclude fast time oscillations from the system of equations for the fields e, $\tilde{u}_{yx}$, and $\tilde{u}_{yy}$, putting that

$$v = \varphi \exp(\mathrm{i}\,\omega_0\,t) + \text{c. c.}, \quad \partial_t^2 v + \omega_0^2 v \cong 2\mathrm{i}\,\omega_0 \partial_t \varphi \exp(\mathrm{i}\,\omega_0 t) + \text{c. c.}$$

Averaging over the fast oscillations yields

$$\begin{gathered} \partial_t^2 e - S^2 \partial_y^2 e + r\,\partial_y^4 e - \partial_y^2(\alpha_1 e^3 + \alpha_2 e|\varphi|^2 + \alpha_3 e\,\tilde{u}_{yy}) = 0, \\ \partial_t^2 \tilde{u}_{yy} - S_{11}^2 \partial_y^2 \tilde{u}_{yy} - \frac{B_{11} - B_{12}}{\rho} \partial_y^2 \left(\frac{b^2}{\omega_0^4} e^2 + 2|\varphi|^2 \right) = 0, \\ \mathrm{i}\,\partial_t \varphi - f \partial_y^2 \varphi + g_1 e^2 \varphi + g_2 \tilde{u}_{yy}\,\varphi - g|\varphi|^2 \varphi = 0, \end{gathered} \tag{8.4.16}$$

where

$$g_1 = \frac{b^2}{\omega_0^3} n_1 n_2, \quad \alpha_1 = \frac{2\,a\,b^3}{\omega_0^6} n_1 \left(4\pi M_0 + 2H_{me3} + \frac{5}{4} H \right), \quad \alpha_2 = \frac{4ab}{\omega_0^2} n_1 n_2,$$

$$\alpha_3 = \frac{2ab}{\omega_0^2} \frac{B_{11} - B_{12}}{M_0} n_1, \quad n_1 = [4\pi M_0 + H_\square + H_{me6} + H]^{-1},$$

$$n_2 = 4\pi M_0 - 6H_{me3} - 2\,(H_{\square} + H_{me6}) + 7H/4.$$

The quantities g and g_2 were previously deduced (see relations (8.4.12)).

The resonance neighborhood $c_g^2 = S_{11}^2$ is an example of the broad range of applications the equation system (8.4.16) can serve.

To ensure that the eqs. (8.4.16) reduce to the mKdV equation (8.4.9) and NLSE (8.4.11) in the limiting cases, we first assume that the activation modes are not excited. Then $\varphi = 0$, and

$$\begin{aligned} \partial_t^2 e - S^2\partial_y^2 e + r\,\partial_y^4 e - \alpha_1\partial_y^2 e^3 - \alpha_3\partial_y^2(\tilde{u}_{yy}e) &= 0,\\ \partial_t^2\tilde{u}_{yy} - S_{11}^2\partial_y^2\tilde{u}_{yy} - \frac{B_{11}-B_{12}}{\rho}\,\frac{b^2}{\omega_0^4}\,\partial_y^2 e^2 &= 0. \end{aligned} \tag{8.4.17}$$

System (8.4.17) is more general than (8.4.9) because it describes the propagation of waves in both directions without limiting their speed. When we focus on waves moving in the same direction with speeds close to the phase velocity S of the Goldstone modes, then

$$e = e(\xi,\tau), \quad \xi = y + St; \quad (\partial_t^2 - S^2\partial_y^2)e \cong 2S\,\partial_\tau\partial_\xi e;$$
$$\tilde{u}_{yy} \cong \frac{B_{11}-B_{12}}{\rho\,(S^2 - S_{11}^2)}\,\frac{b^2}{\omega_0^4}\,e^2,$$

and system (8.4.17) becomes reduced to the mKdV equation (8.4.9).

The soliton-like solution to system (8.4.17) is also more general than (8.4.10):

$$e = \sqrt{\frac{2\,(V^2 - S^2)}{q(V)}}\,\mathrm{sech}\Big(\sqrt{\frac{V^2 - S^2}{2|r|}}\,\zeta\Big), \quad S^2 < V^2;$$
$$\zeta = y + Vt + y_0, \quad q(V) = \alpha_1 + \frac{b^2}{\omega_0^4}\,\frac{B_{11}-B_{12}}{\rho(V^2 - S_{11}^2)}\,\alpha_3.$$

The former coincides with the latter as long as the soliton group velocity V is close to that of transverse sound:

$$V = S + \lambda, \quad 0 < \lambda \ll 1.$$

To achieve reduction to the NLSE, fast space-time oscillations must be excluded by setting

$$\varphi = \Psi(\xi,t)\exp(\mathrm{i}\,k\,y + \mathrm{i}\,\partial_k^2\omega\,k^2 t/2)$$

and passing to a reference frame that moves at the activation waves' group velocity $c_g = \partial_k\omega_a$. If $c_g^2 \neq S_{11}^2$, then

$$e = 0, \quad \tilde{u}_{yy} \equiv \tilde{u}_{yy}(\xi, \tau),$$
$$(\partial_t^2 - S_{11}^2 \partial_y^2)\tilde{u}_{yy} \approx (c_g^2 - S_{11}^2)\partial_\xi^2 \tilde{u}_{yy}.$$

Finally, system (8.4.16) becomes equivalent to the NLSE (8.4.11).

The entire system includes a solution that generalizes the NLSE soliton, which is accurate even when near the resonance ($c_g^2 \approx S_{11}^2$):

$$\tilde{u}_{yy} = \sigma(V)|c|^2 \mathrm{sech}^2\zeta, \quad \tilde{u}_y = \sigma(V)|c|\sqrt{\frac{\partial_k^2\omega_a}{g(V)}}(1 + \tanh\zeta), \quad e = 0,$$
$$\varphi = c\,\mathrm{sech}\,\zeta\,\exp\left[\frac{\mathrm{i}V}{\partial_k^2\omega_a}y + \frac{\mathrm{i}}{2}t\left(\frac{V^2}{\partial_k^2\omega_a} - g(V)|c|^2\right)\right], \tag{8.4.18}$$

where

$$\zeta = (y + Vt)B + \zeta_0, \quad B^2 = \frac{g(V)}{\partial_k^2\omega_a}|c|^2 > 0,$$

V and ζ_0 are real constants, c is a complex parameter, and

$$\sigma(V) = \frac{2(B_{11} - B_{12})}{\rho(V^2 - S_{11}^2)}, \quad g(V) = g - \sigma(V)g_2.$$

Solution (8.4.18) describes a localized excitation moving at a group velocity V (Figure 8.10). Note that the condition $B^2 > 0$ limits the range of admissible speeds V. Accounting for the ever-present self-action g of activation waves expands this diapason. When the soliton speed approaches that of longitudinal sound, S_{11}, the field $\tilde{u}_{yy}$ in eqs. (8.4.18) intensifies indefinitely. The intrinsic anharmonicity and dispersion of the elastic subsystem become important.

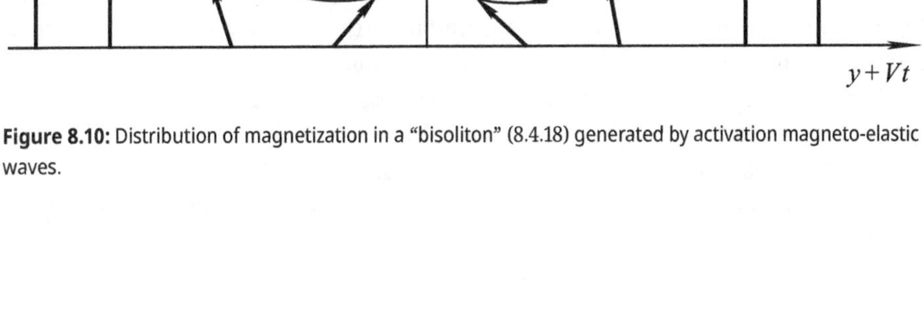

Figure 8.10: Distribution of magnetization in a "bisoliton" (8.4.18) generated by activation magneto-elastic waves.

The vector **M** oscillates rapidly in the "easy plane" of the coordinate system connected with the soliton, reaching its maximum amplitude at the soliton's center. These oscillations induced by magnetostriction generate the medium's longitudinal strains localized on the soliton (see Figure 8.11). Because of the greater inertia, the elastic subsystem is unable to keep up with the fast oscillations of the magnetization vector. As a result, the strains are determined by the averaged-over-fast oscillation square of the amplitude of magnetic oscillations.

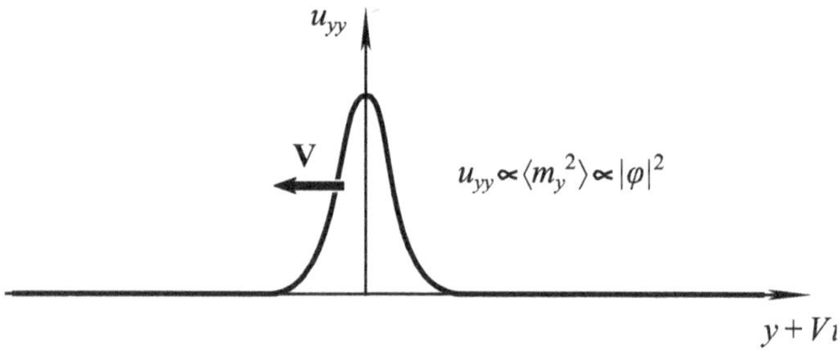

Figure 8.11: Medium's longitudinal strains in a magneto-elastic soliton.

In the neighborhood of resonance, when the group velocity of activated waves meets the inequality

$$\left| \frac{2(B_{11} - B_{12})}{\rho\,(c_g^2 - S_{11}^2)} g_2 \right| \gg g,$$

we can disregard the self-action of activation modes in contrast to their interaction with longitudinal sound. Then system (8.4.16) takes the simpler form ($e = 0$):

$$\begin{aligned} \mathrm{i}\,\partial_\tau \varphi - \frac{1}{2}\,\partial_k^2 \omega_a \partial_y^2 \varphi + g_2 \tilde{u}_{yy} &= 0, \\ (\partial_t^2 - S_{11}^2)\tilde{u}_{yy} - 2\,\frac{B_{11} - B_{12}}{\rho}\,\partial_y^2 |\varphi|^2 &= 0. \end{aligned} \tag{8.4.19}$$

Although the equation system (8.4.19) is not completely integrable, Hirota's method admits multisoliton solutions, which are available in [66]. Numerical solutions to system (8.4.19) are also known. Among these are the most interesting outcomes concerning the formation of a soliton from a preassigned initial distribution ($\varphi \neq 0$, $\tilde{u}_{yy} = 0$), solitons' scattering and "splitting" with emission and absorption of sound waves [20].

If we confine ourselves to looking at quasiacoustic and activation waves propagating unidirectionally at velocities close to the longitudinal sound waves' velocity S_{11}, then system (8.4.19) boils down to a fully integrable Zakharov–Benney model. To deduce appropriate equations, we exclude fast space-time oscillations and pass to a coordinate system traveling at the activation wave's group velocity $c_g = S_{11}$:

$$\varphi = \Psi(\xi, t)\exp\left(\mathrm{i}ky + \frac{\mathrm{i}}{2}\partial_k^2\omega_a k^2 t\right);$$
$$\tilde{u}_{yy} = \tilde{u}_{yy}(\xi, \tau), \quad \xi = y + c_g t, \quad \tau = t. \tag{8.4.20}$$

Transformation (8.4.20) is analogous to that used in passing from system (8.4.16) to the NLSE. However, since in the vicinity of the resonance the sound wave exhibits some freedom, the final equations save the terms containing the first derivatives over the "slow" time:

$$(\partial_t^2 - S_{11}^2\partial_y^2)\tilde{u}_{yy} \cong 2S_{11}\partial_\tau\partial_\xi\tilde{u}_{yy}.$$

The final result is reduced to the Zakharov–Benny equations [22]

$$\mathrm{i}\partial_\tau\Psi - \frac{1}{2}\partial_k^2\omega_a\partial_\xi^2\Psi + g_2\Psi\,\tilde{u}_{yy} = 0,$$
$$\partial_\tau\,\tilde{u}_{yy} - \frac{B_{11} - B_{12}}{\rho\,S_{11}}\,\partial_\xi|\Psi|^2 = 0. \tag{8.4.21}$$

System (8.4.21) highlights the balance between the dispersion effect of an activation wave and its interaction with a sound wave, whereas the sound wave evolves along the self-action of the activation wave.

Solutions to system (8.4.21) can be completely explored using the inverse scattering method [92, 65]. The soliton in model (8.4.21) moves at the group velocity $V < c_g = S_{11}$ and coincides in shape with (8.4.18).

The condition $c_g = S_{11}$ is fulfilled in a region with a certain wave-number k_0. In unbounded ferromagnets, a uniform, alternating magnetic field can excite running spin waves with a predefined value k_0 if their material characteristics are coordinate-dependent. In addition, a nonuniform constant magnetic field can trigger spin waves under the required gradient as well [31].

An analysis of eqs. (8.4.21) shows that the plane activation wave $\Psi = \text{const}$, $\tilde{u}_{yy} = 0$, produced by an external electromagnetic field, is modulationally unstable. As the instability progresses, localized longitudinal sound pulses connected to the soliton's activation mode envelope can arise. The generation of solitons most likely has a threshold nature. Below the threshold energy (amplitude), the activation plane waves spread out due to dispersion.

Interestingly, the field $\tilde{u}_{yy}$ in the soliton solution to system (8.4.19) or (8.4.21) resembles a soliton of the KdV equation, with the same envelope shape φ (Ψ) as that of the NLSE soliton. In this sense the Zakharov–Benney model (8.4.21) is a link between the KdV-type models associated with a Goldstone mode and the NLSE for the activation branch envelope.

Model (8.4.21) describes the so-called long-short wave Benney resonance, sometimes treated as the limiting case of a three-wave resonance. In the limit as $\kappa \to 0$, the conditions for synchronism of three waves,

$$\omega_a(k_1) - \omega_a(k_2) = \omega_G(\kappa), \quad k_1 - k_2 = \kappa,$$

narrow down to the requirement

$$c_g = \partial_k \omega_a(k) = \lim_{\kappa \to 0} \frac{\omega_G(\kappa)}{\kappa} = c_p.$$

In the problem at hand, the beats of activation waves stimulate a dispersion-free sound branch. This broadens a variety of applications for the Benney mechanism and the condition $c_g = c_p$.

Although this section covers a particular problem, the above equations make it possible to analyze the general regularities governing the interaction of low-amplitude magneto-elastic waves in ferro- and antiferromagnets. Accounting for "easy-plane" anisotropy simplifies the analysis but is not principal. The resulting effective equations are strikingly universal and slightly depend on the crystallographic symmetry of the medium. The specificity of a crystal manifests itself mainly through effective interaction constants.

In particular, the methods and models discussed in this section are used for comprehensive study of magneto-elastic solitons and related resonance phenomena in two-sublattice antiferromagnets near orientational phase transitions, both under electric and magnetic fields and under external mechanical deformations that induce such transitions (see [80] and references therein).

The physics of magnetism frequently deals with dispersion relations with double or even multiple branches; therefore a long-short wave resonance is typical not only for magneto-elastic media. Thus, when describing solitons against the background of antiferromagnets' helicoidal magnetic structure, models similar to those presented in this section emerge [14, 44, 53].

9 Nonlinear dynamics of spiral (stripe domain) structure

"The spiral is a spiritualized circle. In the spiral form, the circle, uncoiled, has ceased to be vicious; it has been set free."

V. Nabokov. Other shores

Plane-parallel domain and quasi-one-dimensional helical structures are theoretically described in terms of a lattice of kinks (domain walls). The kink lattice itself is a highly nonlinear state of a magnetically ordered medium. In the context of such a structure, it is challenging to provide an analytical description of collective excitations, particularly soliton ones. Similar tasks can be constructively tackled within simplified models that properly address the main interactions and, simultaneously, allow for exact solutions. With physically justified approximations, the Landau–Lifshitz equations for ferro- and antiferromagnets, as well as multiferroics, are often reduced to the integrable hyperbolic sine-Gordon equation [41, 2, 45, 16, 17, 12]

$$\partial_t^2\Phi - \partial_z^2\Phi + \sin\Phi = 0. \tag{9.1}$$

A field $\Phi(z, t)$ specifies the distribution of the ferro- or antiferromagnetism vector in a structure; z is a spatial coordinate, and t is time.

All solutions to eq. (9.1) with uniform asymptotic behavior of the field $\Phi(z, t)$ as $z \to \pm\infty$ have been examined using the inverse scattering method or its modifications [14, 36]. A significantly nonlinear helical (stripe domain) structure complicates the technique of integrating the model. Although the $U - V$ pair of eq. (9.1) rationally depends on a spectral parameter, the inverse scattering method establishes the conjugation condition for analytic functions on the Riemannian surfaces rather than in the complex plane of the spectral parameter, as in the case of a uniform ground state of the medium. The Riemann surface is connected to the background structure and turns out to be topologically equivalent to a torus.

Solutions to model (9.1) we are interested in have essentially never been explored. Algebro-geometric finite-gap integration methods are ineffective in analyzing the desired solutions. This is because these methods lead to poorly studied transcendental relations and multidimensional theta functions [11] and, as a rule, give complex solutions to an original model. To encompass a variety of real and limited solutions is a labor-intensive mathematical issue. In addition, the finite-gap technique is not suitable for investigating nonlinear waves associated with the continuous spectrum of the conventional scattering method.

References [8, 18, 6] use the Gel'fand–Levitan–Marchenko-type integral transformations within the elliptic sine-Gordon equation

$$\partial_x^2\Phi + \partial_y^2\Phi = \sin\Phi \tag{9.2}$$

https://doi.org/10.1515/9783112235973-009

for researching two-dimensional defects in an incommensurable (stripe domain) structure of magnets, such as soliton vortices and nonsoliton string configurations of domain wall segments in a stripe structure. Although eqs. (9.1) and (9.2) are similar in form, the integration technique and physical findings for them differ.

The "dressing" method involving the Shabat integral operators has been carried out for the first time in [29] for analyzing nonlinear excitations on a pedestal of a traveling cnoidal wave in condensed media described by the universal Korteweg–de Vries model. The present chapter tests the technique for studying the nonlinear dynamics of a spiral structure by "dressing" partial solutions to eq. (9.1) using the Riemann problem on a torus [20, 21]. With this approach, the procedure for constructing the required solutions turns out to be local (the locality is lost under integral transformations). This circumstance greatly facilitates an analytical description of excitations in a helical structure. Obtaining real solutions faces no difficulties; the final formulas are written in terms of well-studied elliptic functions.

Section 9.1 includes a justification for the model used and a formulation of the Riemann problem, necessary for analyzing solitons and spin waves in a spiral structure. Section 9.2 outlines an algorithm for constructing soliton solutions to the model with helical structure. There are other possibilities for constructing soliton solutions, for example, using the Darboux [26–28] or Bäcklund [9] transformations. However, the procedure proposed discloses more deeply the spectral nature of this problem. It enables us to come up with new outcomes and, ultimately, to formulate a general scheme for describing solitons and spin waves under given boundary conditions and arbitrary initial distributions of magnetization in a spiral structure. In other words, an initial-boundary value problem for the sine-Gordon model (9.1) with periodic structure can be solved. Shaping soliton matrices for the Riemann problem requires overcoming the difficulties set forth below.

Helicoidal structure translations inexorably accompany the emergence of solitons. In Section 9.2, we find formulas to relate the shifts in the boundary conditions for the stripe structure with zeros of the Riemann problem, which determine the structure and velocity of solitons.

Due to a periodic background, the Riemann problem's functions become "Bloch" ones. They are multiplied by additional factors when a spectral parameter changes by a period. The Riemann problem used for integrating the elliptic sine-Gordon equation (9.2) [8, 18, 6] lacks the noted feature.

Section 9.2 examines solitons in a spiral structure subjected to an external magnetic field perpendicular to the magnetic spiral axis and elicits the possibilities of their diagnostics. Section 9.3 reduces an analytical description of the nonlinear dispersive waves in a spiral structure to solving linear integral equations on a torus. Section 9.4 finds spectral expansions of integrals of motion for collective excitations in a spiral structure, with separated contributions of solitons and spin waves.

9.1 Basic equations

9.1.1 The sine-Gordon model for helicoidal structure

> Equations are more important to me, because politics is for the present, but an equation is something for eternity.
>
> Albert Einstein

Let us consider a quasi-one-dimensional (along the Oz-axis) helical structure. We describe the distribution of magnetization in the structure by a vector field $\mathbf{M}(z,t)$, where $\mathbf{M}^2 = M_0^2 = \text{const}$, and z and t are a spatial coordinate and time, respectively. For a quasi-one-dimensional inversion-center-free ferromagnet with "easy-plane" anisotropy (the xOy-plane), the energy density in a constant external magnetic field $\mathbf{H} = (H,0,0)$ $(II > 0)$ is written as

$$\tilde{w} = \frac{\alpha}{2}(\partial_z \mathbf{M})^2 + \kappa(M_1\partial_z M_2 - M_2\partial_z M_1) + \frac{\beta}{2}M_3^2 - M_1 H. \tag{9.1.1}$$

Here $\alpha, \beta > 0$ and κ are the constants of the exchange interaction, magnetic anisotropy, and the Dzyaloshinskii interaction, respectively [13, 3]. In easy-plane magnets at $\mathbf{H} = 0$, the Dzyaloshinskii interaction (Lifshitz' invariants) in the energy of eq. (9.1.1) provokes a perfect magnetic ordering of the atoms. The $\mathbf{M}(z,t)$-field vector located in the xOy-plane rotates, when displaced along the Oz-axis, forming a spiral structure with period l_0 significantly greater than and incommensurable with the crystal lattice constant a: $l_0 \backsim \alpha/\kappa \gg a$. In the xOy-plane the external field $\mathbf{H}$ tends to line up the magnetic moments of the atoms along the Ox-direction. The competition of the opposite tendencies shapes extended regions (domains) of width L_0 along the Oz-axis, with the magnetization distribution remaining almost uniform. Near a certain critical field $H < H_c$, a narrow transition layer of width $l_0 \backsim \alpha/\kappa \backsim \sqrt{\alpha M_0/H_c} \ll L_0$ separates neighboring regions where a spiral turn of the magnetization is preserved. At $H > H_c$ the system has a commensurate ferromagnetic ordering.

For strong easy-plane anisotropy, in the parameterization of the field $\mathbf{M}(z,t)$ by angles Θ and Φ,

$$\mathbf{M} = M_0(\sin\Theta\cos\Phi,\ \sin\Theta\sin\Phi,\ \cos\Theta),$$

the angle Θ is close to the value $\Theta = \pi/2$, and $\partial_z\Phi \backsim O(1/l_0)$ even in excited states of the helicoidal structure. Let us use this circumstance to simplify the problem. In terms of the angles Θ and Φ, the Landau–Lifshitz equations for a quasi-one-dimensional ferromagnet have the form

$$\begin{aligned}&\frac{1}{\gamma M_0}\sin\Theta\partial_t\Phi + \alpha\partial_z^2\Theta + \sin\Theta\cos\Theta[\beta - \alpha(\partial_z\Phi)^2 - 2\kappa\partial_z\Phi]\\ &\quad + \frac{H}{M_0}\cos\Theta\cos\Phi = 0,\end{aligned}$$

$$-\frac{1}{\gamma M_0}\sin\Theta\partial_t\Theta + \alpha\partial_z(\sin^2\Theta\partial_z\Phi) + \kappa\partial_z\sin^2\Theta - \frac{H}{M_0}\sin\Theta\sin\Phi = 0, \tag{9.1.2}$$

where γ is the magneto-mechanical ratio. For real materials with helical structure, the problem parameters satisfy the inequality [9]

$$\frac{H}{M_0} \leq \frac{\kappa^2}{\alpha} \ll \beta.$$

Then, assuming that $\Theta = \pi/2 + \theta$ ($\theta \ll 1$), from the first equation of (9.1.2) we find

$$\theta \approx (\gamma M_0\beta)^{-1}\partial_t\Phi.$$

After that, the second equation of (9.1.2) is reduced to the sine-Gordon model [6, 9, 7] in the main approximation to calculate Φ:

$$-\frac{1}{(\gamma M_0)^2\beta}\partial_t^2\Phi + \alpha\partial_z^2\Phi - \frac{H}{M_0}\sin\Phi = 0. \tag{9.1.3}$$

In dimensionless variables $z' = z\sqrt{H/(\alpha M_0)}$ and $t' = \gamma\sqrt{\beta H M_0}t$, the sine-Gordon equation (9.1.3) takes the standard form

$$\partial_{t'}^2\Phi - \partial_{z'}^2\Phi + \sin\Phi = 0. \tag{9.1.4}$$

The prime marks over the new variables are further omitted. In the main approximation the distribution of magnetization is given by the formula

$$\mathbf{M} \approx M_0(\cos\Phi, \sin\Phi, 0).$$

For such configurations, $\operatorname{div}\mathbf{M} = 0$, and therefore magneto-static interactions can be left aside. In dimensionless variables, the energy density for a quasi-one-dimensional ferromagnet is described by the expression

$$w = \frac{1}{2}[(\partial_z\Phi)^2 + (\partial_t\Phi)^2] + q\partial_z\Phi + (1 - \cos\Phi), \tag{9.1.5}$$

where $q = \kappa\sqrt{M_0/(\alpha H)}$.

Not only a magnetic field but also residual quadratic or cubic anisotropy can point some directions for the magnetic moments of the atoms in the basal xOy-plane. Then we replace the sine argument in eq. (9.1.4) by 2Φ or 4Φ, respectively. The residual anisotropy constants vary with temperature; therefore the parameter q in expression (9.1.5) depends on temperature rather than on a magnetic field. Regardless, after scaling transformations, the theoretical description of a quasi-one-dimensional helical structure's nonlinear dynamics can be reduced to solving the sine-Gordon wave equation (9.1.4), which appears not only in describing incommensurable structures. It proves out to be useful when examining the nonlinear dynamics in stripe structures of various magnets [41, 2, 45, 16]. In weak ferromagnets and antiferromagnets [2, 45] the effects of magnetostatics are small. In this case the model's applicability is even broader since the nonlinear dynamics of modulated structures can be also described. The latter manifest themselves so that the end of the antiferromagnetism vector **L**, when displaced along a direction distinguished in space, can move not only along a helix $\operatorname{div}\mathbf{L} = 0$ but also along a cycloid $\operatorname{div}\mathbf{L} \neq 0$.

In addition, such a model can be used to analyze the dynamics of the helical structure of cholesteric liquid crystals under a constant electric or magnetic field [10].

Depending on the value of q, the energy minimum (9.1.5) corresponds to either a homogeneous distribution of the order parameter $\Phi = 0 \pmod{2\pi}$ or a periodic structure, for which

$$\begin{gathered}\Phi = \varphi_0(\chi) = \pi - 2\,\mathrm{am}(\chi, k), \quad \chi = z/k; \\ \cos\frac{\varphi_0}{2} = \mathrm{sn}(\chi, k), \quad \sin\frac{\varphi_0}{2} = \mathrm{cn}(\chi, k), \quad \frac{\partial_\chi \varphi_0}{2} = -\mathrm{dn}(\chi, k),\end{gathered} \tag{9.1.6}$$

where $\mathrm{sn}(\chi, k)$, etc. are Jacobi elliptic functions with modulus k ($k^2 \leq 1$) [4, 1]. The average energy per period of the spiral structure is

$$\bar{W} = \frac{1}{L_0}\int_0^{L_0} \mathrm{d}z = -\frac{\pi q}{Kk} - 2\left(\frac{k'}{k}\right)^3 + \frac{4E}{Kk^2},$$

where $E = E(k)$ is the complete elliptic integral of the second kind. Now we minimize $\bar{W}$ over the variable k. Hence we come to the condition that determines k [17, 7]:

$$\pi q k - 4E = 0. \tag{9.1.7}$$

When $q < 4/\pi$, eq. (9.1.7) has no solutions. The ground state of system (9.1.5) is a commensurate phase: $u = 0 \pmod{2\pi}$.

When $q > 4/\pi$, eq. (9.1.7) is solvable. The energy matching to the spiral structure (9.1.6) is lower than that of the homogeneous order-parameter distribution. The function $\varphi_0(z)$ changes by 2π with period $L_0 = 2Kk$, where $K = K(k)$ is the complete elliptic integral of the first kind. According to the expansion [7],

$$\partial_\chi \varphi_0 = -2\,\mathrm{dn}\chi = -\frac{\pi}{K'}\sum_{p=-\infty}^{+\infty} \mathrm{sech}\left[\frac{\pi}{2K'}(\chi - 2Kp)\right],$$

where p is an integer, $K' = K(k')$, and $k' = \sqrt{1-k^2}$. Changes in the field φ_0 are concentrated near the points $z = 2Kkp$. In the vicinity of each of them with characteristic size $l_0 \sim 2K'k/\pi$, the function φ_0 can be approximated by a 2π-kink, since

$$-\frac{\pi}{K'}\int \mathrm{d}\chi' \,\mathrm{sech}\left(\frac{\pi\chi'}{2K'}\right) = 4\arctan\left[\exp\left(-\frac{\pi}{2K'k}x\right)\right] + \text{const.} \qquad (9.1.8)$$

Thus solution (9.1.6) constitutes a one-dimensional lattice composed of 2π–kinks (9.1.8) separated by the extended regions L_0, within which $\varphi_0(\chi) \approx 2\pi s$, where s is an integer (see Figure 9.1).

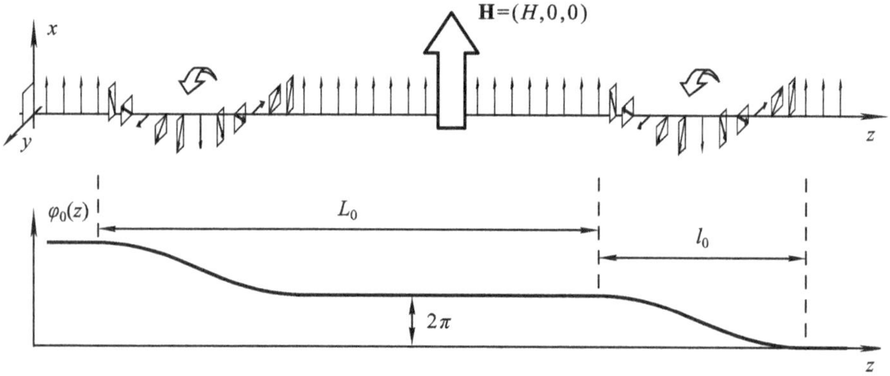

Figure 9.1: Helicoidal structure and lattice $\varphi_0(z)$ of field kinks.

By varying the external field or temperature we can change the period of the spiral structure. Under a magnetic field perpendicular to the axis of the magnetic spiral, near the point $H_c = (k\pi/4)^2 M_0/\alpha$ for transitioning along the field from an incommensurate phase to a commensurate one, the period of the helicoidal structure in dimensional variables has the form

$$L_0 \approx \sqrt{\frac{\alpha M_0}{H_c}} \ln\left[\frac{1}{\sqrt{H_c/H} - 1}\right], \quad H < H_c.$$

It is demonstrated in [43] that model (9.1.14) for a compound $Cr_{1/3}NbS_2$ correctly reflects the data of the Lorenz microscopy and small-angle electron diffraction on the dependence of the helical structure period on an external magnetic field perpendicular to the magnetic helix axis.

9.1.2 Integration procedure

The formation of solitons in a spiral structure is always accompanied by local translations of the structure. Next, we consider boundary value problems of two types. The first problem is characterized by the conditions

$$\begin{aligned}\Phi(z,t) &\to \varphi_2^{(0)}(\chi) \equiv \varphi_0(\chi) \quad \text{as} \quad z \to +\infty,\\ \Phi(z,t) &\to \varphi_1^{(0)}(\chi) = 2\pi\sigma + \varphi_0(\chi+\Delta) \quad \text{as} \quad z \to -\infty,\end{aligned} \tag{9.1.9}$$

where $\sigma = \pm 1$. For $\sigma = 1$, the helical structure experiences compression per period and an additional displacement along the Oz-axis by a distance $k\Delta$. In the end, this necessitates arising a soliton as an extra 2π-kink of the field Φ (a domain wall) in one of the cells of the kink lattice. Its topological charge (chirality) is the same as that of the remaining kinks (domain walls) of the structure. When $\sigma = -1$, the helicoidal structure locally expands for the period. With that, an extra domain wall appears with a chirality opposite to that of the remaining walls of the structure in one of its domains. Under the same boundary conditions (9.1.9), other solitons may also exist in the spiral structure, an even number of extra domain walls with pairwise opposite chiralities and an arbitrary number of breathers. A breather is a pulsating soliton, which can be treated as a bound state of two domain walls with opposite chiralities. In addition, kink-lattice-translation-free spin-wave packets generated by magnetization oscillations near the equilibrium positions of the helicoidal structure can travel along the structure.

The second type of the boundary conditions

$$\begin{aligned}\Phi(z,t) &\to \varphi_2^{(0)}(\chi) \equiv \varphi_0(\chi) \quad \text{as} \quad z \to +\infty,\\ \Phi(z,t) &\to \varphi_1^{(0)}(\chi) \equiv \varphi_0(\chi+\Delta) \quad \text{as} \quad z \to -\infty\end{aligned} \tag{9.1.10}$$

implies the presence of spin waves, an even number of kinks with pairwise opposite charges, and an arbitrary number of breathers. The parameter Δ controls an additional shift of the superstructure due to the formation of solitons (kinks and breathers) in it.

The sine-Gordon equation (9.1.4) is equivalent to the following compatibility condition for an auxiliary system of linear equations [14]:

$$\begin{aligned}\partial_t\Psi &= \frac{\mathrm{i}}{2}\left[\frac{\partial_z\Phi}{2}\sigma_3 + \sigma_1 w_1\cos\frac{\Phi}{2} + \sigma_2 w_2 \sin\frac{\Phi}{2}\right]\Psi \equiv V\Psi,\\ \partial_z\Psi &= \frac{\mathrm{i}}{2}\left[\frac{\partial_t\Phi}{2}\sigma_3 + \sigma_1 w_2\cos\frac{\Phi}{2} + \sigma_2 w_1 \sin\frac{\Phi}{2}\right]\Psi \equiv U\Psi,\end{aligned} \tag{9.1.11}$$

where $w_1^2 - w_2^2 = 1$, and σ_i are Pauli matrices. Under the parameterization $w_1 = \mathrm{cn}(u,k)$ and $w_2 = \mathrm{i}\,\mathrm{sn}(u,k)$, the particular solution to system (9.1.11) corresponding to the background structure $\varphi_0(\chi+\delta)$ has the form

$$\begin{aligned}\Psi_\delta(u,\chi,t) &= M(u,\chi+\delta)\exp\left(A(u,\chi,t)\sigma_3 + \frac{\eta_1\delta}{K}u\,\sigma_3\right);\\ A(u,\chi,t) &= \mathrm{i}\,p(u)\chi + \frac{\mathrm{i}\,t}{2k}\mathrm{dn}(u,k),\\ p(u) &= -\mathrm{i}\left(\frac{\eta_1 u}{K} - \frac{1}{2}[\zeta(u-\mathrm{i}K') + \zeta(u+\mathrm{i}K')]\right) = \frac{\mathrm{i}}{2}Z(u);\end{aligned} \tag{9.1.12}$$

$$M(u,\chi) = \frac{m(\chi)}{\sigma(u)} \begin{pmatrix} \sigma(\chi+u) & i\sigma(\chi+2iK'-u)e^{-\eta_3(\chi+iK'-u)} \\ -i\sigma(\chi+2iK'+u)e^{-\eta_3(\chi+iK'+u)} & \sigma(\chi-u) \end{pmatrix}$$
$$\times \exp\left[-\frac{\eta_1 u}{k}\chi\sigma_3\right];$$
$$m(\chi) = \frac{1}{\sqrt{2}}\left|\frac{\sigma(iK')}{\sigma(\chi+iK')}\right|, \qquad \det M = \frac{\mathrm{dn}u}{1-\mathrm{dn}u}.$$

The parameter δ shifts the solution along the Oz-axis, $\sigma(u)$ and $\zeta(u)$ are the Weierstrass sigma and zeta functions with periods $2K$ and $4iK'$, where $K = K(k)$ and $K' = K(k')$ are the complete elliptic integrals of the first kind of modulus k and additional modulus $k' = \sqrt{1-k^2}$, respectively, and $Z(u)$ is the Jacobi zeta function [4, 1]. The branch of the square root is chosen so that the following condition is fulfilled:

$$\sqrt{\sigma(\chi \pm 2K + iK')\sigma(\chi \pm 2K - iK')} = -|\sigma(\chi+iK')| \exp[\pm 2\eta_1(\chi \pm K)],$$

where

$$|\sigma(\chi+iK')| = \sqrt{\sigma(\chi+iK')\sigma(\chi-iK')}.$$

For the function $\Psi_\delta(u,\chi)$, the quantity $p(u)$ plays the role of the Bloch quasimomentum:

$$\Psi_\delta(\chi \pm 4K, t, u) = \Psi_\delta(\chi, t, u)\exp[\pm 4K i\, p(u)\sigma_3].$$

Under the asymptotic conditions as $z \to \pm\infty$, the sine-Gordon model integration procedure uses the analytic properties of the fundamental solutions to the auxiliary system (9.1.11). In this case the asymptotic conditions must take into account not only the presence of a superstructure, but also its translation when solitons emerge. Although the original nonlinear equation (9.1) is invariant under the transformation $\Phi \to \Phi \pm 2\pi$, the auxiliary linear system (9.1.11) does not have this property and distinguishes the states Φ and $\Phi \pm 2\pi$. The boundary conditions (9.1.9) and (9.1.10) can be determined by mod 4π (the quantity Δ is given by mod $4K$). Such boundary conditions do not exclude the appearance of breathers as weakly bound states of domain walls of the same chirality and/or as the formation of an even number of additional walls both with pairwise opposed and identical chiralities.

The basic Jost solutions to system (9.1.11) under the asymptotic conditions (9.1.9) for the sine-Gordon model are defined as follows:

$$\begin{aligned} \Psi_1(z,t,u) &\to \Psi_1^{(0)} = \sigma_3\Psi_\Delta(z,t,u)\sigma_3 \quad \text{as} \quad z \to -\infty, \\ \Psi_2(z,t,u) &\to \Psi_2^{(0)} = \Psi_0(z,t,u) \quad \text{as} \quad z \to +\infty. \end{aligned} \tag{9.1.13}$$

The boundary conditions (9.1.10) change the definitions of the Jost solutions:

$$\begin{aligned}\Psi_1(z,t,u) &\to \Psi_1^{(0)} = \Psi_\Delta(z,t,u) \quad \text{as} \quad z \to -\infty,\\ \Psi_2(z,t,u) &\to \Psi_2^{(0)} = \Psi_0(z,t,u) \quad \text{as} \quad z \to +\infty.\end{aligned} \tag{9.1.14}$$

The set

$$\gamma = \{u : \operatorname{Im} p(u) = 0\} = \{\operatorname{Re} u = 0, K; \quad |\operatorname{Im} u| \le 2K'\}, \quad \operatorname{mod}(2K, 4\mathrm{i}K')$$

corresponds to the continuous spectrum of system (9.1.11).

On the contour γ, the fundamental solutions $\Psi_{1,2}$ are determined simultaneously and therefore are interconnected:

$$\Psi_1(u) = \Psi_2(u)T(u), \quad u \in \gamma. \tag{9.1.15}$$

The transition matrix $T(u)$ depends only on the spectral parameter u and possesses the following properties:

$$\begin{aligned}&T(u) = \begin{pmatrix} a(u) & -b^*(-u^*) \\ b(u) & a^*(-u^*) \end{pmatrix},\\ &a(u)a^*(-u^*) + b(u)b^*(-u^*) = 1,\\ &a(u \pm 2K) = a(u)e^{\pm 2\eta_1\Delta}, \quad a^*[(u \pm 2\mathrm{i}K')^*] = -\theta a(u)e^{\pm\eta_3\Delta},\\ &b(u \pm 2K) = -b(u)e^{\pm 2\eta_1\Delta}, \quad b^*[(u \pm 2\mathrm{i}K')^*] = -\theta b(u)e^{\pm\eta_3\Delta},\end{aligned} \tag{9.1.16}$$

where $\theta = 1$ and $\theta = -1$ under the boundary conditions (9.1.9) and (9.1.10), respectively.

The Jost functions $\Psi_{1,2}$ satisfy the equations

$$\begin{aligned}\Psi_1(z) &= \Psi_1^{(0)}(z)\\ &\quad\times\left(1 + \int_{-\infty}^{z} \mathrm{d}z'[\Psi_1^{(0)}(z')]^{-1}(U(z') - U_1^{(0)}(z'))\Psi_1(z')\right),\\ \Psi_2(z) &= \Psi_2^{(0)}(z)\\ &\quad\times\left(1 - \int_{z}^{+\infty} \mathrm{d}z'[\Psi_2^{(0)}(z')]^{-1}(U(z') - U_2^{(0)}(z'))\Psi_2(z')\right).\end{aligned} \tag{9.1.17}$$

Here $U_{1,2}^{(0)}$ are the values of the matrix U for $\Phi = \varphi_{1,2}^{(0)}$. Successive iterations of eqs. (9.1.17) lead to the representation of their solutions $\Psi_{1,2}$ as series in powers of $\Psi_{1,2}^{(0)}$. The series converge when the sine-Gordon model solutions approach the asymptotic values of (9.1.9) and (9.1.10) fast enough (faster than any power of z). Both the arrangement of the over-z-integration limits in the series for $\Psi_{1,2}(u)$ and the sign of the quantity $\operatorname{Im} p(u)$ determine the analytical properties of the columns of the matrix $\Psi_{1,2}(u)$. The matrix functions

$$\Psi_-(u) = (\Psi_1^{(1)}(u), \Psi_2^{(2)}(u)), \quad \Psi_+(u) = (\Psi_2^{(1)}(u), \Psi_1^{(2)}(u)) \tag{9.1.18}$$

turn out to be analytic, respectively, in the domains

$$D_- = \{-K < \operatorname{Re} u < 0, \quad |\operatorname{Im} u| \le 2K'\}, \quad \operatorname{mod}(2K, 4iK'),$$
$$D_+ = \{0 < \operatorname{Re} u < K, \quad |\operatorname{Im} u| \le 2K'\}, \quad \operatorname{mod}(2K, 4iK').$$

Here we have denoted the ith column of the matrix Ψ as $\Psi^{(i)}$ $(i = 1, 2)$.

Equation (9.1.15) implies the relations

$$a(u) = \frac{\det \Psi_-(u)}{\det M}, \quad a^*(-u^*) = \frac{\det \Psi_+(u)}{\det M}. \tag{9.1.19}$$

According to (9.1.19), the functions $a(u)$ and $a^*(-u^*)$ continue analytically from the contour γ to the domains D_- and D_+, respectively (with an exception for, perhaps, the points where $\det M = \operatorname{dn} u/(1 - \operatorname{dn} u) = 0$).

The zeros of the functions $a(u)$ and $a^*(-u^*)$ in the domains D_- and D_+ determine the discrete spectrum of system (9.1.11). Further, for definiteness, the zeros are thought to be simple, if any. Due to reductions (9.1.16), the zeros of the function $a(u)$ are divided into two groups:

$$\begin{aligned} &\text{a)} \quad u = \nu_p + i\varepsilon_p K' \equiv \nu_p, \quad -K < \nu_p < 0, \quad \varepsilon_p = \pm 1, \quad p = 1, 2, \ldots, M, \\ &\text{b)} \quad u = \mu_s,\ \mu_s^* - 2iK', \quad -K < \operatorname{Re} \mu_s < 0, \quad s = 1, 2, \ldots, N. \end{aligned} \tag{9.1.20}$$

Each specific zero of ν_p ensures the appearance of an extra kink in the structure. Each group of zeros of μ_s and $\mu_s^* - 2iK'$ parameterizes a pulsating soliton-breather against the background of a spiral structure.

The function $a(u)$ accepts the representation

$$a(u) = \lim_{z \to +\infty} [(\Psi_2^{(0)}(u))^{-1} \Psi_1(u)]_{11}. \tag{9.1.21}$$

Using the expansions of the functions $\Psi_2^{(0)}(u)$ and $\Psi_1(u)$ near the point $u = iK'$ (see Section 9.4), we find the normalization condition

$$a(u = iK') = s \exp(\eta_3 \Delta/2), \tag{9.1.22}$$

where $s = -i\sigma$ and $s = 1$ under the boundary conditions (9.1.9) and (9.1.10), respectively.

The problem of integrating eqs. (9.1.11), and hence the sine-Gordon model (9.1), reduces to solving the Riemann problem on a torus, which is formulated as follows. It is necessary to find the functions $\Psi_+(u)$ and $\Psi_-(u)$ (9.1.18) that are analytic in the domains D_+ and D_- and satisfy simultaneously the conjugation condition on the contour γ

$$\Psi_-(u) = \frac{\Psi_+(u)}{a^*(-u^*)} \begin{pmatrix} 1 & b^*(-u^*) \\ b(u) & 1 \end{pmatrix}, \quad u \in \gamma, \tag{9.1.23}$$

the reductions

$$\Psi_+(u \pm 2K) = \sigma_3 \Psi_+(u) \sigma_3 \begin{pmatrix} 1 & 0 \\ 0 & e^{\mp 2\eta_1 \Delta} \end{pmatrix},$$

$$\Psi_-(u \pm 2K) = \sigma_3 \Psi_-(u) \sigma_3 \begin{pmatrix} e^{\pm 2\eta_1 \Delta} & 0 \\ 0 & 1 \end{pmatrix},$$

$$\Psi_+^*[(u \pm 2iK')^*] = q(u) \sigma_1 \Psi_+(u) \sigma_3 \begin{pmatrix} 1 & 0 \\ 0 & e^{\mp \eta_3 \Delta} \end{pmatrix}, \tag{9.1.24}$$

$$\Psi_-^*[(u \pm 2iK')^*] = -\theta q(u) \sigma_1 \Psi_-(u) \sigma_3 \begin{pmatrix} e^{\pm \eta_3 \Delta} & 0 \\ 0 & 1 \end{pmatrix},$$

$$i \frac{\sigma(u)}{\sigma(u + 2iK')} \exp[\eta_3(u + iK')] = \frac{k\, \mathrm{sn} u}{1 + \mathrm{dn} u} \equiv q(u),$$

and the constraint

$$\Psi_+(u) = -\sigma_2 \Psi^*(-u^*) \sigma_2, \quad u \in \gamma. \tag{9.1.25}$$

The conjugation condition for the analytic functions (9.1.23) is another way to write connection (9.1.15) of the Jost solutions on the contour γ. The symmetry properties (9.1.24) and (9.1.25) for the functions $\Psi_\pm(u)$ follow from those for the Jost functions.

Once the Riemann problem is solved, the exact solution Φ of the original essentially nonlinear model (9.1) can be restored from it along with the solution of the linear system (9.1.11). To do this, it suffices to use the asymptotic expansion of the function $\Psi_-(u)$ near the singular point $u = iK'$ of system (9.1.11):

$$\Psi_-(u = iK' + \varepsilon) = \frac{i}{\sqrt{2}} \exp\left[\frac{i}{4}(\Phi + \pi)\sigma_3\right](1 + i\sigma_1)(1 + O(\varepsilon))$$
$$\times \exp\left[-\frac{(z - t)}{2k\varepsilon}\sigma_3 + O(\varepsilon)\right] \mathrm{diag}\left(s \exp\left[\frac{\eta_3 \Delta}{2}\right], 1\right), \quad |\varepsilon| \ll 1, \tag{9.1.26}$$

where I is then identity matrix, and $s = -i\sigma$ and $s = 1$ under the boundary conditions (9.1.9) and (9.1.10), respectively.

9.2 Solitons in a spiral structure

These imperfections of periodicity are all of great importance. They are, for example, ultimately responsible for... the properties of crystals

N. Ashcroft, N. Mermin. Solid state physics

Solitons in the helicoidal structure correspond to solutions to the Riemann problem (9.1.23) with zeros of the coefficients $a(u)$ and $a^*(-u^*)$ in the domains D_- and D_+, respectively, when $b = b^* \equiv 0$. For soliton states, the unimodularity condition for a transition matrix takes the form $a(u)a^*(-u^*) = 1$. This implies that the function $a(u)$ is meromorphic inside a fundamental rectangle with sides $2K$ and $4iK'$. Moreover, if the function

$a(u)$ has a zero at the point $u = u_p$ in the domain D_-, then it necessarily has a pole at the point $u = -u_p^*$ in the domain D_+. For soliton excitations, the quasiperiodic function $a(u)$ is completely determined by its zeros (9.1.20), its poles, and the normalization condition (9.1.22) (see [1] and Appendix):

$$a_{\text{sol}}(u) = c \prod_{p=1}^{M} \varepsilon_p \frac{\sigma(u - \nu_p)}{\sigma(u + \nu_p^*)} e^{-\eta_3 \varepsilon_p \nu_p} \times \prod_{s=1}^{N} \frac{\sigma(u - \mu_s)\sigma(u - \mu_s^* + 2iK')}{\sigma(u + \mu_s^*)\sigma(u + \mu_s + 2iK')} e^{\eta_3(\mu_s + \mu_s^*)}, \tag{9.2.1}$$

where $2K$ and $4iK'$ are the periods of the Weierstrass functions. Under the boundary conditions (9.1.9), the number M of kinks is odd ($c = i\sigma$), whereas under the boundary conditions (9.1.10), M is an even number ($c = 1$). There are no limits on the number N of breathers. Representation (9.2.1) leads to the correct transformation properties of the function $a(u)$, provided that the positions of its zeros are connected to the superstructure shift Δ:

$$\sum_{p=1}^{M} \nu_p + \sum_{s=1}^{N} (\mu_s + \mu_s^*) = -\Delta/2 \qquad \text{mod } (2K). \tag{9.2.2}$$

Recall that mod $(2K, 4iK')$ defines the zeros of the function $a(u)$, and mod $4K$ controls the shift Δ. Further, for definiteness, the zeros are assumed to lie in the domain D_-. Then mod $2K$ makes eq. (9.2.2) valid.

Bearing in mind eq. (9.1.19), constraint (9.1.25) can be rewritten as

$$\Psi_+^{-1}(u) = -\frac{\Psi_-^\dagger(-u^*)}{a^*(-u^*) \det M}, \qquad u \in \gamma, \tag{9.2.3}$$

where the superscript "†" stands for the Hermitian conjugation. Hence, for soliton states, the problem of finding a solution $\Psi_-(u)$ is reduced to constructing a meromorphic quasiperiodic function with periods $2K$ and $4iK'$ and reductions (9.1.24) satisfying the following condition on the contour γ:

$$\Psi_-^\dagger(-u^*)\Psi_-(u) = \Psi_-(u)\Psi_-^\dagger(-u^*) = \left(\frac{\text{dn}u}{\text{dn}u - 1}\right) I, \qquad u \in \gamma, \tag{9.2.4}$$

where I is the identity matrix. If $\Psi_-(u)$ has a zero at the point $u = u_p$ in the domain D_-, then it has a pole at the point $u = -u_p^*$ in the domain D_+. Due to the specifics of $U - V$ operators, the function $\Psi_-(u)$ cannot be normalized to I for any value of u. Therefore, for the Riemann problem to be completely determined, the asymptotic values for $\Psi_-(u)$ are exploited as $z \to \pm\infty$.

Unlike the Riemann problem functions applied for integrating the static sine-Gordon model with a periodic background [8, 18, 6], the functions $\Psi_-(z, t, u)$ used for describing solitons in the wave model (9.1) are not doubly periodic in the parameter u. This circumstance complicates the problem. The procedure for constructing the soliton functions u boils down to algebraic calculations and, in many instances, resembles that [19] dealt with (see also Sections 3.7 and 3.8). Below, its features will be explained with examples.

9.2.1 Extra domain wall (kink) in structure

Under the boundary conditions (9.1.9), the function $\Psi_-(z,t,u)$ is completely defined by its asymptotics to be extracted from (9.1.13) and (9.1.15):

$$\begin{aligned} \Psi_- &\to \Psi_0(z,t,u)\begin{pmatrix} a(u) & 0 \\ 0 & 1 \end{pmatrix} \quad \text{as} \quad z\to+\infty, \\ \Psi_- &\to \sigma_3\Psi_\Delta(z,t,u)\sigma_3\begin{pmatrix} 1 & 0 \\ 0 & a(u) \end{pmatrix} \quad \text{as} \quad z\to-\infty. \end{aligned} \tag{9.2.5}$$

Formulas (9.2.1), (9.2.2), and (9.2.5) and reductions (9.1.24) reveal the algebraic structure of the “one-kink” function Ψ_-:

$$\Psi_-(u) = r\begin{pmatrix} (\Psi_0)_{11}a(u|v_1)\alpha + (\Psi_\Delta)_{11}\beta & (\Psi_0)_{12}\alpha - (\Psi_\Delta)_{12}a(u|v_1)\beta \\ (\Psi_0)_{21}a(u|v_1)\alpha^* - (\Psi_\Delta)_{21}\beta^* & (\Psi_0)_{22}\alpha^* + (\Psi_\Delta)_{22}a(u|v_1)\beta^* \end{pmatrix}. \tag{9.2.6}$$

Here

$$a(u|v_1) = \mathrm{i}\sigma\varepsilon_1\frac{\sigma(u-v_1)}{\sigma(u+v_1^*)}\exp\left(\frac{\varepsilon_1\eta_3\Delta}{2}\right), \quad v_1 = \mathrm{i}K'\varepsilon_1 - \Delta/2, \quad 0<\Delta<2K,$$

and α, β and r are complex and real functions of the variables z and t.

After substituting representation (9.2.6) into eq. (9.2.4), the left-hand side of eq. (9.2.4) contains poles at the points $u = v_1$ and $u = -v_1^* \bmod(2K, 4\mathrm{i}K')$, whereas the right-hand side has no such poles. The requirement that residues in extra poles vanish leads to the constraints

$$\alpha = p\,e^{\mathrm{i}\gamma}, \quad \beta = h\,e^{\mathrm{i}\gamma}, \tag{9.2.7}$$

where p, h, and γ are yet unknown real functions of the variables z and t.

By direct verification we can make sure that the left-and right-hand sides of formula (9.2.4) have the same periods, zeros, and poles. The zeros reside at the points $u = \pm(K + \mathrm{i}K')$, whereas the double pole is at the point $u = 0 \bmod(2K, 4\mathrm{i}K')$. Therefore, as Liouville's theorem claims (see Appendix), the left- and right-hand sides of eq. (9.2.4) can differ only by a u-independent coefficient. The requirement that eq. (9.2.4) is valid for $u = \mathrm{i}K'$ gives an equation for calculating the coefficient r^2. After transformations using the identity

$$-\frac{\mathrm{i}\,\sigma(\chi - \mathrm{i}K')}{\sigma(\chi + \mathrm{i}K')} \exp(\eta_3 \chi) = -\mathrm{sn}\chi + \mathrm{i}\,\mathrm{cn}\chi, \tag{9.2.8}$$

we arrive at

$$r^2\left(p^2 + h^2 + 2ph\cos\left[\frac{\varphi_2^{(0)} - \varphi_1^{(0)}}{4}\right]\right) = 1. \tag{9.2.9}$$

The matrix function Ψ_- in eqs. (9.2.6), (9.2.7), and (9.2.9) allows us to solve the Riemann problem on a torus (9.2.4) with the specific zero v_1 and reductions (9.1.24).

An additional differential constraint on the choice of the matrices $\Psi_-(z,t,u)$ consists in fulfilling the ratios in the domain D_- (see (9.1.11)):

$$\partial_t \Psi_- \Psi_-^{-1} = V, \quad \partial_z \Psi_- \Psi_-^{-1} = U. \tag{9.2.10}$$

Plugging eq. (9.2.6) into eq. (9.2.10) again gives extra poles at the points $u = v_1$ and $u = v_1^* \bmod(2K, 4\mathrm{i}K')$ in the left-hand sides of eq. (9.2.10). The no-extra-poles requirement comes down to differential equations for calculating the quantity p/h. Their solution appears as

$$\frac{p}{h} = -\frac{(\Psi_\Delta(u))_{21}}{(\Psi_0(u))_{22}}\bigg|_{u=v_1} c_0 = \left|\frac{\sigma(\chi + \mathrm{i}K')}{\sigma(\chi + \Delta + \mathrm{i}K')}\right| \times \exp\left[\frac{\eta_1 \Delta}{K}\chi + A(v_1, \chi, t) + A^*(v_1, \chi, t)\right] c_0, \tag{9.2.11}$$

where c_0 is a real constant of integration.

When calculating $\partial_t \Psi_- \Psi_-^{-1}$ and $\partial_z \Psi_- \Psi_-^{-1}$, the zeros and poles associated with the function $\mathrm{dn}u/(\mathrm{dn}u - 1)$ arise and are completely eliminated (see (9.2.4)). It is worth emphasizing that there are no equations for calculating the function $\gamma(z,t)$ (9.2.7). Conditions (9.2.5) fix the quantity γ, which turns out to be equal to zero. In this case the constant c_0 in formula (9.2.11) and the root r of eq. (9.2.9) should be chosen positive.

The conditions formulated completely determine the solution $\Psi_-(z,t,u)$ to system (9.1.11). From eq. (9.1.26) we can restore the solution of the sine-Gordon model:

$$\Phi(z,t) = 4\arg\left[\sqrt{\frac{p}{h}}\exp\left(\frac{\mathrm{i}\varphi_2^{(0)}}{4}\right) + \sqrt{\frac{h}{p}}\exp\left(\frac{\mathrm{i}\varphi_1^{(0)}}{4}\right)\right]. \tag{9.2.12}$$

It describes an “extra” kink of width $d \backsim [2\,\mathrm{Re}\,\partial_z A(v_1, \chi, t)]^{-1}$, which travels at velocity

$$V = -\frac{\mathrm{Re}[\partial_t A(v_1, \chi, t)]}{\mathrm{Re}[\partial_z A(v_1, \chi, t)]}$$

in a spiral structure.

The paper [26] affirms that a periodic ground state imposes upper and lower constraints on the velocity of moving kinks:

$$S_\mathrm{a} < |V| < S_\mathrm{opt},$$

where $S_a = k'K/E$ and $S_{opt} = 1$ are the maximum group velocities of the gapless and activation spin-wave modes against the background of the helical structure, respectively. We note that the equation for a one-kink state (9.2.12) is simpler than that found by the Darboux and Bäcklund transformations in terms of elliptic integrals of the third kind [9, 26–28]. According to the findings secured, two scenarios for the emergence of such a soliton are possible in a spiral structure. The former produces an extra domain wall with the same chirality as the domain walls of the background structure have after local contraction of the helical structure per period. The extra domain wall (2π-kink of the field Φ) is stable and deforms the helical structure. The latter locally expands the helical structure per period and generates an extra domain wall in one of its domains, the chirality of which is opposite to that of the helical structure walls.

In both cases the helicoidal structure tends to get rid of new formations as the soliton moves along the helicoidal structure. Macroscopic local translations of the helicoidal structure accompany the soliton's motion over a distance of $k\Delta$ in the direction opposite to its motion, followed by restoring the structure periodicity. Figure 9.2a illustrates a segment of a periodic helical structure before passing the soliton through it. The domain walls of the structure are depicted as hatched stripes; the darker parts of the stripes correspond to larger field values. Figure 9.2b displays a local shift and deformation of the structure after passing an "extra" domain wall of the same chirality as that of the structure walls through it. The field Φ undergoes local oscillations when the soliton passes through each region that contains the next kink of the background structure.

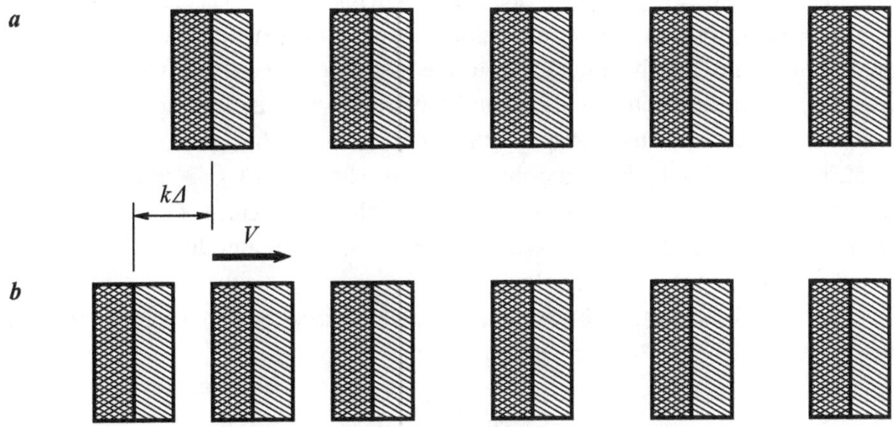

Figure 9.2: Motion of an "extra" domain wall in a spiral structure.

An analysis shows that "extra" domain walls whose chirality coincides with that of the domain walls of the helical structure have less energy. In turn, among such solitons, there are distinguished ones with the lowest energy. For them, we have the parameter

$v_1 = \mathrm{i}K'\varepsilon_1 - K/2$ ($\Delta = K$), and the quantity p/h is expressed in terms of Jacobi elliptic functions:

$$\frac{p}{h} = \frac{1}{\sqrt{\mathrm{dn}\chi}} \exp\left[\frac{(1+k')\chi}{2} - \frac{\varepsilon_1\sqrt{k'}}{k}t\right]c_0. \tag{9.2.13}$$

In dimensionless variables, the width of solitons (9.2.12) and (9.2.13) is equal to $d \backsim 2k/(1+k')$, and the average velocity V of its motion in the superstructure depends on the sign of ε_1: $V = 2\varepsilon_1\sqrt{k'}/(1+k')$.

The real width and velocity of solitons are computed as the products of the dimensionless values of these quantities by the factors $\sqrt{\alpha M_0/H}$ and $\gamma M_0\sqrt{\alpha\beta}$, respectively. Here α and β are the constants of the exchange interaction and anisotropy; γ is the magneto-mechanical ratio, and M_0 is the nominal magnetization. In dimensional variables, the width of solitons (9.2.12) and (9.2.13) weakly depends on the magnetic field in the range $0 < H \lesssim 0.6H_c$ and in dimensional variables is estimated as $d_{\mathrm{dim}} \backsim \pi\sqrt{\alpha M_0/H_c}$. The dimensional velocity noticeably changes only near a phase transition point, where it decreases according to the law

$$V_{\mathrm{dim}} \backsim \gamma M_0\sqrt{\alpha\beta}[\sqrt{H_c/H} - 1]^{1/4}, \quad H < H_c.$$

After passing solitons (9.2.12) and (9.2.13), the helical structure shifts for half a period in the direction opposite to the soliton velocity. In optically transparent materials, macroscopic shifts of the helicoidal structure as kinks travel inside it can be directly observed using the magneto-optical Faraday effect. Similar experiments have been used in [37, 38] to visualize the drift of stripe domains in ferrite-garnet crystals.

In recent years, many molecular chiral magnets with a significant contribution of the Dzyaloshinskii–Moriya interaction to the magnetic properties have been synthesized [40, 23, 31]. Such compounds are optically transparent. By varying their chemical composition we can change the spin–spin interactions and the anisotropy energy, and create conditions favorable for the observation of solitons [33, 32].

We also note that the solitons arising in the cycloidal structure of multiferroics necessitate the emergence of local electric fields [34]. This facilitates the detection and diagnosis of solitons in such materials.

Near an incommensurability–commensurability phase-transition point (in the limit as $k \to 1$), solution (9.2.12) takes the simple form

$$\Phi = 4\arctan\left[\frac{e^{\rho}(e^{\xi} + \sigma e^{y})}{\sigma e^{\xi}e^{y} - 1}\right], \tag{9.2.14}$$

where $\rho = \Delta/2$, $\sigma = \pm 1$, $\xi = z + \rho$, $y = z\coth\rho - \varepsilon t/\sinh\rho + \ln|c_0|$, $\varepsilon = \pm 1$, $c_0 = \mathrm{const}$. In easy-plane magnets with residual quadratic-in-magnetization anisotropy in the basal plane, expression (9.2.14) describes how a moving 180°th domain wall passes through a motionless one. In dimensionless variables the width of the motionless domain wall is about unity, the velocity of the moving wall is $V = \varepsilon/\cosh\rho$, and its width is $\backsim \tanh\rho$.

For $\sigma = 1$, the domain walls have the same chirality, and for $\sigma = -1$, they have opposite chirality. In both cases, they restore their original shape and velocity after interacting with each other (Figure 9.3). Both their partial annihilation and Δ-dependent magnetization oscillations accompany the meeting of domain walls of different chiralities. When $\Delta \ll 1$, the domain walls almost completely compensate for each other: during collision, the oscillation amplitude of the field Φ in Figure 9.3(1b) is the order of Δ. When $\Delta \gg 1$, the domain walls oscillate "in antiphase". When the values of Φ in the left wall are close to zero, whereas in the right wall, its value is $\Phi \backsim \pm 4 \arctan e^{2\Delta} \approx \pm 2\pi$ (and vice versa).

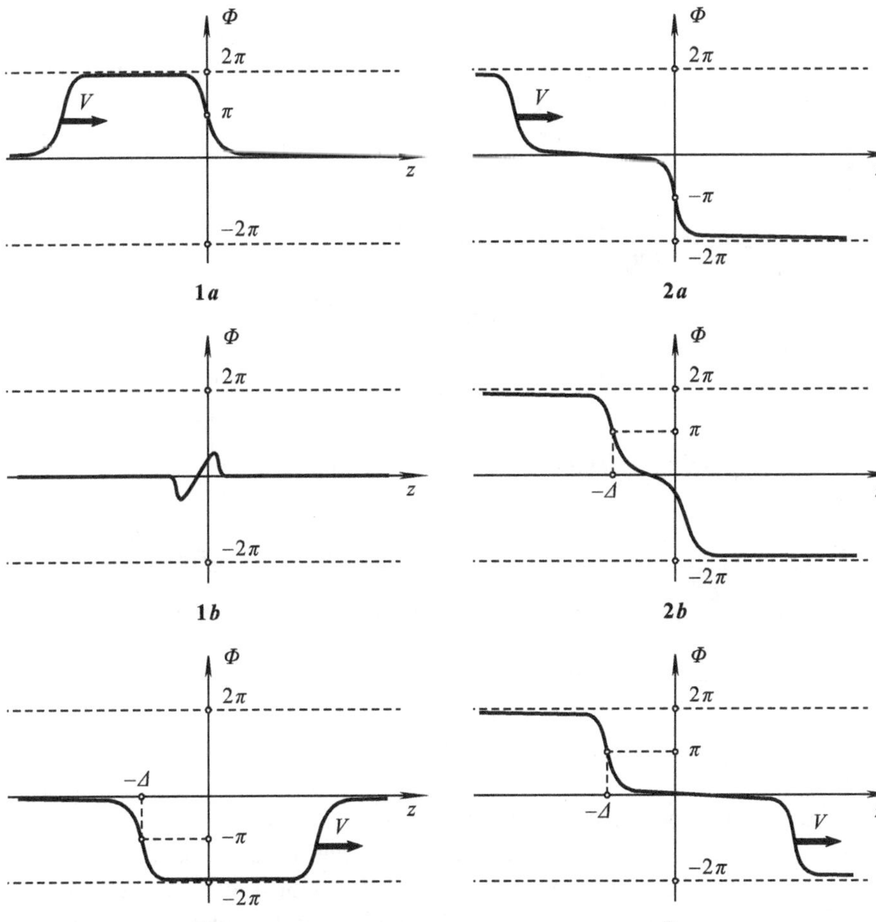

Figure 9.3: Passage of a moving domain wall through a motionless one for $\sigma = -1$ (1a, 1b, 1c) and for $\sigma = 1$ (2a, 2b, 2c).

We can imagine the motion of a kink through a long-period helical structure as repeatable interaction processes between the latter and each of its 2π-kinks.

9.2.2 Breathers (bound states of domain walls) in structure

Under the boundary conditions (9.1.10), the Riemann problem is completely defined by the conditions prescribed by constraint (9.1.15):

$$\begin{aligned}\Psi_-(z,t,u) &\to \Psi_0(z,t,u)\begin{pmatrix} a(u) & 0\\ 0 & 1\end{pmatrix} \quad \text{as} \quad z\to+\infty,\\ \Psi_-(z,t,u) &\to \Psi_\Delta(z,t,u)\begin{pmatrix} 1 & 0\\ 0 & a(u)\end{pmatrix} \quad \text{as} \quad z\to-\infty.\end{aligned} \tag{9.2.15}$$

When shifted by periods in the parameter u, the breather functions $\Psi_-(z,t,u)$ acquire additional multipliers. Therefore a conventional algorithm of constructing them by expanding them into "simple fractions" in terms of the Weierstrass zeta functions [30, 5, 39] (see Chapters 5 and 7) turns out to be inefficient. Instead of the zeta functions, it is more convenient to use the functions

$$\begin{aligned}g_1(u) &= \frac{\sigma(u-\mu_1)}{\sigma(u+\mu_1^*)}\exp\left[-\frac{\eta_3(\mu_1+\mu_1^*)}{2}\right],\\ g_2(u) &= \frac{\sigma(u-\mu_1^*-2\mathrm{i}K')}{\sigma(u+\mu_1-2\mathrm{i}K')}\exp\left[-\frac{\eta_3(\mu_1+\mu_1^*)}{2}\right],\\ g_3(u) &= \mathrm{i}\frac{\sigma(u+\mathrm{i}K')\sigma(u-\mathrm{i}K')}{\sigma(u+\mu_1^*)\sigma(u+\mu_1-2\mathrm{i}K')}\exp\left[-\eta_3\left(u+\frac{\mu_1+\mu_1^*}{2}\right)\right]\end{aligned}$$

to define the representation of the reduction group of the problem at hand up to a Bloch factor:

$$\begin{aligned}g_{1,2}(u\pm 2K) &= g_{1,2}(u)e^{\mp 2\eta_1(\mu_1+\mu_1^*)}, \quad g_3(u\pm 2K) = -g_3(u)e^{\mp 2\eta_1(\mu_1+\mu_1^*)},\\ g_{1,2}^*[(u\pm 2\mathrm{i}K')^*] &= g_{2,1}(u)e^{\mp\eta_3(\mu_1+\mu_1^*)}, \quad g_3^*[(u\pm 2\mathrm{i}K')^*] = g_3(u)e^{\mp\eta_3(\mu_1+\mu_1^*)}.\end{aligned}$$

Moreover, they satisfy the additional relations

$$g_{1,2}^*(-u^*) = \frac{1}{g_{1,2}(u)}, \quad g_3^*(-u^*) = -\frac{g_3(u)}{g_1(u)g_2(u)}$$

for providing transition (9.2.3) from the solution $\Psi_-(u)$ to $\Psi_+(u)$.

According to eqs. (9.2.1) and (9.2.2), for one breather in a spiral structure, the zero of $\mu_1 = \mathrm{i}\,\theta_1 - \Delta/4$ in the Riemann problem parameterizes the transition matrix element $a(u) = g_1(u)g_2(u)$, where $0 < \Delta < 4K$ and $-2K' < \theta_1 < 2K'$. We will look for a breather solution of the Riemann problem in the form

$$\Psi_-(z,t,u) = \Psi(\chi,t,u)\Psi_{\Delta/2}(\chi,t,u). \tag{9.2.16}$$

The scheme for constructing the function $\Psi(u)$ coincides with that given in Section 3.8: reductions and the no-extra-poles requirement in the relation (9.2.4) determine it up to a phase factor, which turns out to be z- and t-independent and is fixed by condition (9.2.15). Here is the final result:

$$\Psi(\chi, t, u|\mu_1, \mathbf{m}) = \mathrm{i}\left|\frac{r}{\kappa}\right| \begin{pmatrix} |m_2|^2 g_2(u)\kappa^* - |m_1|^2 g_1(u)\kappa & \mathrm{i}|\kappa|^2 g_3(u) m_1 m_2^* \\ \mathrm{i}|\kappa|^2 g_3(u) m_1^* m_2 & |m_1|^2 \kappa^* g_2(u) - |m_2|^2 g_1(u)\kappa \end{pmatrix},$$

where $g_i(u) \equiv g_i(u|\mu_1)$ ($i = 1, 2, 3$), and

$$\begin{aligned} \kappa &= \frac{\sigma(\mu_1 + \mu_1^*)\sigma(\mu_1 - \mu_1^* - 2\mathrm{i}K')}{\sigma(\mu_1 + \mathrm{i}K')\sigma(\mu_1 - \mathrm{i}K')} \exp(\eta_3 \mu_1), \\ |r|^{-2} &= |m_1|^4 + |m_2|^4 + 2|m_1 m_2|^2 \cos\theta, \\ \theta &= \mathrm{am}(\mu_1, k) + \mathrm{am}(\mu_1^*, k). \end{aligned} \tag{9.2.17}$$

The residues at the poles of the function Ψ are projection matrices. Therefore, as in [36] and the previous chapters, the no-extra-poles requirement in formulas (9.2.10) for the operators U and V establishes χ- and t-dependencies of the vector $\mathbf{m} = (m_1, m_2) = \Psi_{\Delta/2}(\chi, t, \mu_1)\mathbf{c}$, where $\mathbf{c}$ is an arbitrary constant complex vector. The complex parameter μ_1 and vector $\mathbf{m}$ specify the form of the function Ψ, so they are indicated in the notation after the vertical bar.

From eqs. (9.1.26) and (9.2.16) we find a solution of the nonlinear sine-Gordon model (9.1), corresponding to a breather against the background of a spiral structure:

$$\Phi(z, t) = \varphi_0(\tilde{\chi}) + 4\,\mathrm{Arctan}\left[\tan\frac{\theta}{2}\left(\frac{|m_1|^2 - |m_2|^2}{|m_1|^2 + |m_2|^2}\right)\right], \tag{9.2.18}$$

where $\tilde{\chi} = \chi + \Delta/2$, $\Delta = -4\,\mathrm{Re}\,\mu_1$, and $\mathrm{Arctan}(x)$ is a complete quasiperiodic function.

Using eqs. (9.1.12) and (9.1.24), we arrive at

$$\lim_{z\to+\infty} \frac{m_1}{m_2} = \frac{k\,\mathrm{sn}(\tilde{\chi} + \mu_1)}{1 + \mathrm{dn}(\tilde{\chi} + \mu_1)} \equiv q(\tilde{\chi} + \mu_1), \quad \lim_{z\to-\infty} \frac{m_1}{m_2} = -\frac{1}{q(\tilde{\chi} - \mu_1)}.$$

At a first glance, it seems that for solution (9.2.18), the boundary conditions (9.1.10) are violated. However, this is not the case due to the identities

$$\tan\frac{\theta}{2} = \mathrm{dn}\nu \frac{\mathrm{sn}R}{\mathrm{cn}R}, \quad |q(\tilde{\chi} \pm \mu_1)|^2 = \frac{\mathrm{dn}\nu - \mathrm{dn}(\tilde{\chi} \pm R)}{\mathrm{dn}\nu + \mathrm{dn}(\tilde{\chi} \pm R)},$$

$$R = \mathrm{Re}\,\mu_1, \quad \nu = \mathrm{i}\,\mathrm{Im}\,\mu_1;$$

$$\mathrm{Arctan}\left(\frac{\mathrm{cn}\alpha}{1 + \mathrm{sn}\alpha}\right) - \mathrm{Arctan}\left(\frac{\mathrm{sn}\beta}{\mathrm{cn}\beta}\mathrm{dn}(\alpha + \beta)\right) = \mathrm{Arctan}\left(\frac{\mathrm{cn}(\alpha + 2\beta)}{1 + \mathrm{sn}(\alpha + 2\beta)}\right).$$

The last of the equalities allows us to convert the difference of two functions of the arc tangent type into a function of the same type but of a different argument. This leads to the correct boundary conditions for solution (9.2.18).

The extent and dynamic properties of breather (9.2.18) are ultimately determined by the dependence of the vector-function **m** on the variables χ and t. Since the matrix $M(\mu_1, \chi)$, included in $\Psi_{\Delta/2}(\chi, t, \mu_1)$, is periodic in χ, the main $\chi - t$-dependence of **m** is related to the exponential factors $e^{\pm A(\mu_1, \chi, t)}$ and $e^{\pm A^*(\mu_1, \chi, t)}$. Let us represent the function A (9.1.12), which is linear in χ and t, in the form

$$2A(\mu_1, \chi, t) = q^{(1)}(z - V^{(1)}t) + \mathrm{i}(k^{(1)}z + \omega^{(1)}t),$$

where

$$\begin{aligned} q^{(1)} &= -\frac{2}{k}\operatorname{Im} p(\mu_1) > 0, \quad V^{(1)} = \frac{\operatorname{Im}(\mathrm{dn}(\mu_1))}{kq^{(1)}}, \\ k^{(1)} &= \frac{2}{k}\operatorname{Re} p(\mu_1), \quad \omega^{(1)} = \frac{\operatorname{Re}(\mathrm{dn}(\mu_1))}{k}. \end{aligned} \tag{9.2.19}$$

Then the parameter $q^{(1)}$ features the breather size: $d \backsim 1/q^{(1)}$, $V^{(1)}$ is the velocity of its motion as a whole, and $k^{(1)}$ and $\omega^{(1)}$ are the wave vector and the frequency of magnetization pulsations, respectively, in the region of the breather localization (Figure 9.4). If we write two-parameter soliton-breather in terms of its motion velocity V and pulsation frequency Ω in the comoving reference frame, then the values of these parameters inside the plane $V - \Omega$ fall into the energy gap between the Goldstone and activation branches of the spectrum of spin-wave modes [26].

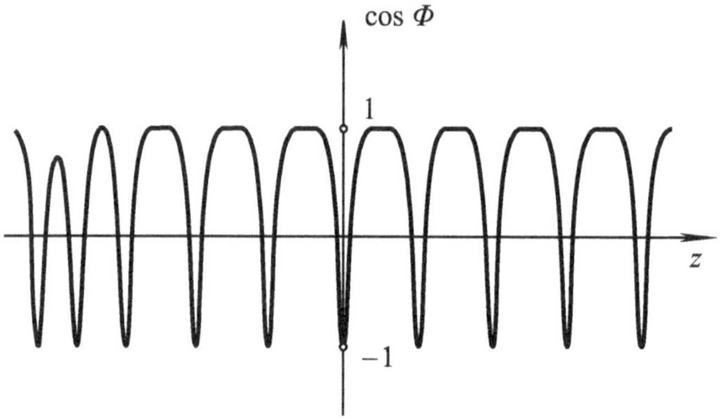

Figure 9.4: Projection of magnetization $M_x \backsim \cos\Phi$ for a breather in a spiral structure.

When a spiral structure contains pulsating solitons-breathers, an external pumping power may be absorbed at the frequencies of the internal oscillations of the breathers. Unlike kinks, breathers can be both moving and motionless. The latter are the most convenient for observations. Equation (9.2.19) implies that the breather velocity vanishes if

$\mu_1 = -\Delta/4$ or $\mu_1 = \pm 2iK' - \Delta/4$. The dimensional frequency of internal oscillations of a motionless breather is given by the formula

$$\tilde{\omega}(\Delta) = \gamma\sqrt{\beta H_c M_0}\,\mathrm{dn}(\Delta/4, k)/E(k). \tag{9.2.20}$$

When composing eq. (9.2.20), we have taken relation (9.1.7) into account. Recall that the parameter Δ in dimensionless variables characterizes a local shift $k\Delta$ of the spiral structure as a breather emerges.

Further in Section 9.3, we show that the spectrum of spin-wave modes against the background of a spiral (stripe domain) structure has two branches, gapless and activation ones. The former characterizes intraboundary oscillations of the magnetization in the stripe structure, whereas the latter is responsible for intradomain oscillations [15]. With uniform external pumping, only activation modes are excited, running in opposite directions along the structure. Ultimately, their interference generates standing waves with frequencies $\tilde{\Omega}_n \backsim 2\pi n/L_0$, where L_0 is a period of the spiral structure, and n are natural numbers. When the pumping frequency coincides with that of the standing wave frequencies, the pumping field energy is resonantly absorbed [22, 24, 35]. It is worth noting that admissible frequencies of the breather pulsations lie almost everywhere below the standing spin-wave frequencies $\tilde{\Omega}_n$ in the spiral structure and continuously fill the band shaded in Figure 9.5. Only a slight overlap of soliton and spin-wave frequencies is possible near a phase transition point across the field from an incommensurate phase to a commensurate one. This fact can be used to detect breathers in a spiral structure.

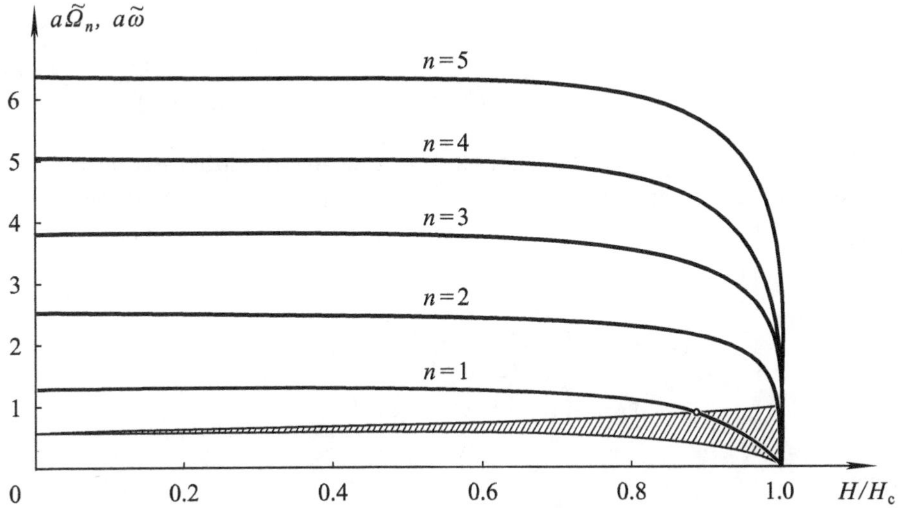

Figure 9.5: A frequency band of a motionless breather and a spectrum of resonant frequencies of standing spin waves in a helicoidal structure, $a^{-1} = \gamma\sqrt{\beta M_0 H_c}$.

Next, we examine the obtained solution near the phase transition point in the helicoidal structure as $k \to 1$ and the period of the 2π-kink lattice also increases unlimitedly. In the limit as $k \to 1$ and $\mu_1 = -\rho/2 + \mathrm{i}\,\theta = \text{const}$, solution (9.2.18) takes the simple form

$$\begin{aligned}\Phi &= 4\,\mathrm{Arctan}\left[\frac{e^{\xi}\sinh(\rho/2)\cos(\gamma+\theta)+\cosh(y+\rho/2)\cos\theta}{e^{\xi}\cos\theta\cosh(y-\rho/2)+\sinh(\rho/2)\cos(\gamma-\theta)}\right],\\ y &= \frac{z\sinh\rho - 2t\,\sin\theta\sinh(\rho/2)}{\cosh\rho+\cos 2\theta}+y_0,\\ \gamma &= \frac{-z\,\sin 2\theta + 2t\,\cos\theta\cosh(\rho/2)}{\cosh\rho+\cos 2\theta}+\gamma_0,\end{aligned} \tag{9.2.21}$$

where $\xi = z + \rho$, and y_0 and γ_0 are real constants. Formula (9.2.21) controls the interaction between the breather and one motionless 2π-kink of the structure. The breather moves at a speed of $V = \sin\theta/\cosh(\rho/2)$ and pulsates at a frequency of $\omega = 2|\cos\theta|\cosh(\rho/2)[\cosh\rho + \cos(2\theta)]^{-1}$. Upon passing the breather through 2π-kink, the latter displaces along the Oz-axis by a distance of $2\rho = \Delta$ toward the opposite breather motion.

Now we discuss a special case where $\gamma_0 = 0$, $\theta \to \pi\varepsilon/2 + \delta$, $\varepsilon = \pm 1$, $\delta \to 0$. In this limit, solution (9.2.21) degenerates into the exponential-polynomial solution

$$\Phi = 4\,\mathrm{Arctan}\left[\frac{(s-1)\sinh(\rho/2)-\cosh(y-\rho/2)e^{\xi}}{(s+1)\sinh(\rho/2)e^{\xi}+\cosh(y+\rho/2)}\right], \tag{9.2.22}$$

where

$$\xi = z+\rho,\quad y = z\,\coth(\rho/2) - \frac{\varepsilon\, t}{\sinh(\rho/2)}+y_0,\quad s = \frac{z-\varepsilon\, t\cosh(\rho/2)}{\sinh^2(\rho/2)}.$$

The breather is localized in the region where $y \approx \text{const}$. At large distances from 2π-kink, "the kernel" of the breather acquires the same form as in the case of a uniform ground state of a medium. So, in the limit as $y = \text{const}$ and $z \to +\infty$, the condition $\varepsilon t \to +\infty$ must be met, so we get

$$\begin{aligned}\Phi &\approx 4\,\mathrm{Arctan}\left(\frac{\cosh(y-\rho/2)}{\varepsilon t\coth(\rho/2)}\right)\\ &= 4\,\mathrm{Arctan}\left(e^{\coth(\rho/2)(z_0-z_2)}+e^{-\coth(\rho/2)(z_0-z_1)}\right),\end{aligned}$$

where

$$\begin{aligned}z_0 &= z - Vt,\quad V = \frac{\varepsilon}{\cosh(\rho/2)},\\ z_{1,2} &= \tanh\left(\frac{\rho}{2}\right)\left[\mp\ln\left(2\,\varepsilon\, t\coth\left(\frac{\rho}{2}\right)\right)+\frac{\rho}{2}-y_0\right].\end{aligned}$$

Hence it follows that the breather consists of two 2π-kinks with opposite topological charges and the same width $\tanh(\rho/2)$.

In easy-plane ferromagnets with residual anisotropy $\propto M_1^2$ in the basal plane, such an object corresponds to a magnetization reversal nucleus in one of two domains separated by a motionless domain wall. When $\varepsilon = 1$ and $y_0 = 0$, the nucleus initially (as $t \to -\infty$) located in the left domain moves from left to right. Moreover, its right (left) domain wall has the same (opposite) chirality as the motionless domain boundary. Domain walls of the same chirality repel, whereas those of opposite chirality attract. Therefore the nucleus approaching the domain boundary downsizes and, moving as a whole at a speed of V, easily passes through the domain boundary (Figure 9.6).

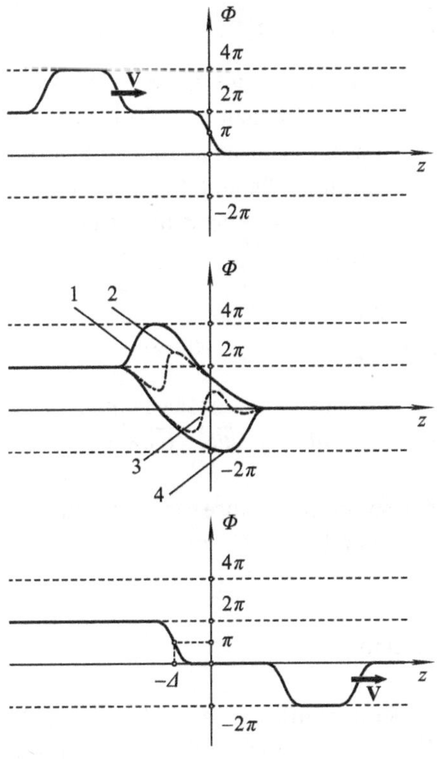

Figure 9.6: A magnetization reversal nucleus and a motionless domain wall before, during, and after the collision. The digits 1, 2, 3, and 4 enumerate the distribution of the field Φ at different moments of time as the nucleus passes through the wall.

Upon the "collision" of the nucleus with the domain boundary, the magnetization field is significantly distorted. However, once collided, the right domain produces a new magnetization reversal nucleus moving at the earlier speed. With that, the domain boundary restores its shape, shifts to the left along the Oz-axis by a distance of

$\Delta = 2\rho$, and then remains immobile. Now the left (right) domain wall of the new nucleus has the same (opposite) chirality as the motionless domain boundary. At small distances from the domain boundary, the nucleus size is determined by the interactions between the domain walls of these objects. As the nucleus moves away from the domain boundary, its width $z_2 - z_1$ increases with time according to the logarithmic law: $z_2 - z_1 = 2\tanh(\rho/2)\ln[2t\coth(\rho/2)]$. In the limit as $t \to +\infty$, the nucleus involves an extended region of an overmagnetized material. The energy of such a formation is equal to the sum of the energies of its bounding domain walls.

For motionless breathers, the parameter μ_1 is either real, $\mu_1 = -\rho/2$, or equal to $\mu_1 = -\rho/2 \pm \pi\mathrm{i}$. In these cases the field Φ appears as

$$\Phi = 4\mathrm{Arctg}\left[\frac{e^{\xi}\sinh(\rho/2)\cos\gamma + \cosh(y+\rho/2)}{\sinh(\rho/2)\cos\gamma + e^{\xi}\cosh(y-\rho/2)}\right],$$

where $y = z\tanh(\rho/2) + y_0$, $\gamma = t/\cosh(\rho/2) + \gamma_0$ for real μ_1, and $\gamma = t/\cosh(\rho/2) - \gamma_0$ for $\mu_1 = -\rho/2 \pm \pi\mathrm{i}$ ($\gamma_0 = \mathrm{const}$). The breather "breathes" with a frequency of $\omega = 1/\cosh(\rho/2)$ (in dimensionless variables) but does not move as a whole. A distance from the breather to the 2π-kink depends on the parameter y_0. Far from the 2π-kink (for $z \gg 1$, $y_0 \gg 1$) in the region where $y \approx \mathrm{const}$, we have an ordinary pulsating soliton against the background of a uniform ground state of the medium [14, 36]:

$$\Phi = 4\,\mathrm{Arctan}\left[\frac{\sinh(\rho/2)\cos\gamma}{\cosh(y-\rho/2)}\right].$$

The real frequency, width, and velocity of solitons can be obtained by multiplying dimensionless values of these quantities by the factors $\gamma\sqrt{\beta H_c M_0}$, $\sqrt{\alpha M_0/H_c}$, and $\gamma M_0\sqrt{\alpha\beta}$, respectively.

The motion of a breather through a long-period spiral structure can be imagined as repeatable interaction processes between the latter and each of its domain walls.

9.2.3 Structure translations under soliton collisions

Multibreathers against the background of the superstructure are expressed through the matrix functions $\Psi(\chi, t, u|\mu_s, \mathbf{m}^{(s)})$, where $\mathbf{m}^{(s)} = \Psi_{\rho_s}(\chi, t, \mu_s)\mathbf{c}^{(s)}$, $\mathbf{c}^{(s)} = (c_1^{(s)}, c_2^{(s)})$ is an arbitrary constant complex vector, $\mu_s = -\rho_s/2 + \mathrm{i}\,\theta_s$, $0 < \rho_s < 2K$, $|\theta_s| < 2K'$, and

$$\Psi_{\rho_s}(\chi, t, u) = M(u, \chi + \rho_s)\exp\left(A(u, \chi, t)\sigma_3 + \frac{\eta_1 \rho_s u}{K}\sigma_3\right).$$

To construct a two-breather solution, we recall that the solution to system (9.1.11) remains unchanged after multiplying it from the right by a $\chi - t$-independent matrix and shifting the argument χ: $\chi \to \chi + \rho_2$. The above transformations, instead of the one-breather solution (9.2.16), bring to a new one-breather solution of system (9.1.11):

$$\Psi(\chi, t, u|\mu_1, \mathbf{n}^{(1)})\Psi_{\Delta/2}(\chi, t, u), \tag{9.2.23}$$

where $\Delta/2 = \rho_1 + \rho_2$, $\mathbf{n}^{(1)} = \Psi_{\Delta/2}(\chi, t, \mu_1)\mathbf{d}^{(1)}$, and $\mathbf{d}^{(1)}$ is an arbitrary constant complex vector. This solution may be treated as a "seeding" solution and "dressed" along the outlined algorithm. Then we come up with a two-breather solution

$$\begin{aligned}&\tilde{\Psi}_-(\chi, t, u|\mu_1, \mu_2, \mathbf{n}^{(1)}, \mathbf{n}^{(2)})\\&\quad = \Psi(\chi, t, u|\mu_2, \mathbf{n}^{(2)})\Psi(\chi, t, u|\mu_1, \mathbf{n}^{(1)})\Psi_{\Delta/2}(\chi, t, u),\end{aligned} \tag{9.2.24}$$

where

$$\mathbf{n}^{(2)} \equiv \Psi(\chi, t, \mu_2|\mu_1, \mathbf{n}^{(1)})\Psi_{\Delta/2}(\chi, t, \mu_2)\mathbf{d}^{(2)}$$

with arbitrary constant complex vector $\mathbf{d}^{(2)}$ and

$$\mu_s = -\rho_s/2 + \mathrm{i}\,\theta_s \ (s = 1, 2), \quad \rho_2 + \rho_1 = \Delta/2.$$

We can make sure by direct verification that eq. (9.2.24) satisfies the differential constraints and transformation properties of the solution to system (9.1.11) under the boundary conditions (9.1.10).

Thus we have outlined a recurrent breeding scheme for breather solutions to system (9.1.11). Consequently, this means that the sine-Gordon model offers a description of N breathers in a spiral structure at its $(N-1)$th step.

Under the boundary conditions (9.1.10), another "elementary" solution to the model corresponds to two "extra" domain walls (kinks) with opposite chiralities (charges) against the background of a spiral structure (a kink lattice). Factorization (9.2.24) cannot describe such solitons, but a two-kink solution to the Riemann problem with zeros at the points $v_s = -\Delta_s/2 + \mathrm{i}\,\varepsilon_s K'$ $(s = 1, 2)$, $\Delta_1 + \Delta_2 = \Delta$, is obtained from relation (9.2.6) after formal replacements:

$$\begin{aligned}&\Psi_0 \to \Psi_-^{(2)}, \ \Psi_\Delta \to \Psi_-^{(1)},\\&a(u|v_1) = \mathrm{i}\sigma\varepsilon_1 \frac{\sigma(u - v_1)}{\sigma(u + v_1^*)} \exp\left(\frac{\varepsilon_1 \eta_3 \Delta_1}{2}\right).\end{aligned} \tag{9.2.25}$$

Here, $\Psi_-^{(s)}$ $(s = 1, 2)$ are the solutions of the Riemann problem with zero at the point $u = v_2$, which correspond to two one-kink states $\Phi^{(s)}$ of the sine-Gordon model with different conditions at infinity

$$\begin{aligned}\Phi^{(s)} &\to \varphi_2^{(0)}(\chi + \delta_s) \equiv \varphi_0(\chi + \delta_s, k) \quad \text{as} \quad z \to +\infty,\\\Phi^{(s)} &\to \varphi_1^{(0)}(\chi + \delta_s) \equiv -2\pi\sigma + \varphi_0(\chi + \Delta_2 + \delta_s, k) \quad \text{as} \quad z \to -\infty,\end{aligned}$$

where $\sigma = \pm 1$, $\delta_1 = \Delta_1$, $\delta_2 = 0$, and $0 < \Delta_{1,2} < K$. The expressions for $\Psi_-^{(s)}$ can be deduced from formulas (9.2.6), (9.2.9), and (9.2.11) after obvious replacements related to a change in the asymptotic conditions and the zero of the Riemann problem. We list important limitations.

Constructing the matrices $\Psi_-^{(1)}$ and $\Psi_-^{(2)}$ requires the use of a function that differs in sign from $a(u|v_1)$ (9.2.25):

$$\tilde{a}(u|v_2) = -\mathrm{i}\sigma\varepsilon_2 \frac{\sigma(u - v_2)}{\sigma(u + v_2^*)} \exp\left(\frac{\varepsilon_2}{2}\eta_3\Delta_2\right), \quad v_2 = -\frac{\Delta_2}{2} + \mathrm{i}\varepsilon_2 K'.$$

Only then a two-soliton solution to model (9.1) can describe two domain walls of different chirality in a spiral structure with boundary conditions (9.1.10). The use of the identical functions ($a(u|v) = \tilde{a}(u|v)$) leads to the solution to model (9.2) with other boundary conditions, which differ in mod 4π from (9.1.10) and take into account the presence of two walls of the same chirality in the spiral structure.

According to relation (9.2.11), each of the matrix functions $\Psi_-^{(s)}$ depends on the integration constant $c_s > 0$ ($s = 1, 2$). These parameters should be subject to the constraint $c_2/c_1 = |a(v_2|v_1)|^2$. Then eq. (9.2.4) contains no extra poles at the points $u = v_2$ and $-v_2^*$ mod $(2K, 4\mathrm{i}K')$. In the process the functions α and β of the two-soliton solution go positive. Finally, we have

$$\frac{\alpha}{\beta} = \left|\frac{(\Psi_-^{(1)}(v_1))_{21}}{(\Psi_-^{(2)}(v_1))_{22}}\right| c, \quad r = \left(\alpha^2 + \beta^2 + 2\sigma\alpha\beta \sin\left[\frac{\Phi^{(2)} - \Phi^{(1)}}{4}\right]\right)^{-1/2},$$

where $c > 0$ is another constant of integration.

The solution of the sine-Gordon model with asymptotics (9.1.10), which describes an elastic collision between a kink and an antikink traveling in a kink lattice, has the form

$$\Phi = 4 \arg(\alpha e^{\mathrm{i}\Phi^{(2)}/4} + \mathrm{i}\sigma\beta e^{\mathrm{i}\Phi^{(1)}/4}).$$

Interestingly, collisions of solitons in a spiral structure are accompanied by local translations of the structure; the latter can be visualized in optically transparent crystals. The formulas derived for multisolitons make it feasible to completely evaluate mutual interaction processes both between solitons and between solitons and a background structure.

The two-breather solution (9.2.24) can serve as an example to be discussed. As before, the propagation of breathers obeys the exponential factors $\exp[\pm A(\mu_s, \chi, t)]$ and $\exp[\pm A^*(\mu_s, \chi, t)]$ ($s = 1, 2$), where

$$2A(\mu_s, \chi, t) = q^{(s)}(z - V^{(s)}t) + \mathrm{i}(k^{(s)}z + \omega^{(s)}t).$$

The parameters $q^{(s)}$, etc. are obtained by replacing the index "1" by the index "s" in formulas (9.2.19). Further, we go over to a frame of reference associated with breather "2" and analyze its internal structure before and after the collision with breather "1". This problem can be simplified mathematically to finding the limits of the function (9.2.24) as

$t \to \mp\infty$, given the constraint $\xi_2 \equiv z - V^{(2)}t = \text{const}$. We represent $A(\mu_s, \chi, t)$ in a format that makes computations easy:

$$\begin{aligned} 2A(\mu_2, \chi, t) &= q^{(2)}\xi_2 + \mathrm{i}[k^{(2)}\xi_2 + (\omega^{(2)} + k^{(2)}V^{(2)})\,t], \\ 2A(\mu_1, \chi, t) &= q^{(1)}[\xi_2 + (V^{(2)} - V^{(1)})\,t] \\ &\quad + \mathrm{i}[k^{(1)}\xi_2 + (\omega^{(1)} + k^{(1)}V^{(2)})\,t]. \end{aligned} \tag{9.2.26}$$

Let, for definiteness, the velocity of breather "2" be greater than that of breather "1": $(V^{(2)} - V^{(1)}) > 0$. Applying the second relation of (9.2.26) and formulas (9.2.15), we arrive at

$$\Psi(\chi, t, u|\mu_1, \mathbf{n}^{(1)})\Psi_{\Delta/2}(\chi, t, u)\Big|_{\substack{\xi_2=\text{const}\\ t\to+\infty}} \to \Psi_{\Delta/2-\rho_1=\rho_2}(\chi, t, u)\begin{pmatrix} a^{(1)} & 0 \\ 0 & 1 \end{pmatrix}, \tag{9.2.27}$$

$$\Psi(\chi, t, u|\mu_1, \mathbf{n}^{(1)})\Psi_{\Delta/2}(\chi, t, u)\Big|_{\substack{\xi_2=\text{const}\\ t\to-\infty}} \to \Psi_{\Delta/2+\rho_1}(\chi, t, u)\begin{pmatrix} 1 & 0 \\ 0 & a^{(1)} \end{pmatrix}, \tag{9.2.28}$$

where $a^{(1)}(u) = g_1(u|\mu_1)g_2(u|\mu_1)$.

Relations (9.2.27) and (9.2.28) immediately give the limit values of the vector $\mathbf{n}^{(2)}$:

$$\begin{aligned} \mathbf{n}^{(2)} &\to \mathbf{p} = \Psi_{\rho_2}(\chi, t, \mu_2)\begin{pmatrix} a^{(1)}(\mu_2)d_1^{(2)} \\ d_2^{(2)} \end{pmatrix} \quad \text{as} \quad t \to +\infty, \\ \mathbf{n}^{(2)} &\to \mathbf{q} = \Psi_{\Delta/2+\rho_1}(\chi, t, \mu_2)\begin{pmatrix} d_1^{(2)} \\ a^{(1)}(\mu_2)d_2^{(2)} \end{pmatrix} \quad \text{as} \quad t \to -\infty. \end{aligned} \tag{9.2.29}$$

In the limit as $t \to \pm\infty$, the exponential factors with index $s = 2$ remain bounded according to (9.2.26). Therefore the desired expressions for $\tilde{\Psi}_-$ have the form

$$\tilde{\Psi}_-\Big|_{\substack{\xi_2=\text{const}\\ t\to+\infty}} \to \Psi(\chi, t, u|\mu_2, \mathbf{p})\Psi_{\rho_2}(\chi, t, u)\begin{pmatrix} a^{(1)}(u) & 0 \\ 0 & 1 \end{pmatrix}, \tag{9.2.30}$$

$$\tilde{\Psi}_-\Big|_{\substack{\xi_2=\text{const}\\ t\to-\infty}} \to \Psi(\chi, t, u|\mu_2, \mathbf{q})\Psi_{\Delta/2+\rho_1}(\chi, t, u)\begin{pmatrix} 1 & 0 \\ 0 & a^{(1)}(u) \end{pmatrix}. \tag{9.2.31}$$

As we can see, the functions (9.2.16) and (9.2.17) typical of an isolated breather govern the second breather moving away from the first one. Since $V^{(2)} > V^{(1)}$, expressions (9.2.30) and (9.2.31) match situations where the second breather is, respectively, at a large distance to the right or left of the first one. As $\xi_2 \to +\infty$ and $\xi_2 \to -\infty$, the asymptotic values of the function (9.2.30) provide a glimpse of the change in the helicoidal structure due to the influence of the second breather. The appropriate limits are easily calculated using (9.2.15):

$$\tilde{\Psi}_-(t \to +\infty)|_{\xi_2 \to +\infty} \to \Psi_0(\chi, t, u)\begin{pmatrix} a(u) & 0 \\ 0 & 1 \end{pmatrix},$$
$$\tilde{\Psi}_-(t \to +\infty)|_{\xi_2 \to -\infty} \to \Psi_{2\rho_2}(\chi, t, u)\begin{pmatrix} a^{(1)}(u) & 0 \\ 0 & a^{(2)}(u) \end{pmatrix}, \tag{9.2.32}$$

where $a^{(2)}(u) = g_1(u|\mu_2)g_2(u|\mu_2)$, and $a(u) = a^{(1)}(u)a^{(2)}(u)$. Consequently, the motion of the second breather is accompanied by "compression" of the helicoidal structure along the Oz-axis by a magnitude of $2k\rho_2$. Analysis of similar limits of expression (9.2.31) leads to the same conclusion. At the same time the first breather causes the superstructures where the second breather moves as $t \to +\infty$ and as $t \to -\infty$ to shift along the Oz-axis relative to each other by a magnitude of $-2k\rho_1$.

The second difference of expressions (9.2.30) and (9.2.31) from the isolated breather functions (9.2.16) and (9.2.17) is renormalizing the constant vector $\mathbf{d}^{(2)}$. For example, in the limit as $t \to +\infty$, the vector components $\mathbf{d}^{(2)} = (d_1^{(2)}, d_2^{(2)})$ acquire additional factors (see (9.2.29)): $d_1^{(2)} \to a^{(1)}(\mu_2)d_1^{(2)}$, $d_2^{(2)} \to d_2^{(2)}$. The internal structure of the solitary second breather is described by an analog of expression (9.2.18) and depends on the component ratio $n_1^{(2)}/n_2^{(2)}$. The explicit form of the vectors $\mathbf{n}^{(2)}$ and $\mathbf{p}$ stands for a shift of the second soliton's center coordinate in the limit as $t \to +\infty$,

$$z = V^{(2)}t - \frac{1}{q^{(2)}} \ln\left|\frac{d_1^{(2)}}{d_2^{(2)}}\right|,$$

by a magnitude of

$$\delta z(t \to +\infty) = -\frac{\ln |a^{(1)}(\mu_2)|}{q^{(2)}}. \tag{9.2.33}$$

With that, the initial phase of pulsations in the breather's localization region acquires the increment

$$\delta\varphi(t \to +\infty) = \arg a^{(1)}(\mu_2). \tag{9.2.34}$$

Before colliding with the first breather ($t \to -\infty$), the above parameters for the second breather differ in sign from the values of eqs. (9.2.33) and (9.2.34).

To trace the motion of the first soliton as it scatters by the second one, factorization (9.2.24) should be carried out in reverse order. A new notation of the same solution is convenient for studying its asymptotics in another limiting case: $z - V^{(1)}t = \text{const}$ as $t \to \pm\infty$. Calculations give close results.

Thus multisolitons in a helicoidal structure, in contrast to solitons on a homogeneous background, are characterized not only by increments of coordinates and phases after each pair of collisions but also by macroscopic local translations of the background structure accompanying the motion of each of the solitons.

9.3 Nonlinear dynamics of spin waves

> But here is the mathematics of waves: It proved to be incredibly intricate right from the start, and it still keeps getting more convoluted. It has become so difficult that one has to use all the power of a computing apparatus in order to at least partially dodge these difficulties and bring solving problems to an end, to a number, as physicists say.
>
> B. Kadomtsev, V. Rydnik. "Waves around us"

When slightly exposed to external impacts insufficient for macroscopic shifts and the formation of solitons, a structure carries spin waves in which magnetization fluctuations occur near the equilibrium positions of the spiral structure. Spin-wave states initially localized against the background of the superstructure are spread out over time due to dispersion effects. A theoretical description of the nonlinear dynamics of spin waves is connected with solving the so-called regular Riemann problem, for which the functions $a(u)$ and $a^*(-u^*)$ have no zeros in their analyticity domains. The boundary conditions for such a problem coincide with those of (9.1.10) for $\Delta = 0$: $\Phi(z,t) \to \varphi_0(\chi)$ as $|z| \to \infty$. Since $\Delta = 0$, the functions $\Psi_\pm(u)$ and $\Psi_0(u)$ have periods $[4K, 4iK']$ and are uniquely defined on a Riemann surface that is topologically equivalent to a torus.

The functions

$$F_+(u) = \exp\left[\frac{i}{4}(\Phi - \varphi_0)\sigma_3\right]\Psi_+(u)\Psi_0^{-1}(u), \quad \det F_+ = a^*(-u^*),$$
$$F_-(u) = \exp\left[\frac{i}{4}(\Phi - \varphi_0)\sigma_3\right]\frac{\Psi_-(u)}{a(u)}\Psi_0^{-1}(u), \quad \det F_- = \frac{1}{a(u)},$$

are useful to solve the regular Riemann problem on a torus; they are not only doubly periodic and analytic in the domains $D_\pm$ but also continuous in these domains up to the boundary. A transition from $\Psi_\pm(u)$ to $F_\pm(u)$ is necessary to eliminate the former's features generated by the background structure.

Using asymptotic expansions of the functions $\Psi_\pm(u)$ and $\Psi_0(u)$ near the points $u = \pm iK'$ (see Section 9.4), we obtain the formulas

$$\begin{aligned} a(\pm iK') &= a(\pm iK' + 2K) = 1, \\ F_\pm(-iK') &= F_\pm(-iK' + 2K) = I, \\ F_\pm(iK') &= \exp\left[\frac{i}{2}(\Phi - \varphi_0)\sigma_3\right]. \end{aligned} \tag{9.3.1}$$

Reductions to functions $F_\pm(u)$ in their domains of analyticity follow from those for $\Psi_\pm(u)$ and $\Psi_0(u)$:

$$F_\pm(u + 2K) = \sigma_3 F_\pm(u)\sigma_3, \quad F_\pm^*[(u^* + 2iK')^*] = \sigma_1 F_\pm(u)\sigma_1. \tag{9.3.2}$$

In the cell $[4K, 4iK']$ of their periodicity at the borders $\gamma = \gamma_1 \cup \gamma_2 \cup \tilde{\gamma}_1 \cup \tilde{\gamma}_2$ between the domains D_+ and D_- (Figure 9.7),

$$\gamma_1 = \{\operatorname{Re} u = 0, \quad |\operatorname{Im} u| \le 2K'\}, \quad \tilde{\gamma}_1 = \{\operatorname{Re} u = 2K, \quad |\operatorname{Im} u| \le 2K'\},$$
$$\gamma_2 = \{\operatorname{Re} u = -K, \quad |\operatorname{Im} u| \le 2K'\}, \quad \tilde{\gamma}_2 = \{\operatorname{Re} u = K, \quad |\operatorname{Im} u| \le 2K'\},$$

and the matrix functions $F_\pm(u)$ are related by condition (9.1.23) as follows:

$$F_+(u) = F_-(u)(I - G(u)),$$
$$G(u) = \Psi_0(u) \begin{pmatrix} 0 & b^*(-u^*) \\ b(u) & 0 \end{pmatrix} \Psi_0^{-1}(u). \tag{9.3.3}$$

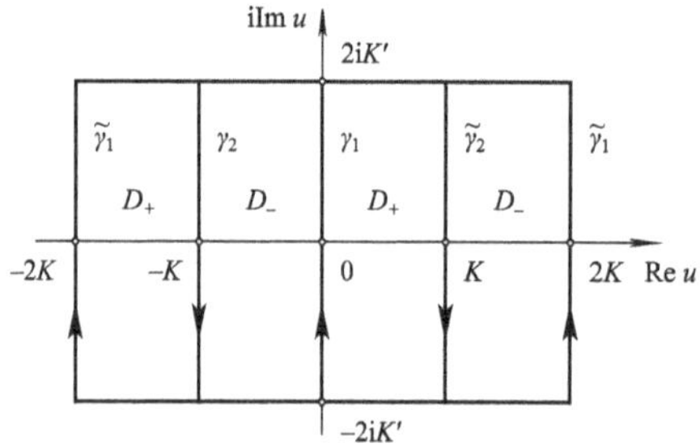

Figure 9.7: A border $\gamma = \gamma_1 \cup \gamma_2 \cup \tilde{\gamma}_1 \cup \tilde{\gamma}_2$ between the domains D_+ and D_-.

Now we show that solving the regular Riemann problem (9.3.3) reduces to searching for a solution to linear integral equations. To do this, we define a piecewise analytic function $\tilde{F}(u) = F_\pm(u)$ for $u \in D_\pm$ and introduce a contour $\Gamma = \Gamma_1 \cup \Gamma_2 \cup \Gamma_3 \cup \Gamma_4$. Here Γ_i are contours bounding the domains D_+ and D_- in the elementary cell $[4K, 4iK']$ (Figure 9.8). We employ positive directions of traversals so that each of the domains D_+ and D_- should stay to the left of the surrounding contour.

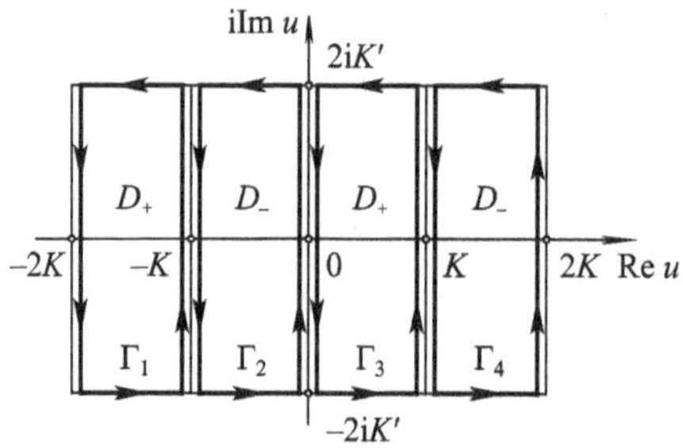

Figure 9.8: A contour $\Gamma = \Gamma_1 \cup \Gamma_2 \cup \Gamma_3 \cup \Gamma_4$ and domains D_+ *and* D_- in the cell $[4K, 4iK']$.

Since the quantities $\tilde{F}(u')$ and $\zeta(u' + iK')[\tilde{F}(u') - I]$ for $u' \in \Gamma$ are the boundary values of functions that are analytic inside the contour Γ, we have the following identities:

$$\oint_\Gamma du'\tilde{F}(u') = 0, \quad \oint_\Gamma du'\zeta(u' + iK')[\tilde{F}(u') - I] = 0, \tag{9.3.4}$$

where $[4K, 4iK']$ are the periods of the Weierstrass functions.

Solving the regular Riemann problem on a Riemann surface is difficult due to the choice of an analog of the Cauchy kernel [44, 25]. The problem at hand contains an appropriate Cauchy kernel on a torus in the form

$$Y(u', u) = \zeta(u' - u) - \zeta(u' + iK').$$

This kernel is doubly periodic in the variable u' with periods $[4K, 4iK']$ but a quasi-periodic-in-u one:

$$\zeta(u \pm 2K) = \zeta(u) \pm 2\eta_1, \quad \zeta(u \pm 4iK') = \zeta(u) \pm 2\eta_3.$$

Nevertheless, by (9.3.4) the transformation

$$\frac{1}{2\pi i}\oint_\Gamma du' Y(u', u)[\tilde{F}(u') - I] \tag{9.3.5}$$

defines a doubly periodic function of the spectral parameter u. Applying the Cauchy theorem, calculation of the integral (9.3.5) gives the integral representation

$$\tilde{F}(u) = I + \frac{1}{2\pi i}\oint_\Gamma du' Y(u', u)[\tilde{F}(u') - I]. \tag{9.3.6}$$

The integration (9.3.6) covers horizontal segments of the contour Γ in opposite directions, and the integrand takes the same values inside them. Therefore the integration (9.3.6) includes only integrals along vertical segments of the contour Γ, which by eqs. (9.3.3) can be expressed in terms of jumps of the function $\tilde{F}(u')$ on γ:

$$\begin{aligned}\tilde{F}(u) &= I + \frac{1}{2\pi i}\int_\gamma du' Y(u', u)[F_-(u') - F_+(u')] \\ &= I + \frac{1}{2\pi i}\int_\gamma du' Y(u', u) F_-(u') G(u').\end{aligned} \tag{9.3.7}$$

In eq. (9.3.7) the integration is carried out over the segments γ in the direction of bypassing the domain D_-.

Suppose u tends to a point $v \in \gamma$ from the domain D_-. Then by the Sokhotski–Plemelj formula relation (9.3.7) leads to a singular integral equation for calculating $F_-(v)$:

$$F_-(v)\left(I - \frac{G(v)}{2}\right) = I + \frac{\text{V.p.}}{2\pi i}\int_\gamma du' Y(u', v) F_-(u') G(u'). \tag{9.3.8}$$

The letters "V. p." before the integral stands for a calculation in the sense of the Cauchy principal value. It suffices to find a solution to eq. (9.3.8) for $\nu = \mathrm{i}\,v$, where $|v| \le K'$, because reductions (9.3.2) restore this solution in other segments γ. Using these reductions and relations like

$$\int_{-2K'\mathrm{i}}^{2K'\mathrm{i}} \mathrm{d}u' Y(u',\nu)F_-(u')G(u') = \mathrm{i}\int_{-K'}^{K'} \mathrm{d}v\big(Y(\mathrm{i}v,\nu)F_-(\mathrm{i}v)G(\mathrm{i}v)$$
$$+ Y(-2K'\mathrm{i}-\mathrm{i}v,\nu)\sigma_1[F_-(\mathrm{i}v)G(\mathrm{i}v)]^*\sigma_1\big),$$
$$G^*(\mathrm{i}v) = -\sigma_2 G(\mathrm{i}v)\sigma_2,$$

we get

$$F_-(\nu)\left(I - \frac{G(\nu)}{2}\right) = I + \frac{\mathrm{V.p.}}{2\pi}\int_{-K'}^{K'} \mathrm{d}v[M(\mathrm{i}v,\nu) - M(\mathrm{i}v - K,\nu)], \tag{9.3.9}$$
$$M(\mathrm{i}v,\nu) = \big[Y(\mathrm{i}v,\nu)F_-(\mathrm{i}v) - Y(2K - 2K'\mathrm{i} - \mathrm{i}v,\nu)\sigma_2 F_-^*(\mathrm{i}v)\sigma_2\big]G(\mathrm{i}v)$$
$$+ \big[Y(2K + \mathrm{i}v,\nu)\sigma_3 F_-(\mathrm{i}v)\sigma_3 - Y(-2K'\mathrm{i} - \mathrm{i}v,\nu)\sigma_1 F_-^*(\mathrm{i}v)\sigma_1\big]\sigma_3 G(\mathrm{i}v)\sigma_3.$$

The known values of the function $F_-(u')$ in the segments γ allow us to compute the values of $F_+(u)$ and $F_-(u)$ in the domains D_+ and D_-, respectively, exploiting the Cauchy formula (9.3.7). Once derived, the functions $F_\pm(u)$ solve the sine-Gordon model (see the last of formulas (9.3.1)) to describe spin waves in a helicoidal structure with dispersion effects dominating over nonlinearity effects.

For clarity, we should discuss a simple example. Let $|b(u)| \ll 1$ ($u \in \gamma$). Then the method of successive approximations enables us to find an approximate solution to eq. (9.3.9): $F_-(\nu) = I + O(b)$. In a linear-in-$|b|$ approximation, we have

$$\tilde{F}(u) = I + \frac{1}{2\pi\mathrm{i}}\int_\gamma \mathrm{d}u' Y(u',u)G(u') + O(b^2).$$

Given that

$$Y(\mathrm{i}v,\mathrm{i}K') - Y(2K - 2K'\mathrm{i} - \mathrm{i}v,\mathrm{i}K') = \mathrm{i}\,\mathrm{dn}(\mathrm{i}v,k),$$
$$G_{11}(\mathrm{i}v) = \frac{(\mathrm{dn}(\mathrm{i}v) - 1)}{\mathrm{dn}(\mathrm{i}v)}\,\mathrm{Re}\big[(\Psi_0(\mathrm{i}v))_{11}(\Psi_0(\mathrm{i}v))_{21}b^*(\mathrm{i}v)\big],$$

from (9.3.1) we arrive at

$$\Phi - \varphi_0 \approx \frac{4}{\pi}\int_{-K'}^{K'} \mathrm{d}v\,\big(\mathrm{Re}\big[(\Psi_0(\mathrm{i}v))_{11}(\Psi_0(\mathrm{i}v))_{21}b^*(\mathrm{i}v)\big](\mathrm{dn}(\mathrm{i}v) - 1)$$
$$+ \mathrm{Re}\big[(\Psi_0(\mathrm{i}v - K))_{11}(\Psi_0(\mathrm{i}v - K))_{21}b^*(\mathrm{i}v - K)\big]\,(\mathrm{dn}(\mathrm{i}v - K) - 1)\big). \tag{9.3.10}$$

It is interesting that

$$(\mathrm{dn}u - 1)(\Psi_0(u))_{11}(\Psi_0(u))_{21} = \Lambda(\chi, u)\exp(\mathrm{i}\,t\,\mathrm{dn}(u,k)/k),$$

where

$$\begin{aligned}\Lambda(\chi, u) = {} & \frac{\mathrm{i}}{2}\frac{|\sigma(\mathrm{i}K')|^2\sigma(K)\sigma(K+2\mathrm{i}K')}{\sigma^2(K+\mathrm{i}K')\sigma(u-\mathrm{i}K')\sigma(u+\mathrm{i}K')}\frac{\sigma(\chi+u)\sigma(\chi+2\mathrm{i}K'+u)}{\sigma(\chi-\mathrm{i}K')\sigma(\chi+\mathrm{i}K')} \\ & \times\exp(-\eta_3(\chi+\mathrm{i}K'+u)-\chi[\zeta(u+\mathrm{i}K')+\zeta(u-\mathrm{i}K')])\end{aligned}$$

is the solution to the Lame equation [1] written in terms of the Weierstrass sigma functions with periods $[2K, 4\mathrm{i}K']$. Linearizing the sine-Gordon equation regarding the background structure $\varphi_0(\chi, k)$ generates Lamé's equation

$$\begin{aligned}\left(\partial_t^2 - \frac{1}{k^2}[\partial_\chi^2 - 2k^2\mathrm{sn}^2\chi + k^2]\right)(\Phi - \varphi_0) = 0, \\ [\partial_\chi^2 - 2k^2\mathrm{sn}^2\chi + k^2]\Lambda = -\varepsilon\Lambda.\end{aligned}$$

The two terms on the right-hand side of (9.3.10) correspond to two branches of the spectrum of spin waves in the helicoidal structure [42]. The quasimomentum $p_\Lambda(u)$ of the Lamé function differs from the quasimomentum $p(u)$ of the Jost solutions:

$$p_\Lambda(u) = 2p(u) + \pi/(2K).$$

The frequency ω of the spin-wave modes correlates to the spectrum ε of the Lamé equation as follows:

$$\omega = \sqrt{\varepsilon}/k.$$

It is worth emphasizing that this relation has lost the multiplier k^{-1} in [17]. For the activation and gapless branches, we have

$$\sqrt{\varepsilon_{\mathrm{opt}}} = \mathrm{dn}(\mathrm{i}v, k) = \mathrm{sn}^{-1}(v + K', k')$$

and

$$\sqrt{\varepsilon_{\mathrm{a}}} = \mathrm{dn}(\mathrm{i}v - K, k') = k'\mathrm{sn}(v + K', k'),$$

respectively.

The periodicity of a spiral structure brings some features into spectra of spin waves, similar to those observed in the spectrum of conduction electrons in the periodic potential of a crystal lattice. As for electrons, the spectrum of spin waves turns out to be "striped". However, unlike the spectrum of electrons, the spectrum of spin waves in the helicoidal structure has only two bands. This is related to a specific form of the kink lattice potential.

The squared modulus of the Lamé functions describing low-amplitude gapless modes is large in regions where the domain walls of the structure are localized. In this connection, such types of excitations are called "intraboundary" spin waves in the theory of stripe domain structures. On the contrary, the squared modulus of the Lamé functions for activation spin excitations is large inside the domains. Therefore such excitations are often called "intradomain" spin waves [15].

The considered problems, in principle, make it possible to trace the evolution of any initial distribution of magnetization in a helicoidal structure under preassigned boundary conditions. In this case an algorithm for the sine-Gordon model to be integrated consists in the following:

1. Relying on a preset initial distribution of magnetization, we find the coefficients $a(u)$ and $b(u)$ of a transition matrix.
2. We extract a soliton part from the function $a(u) = a_{\mathrm{sol}}(u)a_0(u)$ Using $a_{\mathrm{sol}}(u)$ of eq. (9.2.1) and the Riemann problem with zeros, we construct soliton solutions $\Psi_{\pm}^{(\mathrm{sol})}(u,\chi,t)$ and $\Phi_{\mathrm{sol}}(z,t)$ to systems (9.1.11) and sine-Gordon models.
3. A complete solution to system (9.1.11) needs to be sought in the form

$$\begin{aligned}\Psi_+(u) &= \exp\left(-\frac{\mathrm{i}\sigma_3}{4}(\Phi - \Phi_{\mathrm{sol}})\right)F_+(u)\Psi_+^{(\mathrm{sol})}(u),\\ \Psi_-(u) &= \exp\left(-\frac{\mathrm{i}\sigma_3}{4}(\Phi - \Phi_{\mathrm{sol}})\right)F_-(u)\Psi_-^{(\mathrm{sol})}(u)a_0(u).\end{aligned}$$

Then the functions $F_{\pm}(u)$ will be doubly periodic with periods $[4K, 4\mathrm{i}K']$, analytic in the domains $D_{\pm}$, and continuous in them up to the boundary. The functions $F_{\pm}(u)$ contain no peculiarities associated with solitons and a helicoidal structure:

$$\det F_+(u) = a_0^*(-u^*), \quad \det F_-(u) = 1/a_0(u).$$

They satisfy the normalization conditions

$$\begin{aligned}F_{\pm}(-\mathrm{i}K') &= F_{\pm}(-\mathrm{i}K' + 2K) = I,\\ F_{\pm}(\mathrm{i}K') &= F_{\pm}(\mathrm{i}K' + 2K) = \exp\left(\frac{\mathrm{i}\sigma_3}{2}(\Phi - \Phi_{\mathrm{sol}})\right).\end{aligned} \tag{9.3.11}$$

In view of the ratios

$$\left(\Psi_+^{(\mathrm{sol})}(u)\right)^{-1} = -\frac{(\Psi_-^{(\mathrm{sol})}(-u^*))^\dagger}{a_{\mathrm{sol}}^*(-u^*)\det M}, \quad a_{\mathrm{sol}}^*(-u^*)a_{\mathrm{sol}}^*(u) = 1,$$

calculation of these functions is boiled down to solving the regular Riemann problem

$$F_+(u) = F_-(u)(I - G(u)), \quad u \in \gamma,$$

where

$$G(u) = \Psi_-^{(\mathrm{sol})}(u)\begin{pmatrix} 0 & b^*(-u^*) \\ b(u) & 0 \end{pmatrix}(\Psi_-^{(\mathrm{sol})}(-u^*))^\dagger \frac{(\mathrm{dn}u - 1)}{\mathrm{dn}u}.$$

Thus we have deduced a solution to this problem. Using the second relation of (9.3.11) and the function $F_\pm(u)$, we find a solution to the initial-boundary value problem for the original sine-Gordon model.

9.4 Dispersion relations on a torus and a spectrum of integrals of motion

Whenever a new particle or fact is found, I notice that all theorists are either united or undergone "dispersion".

R. Feynman

The element $a(u)$ of the transition matrix is time-independent and therefore can be used as a generating function of integrals of motion for nonlinear excitations against the background of a superstructure. The most informative notation of the integrals of motion follows from the dispersion relations for the function $a(u)$.

The function $a(u)$ is analytic in the domain D_-. Let us extract its soliton part:

$$a(u) = a_{\mathrm{sol}}(u)a_0(u). \tag{9.4.1}$$

The multiplier a_{sol} (9.2.1) depends on the boundary conditions (9.1.9) and (9.1.10) for the sine-Gordon model. In both cases the function $a_0(u)$ in eq. (9.4.1) is not only analytic in the domain D_- but also does not vanish at any point within it. Transformation properties for the functions $a(u)$ and $a_{\mathrm{sol}}(u)$ dictate those of $a_0(u)$:

$$a_0(u \pm 2K) = a_0(u), \quad a_0^*[(u \pm 2\mathrm{i}K')^*] = a_0(u), \quad a_0(\pm \mathrm{i}K') = 1. \tag{9.4.2}$$

The function $\bar{a}_0(u) = a_0^*(-u^*)$ is analytic in the domain D_+. Constructing the functions $\ln a_0(u)$ and $\ln \bar{a}_0(u)$ is reduced to solving the following scalar Riemann problem on a torus: Find the doubly periodic-in-mod $4\pi\mathrm{i}$ functions $\ln a_0(u)$ and $\ln \bar{a}_0(u)$ with periods $[2K, 4\mathrm{i}K']$ that are analytic in the domains D_- and D_+, respectively (see Figure 9.9) and on the contour $\Gamma = \Gamma_2 \cup \Gamma_3$ satisfy (see Figure 9.9)

(1) the conjugation conditions

$$\ln a_0(u) - \ln[\bar{a}_0(u)]^{-1} = \ln[1 - b(u)\bar{b}(u)], \quad u \in \Gamma, \tag{9.4.3}$$

where $\bar{b}(u) = b^*(-u^*)$;

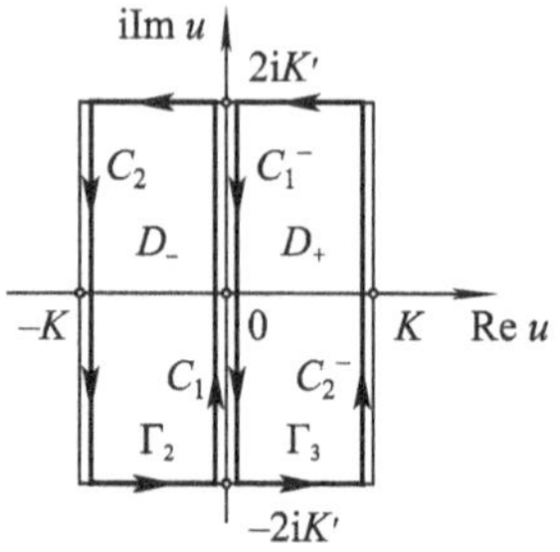

Figure 9.9: Domains D_- and D_+ and a contour $\Gamma = \Gamma_2 \cup \Gamma_3$ bounding them in the complex u-plane.

(2) the reductions

$$\begin{aligned} &\ln a_0(u \pm 2K) = \ln a_0(u), \\ &\ln a_0^*[(u \pm 2\mathrm{i}\,K)^*] = \ln a_0(u), \quad \bmod 4\pi\mathrm{i}, \quad u \in D_-; \\ &\bar{a}_0(u) = a_0^*(-u^*), \quad u \in D_+; \end{aligned} \tag{9.4.4}$$

and

(3) the normalization conditions

$$\ln a_0(u = \pm\mathrm{i}\,K') = 0, \quad \bmod 2\pi\mathrm{i}. \tag{9.4.5}$$

Next, we incorporate the piecewise analytic function

$$\Phi(u) = \begin{cases} \ln a(u), & u \in D_-, \\ \ln[\bar{a}(u)]^{-1}, & u \in D_+. \end{cases}$$

Its values at the internal points of the domains D_- and D_+ are expressed in terms of conditions (9.4.3) on the contour $\Gamma = \Gamma_2 \cup \Gamma_3$. As a Cauchy kernel, we use a combination of Weierstrass zeta functions with periods $2K$ and $4\mathrm{i}\,K'$:

$$S(u', u) = \zeta(u' - u) - \zeta(u' + \mathrm{i}\,K').$$

This combination is periodic in the variable u' with periods $2K$ and $4\mathrm{i}\,K'$ and satisfies the condition $S(u', u = -\mathrm{i}\,K') = 0$. Although $S(u', u)$ is not a doubly periodic-in-u function, the integral representation

$$\Phi(u) = \frac{1}{2\pi\mathrm{i}} \int_\Gamma \mathrm{d}u'\,\Phi(u')S(u', u) - \frac{2\eta_1}{K}(u + \mathrm{i}\,K')Q \tag{9.4.6}$$

provides a solution to the Riemann problems (9.4.3)–(9.4.5) under an appropriate choice of constant Q. Let us justify this statement.

The integrand in (9.4.6) has the same values in the horizontal segments of the contour Γ. These segments' contributions to the integral are compensated for, since they are traversed in opposite directions when calculating the integral over the parameter u. Therefore an integration only over the vertical segments $C_1 = \{u' = \mathrm{i}\,v, -2K' \le v \le 2K'\}$ and $C_2 = \{u' = \mathrm{i}\,v - K, -2K' \le v \le 2K'\}$ (Figure 9.9) can be performed; their banks are traversed in opposite directions. Appropriate contributions adopt a shape of integrals along the lines $C_{1,2}$ of the function $\Phi(u)$ jumps (9.4.3).

On the segments C_1 and C_2, the identities

$$\begin{aligned}
&\ln a_0(\mathrm{i}\,v) - \ln[\bar{a}_0(\mathrm{i}\,v)]^{-1} = \ln|a_0(\mathrm{i}\,v)|^2 \\
&\quad = \ln[1 - |b(\mathrm{i}\,v)|^2] \equiv -\frac{\pi k}{4} n_{\mathrm{opt}}(v) < 0, \\
&\ln a_0(\mathrm{i}\,v - K) - \ln[\bar{a}_0(\mathrm{i}\,v - K)]^{-1} = \ln|a_0(\mathrm{i}\,v - K)|^2 \\
&\quad = \ln[1 + |b(\mathrm{i}\,v - K)|^2 \exp 2\eta_1\Delta] \equiv \frac{\pi k}{4} n_a(v) > 0, \\
&\int\limits_{-2K'}^{+2K'} n_{\mathrm{opt}}(v) S(\mathrm{i}\,v - u)\mathrm{d}v = \int\limits_{-K'}^{+K'} n_{\mathrm{opt}}(v)[\zeta(\mathrm{i}\,v - u) - \zeta(2\mathrm{i}\,K' + \mathrm{i}\,v + u)]\mathrm{d}v, \\
&\int\limits_{-2K'}^{+2K'} n_a(v) S(\mathrm{i}\,v - K - u)\mathrm{d}v = \int\limits_{-K'}^{+K'} n_a(v)[\zeta(\mathrm{i}\,v - K - u) - \zeta(2\mathrm{i}\,K' + \mathrm{i}\,v - K + u)]\mathrm{d}v
\end{aligned}$$

are consequences of relations (9.4.3), (9.4.4), and (9.1.16). Given the above remarks, we specify representation (9.4.6):

$$\begin{aligned}
\Phi(u) = \frac{k}{8} \int\limits_{-K'}^{+K'} &\{n_{\mathrm{opt}}(v)[\zeta(2\mathrm{i}\,K' + \mathrm{i}\,v + u) - \zeta(\mathrm{i}\,v - u)] \\
&+ n_a(v)[\zeta(2\mathrm{i}\,K' + \mathrm{i}\,v - K + u) - \zeta(\mathrm{i}\,v - K - u)]\}\mathrm{d}v \\
&- \frac{2\eta_1}{K}(u + \mathrm{i}\,K')Q.
\end{aligned} \tag{9.4.7}$$

Under translations over periods $[2K, 4iK']$, the Weierstrass zeta function is quasiperiodic:

$$\zeta(u \pm 2K) = \zeta(u) \pm 2\eta_1, \quad \zeta(u \pm 4\mathrm{i}\,K') = \zeta(u) \pm 2\eta_3.$$

Its periods and the parameters $\eta_{1,3}$ are related by the Legendre identity [4, 1]

$$4\mathrm{i}K'\eta_1 - 2K\eta_3 = \pi\,\mathrm{i}.$$

This means that for the second of conditions (9.4.5) to be fulfilled, the spectral data of the continuous spectrum and the constant Q must satisfy the condition

$$\frac{k}{8}\int_{-K'}^{+K'}[n_{\text{opt}}(v)+n_a(v)]\,\mathrm{d}v = 2mK = Q, \tag{9.4.8}$$

where $m = 1, 2, \dots$.

Next, we demonstrate that the left-hand side of eq. (9.4.8) determines the total number of spin-wave modes in the domain structure. By direct verification using the Sokhotski–Plemelj formulas it is easy to make sure that formulas (9.4.7) and (9.4.8) furnish a solution to the Riemann problems (9.4.3)–(9.4.5). The above calculation is more accurate than those in [20]. The right-hand side of formula (9.4.7) differs from that given in [20] by an additional summand with parameter m. This circumstance, although important, does not change the spectral expansions of the integrals of motion found in [20] for solitons and waves in the domain structure.

Ultimately, we get an integral representation for the function $a(u)$:

$$\begin{aligned} a(u) = a_{\text{sol}}(u)\exp\Bigg(&\frac{k}{8}\int_{-K'}^{+K'}\{n_{\text{opt}}(v)[\zeta(2\mathrm{i}K'+\mathrm{i}v+u)-\zeta(\mathrm{i}v-u)] \\ &+ n_a(v)[\zeta(2\mathrm{i}K'+\mathrm{i}v-K+u)-\zeta(\mathrm{i}v-K-u)]\}\,\mathrm{d}v \\ &- 4m\eta_1(u+\mathrm{i}K')\Bigg), \end{aligned} \tag{9.4.9}$$

$$\frac{k}{8}\int_{-K'}^{+K'}[n_{\text{opt}}(v)+n_a(v)]\,\mathrm{d}v = 2mK, \quad m = 1, 2, \dots.$$

The expressions for the system densities of field momentum p and energy w have the form

$$p = \partial_t\Phi\,\partial_z\Phi, \quad w = \frac{1}{2}[(\partial_t\Phi)^2 + (\partial_z\Phi)^2] + q\partial_z\Phi + (1-\cos\Phi).$$

When writing the integrals of motion, we should separate the contribution of the inhomogeneous ground state of the medium (it is infinite for an infinite spiral structure). The stability conditions for the ground state usually require calculating the period-averaged energy per unit length of the structure. However, when excited, the helicoidal structure undergoes translations and loses the periodicity. Therefore, correctly defining the integrals of motion for excitations against the background of the structure is too involved.

Let us show that the usage of the function $a(u)$ as a generating function of the integrals of motion makes it possible to overcome this difficulty. The most interesting integrals of motion are obtained by expanding $\ln a(u)$ near the singular points $u = \pm\mathrm{i}K'$ of system (9.1.11). According to representation (9.1.21), series for $a(u)$ can be found using asymptotic expansions of the Jost solution $\Psi_1(u)$ near the indicated points. We first obtain an asymptotic series for $\Psi_1(u)$ at $u = \mathrm{i}K' + \varepsilon$ ($|\varepsilon| \ll 1$) in the problem with boundary conditions (9.1.10) when

$$\Psi_1 \to \Psi_1^{(0)} = \Psi_\Delta(u, \chi, t) \quad \text{as} \quad z \to -\infty.$$

Both the definition of the function $\varphi_0(\chi)$,

$$\begin{aligned}&\exp\left[\frac{\mathrm{i}\sigma_3}{4}(\pi + \varphi_0(\chi))\right]\\ &= \operatorname{diag}\left(\sqrt{\frac{\sigma(\chi + \mathrm{i}K')}{\sigma(\chi - \mathrm{i}K')}}\exp\left(-\frac{\eta_3\chi}{2}\right), \sqrt{\frac{\sigma(\chi - \mathrm{i}K')}{\sigma(\chi + \mathrm{i}K')}}\exp\left(\frac{\eta_3\chi}{2}\right)\right),\end{aligned}$$

and identities (9.2.8) imply the relation

$$\begin{aligned}\Psi_\delta = {}&\exp\left[\frac{\mathrm{i}}{4}(\varphi_0(\chi + \delta) + \pi)\sigma_3\right]\\ &\times\begin{pmatrix}\frac{\sigma(\chi+\delta+u)}{\sigma(\chi+\delta+\mathrm{i}K')} & \mathrm{i}\,\frac{\sigma(\chi+\delta+2\mathrm{i}K'-u)}{\sigma(\chi+\delta+\mathrm{i}K')}\,e^{\eta_3(u-\mathrm{i}K')}\\ \mathrm{i}\,\frac{\sigma(\chi+\delta+u-2\mathrm{i}K')}{\sigma(\chi+\delta-\mathrm{i}K')}\,e^{\eta_3(u-\mathrm{i}K')} & \frac{\sigma(\chi+\delta-u)}{\sigma(\chi+\delta-\mathrm{i}K')}\end{pmatrix}\\ &\times\frac{|\sigma(\mathrm{i}K')|}{\sqrt{2}\sigma(u)}\exp\left(-\frac{\chi\sigma_3}{2}[\zeta(u + \mathrm{i}K') + \zeta(u - \mathrm{i}K') - \eta_3] + \frac{\eta_3\delta}{2}\sigma_3 + \frac{\mathrm{i}\,t\,\mathrm{dn}u}{2k}\sigma_3\right).\end{aligned}$$

The latter is convenient for the function Ψ_δ to be further calculated: The function Ψ_δ ($\varphi_0(\chi + \delta)$) coincides with $\Psi_1^{(0)}$ ($\varphi_1^{(0)}$) for $\delta = \Delta$ and $\Psi_2^{(0)}$ ($\varphi_2^{(0)}$) for $\delta = 0$.

Next, we asymptotically expand $\Psi_1(u)$ so that, when constructed, series in powers of ε for $a(u)$ should leave out the extra factors in representation (9.1.21). In doing so, we seek Ψ_1 in the form

$$\Psi_1 = \exp\left(\frac{\mathrm{i}\,\Phi\,\sigma_3}{4}\right)g\tilde{\Psi}_1^{(0)}, \quad \tilde{\Psi}_1^{(0)} \equiv \exp\left(-\frac{\mathrm{i}\varphi_1^{(0)}\sigma_3}{4}\right)\Psi_1^{(0)}.$$

Then from eq. (9.1.11) an equation for calculating g follows:

$$\partial_z g = Lg - gL_0. \tag{9.4.10}$$

Here L_0 comes from L by the replacement $\Phi \to \varphi_1^{(0)}$ ($\partial_t\varphi_1^{(0)} = 0$),

$$L = \frac{\mathrm{i}}{4}[(\partial_t\Phi - \partial_z\Phi)\sigma_3 + (w_2 - w_1)\sigma_1 + (w_1 + w_2)(\cos\Phi\,\sigma_1 + \sin\Phi\,\sigma_2)].$$

The matrix function g must satisfy the condition $g \to I$ as $z \to -\infty$. A solution to eq. (9.4.10) should be sought in the form of a series in powers of $\varepsilon = u - \mathrm{i}K'$:

$$g = I + \sum_{n=1}^{+\infty} g_n \varepsilon^n,$$

where $g_n \to 0$ as $z \to -\infty$. Using an expansion of g_n in the Pauli matrix, we find

$$\Psi_1 = \exp\left(\frac{\mathrm{i}\Phi\sigma_3}{4}\right)[I + \varepsilon(b_1\sigma_1 + c_1\sigma_2) + O(\varepsilon^2)]\tilde{\Psi}_1^{(0)},$$

where

$$b_1 = \frac{k}{8}\int_{-\infty}^{z} \mathrm{d}z'\left(\cos\varphi_1^{(0)} - \cos\Phi + \frac{1}{2}[(\partial_t\Phi - \partial_{z'}\Phi)^2 - (\partial_{z'}\varphi_1^{(0)})^2]\right),$$

$$c_1 = \frac{k}{4}[\partial_t\Phi - \partial_z(\Phi - \varphi_1^{(0)})].$$

To expand the function $a(u)$ in powers of ε, it suffices to expand the pre-exponential factors in the expressions for $(\Psi_2^{(0)})^{-1}$ and $\Psi_1^{(0)}$ in eq. (9.1.21). This is achieved through the formula

$$\sigma(a + \varepsilon) = \sigma(a)[1 + \zeta(a)\varepsilon + O(\varepsilon^2)].$$

Simple calculations yield

$$\begin{aligned} a(\mathrm{i}K' + \varepsilon) &\approx \exp\left(\frac{\eta_3\Delta}{2}\right)\left(1 + \varepsilon \lim_{z\to+\infty}[f(\chi) + b_1(z)]\right), \\ f(\chi) &= \frac{1}{2}\,[\zeta(\chi + \Delta + \mathrm{i}K') + \zeta(\chi + \Delta - \mathrm{i}K') \\ &\quad - \zeta(\chi + \mathrm{i}K') - \zeta(\chi - \mathrm{i}K')]\,. \end{aligned} \tag{9.4.11}$$

The above procedure ensures the convergence of expression (9.4.11) because

$$\begin{aligned} &\frac{k}{8}\int_{-\infty}^{z} \mathrm{d}z'\left(\cos\varphi_1^{(0)} - \cos\varphi_2^{(0)} + \frac{1}{2}[(\partial_{z'}\varphi_2^{(0)})^2 - (\partial_{z'}\varphi_1^{(0)})^2]\right) \\ &= -f(\chi) + \text{const}. \end{aligned}$$

In further calculations, we use the identity

$$\zeta(u) = \frac{Z(u)}{2} + \frac{\mathrm{cn}u}{2\,\mathrm{sn}u}(1 + \mathrm{dn}u) + \frac{\eta_1 u}{K}, \tag{9.4.12}$$

which is proved by expanding in the poles of its right-hand side. The condition $\zeta(K) = \eta_1$ fixes an additive constant, and we get

$$f(\chi) = \frac{1}{2}\left[Z(\chi + \Delta) - Z(\chi) + \frac{2\eta_1\Delta}{K}\right].$$

Let the length of the magnetic spiral be equal to $|z| \leq R = 2kKn$ ($n \gg 1$ is an integer). Then eq. (9.4.11) takes the form

$$a(\mathrm{i}K' + \varepsilon) = \exp\Bigg(\frac{\eta_3\Delta}{2} + \frac{\varepsilon}{2}\bigg[Z(\Delta) + \frac{2\eta_1\Delta}{K}\bigg] + \frac{k\,\varepsilon}{8}\int_{-R}^{R} \mathrm{d}z\bigg[\cos\varphi_1^{(0)} - \cos\Phi + \frac{1}{2}\big((\partial_t\Phi - \partial_z\Phi)^2 - (\partial_z\varphi_1^{(0)})^2\big)\bigg] + O(\varepsilon^2)\Bigg).$$

In the vicinity of the point $u = -\mathrm{i}K'$, the asymptotic expansion of the function Ψ_1 should be sought in the form

$$\Psi_1 = \exp\bigg(-\frac{\mathrm{i}\Phi\sigma_3}{4}\bigg) g \exp\bigg(\frac{\mathrm{i}\varphi_1^{(0)}\sigma_3}{4}\bigg)\Psi_1^{(0)}.$$

Similar calculations give

$$a(-\mathrm{i}K' + \varepsilon) = \exp\Bigg(-\frac{\eta_3\Delta}{2} + \frac{\varepsilon}{2}\bigg[Z(\Delta) + \frac{2\eta_1\Delta}{K}\bigg] + \frac{k\,\varepsilon}{8}\int_{-R}^{R} \mathrm{d}z\bigg[\cos\varphi_1^{(0)} - \cos\Phi + \frac{1}{2}\big((\partial_t\Phi + \partial_z\Phi)^2 - (\partial_z\varphi_1^{(0)})^2\big)\bigg] + O(\varepsilon^2)\Bigg).$$

The energy of localized nonlinear excitations is thought to be the difference between the total system energy and the background structure energy:

$$\begin{aligned} H &= \frac{4}{k}\bigg(Z(\Delta) + \frac{2\eta_1\Delta}{K}\bigg) + \int_{-R}^{R} \mathrm{d}z(\cos\varphi_1^{(0)} - \cos\Phi \\ &\quad + \frac{1}{2}[(\partial_t\Phi)^2 + (\partial_z\Phi)^2 - (\partial_z\varphi_1^{(0)})^2] + q(\partial_z\Phi - \partial_z\varphi_1^{(0)})) \\ &= \frac{4}{k}\frac{\mathrm{d}}{\mathrm{d}\varepsilon}\ln[a(\mathrm{i}K' + \varepsilon)a(-\mathrm{i}K' + \varepsilon)]\big|_{\varepsilon=0} + \frac{8E}{\pi k}\,\mathrm{am}(\Delta, k). \end{aligned} \tag{9.4.13}$$

The last summand in eq. (9.4.13) accounts for the explicit form of the functions $\varphi_{1,2}^{(0)}$ and relation (9.1.7): $q = 4E/(\pi k)$. Recall that mod$(4K)$ defines the dilatation Δ of the domain structure; therefore the right-hand side of formula (9.4.13) omits the summand proportional to m henceforth. Then we arrive at

$$\begin{aligned} &\frac{4}{k}\frac{\mathrm{d}}{\mathrm{d}\varepsilon}\ln[a(\mathrm{i}K' + \varepsilon)a(-\mathrm{i}K' + \varepsilon)] \\ &\quad = \int_{-K'}^{K'} \mathrm{d}v[\varepsilon_{\mathrm{a}}(v)n_{\mathrm{a}}(v) + \varepsilon_{\mathrm{opt}}(v)n_{\mathrm{opt}}(v)] \end{aligned}$$

$$- \frac{8}{k}\left(\sum_{p=1}^{2M} \mathrm{Re}\left[Z(\nu_p) + \frac{2\eta_1 \nu_p}{K}\right] + 2\sum_{s=1}^{N} \mathrm{Re}\left[Z(\mu_s) + \frac{2\eta_1 \mu_s}{K}\right]\right). \tag{9.4.14}$$

The quantities ε_{a} (n_{a}) and $\varepsilon_{\mathrm{opt}}$ (n_{opt}) handle the spectrum (density) of quasiacoustic and quasioptical spin waves in the spiral structure:

$$\begin{aligned} \varepsilon_{\mathrm{a}}(\nu) &= k'^2 \mathrm{sn}^2(\nu + K', k'), \\ n_{\mathrm{a}}(\nu) &= \frac{4}{\pi k} \ln(1 + |b(-K + \mathrm{i}\nu)|^2 e^{2\eta_1 \Delta}) > 0, \\ \varepsilon_{\mathrm{opt}}(\nu) &= \frac{1}{\mathrm{sn}^2(\nu + K', k')}, \\ n_{\mathrm{opt}}(\nu) &= -\frac{4}{\pi k} \ln(1 - |b(\mathrm{i}\nu)|^2) > 0. \end{aligned} \tag{9.4.15}$$

When solitons arise, the superstructure is compressed along the Oz-axis by a magnitude of $k\Delta$. This circumstance manifests itself in changes in the density of quasiacoustic magnons and in the energy of the superstructure. The last summand in eq. (9.4.13) takes it into account. The discrete terms in eq. (9.4.14) correspond to the energies of kinks and breathers in the superstructure.

The field pulse of solitons and spin-wave modes in the problem with boundary conditions (9.1.10) is determined by the expression

$$P = \int_{-R}^{R} \mathrm{d}z\, \partial_t \Phi\, \partial_z \Phi = \frac{4}{k} \frac{\mathrm{d}}{\mathrm{d}\varepsilon} \ln \left. \frac{a(-\mathrm{i}K' + \varepsilon)}{a(\mathrm{i}K' + \varepsilon)} \right|_{\varepsilon=0}.$$

Hence, resorting to eq. (9.4.12) and the identity

$$\zeta(\mu - \mathrm{i}K') - \zeta(\mu + \mathrm{i}K') + \eta_3 = \mathrm{i}\, \mathrm{dn}\mu,$$

we find

$$\begin{aligned} P = &\int_{-K'}^{K'} \mathrm{d}\nu [p_{\mathrm{a}}(\nu) n_{\mathrm{a}}(\nu) + p_{\mathrm{opt}}(\nu) n_{\mathrm{opt}}(\nu)] \\ &- \frac{8}{k}\left(\sum_{p=1}^{2M} \mathrm{Im}(\mathrm{dn}\nu_p) + 2\sum_{s=1}^{N} \mathrm{Im}(\mathrm{dn}\mu_s)\right), \end{aligned} \tag{9.4.16}$$

where p_{a} and p_{opt} are the momenta of the gapless and activation spin-wave modes, respectively:

$$\begin{aligned} p_{\mathrm{a}}(\nu) &= k' \mathrm{dn}(\nu + K', k') \mathrm{cn}(\nu + K', k'), \\ p_{\mathrm{opt}}(\nu) &= -\frac{\mathrm{dn}(\nu + K', k') \mathrm{cn}(\nu + K', k')}{\mathrm{sn}^2(\nu + K', k')}. \end{aligned}$$

In the spectral expansions of the integrals of motion, we have succeeded in completely separating the contributions of two independent types of collective excitations against the background of a helicoidal structure, such as solitons and spin waves. The dispersion laws for spin waves in helicoidal structures are usually written in terms of the Bloch quasimomentum of the Lamé functions [17]. The appropriate dependencies turn out to be transcendental. The findings secured show that the quasimomentum of solutions to the Lamé equation differs from the field momentum of spin waves. In terms of the field momentum of spin waves, the dispersion laws for spin-wave modes take on a simple form and turn out to be algebraic (Figure 9.10):

$$\varepsilon_a(p_a) = \frac{1}{2}(1 + k'^2 - \sqrt{k^4 + 4p_a^2}), \quad |p_a| \le k',$$
$$\varepsilon_{opt}(p_{opt}) = \frac{1}{2}(1 + k'^2 + \sqrt{k^4 + 4p_{opt}^2}), \quad -\infty \le p_{opt} \le +\infty.$$

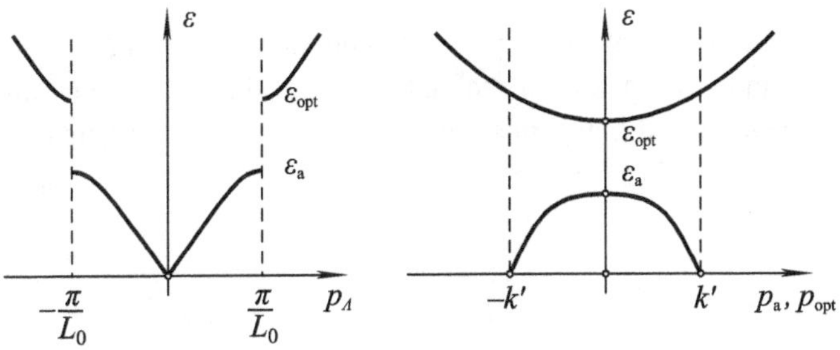

Figure 9.10: Dispersion laws for spin waves in a helicoidal structure in terms of the quasimomentum of the Lamé functions (left) and the magnon momentum (right).

Evaluating the integrals of motion in the problem with boundary conditions (9.1.9) differs little from the above calculations. In this case the helical structure contains an odd number of kinks; therefore, after the substitutions $2M \to 2M + 1$ and $\mathrm{am}\Delta \to \mathrm{am}\Delta - \pi\sigma$, eqs. (9.4.13), (9.4.14), and (9.4.16) produce the final formulas for the energy and momentum of collective excitations.

At the end of the chapter, let us summarize its results so that the necessary but rather tedious calculation part does not obscure the results obtained.

Having solved the initial-boundary value problem for model (9.1), we have discovered that there are two fundamentally different classes, into which potential nonlinear excitations of a helical magnetic structure can fall. The former includes previously unstudied structurally stable solitons, superfluous kinks and breathers in the superstructure kink lattice. Local translations of the helical (strip domain) structure always accompany their formation and motion. Formulas have been found to relate the parameters of solitons to shifts in the helical structure. These outcomes can be used to diagnose solitons or to design switching devices.

Magneto-optical techniques can be exploited to visualize macroscopic shifts of a spiral structure caused by solitons traveling through it. Near the incommensurability/commensurability phase transition, we can regulate the velocity of kinks in the spiral structure by manipulating the temperature of the sample or applying an external magnetic field perpendicular to the spiral axis. This allows for the conditions for their observation to be optimized.

A resonant absorption of microwave power at the characteristic frequencies of the internal modes of breathers can be used to identify them within a spiral structure.

The second class of nonlinear excitations consists of spin-wave pulses that emerge without accompanying macroscopic shifts in the background structure. Regarding the inhomogeneous medium state corresponding to an equilibrium spiral (stripe domain) structure, the magnetization in pulses performs nonlinear oscillations. Even initially localized pulses are spread over time due to dispersion effects, unlike solitons, because the intensity of interactions in pulses is not sufficient for their stability. We have shown that low-amplitude pulses are theoretically described as superpositions of well-known linear spin waves in a spiral structure.

These collective excitations constitute a complete set of nonlinear modes of a modulated structure. Integrals of motion of any localized magnetization distribution in a spiral structure are expressed as sums of independent contributions only of these modes.

A Appendix

A.1 Technique for working with elliptic functions

They all went off to discover the Pole... It's just a thing you discover.

A. Milne, "Winnie the Pooh"

The book [1] sets forth circumstantially the basic principles of the theory of elliptic functions. We roll a number of useful conclusions from this theory.

As Jacobi's theorem states, there is no single-valued nonconstant analytic function with $n \geq 3$ primitive periods. With two preset primitive periods 2ω and $2\omega'$, such a function exists if and only if

$$\operatorname{Im}\left(\frac{\omega'}{\omega}\right) \neq 0.$$

Elliptic functions are thought to be doubly periodic functions $f(u)$ whose poles in any finite part of the u-plane can be the only singularities.

When studying an elliptic function $f(u)$, we can limit ourselves to considering it in some parallelogram of periods; only one of its two parallel sides belongs to the parallelogram.

The sum of the residues of an elliptic function at all poles within its period parallelogram is zero.

It should be emphasized that all the poles in the parallelogram, including those that lie on its boundary, need to be taken.

We say that two points u_1 and u_2 are comparable in a period modulo (or equivalent) if

$$u_1 - u_2 = 2m\omega + 2m'\omega',$$

where m and m' are the integers, and we write

$$u_1 = u_2 \quad \operatorname{mod}(2\omega, 2\omega').$$

Next, we call a-points the points u at which the function $f(u)$ takes the value a, i. e., $f(u) = a$.

The following fundamental Liouville theorems are valid for elliptic functions:

1. A correctly counted (i. e., counted with multiplicity) number of poles of a nonconstant elliptic function inside a period parallelogram is equal to a correctly counted number of a-points, whatever a is.
2. There is no nonconstant elliptic function that is regular inside a period parallelogram.

https://doi.org/10.1515/9783112235973-010

3. The number of poles of an elliptic function inside a period parallelogram, counted with multiplicity (it is called the order of an elliptic function), cannot be less than two.
4. The sum of α-points of a function $f(u)$ for any α is comparable in a period modulo with the sum of the poles of this function if all the α-points belonging to the same period parallelogram are considered.

All theoretical constructions related to elliptic functions are easier to carry out in terms of the Weierstrass functions. Let us give their definitions and basic properties.

The Weierstrass sigma function is defined by the infinite product

$$\sigma(u) = u \prod_{m,m'}{}' \left(1 - \frac{u}{s}\right) \exp\left[\frac{u}{s} + \frac{u^2}{2s^2}\right], \qquad s = 2m\omega + 2m'\omega'.$$

The product applies to all integers m and m' except the pair $m = m' = 0$. The stroke in the sign of the product reflects this circumstance.

As we can see, $\sigma(u)$ is an entire function having only simple zeros lying at the vertices of the period grid:

$$u = 2m\omega + 2m'\omega'. \tag{A.1}$$

The sigma function is odd:

$$\sigma(-u) = -\sigma(u).$$

The Weierstrass ζ-function is an analytic function with simple poles:

$$\zeta(u) = \frac{\mathrm{d}}{\mathrm{d}u} \ln \sigma(u) = \frac{1}{u} + \sum_{m,m'}{}' \left(\frac{1}{u-s} + \frac{1}{s} + \frac{u}{s^2}\right),$$

and it is odd:

$$\zeta(-u) = -\zeta(u).$$

The Weierstrass function $\wp(u)$ is defined by the convergent series

$$\wp(u) = -\frac{\mathrm{d}}{\mathrm{d}u}\zeta(u) = \frac{1}{u^2} + \sum_{m,m'}{}' \left[\frac{1}{(u-s)^2} - \frac{1}{s^2}\right],$$

which shows that the second-order poles at the points of (A.1) are the only singularities of the function $\wp(u)$. The latter is even:

$$\wp(-u) = \wp(u).$$

The transformation properties of the Weierstrass functions are usually written in symmetric form using the notations

$$\omega_1 = \omega, \quad \omega_2 = -\omega - \omega', \quad \omega_3 = \omega'; \quad \eta_\alpha = \zeta(\omega_\alpha), \quad \alpha = 1, 2, 3.$$

The constants η_α satisfy the following identities:

$$\eta_1 + \eta_2 + \eta_3 = 0, \quad \eta_1\omega_3 - \eta_3\omega_1 = \mathrm{i}\pi/2. \tag{A.2}$$

The function $\wp(u)$ turns out to be doubly periodic, whereas the functions $\sigma(u)$ and $\zeta(u)$ are quasiperiodic:

$$\begin{aligned} \wp(u + 2\omega_\alpha) &= \wp(u), \\ \zeta(u + 2\omega_\alpha) &= \zeta(u) + 2\eta_\alpha, \\ \sigma(u + 2\omega_\alpha) &= -\sigma(u)\exp[2\eta_u(u + \omega_u)]. \end{aligned} \tag{A.3}$$

The function $\wp(u)$ satisfies the algebraic differential equation

$$\partial_u\wp(u) = -2\sqrt{(\wp(u) - e_1)(\wp(u) - e_2)(\wp(u) - e_3)}. \tag{A.4}$$

Here $e_\alpha = \wp(\omega_\alpha)$ ($\alpha = 1, 2, 3$), and the square root is chosen so that the main part of the function $\partial_u\wp(u)$ near the point $u = 0$ is equal to $-2u^{-3}$.

The square root in the right-hand side of eq. (A.4) is "extracted", because the multipliers $\wp(u) - e_\alpha$ have second-order poles at all points that are equivalent to the point $u = 0$ and second-order zeros at all points that are equivalent to the points $u = \omega_\alpha$.

The numbers e_α satisfy the relation $e_1 + e_2 + e_3 = 0$.

By eq. (A.4) any derivative of the $\wp$-function is expressed rationally through the meromorphic functions $\wp(u)$ and $\partial_u\wp(u)$.

Recall that every rational function $R(z)$ admits the following two representations:

$$R(z) = c\,\frac{(z - b_1)(z - b_2)\dots(z - b_m)}{(z - a_1)(z - a_2)\dots(z - a_n)}, \quad R(z) = E(z) + \sum_{i,k}\frac{A_k^{(i)}}{(z - a_k)^i},$$

where c, b_i, a_k, and $A_k^{(i)}$ are constants, and $E(z)$ is a polynomial, the so-called integer part of the function $R(z)$. Each of these representations carries specific information regarding the function $R(z)$. It is clear that the former shows which zeros and poles of the function $R(z)$ are, and the latter, as always used in integral calculus, provides the main part of the function $R(z)$ for each of its poles.

It is important that any elliptic function allows for similar representations. An elliptic function can be expanded into "multipliers" as

$$f(u) = c_1\,\frac{\sigma(u - b_1)\sigma(u - b_2)\dots\sigma(u - b_n)}{\sigma(u - a_1)\sigma(u - a_2)\dots\sigma(u - a_n)}. \tag{A.5}$$

In formula (A.5), each zero and pole is counted up to its multiplicity. Liouville's Theorems 1 and 4 claim that

$$a_1 + a_2 + \cdots + a_n = b_1 + b_2 + \cdots + b_n \mod(2\omega,\ 2\omega').$$

The expansion of the elliptic function $f(u)$ into "simple fractions" is

$$f(u) = c_2 + \sum_{k=1}^{n} A_k \zeta(u - a_k) + \sum_{k,r\ (r\geq 2)} A_k^{(r-1)} \wp^{(r-2)}(u - a_k). \tag{A.6}$$

In representation (A.6) the numbers A_k and $A_k^{(r-1)}$ characterize the main part of the function $f(u)$ near the pole $u = a_k$:

$$f \approx \frac{A_k}{u - a_k} + \sum_{r=2}^{m_k} \frac{(-1)^r (r-1)!\, A_k^{(r-1)}}{(u - a_k)^r}.$$

Recall once again that the sum of residues of any elliptic function is always equal to zero:

$$\sum_{k=1}^{n} A_k = 0.$$

Formulas (A.5) and (A.6) compute the constants c_1 and c_2 according to the known value of the function $f(u)$ at some point. Note that by using previously known symmetry properties of the function $f(u)$ we can often find the constant c_2 in expansion (A.6).

Representations (A.5) and (A.6) are extremely useful in performing algebraic calculations with elliptic functions. Let us illustrate their effectiveness with specific examples.

We simplify the expression

$$\Delta(u) = \Delta(\lambda(u), \tau(u)) = \frac{\tau - (w_3 \cos\theta_0 - p/2)}{w_1 \sin\theta_0} \tag{A.7}$$

from formulas (3.7.7) as defining the "seeding" functions $\chi_{1,2}^{(0)}$.

It is worth especially emphasizing that a problem about the interaction between a domain wall and a spin wave uses functions of the spectral parameter u, which have zeros and poles at the points

$$u = \pm \mathrm{i}K' - K/2, \quad \pm \mathrm{i}K' + K/2.$$

In Section 3.7, it was shown that such selected points are associated with the group of reductions of the Riemann problem. To carry out specific calculations, it is useful to know the values of the functions $\lambda = \lambda(u)$ and $\tau = \tau(u)$ at the indicated points:

$$\begin{aligned}
\lambda(\mathrm{i}K' + K/2) &= \lambda(\mathrm{i}K' - K/2) = \mathrm{i},\\
\lambda(-\mathrm{i}K' + K/2) &= \lambda(-\mathrm{i}K' - K/2) = -\mathrm{i},\\
\tau(\mathrm{i}K' + K/2) &= -\tau(\mathrm{i}K' - K/2) = (-p + \mathrm{i}\beta\cos\theta_0)/2,\\
\tau(-\mathrm{i}K' - K/2) &= -\tau(-\mathrm{i}K' + K/2) = (p + \mathrm{i}\beta\cos\theta_0)/2.
\end{aligned} \tag{A.8}$$

Using formulas (A.8), it is easy to verify that the function

$$\tau(u) - [w_3(u)\cos\theta_0 - p/2] \tag{A.9}$$

has double zeros at the points

$$u = \pm iK' + K/2 \quad \mathrm{mod}(2K,\, 4iK'),$$

whereas the function

$$w_1(u) = \beta[\lambda(u) + \lambda^{-1}(u)]/4 \tag{A.10}$$

has simple zeros at the points

$$u = \pm iK' + K/2, \quad \pm iK' - K/2 \quad \mathrm{mod}(2K,\, 4iK').$$

The poles of the functions (A.9) and (A.10) coincide and therefore make no contribution to relation (A.7).

Applying the theorem on the factorization of an elliptic function, we obtain the representation

$$\Delta(u) = c\,\frac{\sigma(u - iK' - K/2)\,\sigma(u + iK' - K/2)}{\sigma(u + iK' + K/2)\,\sigma(u - iK' + K/2)}\exp(2\eta_1 u). \tag{A.11}$$

Here the sigma-functions have periods $2K$ and $4i\,K'$. The value of the constant $c = 1$ is found from the condition that the left and right sides of the equality coincide for $u = 0$.

On the other hand, a meromorphic function $\mathrm{dn}(u+K/2)$ has exactly the same primitive periods, zeros, and poles as the function $\Delta(u)$ (A.11). Hence, in turn, we get the other representation

$$\Delta(u) = c_1\mathrm{dn}(u + K/2).$$

The proportionality coefficient $c_1 = (k')^{-1/2}$ can be derived from the condition of normalizing the equality at the point $u = 0$. Finally, we have

$$\Delta(u) = \frac{\mathrm{dn}(u + K/2)}{\sqrt{k'}}. \tag{A.12}$$

It turns out to be tedious to prove identity (A.12) for the Jacobi functions using the conventional formulas in [2]. At the same time, it is not immediately obvious which form of the Jacobi function will simplify the original expression.

Expansion (A.11) allows us to deduce another representation for the function $\Delta(u)$, which greatly simplifies the "dressing procedure" in Section 3.7:

$$\Delta(u) = \frac{\beta(u)}{\beta(-u)}, \quad \beta(u) = c_1\,\frac{\sigma(u + iK' - K/2)}{\sigma(u + iK' + K/2)}\exp(\eta_1 u).$$

The multiplier c_1 can be arbitrary.

Exploiting the transformation properties of the sigma-functions (A.3) and the second identity (A.2) (Legendre's relation), it is easy to verify that

$$\beta(u + 2K) = \beta(u), \quad \beta(u + 4\mathrm{i}K') = -\beta(u). \tag{A.13}$$

In other words, $2K$ and $8\mathrm{i}\,K'$ are primitive periods of the functions $\beta(u)$ and $\beta(-u)$. Therefore the latter can be factorized more correctly in terms of sigma-functions with periods $2K$ and $8\mathrm{i}\,K'$:

$$\beta(u) = c\,\frac{\sigma(u + \mathrm{i}K' - K/2)\,\sigma(u - 3\mathrm{i}K' - K/2)}{\sigma(u + \mathrm{i}K' + K/2)\,\sigma(u - 3\mathrm{i}K' + K/2)}\exp(2\eta_1 u).$$

Here $\eta_1 = \zeta(K|K, 4\mathrm{i}K')$ and $\sigma(u) = \sigma(u|K, 4\mathrm{i}K')$. After the vertical line, halves of primitive periods are sometimes indicated.

In a fundamental rectangle with sides $2K$ and $8\mathrm{i}\,K'$, the function $\beta(u)$ has only two simple poles. This means that the residues of the function $\beta(u)$ at these poles must differ in sign. Since the function $\beta(u)$ is defined with accuracy up to a constant factor, the latter can be chosen so that the residues should be equal to ± 1. Then by the theorem on the expansion of an elliptic function into simple fractions we have

$$\beta(u) = \zeta(u + \mathrm{i}K' + K/2) - \zeta(u - 3\mathrm{i}K' + K/2) + c,$$

where $2K$ and $8\mathrm{i}\,K'$ are the periods of the zeta functions. The second symmetry condition (A.13) for the function $\beta(u)$ yields the constant. Bearing in mind the quasiperiodicity of the zeta functions (A.3), we obtain

$$\beta(u) = \zeta(u + \mathrm{i}K' + K/2) - \zeta(u - 3\mathrm{i}K' + K/2) - \eta_3. \tag{A.14}$$

Our immediate goal is to rewrite representation (A.14) in terms of Jacobi functions. The primitive periods of all three Jacobi functions, such as $\mathrm{sn}u$, $\mathrm{cn}u$, and $\mathrm{dn}u$, are $4K$ and $4\mathrm{i}\,K'$. At the same time, the rectangle with sides $2K$ and $8\mathrm{i}\,K'$ offers an elementary cell of the function $\beta(u)$. It is possible to reconcile the periodicity properties of functions, for example, within a fundamental rectangle with sides $8K$ and $8\mathrm{i}\,K'$. Therefore we replace representation (A.14) for the function $\beta(u)$ by its expansion in zeta-functions with periods $8K$ and $8\mathrm{i}\,K'$. It can be found similarly:

$$\begin{aligned}\beta(u) = \sum_{\alpha=1}^{4}\big[&\zeta(\nu_\alpha + 4\mathrm{i}K'|4K, 4\mathrm{i}K') \\ &- \zeta(\nu_\alpha|4K, 4\mathrm{i}K') - \zeta(4\mathrm{i}K'|4K, 4\mathrm{i}K')\big].\end{aligned} \tag{A.15}$$

Here $\nu_1 = -u - \mathrm{i}K' - 5K/2$, $\nu_2 = -u - \mathrm{i}K' - K/2$, $\nu_3 = -u - \mathrm{i}K' + 3K/2$, and $\nu_4 = -u - \mathrm{i}K' + 7K/2$

For the zeta-functions with periods $2\omega_\alpha$, we have the following formulas [2]:

$$\zeta(z \pm \omega_\alpha) - \zeta(z) \mp \eta_\alpha = \frac{1}{2}\frac{\partial}{\partial z}\ln[\wp(z) - e_\alpha].$$

Since expansion (A.15) includes only such combinations of zeta-functions, the issue of transition in formula (A.15) to the Jacobi functions is reduced to the problem of transition from the $\wp$-Weierstrass functions to the Jacobi functions. It turns out [2] that

$$e_1 : e_2 : e_3 = (2 - k^2) : (2k^2 - 1) : (-1 - k^2)$$

and

$$\operatorname{sn}(u, k) = \sqrt{\frac{e_1 - e_3}{\wp(z) - e_3}}, \quad \operatorname{cn}(u, k) = \sqrt{\frac{\wp(z) - e_1}{\wp(z) - e_3}}, \quad \operatorname{dn}(u, k) = \sqrt{\frac{\wp(z) - e_2}{\wp(z) - e_3}},$$

where $z = u(e_1 - e_3)^{-1/2}$.

By the homogeneity of the Weierstrass functions we have

$$\partial_{tz}\wp(tz|t\omega, t\omega') = t^{-3}\partial_z\wp(z|\omega, \omega'), \quad \wp(tz|t\omega, t\omega') = t^{-2}\wp(z|\omega, \omega'),$$
$$\zeta(tz|t\omega, t\omega') = t^{-1}\zeta(z|\omega, \omega'), \quad \sigma(tz|t\omega, t\omega') = t\,\sigma(z|\omega, \omega').$$

The coefficient of proportionality between variables z and u can usually be chosen for specific calculations without any difficulties. In our case, we have

$$\begin{aligned} &\zeta(v/2 + 2\mathrm{i}K'|2K, 2\mathrm{i}K') - \zeta(v/2|2K, 2\mathrm{i}K') \\ &\quad - \zeta(2\mathrm{i}K'|2K, 2\mathrm{i}K') = -\frac{1}{2}\frac{\operatorname{cn}(v/4, k)\operatorname{dn}(v/4, k)}{\operatorname{sn}(v/4, k)}. \end{aligned} \tag{A.16}$$

Ultimately, formula (A.16) holds by comparing the properties of symmetry and residues at the poles of the left-hand and right-hand sides of the equality.

Having transformed expansion (A.15) using identity (A.16) and conventional formulas for the Jacobi functions [2], we arrive at

$$\beta(u) = \frac{1}{2}\left[\frac{\operatorname{cn}(v, k)}{\operatorname{sn}(v, k)} - k'\frac{\operatorname{sn}(v, k)}{\operatorname{cn}(v, k)}\right]\Bigg|_{v=(u+\mathrm{i}K'+K/2)/2}.$$

In a similar way, we can prove the following useful relations:

$$\begin{aligned} \beta(u)\beta^*(u^*) &= \frac{(1 + k')}{2}\operatorname{dn}(u + K/2) = \frac{\sqrt{k'}}{2}(1 + k')\frac{\beta(u)}{\beta(-u)}, \\ \beta(-u)\beta^*(u^*) &= \beta^*(-u^*)\beta(u) = \frac{\sqrt{k'}}{2}(1 + k'), \\ \beta(u)\beta(-u) &= \frac{\sqrt{k'}}{2}(1 + k')\left(\frac{\mathrm{i} + 1}{\mathrm{i} - 1}\right)\left(\frac{\operatorname{dn}u - \mathrm{i}\sqrt{k'}}{\operatorname{dn}u + \mathrm{i}\sqrt{k'}}\right) \\ &\equiv -\frac{\mathrm{i}}{2}\sqrt{k'}(1 + k')(\operatorname{sn}v - \mathrm{i}\operatorname{cn}v)(\operatorname{sn}w - \mathrm{i}\operatorname{cn}w)|_{v=u+K/2,\, w=-u+K/2}. \end{aligned} \tag{A.17}$$

Index

https://doi.org/10.1515/9783112235973-011

Bibliography

For Chapter 1

[1] *Akhiezer A. I., Bar'yakhtar V. G. and Peletminskii S. V.* Spin waves. New York: Interscience Publishers, 1968. 369 p.

[2] *Andreev A. F.* Magnetic properties of disordered media. Sov. Phys. JETP. 1978. V. 47, # 2, P. 411–417.

[3] *Andreev A. F., Marchenko V. I.* Macroscopic theory of spin waves. Sov. Phys. JETP. 1976. V. 43, # 4, P. 794–803.

[4] *Andreev A. F., Marchenko V. I.* Symmetry and the macroscopic dynamics of magnetic materials. Sov. Phys. Usp. 1980. V. 23, #1, P. 21–34.

[5] *Arnold V. I.* Mathematical methods of classical mechanics. 2 Ed. Springer–Verlag, 1989. 516 p.

[6] *Bar'yachtar V. G., Chetkin M. V., Ivanov B. A., Gadetskii S. N.* Springer tracts in modern physics 129. New York: Springer, 1994.

[7] *Bar'yakhtar V. G.* Phenomenological description of relaxation processes in magnetic materials. Sov. Phys. JETP. 1984. V. 60, #4, P. 863–867.

[8] *Bar'yakhtar V. G.* Spontaneous ground state degeneracy and relaxation processes in a ferromagnet. Fiz. Tverd. Tela. 1987. V. 29, #5, P. 1317–1322. [In Russian].

[9] *Bar'yakhtar V. G.* The phenomenological theory of relaxation processes in magnets. In the book: Frontiers in magnetism of reduced dimension systems, edited by Bar'yakthar V. G., Wigen P. E., Lesnik N. A., NATO ASI Series, vol. 49. Dordrecht: Springer.

[10] *Baryakhtar V. G., Belykh V. G., Soboleva T. K.* Relaxation of spin wanes in non-collinear magnets: Macroscopic theory. Phys. Lett. A. 1988. V. 132, #6,7, P. 359–362.

[11] *Bar'yakhtar V. G., Ivanov B. A., Soboleva T. K., Sukstanskii A. L.* Theory of dynamical-soliton relaxation in ferromagnets. Sov. Phys. JETP. 1986. V. 64, #4, P. 857–863.

[12] *Bar'yakhtar V. G., Vitebskii I. M., Yablonskii D. A.* Magnetoelastic effects in noncollinear antiferromagnets. Sov. Phys. JETP. 1985. V. 62, #1, P. 108–114.

[13] *Bogoliubov N. N., Shirokov D. V.* Introduction to theory of quantized fields. New York, Chichester, etc: John Wiley & Sons, 1980. xvii, 620 p.

[14] *Borisov A. B.* Nonlinear excitations and 2D topological solitons in magnets. Post-doctoral Thesis. Sverdlovsk, 1986. 303 p. [In Russian].

[15] *Borisov A. B., Ogievetskii V. I.* Theory of dynamical affine and conformal symmetries as the theory of the gravitational field. Theor. Math. Phys. 1974. V. 21, #3, P. 1179–1188.

[16] *Brown W. F.* Micromagnetics. New York, London: Interscience Publishers a division of John Wiley and Sons, 1963.

[17] *Buchgoltz L. J., Fetter A. L.* Static and dynamic textures in superfluid HeA. In the book: Quantum fluids and solids, edited by Trickey, S. B., Adams, E. D., Dufty, J. W., New York–London: Plenum Press, 1977, 473 p.

[18] *Callan C., Coleman S., Wess J., Zumino B.* Structure of phenomenological Lagrangians. II. Phys. Rev. 1969. V. 177, #5, P. 2247–2250.

[19] *Coleman S., Wess J., Zumino B.* Structure of phenomenological Lagrangians. I. Phys. Rev. 1969. V. 177, #5, P. 2239–2247.

[20] *Dubrovin B. A., Fomenko A. T., Novikov S. P.* Modern geometry — methods and applications. Part I: The geometry of surfaces, transformation groups, and fields. Graduate Texts in Mathematics, 93, 2nd Ed. Springer, 1991. 486 p.

[21] *Dzyaloshinski I. E., Volovik G. E.* On the concept of local invariance in theory of spin glasses. J. de Phys. 1978. V. 39, #6, P. 693–700.

[22] *Dzyaloshinskii I. E., Kukarenko B. G.* Spin waves in UO2. Sov. Phys. JETP. 1978. V. 48, #6, P. 1155–1157.

https://doi.org/10.1515/9783112235973-012

[23] *Dzyaloshinskii I. E., Man'ko V. I.* Nonlinear effects in antiferromagnets. "Latent" antiferromagnetism. Sov. Phys. JETP. 1964. V. 19, #4, P. 915–919.

[24] *Egorushkin V. E.* The gauge dynamic theory of defects in structured media under inhomogeneous deformation. Izv. Vyssh. Uchebn. Zaved. Fiz. 1990. V. 33, #2, P. 51–68. [In Russian].

[25] *Elliot R., Thorpe M.* Orbital effects on exchange interactions. J. Appl. Phys. 1968. V. 39, #2, P. 802–807.

[26] *Frenkel Y. I.* Electrodynamics, vol. I. Leningrad: ONTI–GTTI, 1934. 428 p. [In Russian].

[27] *Galkina E. G., Ivanov B. A., Safaryan K. A.* Theory of retardation of magnetic domain walls in rhombic magnetic materials. J. Exp. Theor. Phys. 1997. V. 84, #1, P. 87–95.

[28] *Gell-Mann M., Levi M.* The axial vector current in beta decay. Nuovo Cimento. 1960. V. 16, P. 705.

[29] *Goldenblatt I. I.* Nonlinear problems of the theory of elasticity. Moscow: Nauka, 1969. [In Russian].

[30] *Goldstein H.* Classical mechanics. Cambridge: Addison-Wesley press, inc. 42, Mass., 1950.

[31] *Grib A. A.* The problem of the vacuum non-invariance in the quantum field theory. Moscow: Atomizdat, 1978. 126 p. [In Russian].

[32] *Hubert A.* Theorie der domanenwande in geordneten medien. Berlin, Heidelberg, New York: Springer Verlag, 1974.

[33] *Ignatchenko V.* A boundary conditions for magnetic and magnetoelastic systems. Phys. Met. Metallogr. 1973. V. 36, #6, P. 87–96.

[34] *Ivanenko D. D.* Elementary particles and compensating fields. Moscow: Mir, 1964. 299 p. [In Russian].

[35] *Ivanov B. A., Safaryan K. A.* Mobility of domain walls of ferromagnets in transversal magnetic field. Fiz. Tverd. Tela. 1990. V. 32, #12, P. 3507–3510. [In Russian].

[36] *Ivanov B. A., Safaryan K. A.* Dynamics of domain walls of an uniaxial ferromagnet in transversal magnetic field. Fiz. Nizk. Temp. 1992. V. 18, #7, P. 722–730. [In Russian].

[37] *Ivanov E. A., Ogievetskii V. I.* Gauge theories as spontaneous breaking theories. JETP Lett. 1976. V. 23, #11, P. 606–609.

[38] *Kadic A., Edelen D. G. B.* A gauge theory of dislocations and disclination. Berlin, Heidelberg, New York: Springer Verlag, 1983.

[39] *Kiselev V. V.* Dynamical solitons in magnets with few magnetic sublattices. Ph. D. Thesis. Sverdlovsk, 1984. 168 p. [In Russian].

[40] *Konopleva V. P. and Popov V. N.* Calibration fields Moscow: Atomizdat, 1972. 236 p. [In Russian].

[41] *Kovalev A. S., Gvozdikova M. V.* Exotic solitons in magnets with strongly anisotropic exchange interaction. Low Temp. Phys. 1999. V. 25, #3, P. 184–191.

[42] *Landau L. D.* Collected works, vol. 1. Moscow: Science, 1969. P. 128–143. [In Russian].

[43] *Landau L. D., Lifshitz E. M.* Mechanics: Volume 1. Course of Theoretical Physics, 3 Ed. Butterworth Heinemann, 1976. 224 p.

[44] *Landau L. D., Lifshitz E. M.* Course of theoretical physics. Vol. 9. Statistical Physics. Part 2. Oxford, New York, Toronto, etc: Pergamon Press, 1980. 397 p.

[45] *Lange R. V.* Nonrelativistic theorem analogous to the Goldstone theorem. Phys. Rev. 1966. V. 146, #1, P. 301–303.

[46] *Maki K.* General gauge invariance and spin waves in the B-phase of super fluid ^{3}He. Phys. Rev. B. 1975. V. 11, #11, P. 4264–4271.

[47] *Malozemoff A. P., Slonczewski J. C.* Magnetic domain walls in bubble materials. New York, London, Toronto, Sydney, San Francisco: Academic Press, 1979.

[48] *Malyshev K. L., Romanov A. E.* Aspects of the T(3) $\otimes$ SO(3) – gauge theory of dislocations and disclinations, Pt. 1. LOMI Preprint, R-2-90. Leningrad, 1990. 47 p. [In Russian].

[49] *Malyshev K. L., Romanov A. E.* Aspects of the T(3) $\otimes$ SO(3) – gauge theory of dislocations and disclinations, Pt. 2. Preprint LOMI R-5-90. Leningrad 1990. 52 p. [In Russian].

[50] *Mc Connell A. J.* Application of tensor analysis. New York: Dover Publications, inc., 1963. 318 p.

[51] *Men'shenin V. V., Mirsayev I. F.* Non-linear generation of longitudinal magnetoelastic waves in disordered ferromagnets. Phys. Met. Metallogr. 1989. V. 68, #1, P. 9–16.

[52] *Men'shenin V. V., Mirsayev I. F., Taluts G. G.* Magnetoelastic interactions in disordered ferromagnets. Phys. Met. Metallogr. 1982. V. 54, #3, P. 451–458.
[53] *Mielke E. W.* Geometrodynamics of gauge fields. Berlin: Akademie Verlag, 1987. 242 p.
[54] *Murnaghan F. D.* Finite deformation of on elastic solid. New York: John Willy & Sons, inc. London Chapman & Hall, limited, 1951. 140 p.
[55] *Nesterov A. I., Ovchinnikov S. G.* Gauge theory of amorphous magnets. Theor. Math. Phys. 1988. V. 76, #1, P. 704–709.
[56] *Panin V. E., Grinyaev Y. V., Egorushkin V. E., Buchbinder I. L., Kul'Kov S. N.* Spectrum of excited states and the rotational mechanical field in a deformed crystal. Sov. Phys. J. 1987. V. 30, #1, P. 24–38.
[57] *Patashinskii A. Z., Shumilo B. I.* Theory of condensed matter based on the hypothesis of a local crystalline order. Sov. Phys. JETP. 1985. V. 62, #1, P. 177–184.
[58] *Petrashen M. I., Trifonov E. D.* Applications of group theory in quantum mechanics. 2 Ed. Moscow: URSS, 2002. 321 p.
[59] *Pshisukha A. M., Chernyi A. S., Zvyagin A. I.* Effect of the low symmetry in the EPR spectrum of Er^{3+} ion in $KY(MoO_4)_2$. Sov. J. Low Temp. Phys. 1975. V. 1, P. 233.
[60] *Pshisukha A. M., Zvyagin A. I., Stepanov A. A.* Paramagnetic absorption in $KDy(MoO_4)_2$, in parallel fields. Sov. Phys., Solid State. 1971. V. 13, P. 2630.
[61] *Rashevsky P. K.* Riemannian geometry and tensor analysis. Moscow: Nauka, 1967. [In Russian].
[62] *Rumer Y. B., Fet A. I.* Theory of unitary symmetry. Moscow: Nauka, 1970. 400 p. [In Russian].
[63] *Schwinger J.* Chiral dynamics. Phys. Lett. B. 1967. V. 24, P. 473–475.
[64] *Streitwolf H.-W.* Group theory in solid-state physics. Macdonald and company, 1971. 246 p. *Streitwolf H.-W.* Gruppentheorie in der festkorperphysik. Akademische Verlagsgesellshaft Geest & Porting K.-G.: Leipzig, 1967.
[65] *Tiersden H. F.* Variational principle for saturated magnetoelastic insulators. J. Math. Phys. 1965. V. 6, #5, P. 779–788.
[66] *Truesdell C.* A first course in rational continuum mechanics. Baltimore, Maryland: The Johns Hopkins University, 1972.
[67] *Turov E. A.* Physical properties of magnetically ordered crystals. New York London: Academic Press, 1965. 222 p.
[68] Turov E. A., Kolchanov A. V., Men'shenin V. V., Mirsaev I. F., and Nikolaev V. V. Symmetry and physical properties of antiferromagnets. Moscow: Fizmatlit, 2001. 560 p. [In Russian].
[69] *Turov E. A., Shavrov V. G.* On an energy gap in ferro- and antiferromagnet related to the magnetoelastic energy. Fiz. Tverd. Tela. 1965. V. 7, #1, P. 217–226. [In Russian].
[70] *Turov E. A., Shavrov V. G.* Broken symmetry and magnetoacoustic effects in ferro and antiferromagnetics. Sov. Phys. Usp. 1983. V. 26, #3, P. 593–611.
[71] *Vilenkin N. Y.* Special functions and the theory of group representations. Transactions of Mathematical Monographs, Vol. 22. American Mathematical Society, 1968. 613 p.
[72] *Volkov D. V.* Phenomenological Lagrangian of interraction of the Goldstone particles. Preprint ITF-69-75. Kyiv, 1969. 51 p. [In Russian].
[73] *Volkov D. V.* Phenomenological Lagrangians. Fiz. Elem. Chast. Atom. Yadra. 1973. V. 4, P. 3–41.
[74] *Volkov D. V., Akulov V. P.* Possible universal neutrino interaction. JETP Lett. 1972. V. 16, #11, P. 438–440.
[75] *Volkov D. V., Zheltuhin A. A.* Phenomenological Lagrangian of spin waves in spatially disordered media. Fiz. Nizk. Temp. 1979. V. 5, #11, P. 1359–1363. [In Russian].
[76] *Volkov D. V., Zheltukhin A. A.* Spin-wave propagation in spatially disordered media. Sov. Phys. JETP. 1980. V. 51, # 5, P. 937–942.
[77] *Volkov D. V., Zheltukhin A. A., Bliokh Yu. P.* Phenomenological Lagrangian for spin waves. In the book: Supersymmetry and quantum field theory, edited by Wess J., Akulov V. P., Lecture Notes in Physics, vol. 509. Springer, Berlin, Heidelberg, 1998, P. 370–382.
[78] *Volkov M. K., Pervushin V. N.* Essentially nonlinear quantum theories, dynamical symmetries, and meson physics. Moscow: Atomizdat, 1978. 240 p. [In Russian].

[79] *Volovik G. E., Dzyaloshinskii I. E.* Additional localized degrees of freedom in spin glasses. Sov. Phys. JETP. 1978. V. 48, # 3, P. 555–559.
[80] *Vonsovskii S. V.* Magnetism. New York: J. Wiley, 1974.
[81] *Wagner H.* Long-wavelength excitations and the Goldstone theorem in many-particle system with "broken symmetries". Z. Phys. 1966. V. 195, P. 273–299.

For Chapter 2

[1] *Ablowitz M., Segur H.* Solitons and the inverse scattering transform. Philadelphia: SIAM, 1981. 425 p.
[2] *Ablowitz M. J.* Remarks nonlinear evolution equations and ordinary differential equations Painlevé type. Physica D. 1981. V. 3, #1–2, P. 129–141.
[3] *Ablowitz M. J., Kaup D. J., Newell A. C., Segur H.* The inverse scattering transform. Fourier analysis for non-linear problems. Stud. Appl. Math. 1974. V. 53, #4, P. 249–317.
[4] *Ablowitz M. J., Ramani A. Segur H.* A connection between non-linear evolution equations and ordinary differential equations P-type. I. J. Math. Phys. 1980. V. 21, #4, P. 715–721.
[5] *Ablowitz M. J., Ramani A., Segur H.* A connection between non-linear evolution equations and ordinary differential equations P-type. II. J. Math. Phys. 1980. V. 21, #4, P. 1006–1015.
[6] *Akhiezer I. A., Borovik A. E.* Theory of finite-amplitude spin waves. Sov. Phys. JETP. 1967. V. 35, #2, P. 332–335.
[7] *Akhiezer I. A., Borovik A. E.* Nonlinear spin waves in ferromagnets and antiferromagnets. Sov. Phys. JETP. V. 25, #5, P. 885–893.
[8] *Arnold V. I.* Mathematical methods of classical mechanics. 2 Ed. Springer–Verlag, 1989. 516 p.
[9] *Babich I. M., Kosevich A. M., Manzhos I. V.* Magnetic excitations localized in space and periodic in time. Fiz. Nizk. Temp. 1983. V. 9, #6, P. 636–644. [In Russian].
[10] *Bar'yakhtar I. V., Ivanov B. A.* Nonlinear magnetization waves in antiferromagnets. Sov. J. Low Temp. Phys. 1979. V. 5, P. 361.
[11] *Bar'yakhtar I. V., Ivanov B. A.* On the nonlinear waves of magnetization of antiferromagnets. Preprint #80-4 DonPhTI Acad. Sci. Ukr. SSR Donetsk, 1980. 61 p. [In Russian].
[12] *Bar'yakhtar V. G., Ivanov B. A., Sukstanskii A. L.* Nonlinear waves and the dynamics of domain walls in weak ferromagnets. Sov. Phys. JETP. 1980. V. 51, #4, P. 757–764.
[13] *Baxter R. J.* Exactly solved models in statistical mechanics. London, New York: Academic Press, 1982.
[14] *Belokolos E. D., Bobenko A. I., Enolskii V. Z., Its A. R., Matveev V. B.* Algebro-geometric approach to nonlinear integrable equations. Springer Series in Nonlinear Dynamics, Berlin: Springer - Verlag, 1994, XII. 320 p.
[15] *Bethe H.* Zur theore der metalle. I. Eigenwerte und eigen-funktion der linearen atomkette. Z. Phys. 1931. Bd. 71, H. 3–4. S. 205–226.
[16] *Bezmaternih G. V.* Exact solutions of the sine–Gordon and Landau–Lifshitz equations: Rational–exponential solutions. Phys. Lett. A. 1985. V. 146, #9, P. 492–495.
[17] *Bezmaternih G. V., Borisov A. B.* Rational–exponential solutions of nonlinear equations. Lett. Math. Phys. 1985. V. 18, P. 1–8.
[18] *Bikbaev R. F., Bobenko A. I.* On finite-gap integration of the Landau–Lifshitz equation: XYZ case. Preprint LOMI E-8-83. Leningrad, 1983. 27 p. [In Russian].
[19] *Bikbaev R. F., Bobenko A. I., Its A. R.* Finite-zone integration of the Landau–Lifshits equations. Sov. Math. Dokl. 1983. V. 28, P. 512–516.
[20] *Bikbaev R. F., Bobenko A. I., Its R. A.* The Landau–Lifshitz equations. Theory of exact solutions. Part 1. Preprint FTI-84-6 (81). Donetsk, 1984. 46 p. [In Russian].
[21] *Bikbaev R. F., Bobenko A. I., Its R. A.* The Landau–Lifshitz equations. Theory of exact solutions. Part II. Preprint FTI-84-2. Donetsk, 1984. 51 p. [In Russian].

[22] *Bikbaev R. F., Its A. R.* Asymptotics at $t \to \infty$ of the solution of the Cauchy problem for the Landau–Lifshitz equation. Theor. Math. Phys. 1988. V. 76, #1, P. 665–675.
[23] *Bishop A.* Solitons and physical perturbations. In the book: Solitons in action, edited by Lonngren K., Scott A., New York, San Francisco, London: Academic Press, 1978, P. 61–88.
[24] *Bishop A. R., Krumhansl J. A., Trullinger S. E.* Solitons in condensed matter: A paradigm. Physica D. 1980. V. 1, #1, P. 1–14.
[25] *Bobenko A. I.* Landau–Lifshitz equation. Dressing up of the solutions and the elementary excitations. Differential geometry, Lie groups and mechanics. Part V, Zap. Nauchn. Sem. LOMI. V. 123. Leningrad: "Nauka", Leningrad. Otdel., 1983, P. 58–66.
[26] *Bobenko A. I.* Real algebrogeometric solutions of the Landau–Lifshits equation in prym theta functions. Funct. Anal. Appl. 1985. V. 19, #1, P. 5–17.
[27] *Bogdan M. M., Kovalev A. S.* Exact multisoliton solutions of the Landau–Lifshitz equations for an one dimensional non-izotropic magnet. Preprint IFM UNC-80-4. Sverdlovsk, 1980. 36 p. [In Russian].
[28] *Bogdan M. M., Kovalev A. S.* Exact multisoliton solution of one-dimensional Landau-Lifshitz equations for an anisotropic ferromagnet. JETP Lett. 1980. V. 31, #8, P. 424–427.
[29] *Bogoliubov N. M., Izergin A. G., Korepin V. E.* Quantum inverse scattering method and correlation functions. Cambridge university press, 1993. 554 p.
[30] *Borisov A. B.* The Hilbert problem for matrices and a new class of integrable equation. Lett. Math. Phys. 1983. V. 7, P. 195–199.
[31] *Borisov A. B.* Multisoliton solutions of the equations of a non-isotropic magnet. Phys. Met. Metallogr. 1983. V. 55, #2, P. 17–21.
[32] *Borisov A. B.* Direct and inverse scattering problem for the Landau–Lifshits equation. Sov. Phys. Dokl. 1986. V. 31, P. 482.
[33] *Borisov A. B.* Bäcklund transformation and dressing chains for the Landau–Lifshitz equation. Theor. Math. Phys. 2001. V. 128, #2, P. 1025–1033.
[34] *Borisov A. B.* A new class of exact solutions to the n-field equations and their geometric interpretation. Dokl. Math. 2003. V. 67, #2, P. 281–282.
[35] *Borisov A. B.* Differential and geometric approach and a new class of exact solutions for the n-field equations. J. Math. Phys. Anal. Geom. 2004. V. 10, P. 326–331. [In Russian].
[36] *Borisov A. B., Kiselev V. V.* Many-soliton solutions of asymmetric chiral SU(2) and SL(2, R) theories ($d = 1$). Theor. Math. Phys. 1983. V. 54, #2, P. 160–167.
[37] *Borisov A. B., Kiselev V. V.* Dynamics of a quasi-one-dimensional ferromagnet with easy-plane anisotropy. Phys. Met. Metallogr. 1984. V. 58, #2, P. 23–35.
[38] *Borisov A. B., Kiselev V. V., Taluts G. G.* Soliton solutions of nonlinear dynamics of magnets. Fiz. Nizk. Temp. 1983. V. 9, #2, P. 170–178. [In Russian].
[39] *Borisov A. B., Kiselev V. V., Talutz G. G.* Disclination in isotropic antiferromagnet. In the book: Research on physics of kinetic phenomena. Sverdlovsk: Acad. Sci. USSR, The Urals Branch, 1984. P. 118–126. [In Russian].
[40] *Borisov A. B., Kiseliev V. V.* Two-dimensional solutions of the Landau–Lifshitz equation. Phys. Lett. A. 1985. V. 107, #4, P. 161–163.
[41] *Borisov A. B., Kiseliev V. V.* Dynamical solitons in a quasi-one-dimensional ferromagnet with easy plane type anisotropy. Physica D. 1986. V. 19, P. 411–422.
[42] *Borisov A. B., Kiseliev V. V.* Inverse problem for an elliptic sine–Gordon equation with asymptotic behavior of the cnoidal-wave type. Inverse Probl. 1989. V. 5, P. 952–982.
[43] *Borisov A. B., Kiseliev V. V.* Vortex dipoles on a soliton lattice background: Solution of the boundary-value problem by inverse spectral transform. Physica D. 1998. V. 111, P. 96–128.
[44] *Borisov A. B., Kiseliev V. V., Taluts G. G.* Solitons in a ferrimagnet. Solid State Commun. 1982. V. 44, #3, P. 441–442.
[45] *Borisov A. B., Kiseliev V. V., Talutz G. G.* N-soliton solutions for asymmetric chiral SU(2) theory. In the book: Nonlinear and turbulent processes in physics, vol. 3, edited by Sagdeev R. S., New York: Gordon and Breach, Harwood Academic Publishers, 1984. P. 1211–1217.

[46] *Borisov A. B., Zykov S. A.* The dressing chain of discrete symmetries and proliferation of nonlinear equations. Theor. Math. Phys. 1998. V. 115, #2, P. 530–541.

[47] *Borisov A. B., Zykov S. A.* Proliferation scheme for Kaup–Boussinesq equation. Physica D. 2001. V. 152–153, P. 104–106.

[48] *Borisov A. B., Zykov S. A., Pavlov M. V.* Tzitzéica equation and proliferation of nonlinear integrable equations. Theor. Math. Phys. 2002. V. 131, #1, P. 550–557.

[49] *Borovik A. E.* N-soliton solutions of the nonlinear Landau–Lifshitz equation. JETP Lett. 1978. V. 28, #10, P. 581–584.

[50] *Borovik A. E.* Solitons investigations in nonlinear dynamics – new concept. Poznan Institute of molecular physics, Polish Academy of Sciences, 1980. 113 p.

[51] *Borovik A. E.* Exact integration on the nonlinear Landau–Lifshitz equation. Solid State Commun. 1980. V. 34, #8, P. 721–722.

[52] *Borovik A. E.* Solitons investigations in nonlinear dynamics – new concept. Poznan Institute of molecular physics, Polish Academy of Sciences, 1980. 113 p.

[53] *Borovik A. E., Klama S., Kulinich S. I.* Integration of the Landau–Lifshitz equation with preferred-axis anisotropy by the method of the inverse scattering problem. Physica D. 1988. V. 32, P. 107–134.

[54] *Borovik A. E., Kulinich S. I.* Integration of the nonlinear dynamics of a uniaxial ferromagnet by the method of the inverse scattering problem. JETP Lett. 1984. V. 39, #7, P. 384–388.

[55] *Borovik A. E., Robuk V. N.* Linear pseudopotentials and conservation laws for the Landau–Lifshits equation describing the nonlinear dynamics of a ferromagnet with uniaxial anisotropy. Theor. Math. Phys. 1981. V. 46, #3, P. 242–248.

[56] *Burtsev S. P., Mukhailov A. V., Zakharov V. E.* The inverse scattering transform with a movable spectral parameter. Preprint. The Academy of Sciences of the USSR, L. D. Landau institute for theoretical physics. Chernogolovka, 1987. 37 p.

[57] *Burtsev S. P., Zakharov V. E., Mikhailov A. V.* Inverse scattering method with variable spectral parameter. Theor. Math. Phys. 1987. V. 70, #3, P. 227–240.

[58] *Cherednik I. V.* Relativistically invariant quasiclassical limits of integrable two-dimensional quantum models. Theor. Math. Phys. 1981. V. 47, #2, P. 422–425.

[59] *Cherednik I. V.* Integrability of the equation of a two–dimensional asymmetric chiral $O(3)$ field and its quantum analog. Sov. J. Nucl. Phys. 1981. V. 33, #1.

[60] *Deift P., Zhou X.* A steepest descent method for oscillatory Riemann–Hilbert problems. Bull. Am. Math. Soc. 1992. V. 26, #1, P. 119–123.

[61] *Deift P., Zhou X.* A steepest descent method for oscillatory Riemann–Hilbert problems. Asymptotic for the MKdV equation. Ann. Math. 1993. V. 137, P. 295–368.

[62] *Dodd R. K., Eilbeck J. C., Gibbon J. D., Morris H. C.* Solitons and nonlinear wave equation. London, Orlando, San Diego, New York, Toronto, Montreal, Sydney, Tokyo: Academic Press, Inc. (Harcourt Brace Jovanovich, Publishers), 1984.

[63] *Dubrovin B. A.* Theta functions and non-linear equations. Russ. Math. Surv. 1981. V. 36, #2, P. 11–92.

[64] *Dubrovin B. A.* Riemann surfaces and nonlinear equations. Izhevsk: NIC “Regularnaya i chaotichnaya dinamica”, 2001. 152 p. [In Russian].

[65] *Dubrovin B. A., Matveev V. B., Novikov S. P.* Non-linear equations of Korteweg–de Vries type, finite-zone linear operators, and Abelian varieties. Russ. Math. Surv. 1976. V. 31, #1, P. 59–146.

[66] *Eleonskii V. M., Kirova N. N., Kulagin N. E.* Exact solutions of the Landau–Lifshitz equations for weak ferromagnets. Sov. Phys. JETP. 1980. V. 52, #1, P. 162–167.

[67] *Eleonskii V. M., Kirova N. N., Kulagin N. E.* Exactly solvable models for two-sublattice magnets. Sov. Phys. JETP. 1981. V. 53, #1, P. 182–185.

[68] *Faddeev L. D., Takhtadzhyan L. A.* Hamiltonian approach in the theory of solitons, classics in mathematics. Berlin: Springer, 2017. 601 p.

[69] *Faddeev L. D., Takhtajan L. A.* What is the spin of a spin wave? Phys. Lett. A. 1981. V. 85, #6–7, P. 375–377.

[70] *Flashka H., Newell A. C.* Monodromy and spectrum-preserving deformations. I. Commun. Math. Phys. 1980. V. 76, #1, P. 65–116.

[71] *Fokas A. S., Its A. R., Kapaev A. A. Novokshenov V. Yu.* Painlevé transcendents: The Riemann–Hilbert approach. American Mathematical Society, 2023. 553 p.

[72] *Gardner G., Green J., Kruskal M., Miura R.* A method for solving of the Korteweg–de Vries equation. Phys. Rev. Lett. 1967. V. 19, #19, P. 1095–1098.

[73] *Gaudin M.* La function d' onde de Bethe. Masson, 1983.

[74] *Hirota R.* Direct methods in soliton theory. In the book: Solitons, edited by Bullough R. K., Caudrey P. J., Berlin, Heidelberg, New York: Springer Verlag, 1980. P. 157–176.

[75] *Hirota R.* Bilinearization of soliton equation. J. Phys. Soc. Jpn. 1982. V. 51, #1, P. 323–331.

[76] *Its A. R.* Asymptotics of solutions of the nonlinear Schrodinger equation and isomonodromic deformations of systems of linear differential equations. Sov. Math. Dokl. 1981. V. 24, #1, P. 452–456.

[77] *Its A. R.* "Isomonodromy" solutions of equations of zero curvature. Math. USSR, Izv. 1986. V. 26, #3, P. 497–529.

[78] *Its A. R., Novokshenov V. Yu.* The isomonodromy deformation method in theory of Painlevé equations. Lect. Notes in Math. V. 1191 Berlin, Heidelberg, New York: Springer Verlag, 1986. 313 p.

[79] *Ivanov B. A., Kosevich A. M., Babich I. M.* Localized nonlinear oscillations in ferromagnetic substances. JETP Lett. 1979. V. 29, #12, P. 714–717.

[80] *Ivanov B. A., Kosevich A. M., Kovalev A. S.* Nonlinear magnetization waves. Dynamic and topological solitons. Kiev: Naukova Dumka, 1983. 294 p. [In Russian].

[81] *Ivanov B. A., Kosevich A. M., Manshos I. V.* Algebraic soliton in a ferromagnet in the presence of the magnetic field directed along the anisotropy axis. Solid State Commun. 1980. V. 34, #6, P. 417–418.

[82] *Izyumov Y. A., Skryabin Y. N.* Statistical mechanics of magnetically ordered systems. Moscow: Nauka, 1987. 264 p. [In Russian].

[83] *Jimbo M., Miwa T.* Monodromy preserving deformation of linear ordinary differential equations with rational coefficients. II. Physica D. 1981. V. 2, #3, P. 407–448.

[84] *Jimbo M., Miwa T.* Monodromy preserving deformation of linear ordinary differential equations with rational coefficients. III. Physica D. 1981. V. 4, #1, P. 26–46.

[85] *Jimbo M., Miwa T., Ueno K.* Monodromy preserving deformation of linear ordinary differential equations with rational coefficients. I. General theory and τ-function. Physica D. 1981. V. 2, #2, P. 306–352.

[86] *Karpman V. I., Maslov E. M.* Perturbation theory for solitons. Sov. Phys. JETP. 1977. V. 46, #2, P. 281–291.

[87] *Kaup D. J.* The method of solution for stimulated Raman scattering and two-photon absorption. Preprint Clarkson college of Technology, April, 1980.

[88] *Kaup D. J.* The method of solution for stimulated Raman scattering and two-photon absorption. Physica D. 1983. V. 6, #2, P. 143–154.

[89] *Kivshar Y. S., Malomed B. A.* Dynamics of solitons in nearly integrable systems. Rev. Mod. Phys. 1989. V. 61, #4, P. 763–916.

[90] *Konopelchenko B. G.* Introduction to multidimensional integrable equations: The inverse spectral transform in 2 + 1 dimensions. New York and London: Plenum Press, 1992. 292 p.

[91] *Konopelchenko B. G.* Solitons in multidimensions: Inverse spectral transform method. World Scientific Publishing Co.Pte.Ltd., 1993. 294 p.

[92] *Kosevich A. M., Voronov V. P., Manzhos I. V.* Nonlinear collective excitations in an easy plane magnet. Sov. Phys. JETP. 1983. V. 57, #1, P. 86–92.

[93] *Krichever M. I.* Methods of algebraic geometry in the theory of non-linear equations. Russ. Math. Surv. 1977. V. 32, #6, P. 185–213.

[94] *Kudryashov N. A.* Analytical theory of nonlinear differential equations. Moscow: Institute of Computer Investigations, 2004. 360 p. [In Russian].

[95] *Lakshamanan M.* Continuum spin system as an exactly soluble dynamical system. Phys. Lett. A. 1977. V. 61, #1, P. 53–54.

[96] *Lamb G. L.* Elements of soliton theory. New York, Chichester, Brisbane, Toronto: Wiley, 1980. xii, 289 p.
[97] *Lax P. D.* Integrals of non-linear equations of evolution and solitary waves. Commun. Pure Appl. Math. 1968. V. 21, #5, P. 467–490.
Lax P. D. Invariant functionals of nonlinear evolution equations. In the book: Nonlinear waves, edited by Leibovich S., Seebass A. R., Ithaca and London: Cornell University Press, 1974.
[98] *Marchenko V. A.* Sturm-Liouville operators and applications. Springer Basel AG, 1986. XI, 367 p.
[99] *Matveev V. B., Salle M. A.* Darboux transformation and solitons. Springer Verlag, 1991. 129 p.
[100] *McLaughlin D., Scott A.* Multisoliton theory of excitations. In the book: Solitons in action, edited by Lonngren K., Scott A., New York, San Francisco, London: Academic Press, 1978. P. 201–256.
[101] *Mikhailov A. V.* The reduction problem and the inverse scattering method. Physica D. 1981. V. 3, #1–2, P. 73–117.
[102] *Mikhailov A. V.* The Landau–Lifshitz equation and the Riemann-boundary problem on a torus. Phys. Lett. A. 1982. V. 92, #2, P. 51–55.
[103] *Mikhailov A. V.* Integrable magnetic models. In the book: Solitons, edited by Trullinger S. E., Zakharov V. E., Pokrovsky V. L., Elsevier Science Publishers B. V., 1986. P. 625–690.
[104] *Mikhailov A. V., Yaremchuk A. I.* Axially symmetrical solutions of the two-dimensional Heisenberg model. JETP Lett. 1982. V. 36, #3, P. 95–100.
[105] *Mikhailov A. V., Yaremchuk A. I.* Cylindrically symmetric solutions of the nonlinear chiral field model (σ-model). Nucl. Phys. B. 1982. V. 202, P. 508–522.
[106] *Mitropolskii Y. F., Bogoliubov N. N. (Jr), Prikarpatskii A. K., Samoilenko V. G.* Integrable dynamical systems: Spectral and differential-geometric aspects. Kyiv: Naukova Dumka, 1987. 296 p. [In Russian].
[107] *Newell A.* Solitons in mathematics and physics. Philadelphia, PA: Society for Industrial and Applied Mathematics, 1985. xvi, 244 p.
[108] *Novikov S., Manakov S. V., Pitaevskii L. P., Zakharov V. E.* Theory of solitons: The inverse scattering method. N. York and London: Consultants Bureau, 1984. 276 p.
[109] *Rodin Yu. L.* The Riemann boundary problem on a torus and the inverse scattering problem for the Landau–Lifshitz equation. Lett. Math. Phys. 1983. V. 7, P. 3–8.
[110] *Rodin Yu. L.* The Riemann boundary problem on Riemann surfaces and the inverse scattering problem for the Landau–Lifschitz equation. Physica D. 1984. V. 11, P. 90–108.
[111] *Sasada T. J.* Magnons, solitons and a critical field in the Heisenberg ferromagnetic chain with easy-plane anisotropy. J. Phys. Soc. Jpn. 1982. V. 51, #8, P. 2446–2449.
[112] *Scott A.* Nonlinear science: Emergence and dynamics of coherent structures. Oxford University press, 2003. 480 p.
[113] *Scott A. C., Chu F. Y., Mc Iaughlin D. W.* The soliton: A new concept in applied science. Proc. IEEE 1973. V. 61, P. 1443–1483.
[114] *Sklyanin E. K.* On complete integrability of the Landau–Lifshitz equation. Preprint LOMI No 79-E-3. Leningrad, 1979. 32 p. [In Russian].
[115] *Sklyanin E. K., Takhtadzhyan L. A. Faddeev L. D.* Quantum inverse problem method. I. Theor. Math. Phys. 1979. V. 40, #2, P. 688–706.
[116] *Takhtadzhan L. A., Faddeev L. D.* The quantum method of the inverse problem and the Heisenberg XYZ model. Russ. Math. Surv. 1979. V. 34, #5, P. 11–68.
[117] *Takhtadzhyan L. A., Faddeev L. D.* Spectrum and scattering of excitations in the one-dimensional isotropic Heisenberg model. J. Sov. Math. 1984. V. 24, #2, P. 241–267.
[118] *Takhtajan L. A.* Integration of the continuous Heisenberg spin chain through the inverse scattering method. Phys. Lett. A. 1977. V. 64, #2, P. 235–237.
[119] *Ueno K.* Monodromy preserving deformation and its application to soliton theory. Proc. Jpn. Acad., Ser. A, Math. Sci. 1980. V. 56, #3, P. 103–108.
[120] *Ueno K.* Monodromy preserving deformation and its application to soliton theory. II. Proc. Jpn. Acad., Ser. A, Math. Sci. 1980. V. 56, #5, P. 210–211.

[121] *Veselov A. P.* Cnoidal solutions of the Landau–Lifshits equation for a bisublattice magnet. Dokl. Akad. Nauk SSSR. 1984. V. 276, #3, P. 590–593. [In Russian].

[122] *Volzhan E. B., Giorgadze N. P., Pataraya A. D.* On weakly magnetization density waves in magnetically ordered crystals. Sov. Phys., Solid State. 1976. V. 18, P. 1487.

[123] *Yang C. N., Yang C. P.* One-dimensional chain of anisotropic spin-spin interaction. I. Proof of Bethe's hypothesis for ground state in a finite system. Phys. Rev. 1966. V. 150, #1, P. 321–327.

[124] *Yang C. N., Yang C. P.* One-dimensional chain of anisotropic spin-spin interaction. II. Properties of the ground state energy per lattice site for a infinite system. Phys. Rev. 1966. V. 150, #1, P. 327–339.

[125] *Yang C. N., Yang C. P.* One-dimensional chain of anisotropic spin-spin interaction. III. Applications. Phys. Rev. 1966. V. 151, #1, P. 258–264.

[126] *Yaremchuk A. I.* Interaction of domain wall with a spin wave in the framework of an integrable case of the Landau–Lifshitz equations. Theor. Math. Phys. 1985. V. 62, #1, P. 101–105.

[127] *Zabusky N., Kruskal M.* Interaction of "solitons" in a collisionless plasma and the recurrence of initial states. Phys. Rev. Lett. 1965. V. 15, #6, P. 240–243.

[128] *Zakharov V. E., Faddeev L. D.* Korteweg–de Vries equation: A completely integrable Hamiltonian system. Funct. Anal. Appl. 1971. V. 5, #4, P. 280–287.

[129] *Zakharov V. E., Mikhailov A. V.* Relativistically invariant two-dimensional models of field theory which are integrable by means of the inverse scattering problem method. Sov. Phys. JETP. 1978. V. 47, #6, P. 1017–1027.

[130] *Zakharov V. E., Shabat A. B.* A scheme for integrating the nonlinear equations of mathematical physics by the method of the inverse scattering problem. I. Funct. Anal. Appl. 1974. V. 8, #3, P. 226–235.

[131] *Zakharov V. E., Shabat A. B.* Integration of nonlinear equations of mathematical physics by the method of inverse scattering. II. Funct. Anal. Appl. 1979. V. 13, #3, P. 166–174.

[132] *Zvezdin A. K.* Dynamics of domain walls in weak ferromagnets. JETP Lett. 1979. V. 29, #10, P. 553–557.

For Chapter 3

[1] *Akhiezer N. I.* Elements of the theory of elliptic functions. Transactions of mathematical monographs, volume 79. Providence, Rhod Island: American Mathematical Society, 1990. 237 p.

[2] *Batalov S. V., Maslov E. M., Shagalov A. G.* Autophasing of solitons. J. Exp. Theor. Phys. 2009. V. 108, #5, P. 890–897.

[3] *Bateman H., Erdélyi A.* Higher transcendental functions, vol. 3. New York, Toronto, London: McGraw–Hill Book Company, Inc., 1955. 310 p.

[4] *Borisov A. B., Kiselev V. V.* Many-soliton solutions of asymmetric chiral SU(2) and SL(2, *R*) theories ($d = 1$). Theor. Math. Phys. 1983. V. 54, #2, P. 160–167.

[5] *Borisov A. B., Kiselev V. V.* Dynamics of a quasi-one-dimensional ferromagnet with easy-plane anisotropy. Phys. Met. Metallogr. 1984. V. 58, #2, P. 23–35.

[6] *Borisov A. B., Kiselev V. V., Taluts G. G.* Soliton solutions of nonlinear dynamics of magnets. Fiz. Nizk. Temp. 1983. V. 9, #2, P. 170–178. [In Russian].

[7] *Borovik A. E., Klama S., Kulinich S. I.* Integration of the Landau–Lifshitz equation with preferred-axis anisotropy by the method of the inverse scattering problem. Physica D. 1988. V. 32, P. 107–134.

[8] *Borovik A. E., Kulinich S. I.* Integration of the nonlinear dynamics of a uniaxial ferromagnet by the method of the inverse scattering problem. JETP Lett. 1984. V. 39, #7, P. 384–388.

[9] *Borovik A. E., Robuk V. N.* Linear pseudopotentials and conservation laws for the Landau–Lifshits equation describing the nonlinear dynamics of a ferromagnet with uniaxial anisotropy. Theor. Math. Phys. 1981. V. 46, #3, P. 242–248.

[10] *Eleonskii V. M., Kulagin N. E., Novozhilova N. S.* New examples of topological solitons in magnetically disordered media. Sov. Phys. JETP. 1985. V. 62, #6, P. 1255–1258.

[11] *Filippov B. N., Tankeev A. P.* Dynamic effects in ferromagnets with a domain structure. Moscow: Nauka, 1985. 217 p. [In Russian].

[12] *Ivanov B. A., Kosevich A. M., Kovalev A. S.* Nonlinear magnetization waves. Dynamic and topological solitons. Kiev: Naukova Dumka, 1983. 294 p. [In Russian].

[13] *Karpman V. I., Maslov E. M.* Perturbation theory for solitons. Sov. Phys. JETP. 1977. V. 46, #2, P. 281–291.

[14] *Kiselev V. V., Rascovalov A. A.* Forced motion of breathers and domain boundaries against the background of nonlinear magnetization wave. Chaos Solitons Fractals. 2012. V. 45, P. 1551–1565.

[15] *Kiselev V. V., Raskovalov A. A.* Interaction of a breather with a magnetization wave in a ferromagnet with light-axis anisotropy. Theor. Math. Phys. 2010. V. 163, #1, P. 479–495.

[16] *Kiselev V. V., Raskovalov A. A.* Forced motion of solitary domains and domain walls in the field of a nonlinear magnetization wave. Phys. Met. Metallogr. 2010. V. 109, #6, P. 585–598.

[17] *Kiselev V. V., Raskovalov A. A.* Solitons against a pump wave background in an easy-axis ferromagnet. Low Temp. Phys. 2010. V. 36, #8/9, P. 656–664.

[18] *Kittel C.* Note on the inertia and damping constant of ferromagnetic domain boundaries. Phys. Rev. 1950. V. 80, P. 918.

[19] *Kivshar Y. S., Malomed B. A.* Dynamics of solitons in nearly integrable systems. Rev. Mod. Phys. 1989. V. 61, #4, P. 763–916.

[20] *Kovalev A. S., Kulagin N. E.* Dynamics of hole-like solitons in stable condensate. In the book: Nonlinear and turbulent processes in physics, vol. 2, Kiev: Naukova Dumka, 1988, P. 273–276.

[21] *Kovalev A. S., Kulagin N. E.* Dynamics of the dark solitons of the non zero vacuum in an one-component system. Preprint. Kharkiv: FTINT, 1987, #14. 23 p. [In Russian].

[22] *Kuznetsov E. A., Mikhailov A. V.* Stability of stationary waves in nonlinear weakly dispersive media. Sov. Phys. JETP. 1974. V. 40, #5, P. 855–859.

[23] *Kuznetsov E. A., Mikhailov A. V., Shimokhin I. A.* Nonlinear interaction of solitons and radiation. Physica D. 1995. V. 87, P. 201–215.

[24] *McLaughlin D., Scott A.* Multisoliton theory of excitations. In the book: Solitons in action, edited by Lonngren K., Scott A., New York, San Francisco, London: Academic Press, 1978. P. 201–256.

[25] *Mikhailov A. V.* The Landau–Lifshitz equation and the Riemann-boundary problem on a torus. Phys. Lett. A. 1982. V. 92, #2, P. 51–55.

[26] *Mikhailov A. V., Yaremchuk A. I.* Forced motion of a domain wall in the field of a spin wave. JETP Lett. 1984. V. 39, #7, P. 354–357.

[27] *Novikov S., Manakov S. V., Pitaevskii L. P., Zakharov V. E.* Theory of solitons: The inverse scattering method. N. York and London: Consultants Bureau, 1984. 276 p.

[28] *Rodin Yu. L.* The Riemann boundary problem on a torus and the inverse scattering problem for the Landau–Lifshitz equation. Lett. Math. Phys. 1983. V. 7, P. 3–8.

[29] *Sklyanin E. K.* On complete integrability of the Landau–Lifshitz equation. Preprint LOMI No 79-E-3. Leningrad, 1979. 32 p. [In Russian].

[30] *Yaremchuk A. I.* Interaction of domain wall with a spin wave in the framework of an integrable case of the Landau–Lifshitz equations. Theor. Math. Phys. 1985. V. 62, #1, P. 101–105.

For Chapter 4

[1] *Akhiezer I. A., Borovik A. E.* Nonlinear spin waves in ferromagnets and antiferromagnets. Sov. Phys. JETP. V. 25, #5, P. 885–893.

[2] *Babich I. M., Kosevich A. M., Manzhos I. V.* Magnetic excitations localized in space and periodic in time. Fiz. Nizk. Temp. 1983. V. 9, #6, P. 636–644. [In Russian].

[3] *Bar'yakhtar V. G., Stefanovskii E. P.* Spectrum of spin waves in antiferromagnets with spiral magnetic structure. Sov. Phys., Solid State. 1970. V. 11, P. 1566.

[4] *Borisov A. B., Izyumov Yu. A.* Nonlinear excitation in spiral magnetic structures. Dokl. Akad. Nauk SSSR. 1985. V. 283, #3, P. 859–861. [In Russian].

[5] *Borisov A. B., Kiselev V. V.* Dynamics of a quasi-one-dimensional ferromagnet with easy-plane anisotropy. Phys. Met. Metallogr. 1984. V. 58, #2, P. 23–35.

[6] *Borisov A. B., Kiseliev V. V.* Dynamical solitons in a quasi-one-dimensional ferromagnet with easy plane type anisotropy. Physica D. 1986. V. 19, P. 411–422.

[7] *Dzyaloshinskii I. E.* Theory of helicoidal structures in antiferromagnets. III. Sov. Phys. JETP. 1965. V. 20, #3, P. 665–668.

[8] *Ivanov B. A., Kosevich A. M., Kovalev A. S.* Nonlinear magnetization waves. Dynamic and topological solitons. Kiev: Naukova Dumka, 1983. 294 p. [In Russian].

[9] *Ivanov B. A., Kosevich A. M., Manshos I. V.* Algebraic soliton in a ferromagnet in the presence of the magnetic field directed along the anisotropy axis. Solid State Commun. 1980. V. 34, #6, P. 417–418.

[10] *Izyumov Y. A.* Neutron diffraction on the long periodic structures. Moscow: EnergoAtomIzdat, 1987. 200 p. [In Russian].

[11] *Izyumov Yu. A.* Solitons in quasi-one-dimensional magnetic materials and their study by neutron scattering. Sov. Phys. Usp. 1988. V. 31, #4, P. 689–712.

[12] *Kjems I. K., Stelner M.* Evidence for soliton modes in the one-dimensional ferromagnet $CsNiF_3$. Phys. Rev. Lett. 1978. V. 41, #16, P. 1137–1140.

[13] *Kosevich A. M., Voronov V. P., Manzhos I. V.* Nonlinear collective excitations in an easy plane magnet. Sov. Phys. JETP. 1983. V. 57, #1, P. 86–92.

[14] *Kovalev A. S.* Influence of field on magnetization dynamics of a light–plane ferromagnet. In the book: Modern problems of magnetism. Kyiv: Naukova Dumka, 1986. P. 72–83. [In Russian].

[15] *Landau L. D., Lifshitz E. M.* Electrodynamics of continuous media. Course of Theoretical Physics. Vol. 8. 2 Ed. Pergamon press, 1984. 460 p.

[16] *Reiter G.* Have solitons been observed in $CsNiF_3$. Phys. Rev. Lett. 1981. V. 46, #3, P. 202–205.

[17] *Sasada T. J.* Magnons, solitons and a critical field in the Heisenberg ferromagnetic chain with easy-plane anisotropy. J. Phys. Soc. Jpn. 1982. V. 51, #8, P. 2446–2449.

[18] *Schlömann E.* Generation of spin waves in uniform magnetic fields. 1. Conversion of electromagnetic power into spin wave power and vice versa. J. Appl. Phys. 1964. V. 35, P. 159–166.

[19] *Schlömann E., Joseph R. I.* Generation of spin waves in nonuniform magnetic fields. 2. Calculation of the coupling length. J. Appl. Phys. 1964. V. 35, #1, P. 167–170.

[20] *Steiner M., Kakarai K., Knop W.* Neutron inelastic scattering study of transverse spin fluctuation in $CsNiF_3$: A soliton only central peak. Solid State Commun. 1982. V. 41, #4, P. 329–332.

[21] *Steiner M., Kakurai K., Kjems J. K.* Experimental study of the spindynamics in the 1-d-ferromagnet with planar anisotropy, $CsNiF_3$, in an external magnetic field. Z. Phys. B, Condens. Matter. 1983. V. 53, #2, P. 117–143.

[22] *Steiner M., Villain J., Windsor C.* The theoretical and experimental studies on one-dimensional magnetic systems. Advances in Physics 1976. V. 25, #2, P. 87–209.

[23] *Thiele A. A.* Excitation spectrum of magnetic domain walls. Phys. Rev. B. 1973 V. 7, #1, P. 391–397.

[24] *Turov E. A.* Physical properties of magnetically ordered crystals. New York, London: Academic Press, 1965. 222 p.

[25] *Villain J.* Quantum theory of one and two-dimensional ferro- and antiferromagnets with easy magnetization plane, I. Ideal 1-D or 2-D lattices without in-plane anisotropy. J. Phys. 1974. V. 35, #1, P. 27–48.

For Chapter 5

[1] *Bikbaev R. F., Bobenko A. I.* On finite–gap integration of the Landau–Lifshitz equation: XYZ case. Preprint LOMI E-8-83. Leningrad, 1983. 27 p. [In Russian].

[2] *Bikbaev R. F., Bobenko A. I., Its A. R.* Finite-zone integration of the Landau-Lifschitz equations. Sov. Math. Dokl. 1983. V. 28, P. 512–516.
[3] *Bikbaev R. F., Bobenko A. I., Its A. R.* Landau–Lifshitz equation, uniaxial anisotropy case: Theory of exact solutions. Theor. Math. Phys. 2014. V. 178, #2, P. 143–193.
[4] *Bikbaev R. F., Bobenko A. I., Its R. A.* The Landau–Lifshitz equations. Theory of exact solutions. Part 1. Preprint FTI-84-6 (81). Donetsk, 1984. 46 p. [In Russian].
[5] *Bikbaev R. F., Bobenko A. I., Its R. A.* The Landau–Lifshitz equations. Theory of exact solutions. Part II. Preprint FTI-84-2.Donetsk, 1984. 51 p. [In Russian].
[6] *Bobenko A. I.* Landau–Lifshits equation. "Dressing up" procedure. Elementary excitations. J. Sov. Math. 1985. V. 28, P. 499–505.
[7] *Bobenko A. I.* Real algebrogeometric solutions of the Landau–Lifshits equation in prym theta functions. Funct. Anal. Appl. 1985. V. 19, #1, P. 5–17.
[8] *Bogdan M. M., Kovalev A. S.* Exact multisoliton solutions of the Landau–Lifshitz equations for an one dimensional non-izotropic magnet. Preprint IFM UNC-80-4. Sverdlovsk, 1980. 36 p. [In Russian].
[9] *Bogdan M. M., Kovalev A. S.* Exact multisoliton solution of one-dimensional Landau–Lifshitz equations for an anisotropic ferromagnet. JETP Lett. 1980. V. 31, #8, P. 424–427.
[10] *Bogoliubov N. N. Jr., Prikarpatskii A. K.* On finite-zone solutions of a Heisenberg-type equation In the book: Mathematical methods and physical mechanical fields. Kyiv: Naukova Dumka, 1983 #18, P. 7–13. [In Russian].
[11] *Borisov A. B.* Multisoliton solutions of the equations of a non-isotropic magnet. Phys. Met. Metallogr. 1983. V. 55, #2, P. 17–21.
[12] *Borisov A. B.* Direct and inverse scattering problem for the Landau–Lifshits equation. Sov. Phys. Dokl. 1986. V. 31, P. 482.
[13] *Borisov A. B.* Nonlinear excitations and 2D topological solitons in magnets. Post-doctoral Thesis. Sverdlovsk, 1986. 303 p. [In Russian].
[14] *Borisov A. B., Kiselev V. V.* Many-soliton solutions of asymmetric chiral SU(2) and SL(2, R) theories ($d = 1$). Theor. Math. Phys. 1983. V. 54, #2, P. 160–167.
[15] *Borisov A. B., Kiseliev V. V., Talutz G. G.* N-soliton solutions for asymmetric chiral SU(2) theory. In the book: Nonlinear and turbulent processes in physics, V. 3, edited by Sagdeev R. S., New York: Gordon and Breach, Harwood Academic Publishers, 1984. P. 1211–1217.
[16] *Cherednik I. V.* Integrable differential equations and coverings of elliptic curves. Math. USSR, Izv. 1984. V. 22, #2, P. 357–377.
[17] *Courant R.* Partial differential equations. Methods of Mathematical Physics by R. Courant and D. Hilbert. Vol. II. New York, London: Wiley-VCH, 1989. 856 p.
[18] *Eleonskii V. M., Kulagin N. E.* Integrable models in the problem for particle motion in a two-dimensional potential well. Sov. Phys. JETP. 1983. V. 58, #4, P. 833–837.
[19] *Eleonskii V. M., Kulagin N. E.* New cases of integrability of the Landau-Lifshitz equations. Sov. Phys. JETP. V. 57, #2, P. 356–362.
[20] *Faddeev L. D., Takhtadzhyan L. A.* Hamiltonian approach in the theory of solitons, classics in mathematics. Berlin: Springer, 2017. 601 p.
[21] *Hirota R.* Bilinearization of soliton equations. J. Phys. Soc. Jpn. 1982. V. 51, #1, P. 323–331.
[22] *Ivanov B. A., Kosevich A. M., Kovalev A. S.* Nonlinear magnetization waves. Dynamic and topological solitons. Kiev: Naukova Dumka, 1983. 294 p. [In Russian].
[23] *Jimbo M., Miwa T.* Monodromy preserving deformations of the linear differential equation with the rational coefficients II. Preprint/RIMS-327. Kyoto, 1930. 34 p.
[24] *Jimbo M., Miwa T., Veno K.* Monodromy preserving deformations of the linear differential equations with the rational coefficients I. Preprint/RIMS-319. Kyoto, 1980. 31 p.
[25] *Kolmogorov A. N., Fomin S. V.* Elements of the theory of functions and functional analysis. Dover Books on Mathematics. Dover Publications, 1999. 288 p.

[26] *Mikhailov A. V.* The Landau–Lifshitz equation and the Riemann-boundary problem on a torus. Phys. Lett. A. 1982. V. 92, #2, P. 51–55.
[27] *Mikhailov A. V.* Integrable magnetic models. In the book: Solitons. edited by Trullinger S. E., Zakharov V. E., Pokrovsky V. L., Elsevier Science Publishers B. V., 1986. P. 625–690.
[28] *Petrovsky I. G.* Lectures on partial differential equations. Dover Books on Mathematics. Dover Publications. 272 p.
[29] *Rodin Yu. L.* The Riemann boundary problem on a torus and the inverse scattering problem for the Landau–Lifshitz equation. Lett. Math. Phys. 1983. V. 7, P. 3–8.
[30] *Rodin Yu. L.* The Riemann boundary problem on Riemann surfaces and the inverse scattering problem for the Landau–Lifschitz equation. Physica D. 1984. V. 11, P. 90–108.
[31] *Sklyanin E. K.* On complete integrability of the Landau–Lifshitz equation. Preprint LOMI #79-E-3. Leningrad, 1979. 32 p. [In Russian].
[32] *Tjon J., Wright J.* Solitons in the continuous Heisenberg spin chain. Phys. Rev. 1977. V. 15, #7, P. 3470–3476.
[33] *Veselov A. P.* The Landau–Lifshits equation and integrable systems of classical mechanics. Sov. Phys. Dokl. 1983. V. 28, #6, P. 458–459.

For Chapter 6

[1] *Bar'yakhtar I. V., Ivanov B. A.* Nonlinear magnetization waves in antiferromagnets. Sov. J. Low Temp. Phys. 1979. V. 5, P. 361.
[2] *Bar'yakhtar I. V., Ivanov B. A.* On the nonlinear waves of magnetization of antiferromagnets. Preprint #80-4 DonPhTI Acad. Sci. Ukr. SSR Donetsk, 1980. 61 p. [In Russian].
[3] *Bogoliubov N. N., Shirokov D. V.* Introduction to theory of quantized fields. New York, Chichester, etc.: John Wiley & Sons, 1980. xvii, 620 p.
[4] *Borisov A. B., Kiselev V. V., Taluts G. G.* Soliton solutions of nonlinear dynamics of magnets. Fiz. Nizk. Temp. 1983. V. 9, #2, P. 170–178. [In Russian].
[5] *Borisov A. B., Kiselev V. V., Talutz G. G.* Disclination in isotropic antiferromagnet. In the book: Research on physics of kinetic phenomena. Sverdlovsk: Acad. Sci. USSR, The Urals Branch, 1984. P. 118–126. [In Russian].
[6] *Borisov A. B., Kiseliev V. V.* Two-dimensional solutions of the Landau–Lifshitz equation. Phys. Lett. A. 1985. V. 107, #4, P. 161–163.
[7] *Borisov A. B., Kiseliev V. V., Taluts G. G.* Solitons in a ferrimagnet. Solid State Commun. 1982. V. 44, #3, P. 441–442.
[8] *Chikazumi S.* Physics of ferromagnetism, 2 ed. Oxford: Oxford science publications, 2009. 672 p.
[9] *Courant R.* Partial differential equations. Methods of Mathematical Physics by R. Courant and D. Hilbert. Vol. II. New York, London: Wiley-VCH, 1989. 856 p.
[10] *Ivanov B. A., Sukstanskii A. L.* Nonlinear magnetization waves in ferrites. Sov. Phys. JETP. 1983. V. 57, #1, P. 214–219.
[11] *Izyumov Y. A., Ozerov R. P.* Magnetic neutron diffraction. New York: Plenum Press, 1970. xviii, 598 p.
[12] *Kuznetsov E. A., Mikhailov A. V.* On the complete integrability of the two-dimensional classical Thirring model. Theor. Math. Phys. 1977. V. 30, #3, P. 193–200.
[13] *Lakshamanan M.* Continuum spin system as an exactly soluble dynamical system. Phys. Lett. A. 1977. V. 61, #1, P. 53–54.
[14] *Landau L. D., Lifshitz E. M.* The classical theory of fields. Course of Theoretical Physics, Vol. 2. 4 Ed. Amsterdam, Boston, etc.: Butterworth Heinemann, 1994. 444 p.
[15] *Mikhailov A. V., Yaremchuk A. I.* Axially symmetrical solutions of the two-dimensional Heisenberg model. JETP Lett. 1982. V. 36, #3, P. 95–100.

[16] *Novikov S., Manakov S. V., Pitaevskii L. P., Zakharov V. E.* Theory of solitons: The inverse scattering method. N. York and London: Consultants Bureau, 1984. 276 p.
[17] *Papanicolaou N.* Duality rotation for 2d classical ferromagnets. Phys. Lett. A. 1985. V. 84, #3, P. 151–154.
[18] *Petrovsky I. G.* Lectures on partial differential equations. Dover Books on Mathematics. Dover Publications. 272 p.
[19] *Pohlmeyer K.* Integrable Hamiltonian systems and interaction through quadratic constraints. Commun. Math. Phys. 1976. V. 46, P. 207–221.
[20] *Poluektov Y. M.* Nonlinear magnetization waves in superfluid ^{3}He $- A$ Sov. J. Low Temp. Phys. 1983. V. 9, #5.
[21] *Takhtajan L. A.* Integration of the continuous Heisenberg spin chain through the inverse scattering method. Phys. Lett. A. 1977. V. 64, #2, P. 235–237.
[22] *Turov E. A.* Physical properties of magnetically ordered crystals. New York, London: Academic Press, 1965. 222 p.
[23] *Walquest H. D., Estabrook F. B.* Prolongation structure of nonlinear evolution equations. J. Math. Phys. 1975. V. 16, #1, P. 1–7.

For Chapter 7

[1] *Akhiezer N. I.* Elements of the theory of elliptic functions. Transactions of Mathematical Monographs, volume 79. Providence, Rhod Island: American Mathematical Society, 1990. 237 p.
[2] *Andreev A. F., Marchenko V. I.* Symmetry and the macroscopic dynamics of magnetic materials. Sov. Phys. Usp. 1980. V. 23, #1, P. 21–34.
[3] *Bateman H., Erdélyi A.* Higher transcendental functions, vol. 3. New York, Toronto, London: McGraw–Hill Book Company, Inc., 1955. 310 p.
[4] *Borisov A. B.* Nonlinear excitations and 2D topological solitons in magnets. Post-doctoral Thesis. Sverdlovsk, 1986. 303 p. [In Russian].
[5] *Borisov A. B., Kiselev V. V.* Many-soliton solutions of asymmetric chiral SU(2) and SL(2, R) theories ($d = 1$). Theor. Math. Phys. 1983. V. 54, #2, P. 160–167.
[6] *Borisov A. B., Kiselev V. V.* Solitons in the modulated magnetic structure of MnOOH and of isomorphic compounds. Sov. Phys., Solid State. 1990. V. 32, P. 119.
[7] *Borisov A. B., Kiselev V. V., Taluts G. G.* Soliton solutions of nonlinear dynamics of magnets. Fiz. Nizk. Temp. 1983. V. 9, #2, P. 170–178. [In Russian].
[8] *Borisov A. B., Kiseliev V. V., Talutz G. G.* N-soliton solutions for asymmetric chiral SU(2) theory. In the book: Nonlinear and turbulent processes in physics, V. 3, edited by Sagdeev R. S., New York: Gordon and Breach, Harwood Academic Publishers, 1984. P. 1211–1217.
[9] *Borisov A. B., Kiseliev V. V., Talutz G. G.* N-soliton solutions for asymmetric chiral SU(2) theory. In the book: Nonlinear and turbulent processes in physics, vol. 3, edited by Sagdeev R. S., New York: Gordon and Breach, Harwood Academic Publishers, 1984. P. 1211–1217.
[10] *Izyumov Y. A., Ozerov R. P.* Magnetic neutron diffraction. New York: Plenum Press, 1970. xviii, 598 p.
[11] *Kaup D. J.* The method of solution for stimulated Raman scattering and two photon absorption. Preprint: Clarkson College of Technology, April, 1980.
[12] *Kiselev V. V.* Dynamical solitons in magnets with few magnetic sublattices. Ph. D. Thesis. Sverdlovsk, 1984. 168 p. [In Russian].
[13] *Mikhalev V. G.* On complete integrability of the $O(3)$–field in the class of rapidly decreasing functions. J. Sov. Math. 1992. V. 59, #5, P. 1092–1096.
[14] *Novikov S., Manakov S. V., Pitaevskii L. P., Zakharov V. E.* Theory of solitons: The inverse scattering method. N. York and London: Consultants Bureau, 1984. 276 p.

[15] *Pohlmeyer K.* Integrable Hamiltonian systems and interaction through quadratic constraints. Commun. Math. Phys. 1976. V. 46, P. 207–221.
[16] *Smolenskii G. A., Lemanov V. V., Nedlin G. M., Petrov M. P., and Pisarev R. V.* Physics of magnetic dielectrics. Leningrad: Nauka, 1974. 454 p. [In Russian].
[17] *Stefanoskii E. P.* Modulated magnetic structures in monoclinic systems (MnOOH and isomorphic compounds). Fiz. Nizk. Temp. 1987. V. 13, #7, P. 740–746. [In Russian].
[18] *Steudel H.* Solitons in stimulated Raman scattering. Ann. Phys. 1977. V. 7, #3, s. 188–200.
[19] *Zakharov V. E., Mikhailov A. V.* Example of nontrivial interaction of solitons in two-dimensional classical field theory. JETP Lett. 1977. V. 27, #1, P. 42–46.

For Chapter 8

[1] *Ablowitz M., Fokas A. S., Anderson R. L.* The direct linearising transform and the Benjamin–Ono equation. Phys. Lett. A. 1983. V. 93, #8, P. 375–378.
[2] *Ablowitz M., Segur H.* Solitons and the inverse scattering transform. Philadelphia: SIAM, 1981.
[3] *Ablowitz M. J., Segur H.* Asymptotic solutions of the Korteweg–de Vries equation. Stud. Appl. Math. 1977. V. 57, P. 13–44.
[4] *Akhiezer A. I., Bar'yakhtar V. G. and Peletminskii S. V.* Spin waves. New York: Interscience Publishers, 1968. 369 p.
[5] *Bar'yakhtar V. G., Gorobets Y. I.* Bubble domains and their lattices. Kyiv, Ukraine: Naukova Dumka, 1988. 168 p. [In Russian].
[6] *Benjamin T. V.* Internal waves of permanent form in fluids of great depth. J. Fluid Mech. 1967. V. 59, part 3, P. 559–592.
[7] *Benney D. J.* Significant interaction between long and short waves. Stud. Appl. Math. 1976. V. 55, P. 93–106.
[8] *Benney D. J.* A general theory for interactions between long and short waves. Stud. Appl. Math. 1977. V. 56, P. 81–94.
[9] *Berezhnov V. V., Evtikhiev N. N., Preobrazhenskiy V. L., Ekonomov N. A.* None resonant interraction of acoustic waves and the correlation treatment in antiferromagnets. Akusticheskij Zhurnal 1980. V. 26, #3, P. 328–335. [In Russian].
[10] *Bibikov P. N., Tarasov V. O.* Boundary-value problem for nonlinear Schrödinger equation. Theor. Math. Phys. 1989. V. 79, #3, P. 570–579.
[11] *Bikbaev R. F., Its A. R.* Algebrogeometric solutions of a boundary-value problem for the nonlinear Schrödinger equation. Math. Notes. 1989. V. 45, #5, P. 349–354.
[12] *Bikbaev R. F., Tarasov V. O.* Initial-boundary problem for the nonlinear Schrodinger equation. J. Phys. A, Math. Gen. 1991. V. 24, P. 2507–2518.
[13] *Boardman A. D., Nikitov S. A., Waby N. A.* Existence of spin-wave solitons in an antiferromagnetic film. Phys. Rev. B. 1993. V. 48, #18, P. 13602–13606.
[14] *Borisov A. B., Kiselev V. V.* Solitons in the modulated magnetic structure of MnOOH and of isomorphic compounds. Sov. Phys., Solid State. 1990. V. 32, P. 119.
[15] *Chen M., Nash J. M., Patton C. E.* A numerical study of nonlinear Schrodinger equation solutions for microwave solitons in magnetic thin films. J. Appl. Phys. 1993. V. 73, P. 3906–3909.
[16] *Chen M., Tzankov M. A., Nash J. M., Patton C. E.* Microwave magnetic-envelope dark solitons in yttrium iron garnet films. Phys. Rev. Lett. 1993. V. 67, #11, P. 1707–1710.
[17] *Chetkin M. V., Lykov V. V.* Solitary magnetoelastic waves in iron borate. Fiz. Tverd. Tela. 1990. V. 32, #9, P. 2848–2849. [In Russian].
[18] *Davydov A. S.* Solitons in molecular systems. Springer Dordrecht, 1985. XVIII, 319 p.

[19] *De Wames R. E., Wolfram T.* Dipole-exchange spin waves in ferromagnetic films. J. Appl. Phys. 1970. V. 41, #3, P. 987–993.

[20] *Degtyarev L. M., Nakhan'kov V. G., Rudakov L. I.* Dynamics of the formation and interaction of Langmuir solitons and strong turbulence. Sov. Phys. JETP. 1975. V. 40, #2, P. 264–268.

[21] *Dikshtein I. E., Turov E. A., Shavrov V. G.* Magnetoacoustic phenomena and soft modes near the magnetic orientational phase transitions. In the book: Dynamic and kinetic properties of magnets, edited by Vonsovsky S. V., Turov E. A., Moscow: Nauka, 1986. P. 68–103. [In Russian].

[22] *Dodd R. K., Eilbeck J. C., Gibbon J. D., Morris H. C.* Solitons and nonlinear wave equation. London, Orlando, San Diego, New York, Toronto, Montreal, Sydney, Tokyo: Academic Press, Inc. (Harcourt Brace Jovanovich, Publishers), 1982.

[23] *Filippov B. N.* Vibrations of magnetization in ferromagnetic plates. I. Phys. Met. Metallogr. 1971. V. 32, #5, P. 12–26.

[24] *Filippov B. N., Tityakov I. G.* Concerning oscillations of the magnetization in ferromagnetic plates. II. Phys. Met. Metallogr. 1973. V. 35,#1, P. 21–30.

[25] *Fokas A. S.* An initial-boundary value problem for nonlinear Schrodinger equation. Physica D. 1989. V. 35, P. 167–185.

[26] *Fokas A. S., Ablowitz M.* The inverse scattering transform for Benjamin–Ono equation – a pivot to multidimensional problems. Stud. Appl. Math. 1983. V. 68, P. 1–10.

[27] *Fokas A. S., Ablowitz M. J.* Forced nonlinear evolution equations and the inverse scattering transform. Stud. Appl. Math. 1989. V. 80, #3, P. 253–272.

[28] *Gann V. V.* Inhomogeneous resonance in a ferromagnetic film. Sov. Phys., Solid State. 1966. V. 6, P. 2537–2540.

[29] *Gromov E. M.* Propagation of short nonlinear wave packets and solitons in smoothly inhomogeneous media. Phys. Lett. A. 1997. V. 227, P. 67–71.

[30] *Gromov E. M., Talanov V. I.* Nonlinear dynamics of short wave trains in dispersive media. J. Exp. Theor. Phys. 1996. V. 83, #1, P. 73–79.

[31] *Gurevich A. G.* Magnetic resonance in ferrites and antiferromagnets. Moscow: Nauka, 1973. 591 p. [In Russian].

[32] *Harris A. B., Kumar D., Halperin B. I., Hohenberg P. C.* Dynamics of an antiferromagnet at low temperatures: Spin-wave damping and hydrodinamics. Phys. Rev. B. 1971. V. 3, #3, P. 961–1024.

[33] *Izyumov Yu. A., Laptev V. M.* Soliton magnetoelastic excitation in the Heisenberg ferromagnetic chain. Phys. Status Solidi (b). 1982. V. 112, #1, P. 155–159.

[34] *Kaganov M. I., Tsukernik V. M.* Phenomenological theory of kinetic processes in ferromagnetic dielectrics. II. Interaction of spin waves with phonons. Sov. Phys. JETP. 1959. V. 9, #1, P. 151–156.

[35] *Kalinikos B. A.* Spin wave spectrum and linear excitation of spin waves in ferromagnetic films. Izv. Vyssh. Uchebn. Zaved. Fiz. 1981. V. 24, #8, P. 42–56. [In Russian].
Kalinikos B. A. Spectrum and linear excitation of spin waves in ferromagnetic films. Soviet Physics Journal. 1981. V. 24, #8, P. 718–731.

[36] *Kalinikos B. A., Kovshikov N. G., Patton C. E.* Observation of self-generation of dark envelope solitons for spin waves in ferromagnetic films. JETP Lett. 1998. V. 68, #3, P. 243–247.
Kalinikos B. A., Kovshikov N. G., Patton C. E. Excitation of bright and dark microwave magnetic envelope solitons in a resonant ring. Appl. Phys. Lett. 1999. V. 75, #2, P. 265–267.

[37] *Kalinikos B. A., Kovshikov N. G., Slavin A. N.* Observation of spin-wave solitons in ferromagnetic films. JETP Lett. 1983. V. 38, #7, P. 413–417.

[38] *Kalinikos B. A., Kovshikov N. G., Slavin A. N.* Envelope solitons and modulation instability of dipole-exchange magnetization waves in yttrium iron garnet films. Sov. Phys. JETP. 1988. V. 67, #2, P. 303–312.

[39] *Kalinikos B. A., Kovshikov N. G., Slavin A. N.* Experimental observation of magnetostatic wave envelope solitons in yttrium iron garnet. Phys. Rev. B. 1990. V. 42, #13, P. 8658–8660.

[40] *Kalinikos B. A., Slavin A. N.* Theory of dipole-exchange spin wave spectrum for ferromagnetic films with mixed exchange boundary conditions. J. Phys. C, Solid State Phys. 1986. V. 19, #35, P. 7013–7033.
[41] *Karpman V. I.* Radiation by solitons due to higher-order dispersion. Phys. Rev. E. 1993. V. 47, #3, P. 2073–2083.
[42] *Karpman V. I.* Stationary and radiating dark solitons of the third order nonlinear Schrodinger equation. Phys. Lett. A. 1993. V. 181, P. 211–215.
[43] *Karpman V. I.* Evolution of the solitons described by higher-order nonlinear Schrodinger equation. Jerusalem (Israel), Racah Institute of Physics, Hebrew University, preprint, 1998. 13 p.
[44] *Kiselev V. V.* Solitons and three-wave resonance against a modulated magnetic structure of antiferromagnet. Preprint of Ural Div. Acad. Sci. Sverdlovsk, 1991. 25 p.
[45] *Kiselev V. V., Tankeev A. P.* Magnetoelastic resonance of long and short waves in magnets. Phys. Met. Metallogr. 1993. V. 75, #1, P. 26–35.
[46] *Kiselev V. V., Tankeev A. P.* Solitons in an antiferromagnetic film. Fiz. Tverd. Tela. 1994. V. 36, #10, P. 3055–3066. [In Russian].
[47] *Kiselev V. V., Tankeev A. P.* Weakly nonlinear long-wavelength spin excitations in thin ferromagnetic films. Phys. Met. Metallogr. 1996. V. 82, #3, P. 233–240.
[48] *Kiselev V. V., Tankeev A. P., Kobelev A. V.* Weakly linear dynamics of dipole-exchange spin waves in ferromagnetic plates of finite thickness. Phys. Met. Metallogr. 1996. V. 82, #5, P. 458–469.
[49] *Kiselev V. V., Tankeyev A. P.* Weakly non-linear excitations in an antiferromagnetic film. J. Phys. Condens. Matter. 1995. V. 7, P. 2087–2094.
[50] *Kiselev V. V., Tankeyev A. P.* Non-local dynamics of weakly nonlinear spin excitations in thin ferromagnetic films. J. Phys. Condens. Matter. 1996. V. 8, P. 10219–10229.
[51] *Kiselev V. V., Tankeyev A. P.* Non-local dynamics of exchange-dipole spin waves in ferromagnetic films. In the book: Studies in applied electromagnetics and mechanics, V. 13: Non-linear electromagnetic systems. Advanced Techniques and Mathematical Methods. Amsterdam: IOS Press, 1998. P. 131–135.
[52] *Kiselev V. V., Tankeyev A. P.* Non-local dynamics and soliton regimes in an antiferromagnetic film. In the book: Studies in applied electromagnetics and mechanics, V. 13: Non-linear electromagnetic systems. Advanced techniques and mathematical methods edited by Kose V., Sievert I., Amsterdam: IOS Press, 1998.
[53] *Kiseliev V. V.* Solitons and three-wave resonance against the background of the modulated magnetic structure of antiferromagnets. Fiz. Tverd. Tela. 1992. V. 34, #3, P. 769–778. [In Russian].
[54] *Kiseliev V. V., Tankeyev A. P., Kobelev A. V., Shagalov A. G.* Non-linear spin dynamics in ferromagnetic films and Schrodinger equation in the vicinity of the zero-dispersion point. J. Phys. Condens. Matter. 1999. V. 11, P. 3461–3474.
[55] *Kolmogorov A. N., Fomin S. V.* Elements of the theory of functions and functional analysis. Dover Books on Mathematics. Dover Publications, 1999. 288 p.
[56] *Kovalev A. S., Kosevich A. M., Manzhos I. V., Maslov K. V.* Precession soliton in a ferromagnetic film. JETP Lett. 1986. V. 44, #4, P. 222–225.
[57] *Kuehl H. H., Zhang C. Y.* Effects of higher-order dispersion on envelope solitons. Phys. Fluids B. 1990. V. 2, #5, P. 889–900.
[58] *Lamb G. L.* Elements of soliton theory. New York, Chichester, Brisbane, Toronto: Wiley, 1980. xii, 289 p.
[59] *Lebedev A. Yu., Ozhogin V. I., Yakubovskii A. Yu.* Stimulated acoustic Raman scattering in an antiferromagnet. JETP Lett. 1981. V. 34, #1, P. 19–22.
[60] *Lee Y. C., Chen H. H.* Nonlinear dynamical models of plasma turbulence. Phys. Scr. 1982. V. 2/1, P. 41–47.
[61] *Lemanov V. V.* Magnetoelastic interactions. In the book: Physics of magnetic dielectrics, edited by Smolenskii G. A., Leningrad: Nauka, 1974. P. 284–355. [In Russian].
[62] *Lui M., Drucker A. R., King A. R., Kotthaus J. R., Hansma R. K., Jaccarino V.* Observation of antiferromagnetic resonance in epitaxial films of MnF_2. Phys. Rev. B. 1986. V. 33, #11, P. 7720–7723.
[63] *Lui M., Ramos C A., King A. R., Jaccarino V.* Antiferromagnetic standing-spin-wave resonance in epitaxial films of MnF_2. J. Appl. Phys. 1990. V. 67, #9, P. 5518–5523.

[64] *Lukomskiy V. P.* Nonlinear magnitostatic waves in ferromagnetic plates. Ukr. J. Phys. 1978. V. 23, #1, P. 134–139. [In Russian].

[65] *Ma Y. G.* The complete solution of the long wave – short wave resonance equations. Stud. Appl. Math. 1978. V. 59, #2, P. 201–221.

[66] *Ma Y. G.* On the multi-soliton solutions of some nonlinear evolution equations. Stud. Appl. Math. 1979. V. 60, #1, P. 73–82.

[67] *Matsuno Y.* Exact multi-soliton solutions of the Benjamin–Ono equation. J. Phys. A, Math. Gen. 1979. V. 12, #4, P. 619–621.

[68] *Men'shenin V. V.* Non-linear excitation of non-activational magnetoelastic waves near orientation phase transitions in magnets. Phys. Met. Metallogr. 1990. V. 70, #5, P. 18–25.

[69] *Mikhailovskaya L. V., Khlebopros R. G.* Magnetostatic spectrum of ferromagnetic film. Sov. Phys., Solid State. 1969. V. 11, P. 1381.

[70] *Mikhailovskaya L. V., Khlebopros R. G.* Influence of surface spin pinning on the magnetostatic spectrum of a ferromagnetic layer. Sov. Phys., Solid State. 1974. V. 16, P. 46.

[71] *Miles J. W.* The asymptotic solutions of the Korteweg–de Vries equation in the absence of solitons. Stud. Appl. Math. 1979. V. 60, P. 59–72.

[72] *Nayyar A. H., Murtaza G.* Motion of a magnetic soliton about a lattice soliton in Heisenberg chain. Phys. Rev. 1982. V. 26, #7, P. 3904–3909.

[73] *Novikov S., Manakov S. V., Pitaevskii L. P., Zakharov V. E.* Theory of solitons: The inverse scattering method. N. York and London: Consultants Bureau, 1984. 276 p.

[74] *Ono H.* Algebraic solitary waves in stratified fluids. J. Phys. Soc. Jpn. 1975. V. 39, #4, P. 1082–1091.

[75] *Ozhogin V. N., Lebedev A. Yu., Yakubovskii A. Yu.* Doubling of the sound frequency and acoustic detection in hematite. JETP Lett. 1978. V. 27, #6, P. 313–315.

[76] *Preobrazhenskii V. L., Savchenko M. A., Ekonomov N. A.* Nonlinear self-action of sound waves in an antiferromagnet with easy-plane anisotropy. JETP Lett. 1978. V. 28, #2, P. 87–90.

[77] *Pushkarov D. I., Pushkarov Kh. I.* Solitons in one-dimensional ferromagnetic systems. Phys. Status Solidi (b). 1977. V. 81, #2, P. 703–707.

[78] *Raevskii V. Y.* Mathematical modelling of domain moving systems for memory devices. Ph. D. Thesis. Sverdlovsk: Institute of Metal Physics, Ural Div. Acad. Sci., 1986. 146 p. [In Russian].

[79] *Ramani A., Grammaticos B, Bountis T.* The Painlevé property and singularity analysis of integrable and non-integrable systems. Phys. Rep. 1989. V. 180, #3, P. 159–245.

[80] *Shamsutdinov M. A., Lomakina I. Y., Nazarov V. N., Kharisov A. T., Shamsutdinov D. M.* Ferro- and antiferromagnetics. Nonlinear oscillation, waves and solitons. Moscow: Nauka, 2009. 456 p. [In Russian].

[81] *Slavin A. N., Dudko G. M.* Numerical modeling of spin wave soliton propagation in ferromagnetic films. J. Magn. Magn. Mater. 1990. V. 86, #1, P. 115–123.

[82] *Taniuti T., Yajima N. J.* Perturbation method a non-linear wave modulation. I. J. Math. Phys. 1969. V. 10, #8, P. 1369–1372.

[83] *Taniuti T., Yajima N. J.* Perturbation method a non-linear wave modulation. II. J. Math. Phys. 1969. V. 10, #11, P. 2020–2024.

[84] *Tarasov V. O.* The boundary value problem for the nonlinear Schrödinger equation In the book: Questions of quantum field theory and statistical physics. Part 8, Zap. Nauchn. Sem. LOMI. V. 169. Leningrad: "Nauka", Leningrad. Otdel., 1988, P. 151–165. [In Russian].

[85] *Toupin R. A., Gazis D. S.* Lattice dynamics. Oxford: Pergamon Press, U.K., 1965. 597 p.

[86] *Turitsyn S. K., Fai'kovich G. E.* Stability of magnetoelastic solitons and self-focusing of sound in antiferromagnets. Sov. Phys. JETP. 1985. V. 62, #1, P. 146–152.

[87] *Turov E. A.* Kinetic, optical an acoustical properties of antiferromagnets. Sverdlovsk: Ural Div. Acad. Sci. Press, 1990. 134 p. [In Russian].

[88] Turov E. A., Kolchanov A. V., Men'shenin V. V., Mirsaev I. F., and Nikolaev V. V. Symmetry and physical properties of antiferromagnets. Moscow: Fizmatlit, 2001. 560 p. [In Russian].

[89] *Volzhan E. B., Giorgadze N. P., Pataraya A. D.* On weakly nonlinear magnetoelastic oscillations in ferromagnets. Sov. Phys. JETP. 1976. V. 43, #4, P. 693–698.
[90] *Wai P. K. A., Chen H. H., Lee Y. C.* Radiations by "solitons" at the zero group-dispersion wavelength of single-mode optical fibers. Phys. Rev. A. 1990. V. 41, #1, P. 426–438.
[91] *Whitham G. B.* Linear and nonlinear waves. A Wiley – Interscience Publication, 1974.
[92] *Yajima N., Oikava M.* Formation and interaction of sonic-Langmuir solitons. Prog. Theor. Phys. 1976. V. 56, #5, P. 1719–1739.
[93] *Yajima N., Satsuma J.* Soliton solution in a diatomic lattice system. Prog. Theor. Phys. 1972. V. 62, #2, P. 370–378.
[94] *Zvezdin A. K., Popkov A. F.* Contribution to the nonlinear theory of magnetostatic spin waves. Sov. Phys. JETP. 1983. V. 57, #2, P. 350–355.

For Chapter 9

[1] *Akhiezer N. I.* Elements of the theory of elliptic functions. Transactions of Mathematical Monographs, volume 79. Providence, Rhod Island: American Mathematical Society, 1990. 237 p.
[2] *Bar'yakhtar I. V., Ivanov B. A.* Nonlinear magnetization waves in antiferromagnets. Sov. J. Low Temp. Phys. 1979. V. 5, P. 361.
[3] *Bar'yakhtar V. G., Stefanovski E. P.* Spectrum of spin waves in antiferromagnets with a spiral magnetic structure. Sov. Phys., Solid State. 1970. V. 11, P. 1566.
[4] *Bateman H., Erdélyi A.* Higher transcendental functions, vol. 3. New York, Toronto, London: McGraw–Hill Book Company, Inc., 1955. 310 p.
[5] *Borisov A. B., Kiselev V. V.* Many-soliton solutions of asymmetric chiral SU(2) and SL(2, R) theories ($d = 1$). Theor. Math. Phys. 1983. V. 54, #2, P. 160–167.
[6] *Borisov A. B., Kiselev V. V.* Two and three dimensional topological defects, solitons and textures in magnets. Moscow: Physmatlit, 2022. 456 p. [In Russian].
[7] *Borisov A. B., Kiseliev V. V.* Topological defects in incommensurate magnetic and crystal structures and quasi-periodic solutions of the elliptic sine–Gordon equation. Physica D. 1988. V. 31, P. 49–64.
[8] *Borisov A. B., Kiseliev V. V.* Vortex dipoles on a solution lattice background: Solution of the boundary-value problem by inverse spectral transform. Physica D. 1998. T. 111, C. 96–128.
[9] *Borisov A. B., J. Kishine, Bostrem Y. G., Ovchinnikov A. S.* Magnetic soliton transport over topological spin texture in chiral helimagnet with strong easy-plane anisotropy. Phys. Rev. B. 2009. V. 79, P. 134436–134446.
[10] *de Gennes P. G.* The physics of liquid crystals. Oxford: Clarendon Press, 1974. 597 p.
[11] *Dubrovin B. A.* Riemann surfaces and nonlinear equations. Izhevsk: NIC "Regularnaya i chaotichnaya dinamica", 2001. 152 p. [In Russian].
[12] *Dzyaloshinskii I.* Magnetoelectricity in ferromagnets. EPL 2008. V. 83, 67001.
[13] *Dzyaloshinskii I. E.* Theory of helicoidal structures in antiferromagnets. III. Sov. Phys. JETP. 1965. V. 20, #3, P. 665–668.
[14] *Faddeev L. D., Takhtadzhyan L. A.* Hamiltonian approach in the theory of solitons, Classics in Mathematics. Berlin: Springer, 2017. 601 p.
[15] *Filippov B. N., Tankeev A. P.* Dynamic effects in ferromagnets with a domain structure. Moscow: Nauka, 1985. 217 p. [In Russian].
[16] *Ivanov B. A., Kosevich A. M., Kovalev A. S.* Nonlinear magnetization waves. Dynamic and topological solitons. Kiev: Naukova Dumka, 1983. 294 p. [In Russian].
[17] *Izyumov Y. A.* Neutron diffraction on the long periodic structures. Moscow: EnergoAtomIzdat, 1987. 200 p. [In Russian].

[18] *Kiselev V. V.* 2D vortices in incommensurate (stripe-domain) magnetic structures. Phys. Met. Metallogr. 2003. T. 95, Suppl. 1, C. 28–34.

[19] *Kiselev V. V., Raskovalov A. A.* Interaction of a breather with a magnetization wave in a ferromagnet with light-axis anisotropy. Theor. Math. Phys. 2010. V. 163, #1, P. 479–495.

[20] *Kiselev V. V., Raskovalov A. A.* Nonlinear dynamics of a quasi-one-dimensional helicoidal structure. Theor. Math. Phys. 2012. V. 173, #2, P. 1565–1586.

[21] *Kiselev V. V., Raskovalov A. A.* Nonlinear collective excitations in helical magnetic structures. Phys. Met. Metallogr. 2012. V. 113, #12, P. 1114–1126.

[22] *Kiselev V. V., Raskovalov A. A.* Standing spin waves and solitons in a quasi-one-dimensional spiral structure. J. Exp. Theor. Phys. 2013. V. 116, #2, P. 272–279.

[23] *Kishine J., Inoue K., and Yoshida Y.* Synthesis structure and magnetic properties of chiral molecule-based magnets. Prog. Theor. Phys. Suppl. 2005. V. 159, P. 82–85.

[24] *Kishine J., Ovchinnikov A. S.* Theory of spin resonance in a chiral helimagnet. Phys. Rev. B. 2009. V. 79, #22, P. 220405 (1–4).

[25] *Koppelman W.* Singular integral equations, boundary value problems and Riemann–Roch theorem. J. Math. Mech. 1961. V. 10, #2, P. 13–35.

[26] *Kovalev A. S., Gerasimchuk I. V.* Nonlinear localization of excitations and the dynamics of solitons in self-modulated systems. J. Exp. Theor. Phys. 2002. V. 95, #5, P. 965–972.

[27] *Kovalev A. S., Gerasimchuk I. V., Maugin G. A.* Nonlinear excitations of incommensurate surface structures. Proc. Est. Acad. Sci., Phys. Math. 2003. V. 52, #1, P. 94–102.

[28] *Kovalev A. S., Gerasimchuk I. V., Maugin G. A.* Nonlinear dynamics of incommensurate surface layers. Phys. Rev. Lett. 2004. V. 92, #24, P. 244101-1(4).

[29] *Kuznetsov E. A., Mikhailov A. V.* Stability of stationary waves in nonlinear weakly dispersive media. Sov. Phys. JETP. 1975. V. 40, #5, P. 855–859.

[30] *Mikhailov A. V.* The Landau–Lifshitz equation and the Riemann-boundary problem on a torus. Phys. Lett. A. 1982. V. 92, #2, P. 51–55.

[31] Magnetism: Molecules to materials. V, edited by Miller J. S. and Drillon M., 2005. Weinhein: Wiley–VCH Verlag GmbH and Co. KGaA.

[32] *Morgunov R., Kirman M. V., Inoue K., Tanimoto Y., Kishine J., Ovchinnikov A. S., and Kazakova O.* Spin solitons and spin waves in chiral and racemic molecular based ferrimagnets. Phys. Rev. B. 2008 V. 77, P. 184419 (1–9).

[33] *Morgunov R. B., Berdinski V. L., Kirman M. V., Inoue K., Kishine J., Yoshida Y. and Tanimoto Y.* Spin solitons in molecular magnetic materials with the chiral structure. JETP Lett. 2006. V. 84, #8, P. 446–450.

[34] *Mostovoy M.* Ferroelectricity in spiral magnets. Phys. Rev. Lett. 2006. V. 96, P. 067601-1(4).

[35] *Mushenok F. B., Kirman M. V., Koplak O. V. et al.* FMR determination of the period of an incommensurate magnetic structure in chiral organometallic crystals. Phys. Solid State. 2012. V. 54, #7, P. 1358–1362.

[36] *Novikov S., Manakov S. V., Pitaevskii L. P., Zakharov V. E.* Theory of solitons: The inverse scattering method N. York and London: Consultants Bureau, 1984. 276 p.

[37] *Pamyatnykh L. A., Lysov M. S., Kandaurova G. S.* Mechanism of drift of stripe magnetic domains in ferrite-garnet single crystals. Bulletin of the Russian Academy of Sciences: Physics. 2007. V. 71, #11, P. 1497–1499.

[38] *Pamyatnykh L. A., Lysov M. S., Shmatov G. A., Kandaurova G. S., Druzhinin A. V.* Dynamic drift of the magnetic domains in ferrite–garnet crystals. Bull. Russ. Acad. Sci., Phys. 2010. V. 74, #10, P. 1417–1419.

[39] *Rodin Yu. L.* The Riemann boundary problem on a torus and the inverse scattering problem for the Landau–Lifshitz equation. Lett. Math. Phys. 1983. V. 7, P. 3–8.

[40] *Roessli D., Schefer J., Petrakovskii G. A., Ouladdiaf B., Boehm M., Staub U., Vorotinov A., and Bezmaternikh L.* Formation of a magnetic soliton lattice in copper metaborate. Phys. Rev. Lett. 2001. V. 86, P. 1885–1886.

[41] *Shirobokov M. K.* On the theory of the magnetization mechanism of ferromagnets. Zh. Eksp. Teor. Fiz. 1945. V. 15, #1–2, P. 57–76. [In Russian].

[42] *Soboleva T. K., Stefanovskii E. P., Tarasenko V. V.* Equilibrium state and elementary excitation spectrum in magnets with inhomogeneous exchange-relativistic interaction in an external magnetic field. Sov. Phys., Solid State. 1980. V. 22, #8, P. 1370.

[43] *Togawa Y., Koyama T., Takayangi K., Mori S., Kousaka Y., Akimitsu J., Nishihara S., Inoue K., Ovchinnikov A. S. and Kishine J.* Chiral magnetic soliton lattice on a chiral helimagnet. Phys. Rev. Lett. 2012. V. 108, P. 107202(1–5).

[44] *Zverovich E. I.* Boundary value problems in the theory of analytic functions in Hölder classes on Riemann surfaces. Russ. Math. Surv. 1971. V. 26, #1, P. 117–192.

[45] *Zvezdin A. K.* Dynamics of domain walls in weak ferromagnets. JETP Lett. 1979. V. 29, #10, P. 553–557.

For Appendix A.1

[1] *Akhiezer N. I.* Elements of the theory of elliptic functions. Transactions of Mathematical Monographs, volume 79. Providence, Rhod Island: American Mathematical Society, 1990. 237 p.

[2] *Bateman H., Erdélyi A.* Higher transcendental functions, vol. 3. New York, Toronto, London: McGraw–Hill Book Company, Inc., 1955. 310 p.